The Most Successful Way of JEE Preparation

DPP

Daily Practice Problems

JEE Main & Advanced

Electrostatics & Current Electricity

with Heating & Chemical Effect of Current

PHYSICS Vol-6

DEEPAK PALIWAL

arihant
ARIHANT PRAKASHAN (SERIES), MEERUT

DPP
Daily Practice Problems

arihant
ARIHANT PRAKASHAN (SERIES), MEERUT

Administrative & Production Offices

Corporate Office 'Ramchhaya' 4577/15, Agarwal Road, Darya Ganj, New Delhi -110002
Tele: 011- 47630600, 43518550; Fax: 011- 23280316

Head Office Kalindi, TP Nagar, Meerut (UP) - 250002
Tel: 0121-2401479, 2512970, 4004199; Fax: 0121-2401648

Sales & Support Offices

Agra, Ahmedabad, Bengaluru, Bareilly, Chennai, Delhi, Guwahati, Hyderabad, Jaipur, Jhansi, Kolkata, Lucknow, Meerut, Nagpur & Pune

ISBN : 978-93-13193-36-4

Price : ₹260

Published by Arihant Publications (I) Ltd.

Production Team

Publishing Managers Mahendra Singh Rawat & Keshav Mohan
Project Head Sorabh Chaudhary
Cover Designer Shanu Mahad
Inner Designer Mazher Chaudhary
Page Layouting Sunil Sharma, Hemant
Proof Reading Ashish

For further information about the products from Arihant, log on to www.arihantbooks.com or email to info@arihantbooks.com

/arihantpub /@arihantpub Arihant Publications /arihantpub

ABOUT DPP

Daily Practice Problems

Indian Institutes of Technology better known as IITs have always been the dream destination of each science group students who want to pursue their career as Engineers. These are the institutes which have got very high value not even in India but in the whole world. To get the admission in such institutes one need to face entrance Examinations; JEE Main & JEE Advanced, out of which JEE Main is considered as the preliminary exam & JEE Advanced is the final exam to get into the IITs. Approximately per year 12 lakh students enroll themselves for JEE Main and only the top 2.25 lakh in JEE Main merit can attempt for JEE Advanced, the fact clearly express the degree of toughness of JEE Main & JEE Advanced.

To have success in such an exam one needs to be very planned & systematic in its preparation. A very popular wise saying 'Rome was not built in a day' suggests that to achieve a big target like IITs we need to work continuously. For continuous work, one need to have proper planning that how much work he/she has to do on regular basis, i.e., on daily basis. Each student preparing for JEE has many books but unfortunately none of them gives the students a plan for daily studies, which is the most effective success key for JEE.

Keeping the daily practice needs of the students in mind we have come up with DPP Series (Daily Practice Problems) having approximately 26 books. The primary focus of this series is on achieving success through practice & proper planning. For that we have divided the whole syllabus of JEE Main & JEE Advanced; Physics, Chemistry & Mathematics, into different volumes (around 8 volumes of Mathematics & 9 volumes each of Physics & Chemistry), each volume has Daily Practice Problems Sheets having specific questions on various Topics of the individual chapter of the volume ensuring the complete Practice & Assessment of the topic.

DPP

Daily Practice Problems

In this way we have tried to give you an effective daily planning for JEE keeping in mind the daily hours that a student can devote to his practice, you are just required to practice one sheet of each subject on daily basis. This habit of daily practice will keep you away from exam stress and thoughts like; how much I have prepared/am I preparing properly or not/ do I need to study more on daily basis or what I am doing is ok, etc., which regularly tease your mind and can give you panic at any stage of your preparation.

Some special features that we have tried to incorporate in our Daily Practice Problems (DPP) series are discussed below

1. Micro Level coverage of all the topics of a Chapter/ Unit given in JEE Main & Advanced Syllabus.
2. Each DPP has various types of questions; Subjective Questions, Single/ Multiple Correct Options, Statements type, Comprehension based, Integer Value Answer type, Matching, etc., making you practice & ready for all the question formats of JEE.
3. New & innovative questions in each DPP challenging & improving your problem solving skills.
4. Along with Topical Coverage, Revisal DPPs for JEE Main & JEE Advanced with each Chapter, covering multidisciplinary questions proving comprehensive practice.
5. JEE Main & JEE Advanced Archive (collection of last 13 years' exams questions) with each chapter.
6. Complete Solutions for each DPP.

We hope that DPP series will prove to be very beneficial for the students preparing for JEE, by giving them daily practice in a systematic & planned manner. The suggestions/criticism from the students, teachers & parents about DPP series will be welcomed whole heartedly and we promise to update this series accordingly from time to time.

DPP
Daily Practice Problems

DAILY TASKS...

1 ELECTROSTATICS

2 CAPACITANCE

DPP

Daily Practice Problems

3 CURRENT ELECTRICITY

1. Electrostatics

DPP

DAILY PRACTICE PROBLEMS

DPP-1 Charge and Properties of Charges

Subjective Questions

Working Space

Direction (Q. Nos. 1-5) *These questions are subjective in nature, need to be solved completely on notebook.*

1 Two identical metal spheres *A* and *B* have charges of +12 nC and −6 nC respectively. A neutral identical metal sphere *C* is touched to *A*, then to *B* and finally removed.

i. What will be the charge on sphere *C* finally?
ii. Have electrons been added to or removed from sphere *C*?
iii. How many electrons are removed from or given to sphere *C*?

2 In each of the following processes that identify the nature of charge body (which is being charged) acquires.

i. Charging by friction using another body with more work function value.
ii. Charging by induction using a negatively charged body.
iii. Charging by conduction using a positively charged body.
iv. Charging by excessively heating the body.
v. Charging by radiating the body with an electromagnetic wave of sufficiently high frequency.
vi. Charging by placing the body in the region of a very high intensity electric field.

3 Rajesh, Ena, Ayan, Julia and Lasit are of class XII students. Rajesh inflate a rubber balloon and rub it with his rabbit fur glove and then stuck it to room's painted wall.

The balloon remains stucked for some time to the wall and eventually it will fall.

All of them present different justifications for the incident which are given below

i. Ena said, inflating the balloon comes polarisation of rubber molecules and it got stuck to the wall.
ii. Ayan said, rubbing the balloon with fur causes charging of balloon due to which it got stuck to the wall. But, the gravity finally pulls the balloon down.
iii. Julia said, rubbing balloon causes its charging and it got stuck to the wall which conducts the excess charge and the balloon will fall.
iv. Lasit said, rubbed balloon is charged which causes polarisation of wall molecules and it stuck to wall. The excess charge of balloon is carried away by air molecules, so eventually it falls.

How does the charge effectfully work. Explain.

4 A conductor has been given a charge of $+3\times10^{-7}$ C by conduction process.

i. How many number of electrons are removed from the conductor?
ii. How much mass of conductor decrease?

5 Charge produced by induction process is either less than or equal to charge on the body used for inducing the charge. Does this against the conservation of charge? Explain.

Only One Option Correct Type

Working Space

Direction (Q. Nos. 6-15) This section contains 10 multiple choice questions. Each question has four choices (a), (b), (c) and (d), out of which ONLY ONE option is correct.

6 A body has a negative charge of $2\mu C$. If body has 2.25×10^{12} protons, then number of electrons present in the body will be

a. 1.25×10^{12}

b. 2.25×10^{12}

c. 1×10^{12}

d. 3.50×10^{12}

7 A common concept on charge is given. Select the odd statement.

a. Charge gained by an uncharged body by conduction from a charged body is equal to half of the total charge initially present

b. Magnitude of charge does not increase with its velocity

c. Charge cannot be coexisted without matter although matter can exist without charge

d. Between two substances repulsion is true test of presence of charge

8 A plastic scale is positively charged by rubbing against a woolen shirt. That means

a. all molecules of plastic become ions

b. few of bonds are broken to release electrons

c. few of atoms are ionised to release electrons

d. few of molecules are polarised to show presence of charge

9 A conductor has 14.4×10^{-19}C positive charge. The conductor has

a. 9 electrons in excess

b. 27 electrons in excess

c. 9 electrons in short

d. 27 electrons in short

10 A glass rod rubbed with fur is used to charge a gold leaf electroscope and the leaves are observed to diverge. The electroscope disc is then exposed to X-rays for a short period. Then,

a. the divergence of leaves is not affected

b. the leaves diverge further

c. the leaves will collapse

d. the leaves will melt

11 Which of the following charge is not possible?

a. 1.6×10^{-18}C

b. 1.6×10^{-19}C

c. 1.6×10^{-20}C

d. 1.6×10^{-12}C

12 Five balls numbered 1 to 5 are checked pairwise. Pair (1, 2), (2, 4) and (4, 1) show electrostatic attraction while pairs (2, 3) and (4, 5) show repulsion.

Ball 1 must be

a. positively charged

b. negatively charged

c. neutral

d. either positively or negatively charged

13 Charging by rubbing (friction) does not create electricity because

a. charge produced is of very small magnitude

b. charge produced is not free charge

c. charge produced cannot made to flow through a conductor

d. charge is conservative in nature

Working Space

14 Why does a charged comb attract bits of paper?

a. Paper bits remain stuck to the comb forever
b. Paper bits stuck to the comb and then fall down
c. Paper bits are initially attracted to the comb but they remain at their places
d. Paper bits 'fly away' when they touch the comb

15 A foreigner student who grow up in a tropical country but is studying in United States may not have experience with static electricity sparks or shocks. The student most probably gets his/her first shock in

a. during a thunderstorm
b. while having a stroll in rain
c. in summer
d. in winter

Statement Type

Direction (Q. Nos. 16 and 17) *This section is based on Statement I and Statement II. Select the correct answer from the codes given below.*

a. Both Statement I and Statement II are correct and Statement II is the correct explanation of Statement I
b. Both Statement I and Statement II are correct and Statement II is not the correct explanation of Statement I
c. Statement I is correct but Statement II is incorrect
d. Statement II is correct but Statement I is incorrect

16 **Statement I** The whole charge of conductor cannot be transferred to an isolated conductor (by touching charged one to isolated one).

Statement II Charge flows from a conductor to another due to difference of potentials of conductors and flow stops as soon as both conductors reaches same potential.

17 **Statement I** Tyres of aircraft's are slightly conducting.

Statement II When a conductor is connected to ground, the extra charge induced on the conductor will flow to ground.

One or More than One Options Correct Type

Direction (Q. Nos. 18-21) *This section contains 4 multiple choice questions. Each question has four choices (a), (b), (c) and (d), out of which ONE or MORE THAN ONE are correct.*

18 Consider a neutral conducting sphere. A positive point charge is placed outside it. The net charge on the sphere is then **[IIT JEE 2007]**

a. negative and uniformly distributed over the sphere
b. negative and appears only on the point closest to the outside placed charge
c. negative and non-uniformly distributed over the entire volume of sphere
d. zero

19 When an object A is charged by induction process, another charged object B is required to induce charge. If q_B is the charge on B in such an experiment, then

a. $q_A + q_B = 0$ in every such experiment
b. q_A and q_B are of dissimilar nature in all such experiments
c. $q_B = q_A\left(1 - \frac{1}{K_A}\right)$, where, K_A = dielectric constant of object A
d. Charge can be induced on A only when A contains few free electrons

Working Space

20 A small tennis ball is negatively charged. Then, which of these are true?

a. Its mass increases
b. Its charge is $q = ne$
c. It moves with a larger velocity through air
d. Charge is found uniformly distributed over its entire surface

21 When a plastic comb is charged by rubbing it against silk cloth, the net charge acquired by comb is typically a few micro coulombs. If that charge is 3.0μC and mass of comb is 35 g, then

a. comb acquires negative charge
b. comb is warmed by rubbing
c. comb shows a very little change of mass
d. combing is difficult when comb is charged

Matching List Type

Direction *(Q. Nos. 22 and 23) Choices for the correct combination of elements from Column I and Column II are given as options (a), (b), (c) and (d), out of which one is correct.*

22 Two objects A and B are charged.
Let q_A = charge on A and q_B = charge on B, then match Column I and Column II.

	Column I		Column II
i.	$q_A = q_B$	p.	Charge transfer by conduction
ii.	$q_A \geq q_B$	q.	Charging by friction
iii.	$q_B = \frac{q_A}{2}$	r.	Charging by induction

Codes

	i	ii	iii
a.	q	r	p
b.	r	q	p
c.	p	r	q
d.	q	p	r

23 Let A and B are two identical objects of same mass initially are charged electrostatically by rubbing against each other. Let A has got a positive charge and B has got a negative charge, then match Column I and Column II.

	Column I		Column II
i.	Mass of A	p.	increases
ii.	Mass of B	q.	decreases
iii.	Work function of A	r.	is less
iv.	Work function of B	s.	is more

Codes

	i	ii	iii	iv
a.	p	q	r	s
b.	r	s	q	p
c.	q	p	r	s
d.	s	r	q	p

Working Space

One Integer Value Correct Type

Direction *(Q. No. 24 and 25) This section contains 2 questions. When worked out will result in an integer from 0 to 9 (both inclusive).*

24 A monochromatic light source emitting photon of energy 2×10^{-19} J having a power of 2 W is located at the centre of curvature of a concave metal disc of radius $\left(\frac{1}{\sqrt{\pi}}\right)$ m.

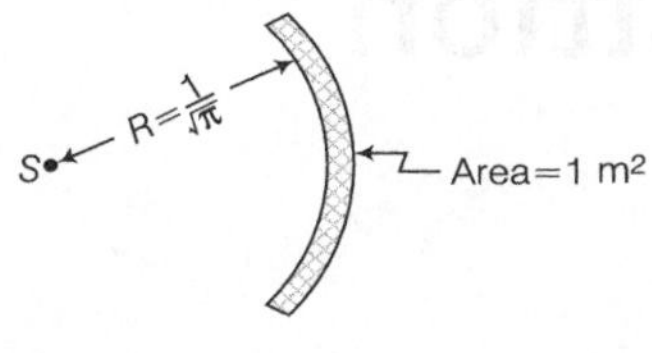

Assume that

i. Each photon gives sufficient energy to a surface electron, so that it leaves the metal surface.

ii. Electron once leave the metal surface does not come back.

iii. Work function of metal remains constant.

If, area of metal disc is 1.0 m^2 and the charge on the disc after 1 s the light source is q.

Find value of 10 q.

25 In an millikans oil-drop experiment following are the observations made.

Drop	Charge for going up $q_1 \times 10^{-19}$ C	Charge for going down $q_2 \times 10^{-19}$ C	Average charge on drop $=\left(\frac{q_1+q_2}{2}\right)\times 10^{19}$ C
1A	4.72	4.71	4.72
1B	6.18	6.35	6.27
1C	1.51	1.68	1.60
1D	3.09	3.06	3.08
1E	1.52	1.52	1.52

To calculate the value of elementary charge, difference of charges observed on drops is found and then, difference of average value is taken.

Difference between letter charges	Value of difference ($\times 10^{19}$ C)
D-E	1.56
( )	--
( )	--
( )	--

We just forget to put some values in above table. If we take average difference value from IInd table as our elementary charge value, then the true value of elementary charge ($e = 1.6 \times 10^{-19}$ C) is about x% more than experimental value. Find the value of $2x$.

DPP-2 Coulomb's Law and Principle of Superposition

Subjective Questions

Working Space

Direction (Q. Nos. 1-5) *These questions are subjective in nature, need to be solved completely on notebook.*

1 Two charged particles having charges of different magnitude when placed at a distance d from each other experiences a force of attraction F. These two particles are put in contact and again placed at same distance from each other.

i. What is nature of new force between them?

ii. Is the magnitude of the force of interaction between them, now more or less than F?

2 Draw a graph showing variation of Coulomb force F *versus* $\left(\frac{1}{r^2}\right)$, where, r is distance between charges of each pair of charges, $(1\,\mu C, 2\,\mu C)$, $(2\,\mu C, -3\,\mu C)$.

Which quantity represents slope of graph?

3 Two identical conducting spheres are separated by a fixed centre to centre distance of 25 cm and have different charges.

Initially spheres attract each other with a force of 1.728 N. They are connected with a wire when the spheres are found positvely charged and repel each other with a force of 0.576 N when the wire is removed. Find

i. the final and

ii. the initial charges on the spheres.

4 A small sphere with a charge of $2.44\,\mu C$ is attached to a released horizontal spring whose force constant is 89.2 N / m. The spring extends along the X-axis and the sphere rests on a frictionless surface with its centre at origin. A point charge of charge $-8.55\,\mu C$ is now moved slowly from infinity to a point $x = d$ $\{d > 0\}$ on the X-axis. This causes small sphere to move to position $x = 0.124$ m. Find the value of d.

5 Give the answer of the following with explanation.

i. An electrically neutral object is given a positive charge. Does the mass of object increase, decrease or remain same.

ii. An insulating sphere of radius R has a uniform surface charge density σ. A point charge of magnitude q is at a distance of $4R$ from centre. Force on q exerted by sphere is F. If insulating sphere is replaced by a conducting sphere of same size and same charge. Does the force increase, decrease or remain same.

iii. An electron and a proton are released from rest in space.The particles move towards each other, due to their mutual electrical attraction. The kinetic energy of the electron at greater than, less than or equal to the kinetic energy of the proton, when they collide?

Working Space

Only One Option Correct Type

Direction *(Q. Nos. 6-15) This section contains 10 multiple choice questions. Each question has four choices (a), (b), (c) and (d), out of which ONLY ONE option is correct.*

6 Four charges each with charge $+q$ are placed at the four corners of a square of side l, a charge $\frac{-q}{4}$ is at centre of the square. Force on the charge at the centre due to other charges is

a. $\frac{Kq^2}{l^2}$ b. $K\frac{\sqrt{2}q^2}{l^2}$

c. zero d. $K\frac{2\sqrt{2}q^2}{l^2}$

7 Three point charges q_1, q_2 and q_3 are taken such that when q_1 and q_2 are placed close together to form a single point charge, the force on q_3 at distance L from this combination is a repulsion of 2 unit magnitude. When q_2 and q_3 are combined, the force on q_1, from same distance L is of magnitude 4 unit and is attractive. Also, q_3 and q_1 combining exerts a force of magnitude 18 unit in q_2 from same distance L. Algebraic ratio of charges q_1, q_2 and q_3 is

a. 1:2:3 b. 2:−3:4

c. 4:−3:1 d. 4:−3:2

8 Two equal charges are placed at d distance apart. A third charge placed at x distance from line joining charges on a perpendicular bisector. Third charge experiences maximum force when

a. $x = \frac{d}{\sqrt{2}}$ b. $x = \frac{d}{2}$

c. $x = \frac{d}{2\sqrt{2}}$ d. $x = \frac{d}{2\sqrt{3}}$

9 Three charges are placed as shown.

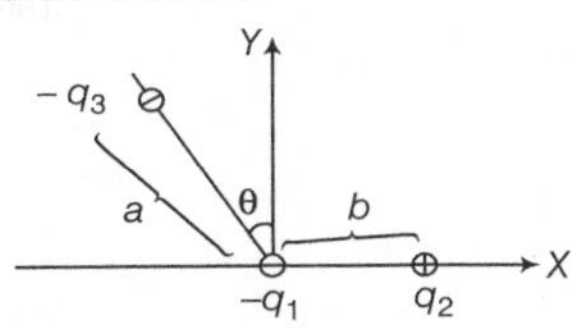

Components of force on $-q_1$ are

a. $K\left(\frac{q_1q_2}{b^2} + \frac{q_1q_3}{a^2}\sin\theta\right)\hat{i}, \frac{-Kq_1q_3}{a^2}\cos\theta\,\hat{j}$

b. $K\left(\frac{q_1q_2}{b^2} + \frac{q_1q_3}{a^2}\right)\hat{i}, \frac{Kq_1q_3}{a^3}\hat{j}$

c. $K\left(\frac{q_1q_2}{b^2} + \frac{q_1q_3}{a^2}\cos\theta\right)\hat{i}, \frac{Kq_1q_3}{a^2}\sin\theta\,\hat{j}$

d. $K\left(\frac{q_1q_3}{b^2} + \frac{q_1q_3}{a_2}\right)\hat{i}, \frac{Kq_1q_3}{a^2}\cos\theta\,\hat{j}$

10 Two identical charged spheres are suspended in air by equal length strings pivoted at same point, makes an angle of 30° with each other.
When suspended in a liquid of density 800 kg / m^3, the angle remains same. Dielectric constant of the liquid is
(density of material of spheres is 1600 kg / m^3)

a. $K = 6$ b. $K = 2$

c. $K = 7$ d. $K = 3$

Working Space

11 Two identical positive point charges Q each are fixed at a distance $d = 2a$ apart in air. A point charge q lies at mid-point on the line joining the charges. Now q executes SHM, if

i. q is positive and it is given a small displacement along the line joining the charges.

ii. q is negative and it is given a small lateral displacement. If frequencies of oscillations are n_a and n_b, in case (i) and case (ii) then the ratio $\frac{\sqrt{2}n_a}{n_b}$ is

a. 2 b. 4 c. 6 d. 8

12 A point charge Q is situated at a distance d from one end of a non-conducting rod of length L, carrying a charge q as shown.

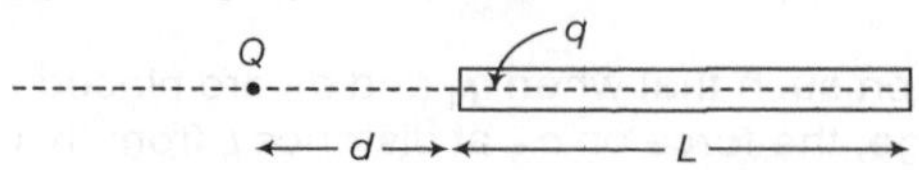

The force experienced by Q charge is

a. $\frac{KqQ}{d(d+L)}$ b. $\frac{4KqQ}{d(d+L)}$

c. $\frac{2KqQ}{(d+L)^2}$ d. $\frac{4KqQ}{(d+L)^2}$

13 Force between two point charges q_1 and q_2 is F at a separation of distance r. When a dielectric of relative permittivity 4 and of thickness $r/2$ is put between the charges, force is reduced by a factor of

a. $\frac{1}{4}$ b. $\frac{3}{4}$ c. $\frac{4}{9}$ d. $\frac{2}{9}$

14 Consider a system of three charges $\frac{q}{3}, \frac{q}{3}$ and $\frac{-2q}{3}$ placed at point A, B and C respectively, as shown in the figure. Take O to be the centre of the circle of radius R and $\angle CAB = 60°$. **[AIEEE 2008]**

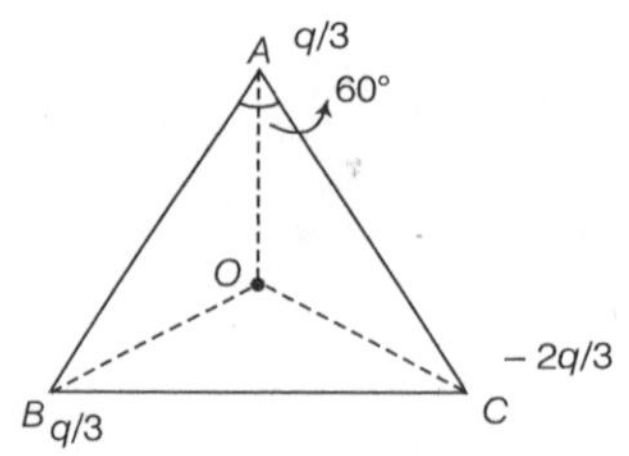

Which of the following statements is correct?

a. The electric field at point O is $\frac{q}{8\pi\varepsilon_0 R^2}$ directed along the negative X-axis

b. The potential energy of the system is zero

c. The magnitude of the force between the charges at C and B is $\frac{q^2}{54\pi\varepsilon_0 R^2}$

d. The potential at point O is $\frac{q}{12\pi\varepsilon_0 R}$

15 Two charges, each equal to q, are kept at $x = -a$ on the X-axis. A particle of mass m and charge $q_0 = \frac{q}{2}$ is placed at the origin. If charge q_0 is given a small displacement $y (y << a)$ along the Y-axis, the net force acting on the particle is proportional to **[JEE Main 2013]**

a. y b. $-y$ c. $\frac{1}{y}$ d. $-\frac{1}{y}$

One or More than One Options Correct Type

Direction (Q. Nos. 16-21) *This section contains 6 multiple choice questions. Each question has four choices (a), (b), (c) and (d), out of which ONE or MORE THAN ONE are correct.*

16 Two small charged spheres are suspended on cords of equal length.

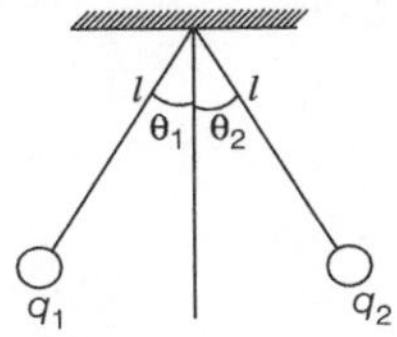

Cords makes angle θ_1 and θ_2 with the vertical.
Which of the following statements is/are correct?

a. If $q_1 = q, q_2 = 2q$ and balls have equal masses $m_1 = m_2 = m$, then $\frac{\theta_1}{\theta_2} = 1$
b. If $q_1 = q, q_2 = 2q$ and $m_1 = m, m_2 = 2m$, then $\frac{\theta_1}{\theta_2} = 2$
c. $\theta_1 = \theta_2$ when spheres have equal mass
d. $\theta_1 = \theta_2$ when spheres have equal charge on them

17 Two positive charges $+Q$ are fixed on X-axis at $(+d, 0)$ and $(-d, 0)$. A third charge $+q$ of mass m is displaced from origin by a small displacement $s << d$ and then released from rest. Then,

a. particle will execute SHM
b. if T = period of oscillation, then $T^2 \propto d^3$
c. motion of particle is periodic but not simple harmonic
d. particle comes quickly to rest at origin when released

18 Three charges are placed as shown.

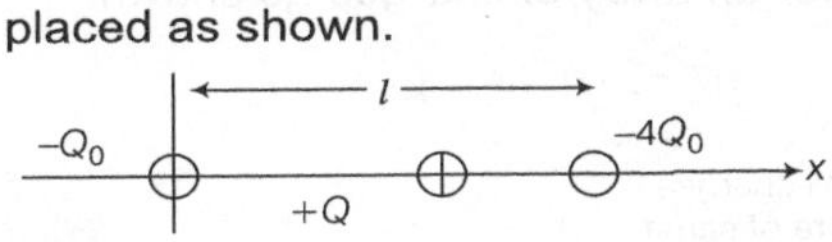

Charges are free to move

a. system is never in stable equilibrium
b. system is in equilibrium, when $Q = \frac{4}{9}Q_0$
c. system is in equilibrium when distance between $-Q_0$ and Q is $\frac{1}{3}l$
d. system is in equilibrium for more than one position of Q

19 A ring of radius 0.1 m is made of thin metallic wire of area of cross-section $10^{-6} m^2$. The ring has a uniform charge of π coulomb. Young's modulus of metal of wire in 2×10^{11} N/m^2 and $E_0^{-1} = 36\pi \times 10^9$ Vm/As. When a charge of 10^{-8} C is placed at the centre of the ring, then

a. tension in the wire is 4.5×10^3 N
b. tension in the wire is 4.5×10^5 N
c. change in radius is 2.25×10^{-3} m
d. change in radius is 2.25×10^{-5} m

20 A small ball of mass 2×10^{-3} kg having a charge of 1 μC is suspended by means of a string of length 0.8 m. Another identical ball with same charge is kept at the point of suspension. Then, correct options are

a. lower ball makes a complete revolution when it is given a horizontal velocity of 2 m/s
b. lower ball makes a complete revolution when it is given a horizontal velocity 6 m/s
c. minimum horizontal velocity for complete revolution required by lower ball is independent of its charge
d. minimum horizontal velocity for complete revolution required by lower ball depends on its charge

Working Space

21 A 2 μC charge is placed at every corner of a cube of side 5 cm.

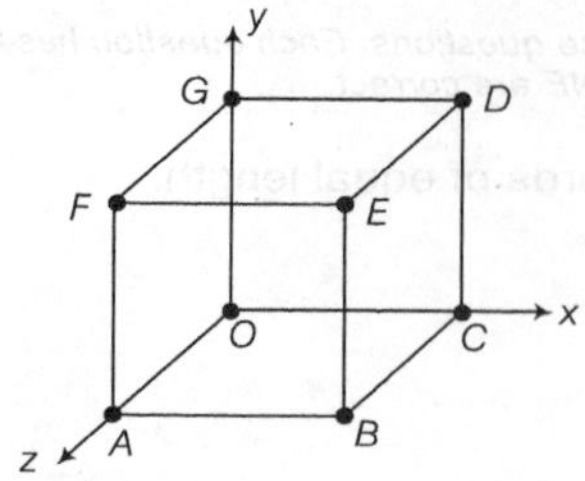

a. Total force on charge at origin is

$$\mathbf{F} = \frac{36 \times 10^{-3}}{125}\left(5\hat{\mathbf{k}} + \frac{5\hat{\mathbf{i}} + 5\hat{\mathbf{j}}}{2^{3/2}} + \frac{5\hat{\mathbf{i}} + 5\hat{\mathbf{j}} + 5\hat{\mathbf{k}}}{3^{3/2}} + \frac{5\hat{\mathbf{j}} + 5\hat{\mathbf{k}}}{2^{3/2}} + 5\hat{\mathbf{j}} + \frac{5\hat{\mathbf{j}} + 5\hat{\mathbf{k}}}{2^{3/2}} + 5\hat{\mathbf{i}}\right)$$

b. Force on charge at origin is

$$\mathbf{F} = -\frac{36 \times 10^{-3}}{125}\left\{5(\hat{\mathbf{i}} + \hat{\mathbf{j}} + \hat{\mathbf{k}}) + \left(\frac{5}{2^{3/2}} + \frac{5}{3^{3/2}}\right)(\hat{\mathbf{i}} + \hat{\mathbf{j}} + \hat{\mathbf{k}})\right\}$$

c. Force on charge at origin is $\mathbf{F} = -\left(\frac{36 \times 10}{25}\right)\left[1 + \frac{1}{2^{3/2}} + \frac{1}{3^{3/2}}\right](\hat{\mathbf{i}} + \hat{\mathbf{j}} + \hat{\mathbf{k}})$

d. Magnitude of force on each charge is same

Matching List Type

Direction (Q. No. 22) *Choices for the correct combination of elements from Column I and Column II are given as options (a), (b), (c) and (d), out of which one is correct.*

22 To write vector expression of Coulomb's force between two charges, we have to find a vector in the direction of force. Consider an array of charges as shown.

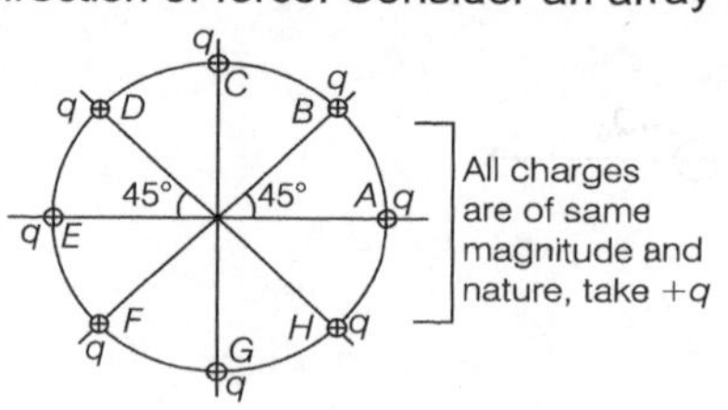

Charges are named A, B, C, D, E, F, G and H. Radius of circle is 1 unit. Lines FB and DH are inclined with X-axis at 45°. Now, match Column I and Column II.

	Column I		Column II
i.	Force on charge A is along	p.	$-\frac{\hat{\mathbf{i}}}{\sqrt{2}} + \frac{\hat{\mathbf{j}}}{\sqrt{2}}$
ii.	Force on charge B is along	q.	$-\frac{\hat{\mathbf{i}}}{\sqrt{2}} - \frac{\hat{\mathbf{j}}}{\sqrt{2}}$
iii.	Force on charge H is along	r.	$\frac{\hat{\mathbf{i}}}{\sqrt{2}} + \frac{\hat{\mathbf{j}}}{\sqrt{2}}$
iv.	Force on charge D is along	s.	$\hat{\mathbf{i}}$

Codes

	i	ii	iii	iv		i	ii	iii	iv
a.	p	r	q	s	b.	s	r	q	p
c.	r	s	q	q	d.	s	q	r	p

Working Space

Comprehension Type

Direction *(Q. Nos. 23-25) This section contains a paragraph, describing theory, experiments, data etc. Three questions related to the paragraph have been given. Each question has only one correct answer among the four given options (a), (b), (c) and (d).*

Passage

Two equal point charges of 8 μC are fixed at points (2 m, 0), (– 2 m, 0). Another charge q with a mass of 91 mg is released at point (0, 0.1 m) at $t = 0$. It is observed that q oscillates in simple harmonic motion about the origin. At $t = 0$, force experienced by q is 9×10^{-3} N.

23 Charge q is

a. – 8 μC
b. – 6.5 μC
c. – 5 μC
d. + 6.5 μC

24 Amplitude of simple harmonic motion is

a. 10 cm
b. 20 cm
c. 30 cm
d. 40 cm

25 Frequency of oscillation is

a. 8
b. 10
c. 5
d. 2

One Integer Value Correct Type

Direction *(Q. No. 26 and 27) This section contains 2 questions. When worked out will result in an integer from 0 to 9 (both inclusive).*

26 Two small metal sphere having equal charge and mass are suspended from some point on the ceiling of a damp room with silk threads of equal length. Let centre to centre distance between sphere be x, $x << l$, l is length of silk thread. Due to ionisation of medium, charge leaks off from each sphere and they keep on coming closer to each other at a constant rate.

Let their approach velocity v varies as $v \propto x^{-1/2}$.

If mass of each sphere is m then the rate at which charge varies with respect to time is

$$\frac{dq}{dt} \propto \frac{N}{2}\sqrt{\frac{2\pi\varepsilon_0 mg}{l}}$$

Find the value of N.

27 In Bohr's theory of the hydrogen atom, an electron moves in a circular orbit about a proton of radius 0.53 Å. Speed of electron in Bohr's orbit is about

$$\frac{1}{N \times 69} \times \text{speed of light}$$

Find the value of N.

DPP-3 Electric Field, Field Intensity and Superposition of Electric Field

Subjective Questions

Working Space

Direction *(Q. Nos. 1-4) These questions are subjective in nature, need to be solved completely on notebook.*

1 Electric field at a point due to a point charge q is defined as, $\mathbf{E} = \lim_{q_0 \to 0} \frac{\mathbf{E}}{q_0}$

where, q_0 is test charge and field exists due to charge q.

i. What is the significance of electric field?

ii. What is the purpose of taking limit, $\lim_{q_0 \to 0}$?

2 A thin charged wire ring of radius R has a very long charged thread oriented along the axis of ring, with one end at the centre of the ring.
Find force on the ring by the wire.

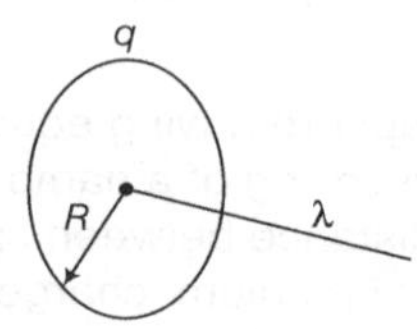

3 Draw field lines for

i. two positive charges separated by a small distance.

ii. a small positively charged sphere is placed electric to a large spherical shell.

iii. for a system of charges consisting of a $+ 4\,\mu C$ charge at origin and $-2\,\mu C$ charge at (5 cm, 0) in XOY-plane.

iv. a positively charged plate when

a. plate is plane (straight). b. plate is convex (curved).

v. a. a uniform field and

b. a field which follows inverse square law.

4 Given figure shows electric field lines in a region.

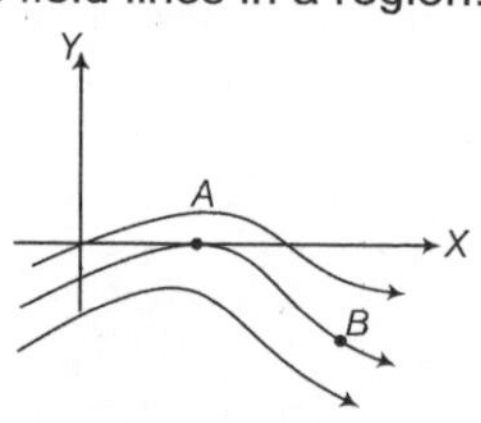

i. What is the direction of force on a positive test charge placed at A?

ii. Acceleration of a positive test charge q_0 is

$$\alpha = \frac{F}{m} = \frac{q_0 E}{m}$$

where, m = mass of q_0
At which point A or B, the acceleration of q_0 is higher?

Only One Option Correct Type

Direction *(Q. Nos. 5-13) This section contains 9 multiple choice questions. Each question has four choices (a), (b), (c) and (d), out of which ONLY ONE option is correct.*

5 Two point charges $q_1 = 2.1\times10^{-8}$ C and $q_2 = -4.0\,q_1$ are fixed on X-axis at 50 cm apart (q_1 at origin) coordinates of point where electric field is zero, are (in cm)

a. (−50, 0) b. (50, 0)
c. (−100, 0) d. (100, 0)

6 Two charges of equal magnitude Q are placed at two vertices of an equilateral triangle of side a. Electric field at the third vertex is

a. $\dfrac{Q}{4\pi\varepsilon_0 a^2}$ b. $\dfrac{\sqrt{2}Q}{4\pi\varepsilon_0 a^2}$
c. $\dfrac{\sqrt{3}Q}{4\pi\varepsilon_0 a^2}$ d. $\dfrac{Q}{2\pi\varepsilon_0 a^2}$

7 Four charges are placed on corners of a square of side 5 cm as shown. If $Q = 1\,\mu$C, then field at centre of square is

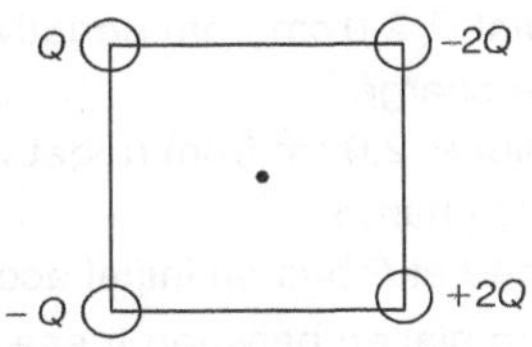

a. 1.02×10^7 N/C upwards
b. 2.04×10^7 N/C downwards
c. 2.04×10^7 N/C upwards
d. 1.02×10^7 N/C downwards

8 Two charges of magnitudes q and $2\,q$ are situated at points (0, 0) and (3, 0) (coordinates are in cm). The net electric field due to the charge system at point (3 cm, 4 cm) is

a. $kq\left(\dfrac{4}{3}\hat{\mathbf{i}} + 4\hat{\mathbf{j}}\right)$ N/C b. $kq\left(4\hat{\mathbf{i}} + \dfrac{141}{3}\hat{\mathbf{j}}\right)\times10^2$ N/C
c. $kq\,(3\hat{\mathbf{i}} + 7\hat{\mathbf{j}})\times10^2$ N/C d. $kq\,(4\hat{\mathbf{i}} + 16\hat{\mathbf{j}})\times10^2$ N/C

9 Two small spheres are equally charged and then joined by a wire. If radii of spheres are R and $3R$ respectively, then ratio of electric field on the surface of spheres after stabilisation will be

a. 1 : 1 b. 3 : 1
c. $1:\sqrt{2}$ d. $\sqrt{3}:1$

10 Electric field at point (30 , 30 , 0) due to a charge of 0.008 μC placed at origin will be, (coordinates are in cm)

a. 8000 NC^{-1} b. $4000(\hat{\mathbf{i}} + \hat{\mathbf{j}})$NC^{-1}
c. $200\sqrt{2}\,(\hat{\mathbf{i}} + \hat{\mathbf{j}})$NC^{-1} d. $400\sqrt{2}\,(\hat{\mathbf{i}} + \hat{\mathbf{j}})$NC^{-1}

11 Three metallic spheres of radii R, $2R$ and $3R$ are taken. First sphere is initially given a charge Q and other two are uncharged. Third sphere (radius = $3R$) is touched with first sphere then with second sphere and finally removed. Ratio of electric field intensity on surfaces of I and II is

a. 3 : 2 b. 2 : 3
c. 27 : 4 d. 10 : 3

12 A metallic shell has a point charge q kept inside its cavity. Which one of the following diagrams correctly represents the electric lines of forces? **[IIT JEE 2003]**

a. 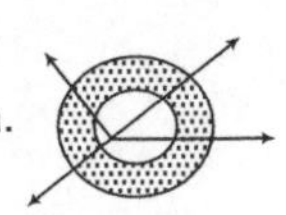 b. 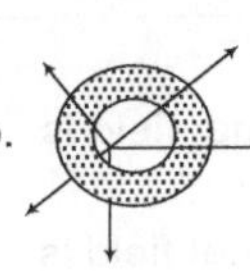 c. 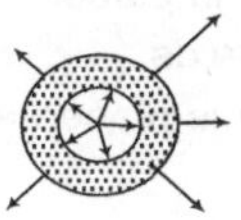 d. 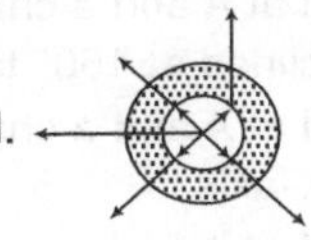

Working Space

13 Charges Q, $2Q$ and $4Q$ are uniformly distributed in three dielectric solid spheres 1, 2 and 3 of radii $R/2$, R and $2R$ respectively, as shown in figure. If magnitudes of the electric fields at point P at a distance R from the centre of spheres 1, 2 and 3 are E_1, E_2 and E_3 respectively, then **[JEE Advanced 2014]**

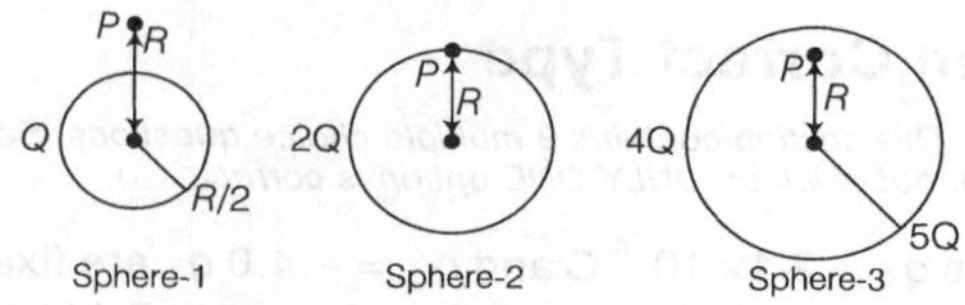

a. $E_1 > E_2 > E_3$ b. $E_3 > E_1 > E_2$ c. $E_2 > E_1 > E_3$ d. $E_3 > E_2 > E_1$

One or More than One Options Correct Type

***Direction** (Q. Nos. 14-18) This section contains 5 multiple choice questions. Each question has four choices (a), (b), (c) and (d), out of which ONE or MORE THAN ONE are correct.*

14 Two point charges are separated by a distance of 10.0 cm. One has a charge of $-25\ \mu C$ and other $+50\ \mu C$. Then,

a. net field at a point P, 2.0 cm from negative charge is 6.3×10^8 N/C and it points towards positive charge
b. net field at a point P, 2.0 cm from negative charge is 6.3×10^8 N/C and it points towards negative charge
c. an electron placed at P has an initial acceleration of $1.1 \times 10^{20}\ m/s^2$ when released
d. an electron when placed between these charges has an acceleration of $1.1 \times 10^{20}\ m/s^2$

15 Two charges placed as shown. Then,

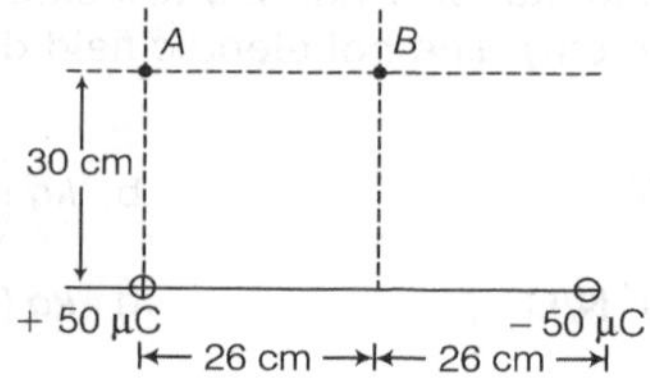

a. field at $A = 4.5 \times 10^{-6}$ N/C
b. field at $B = 3.6 \times 10^{6}$ N/C
c. ratio of magnitude of field at A to that of B is 5/4
d. field at $B = 3.6 \times 10^{-6}$ N/C

16 An electron is released from rest in an uniform electric field of 576 N/C. Then,

a. it does not experience any force
b. its initial acceleration is $1.01 \times 10^{14}\ m/s^2$
c. its acceleration increases as it moves along the field
d. acceleration is directed opposite to field

17 Two charges form a field pattern as shown. The relation between q_1 and q_2 satisfy,

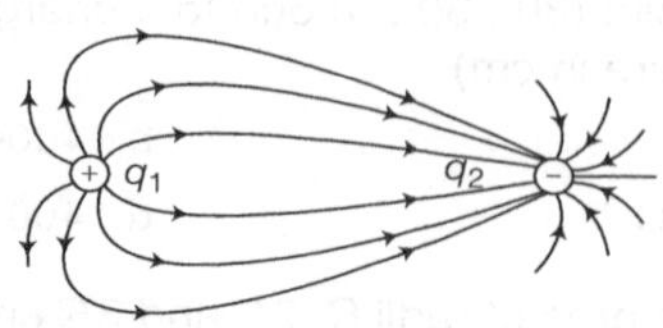

a. $|q_1| = |q_2|$
b. $|q_1| < |q_2|$
c. $|q_1| > |q_2|$
d. A negative test charge can be placed on left side of positive charge such that it experiences zero force

18 Consider an equilateral triangle as shown, let side length is l (m).
Which of the following statements is/are correct?

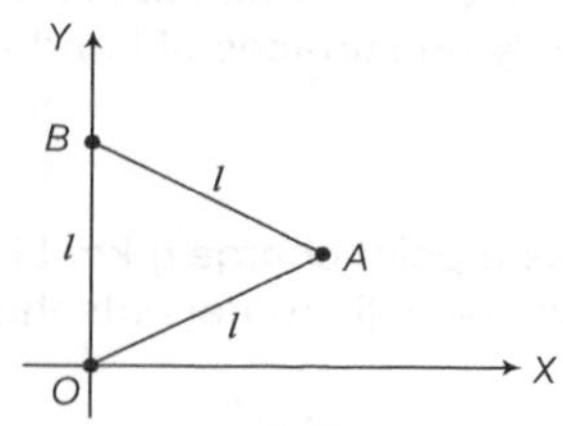

a. If a charge $+q$ is placed at A and a charge $-q$ is placed at B, then net field is inclined at 60° to the X-axis
b. If a charge $+q$ is placed at A and a charge $-q$ is placed at B, then net field is inclined at 150° to the X-axis
c. If a charge $-q$ is placed at A and a charge $+q$ is placed at B, then net field is inclined at 150° to the X-axis
d. If a charge $-q$ is placed at A and a charge $+q$ is placed at B, then net field is inclined at 330° to the X-axis

Working Space

Statement Type

Direction *(Q. Nos. 19-21) This section is based on Statement I and Statement II. Select the correct answer from the codes given below.*

a. Both Statement I and Statement II are correct and Statement II is the correct explanation of Statement I
b. Both Statement I and Statement II are correct and Statement II is not the correct explanation of Statement I
c. Statement I is correct but Statement II is incorrect
d. Statement II is correct but Statement I is incorrect

19 **Statement I** A metallic shield in form of a hollow shell may be build to block an electric field.

Statement II Field lines are reflected from the metallic surface.

20 **Statement I** A proton and an electron when placed in same uniform field are accelerated with same acceleration.

Statement II Electric force is not mass dependent.

21 **Statement I** Three equal charges are placed equidistant from each other on a circle. Net field of charges at centre is zero.

Statement II Force on unit positive charge at centre can be represented by three sides of a triangle, taken in same order.

Matching List Type

Direction *(Q. No. 22) Choices for the correct combination of elements from Column I and Column II are given as options (a), (b), (c) and (d), out of which one is correct.*

22 On Y-axis, magnitude of electric field E is taken and on X-axis, $\frac{1}{r^2}$ is taken. Then, match Column I and Column II.

	Column I		Column II
i.	Field between two positive charges	p.	Curve P
ii.	Field between a positive and a negative charge	q.	Curve Q
iii.	Field due to a positive point charge	r.	Curve R

Codes

	i	ii	iii		i	ii	iii
a.	q	r	p	b.	q	p	r
c.	p	q	r	d.	r	p	q

Comprehension Type

Direction *(Q. Nos. 23-29) This section contains 2 paragraphs, each describing theory, experiments, data, etc. Seven questions related to the paragraph have been given. Each question has only one correct answer among the four given options (a), (b), (c) and (d).*

Passage I

Electric field in an XY-plane produced by a charged particle is $7.2\,(4\hat{i} + 3\hat{j})$ N/C at the point (3 cm, 3 cm) and is $100\,\hat{i}$ N/C at point (2 cm, 0).

23 X-coordinate of the charged particle is

a. −1.0 b. 1.0 c. 2.0 d. −2.0

24 Y-coordinate of the charged particle is

a. 2.0 b. 1.0 c. 0 d. −1.0

Working Space

25 Charge on the particle is

a. 10 mC b. 10 pC c. 10 μC d. 10 nC

Passage II

A thin non-conducting ring is placed in XOY plane with its axis along Z-axis and centre at origin. Ring is charged such that it acquires a charge density of $\lambda = \lambda_0 \cos\theta \left(\frac{\text{C}}{\text{m}}\right)$. λ_0 is constant and θ is angle made with positive direction of X-axis, measured anti-clockwise.

26 Charge on the ring is

a. positive everywhere

b. negative everywhere

c. positive in $\frac{3}{4}$ th part and negative in $\frac{1}{4}$ th part

d. total charge on the ring is zero

27 X-component of electric field at centre of the ring is

a. $\frac{k\lambda_0}{R}$ b. $\frac{2k\lambda_0}{R}$ c. $\frac{\lambda_0}{4\varepsilon_0 R}$ d. $\frac{\lambda_0}{2\varepsilon_0 R}$

28 Y-component of electric field at the centre of the ring is

a. $\frac{k\lambda_0}{R}$ b. $\frac{2k\lambda_0}{R}$ c. $\frac{\lambda_0}{4\varepsilon_0 R}$ d. 0

29 Electric field at a point distant x from centre of ring will be

a. $\frac{\lambda_0 R^2}{4\varepsilon_0 x^3}$ along X-axis b. $\frac{\lambda_0 R^2}{\varepsilon_0 x^3}$ along X-axis

c. $\frac{\lambda_0 R^2}{4\varepsilon_0 x^3}$ along Y-axis d. $\frac{\lambda_0 R^2}{\varepsilon_0 x^3}$ along Y-axis

One Integer Value Correct Type

***Direction** (Q. Nos. 30 and 31) This section contains 2 questions. When worked out will result in an integer from 0 to 9 (both inclusive).*

30 A charged dust particle of radius 5×10^{-7} m is moving in a horizontal electric field of intensity 6.28×10^5 V/m. The surrounding medium is air with coefficient of viscosity $\eta = 1.6 \times 10^{-5}$ N-s/m^2.

If this particle is moving with a uniform horizontal speed of 0.02 m/s, then number of excess electrons on the drop are 6 k. Find the value of k.

31 A uniform field of 10^3 V/m is maintained between two very long plates with a separation of 2 cm.

A particle of mass 1.6×10^{-30} kg with a charge of -1.6×10^{-19} C is projected as shown.

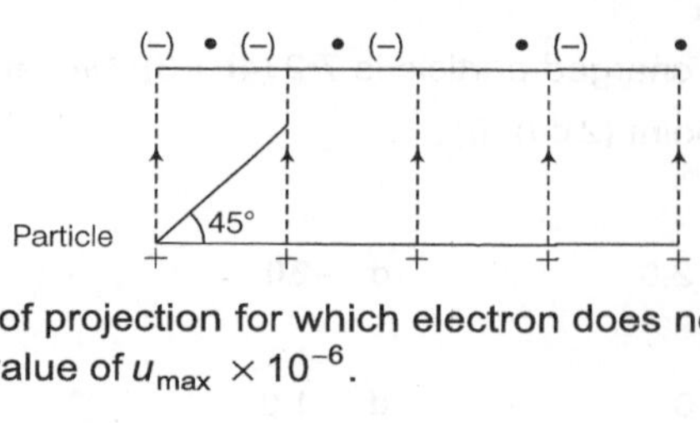

Maximum velocity of projection for which electron does not hit the upper plate is u_{max}. What is the value of $u_{max} \times 10^{-6}$.

DPP-4 Electric Flux and Gauss' Law

Subjective Questions

Working Space

Direction (Q. Nos. 1-4) *These questions are subjective in nature, need to be solved completely on notebook.*

1 Is the electric field in Gauss' law $\phi = \oint \mathbf{E} \cdot d\mathbf{A} = Q_{encl}/\varepsilon_0$ created only by charge enclosed? Can we use Gauss' law to find field of an electric dipole?

2 Consider a disc of radius 0.1 m exists inside a region of an uniform electric field with magnitude 2×10^3 N/C. Disc is located in the field such that its one of diameters is perpendicular to direction of the field.
If the disc is rotated about this diameter, sketch a graph of resulting electric flux with angle turned for one complete rotation.

3 Define gravitational flux in analogy to electric flux. How gravitational flux is different from electric flux? What would be Gauss' law for gravitational flux?

4 Let a spherical surface S encloses charge q. How electric flux changes in following cases?

i. Surface area of S is increased to twice.
ii. Charge is shifted from centre to a point near the surface.
iii. Sphere S is removed by a cubical surface.
iv. Another charge is placed just outside the surface.

Only One Option Correct Type

Direction (Q. No. 5-14) *This section contains 10 multiple choice questions. Each question has four choices (a), (b), (c) and (d), out of which ONLY ONE option is correct.*

5 Flux through an elemental area ds of the Gauss' surface is given by

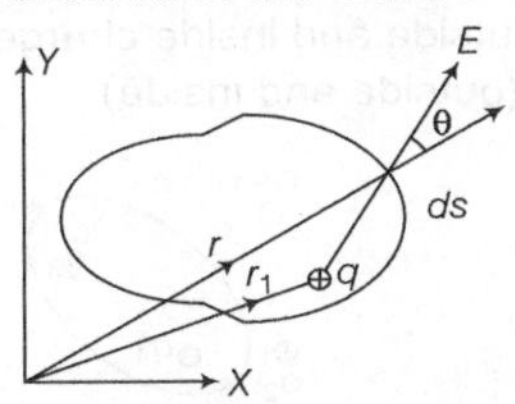

a. $\phi = \dfrac{Kq(\mathbf{r} - \mathbf{r}_1) \cdot ds}{|\mathbf{r} - \mathbf{r}_1|^3}$ b. $\phi = \dfrac{Kq \cdot ds}{|\mathbf{r} - \mathbf{r}_1|^2}$ c. $\phi = \dfrac{Kq\, \mathbf{r} \cdot ds}{|\mathbf{r}|^3}$ d. $\phi = \dfrac{Kq \cdot ds}{|\mathbf{r} - \mathbf{r}_1|^2}$

6 Flux through the rectangular piece of area 10 cm by 20 cm, when placed (as shown) in an uniform electric field of 200 N/C, is

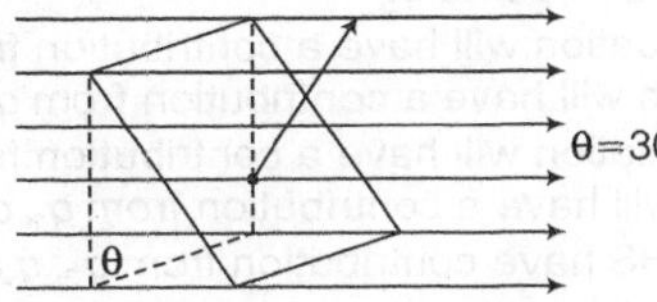

a. 4.0 Nm^2/C b. 3.5 Nm^2/C
c. 2.5 Nm^2/C d. zero

7 A uniform field **E** is parallel to axis of a hollow hemisphere of radius r.

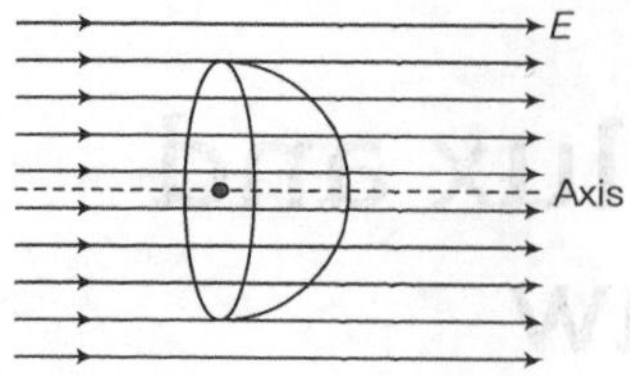

Electric flux through the hemispherical surface is

a. $4\pi r^2 E$ b. $\frac{3}{8}\pi r^3 E$

c. $\pi r^2 E$ d. $2\pi r^2 E$

8 Consider a triangular area with vertices at (2m,0,0) (0, 1m, 0) and (0, 0, 1m) is placed in region of an uniform electric field of $3 \times 10^3\ \hat{\mathbf{i}}$ N/C. Flux passing through the area is

a. 3.0×10^3 Nm²/C b. 1.5×10^3 Nm²/C

c. 6.0×10^3 Nm²/C d. 4.5×10^3 Nm²/C

9 A uniform electric field of magnitude 5.8×10^2 N/C passes through a circular disc of radius 10 cm. Ratio of electric flux through the disc when its face is perpendicular to the field lines and when its face is at 45° to the field lines, is

a. 1 b. $\frac{1}{\sqrt{2}}$ c. $\sqrt{2}$ d. 2

10 In a certain region of space, the electric field always pointing in X-direction but its magnitude decreases from $E = 560$ N/C at $x = 0$ to $E = 410$ N/C at $x = 25$ m

If a cubical box of side length $l = 25$ cm is oriented, such that its four sides are parallel to the field lines, then charge enclosed by the box is

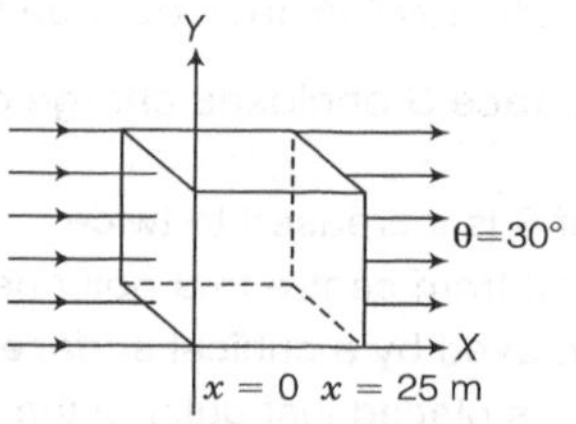

a. 8.3×10^{-7}C b. 8.3×10^5 C

c. -8.3×10^{-5} C d. -8.3×10^{-7}C

11 In the Gauss' law, $\oint \mathbf{E} \cdot d\mathbf{A} = \frac{Q_{encl}}{\varepsilon_0}$

Then,

a. electric field in the expression is created only by enclosed charge
b. electric field in the expression is created only by outside charge
c. electric field in the expression is created by both outside and inside charges
d. electric field created only by the net charge value (outside and inside)

12 Consider a system of charges given as

For the Gaussian surface, Gauss' law states,

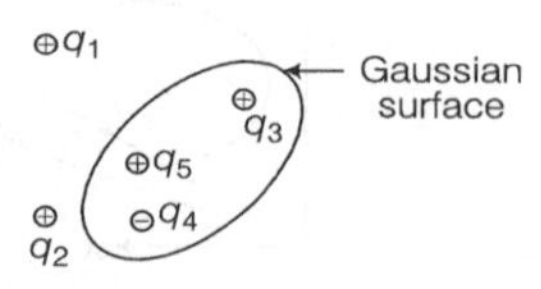

$$\oint \mathbf{E} \cdot d\mathbf{A} = \frac{q}{\varepsilon_0}$$

Which of the following statement(s) is true?

a. LHS of above equation will have a contribution of q_1 and q_2 and RHS will have a contribution from q_3, q_4 and q_5
b. LHS of above equation will have a contribution from q_1, q_2, q_3, q_4 and q_5 and RHS of above equation will have a contribution from q_3, q_4 and q_5
c. LHS of above equation will have a contribution from q_3, q_4 and q_5 and RHS of above equation will have a contribution from q_1, q_2, q_3, q_4 and q_5
d. Both RHS and LHS have contribution from q_3, q_4 and q_5

13 A point charge q is placed at centre of a hemisphere.

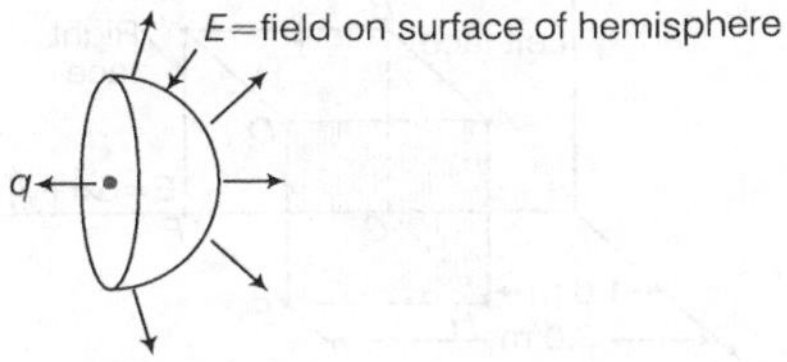

Flux linked with the curved surface is

a. $\frac{q}{\varepsilon_0}$ b. $\pi r^2 E$

c. $2\pi r^2 E$ d. zero

14 A charge q is enclosed by an imaginary Gaussian surface.

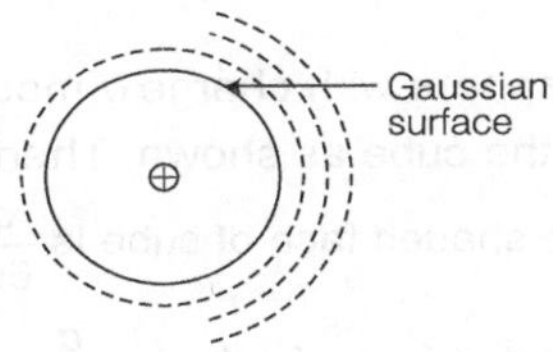

If radius of surface is increasing at a rate $\frac{dr}{dt} = K$, then

a. flux linked with surface is increasing at a rate $\frac{d\phi}{dt} = K$

b. flux linked with surface is decreasing at a rate $\frac{d\phi}{dt} = -K$

c. flux linked with surface is increasing at a rate of $\frac{d\phi}{dt} = \frac{1}{K}$

d. flux linked with surfaces is $\frac{q}{\varepsilon_0}$

One or More than One Options Correct Type

Direction (Q. Nos. 15-20) *This section contains 6 multiple choice questions. Each question has four choices (a), (b), (c) and (d), out of which ONE or MORE THAN ONE are correct.*

15 Consider a cubical surface of side 1 m and let a long wire of linear charge density $\lambda\, C/m^3$ passes through the cube. Then, correct options are

a. maximum flux through the cube is $\sqrt{2}\lambda / \varepsilon_0$

b. minimum flux through the cube is zero

c. maximum flux through the cube is $\sqrt{3}\lambda / \varepsilon_0$

d. minimum flux through the cube is λ / ε_0

16 An imaginary cube of side a is in region of uniform electric field **E**.

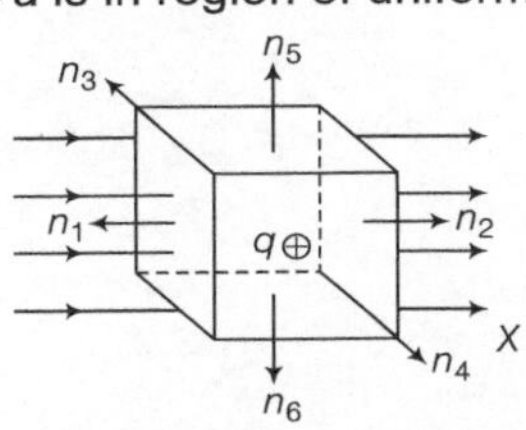

The cube is then turned by angle θ about a vertical axis. Correct options are

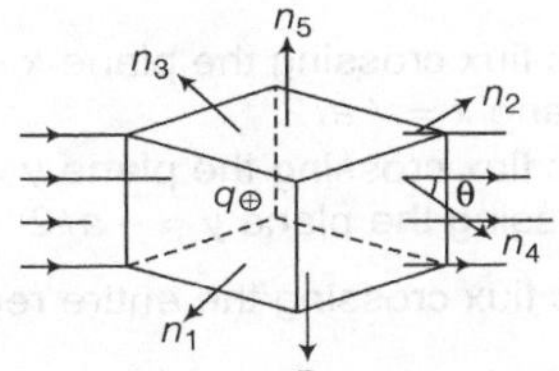

a. flux through surface 1 $\phi_{E_1} = -Ea^2 \cos\theta$

b. flux through surface 2 $\phi_{E_2} = Ea^2 \cos\theta$

c. flux through surface 3 $\phi_{E_3} = 0$

d. flux through surface 4 $\phi_{E_4} = Ea^2 \sin\theta$

Working Space

17 A non-uniform electric field $\mathbf{E} = (5x\hat{\mathbf{i}} + 3\hat{\mathbf{j}})$ N/C goes through a cube of side length 2.0 m, oriented as shown.
Then,

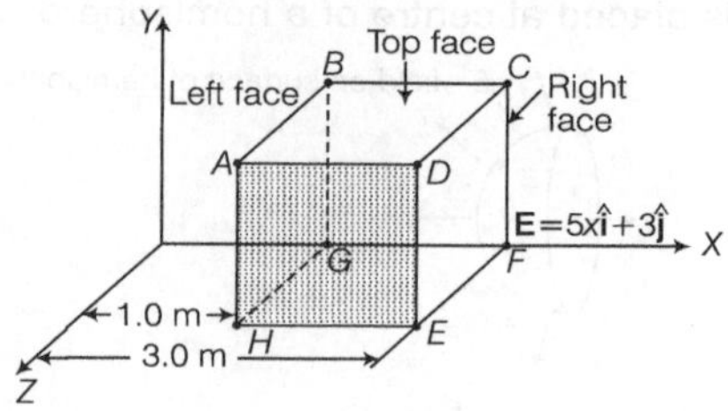

a. flux through face $CDEF = 60 \dfrac{\text{Nm}^2}{\text{C}}$

b. flux through face $ABCD = 60 \dfrac{\text{Nm}^2}{\text{C}}$

c. flux through face $ABGH = 20 \dfrac{\text{Nm}^2}{\text{C}}$

d. flux through face $EFGH = 12 \dfrac{\text{Nm}^2}{\text{C}}$

18 A small charged sphere with charge q mounted over one of the vertices of the cube as shown. Then,

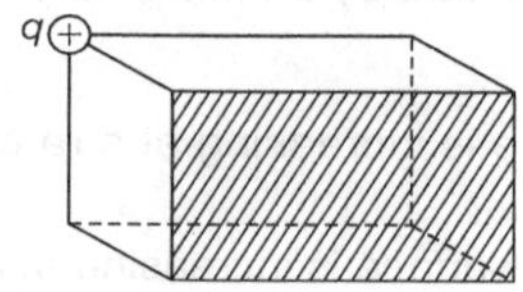

a. flux through the shaded face of cube is $\dfrac{q}{6\varepsilon_0}$

b. flux through shaded face of cube is $\dfrac{q}{24\varepsilon_0}$

c. flux through complete cube is $\dfrac{q}{8\varepsilon_0}$

d. flux through complete cube is $\dfrac{q}{\varepsilon_0}$

19 Which of the following statement (s) is /are correct? **[AIEEE 2011]**

a. If the electric field due to a point charge varies as $r^{-2.5}$ instead of r^{-2}, then the Gauss' law will still be valid

b. The Gauss' law can be used to calculate the field distribution around an electric dipole

c. If the electric field between two point charges is zero somewhere, then the sign of the two charges is same

d. The work done by the external force in moving a unit positive charge from point A at potential V_A to point B at potential V_B is $(V_B - V_A)$

20 A cubical region of side a has its centre at the origin. It encloses three fixed point charges, $-q$ at $(0, -a/4, 0)$, $+3q$ at $(0, 0, 0)$ and $-q$ at $(0, +a/4, 0)$. Choose the correct option(s). **[AIEEE 2012]**

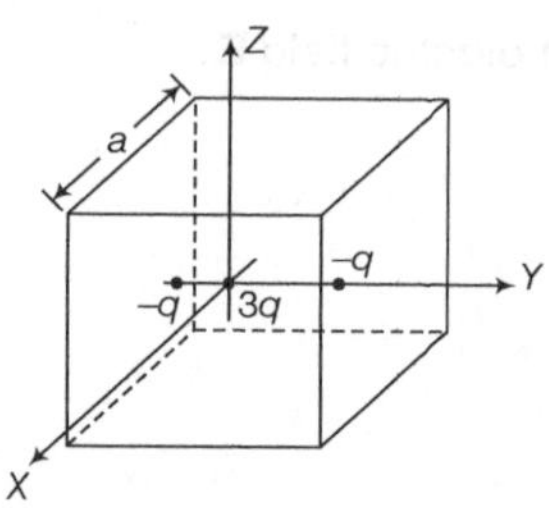

a. The net electric flux crossing the plane $x = +a/2$ is equal to the net electric flux crossing the plane $x = -a/2$

b. The net electric flux crossing the plane $y = +a/2$ is more than the net electric flux electric flux crossing the plane $y = -a/2$

c. The net electric flux crossing the entire region is $\dfrac{q}{\varepsilon_0}$

d. The net electric flux crossing the plane $z = +a/2$ is equal to the net electric flux crossing the plane $x = +a/2$

Matching List Type

Direction (Q. No. 21) Choices for the correct combination of elements from Column I and Column II are given as options (a), (b), (c) and (d), out of which one is correct.

21 Following figure shows for Gaussian surface S_1, S_2, S_3 and S_4.

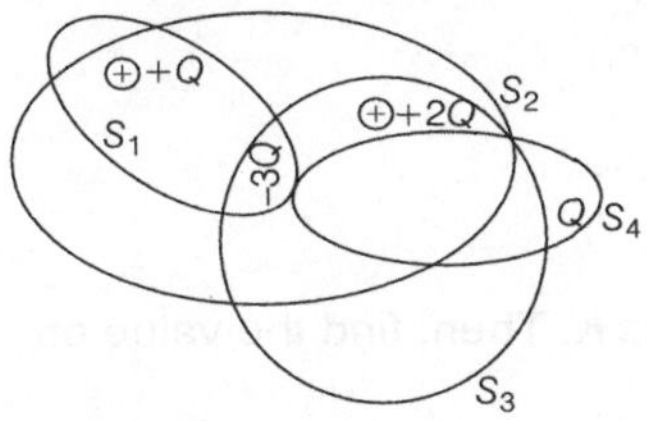

Now, match the following columns.

	Column I		Column II
i.	ϕ_{S_1}	p.	$+Q/\varepsilon_0$
ii.	ϕ_{S_2}	q.	0
iii.	ϕ_{S_3}	r.	$-Q/\varepsilon_0$
iv.	ϕ_{S_4}	s.	$-2Q/\varepsilon_0$

Codes

	i	ii	iii	iv		i	ii	iii	iv
a.	s	q	r	p	b.	p	q	r	s
c.	s	r	q	p	d.	p	r	q	s

One Integer Value Correct Type

Direction (Q. Nos. 22-27) This section contains 6 questions. When worked out will result in an integer from 0 to 9 (both inclusive).

22 A linear charge having linear charge density λ, penetrates a cube diagonally and then it penetrates a sphere diametrically as

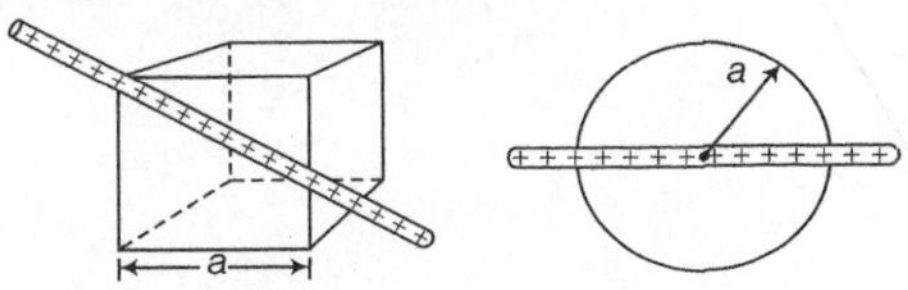

If ratio of flux through cube and through sphere is r then value of $\frac{2}{\sqrt{3}}r$ is

23 Consider a uniform charge distribution with charge density of 2 C/m^3 throughout in space. If a Gaussian sphere has a variable radius which changes at the rate of 2 m/s, then value of rate of change of flux is proportional to r^k, (r = radius of sphere). Then, find the value of k.

24 A point charge q is located inside a cube but not at its centre.

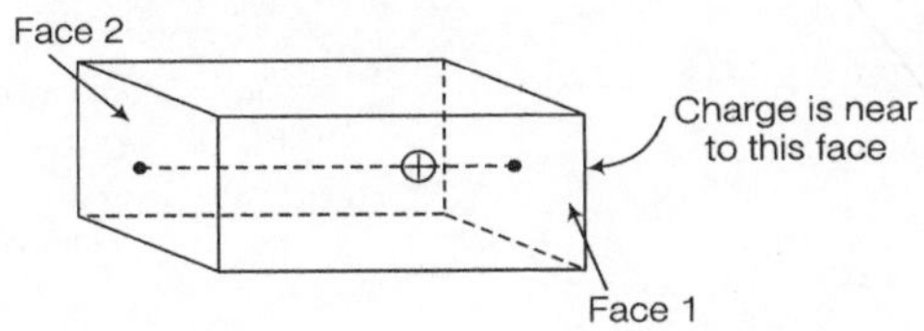

If flux through face 1 is more than $\frac{q/\varepsilon_0}{k}$ and flux through face 2 is less than $\frac{q/\varepsilon_0}{k}$.

Then, find the volue of k.

Working Space

25 S_1 and S_2 are two hollow concentric spheres with charges q and $2q$. Space between S_1 and S_2 is filled with a dielectric of dielectric constants.

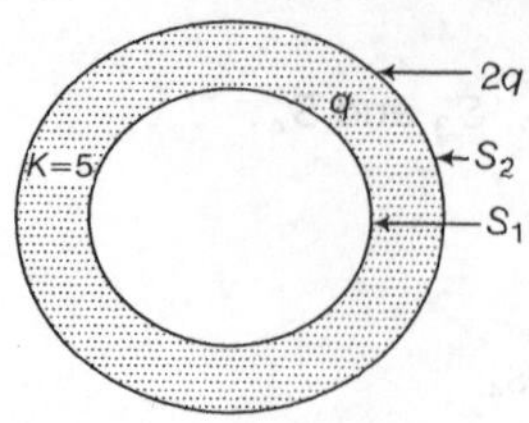

The ratio of flux through S_2 and flux through S_1 is K. Then, find the value of $\frac{9K}{11}$.

26 A charge q is placed just a distance above a square sheet of size $a \times a$ as shown below.

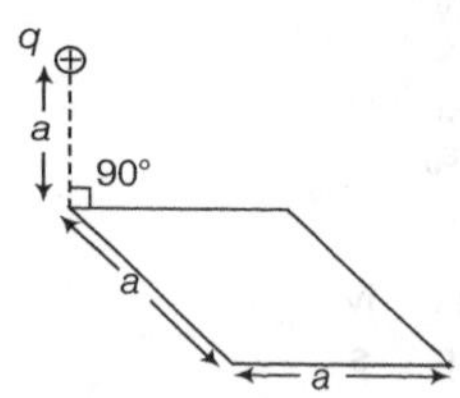

If flux through the square is $\frac{q}{K\varepsilon_0}$, find $\frac{K}{6}$.

27 Flux through the cone when a charge q is placed h distance from the base of centre cone is $\phi = \frac{2}{3}\left(\frac{q}{\varepsilon_0}\right)$.

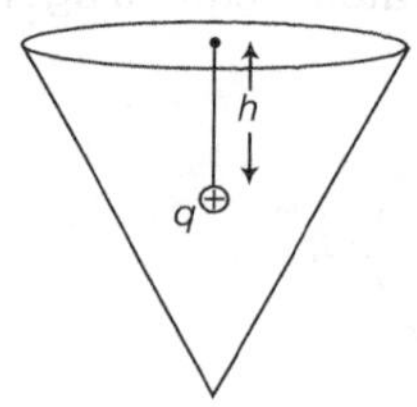

If charge is placed h distance above the base on the centre line, flux through the cone is $\frac{1}{n}\left(\frac{q}{\varepsilon_0}\right)$. Find the value of n.

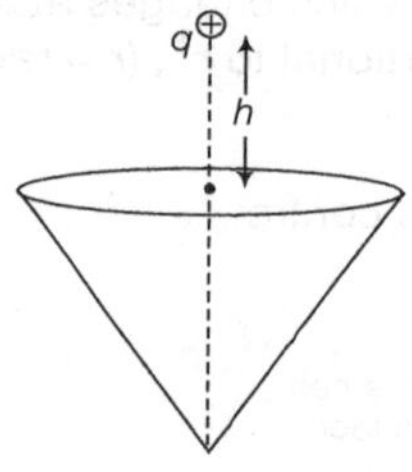

Working Space

Comprehension Type

Direction (Q. Nos. 28-31) *This section contains a paragraph, describing theory, experiments, data, etc. Four questions related to the paragraphs have been given. Each question has only one correct answer among the four given options (a), (b), (c) and (d).*

Passage

In more general cases of finding flux, where the electric field is not uniform and the surface is not flat, we divide the choosen surface into n small areas ΔA_1, ΔA_2, ΔA_3,......, ΔA_n. We choose ΔA_i such that it is small enough to be flat and electric field is constant for this area.

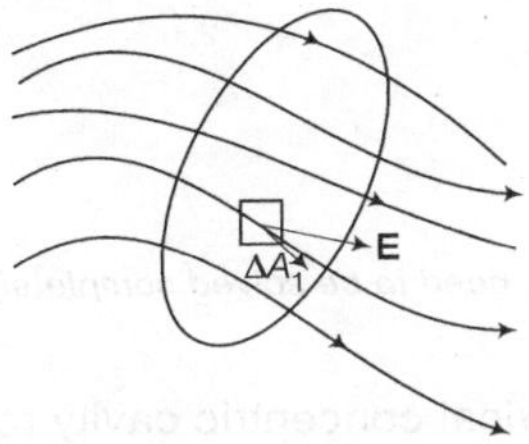

Then, electric flux through the entire surface is approximately

$$\phi_E = \sum_{i=1}^{n} \mathbf{E}_i \cdot \Delta \mathbf{A}_i = \oint \mathbf{E} \cdot d\mathbf{A}$$

28 Flux associated with the cap of a sphere having semi-vertical (semi-vertex) angle α and let charge Q is placed at centre, is

a. $\dfrac{Q}{2\varepsilon_0}$ b. $\dfrac{Q}{\varepsilon_0}$

c. $\dfrac{Q}{\varepsilon_0}(1-\cos\alpha)$ d. $\dfrac{Q}{2\varepsilon_0}(1-\cos\alpha)$

29 Flux associated with a disc subtending semi-vertex (semi-vertical) angle α at a charge Q is

a. $\dfrac{Q}{2\varepsilon_0}(1-\cos\alpha)$ b. $\dfrac{Q}{\varepsilon_0}(1-\cos 2\alpha)$

c. $\dfrac{Q}{\varepsilon_0}\sin^2\left(\dfrac{\alpha}{2}\right)$ d. $\dfrac{Q}{\varepsilon_0}.\cos^2\left(\dfrac{\alpha}{2}\right)$

30 Two point charges q and $-q$ are placed 2λ distance apart. Consider a circle of radius R placed perpendicular to line joining charges, at its mid-point. Electric flux through this circle is

a. $\dfrac{2\sqrt{2}q}{\varepsilon_0}(\sqrt{2}-1)$ b. $\dfrac{q}{\varepsilon_0}(1-\cos 60^\circ)$

c. $\dfrac{q}{\varepsilon_0}\left(\dfrac{\sqrt{2}-1}{\sqrt{2}+1}\right)$ d. $\dfrac{q}{\varepsilon_0}\left(\dfrac{\sqrt{2}-1}{\sqrt{2}}\right)$

31 Consider a very long plastic wire with charge density λ. Flux passing through a circle of radius R having centre at one of ends of wire is

a. $\dfrac{2\pi R\lambda}{\varepsilon_0}$ b. $\dfrac{\lambda R}{\varepsilon_0}$

c. zero d. $\dfrac{4\pi R\lambda}{\varepsilon}$

DPP-5 Applications of Gauss' Law

Subjective Questions

Working Space

Direction *(Q. Nos. 1-4) These questions are subjective in nature, need to be solved completely on notebook.*

1 Let a non-conducting sphere of radius r_0 has spherical concentric cavity radius r_1. Assuming, charge Q is distributed uniformly in the shell (between $r = r_1$ and $r = r_0$). Determine electric field as a function of r for

i. $0 < r < r_1$

ii. $r_1 < r < r_0$

iii. $r > r_0$

If we consider a charge q at centre of cavity, what changes occur in answers (i), (ii) and (iii)?

2 A sphere of radius r_0 carries a volume charge density ρ. A spherical cavity of radius $\frac{r_0}{2}$ is removed out. What is the magnitude and direction of field?

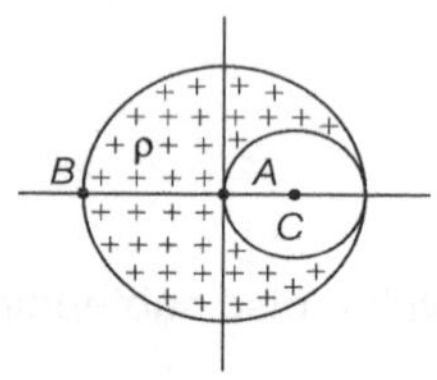

i. At A

ii. At B

3 A cube of side l has one corner at the origin of coordinates and extends along the positive X, Y and Z-axes. If electric field in the region is given by $\mathbf{E} = (ay + b)\hat{\mathbf{j}}$. Determine the charge inside the cube.

4 A spherical charged rubber balloon carries a total charge q on its surface. At $t = 0$, the non-conducting balloon has radius r_0. After that balloon is inflated such that radius increases to $2\,r_0$ in time T second.

Determine the electric field as a function of time.

i. Just outside the balloon.

ii. At $r = 3.3\,r_0$.

Only One Option Correct Type

Direction *(Q. Nos. 5-16) This section contains 12 multiple choice questions. Each question has four choices (a), (b), (c) and (d), out of which ONLY ONE option is correct.*

5 A charged metallic ball is lowered into an insulated metal can.

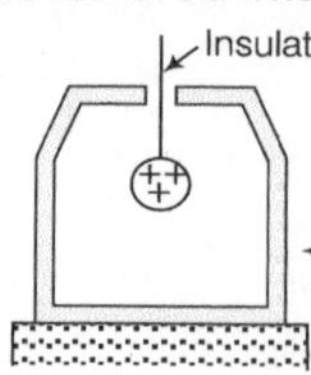

Working Space

The ball is made to touch bottom of the can, then it is placed on the disc of electroscope shown below.

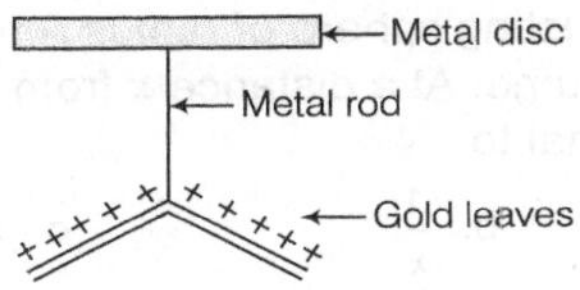

Final observation must be

a. leaves of electroscope diverges
b. leaves of electroscope converges
c. leaves of electroscope remains unaffected
d. leaves of electroscope oscillates

6 Two thin spherical shells of radii r_1 and r_2 contain uniform charge densities of σ_1 and σ_2 respectively. Under which of the given condition, $E = 0$ for $r_1 < r < r_2$?

a. $\sigma_1 > 0$
b. $\sigma_2 < 0$
c. If a charge of $-4\pi\sigma_1 r_1^2$ is placed at centre of the shells
d. If a charge of $-4\pi\sigma_2 r_2^2$ is placed at centre of the shells

7 A point charge Q is placed above a spherical surface as shown. Then,

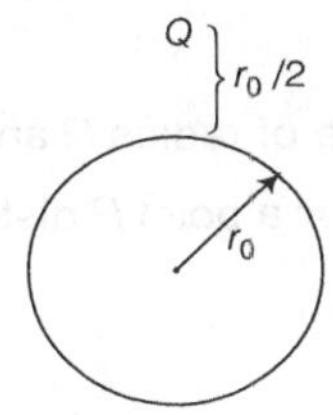

a. E at surface $= 0$

b. $\frac{1}{4\pi\varepsilon_0}\frac{Q}{r^2} \leq E_{\text{surface}} \leq \frac{1}{25\pi\varepsilon_0}\frac{Q}{r^2}$

c. $\frac{1}{\pi\varepsilon_0}\frac{Q}{r^2} \leq E_{\text{surface}} \leq \frac{1}{25\pi\varepsilon_0}\cdot\frac{Q}{r^2}$

d. $0 \leq E_{\text{surface}} \leq \frac{1}{4\pi\varepsilon_0}\frac{Q}{r^2}$

8 Two concentric conducting thin spherical shells A and B having radii r_A and r_B $(r_B > r_A)$ and charges Q_A and Q_B, $(|Q_B| > |Q_A|)$. Variation of electric field along a line through the centre of shells is

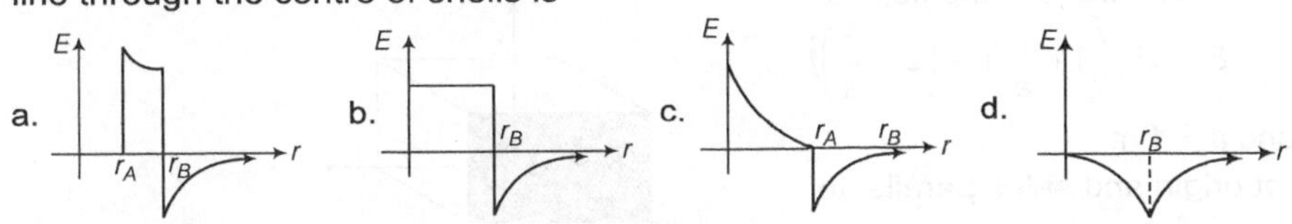

9 Two charged plates with charges Q and $3Q$ are placed d distance apart. Another plate with charge Q is placed between two plates.

Final charges q_3 and q_5 on the surface of plates are

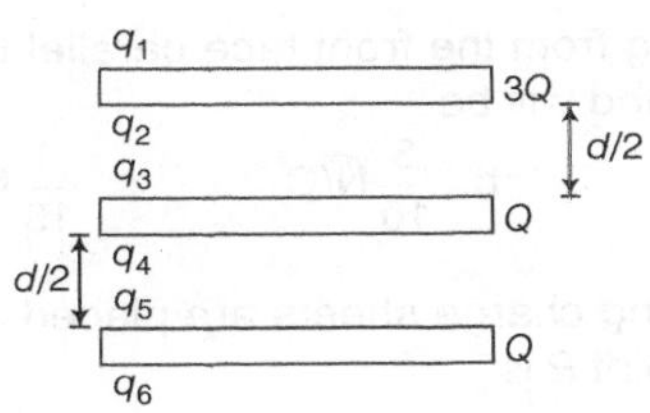

a. $\frac{-Q}{2}, \frac{-3Q}{2}$
b. $\frac{-Q}{2}, \frac{-5Q}{2}$
c. $\frac{-3Q}{2}, \frac{-5Q}{2}$
d. $\frac{-3Q}{2}, \frac{-3Q}{2}$

10 For a uniformly charged conducting solid cylinder of infinite length, which of these graphs shows the variation of **E** and r (distance from the axis).

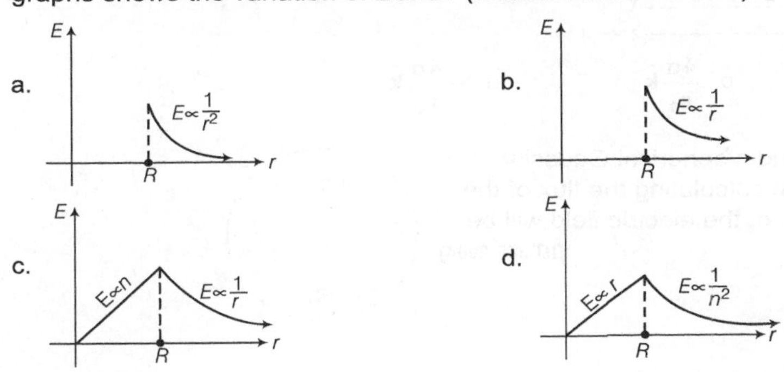

Working Space

11 A solid non-conducting sphere of radius R has a uniform distribution of charge throughout its volume. At a distance x from its centre, for $x < R$, the electric field is directly proportional to

a. $\frac{1}{x^2}$ b. $\frac{1}{x}$ c. x d. x^2

12 Consider a spherical symmetric charge distribution with charge density varying as

$$\rho = \begin{cases} \rho_0\left(1-\frac{r}{R}\right), & r \le R \\ 0, & r > R \end{cases}$$

The electric field at r, $r < R$ will be

a. $\frac{\rho_0}{4\varepsilon_0}\left(\frac{1}{3}-\frac{r}{4R}\right)$ b. $\frac{\rho_0 r}{4\pi\varepsilon_0}\left(1-\frac{r^2}{4R}\right)$

c. $\frac{\rho r}{4\varepsilon_0}\left(1-\frac{r}{R}\right)$ d. $\frac{\rho_0 r}{4\varepsilon_0}\left(\frac{1}{3}-\frac{r}{4R}\right)$

13 For a solid sphere of radius R and total charge Q, let the charge distribution be $\rho = \frac{Qr}{\pi R^4}$. Then, for a point P distant r_1 from centre of sphere ($r_1 < R$) field is

a. 0 b. $Q/4\pi\varepsilon_0 r_1^2$

c. $\frac{Q r_1^2}{4\pi\varepsilon_0 R^4}$ d. $\frac{Q r_1^2}{3\pi\varepsilon_0 R^4}$

14 In a cubical volume of side 1m, the electric field is

$$E = E_0\left(1+\frac{z}{a}\right)\hat{\mathbf{i}} + \left(E_0\,\frac{z}{a}\right)\hat{\mathbf{j}}$$

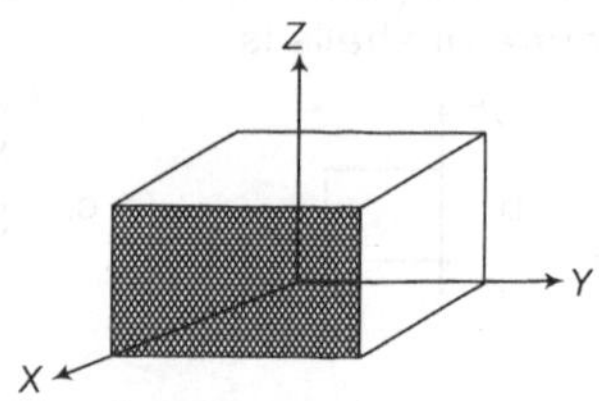

where, $E_0 = 0.2$ N/C and $a = 1$ m.

Cube has one vertex at origin and sides parallel to coordinates axes.

The flux eminating from the front face parallel to *YOZY*-plane shaded will be

a. $\frac{3}{5}$ N/C b. $\frac{3}{10}$ N/C c. $\frac{3}{15}$ N/C d. 0

15 Three infinitely long charge sheets are placed as shown in figure below. The electric field at point P is **[IIT JEE 2005]**

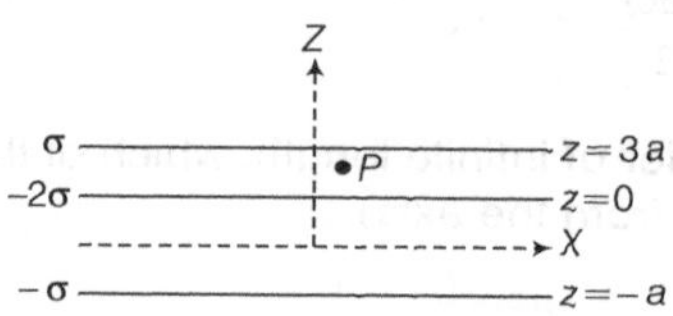

a. $\frac{2\sigma}{\varepsilon_0}\hat{\mathbf{k}}$ b. $-\frac{2\sigma}{\varepsilon_0}\hat{\mathbf{k}}$ c. $\frac{4\sigma}{\varepsilon_0}\hat{\mathbf{k}}$ d. $-\frac{4\sigma}{\varepsilon_0}\hat{\mathbf{k}}$

16 Consider the charge configuration and a spherical Gaussian surface as shown in the figure. When calculating the flux of the electric field over the spherical surface, the electric field will be due to **[IIT JEE 2004]**

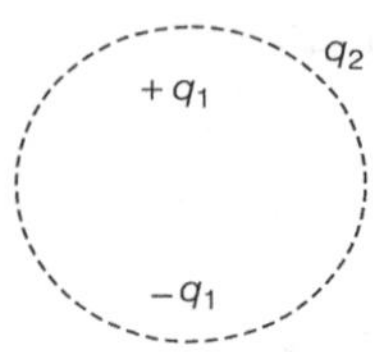

a. q_2
b. only the positive charges
c. all the charges
d. $+q_1$ and $-q_1$

Working Space

One or More than One Options Correct Type

Direction (Q. Nos. 17-21) This section contains 5 multiple choice questions. Each question has four choices (a), (b), (c) and (d), out of which ONE or MORE THAN ONE are correct.

17 A uniformly charged non-conducting sphere of radius r_0 and charge density ρ C/m^3, has

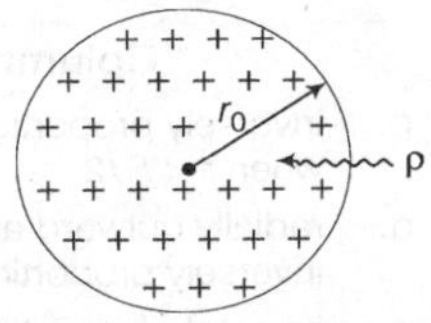

a. a radially outward pointing field of magnitude $E = \dfrac{\rho r_0^3}{3\varepsilon_0 r^2}$, where $r > r_0$

b. a field of magnitude $E = \dfrac{\rho r_0}{3\varepsilon_0}$ at surface

c. a field of magnitude $E = \dfrac{33}{8}\dfrac{\rho R^3}{\varepsilon_0}$ at a distance $r = \dfrac{R}{2}$ from centre

d. no field inside its surface

18 A solid conductor has a cavity. Conductor carries a charge $+Q$ and a point charge $-q$ lies in the centre of cavity. Then,

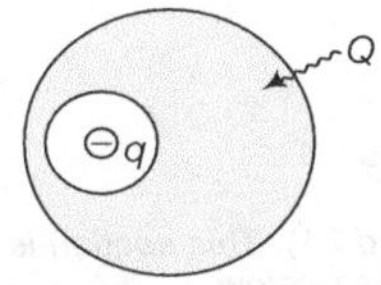

a. charge $(Q - q)$ present on the outer surface of conductor

b. charge $-q$ is present on inner surface of conductor

c. net field inside the conductor is zero

d. net field inside the cavity is zero

19 Three large, thin and charged sheets are placed parallel to each other. Surface I has a charge density of 6.5n C/m^2, surface II has a charge density of -2.0nC/m^2 and surface III has a charge density of 5.0 nC/m^2. Then,

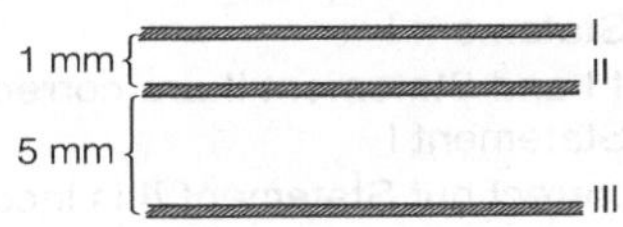

a. the net field between I and II is zero

b. the net field between II and III is zero

c. sheet I experiences a force of 1.1×10^{-6} N/m^2

d. sheet II experiences a force of 1.7×10^{-7} N/m^2

20 An uncharged solid metallic sphere of radius r_0 contains two spherical cavities of radii r_1 and r_2, where $r_0 > r_1 + r_2$.
If a point charge Q_1 is placed in cavity of radius r_1 and another point charge Q_2 is placed in cavity of radius r_2. Then,

a. Net field outside the sphere at distance $r > r_0$ is zero

b. If $Q_1 + Q_2 < 0$, then field outside the conductor points is radially outward direction

c. Net field in region $r < r_0$ is zero

d. Field inside any of the cavity is $\propto \dfrac{1}{r^2}$

21 A spherical metal shell A of radius R_A and a solid metal sphere B of radius R_B $(< R_A)$ are kept far apart and each is given charge $+Q$. Now they are connected by a thin metal wire. Then, **[AIEEE 2011]**

a. $E_A^{\text{inside}} = 0$

b. $Q_A > Q_B$

c. $\dfrac{\sigma_A}{\sigma_B} = \dfrac{R_B}{R_A}$

d. $E_A^{\text{on surface}} < E_B^{\text{on surface}}$

Working Space

Matching List Type

Direction *(Q. No. 22) Choices for the correct combination of elements from Column I and Column II are given as options (a), (b), (c) and (d), out of which one is correct.*

22 Match the information given in Column I with that in Column II and mark the correct code given below.

	Column I		Column II
i.	For an infinitely long line charge, electric field at a distance r from the line is	p.	inversely proportional to r^2 when $r < R/2$.
ii.	For a uniformly charged non-conducting sphere electric field at a distance r from centre is (R = radius of sphere)	q.	radially outward and inversely proportional to r.
iii.	A spherical charged conductor of radius R, with a concentric cavity of radius $R/2$, electric field at a distance r from the centre is	r.	inversely proportional to r^2 when $r > R$.
iv.	A spherical charged conductor of radius R with a cavity of radius $R/2$ has a point charge q ($q > 0$) at its centre. Electric field at a distance r from the centre is	s.	increases with r, reaches a maximum, then decreases.

Codes

	i	ii	iii	iv			i	ii	iii	iv
a.	q	s	r	p		b.	s	q	r	p
c.	q	s	p	r		d.	q	r	s	p

Statement Type

Direction *(Q. Nos. 23 and 24) This section is based on Statement I and Statement II. Select the correct answer from the codes given below.*

a. Both Statement I and Statement II are correct and Statement II is the correct explanation of Statement I
b. Both Statement I and Statement II are correct and Statement II is not the correct explanation of Statement I
c. Statement I is correct but Statement II is incorrect
d. Statement II is correct but Statement I is incorrect

23 **Statement I** Field of a infinite length charged wire can be found using Gauss' law.

Statement II As wire of a cylindrical shape, Gauss' surface chosen will also be cylindrical.

24 **Statement I** To find field due to charge distribution by Gauss' law, Gaussian surface is drawn symmetrical to the charge distribution.

Statement II If Gaussian surface is symmetrical to the charge distribution, then flux integral is easier to evaluate.

One Integer Value Correct Type

Direction *(Q. Nos. 25-27) This section contains 3 questions. When worked out will result in an integer from 0 to 9 (both inclusive).*

25 A solid sphere of radius R has a charge distribution $\rho = kr^a$ with a total charge of Q coulombs. k and a are constants and r is distance from centre.

If electric field at $r = \dfrac{R}{2}$ is $\dfrac{1}{8}$ times of that at $r = R$, then the value of a is

26 In given situation, radius of disc is $\sqrt{3}b$ and distance of point charge from the disc is b.

Ratio of electric flux not going through the disc and electric flux of charge through the disc is k. Then, find value of k which will be an integer from 1 to 10.

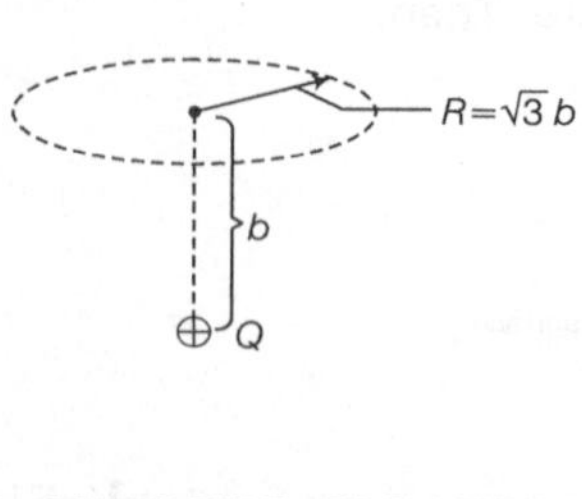

Working Space

27 A point charge of magnitude 1.77×10^{-10} C is placed on the mid-point of axes of a cylinder. Diameter of cylinder is equal to the length of cylinder. If Q is the total flux through curved surface of cylinder then, find the value of $\sqrt{2}\,\phi$.

Comprehension Type

Direction (Q. Nos. 28-34) This section contains 2 paragraphs, each describing theory, experiments, data, etc. Seven questions related to the paragraphs have been given. Each question has only one correct answer among the four given options (a), (b), (c) and (d).

Passage I

A system consists of a ball of radius R carrying a spherically symmetrical charge and the surrounding space is filled with a charge of volume density $\rho = \frac{\alpha}{r}$, where α = constant and r is the distance from the centre of the ball.

28 Electric field around the ball is

a. uniform in nature
b. of increasing magnitude
c. of decreasing magnitude
d. first increases then decreases

29 Magnitude of the electric field as a function of r is

a. $E = \frac{\rho}{\varepsilon_0}$
b. $E = \frac{1}{4\pi\varepsilon_0} \cdot \frac{\left(\frac{4}{3}\pi r^3\right) \cdot \rho}{R^2}$
c. $E = \left(\frac{q}{4\pi\varepsilon_0} - \frac{\alpha R^2}{2\varepsilon_0}\right) \cdot \frac{1}{r^2} + \frac{\alpha}{2\varepsilon_0}$
d. $E = \frac{\alpha R^2}{2\varepsilon_0 r^2} - \frac{\alpha}{2\varepsilon_0}$

30 Electric field in the region outside the ball is independent of distance r when charge on ball is

a. $q = 4\pi\alpha R^2$
b. $q = \frac{4}{3}\alpha R^2$
c. $q = \frac{4}{3}\pi\alpha R^3$
d. $q = 2\pi\alpha R^2$

31 If electric field outside the ball is independent of r, then magnitude of field is

a. $\frac{\alpha}{\varepsilon_0}$
b. $\frac{\alpha}{4\varepsilon_0}$
c. $\frac{2\alpha}{\varepsilon_0}$
d. $\frac{\alpha}{2\varepsilon_0}$

Passage II

Consider a volume charge with density $\rho = \rho_0 e^{-\alpha r^3}$, where ρ_0 and α are constants and r is distance from the centre of the charge density.

32 Electric field strength as a function of r is

a. $\frac{4\rho_0}{3 \times r^2 \varepsilon_0}(1 - e^{-\alpha r^3})$
b. $\frac{\rho_0}{3\varepsilon_0 \times r^2}(e^{-\alpha r^3})$
c. $\frac{\rho_0}{3\varepsilon_0 \times r^2}(1 - e^{-\alpha r^3})$
d. $\frac{\rho_0}{\varepsilon_0 \times r^2} .2e^{-\alpha r^3}$

33 If $\alpha\varepsilon^3 << 1$, then field at a distance r from centre is

a. $\frac{\rho_0 r}{3\varepsilon_0}$
b. $\frac{4\rho_0 \varepsilon}{3\varepsilon_0}$
c. 0
d. $\frac{\rho_0 e^{-\alpha r^3}}{\varepsilon_0}$

34 If $\alpha r^3 >> 1$, then field at a distance r from the centre is

a. $\frac{\rho_0}{3\varepsilon_0 \alpha r^2}$
b. $\frac{\rho_0}{\varepsilon_0 \alpha r^2}$
c. $\frac{\rho e^{-\alpha r^3}}{3\varepsilon_0 \alpha r^2}$
d. $\frac{4\rho e^{-\alpha r^3}}{3\varepsilon_0 \alpha r^2}$

DPP-6 Electric Dipole, Dipole in an External Field

Subjective Questions

Working Space

Direction (Q. Nos. 1-3) *These questions are subjective in nature, need to be solved completely on notebook.*

1 i. In an hypothetical dielectric bar of 5.0 cm length and area of cross-section 1.0 cm^2, all dipole moments are aligned. If dipole moment of one atom is 2.1×10^{-23} cm then, find the net dipole moment associated with dielectric bar. (Given, density of dielectric is 5 g/cm^3, molar mass of dielectric is 3.01g/mol and Avogadro's number is 6.02×10^{23} atoms/mol.)

ii. Do all molecules are polar in nature? What is meant by a polar molecule.

2 Two small identical electrical dipoles *AB* and *CD*, each of dipole moment **p** are kept at an angle of 120° as shown

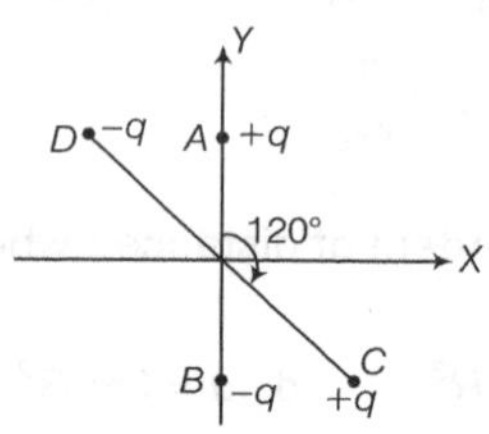

i. What is resultant dipole moment of this system?

ii. If this system is subjected to electric field **E** directed along positive *X*-direction, what will be the magnitude and direction of the torque acting on it?

3 A linear electric quadrapole consists of following charge array.

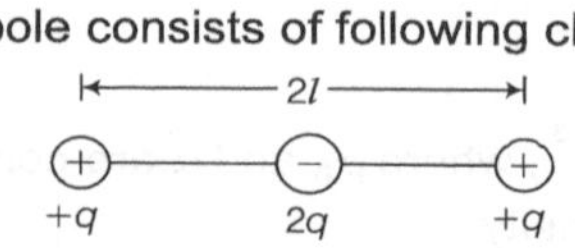

Find an expression of field at a point on axis of quadrapole.

Only One Option Correct Type

Direction (Q. Nos. 4-13) *This section contains 10 multiple choice questions. Each question has four choices (a), (b), (c) and (d), out of which ONLY ONE option is correct.*

4 A given point charge at a certain distance from a dipole on its axis experiences a force *F*. If distance of charge from dipole is increased to double, the force acting on the charge will be

a. $\frac{F}{2}$ b. $\frac{F}{4}$ c. $\frac{F}{8}$ d. $\frac{F}{16}$

5 An electric dipole is in a non-uniform field. When released from rest

a. it shows both rotation and translation
b. it shows only rotation
c. it shows only translation
d. it neither shows rotation nor translation

Working Space

6 An electric dipole is placed along the X-axis at the origin O. A point P is at a distance of 20 cm from the origin such that OP makes an angle $\frac{\pi}{3}$ with the X-axis. If the electric field at P makes an angle θ with the X-axis, the value of θ will be

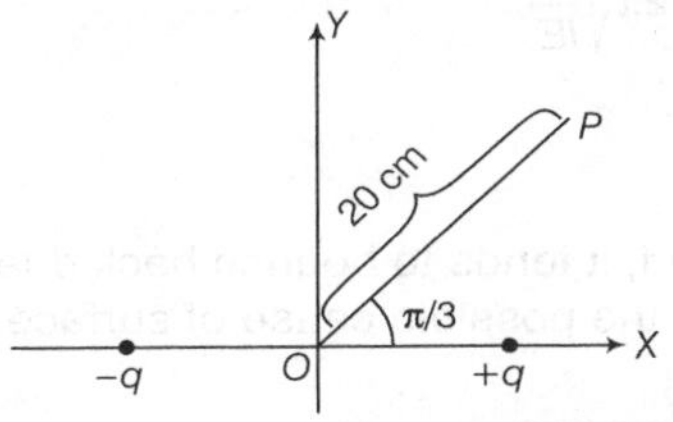

a. $\frac{\pi}{3}$

b. $\frac{\pi}{3} + \tan^{-1}\left(\frac{\sqrt{3}}{2}\right)$

c. $\frac{2\pi}{3}$

d. $\tan^{-1}\frac{\sqrt{3}}{2}$

7 Electric charges q, q and $-2q$ are placed at the corners of an equilateral triangle ABC of side l. The magnitude of electric dipole moment of the system is

a. ql b. $2ql$ c. $\sqrt{3}ql$ d. $\frac{\sqrt{3}ql}{2}$

8 Two charges $+3.2 \times 10^{-19}$C and -3.2×10^{-19}C are placed 2.4 Å apart. Dipole moment of system is

a. 15.36×10^{-29} C-m

b. 15.36×10^{-19} C-m

c. 7.68×10^{-29} C-m

d. 7.68×10^{-19} C-m

9 Consider an electric quadrapole of length $2l$.

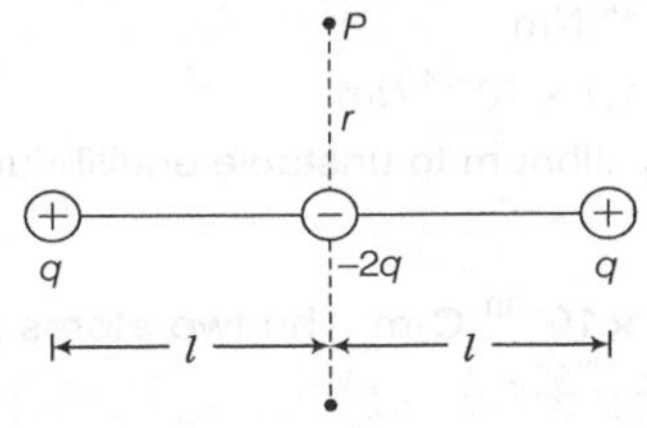

Then, for a point P on perpendicular bisector of quadrapole at a distance r from the centre.

a. Field shows $\frac{1}{r^2}$ dependence and points towards negative charge

b. Field shows $\frac{1}{r^3}$ dependence and points away from negative charge

c. Field shows $\frac{1}{r^4}$ dependence and points towards negative charge

d. Field shows $1/r^2$ dependence and points towards negative charge

10 An electric dipole consists of two charges $+q$ and $-q$ separated by a small distance $2a$. Its centre is on X-axis at $x = x_1$ and it points along the X-axis in positive X-direction. The dipole is in non-uniform field which is given by $\mathbf{E} = Cx\hat{\mathbf{i}}$, where C = a constant. Then,

a. net force on dipole is $Cp\hat{\mathbf{i}}$

b. net force on dipole is $-Cp\hat{\mathbf{i}}$

c. net force on dipole is $CEx_1\hat{\mathbf{i}}$

d. net force on dipole is $-CEx_1\hat{\mathbf{i}}$

Working Space

11 A dipole is placed in an uniform electric field of magnitude E, it is then slightly displaced by giving a small angular displacement θ. Then, time period of oscillation of the dipole will be (I = moment of inertia of dipole).

a. $T = 2\pi\sqrt{\frac{I}{PE}}$ b. $T = 2\pi\sqrt{\frac{P}{IE}}$

c. $T = 2\pi\sqrt{\frac{E}{PI}}$ d. 0

12 When free surface of water is slightly depressed, it tends to bounce back due to surface tension. Which of the following may be the possible cause of surface tension?

a. Gravitational force b. Magnetic force

c. Electrical force d. van der Waals' force

13 A dipole is in stable equilibrium in an uniform external field when

a. dipoles own field opposes the external field for points between the charges

b. dipoles own field points in the direction of external field

c. dipole moment points in the direction opposite to the external field

d. dipole moment points from negative to positive charge

One or More than One Options Correct Type

Direction (Q. Nos. 14-18) *This section contains 5 multiple choice questions. Each question has four choices (a), (b), (c) and (d), out of which ONE or MORE THAN ONE are correct.*

14 A dipole consists of charges of + e and – e separated by 0.68 nm. Dipole is placed in a region of uniform field $E = 2.2 \times 10^4$ N/C. Then

a. dipole moment of system is 1.1×10^{-28} C-m

b. torque when **p** is perpendicular to **E** is 2.4×10^{-24} Nm

c. torque when **p** makes an angle of 45° with **E** is 1.7×10^{-24} Nm

d. work required to rotate dipole from its stable equilibrium to unstable equilibrium position is 4.8×10^{-24} J

15 HCl molecule has a dipole moment of about 3.4×10^{-30} C-m. The two atoms are separated by about 1.0×10^{-10} m.
Mark the correct statement(s).

a. Net charge on each atom is quantised

b. Electrons spend more time near Cl-atom than H-atom

c. Electrons spend more time near H-atom than Cl-atom

d. Energy required to rotate one molecule 45° from its equilibrium position is 2.5×10^{-26} J

16 Suppose both charges + q are separated by distance $2l$. Then

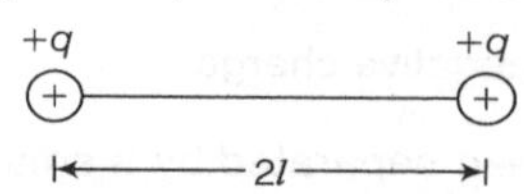

a. field decreases as $\frac{1}{r^2}$

b. field decreases as $\frac{1}{r^3}$

c. at a large distance, it appears as a charge of $2q$

d. at a large distance, it appears as a charge distribution of $\lambda = 2q/d$

Working Space

17 An electric dipole of dipole moment $\mathbf{p}$ and moment of inertia I is placed in an uniform electric field $\mathbf{E}$. It displaced by some angle θ and released, then

a. dipole soon acquires its stable equilibrium position
b. dipole undergoes SHM if θ is small, releasing oscillation energy into heat and acquires stable equilibrium position
c. time period of oscillation does not depends on mass of dipole
d. frequency of dipole is inversely proportional to $\sqrt{I}$

18 Let a dipole is placed in a non-uniform field $\mathbf{E} = E\hat{\mathbf{i}}$ that points in X-direction. If $\mathbf{E}$ only depends on x, then

a. there is no torque on dipole
b. net force on dipole is $\mathbf{F} = \left(\mathbf{p} \cdot \frac{d\mathbf{E}}{dx}\right)\hat{\mathbf{i}}$
c. dipole shows translation and rotation
d. dipole shows no translation

Statement Type

Direction (Q. Nos. 19 and 20) This section is based on Statement I and Statement II. Select the correct answer from the codes given below.

a. Both Statement I and Statement II are correct and Statement II is the correct explanation of Statement I
b. Both Statement I and Statement II are correct and Statement II is not the correct explanation of Statement I
c. Statement I is correct but Statement II is incorrect
d. Statement II is correct but Statement I is incorrect

19 **Statement I** Electric field decays at same rate for a dipole as for a point charge.

Statement II Electric field varies in inverse proportion to square of distance from a charge.

20 **Statement I** When a sample of a polar liquid (like water) is subjected to a changing electric field, it becomes warm.

Statement II An oscillating dipole radiates energy in form of electromagnetic waves.

One Integer Value Correct Type

Direction (Q. Nos. 21 and 22) This section contains 2 questions. When worked out will result in an integer from 0 to 9 (both inclusive).

21 Two opposite and equal charges of 4×10^{-8} C when placed 2×10^{-2} cm apart from a dipole. This dipole is placed in an external field of 4×10^{8} N/C. If the value of maximum torque experienced by dipole is $8k \times 10^{-4}$ Nm, then value of k is

a. 2
b. 4
c. 8
d. 10

22 Force between two short dipoles separated by a small distance is directly proportional to $1/r^n$. The value of n is

a. 1
b. 3
c. 4
d. 2

Comprehension Type

Working Space

Direction *(Q. Nos. 23-29) This section contains 2 paragraphs, each describing theory, experiments, data, etc. Seven questions related to the paragraphs have been given. Each question has only one correct answer among the four given options (a), (b), (c) and (d).*

Passage I

A small dipole with its dipole moment **p**, directed along positive *Z*-axis is located at origin.

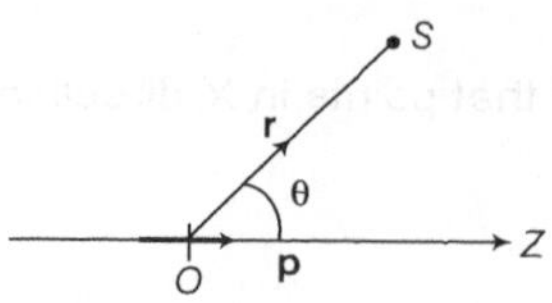

Consider a point *S* with locating vector **r**.

23 Component E_2 of field of dipole parallel to *Z*-axis at point *S* is

a. $\frac{\mathbf{p}}{4\pi\varepsilon_0 r^3}\cdot(3\cos^2\theta-1)$ b. $\frac{\mathbf{p}}{4\pi\varepsilon_0 r^3}\cdot(2\cos^2\theta-1)$

c. $\frac{\mathbf{p}}{4\pi\varepsilon_0 r^3}\cdot(\cos^2\theta-1)$ d. $\frac{\mathbf{p}}{4\pi\varepsilon_0 r^2}\cdot(2\cos^2\theta-1)$

24 Component E_1 of field of dipole perpendicular to *Z*-axis at point *S* is

a. $\frac{3p\sin\theta\cos\theta}{4\pi\varepsilon_0 r^3}$ b. $\frac{3p\sin\theta\cos\theta}{4\pi\varepsilon_0 r^2}$

c. $\frac{2p\sin\theta\cos\theta}{4\pi\varepsilon_0 r^2}$ d. $\frac{2p\sin\theta\cos\theta}{4\pi\varepsilon_0 r^3}$

25 At point *S*, net field is perpendicular to **p** when,

a. $\cos\theta=\frac{\sqrt{3}}{2}$ b. $\cos\theta=\frac{1}{2}$

c. $\cos\theta=\frac{2}{\sqrt{3}}$ d. $\cos\theta=\frac{1}{\sqrt{3}}$

26 If net field at point *S* is perpendicular to **p**, then its magnitude is

a. $\frac{\mathbf{p}}{4\pi\varepsilon_0 r^3}$ b. $\frac{2\mathbf{p}}{4\pi\varepsilon_0 r^3}$ c. $\frac{\sqrt{2}\mathbf{p}}{4\pi\varepsilon_0 r^3}$ d. $\frac{\sqrt{3}\mathbf{p}}{4\pi\varepsilon_0 r^3}$

Passage II

A dipole with an electric dipole moment **p** is located at a distance *r* from a wire with a uniform linear charge density λ.

27 Force on the dipole when **p** is parallel to wire is

a. $\frac{p\lambda}{2\pi\varepsilon_0 r^2}$ b. $\frac{p\lambda}{4\pi\varepsilon_0 r^3}$ c. 0 d. $\frac{\lambda p^2}{4\pi\varepsilon_0 r^3}$

28 Force on the dipole when **p** is along the radius vector *r* is

a. $\frac{p\lambda}{2\pi\varepsilon_0 r^2}$ b. $\frac{p\lambda}{4\pi\varepsilon_0 r^3}$ c. 0 d. $\frac{\lambda p^2}{4\pi\varepsilon_0 r^3}$

29 Force on the dipole when **p** is at right angles to the wire and also at right angles to the radius vector is

a. $\frac{p\lambda}{2\pi\varepsilon_0 r^2}$ b. $\frac{p\lambda}{4\pi\varepsilon_0 r^3}$ c. 0 d. $\frac{\lambda p^2}{4\pi\varepsilon_0 r^3}$

DPP-7 Electric Potential and Potential Energy

Subjective Questions

Working Space

Direction (Q. Nos. 1-4) *These questions are subjective in nature, need to be solved completely on notebook.*

1 Two equal and dissimilar charges $q_1 = +5\,\mu C$ and $q_2 = -5\,\mu C$ are located across ends of a diagonal of a rectangle of sides 10 cm and 20 cm as shown below.

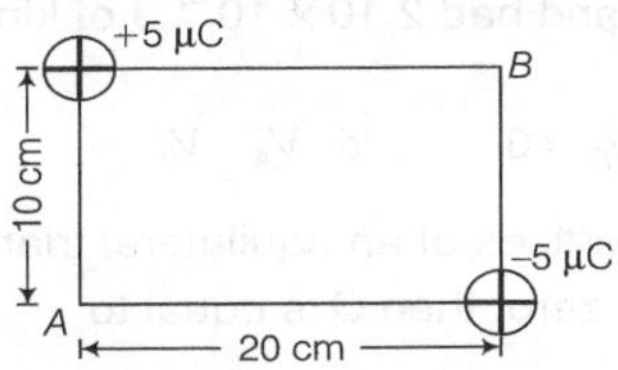

i. What are the potential at points A and B?

ii. How much work is required to move a third charge $q_3 = +2\,\mu C$ from A to B.

iii. Does the work done in moving q_3 from A to B along diagonal is less than or equal to work done in moving q_3 from A to B on a path which lies completely outside the rectangle?

iv. Does the work done in moving q_3 from A to B increases or decreases the elctrical potential energy of three charge system?

2 i. A test charge moves in the region of a uniform electric field. No external force is present. Does the charge always move from a higher potential point to a lower potential point? Explain your answer.

ii. An electron is accelerated by applying a potential difference of 5000 V. Find change in kinetic energy and potential energy of electron. Also, find final speed of electron.

3 i. A charge of $2\,\mu C$ is situated at point (1 cm, 1 cm) near another charge of $4\,\mu C$ at origin. Find work done in moving $2\,\mu C$ charge to (5 cm, 5 cm).

ii. Show that work done in moving $2\,\mu C$ charge from A (1 cm, 1cm) point to B (5 cm, 5 cm) point *via* point C (4 cm, 3 cm,) is same as taking $2\,\mu C$ charge directly from A to B.

4 i. Does the electric potential energy of two unlike charges positive or negative?

ii. Does the electric potential energy of two like charges positive or negative?

iii. What is the significance of sign convention of potential energy in each case.

iv. Give formula for potential energy of two charge system separated by a small distance when potential at infinity is taken as zero.

v. Give the formula for potential energy of two charge systems when potential is zero at some arbitrary chosen point at a finite distance.

Only One Option Correct Type

Working Space

Direction (Q. Nos. 5 -16). *This section contains 12 multiple choice questions. Each question has four choices (a), (b), (c) and (d), out of which ONLY ONE option is correct.*

5 A uniform electric field exists between two opposite charged parallel plates. An electron is released from rest from the surface of the negatively charged plate. The electron strikes the surface of the opposite plate 2.0×10^{-2} m from the first, in 1.5×10^{-8}s. Then, potential difference between plates is

a. 20.2 V b. 2.02 V c. 40.4 V d. 4.04 V

6 A metallic sphere of radius r_1 carries a charge Q, another uncharged metallic sphere of radius r_2 is connected to it using a long wire of negligible resistance. Charge transferred from first sphere to second sphere is

a. zero b. $Q\left(\frac{r_1}{r_1 + r_2}\right)$ c. $Q\left(\frac{r_2}{r_1 + r_2}\right)$ d. $Q\left(\frac{r_1 + r_2}{r_1 r_2}\right)$

7 Work done by an external force to move a negative charge from point A to point B is 7.00×10^{-4} J. If charge was started from rest and had 2.10×10^{-4} J of kinetic energy when it reaches B, then

a. $V_A < V_B$ b. $V_A = V_B$ c. $V_A + V_B = 0$ d. $V_A > V_B$

8 Three charges Q, $+q$ and $+q$ are placed at the vertices of an equilateral triangle of side l. If net electrostatic energy of the system is zero, then Q is equal to

a. $\frac{-q}{2}$ b. $-q$ c. $+q$ d. zero

9 Consider a uniform electric field in a region as shown below.

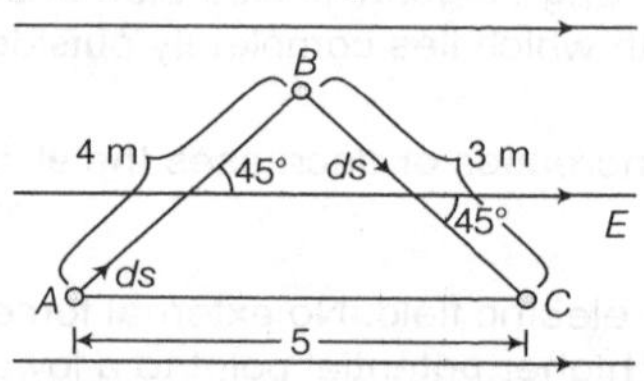

Then, potential difference between points A and C is given by

a. $V_A - V_C = -\int_{AB} Eds \cos 45° - \int_{BC} Eds \cos 45°$

b. $V_A - V_C - V_C = -\int_{AB} Eds \cos 45° - \int_{BC} Eds \cos 45°$

c. $V_A - V_C = 5E$

d. $V_C - V_A = 5E$

10 Consider four charges placed as shown below.

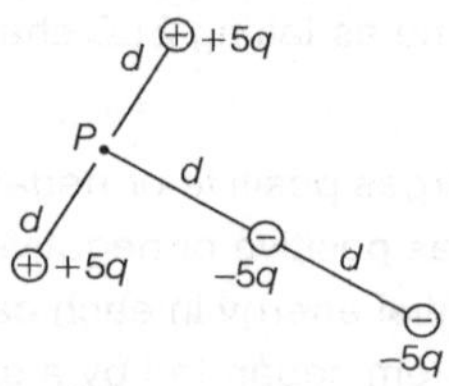

Net electric potential at P is

a. zero b. $\frac{5q}{4\pi\varepsilon_0 d}$ c. $\frac{2.5q}{4\pi\varepsilon_0 d^2}$ d. $\frac{2.5q}{4\pi\varepsilon_0 d}$

11 A spherical charged drop of water carrying a charge of 30 pC has a potential of 400 V at its surface ($V = 0$ at infinity). If eight such drops combined to form a single drop, then potential on the surface of single drop will be

a. 3200 V b. 50 V c. 1600 V d. 1500 V

Working Space

12 Three charges q_1, q_2 and q_3 are fixed on the vertices of an equilateral triangle of side d. If $q_1 = +q$, $q_2 = -4q$ and $q_3 = +2q$, then potential energy of the configuration is

a. $\frac{12q^2}{4\pi\varepsilon_0 d}$ b. $\frac{-12q^2}{4\pi\varepsilon_0 d}$ c. $\frac{10q^2}{4\pi\varepsilon_0 d}$ d. $\frac{-10q^2}{4\pi\varepsilon_0 d}$

13 For two points A and B in an electric field, if ΔW is work done in taking a test charge q' from A to B, $\Delta U = U_B - U_A$ is change in potential energy of a test charge when it is moved from A to B.
$\Delta V = V_B - V_A$ is change in potential due to field if we move along path AB from A to B and $d\mathbf{s}$ is differential displacement along path AB.

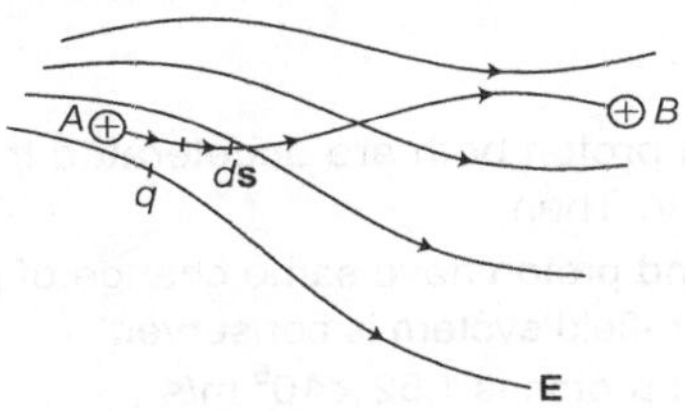

Then, which of the following options is correct?

a. $V_B - V_A = \int_A^B \mathbf{E}\cdot d\mathbf{s}$ b. $\Delta W = -q'\int_A^B \mathbf{E}\cdot d\mathbf{s}$

c. $\Delta V = -\frac{\Delta U}{q'}$ d. $\Delta V = -\frac{\Delta W}{q'} = \frac{\Delta U}{q'}$

14 Two identical thin rings, each of radius R, coaxially placed a distance R apart. If Q_1 and Q_2 are respectively the charges, uniformly spread on the two rings, the work done in moving a charge q from the centre of one ring to that of the other is

a. zero b. $q(Q_1 - Q_2)(\sqrt{2}-1)/\sqrt{2}(4\pi\varepsilon_0 R)$

c. $q\sqrt{2}(Q_1 + Q_2)/(4\pi\varepsilon_0 R)$ d. $q(Q_1/Q_2)(\sqrt{2}+1)\sqrt{2}(4\pi\varepsilon_0 R)$

15 Three charges Q, $+q$ and $+q$ are placed at the vertices of a right angle triangle (isosceles triangle) as shown in the figure. The net electrostatic energy of the configuration is zero, if Q is equal to **[IIT JEE 2000]**

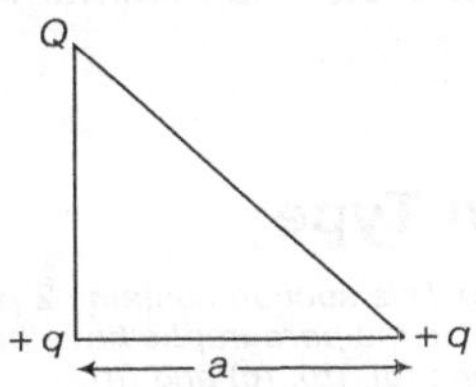

a. $\frac{-q}{1+\sqrt{2}}$ b. $\frac{-2q}{2+\sqrt{2}}$

c. $-2q$ d. $+q$

16 Two equal point charges are fixed at $x = -a$ and $x = +a$ on the X-axis. Another point charge Q is placed at the origin. The change in the electrical potential energy of Q, when it is displaced by a small distance x along the X-axis, is approximately proportional to **[IIT JEE 2002]**

a. x b. x^2 c. x^3 d. $1/x$

One or More than One Options Correct Type

Direction (Q. Nos. 17-19) *This section contains 3 multiple choice questions. Each question has four choices (a), (b), (c) and (d), out of which ONE or MORE THAN ONE are correct.*

17 In a hydrogen atom, an electron revolves around a nucleus containing only one proton. Then,

a. in ground state, when distance of electron and proton is about 0.53 Å, potential energy of system is zero
b. in ground state when distance of electron and proton is 0.53 Å, potential energy of system is – 27.2 eV
c. potential energy of system is – 13.6eV, when potential is taken zero at infinity and kinetic energy of system is 13.6 eV
d. if zero potential is taken at 1.06 Å separation, then total energy of system is zero, if kinetic energy associated with electron is 13.6 eV

Working Space

18 Two points A and B are 2 cm apart and a uniform electric field $\mathbf{E}$ acts along the straight line AB directed from A to B with $E = 200\,NC^{-1}$. A particle of charge 10^{-6} C is taken from A to B along line AB. Then,

a. force on charge $= 2 \times 10^{-4}$ N along B to A
b. $V_A - V_B = 4$ V
c. $W_{A \to B} = 4 \times 10^{-6}$ J
d. $W_{A \to B} = \Delta U_{AB}$

19 An electron and a proton both are accelerated from rest through a potential difference of 120 V. Then,

a. both electron and proton have same change of potential energy
b. energy of proton-field system is conserved
c. gain in speed of proton is 1.52×10^5 m/s
d. gain in speed of electron is 6.49×10^6 m/s

One Integer Value Correct Type

Direction *(Q. Nos. 20 and 21) This section contains 2 questions. When worked out will result in an integer from 0 to 9 (both inclusive).*

20 A uniform electric field of magnitude 250 V/ m is directed in the positive X-direction. A $+12\ \mu$C charge is carried from origin to point (20 cm, 50 cm). If change in potential energy of charge field system is $-k \times 10^{-4}$ J, then the value of k is

21 Three point charges q, $2q$ and $8q$ are to be placed on a 9 cm long straight line. If potential energy of system is minimum possible, then find the distance between $2q$ and q.

Comprehension Type

Direction *(Q. Nos. 22-29) This section contains 2 paragraphs, each describing theory, experiments, data etc. Eight questions related to the paragraphs have been given. Each question has only one correct answer among the four given options (a), (b), (c) and (d).*

Passage I

A block of mass m and with charge $+Q$ is connected to a spring of spring constant k. The block is placed on a frictionless surface and the system is placed in the region of a uniform field E. The block is released from rest when the spring is in unstretched position at $x = 0$.

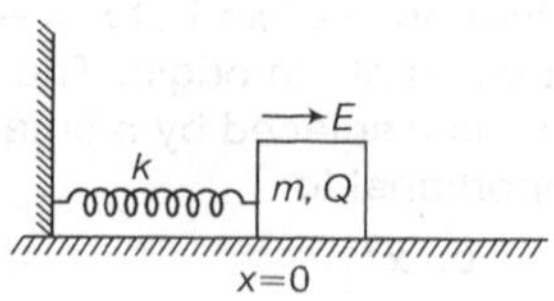

22 Maximum displacement of mass m is

a. $\frac{QE}{k}$
b. $\frac{2QE}{k}$
c. $\frac{QE}{2k}$
d. $\frac{QE}{4k}$

23 Equilibrium position of block is

a. $x = \frac{2QE}{k}$
b. $x = \frac{QE}{4k}$
c. $x = \frac{QE}{k}$
d. $x = \frac{QE}{2k}$

24 Motion of the block is

a. simple harmonic with time period $T = 2\pi\sqrt{\frac{m}{k}}$

b. periodic but not simple harmonic

c. simple harmonic with time period $T = 2\pi\sqrt{\frac{2m}{k}}$

d. periodic with time period $T = 2\pi\sqrt{\frac{m}{2k}}$

25 If coefficient of friction between mass and horizontal surface is μ_k, then

a. $x_{max} = \frac{2(QE) - \mu_k mg}{k}$

b. $x_{max} = \frac{QE - \mu_k mg}{k}$

c. $x_{max} = \frac{QE - \mu_k mg}{2k}$

d. $x_{max} = \frac{2(QE - \mu_k mg)}{k}$

Passage II

Two fixed equal positive charges, each of magnitude $q = 5 \times 10^{-5}$ C are located at points A and B, where AB is 6 m. An equal and opposite charge moves towards AB line along the perpendicular bisector of AB. The moving charge, when reaches at point C at a distance of 4 m from O, has a kinetic energy of 4 J.

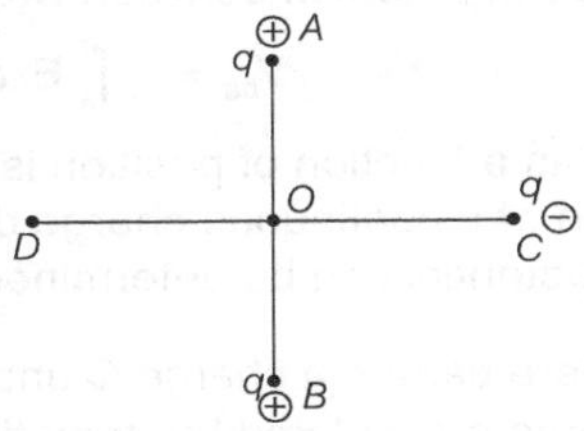

26 Total energy of charge $-q$ at C is

a. 4 J
b. 9 J
c. −9 J
d. −5 J

27 Total energy of charge $-q$ at D is

a. 4 J
b. 9 J
c. −9 J
d. −5 J

28 Total potential energy of charge $-q$ at D is

a. around 4 J
b. around −9 J
c. around −4 J
d. around −5 J

29 The distance of point D at which $-q$ charge will reach before returning towards C, is

a. less than 9 m
b. more than 9 m
c. less than 9.5 m
d. more than 9.5 m

Matching List Type

Direction (Q. No. 30) *Choices for the correct combination of elements from Column I and Column II are given as options (a), (b), (c) and (d), out of which one is correct.*

30 Eight point charges each with charge Q is placed at corners of a cube of side length l. If Q = 1 C and l = 1m. Then, match the Column I with Column II.

	Column I		Column II
i.	Potential at the centre of cube (in volt) (V = 0 at infinity)	p.	$5.70 \times 9 \times 10^9$
ii.	Potential at a corner of cube (in volt) (V = 0 at infinity)	q.	$22.8 \times 9 \times 10^9$
iii.	Potential energy of the system of charges (in joule)	r.	$9.24 \times 9 \times 10^9$

Codes

	i	ii	iii		i	ii	iii
a.	r	p	q	b.	r	q	p
c.	q	p	r	d.	q	r	p

DPP-8 Potential Due to Charge Distributions

Subjective Questions

Working Space

Direction *(Q. Nos. 1-5) These questions are subjective in nature, need to be solved completely on notebook.*

1 To know difference in potential between two points, we use

$$V_{ba} = -\int_a^b \mathbf{E}\cdot d\mathbf{l}$$

In many cases, **E** as a function of position is not known like in case of multiple charges or in case of a continuous charge distribution, **E** is difficult to calculate. Then, state how potential can be determined (In such cases)?

2 A conducting sphere carries a charge Q and a second identical sphere is neutral. Initially both spheres are isolated but then they are put in contact.

i. What is their potential when they are put in contact?
ii. How do you account for the charge flow from Ist sphere to IInd sphere?
iii. If spheres do not have same radius, then explain (i) and (ii)
iv. Show that when a conductor is charged, surface charge density is higher at a pointed end.

3 Two point charges 3.4 μC and $-2.0\ \mu$C are placed 5.0 cm apart on X-axis. At what points on X-axis

i. electric field is zero and
ii. electric potential is zero
iii. what do you infer from these results?

4 Two equal but opposite charges are separated by a distance d. Determine a formula $V_{BA} = V_B - V_A$ for points A and B on the line between the charges.

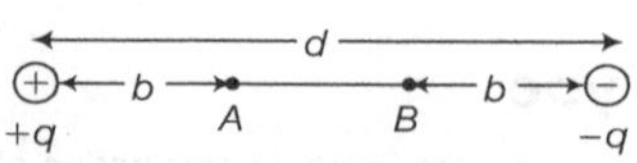

5 A thin circular ring of radius R has charge $+\dfrac{Q}{2}$ uniformly distributed on top half and $-\dfrac{Q}{2}$ on bottom half. Find potential at a point distance x from the centre of ring on the axis of ring.

Only One Option Correct Type

Direction *(Q. Nos. 6-15) This section contains 10 multiple choice questions. Each question has four choices (a), (b), (c) and (d), out of which ONLY ONE option is correct.*

6 In rectangle shown in the figure, the two corners have charges $q_1 = -5\ \mu$C and $q_2 = +2.0\ \mu$C. The work done in moving a charge of $+3.0\ \mu$C from B to A is (take $\dfrac{1}{4\pi\varepsilon_0} = 10^{10}$)

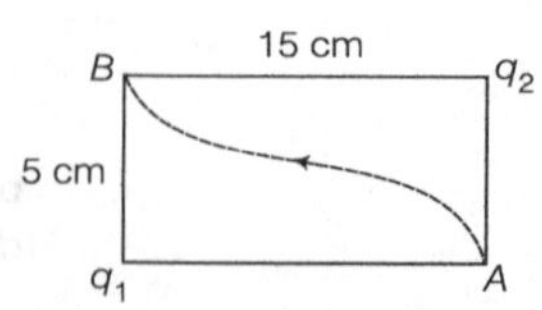

a. 2.8 J
b. 3.5 J
c. 4.5 J
d. 5.5 J

Working Space

7 The radius of a soap bubble whose potential is 16 V is doubled. Then, new potential of the bubble is

a. 2 V
b. 4 V
c. 8 V
d. 16 V

8 Identical charges $(-q)$ are placed at each corner of a cube of side b, then electrical potential energy of charge $+q$ which is placed at centre of cube will be

a. $\dfrac{8\sqrt{2}\,q^2}{4\pi\varepsilon_0 b}$
b. $\dfrac{-8\sqrt{2}\,q^2}{\pi\varepsilon_0 b}$
c. $\dfrac{-4\sqrt{2}\,q^2}{\pi\varepsilon_0 b}$
d. $\dfrac{-4q^2}{\sqrt{3}\pi\varepsilon_0 b}$

9 The charge given to a hollow sphere of radius 10 cm is 3.2×10^{-19}C. At a distance of 4 cm from its centre, the electrical potential is

a. 28.8×10^{-9} V
b. 288 V
c. 28.8 V
d. zero

10 The distance between H^+ and Cl^- ions in HCl molecule is 1.28 Å. What will be potential due to this dipole at a distance of 12 Å on the axis of dipole?

a. 0.26 V
b. 26 V
c. 2.6 V
d. 260 V

11 Consider a ring of radius R with charge Q, uniformly distributed over its perimeter. Potential difference between centre of the ring and a point on the axis at a distance $2R$ from the centre is

a. $\dfrac{\sqrt{5}-1}{\sqrt{5}} \cdot \dfrac{kQ}{R}$
b. $\dfrac{\sqrt{2}-1}{\sqrt{2}} \cdot \dfrac{kQ}{R}$
c. $\dfrac{\sqrt{3}-1}{\sqrt{3}} \cdot \dfrac{kQ}{R}$
d. $\dfrac{kQ}{\sqrt{5}R}$

12 A charged rod of length l and a uniform charge density λ is bent to form a semicircle. Potential at the centre of semicircle of radius r will be

a. $\dfrac{kQ}{r}$
b. $\dfrac{kQ}{2\pi r}$
c. $k\pi\lambda$
d. $\dfrac{k\lambda}{2\pi r}$

13 Electric potential at the point P on axis of the charged disc with a hole is

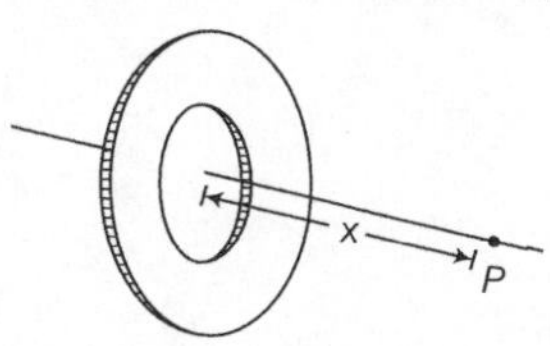

(Given, hole radius = a, outer radius = b and charge density on surface = σ)

a. $k\sigma(\sqrt{x^2+b^2}-\sqrt{x^2+a^2})$
b. $\dfrac{k\pi\sigma}{4}(\sqrt{x^2+b^2}-\sqrt{x^2+a^2})$
c. $4\pi k\sigma(\sqrt{x^2+b^2}-\sqrt{x^2+a^2})$
d. $2\pi k\sigma(\sqrt{x^2+b^2}-\sqrt{x^2+a^2})$

Working Space

14 A non-conducting disc of radius a and uniform positive surface charge density σ is placed on the ground with its vertical axis. A particle of mass m and positive charge q is dropped, along the axis of the disc from a height H with zero initial velocity. The particle has $q/m = 4\varepsilon_0 g/\sigma$. The value of H if the particle just reaches the disc is

a. $\frac{4}{7}a$ b. $\frac{4}{3}a$ c. $\frac{9}{2}a$ d. $\frac{1}{2}a$

15 A conducting bubble of radius a, thickness t $(t << a)$ has potential V. Now, the bubble collapses into a droplet. Find the potential of the droplet. **[JEE Advanced 2005]**

a. $V\left[\frac{a}{3t}\right]^{1/3}$ b. $V\left[\frac{a}{3t}\right]^{1/2}$

c. $V\left[\frac{a}{2t}\right]^{1/3}$ d. $V\left[\frac{a}{4t}\right]^{1/2}$

One or More than One Options Correct Type

Direction (Q. Nos. 16-19) *This section contains 4 multiple choice questions. Each question has four choices (a), (b), (c) and (d), out of which ONE or MORE THAN ONE are correct.*

16 A $+33\,\mu C$ charge is placed 36 cm from an identical charge. A $-1.5\,\mu C$ charge is moved from point A to point B, then

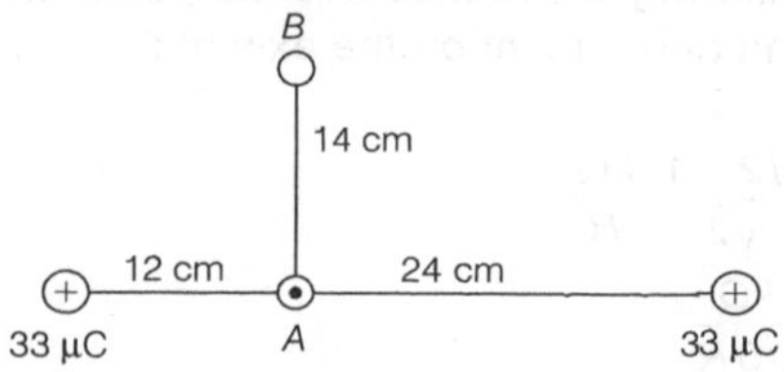

a. $U_B - U_A = q(V_B - V_A)$
b. $U_B - U_A = 1.5\,J$
c. work done by external force is displacing $-1.5\,\mu C$ charge from B to $A = -1.5\,J$
d. $U_B - U_A = 2.0\,J$

17 A cube of side l carries a charge Q at each corner. Then, taking $V = 0$ at $r = \infty$,

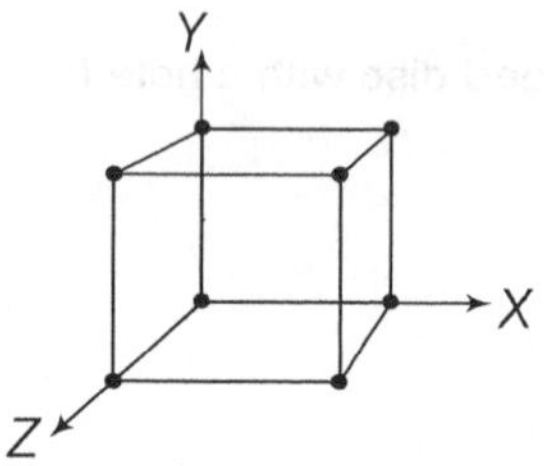

Mark the correct option(s).
a. Potential at centre of cube is zero
b. Potential at centre of cube is $\frac{16}{\sqrt{3}}\frac{kQ}{l}$
c. Potential at each corner of cube due to other seven charge is $\left(3+\frac{3}{\sqrt{2}}+\frac{1}{\sqrt{3}}\right)\frac{kQ}{l}$
d. Potential at each corner of cube due to other seven charges is $\left(1+\frac{3}{\sqrt{2}}+\frac{3}{\sqrt{3}}\right)\frac{kQ}{l}$

18 Three charges are at the corners of an equilateral triangle. A, B and C are mid-points of the sides of triangle.

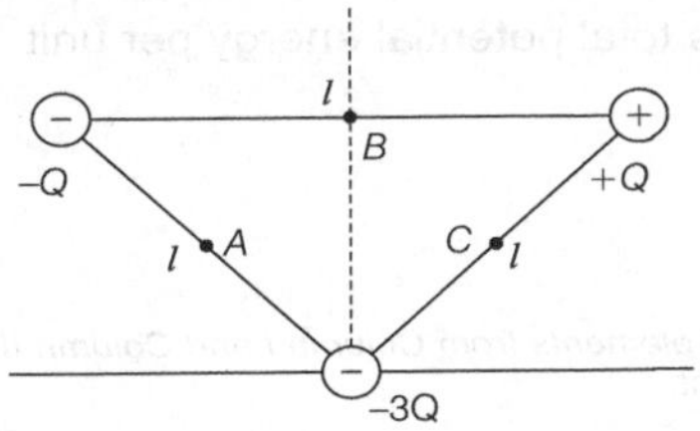

Mark the correct option(s).

a. $V_A = \frac{Q}{\pi\varepsilon_0 l}\left(\frac{\sqrt{3}}{6} - 2\right)$

b. $V_B = \frac{\sqrt{3}\,Q}{2\pi\varepsilon_0 l}$

c. $V_C = -\frac{Q}{\pi\varepsilon_0 l}\left(1 + \frac{\sqrt{3}}{6}\right)$

d. Total potential energy of system of charges is greater than zero

19 A non-conducting sphere of radius r_2 contains a concentric spherical cavity of radius r_1. Material between r_1 and r_2 has a uniform charge density ρ ($V = 0$ at $r = \infty$). Then,

a. potential at the outer surface is greatest
b. potential is same for all points between $r_1 < r < r_2$
c. potential inside the cavity has same value
d. potential varies inversely with r for all exterior points ($V \propto 1/r$)

Statement Type

Direction (Q. Nos. 20 and 21) *This section is based on Statement I and Statement II. Select the correct answer from the codes given below.*

a. Both Statement I and Statement II are correct and Statement II is the correct explanation of Statement I
b. Both Statement I and Statement II are correct and Statement II is not the correct explanation of Statement I
c. Statement I is correct but Statement II is incorrect
d. Statement II is correct but Statement I is incorrect

20 **Statement I** Three charges Q, $+q$ and $+q$ are placed on vertices of a right angled isosceles triangle as,

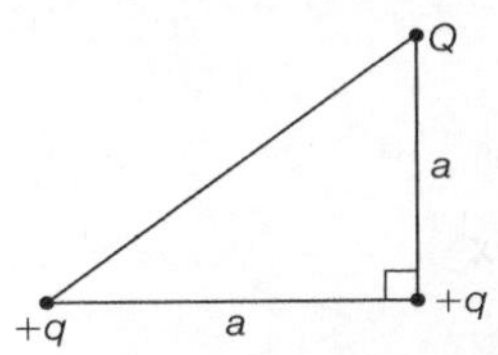

Then, the net electrostatic potential energy of system is zero if Q is equal to $\frac{-2q}{2+\sqrt{2}}$.

Statement II Potential energy of a multiple charge system is $U = \frac{1}{2}\sum_{j=1}^{N}\sum_{k \neq j}\frac{q_j q_k}{r_{jk}}$.

The double notation means for $j = 1$, sum over $k = 2, 3 \cdots N$, then take $j = 2$ and sum over $k = 1, 2, 3, \cdots N$ and so on.

Working Space

21 **Statement I** Conductors having equal charge and volume must also have equal surface potentials.

Statement II Potential of a charged conductor is total potential energy per unit charge.

Matching List Type

Direction (Q. No. 22) Choices for the correct combination of elements from Column I and Column II are given as options (a), (b), (c) and (d), out of which one is correct.

22 If we have N-different charges in space, then potential energy of system can be calculated by summing all pairs. Now, consider a cube of side a with a charge $-e$ at each of its corner and a charge $+2e$ at its centre. Match the Column I with Column II.

	Column I		Column II
i.	PE of charge pairs which are a distance a.	p.	$-18.475\, ke^2/a$
ii.	PE of charge pairs which are at a distance $\sqrt{3/2}\, a$.	q.	$12\, ke^2/a$
iii.	PE of charge pairs which are at a distance $\sqrt{2}\, a$.	r.	$8.485\, ke^2/a$
iv.	PE of charge pairs which are at a distance $\sqrt{3}\, a$.	s.	$2.309\, ke^2/a$

Codes

	i	ii	iii	iv		i	ii	iii	iv
a.	p	q	r	s	b.	q	p	r	s
c.	p	q	s	r	d.	s	r	p	q

One Integer Value Correct Type

Direction (Q. No. 23 and 24) This section contains 2 questions. When worked out will result in an integer from 0 to 9 (both inclusive).

23 The volume charge density of a sphere of radius r_0 is distributed as

$\rho = \rho_0 \left(1 - \frac{r^2}{r_0^2}\right)$ r is distance measured from centre of sphere. Potential at the surface of sphere is $k(\rho_0 r_0^2/15\varepsilon_0)$. Then, find the value of k.

24 Two equal charges are located as shown below.

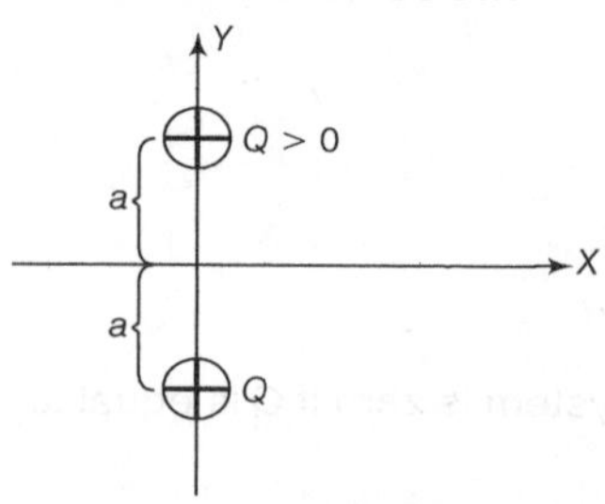

Graph of function $\frac{V(x)}{(kQ/a)}$ *versus* (x/a) has a maxima of p. $\left(k = \frac{1}{4\pi\varepsilon_0}\right)$

Find the value of p when $V(x)$ is potential along X-axis.

Working Space

Comprehension Type

Direction (Q. Nos. 25-30) This section contains 2 paragraphs, each describing theory, experiments, data, etc. Six questions related to the paragraphs have been given. Each question has only one correct answer among the four given options (a), (b), (c) and (d).

Passage I

For a hollow spherical shell, potential at all interior points is same as potential on the surface of shell. Now, consider 3-concentric shells A, B and C with radii a, b and c $(a < b < c)$ with surface charge densities σ, $-\sigma$ and σ, respectively.

25 Potential on the surface of shell A is

a. $\frac{\sigma}{\varepsilon_0}(a+b+c)$

b. $\frac{\sigma}{\varepsilon_0}\left(\frac{a^2}{b}-b+c\right)$

c. $\frac{\sigma}{\varepsilon_0}(a-b+c)$

d. $\frac{\sigma}{\varepsilon_0}\left(\frac{a^2}{c}-\frac{b^2}{c}-c\right)$

26 Potential on the surface of shell B is

a. zero

b. $\frac{\sigma}{\varepsilon_0}(a-b+c)$

c. $\frac{\sigma}{\varepsilon_0}\left(\frac{a^2}{c}-\frac{b^2}{c}+c\right)$

d. $\frac{\sigma}{\varepsilon_0}\left(\frac{a^2}{b}-b+c\right)$

27 In which conditions potential on the surface of shell C is equal to potential on the surface of A

a. for no value of a, b and c

b. for any value of a, b and c

c. only when $a+b=c$

d. only when $\frac{a+b}{2}=c$

Passage II

Two fixed charges $-2Q$ and Q are located at the points $(-3a, 0)$ and $(+3a, 0)$ in the XOY-plane.

28 Locus of all the points with $V=0$ in XY- plane

a. is a straight line of slope $4a$

b. is a circle of radius $4a$

c. is an ellipse of semi-major axis $4a$

d. is a parabola with focal length $4a$

29 If a particle with charge $+q$ is released at point $(5a, 0)$, then it

a. oscillates between points $(a, 0)$ and $(9a, 0)$

b. crosses the point $(9a, 0)$

c. remains stationary at $(5a, 0)$

d. rotates in a circle of radius $2a$

30 Potential function $V(x)$ due to the charges

a. does not have any minima for $x > 9a$

b. does not have any minima at a

c. have a maxima at $x=a$

d. have a minima at $x=a$

DPP-9 Motion of a Charged Particle inside an Electric Field

Subjective Questions

Working Space

Direction (Q. Nos. 1-4) *These questions are subjective in nature, need to be solved completely on notebook.*

1 A point charge moves inside an electric field (non-uniform)

i. Does it move along a field line?

ii. What is the energy balance equation when it moves from some point A to point B?

2 An electron is released from rest on the axis of a uniform positively charged ring 0.1 m from the centre of the ring. If linear charge density of ring is + 0.1 μC/m and radius of ring is 0.2 m, how fast will the electron be moving when it reaches centre of the ring?

3 A small ball of mass 2×10^{-3} kg having a charge of 1 μC is suspended by a string of length 0.8 m. Another identical ball having the same charge is kept at a point of suspension. Determine the minimum horizontal velocity which should be imparted to the lower ball, so that it can make complete revolution?

4 A uniform field of magnitude 2.00×10^3 N/C is set up between horizontal plates by charging the lower plate positively and upper plate negatively. Plates have length $L = 10.0$ cm and separation $d = 2.00$ cm. An electron is shot between the plates from one end of lower plate at $\theta = 45°$ with a velocity of 6.00×10^6 m/s.

i. Did electron strike any of the plates?

ii. If it does, then where?

Only One Option Correct Type

Direction (Q. Nos. 5-12) *This section contains 8 multiple choice questions. Each question has four choices (a), (b), (c) and (d), out of which ONLY ONE option is correct.*

5 An electron and a proton are released from rest in a uniform field. They cover same distance in time t_1 and t_2 in seconds, respectively. Ratio $\frac{t_1}{t_2}$ is

a. 1

b. $\sqrt{m_e/m_p}$

c. $\sqrt{m_p/m_e}$

d. 1836

6 A particle of mass m and charge q_0 is placed at rest in a uniform field of intensity E and then released. The kinetic energy attained by the particle after moving through a distance y is

a. qEy^2

b. qE^2y

c. qEy

d. q^2Ey

Working Space

7 A charged particle of mass m and charge q is accelerated by applying a potential difference V volt, the final speed of the particle will be

a. $V\sqrt{e/m}$
b. $\sqrt{eV/m}$
c. $\sqrt{2eV/m}$
d. $2eV/m$

8 A thin semicircular glass rod of radius R has a non-uniform charge distribution of $\lambda = \lambda_0 \sin\theta$, where λ_0 = a constant and θ is angle with positive X-axis.

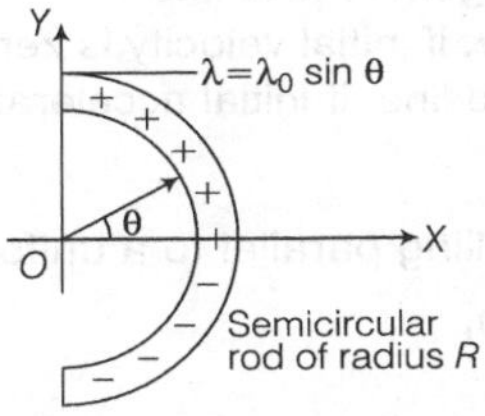

If $R = 1.0$ cm and $\lambda_0 = 1.0\ \mu$C/m, the acceleration of an electron placed at origin O will be

a. 2.5×10^{17} m/s$^2\ \hat{\mathbf{j}}$
b. 2.5×10^{17} m/s$^2\ (-\hat{\mathbf{j}})$
c. 1.5×10^{17} m/s$^2\ (-\hat{\mathbf{j}})$
d. 1.5×10^{17} m/s$^2\ (+\hat{\mathbf{j}})$

9 A 10.0 g block with a charge of 8×10^{-5} C is placed in an electric field $\mathbf{E} = (3\times10^3)\hat{\mathbf{i}} - 600\,\hat{\mathbf{j}}$ N/C. If block is released from rest at the origin at $t = 0$, then its X-coordinates at $t = 3$ s is

a. 100 m
b. 102 m
c. 107 m
d. 108 m

10 An electron starts from rest 3 cm from the centre of a uniformly charged insulating sphere of radius 2 cm and charge 1nC. When electron reaches the surface of sphere, its velocity will be

a. 5.2×10^6 m/s
b. 7.6×10^6 m/s
c. 6.4×10^6 m/s
d. Zero

11 Four identical particles each with charge q and mass m are placed at the vertices of a square of side L.
They are released from rest. Velocity of each particle when their distance from centre of square is twice (of its original value) is

a. $v = \sqrt{\left(1+\frac{1}{\sqrt{5}}\right)\cdot\frac{kq^2}{mL}}$
b. $v = 0$
c. $v = \sqrt{\left(1+\frac{1}{\sqrt{8}}\right)\cdot\frac{kq^2}{mL}}$
d. $v = \sqrt{2}\cdot\frac{kq^2}{mL}$

12 A light unstretched spring has length d. Two identical charged particles each with charge q are connected on each end of the spring. When the particles are released from rest, the system oscillates over a horizontal frictionless table. The spring has a bit of internal kinetic friction or a result, oscillations finally stop when distance between the particles is $3d$. Then,

a. initial PE of system is equal to final PE of system
b. final PE of system is more than its initial PE of system
c. part of energy is lost
d. internal energy of spring increases due to oscillations

Working Space

One or More than One Options Correct Type

Direction *(Q. Nos. 13-17) This section contains 5 multiple choice questions. Each question has four choices (a), (b), (c) and (d), out of which ONE or MORE THAN ONE are correct.*

13 A point positive charge is located in region of an electric field. Choose the correct statement(s).

a. Direction of velocity of charge is along the field line
b. Direction of acceleration of charge is along the tangent to field line
c. Direction of velocity of charge is along the field line, if initial velocity is zero
d. Direction of acceleration of charge is along the field line, if initial acceleration is zero

14 An electron with speed $v_0 = 27.5 \times 10^6$ m/s is travelling parallel to a uniform electric field of magnitude $E = 11.4 \times 10^3$ N/C. Then,

a. electron comes to rest before 0.2 m
b. electron comes to rest after 0.2 m
c. electron comes to starting point after 3×10^{-8} s
d. electron comes to starting point before 3×10^{-8} s

15 An electron has an initial velocity $v_0 = 8 \times 10^4$ m/s $\hat{\mathbf{j}}$. If it enters a region of field $\mathbf{E} = (2\hat{\mathbf{i}} + 8\hat{\mathbf{j}}) \times 10^4$ N/C then

a. acceleration of electron is in XOY-plane
b. acceleration of electron is along X-axis
c. final velocity of electron is in XOY-plane
d. final velocity of electron is along X-axis

16 An electron is travelling through an electric field $\mathbf{E} = (2\,\hat{\mathbf{i}} - 1.20\,\hat{\mathbf{j}}) \times 10^{-11}$ N/C. At $t = 0$, electron is at origin and is travelling along X-axis with a speed of 1.90 m/s. Which of the following statement(s) is/are correct?

a. x-coordinate of e^- after 2s later is 3.2 m
b. x-coordinate of e^- after 2.00 s is – 3.2 m
c. y-coordinate of e^- after 2.00 s is – 4.2 m
d. y-coordinate of e^- after 2.00 s is 4.2 m

17 A positive charge q is placed at centre of a ring of radius R and carrying a uniformly distributed negative charge $-Q$. When q is slightly displaced and released it undergoes SHM. Then,

a. spring constant $\propto \dfrac{1}{R^2}$
b. spring constant $\propto \dfrac{1}{R^3}$
c. time period of oscillation is proportional to R
d. time period of oscillation is proportional to $R^{3/2}$

Statement Type

Direction *(Q. Nos. 18 and 19) This section is based on Statement I and Statement II. Select the correct answer from the codes given below.*

a. Both Statement I and Statement II are correct and Statement II is the correct explanation of Statement I
b. Both Statement I and Statement II are correct and Statement II is not the correct explanation of Statement I
c. Statement I is correct but Statement II is incorrect
d. Statement II is correct but Statement I is incorrect

Working Space

18 **Statement I** An α-particle and a deutron experience same force in electric field.

Statement II Mass of both particles is nearly same.

19 **Statement I** When two charged particles are shot through two parallel plates having different nature of charge but same velocities, the heavier particle bends less than lighter particle.

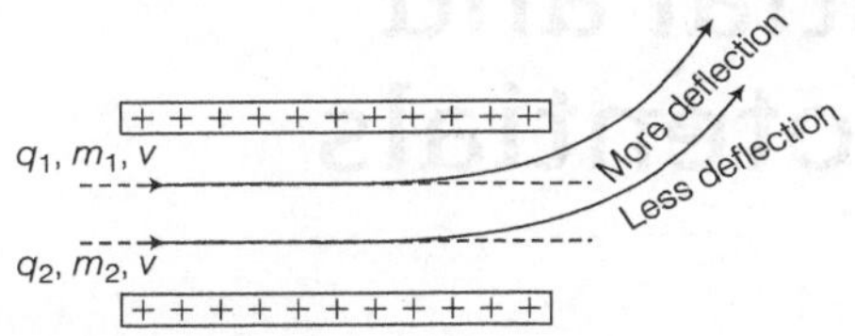

Statement II A heavier particle has more inertia of motion than a lighter one when they are moving with same velocities.

One Integer Value Correct Type

Direction (Q. No. 20) *This section contains 1 question. When worked out will result in an integer from 0 to 9 (both inclusive).*

20 Two parallel plates have opposite charges with equal charge density of 36 *x* C/m^2 are separated by a distance of 12 cm.

A proton is released from positive plate. If speed of proton when it is striking the negative plate is N $\times$ 102 km/s, then value of *N* is

Comprehension Type

Direction (Q. Nos. 21-23) *This section contains a paragraph, describing theory, experiments, data, etc. Three questions related to the paragraph have been given. Each question has only one correct answer among the four given options (a), (b), (c) and (d).*

Passage

A non-conducting disc of radius *a* and uniform positive surface charge density σ is placed on ground with its axis vertical. A particle of mass '*m*' and positive charge *q* is dropped along the axis of disc from height *H* with zero initial velocity. For particle, $\frac{q}{m} = \frac{4\varepsilon_0 g}{\sigma}$.

21 Potential due to charge of disc at height *H* above the axis of disc is

a. $\frac{\sigma}{2\varepsilon_0}(\sqrt{a^2+H^2})$ b. $\frac{\sigma a}{2\varepsilon_0}$ c. $\frac{\sigma}{2\varepsilon_0}$ d. $\frac{\sigma}{2\varepsilon_0}(\sqrt{a^2+H^2}-H)$

22 Potential energy *U versus* height *H* curve for the particle is

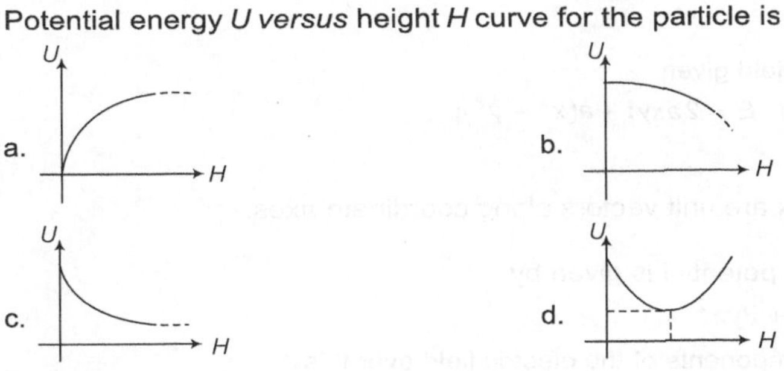

23 Height *H* from which the particle is released, is

a. $H = a$ b. $H = 4a$ c. $H = \frac{2}{3}a$ d. $H = \frac{4}{3}a$

DPP-10 Electric Field, Potential and Equipotentials

Subjective Questions

Working Space

Direction *(Q. Nos. 1-4) These questions are subjective in nature, need to be solved completely on notebook.*

1 i. If you know potential at a point, can you calculate electric field at that point? If you know electric field at a point, can you calculate potential at that point?

ii. If the electric field is uniform in a region, what can be calculated about the electric potential? If potential is uniform in a region of space, what can be concluded about electric field?

iii. In a region of space, electric field points are in North. In what direction the rate of change of potential be

a. greatest b. least c. zero

2 i. For equipotential surfaces,

a. explain why they do not cross each other?

b. does the distance between two equipotentials tell us about the relative strengths of electric field in those regions?

c. prove that field lines are always perpendicular to an equipotential surface.

ii. Draw equipotentials for

a. a negatively charged hollow sphere.

b. a very large uniformly charged plate with charge density $\sigma = 0.75\,\mu C / m^2$, such that equipotentials are at 100 V potential difference.

c. an electric dipole.

d. two dissimilarily charged parallel plates kept nearby.

3 i. Determine magnitude and direction of the electric field if potential function is

a. $V = a(x^2 - y^2)$ b. $V = axy$

c. $V = a(x^2 + y^2) + bz^2$

where, a and b are constants.

ii. Find the potential function of an electric field given

a. $E = a(y\hat{\mathbf{i}} + x\hat{\mathbf{j}})$ b. $E = 2axy\hat{\mathbf{i}} + a(x^2 + y^2)\hat{\mathbf{j}}$

c. $E = ay\hat{\mathbf{i}} + (ax + bz)\hat{\mathbf{j}} + by\hat{\mathbf{k}}$

where, a and b are constants and $\hat{\mathbf{i}}$, $\hat{\mathbf{j}}$ and $\hat{\mathbf{k}}$ are unit vectors along coordinate axes.

4 Over a certain region of space the electric potential is given by

$$V = 5x - 3x^2y + 2yz^2$$

i. Find the expression for the x, y and z components of the electric field over this region.

ii. What is the magnitude of the field at the point P that has coordinates $(1, 0, -2)$ m?

Only One Option Correct Type

Direction *(Q. Nos. 5-17) This section contains 13 multiple choice questions. Each question has four choices (a), (b), (c) and (d), out of which ONLY ONE option is correct.*

5 The electric potential in a region of space is given by $V = 4x^2$ volt. An electron when placed at point (1 m, 0, 2 m), then

a. it moves parallel to X-axis and it reaches point (3m, 0, 2m)
b. it moves parallel to X-axis and it reaches point (–3m, 0, 2m)
c. it moves parallel to Y-axis and it reaches point (1m, 2m, 2m)
d. it moves parallel to Z-axis and it reaches point (1 m, 0, 5m)

6 Points P, Q, R and S are located in region of an electric field as

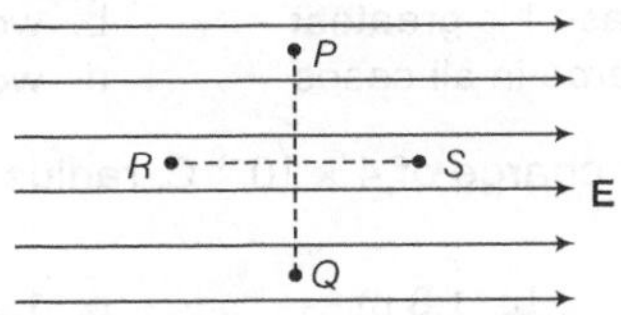

Then,

a. $V_P = V_R$ b. $V_P = V_S$
c. $V_R = V_S$ d. $V_P = V_Q$

7 Points A, B and C exist in region of a uniform field as shown. If V_A, V_B and V_C are potentials of these points respectively, then

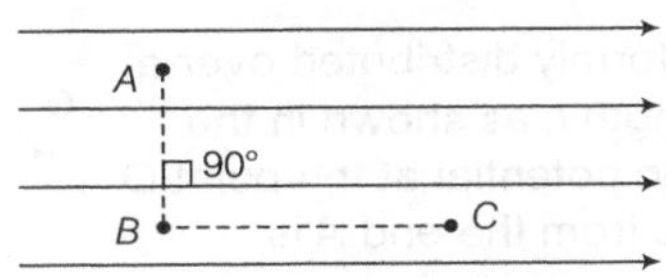

a. $V_A = V_B = V_C$
b. $V_A = V_B > V_C$
c. $V_A = V_B < V_C$
d. $V_A > V_B = V_C$

8 Potential gradient just outside the surface of a uranium nucleus ($Z = 92$) whose diameter is about 15×10^{-15} m, is

a. 2.4×10^{21} V/m b. -2.4×10^{21} V/m
c. 2.4×10^{-21} V/m d. -2.4×10^{21} V/m

9 Electric potential between two parallel plates is given by $V(x) = (8x + 5)$ volt with $x = 0$, at one of plates and x is positive towards the second plate. Charge density on the plates is

a. 7.1×10^{-11} C/m^2 b. 14.2×10^{-11} C/m^2
c. 1.15×10^{-9} C/m^2 d. 2.3×10^{-9} C/m^2

10 In a region of space, potential function is given by $V = y^2 + 2.5\,xy - 3.5\,xyz$ volt Then, **E** at point ($x = 1$m, $y = 1$m, $z = 1$m) is

a. zero b. $-6\hat{\mathbf{i}}$ N/C
c. $\left(-6\hat{\mathbf{i}} - \hat{\mathbf{j}} + \frac{7}{2}\hat{\mathbf{k}}\right)$ N/C d. $(\hat{\mathbf{i}} - \hat{\mathbf{j}} + 3.5\hat{\mathbf{k}})$ N/C

11 For a potential function $V = \phi(x, y, z)$ in space,

a. only a maxima exists in free space
b. only a minima exists in free space
c. can have both maxima and minima in free space
d. can neither have maxima nor minima in free space

12 A negative charge is moved from point A to point B. Charge is at rest at A and it had kinetic energy of 2.10×10^{-4} J at B. Also, work done is moving the charge from A to B by external force is 7×10^{-4} J. Then,

a. $V_B - V_A = 0$ b. $V_B - V_A =$ positive
c. $V_B - V_A =$ negative d. $V_B = 0$

Working Space

13 Figure shows some equipotentials in space. A charged particle is moved from A to B.

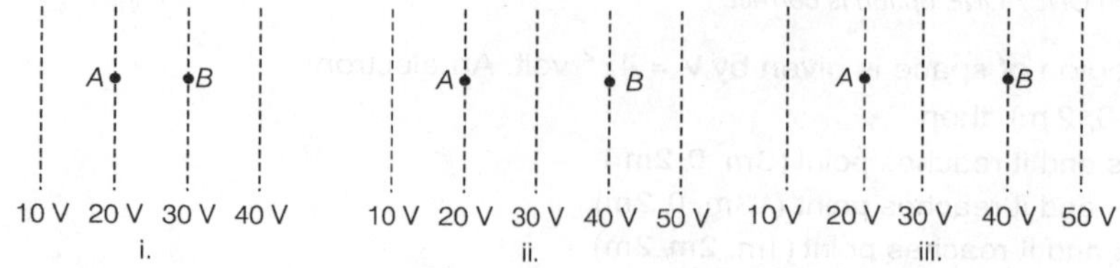

Then, appropriate options are

a. work done in case I is greatest
b. work done in case II is least
c. work done is same in all cases
d. work done in case I is least

14 For a single point charge of 4×10^{-9} C, radius of equipotential surface drawn at 30 V is

a. 3.6 m
b. 1.8 m
c. 1.2 m
d. 0.6 m

15 A charge Q is uniformly distributed over a long rod AB of length L as shown in the figure. The electric potential at the point O lying at distance L from the end A is

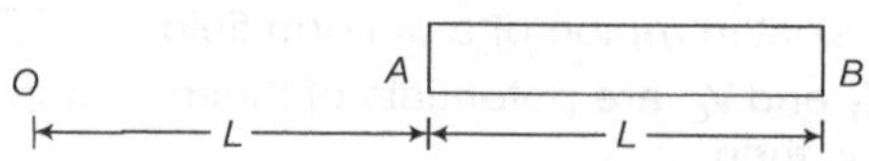

[JEE Main 2013]

a. $\dfrac{Q}{8\pi\varepsilon_0 L}$
b. $\dfrac{3Q}{4\pi\varepsilon_0 L}$
c. $\dfrac{Q}{4\pi\varepsilon_0 L \ln 2}$
d. $\dfrac{Q \ln 2}{4\pi\varepsilon_0 L}$

16 Assume that an electric field $\mathbf{E} = 30x^2\hat{\mathbf{i}}$ exists in space. Then, the potential difference $V_A - V_O$, where V_O is the potential at the origin and V_A is the potential at $x = 2$ m, is **[JEE Main 2014]**

a. 120 J
b. −120J
c. −80 J
d. 80 J

17 Consider a thin spherical shell of radius R with its centre at the origin, carrying uniform positive surface charge density. The variation of the magnitude of the electric field $|\mathbf{E}(r)|$ and the electric potential $V(r)$ with the distance r from the centre, is best represented by which of the graphs given below? **[IIT JEE 2012]**

a.
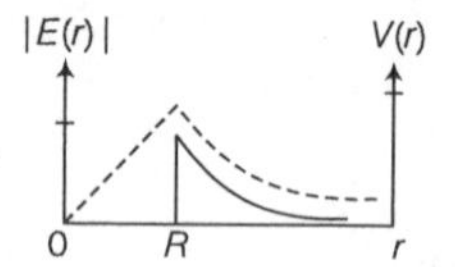

b.
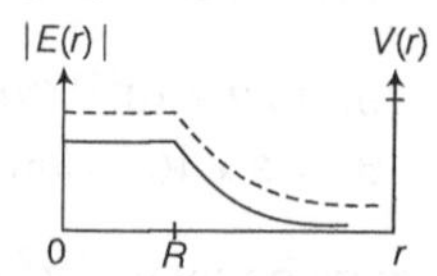

c.
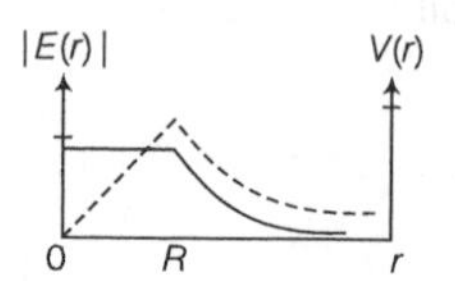

d.
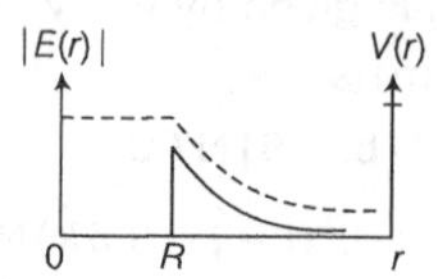

One or More than One Options Correct Type

Direction (Q. Nos. 18-22) This section contains 5 multiple choice questions. Each question has four choices (a), (b), (c) and (d), out of which ONE or MORE THAN ONE are correct.

18 A metal sphere of radius $r_0 = 0.44$ m carries $Q = 0.50\ \mu$C. Consider equipotential surfaces formed due to this sphere with a potential difference of 100 V between two nearby surfaces. Then, which of the following statements is/are correct?

a. Radius of 100th equipotential is around 21 m
b. Equipotentials are equally spaced
c. Radius of equipotentials increases rapidly, when we move away from the surface of sphere
d. Equipotentials become elliptical very far away from sphere

19 A uniform electric field exists in $+X$-direction. The potential in the region

a. is constant in all space
b. is constant for any Y for given X
c. is constant for any Z for given X
d. is constant for any point on a plane parallel to YOZ-plane

20 Equipotentials are surfaces of constant potentials at all points. True options regarding equipotentials are

a. they get closer as we move away from a positively charged conductor
b. they get closer near pointed end of conductor
c. they get closer where charge density is larger
d. they are always equally spaced

21 In certain region, it is observed that potential is constant. The other true observations are

a. electric field is uniform in the region
b. electric field in region is zero
c. charge enclosed in the region is zero
d. electric field is decreasing

22 Six point charges are kept at the vertices of a regular hexagon of side L and centre O as shown in the figure. Given that $K = \frac{1}{4\pi\varepsilon_0}\frac{q}{L^2}$, which of the following statement(s) is/are correct? **[IIT JEE 2012]**

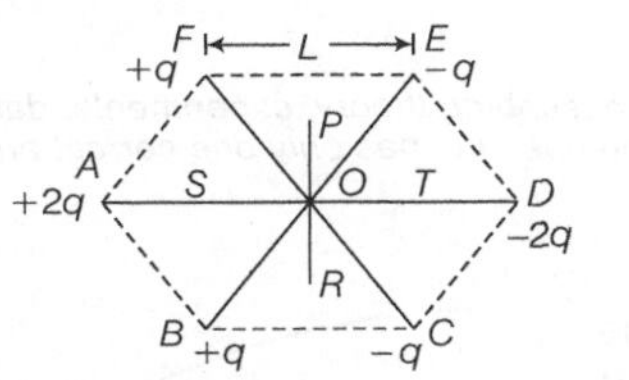

a. The electric field at O is 6 K along OD
b. The potential at O is zero
c. The potential at all points on the line PR is same
d. The potential at all points on the line ST is same

One Integer Value Correct Type

Direction *(Q. Nos. 23 and 24) This section contains 2 questions. When worked out will result in an integer from 0 to 9 (both inclusive).*

23 An insulating rod with linear charge density of 40 μC/m and linear mass density of 0.1kg/m is released from rest in a uniform electric field $E = 100$ V/m directed perpendicular to the rod. If rod is released and after travelling 2 m down the field, speed of rod is formed to be $(v/10)$ m/s, then value of v is

24 Two large metal plates are maintained at potentials $+V_0$ and $-V_0$ at a distance d from each other. A small conducting ball of mass m and radius $R (R << d)$ is suspended between the plates. The thread supporting the ball is a thin copper wire connected to the ground so that potential of ball is fixed at $V = 0$.

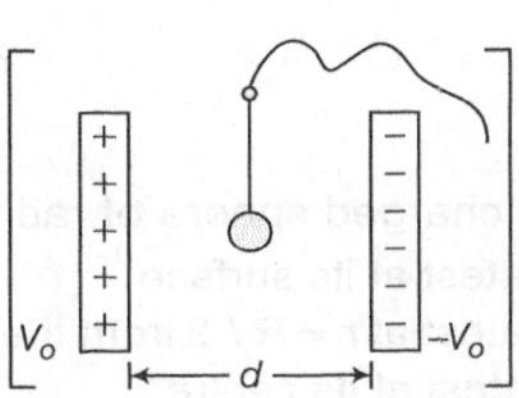

The ball hangs vertically in equilibrium when V_0 is sufficiently small.

The equilibrium of ball is unstable when V_0 exceeds the critical value of

$$V_0 = \frac{d^2 mg}{N(4\pi\varepsilon_0 RL)}$$

Then, the value of N is

Matching List Type

Working Space

Direction *(Q. No. 25) Choices for the correct combination of elements from Column I and Column II are given as options (a), (b), (c) and (d), out of which one is correct.*

25 Consider three spherical metallic shells A, B and C of radii $\frac{3}{2}$ m, $\frac{5}{2}$ m and $\frac{7}{2}$ m respectively with surface charge densities σ, $-\sigma$ and σ, respectively. Now, match the following columns.

	Column I		Column II
i.	Potential of A	p.	$\left(\frac{33}{14}\right)\frac{\sigma}{\varepsilon_0}$
ii.	Potential of B	q.	$\left(\frac{19}{10}\right)\frac{\sigma}{\varepsilon_0}$
iii.	Potential of C	r.	$\left(\frac{5}{2}\right)\frac{\sigma}{\varepsilon_0}$

Codes

	i	ii	iii		i	ii	iii
a.	p	q	r	b.	q	p	r
c.	r	p	q	d.	r	q	p

Comprehension Type

Direction *(Q. Nos. 26-28) This section contains a paragraph, describing theory, experiments, data, etc. Three questions related to the paragraph have been given. Each question has only one correct answer among the four given options (a), (b), (c) and (d).*

Passage

When a charged particle q_0 is moved from some point A to some point B, during the displacement $d\mathbf{r}$, work done on the particle is $dW = \mathbf{F}\cdot d\mathbf{r}$ $= q_0\mathbf{E}\cdot d\mathbf{r}$

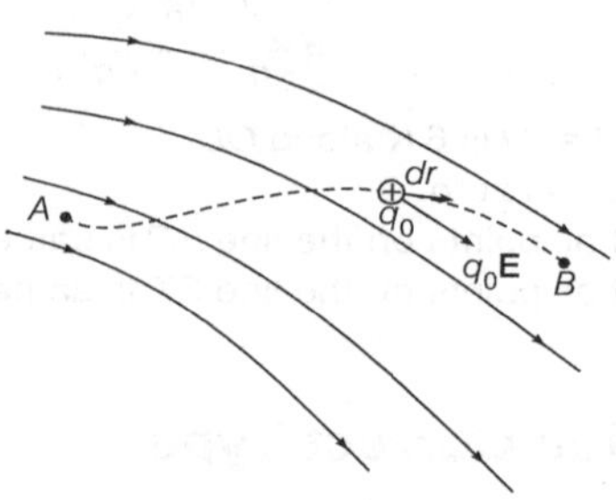

So, total work done $W = q_0\int_A^B \mathbf{E}\cdot d\mathbf{r}$

$$\Rightarrow \quad \frac{W}{q_0} = \Delta V = V_A - V_B = q_0\int_A^B \mathbf{E}\cdot d\mathbf{r}$$

$$\Rightarrow \quad V_B - V_A = -q_0\int_A^B \mathbf{E}\cdot d\mathbf{r}$$

26 For a uniformly charged insulating sphere, potential at a distance r ($r < R$, R = radius of sphere), is

a. $\frac{kQ}{r}$

b. $\frac{kQ}{2R}\left(3 - \frac{r^2}{R^2}\right)$

c. $\frac{kQ}{R}\left(2 - \frac{r^2}{R^2}\right)$

d. $\frac{kQ}{R^2}(2r)$

27 For the insulating charged sphere of radius R,

a. potential is greatest at its surface
b. potential is greatest at $r = R/2$ from the centre
c. potential is greatest at its centre
d. potential has same value at all interior points

28 Potential at centre of a charged insulating sphere is V_0. Then, potential at its surface is

a. $\frac{2}{3}V_0$ b. $\frac{3}{2}V_0$ c. $\frac{3}{2}V_0$ d. 0

DPP-11 Electrical Potential Energy

Subjective Questions

Working Space

Direction (Q. Nos. 1-5) *These questions are subjective in nature, need to be solved completely on notebook.*

1 i. Is the electrical PE of two unlike charges positive or negative? What is the electrical PE of two like charges? What is the significance of the sign of the PE in each case?

ii. Find the energy of a uniformly charged non-conducting sphere having charge q and radius R.

2 A particle of positive charge q is fixed at the centre of a circle of radius r, around the circumference of which another particle of mass m and negative charge $-q$ is moving with a constant speed.
Find work done by an external agent on the second particle to increase the radius of circle to r_2.

3 Two small metal spheres with masses 5 g and 10 g respectively, have equal positive charges of 5 μC. The spheres are initially held at a distance of 1 m from each other.

i. What is the electrical PE of the system?

ii. When spheres are released then, what is the ratio of accelerations of spheres?

iii. What is the ratio of speeds of spheres when they are far away from each other.

4 A plastic rod of length 1 m has a non-uniform linear charge density $\lambda = \left(-3.0 \times 10^{-9} \frac{\text{C}}{\text{m}}\right).x$, where, x is the distance from one end of the rod.

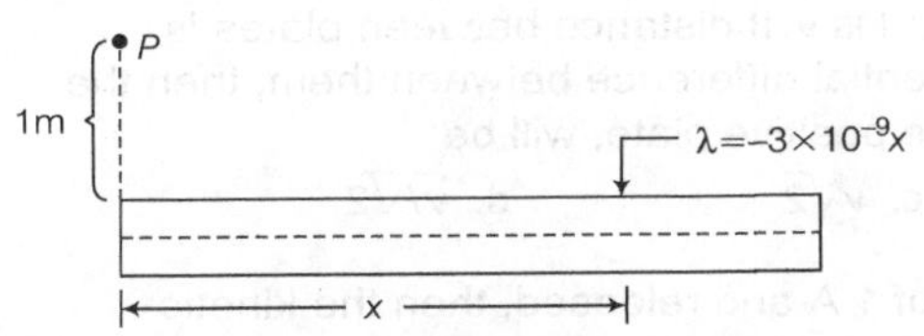

With $V = 0$ at infinity, find electrical potential at point P at a distance of 1 m from one end of the rod.

5 State relation between

i. PE and force

ii. potential and PE

iii. PE and field

Only One Option Correct Type

Working Space

Direction *(Q. Nos. 6-12) This section contains 7 multiple choice questions. Each question has four choices (a), (b), (c) and (d), out of which ONLY ONE option is correct.*

6 A surface will be an equipotential surface, if

a. electric field is tangential to surface at all the points
b. electric potential at all the points is non-zero
c. work done (along any path) in moving a point charge over the surface is non-zero
d. electrical potential energy of a unit positive charge at any point on the surface will be same

7 A point P is equidistant from both point charges as shown, then

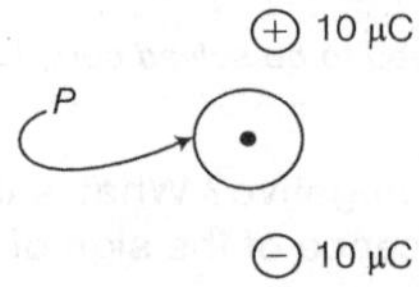

a. electric field at P is zero
b. electric potential at P is not zero
c. work done in moving a charge q from infinity to P is not zero
d. if a test charge of + 1 C is placed at P, its potential energy is zero

8 Three point charges 1 C, – 2 C and – 2 C are placed at the vertices of an equilateral triangle of side 1 m. The work done by an external force to increase the separation of the charges to 2 m (in joule) is

a. $\frac{1}{4\pi\varepsilon_0}$ b. $\frac{1}{8\pi\varepsilon_0}$ c. $\frac{1}{16\pi\varepsilon_0}$ d. zero

9 An electron is released between two plates with charge densities $+\sigma$ and $-\sigma$ respectively as shown.

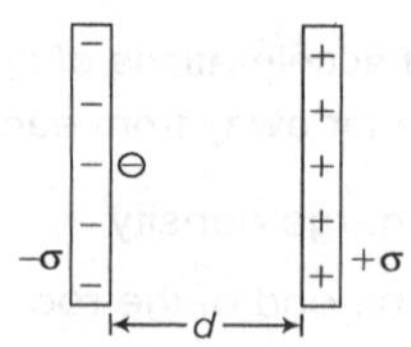

As it just reaches the night plate, its speed is v. If distance between plates is reduced to half without changing the potential difference between them, then the speed of the electron when it reaches the positive plate, will be

a. v b. $2v$ c. $v\sqrt{2}$ d. $v/\sqrt{2}$

10 Two free protons are kept at a distance of 1 Å and released, then the kinetic energy of each proton at infinite separation is

a. 5.6×10^{-12} J b. 11.5×10^{-19} J
c. 23×10^{-19} J d. 46×10^{-19} J

11 Three electrons are brought to corners of an equilateral triangle (side = 1.0×10^{-10} m). The work done in the process is

a. 100 eV b. 37 eV
c. 43 eV d. 54 eV

12 A helium nucleus is accelerated by applying a potential difference of V volt such that it gains a kinetic energy of 125 keV. Then, V is

a. – 62.5 kV b. 62.5 kV
c. – 37.8 kV d. 38.5 kV

One or More than One Options Correct Type

Direction (Q. Nos. 13-18) *This section contains 6 multiple choice questions. Each question has four choices (a), (b), (c) and (d), out of which ONE or MORE THAN ONE are correct.*

13 Which of the statements are true?

a. Speed of a 1.5 keV electron is 2.3×10^{-7} m/s

b. Speed of a 10.5 keV proton is 5.4×10^{5} m/s

c. Speed of a 1.5 keV α-particle is 5.4×10^{5} m/s

d. Speed of a 1.5 keV neutron is 2.3×10^{7} m/s

14 Many chemical reactions release energy and many of them absorb energy. If at the beginning of a reaction, an electron and a proton are separated by 0.110 nm and their final separation is 0.100 nm. Then,

a. reaction is endothermic

b. reaction is exothermic

c. PE is gained by electron proton system

d. PE is lost by electron proton system

15 Four point charges, each $+Q$ are placed at the corners of a square of side a. Choose the correct statements.

a. A fifth charge $+Q$ is placed at centre, then its potential energy is $U = \frac{Q^2}{4\pi\varepsilon_0 a} \cdot (4\sqrt{2})$

b. Potential energy of four charges (at vertices) configuration will be, $\frac{Q^2}{4\pi\varepsilon_0 a} \cdot (4-\sqrt{2})$

c. Fifth charge is in stable equilibrium

d. Fifth charge is in unstable equilibrium

16 A positive charge is situated (as shown) in between two dissimilarity charged plates. If it moves along path I and II (shown in figure).

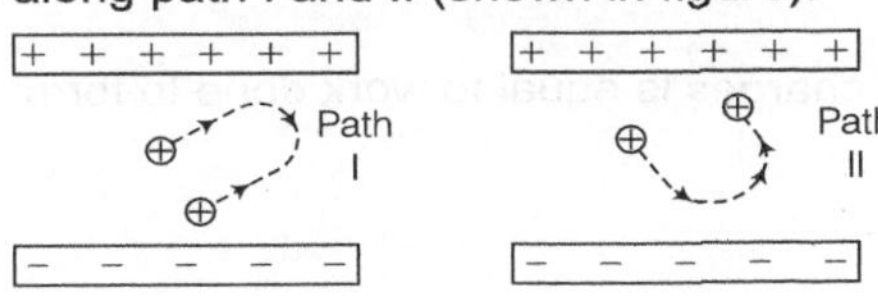

Then,

a. along path I electric field II does negative work on charge

b. along path II electric field does positive work on charge

c. along path II potential energy increases

d. along path II potential energy decreases

17 Following figure shows a series of electric field lines.
A unit positive charge is moved along the paths (marked dotted) shown, then there is no change in PE of the charge when it is being moved along

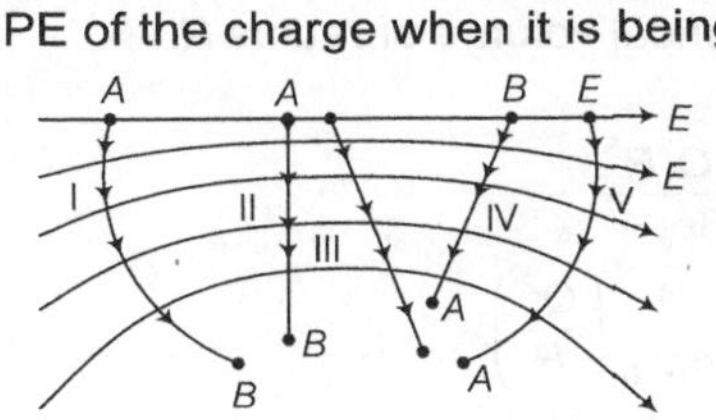

a. path I

b. path III

c. path IV

d. path V

Working Space

18 A set of point charges is held in plane at the vertices of an equilateral triangle of side 10.0 cm. Then,

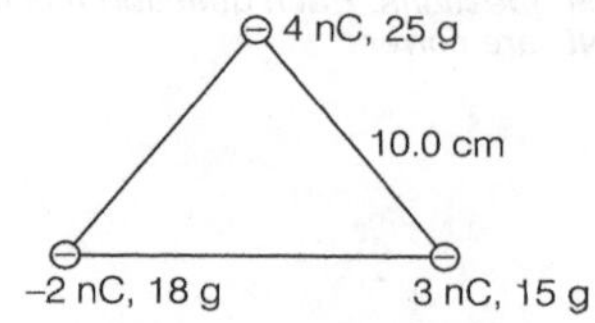

a. kinetic enegy is produced when the charges are released in frictionless void of outer space is equal to potential energy of system
b. kinetic energy is equally shared by all the charges
c. kinetic energy maxima is reached instantaneously when released
d. kinetic energy maxima is reached when charges reach very far off from each other

Statement Type

Direction *(Q. Nos. 19 and 20) This section is based on Statement I and Statement II. Select the correct answer from the codes given below.*

a. Both Statement I and Statement II are correct and Statement II is the correct explanation of Statement I
b. Both Statement I and Statement II are correct and Statement II is not the correct explanation of Statement I
c. Statement I is correct but Statement II is incorrect
d. Statement II is correct but Statement I is incorrect

19 **Statement I** A point charge is placed in a uniform field pointing North. There is no change in potential energy of charge when it is moved along a line pointing East-West.

Statement II Electric field of a point charge is non-uniform.

20 **Statement I** When two positive charges are moved towards each other, potential energy of system decreases.

Statement II Potential energy of a system of charges is equal to work done to form the system.

Comprehension Type

Direction *(Q. Nos. 21-27) This section contains two paragraphs, each describing theory, experiments, data, etc. Seven questions related to the paragraphs have been given. Each question has only one correct answer among the four given options (a), (b), (c) and (d).*

Passage I

Coulomb's force remembers the inverse square law of gravitation with charge playing same role as in mass is gravitation.

So, expressions for potential and potential energy similar in both cases.

21 A charge Q is uniformly distributed over a spherical volume of radius R. Expression for energy of system is

a. $\dfrac{4\pi\, Q^2R^5}{15\varepsilon_0}$
b. $\dfrac{4\pi\, Q^2R^5}{3\varepsilon_0}$
c. $\dfrac{3}{5}\cdot\dfrac{1}{4\pi\varepsilon_0}\left(\dfrac{Q^2}{R}\right)$
d. $\dfrac{4}{3}\cdot\dfrac{1}{4\pi\varepsilon_0}\left(\dfrac{Q^2}{R}\right)$

22 Product of mass and radius of the Earth is 2.5×10^{31} kg-m and $g = 10$ N/kg. Energy required to disassemble the planet earth completely is

a. 1.5×10^{32} J
b. 1.5×10^{31} J
c. 1.5×10^{62} J
d. 2.5×10^{31} J

Working Space

23 If charge Q is given to a spherical conductor of radius R, then energy of the system is

a. $\frac{1}{4\pi\varepsilon_0}\cdot\frac{Q}{2R}$

b. $\frac{1}{4\pi\varepsilon_0}\cdot\frac{Q^2}{2R}$

c. $\frac{3}{15}\cdot\frac{1}{4\pi\varepsilon_0}\cdot\frac{Q^2}{R}$

d. None of these

Passage II

Four point charges $+8\,\mu C$, $-1\,\mu C$, $-1\,\mu C$ and $+8\,\mu C$ are fixed at points $(0, -\sqrt{27/2}\,m)$, $(0, -\sqrt{3/2}\,m)$, $(0, \sqrt{3/2}\,m)$ and $(0, \sqrt{(27/2)}\,m)$ respectively, in XOY-plane. A particle of mass 6×10^{-4} kg and charge $+0.1\,\mu C$ moves along the x-direction towards the origin from a large distance as shown in the figure.

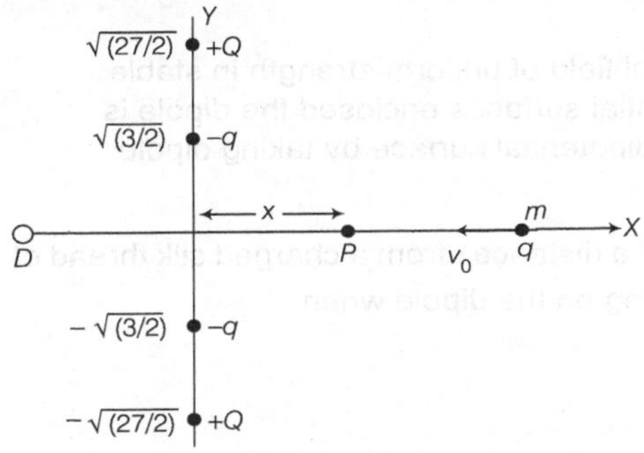

24 Potential at point P, distant x from the origin is

a. $1.8\times10^{-9}\left\{\frac{8}{\sqrt{\left(\frac{27}{2}+x^2\right)}}+\frac{1}{\sqrt{\left(\frac{3}{2}+x^2\right)}}\right\}$

b. $1.8\times10^{4}\left\{\frac{8}{\sqrt{\left(\frac{27}{2}+x^2\right)}}-\frac{1}{\sqrt{\left(\frac{3}{2}+x^2\right)}}\right\}$

c. $1.8\times10^{4}\left\{\frac{1}{\sqrt{\left(\frac{27}{2}+x^2\right)}}-\frac{1}{\sqrt{\left(\frac{3}{2}+x^2\right)}}\right\}$

d. $1.8\times10^{9}\left\{\frac{1}{\sqrt{\left(\frac{27}{2}+x^2\right)}}+\frac{1}{\sqrt{\left(\frac{3}{2}+x^2\right)}}\right\}$

25 Least value of velocity at infinity which must be imparted to the particle so that it crosses the origin is

a. 2 m/s

b. 1 m/s

c. 3 m/s

d. 4 m/s

26 Kinetic energy of the particle at the origin is

a. 2.4×10^{-4} J

b. 2.4×10^{4} J

c. 3×10^{-4} J

d. 3×10^{4} J

27 Position of stable equilibrium of particle on X-axis is

a. $x = 0$

b. $x = +\sqrt{\frac{5}{2}}$ m

c. $x = -\sqrt{\frac{5}{2}}$ m

d. stable equilibrium of particle is not possible

DPP-12 Potential and Potential Energy of Dipole

Subjective Questions

Working Space

***Direction** (Q. Nos. 1-4) These questions are subjective in nature, need to be solved completely on notebook.*

1 A small electric dipole exists in an external field of uniform strength in stable equilibrium position. One of the equipotential surfaces enclosed the dipole is spherical in shape. Find radius of this equipotential surface by taking dipole moment **p** and field strength **E**.

2 A dipole with dipole moment **p** is located at a distance **r** from a charged silk thread of uniform charged density λ. Find force acting on the dipole when

i. **p** is parallel to thread **r**

ii. **p** along radius vector **r**

iii. **p** is perpendicular to **r**

3 Calculate electric potential due to a small dipole of dipole moment 4.8×10^{-30} Cm at a point 4.1×10^{-9} m away, if this point is

i. along axis of dipole nearer to positive charge.

ii. 45° above the axis but nearer to positive charge.

4 Water molecule has two dipole moments $\mathbf{p}_1$ and $\mathbf{p}_2$, both have same magnitudes.

Taking, $V = 0$ at $r = \infty$, show that electric potential due to the molecule (at a very large distance) is $V = \frac{1}{4\pi\varepsilon_0} \frac{p \cos \theta}{r}$.

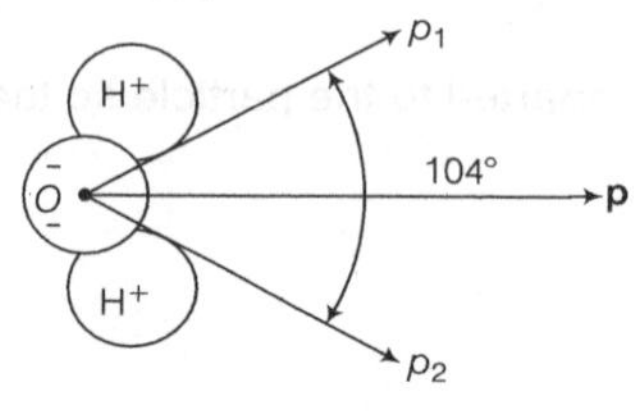

where, θ = angle between line joining point and molecule and dipole moment.

Only One Option Correct Type

***Direction** (Q. Nos. 5-14) This section contains 10 multiple choice questions. Each question has four choices (a), (b), (c) and (d), out of which ONLY ONE option is correct.*

5 Dipole moment is also measured in debye, 1 debye unit is equivalent to

a. 3.34×10^{-30} Cm
b. 3.34×10^{-19} Cm
c. 3.34×10^{-27} Cm
d. 3.34×10^{-12} Cm

Working Space

6 Electric potential due to given charge distribution at P,

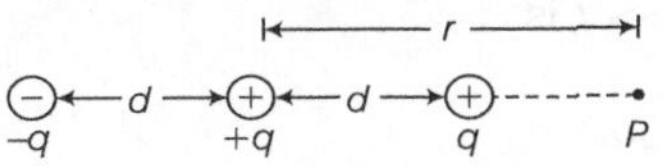

for $r >> d$ is

a. $\frac{kq}{r}\left(1-\frac{2d}{r}\right)$ b. $\frac{kq}{r}\left(1+\frac{2d}{r}\right)$

c. $\frac{kq}{r^2}$ d. $\frac{2kq}{r^2}(1+2d)$

7 A dipole is placed parallel to electric field. If W is the work done in rotating the dipole by 60°, then work done in rotating it by 180 ° is

a. $2W$ b. $3W$

c. $4W$ d. $W/2$

8 Positive and negative charges of equal magnitude are kept at $\left(0,\ 0,\ \frac{a}{2}\right)$ and $\left(0,\ 0,\ -\frac{a}{2}\right)$, respectively. The work done by electric field when another positive point charge is moved from $(-a,\ 0,\ 0)$ to $(0,\ a,\ 0)$, is **[IIT JEE 2007]**

a. positive

b. negative

c. zero

d. depends on the path connecting the initial and final positions

9 An electric dipole with dipole moment 4×10^{-9} Cm makes an angle of 30 ° with a uniform field of 5×10^{4} N/C. Work done in rotating dipole to zero potential energy position is

a. $\sqrt{3}\times10^{-4}$ J b. 2×10^{-4} J

c. 20×10^{-4} J d. $\sqrt{3}/2\times10^{-4}$ J

10 An electric dipole is placed normal to the direction of a uniform field. If p indicates dipole moment and E indicates field intensity, then work done in deflecting the dipole by an angle of 180° is

a. pE b. $2pE$

c. $-2pE$ d. zero

11 Two dipoles are at an angle at 60° to each other. The distance between them is L. Electric field is perpendicular to the first dipole. The dipole moments of both the dipoles are p. Then, the potential energy will be

a. $\left[\frac{Kp^3}{2L^3}-\frac{pE}{2}\right]$ b. $-\left[\frac{Kp^3}{2L^3}+\frac{pE^2}{2}\right]$

c. $-\left[\frac{Kp^2}{2L^3}+\frac{pE}{2}\right]$ d. $\left[\frac{Kp^2}{2L^3}+p\frac{E^2}{2}\right]$

12 An electric dipole is placed with its charges at $(a,\ 0,\ 0)$ and $(-a,\ 0,\ 0)$. The equation of one of its equipotential surface is

a. $\mathbf{r}\cdot\hat{\mathbf{i}}=0$ b. $\mathbf{r}\cdot\hat{\mathbf{j}}=0$

c. $\mathbf{r}\cdot\hat{\mathbf{k}}=0$ d. $y+z=0$

Working Space

13 For the dipole shown in figure, X-components of electric field E_x at some point $P(x, y)$ with position vector $\mathbf{r}$, $|\mathbf{r}| = (x^2 + y^2)^{1/2} >> l$, is

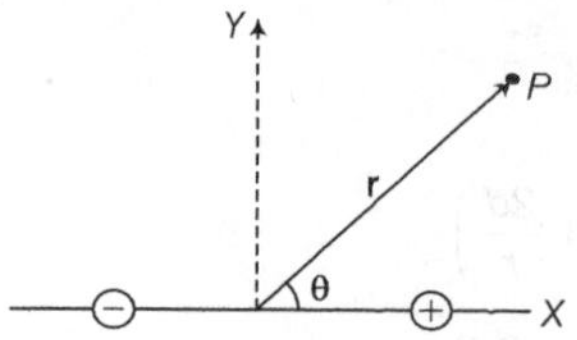

a. $\frac{p}{4\pi\varepsilon_0}\left\{\frac{\cos^2\theta - 2\sin^2\theta}{r^3}\right\}$ b. $\frac{p}{4\pi\varepsilon_0}\left\{\frac{3\cos\theta\sin\theta}{r^3}\right\}$

c. $\frac{p}{4\pi\varepsilon_0}\left\{\frac{\sin^2\theta - \cos^2\theta}{r^3}\right\}$ d. $\frac{p}{4\pi\varepsilon_0}\left\{\frac{2\cos^2\theta - \sin^2\theta}{r^3}\right\}$

14 In a region of space, potential is given by $V = \left(2\frac{\text{V}}{\text{m}^2}\right)x^2$. A dipole of charge 2×10^{-6} C and length $3\overset{\circ}{\text{A}}$ is placed in this region. The net force acting on the dipole when the dipole is placed along the X-axis is

a. 24×10^{-14} N b. 24×10^{-10} N c. 24×10^{-16} N d. 24×10^{-15} N

One or More than One Options Correct Type

***Direction** (Q. Nos. 15-18) This section contains 4 multiple choice questions. Each question has four choices (a), (b), (c) and (d), out of which ONE or MORE THAN ONE are correct.*

15 An electric dipole of dipole moment p is placed in the position of equilibrium in uniform electric field intensity E. This is rotated through an angle θ from the initial position. Then,

a. the work done in deflecting the dipole through the angle θ is $pE\cos\theta$
b. the work done in deflecting the dipole through the angle θ is $pE(1-\cos\theta)$
c. the potential energy of the dipole in the final position is $pE\sin\theta$
d. the potential energy of the dipole in the final position is $pE(1-\cos\theta)$

16 The electric potential due to a tiny dipole whose dipole moment is 4.8×10^{-30} cm at a point 4.1×10^{-9} m away. Which of the following statements is/are correct.

a. If this point is along the axis of the dipole nearer the positive charge, the potential is 2.6×10^3 V
b. If this point is 45° above the axis but nearer the positive charge, the potential is 1.8×10^{-3} V
c. If this point is 45° above the axis but nearer the negative charge, the potential is 1.8×10^3 V
d. If this point is along the axis of the dipole nearer the negative charge, then potential is 2.6×10^3 V

17 An electric dipole consists of two charges of equal magnitude and opposite sign separated by a distance $2a$. The dipole is along the X-axis and is centred at origin. Then,

a. point located at the right of a positive charge, then potential is $\frac{2k_{eq}a}{x^2 - a^2}$
b. point located at the left of a negative charge, then potential is $-\frac{2k_{eq}a}{x^2 - a^2}$
c. electric potential at a point far from the dipole is $\frac{2k_{eq}a}{x^2}$
d. electric potential anywhere between the two charge is $\frac{-2k_{eq}a}{a^2}$

Working Space

18 AB is an electric dipole of length $2a$ and P be any point, where $OP = r$. Let θ be the angle between r and the dipole axis. Choose the correct statement (s).

a. When the point P lies on the axial line of the side of the positive charge, potential will be $\frac{kp}{r^2}$

b. When the point P lies on the axial line of the dipole but on the side of negative charge, potential will be $\frac{kp}{r^3}$

c. When the point P lies on the equatorial line of the dipole, no work is done in bringing a charge from infinity to the dipole along the perpendicular bisector

d. At large distances, $V \propto \frac{1}{r^2}$ and not as $V \propto \frac{1}{r}$

Comprehension Type

Direction (Q. Nos. 19-21) This section contains a paragraph, describing theory, experiments, data, etc. Three questions related to the paragraph have been given. Each question has only one correct answer among the four given options (a), (b), (c) and (d).

Passage

A massless rigid rod carries two small spheres each of mass m and charge q at each of its ends. This rod is held in a region of uniform electric field **E** such that the rod makes a small angle θ (around 3 to 5°) with the direction of field.

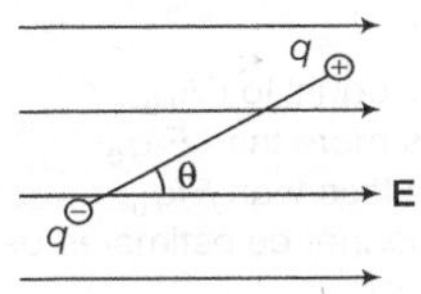

19 When released, initial angular acceleration of rod is

a. $\frac{3qE}{L}\theta$ b. $\frac{4qE}{mL}\theta$ c. $\frac{6qE}{mL}\theta$ d. $\frac{2qE}{mL}\theta$

20 Least time required by rod to become parallel to field is

a. $\frac{\pi}{2}\sqrt{\frac{mL}{2qE}}$ b. $\frac{\pi}{4}\sqrt{\frac{mL}{qE}}$ c. $2\pi\sqrt{\frac{mL}{2qE}}$ d. None of these

21 Change in potential energy of rod when it is aligned with the field w.r.t. initial position is

a. $pE(1-\cos\theta)$ b. $-pE\cos\theta$ c. $pE(\cos\theta - 1)$ d. $-2pE$

Matching List Type

Direction (Q. No. 22) Choices for the correct combination of elements from Column I and Column II are given as options (a), (b), (c) and (d), out of which one is correct.

22 Consider a dipole with dipole moment 4.8×10^{-30} cm. Let a point P lies 4.1×10^{-9}m away from centre of dipole. Match the values of potentials given in column II with column I.

	Column I		Column II
i.	Point is along axis of dipole nearer to positive charge.	p.	-1.8×10^{-3} V
ii.	Point is 45° above the axis but nearer to positive charge.	q.	1.8×10^{-3} V
iii.	Point is 45° above the axis but nearer to negative charge.	r.	2.6×10^{-3} V

Codes

	i	ii	iii			i	ii	iii
a.	r	q	p		b.	p	q	r
c.	r	p	q		d.	q	p	r

Revisal Problems for JEE Main

Only One Option Correct Type

Direction (Q. Nos. 1-20) *This section contains 20 multiple choice questions. Each question has four choices (a), (b), (c) and (d), out of which ONLY ONE option is correct.*

Working Space

1 A positively charged ball with charge q hangs from a rigid support as shown.
An another charged ball with charge q_0 is placed at P for a distance r from q.
Value of force F on charge q_0 due to q is found, then ratio $\frac{F}{q_0}$ is calculated.

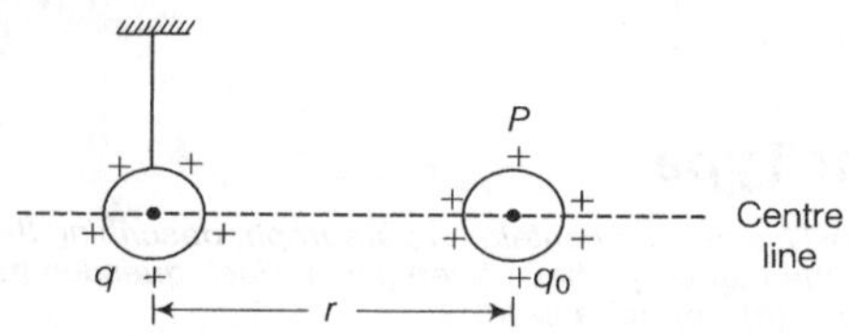

Then, true option is

a. electric field at P is equal to F/q_0
b. electric field at P is more than F/q_0
c. electric field at P is less than F/q_0
d. electric field at P cannot be estimated using such technique

2 A particle of charge $q = 1.6 \times 10^{-19}$ C and mass $m = 12 \times 10^{-24}$ kg is shot between two capacitor plates with a potential difference of 300 V and separation of 2 cm.
If dimensions of plates are as shown in figure, then the greatest initial velocity for which the particle will not be able to come out of plates is

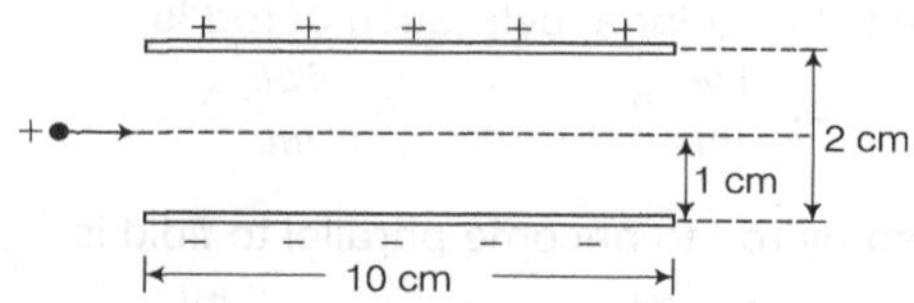

a. 10^3 m/s
b. 10^4 m/s
c. 10^5 m/s
d. particle never comes out of region of plates

3 A small ball of mass m and charge q is rotated in a vertical plane by using a string of length l. An electric field E also exists in the region which points upwards.
If ball is imparted a horizontal velocity v at top of the vertical circle, such that tension in the string in the lower most position of ball is 15 times of the weight of the ball, then v is

a. $\sqrt{\frac{l}{m}(10\,mg + qE)}$ b. $\sqrt{\frac{l}{m}(5\,mg + qE)}$ c. $\sqrt{\frac{l}{m}(10\,mg + 5qE)}$ d. $\sqrt{\frac{l}{m}(mg + qE)}$

4 A positively fixed charged ring is placed on a horizontal surface on the earth. A positively charged particle is released from a height R on its axis. The variation of total potential energy as a function of separation between particle and ring is correctly depicted by which of the following graphs.

a. 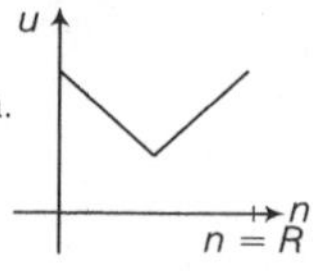

b. 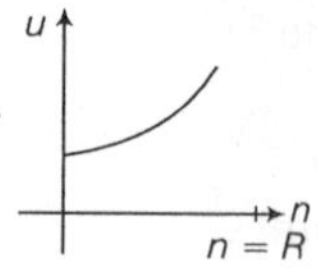

c. 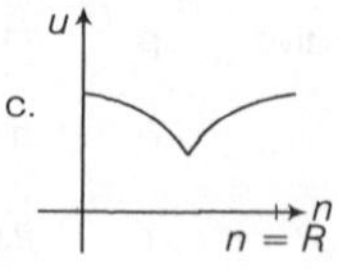

d. 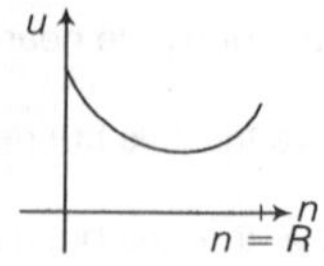

5 A positively charged thin metal ring of radius R is fixed in the XY-plane, with its centre at origin. A negatively charged particle is released from rest at the point $(0, 0, c)$, $c > 0$.

Consider the following statements.

i. The electric field at P due to metallic ring is always directed towards origin.
ii. The negatively charged particle is accelerated towards origin and undergoes periodic motion.

Which of these are correct?

a. Only (i)
b. Only (ii)
c. Both (i) and (ii)
d. Neither (i) nor (ii)

6 A point charge Q is on the axis of a short cylinder at its centre. The diameter of the cylinder is equal to its length l. Then, total flux through the curved sides of the cylinder

a. cannot be found as Gaussian surface drawn is not symmetrical to charge

b. is equal to $\frac{Q}{2\varepsilon_0}\left(1-\frac{1}{\sqrt{2}}\right)$

c. is equal to $\frac{Q}{\varepsilon_0}$

d. is equal to $\frac{Q}{\sqrt{2}\,\varepsilon_0}$

7 A very large thin plate has uniform surface charge density σ. It is touched to a long and wide slab of thickness d with uniform volume charge density ρ_E. Then, field inside the slab is

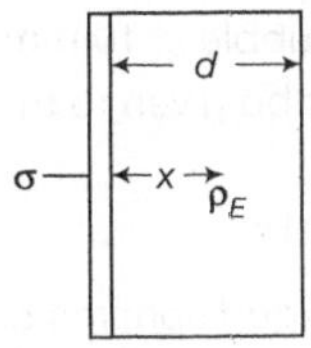

a. $\frac{\sigma+\rho_E(x-d)}{\varepsilon_0}, 0<x<d$
b. $\frac{\sigma+\rho_E(x-d)}{2\varepsilon_0}, 0<x<d$
c. $\frac{\sigma+\rho_E(2x-d)}{2\varepsilon_0}, 0<x<d$
d. $\frac{\sigma+\rho_E(2x-d)}{\varepsilon_0}, 0<x<d$

8 An electron enters a region of a uniform field, midway between two charged plates. $E=3.8\times10^3$ N/C, plate separation is 1.0 cm, length of region of uniform field is nearly 3.464 cm.

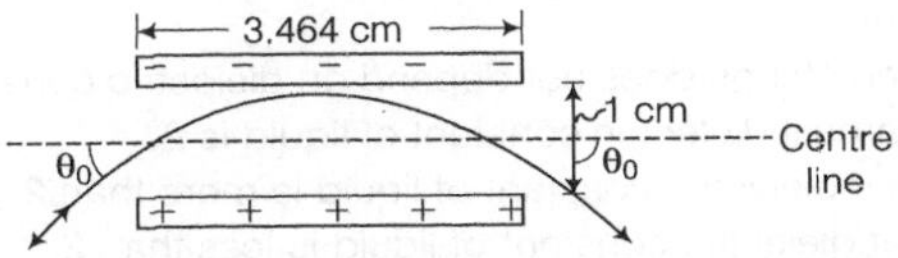

If path of electron is symmetrical, i.e. they enter and leave the region of field at same angle θ_0, just missing the top plate. Then, θ_0 is

a. 45°
b. 30°
c. 60°
d. 18°

9 If a charge q is placed at mid-point of edge of a cube, then total flux through the faces of the cube is

a. $\frac{q}{8\varepsilon_0}$
b. $\frac{q}{4\varepsilon_0}$
c. $\frac{q}{2\varepsilon_0}$
d. $\frac{q}{2\varepsilon_0}$

10 If there is only one type of charge in the universe, then

a. $\oint \mathbf{E}\cdot d\mathbf{s} \neq 0$ for any surface

b. $\oint \mathbf{E}\cdot d\mathbf{s} = 0$ on any surface

c. $\oint \mathbf{E}\cdot d\mathbf{s}$ cannot be defined

d. $\oint \mathbf{E}\cdot d\mathbf{s} = \frac{q}{\varepsilon_0}$, if charges of magnitude q were inside the surface

11 A spherical balloon of 15 cm radius carries a charge of 2 μC. A water droplet at 1 cm distance from the surface of balloon has an induced dipole moment of 6.3 nCm pointing away from the balloon. Force on the water droplet exerted by balloon is

a. 8.78 mN
b. 55.3 mN
c. 71.3 mN
d. 21.56 mN

12 Two equal point charges each of magnitude Q are located at point $(\pm a, 0, 0)$. A third point charge of mass m and charge $-Q$ revolves around the X-axis in a circle of radius r attraction to point A and B. Kinetic energy of rotating particle is given by

a. $\frac{2Q^2r^2}{4\pi\varepsilon_0(r^2+a^2)^{3/2}}$
b. $\frac{Q^2r^2}{4\pi\varepsilon_0(r^2+a^2)^{3/2}}$
c. $\frac{Q^2r}{4\pi\varepsilon_0(r^2+a^2)^{1/2}}$
d. $\frac{2Q^2r}{4\pi\varepsilon_0(r^2+a^2)^{1/2}}$

13 Potential to which a soap bubble of radius R be charged, so that pressure outside and inside is same will be (take, p_0 = atmospheric pressure and T = surface tension of soap film)

a. $V=\sqrt{\frac{Tr}{\varepsilon_0}}$
b. $V=\sqrt{\frac{3Tr}{\varepsilon_0}}$
c. $V=\sqrt{\frac{8Tr}{\varepsilon_0}}$
d. $V=\sqrt{\frac{5Tr}{\varepsilon_0}}$

Working Space

14 A circular ring of radius r is placed at mid-point of a dipole of dipole moment with its plane perpendicular to axis of dipole. Electric flux passing through the ring is (take length of dipole $= 2l$)

a. $\frac{p}{2l\varepsilon_0}\left(1-\frac{1}{\sqrt{1+(r^2/l^2)}}\right)$ b. $\frac{2p}{l\varepsilon_0}\left(1-\frac{1}{\sqrt{1+(r^2/l^2)}}\right)$

c. $\frac{p}{\varepsilon_0}\left(1-\frac{1}{\sqrt{1+(r^2/l^2)}}\right)$ d. 0

15 A spherical metal shell of radius r is given a charge q. Stress developed in the shell due to charging is

a. $\frac{q^2}{4\pi^2\varepsilon_0 r^4}$ b. $\frac{q^2}{16\pi^2\varepsilon_0 r^4}$ c. $\frac{q^2}{32\pi^2\varepsilon_0 r^4}$ d. zero

16 Radius of a soap bubble is 10 mm and surface tension of soap is 3×10^{-2} N/m. Charge which must be given to this bubble, so that it begins to swell is

a. 1.83×10^{-12} C b. 1.83×10^{-20} C
c. 1.83×10^{-6} C d. 1.83×10^{-8} C

17 Two identically charged spheres are suspended by strings of equal length. The strings make 30° angle with each other. When suspended in a liquid of density 0.8 g/cc, the angle remains same. Then,

I. the angle between strings does not depend on dielectric constant of liquid.
II. angle remains same if dielectric constant of liquid is 2.
III. angle decreases if dielectric constant of liquid is more than 2.
IV. angle increases if dielectric constant of liquid is less than 2.

True option is

a. Only I b. II and III c. II and IV d. III and IV

18 A ball of radius $R = 2$ m carries a positive charge whose volume density depends on a distance r from the balls' centre as $\rho = \rho_0\left(1-\frac{r}{R}\right)$, where $\rho_0 =$ constant. Then,

I. electric field is maximum at surface.
II. electric field at a distance $r = 1$ m from centre is $\frac{5}{24}\frac{\rho_0}{\varepsilon_0}$.
III. electric field at a distance $r = 1$ m from centre is $\frac{5}{12}\frac{\rho_0}{\varepsilon_0}$.
IV. electric field is maximum at a distance of $\frac{4}{3}$ m from the centre of ball.

True option is

a. I and II b. I and III c. II and III d. II and IV

19 The variation of electric field between two charges q_1 and q_2 along the line joining charges is as shown below. Then,

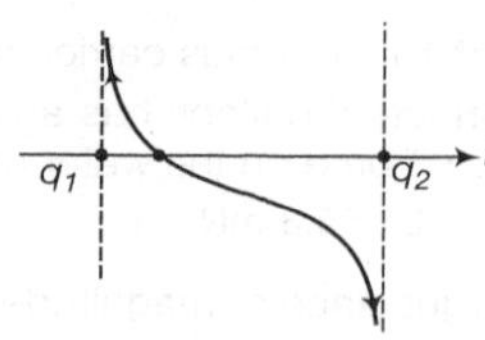

I. q_1 and q_2 are positive and $q_1 < q_2$.
II. q_1 and q_2 are positive and $q_1 > q_2$.
III. q_1 is positive, q_2 is negative and $q_1 < q_2$.
IV. q_1 and q_2 are negative and $q_1 < q_2$.

True option is

a. I and II b. I and IV c. II and III d. II and IV

20 A charge $Q = 8.85$ nC is located at a distance b on the axis of a disc of radius b from the plane of disc. Let radius of disc is started to charge at a rate of 5 cm/s. Then,

a. flux through the disc charges at a rate of $150\sqrt{2}\ b$Vms^{-1}
b. flux through the disc charges at a rate of $\frac{75\sqrt{2}b}{2}$ Vms^{-1}
c. flux through the disc is always zero
d. flux through the disc is always constant

SET - 2

Revisal Problems for JEE Advanced

Only One Option Correct Type

Working Space

Direction *(Q. Nos. 1-10) This section contains 10 multiple choice questions. Each question has four choices (a), (b), (c) and (d), out of which ONLY ONE option is correct.*

1 Two small spheres of masses m_1 and m_2 with charges Q_1 and Q_2 are suspended with insulating threads of lengths L_1 and L_2 such that centres of spheres are in level.

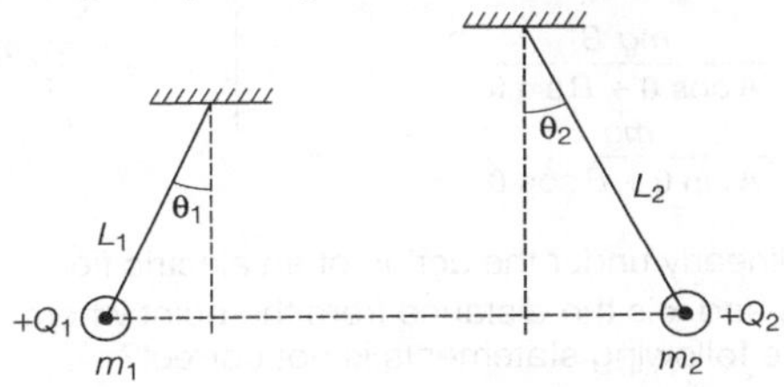

If $\theta_1 = \theta_2$, then which of the following condition is necessary?

a. $m_1 = m_2$

b. $m_1 \neq m_2$ but $Q_1 = Q_2$

c. $Q_1 = Q_2$

d. $L_1 = L_2$

2 Five charges are located in space as shown in figure. If q_A, q_B, q_C, q_D and q_E all are positive q, then force on a negative test charge $-q$ located at origin is given by

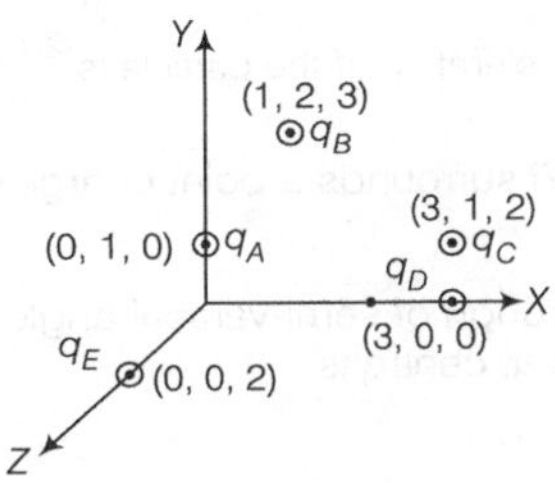

a. $kq^2\left[\hat{i}\left(\frac{1}{3^2}+\frac{4}{14^{3/2}}\right)+\hat{j}\left(1+\frac{3}{14^{3/2}}\right)\right]$

b. $kq^2\left[\hat{i}\left(\frac{1}{3^2}+\frac{4}{14^{3/2}}\right)+\hat{j}\left(1+\frac{3}{14^{3/2}}\right)+\hat{k}\left(\frac{1}{2^2}+\frac{5}{14^{3/2}}\right)\right]$

c. $\frac{8kq^2}{9}$

d. $kq^2\left[\hat{i}\left(\frac{1}{\sqrt{3}}+\frac{1}{\sqrt{2}}\right)+\hat{j}\,\frac{1}{\sqrt{3}}+\hat{k}\,\frac{2}{\sqrt{3}}\right]$

3 In Millikan-oil drop experiment, an oil drop of radius r with charge q falls with a terminal velocity v_1 without presence of an electric field. When an electric field of intensity E is switch on which produces upward motion of oil drop, let a terminal velocity v_2 is observed. If η is coefficient of viscosity of air, then charge on the drop is

a. $\frac{6\pi\eta r(v_1 - v_2)}{E}$

b. $\frac{6\pi\eta\varepsilon v_1 v_2}{E}$

c. $\frac{6\pi\eta\varepsilon(v_1 + v_2)}{E}$

d. $\frac{6\pi\eta\varepsilon(v_1 + v_2)}{2E}$

4 A thin rod of uniform charge per unit length λ lies along the X-axis.

Then, for a point P on the perpendicular bisector of the rod

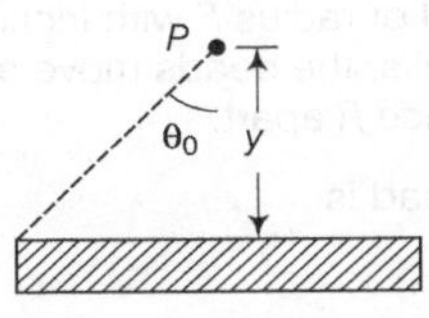

a. x-component of electric field is $E_x = \frac{2k\lambda}{y}$

b. x-component of electric field is $\frac{2k\lambda \sin\theta_0}{y}$

c. y-component of electric field is $\frac{2k\lambda}{y}$

d. y-component of electric field is $\frac{2k\lambda \sin\theta_0}{y}$

5 Using an electric field produced by two plates, a beam of ions each of charge q and mass m with an initial velocity v is turned through an angle of 90° as shown.

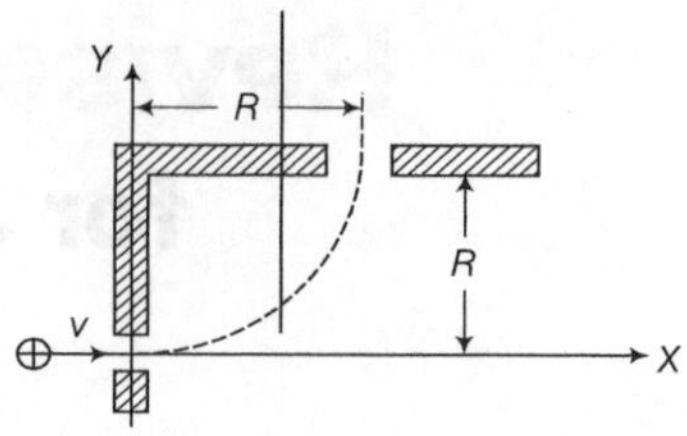

Magnitude of field in the region is

a. $\frac{mv^2}{\sqrt{3}qR}$ b. $\frac{mv^2}{2\sqrt{2}qR}$

c. $\frac{mv^2}{\sqrt{2}qR}$ d. $\frac{mv^2}{2qR}$

6 A charged ball of mass m and charge q is suspended in the region of a uniform electric field $\mathbf{E} = A\hat{\mathbf{i}} + B\hat{\mathbf{j}}$ (N/C).

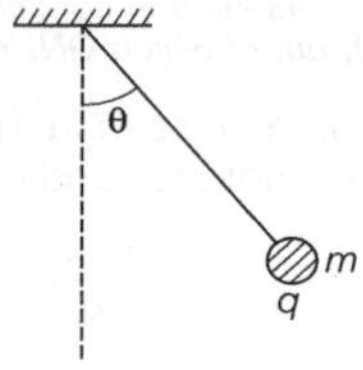

If ball is in the equilibrium at an angle θ with the vertical, then tension in the string is

a. $\frac{mgA}{A\cos\theta + B\sin\theta}$ b. $\frac{mg\,B}{A\cos\theta + B\sin\theta}$

c. $\frac{mg}{A\cot\theta + B}$ d. $\frac{mg}{A\sin\theta + B\cos\theta}$

7 A particle of charge q and mass m moves rectilinearly under the action of an electric field $\mathbf{E} = \alpha - \beta x$. Here, α and β are positive constants and x is the distance from the point where the particle was initial at rest. Then, which of the following statements is not correct?

a. The motion of the particle is oscillatory

b. The amplitude of the particle is $\frac{\alpha}{\beta}$

c. The mean position of the particle is at $x = \frac{\alpha}{\beta}$

d. The maximum acceleration of the particle is $\frac{2q\alpha}{m}$

8 A sphere of radius R surrounds a point charge Q, located at its centre.

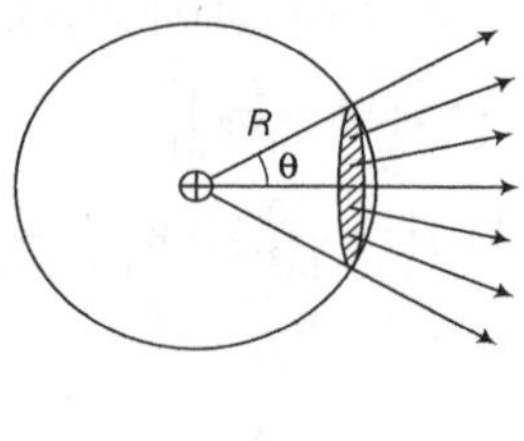

Net outward flux through of semi-vertical angle θ and of slant length R with vertex at centre is

a. $\frac{Q}{2\varepsilon_0(1-\cos\theta)}$ b. $\frac{kQ\cos\theta}{4\varepsilon_0}$

c. $\frac{Q}{\varepsilon_0}(1-\cos\theta)$ d. $\frac{Q}{2\varepsilon_0}(1-\cos\theta)$

9 A rectangular surface in XOY-plane extends from $x = 0$ to $x = w$ and $y = 0$ to $y = h$. If an electric field $\mathbf{E} = ay\hat{\mathbf{i}} + bz\hat{\mathbf{j}} + cx\hat{\mathbf{k}}$ exists in the region, then flux through the surface is

a. $\frac{cw^2b}{2}$ b. $\frac{chw^2}{2}$

c. $\frac{c^2hw}{2}$ d. $\frac{cw^2a}{2}$

10 Two identical beads each of mass m, when placed in a hemispherical bowl of radius R with frictionless non-conducting walls, the beads move and at equilibrium they are at a distance R apart.

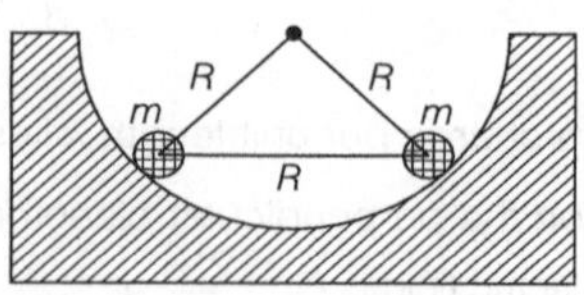

Charge on each bead is

a. $R\left(\frac{mg}{2k}\right)^{1/2}$ b. $R\left(\frac{mg}{3k}\right)^{1/2}$

c. $R\left(\frac{mg}{\sqrt{3}k}\right)^{1/2}$ d. $R\left(\frac{2mg}{\sqrt{3}k}\right)^{1/2}$

Working Space

Working Space

Comprehension Type

***Direction** (Q. Nos. 11-16) This section contains 2 paragraphs, each describing theory, experiments, data, etc. Six questions related to the paragraphs have been given. Each question has only one correct answer among the four given options (a), (b), (c) and (d).*

Passage I

An empty thick conducting shell of inner radius a and outer radius b is shown.

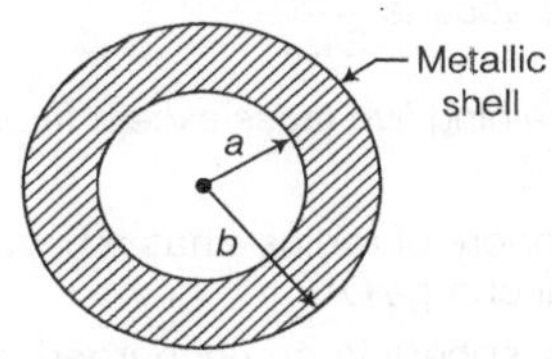

A charge q_A is placed at centre of the shell. It is observed that inner face of shell has a uniform charge density of $-\sigma$ and the outer surface has a charge density of $+\sigma$.

11 Charge q_A must be

a. zero
b. negative
c. having a magnitude $4\pi\sigma(b^2 - a^2)$
d. having a magnitude $4\pi\sigma a^2$

12 If a charge q_B is placed at a distance c, $(c > b)$, then

a. force on charge A is $\frac{\sigma q_A b^2}{\varepsilon_0 c^2}$
b. force on charge A is zero
c. force on charge B is $\frac{\sigma q_A b^2}{\varepsilon_0 c^2}$
d. force on charge B is $\frac{k q_A q_B}{c^2}$

13 If charge q_A is move slowly inside the shell, then

a. charge distribution on inner and outer surfaces of the shell changes charge
b. force on charge B changes
c. charge B also starts moving
d. None of the above

Passage II

A long cylindrical insulating shell of inner radius a and outer radius b has a uniform volumetric charge density ρ. A line of uniform linear charge density λ is placed on the axis of the shell.

14 For $r < a$,

a. $E = \frac{\lambda}{4\pi r\varepsilon_0}$
b. $E = \frac{\lambda}{2\pi r\varepsilon_0}$
c. $E = \frac{\lambda l}{2\pi r\varepsilon_0}$
d. $E = 0$

15 For $a < r < b$,

a. $E = \frac{\lambda + \rho\pi(r^2 - a^2)}{2\pi r\varepsilon_0}$
b. $E = \frac{\lambda}{4\pi r\varepsilon_0}$
c. $E = \frac{\rho\pi(r^2 - a^2)}{2\pi r\varepsilon_0}$
d. $E = \frac{\rho(r^2 - a^2)}{2\pi\varepsilon_0}$

16 For $r > b$,

a. $E = \frac{\lambda}{2\pi r\varepsilon_0}$
b. $E = \frac{\rho}{2\pi r\varepsilon_0}$
c. $E = \frac{\lambda + \rho\pi(b^2 - a^2)}{2\pi r\varepsilon_0}$
d. $E = \frac{\rho\pi(b^2 - a^2)}{2\pi r\varepsilon_0}$

One or More than One Options Correct Type

***Direction** (Q. Nos. 17-20) This section contains 4 multiple choice questions. Each question has four choices (a), (b), (c) and (d), out of which ONE or MORE THAN ONE are correct.*

17 Eight point charges each of magnitude q are located on the corners of a cube of edge a. Then,

a. force on each corner charge is $\frac{kq^2}{a^2}\left(\frac{2}{2\sqrt{2}} + \frac{1}{3\sqrt{3}}\right)$ N
b. force on a charge placed at centre of cube is $\frac{8kq^2}{\left(\frac{\sqrt{3}}{2}a\right)^2}$ N
c. electric field at centre of top face is $\frac{2^{7/2}}{3^{3/2}}\left(\frac{kq}{a^2}\right)$
d. electric field at centre of top face is directed along area vector of the face

Working Space

18 A point charge q exists on the extension line of diagonal ag of cube which is very close to the vertex a of cube. Then,

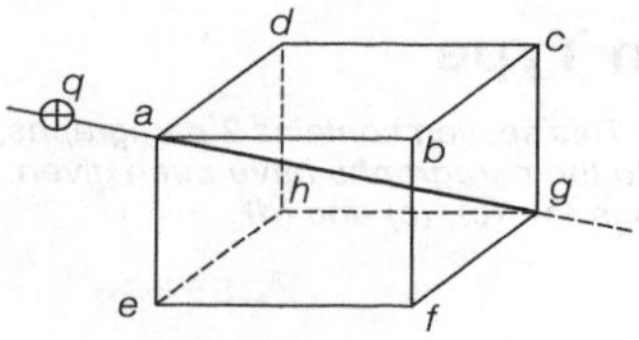

a. net flux through the cube is zero
b. flux through three faces meeting at g is $q/8\varepsilon_0$
c. flux through face $abcd$ is $\frac{-q}{24\varepsilon_0}$
d. flux through remaining five faces except face $abcd$ is $\frac{q}{24\varepsilon_0}$

19 A solid insulating sphere of radius a has a uniform charge density ρ and a total charge Q.

Concentric with this sphere in an uncharged conducting shell whose inner and outer radii are b and c. Then,

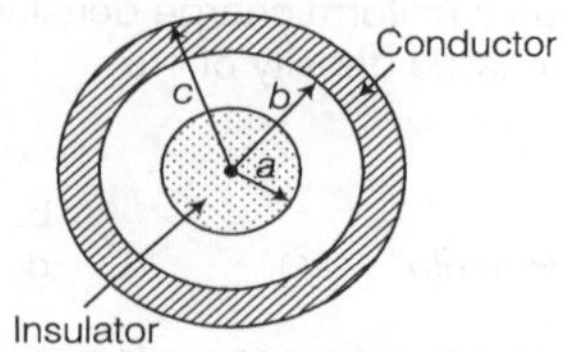

a. for $b \le r \le c, E = 0$
b. for $a < r < b, E = \frac{Q}{4\pi\varepsilon_0 r^2}$
c. induced charge on the inner surface of shell is $-Q/\varepsilon r$
d. induced charge on the outer surface of shell is $+Q$

20 A particle A of mass m and charge q moves at high speed v along the X-axis. It is initially at $x = -\infty$ and it ends up near $x = +\infty$. Another charge Q is placed at point $(0, d)$.

As the moving charge passes through the stationary charge,

a. there is no deflection in the path of A
b. vertical velocity component gained by A is $v_y = \frac{qQ}{mv\, 2\pi d\varepsilon_0}$
c. angle of deflection of A is $\tan^{-1}(qQ/2\pi\varepsilon_0 dmv^2)$
d. angle of deflection of A is $\tan^{-1}(qQ/2\pi d\varepsilon_0)$

Matching List Type

Direction *(Q. No. 21) Choices for the correct combination of elements from Column I and Column II are given as options (a), (b), (c) and (d), out of which one is correct.*

21 Two spherical shells are as shown in figure.

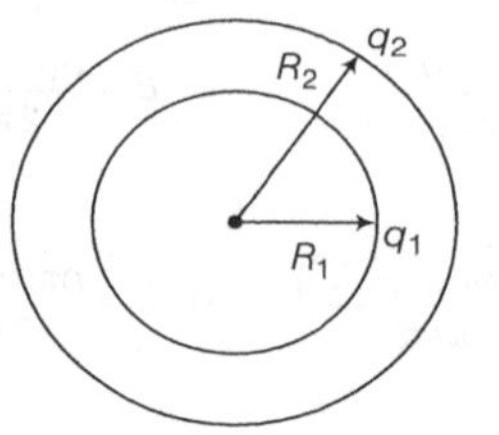

Let r be the distance of a point from their common centre. Then, match the following columns and mark the correct code.

	Column I		Column II
i.	Electric field for $r < R_1$	p.	is constant for q_2 and vary for q_1.
ii.	Electric potential for $r < R_1$	q.	is zero for q_2 and vary for q_1.
iii.	Electric potential for $R_1 < r < R_2$	r.	is constant.
iv.	Electric field for $R_1 < r < R_2$	s.	is zero.

Codes

	i	ii	iii	iv
a.	s	r	q	p
b.	s	r	p	q
c.	r	s	p	q
d.	r	p	s	q

Working Space

Statement Type

Direction *(Q. Nos. 22-24)* ***This section is based on Statement I and Statement II. Select the correct answer from the codes given below.***

a. Both Statement I and Statement II are correct and Statement II is the correct explanation of Statement I
b. Both Statement I and Statement II are correct and Statement II is not the correct explanation of Statement I
c. Statement I is correct but Statement II is incorrect
d. Statement II is correct but Statement I is incorrect

22 **Statement I** Water has much greater dielectric constant (K) than mica.

Statement II Large value of K for water is due to dipole moment associated with a molecule of water.

23 **Statement I** A charged particle in electric field always experiences a force either it is rest or in motion.

Statement II Because $F = qE$.

24 **Statement I** Slope of **E** *versus* **r** (E = field, r = distance) graph for a point charge and for a dipole is same.

Statement II $E \propto \frac{1}{r^2}$.

One Integer Value Correct Type

Direction *(Q. Nos. 25 and 26)* ***This section contains 2 questions. When worked out will result in an integer from 0 to 9 (both inclusive).***

25 A point charge Q is placed at a distance $\frac{r}{2}$ above an imaginary Gaussian sphere of radius r as shown.

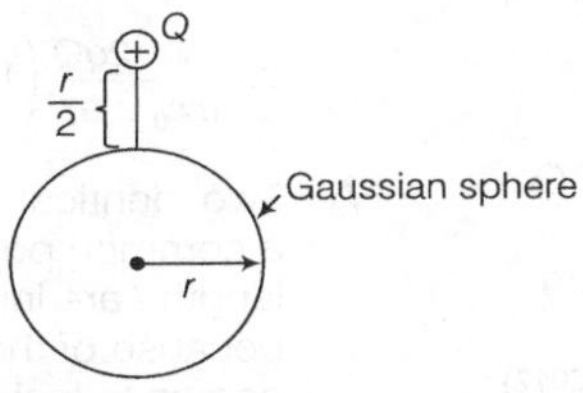

Let E denotes magnitude of electric field over the Gaussian surface.

Value of $E_{max} - E_{min} = N \times \frac{6Q}{25\pi\varepsilon_0 r^2}$

Find the value of N.

26 Consider two flat plates of size $1\text{m} \times 1\text{m}$ with charge densities $\pm 1.5\mu\text{Cm}^{-2}$, placed parallel facing each other. Work done in increasing plate separation from 1 mm to 6 mm is $8N \times 10^{-5}$ J. Find the value of N.

JEE Main & AIEEE Archive

(Compilation of Last 13 Years Questions)

1 Assume that an electric field $\mathbf{E} = 30x^2\,\mathbf{i}$ exists in space. Then, the potential difference $V_A - V_O$, where V_O is the potential at the origin and V_A is the potential at $x = 2$ m is **(JEE Main 2014)**

a. 120 J b. −120 J
c. −80J d. 80J

2 A charge Q is uniformly distributed over a long rod AB of length L as shown in the figure. The electric potential at the point O lying at distance L from the end A is **(JEE Main 2013)**

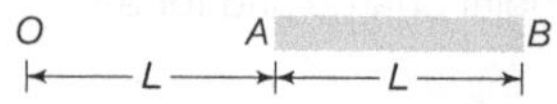

a. $\dfrac{Q}{8\pi\varepsilon_0 L}$ b. $\dfrac{3Q}{4\pi\varepsilon_0 L}$
c. $\dfrac{Q}{4\pi\varepsilon_0 L \ln 2}$ d. $\dfrac{Q \ln 2}{4\pi\varepsilon_0 L}$

3 Two charges each equal to q are kept at $x = -a$ and $x = a$ on the X-axis. A particle of mass m and charge $q_0 = q/2$ is placed at the origin. If charge q_0 is given, a small displacement $(y << a)$ along the Y-axis, then net force acting on the particle is proportional to **(JEE Main 2013)**

a. y b. $-y$
c. $1/y$ d. $-1/y$

4 In a uniformly charged sphere of total charge Q and radius R, the electric field E is plotted as function of distance from the centre. The graph which would correspond to the below will be **(AIEEE 2012)**

a. 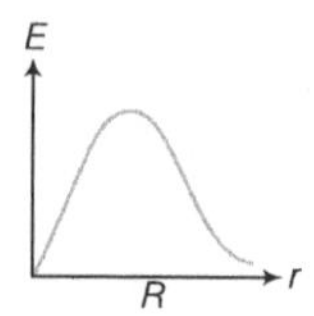

b. 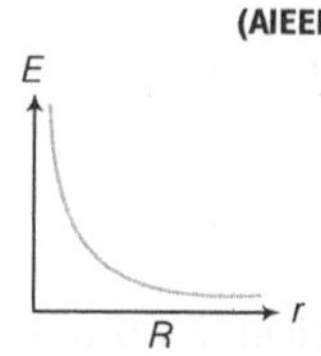

c. 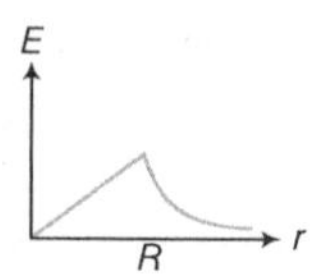

d. 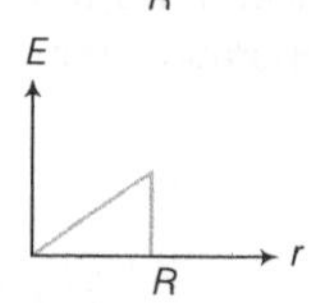

5 An insulating solid sphere of radius R has a uniform positive charge density ρ. As a result of this uniform charge distribution, there is a finite value of electric potential at the centre of the sphere, at the surface of the sphere and also at a point outside the sphere. The electric potential at infinite is zero. **(AIEEE 2012)**

Statement I When a charge q is taken from the centre to the surface of the sphere, its potential energy changes by $\dfrac{q\rho}{3\varepsilon_0}$.

Statement II The electric field at a distance $r\,(r < R)$ from the centre of the sphere is $\dfrac{\rho r}{3\varepsilon_0}$.

a. Both Statement I and Statement II are correct and Statement II is the correct explanation of Statement I
b. Both Statement I and Statement II are correct and Statement II is not the correct explanation of Statement I
c. Statement I is correct but Statement II is incorrect
d. Statement II is correct but Statement I is incorrect

6 Two positive charges of magnitude q are placed at the ends of a side 1 of a square of side $2a$. Two negative charges of the same magnitude are kept at the other corners. Starting from rest, if a charge Q moves from the middle of side 1 to the centre of square, its kinetic energy at the centre of square is **(AIEEE 2011)**

a. $\dfrac{1}{4\pi\varepsilon_0}\dfrac{2qQ}{a}\left(1 - \dfrac{1}{\sqrt{5}}\right)$
b. zero
c. $\dfrac{1}{4\pi\varepsilon_0}\dfrac{2qQ}{a}\left(1 + \dfrac{1}{\sqrt{5}}\right)$
d. $\dfrac{1}{4\pi\varepsilon_0}\dfrac{2qQ}{a}\left(1 - \dfrac{2}{\sqrt{5}}\right)$

7 Two identical charged spheres suspended from a common point by two massless strings of length l are initially a distance $d, (d << l)$ apart because of their mutual repulsion. The charge begins to leak from both the spheres at a constant rate. As a result, charges approach each other with a velocity v. Then, as a function of distance x between them is **(AIEEE 2011)**

a. $v \propto x^{-1}$ b. $v \propto x^{1/2}$
c. $v \propto x$ d. $v \propto x^{-1/2}$

8 A fully charged capacitor C with initial charge q_0 is connected to a coil of self inductance L at $t = 0$. The time at which the energy is stored equally between the electric and the magnetic fields is **(AIEEE 2011)**

a. $\dfrac{\pi}{4}\sqrt{LC}$ b. $2\pi\sqrt{LC}$
c. $\sqrt{LC}$ d. $\pi\sqrt{LC}$

9 The electrostatic potential inside a charged spherical ball is given by $\phi = ar^2 + b$, where r is the distance from the centre and a, b are constants. Then, the charge density inside the ball is **(AIEEE 2011)**

a. $-6a\,\varepsilon_0 r$ b. $-24\pi a\,\varepsilon_0$
c. $-6a\,\varepsilon_0$ d. $-24\pi a\,\varepsilon_0 r$

JEE Archive

10 A thin semi-circular ring of radius r has a positive charge q distributed uniformly over it. The net field E at the centre O is **(AIEEE 2010)**

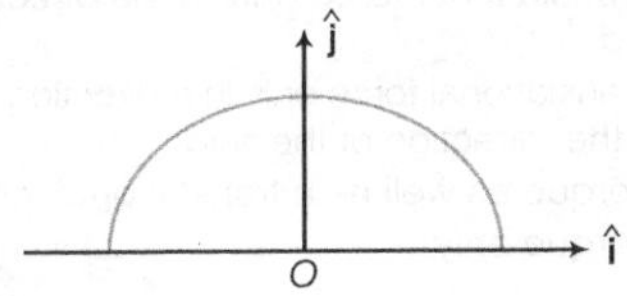

a. $\frac{q}{4\pi^2 \varepsilon_0 r^2}\hat{j}$ b. $-\frac{q}{4\pi^2 \varepsilon_0 r^2}\hat{j}$

c. $-\frac{q}{2\pi^2 \varepsilon_0 r^2}\hat{j}$ d. $\frac{q}{2\pi^2 \varepsilon_0 r^2}\hat{j}$

11 Let there be a spherically symmetric charge distribution with charge density varying as $\rho(r) = \rho_0\left(\frac{5}{4} - \frac{r}{R}\right)$ up to $r = R$ and $\rho(r) = 0$ for $r > R$, where r is the distance from the origin. The electric field at a distance r, $(r < R)$ from the origin is given by **(AIEEE 2010)**

a. $\frac{4\pi\rho_0 r}{3\varepsilon_0}\left(\frac{5}{3} - \frac{r}{R}\right)$ b. $\frac{\rho_0 r}{4\varepsilon_0}\left(\frac{5}{3} - \frac{r}{R}\right)$

c. $\frac{4\rho_0 r}{3\varepsilon_0}\left(\frac{5}{4} - \frac{r}{R}\right)$ d. $\frac{\rho_0 r}{3\varepsilon_0}\left(\frac{5}{4} - \frac{r}{R}\right)$

12 **Statement I** For a charged particle moving from point P to point Q, the net work done by an electrostatic field on the particle is independent of the path connecting point P to point Q.

Statement II The net work done by a conservative force on an object moving along a closed loop is zero. **(AIEEE 2009)**

a. Both Statement I and Statement II are correct and Statement II is the correct explanation of Statement I
b. Both Statement I and Statement II are correct and Statement II is not the correct explanation of Statement I
c. Statement I is correct but Statement II is incorrect
d. Statement II is correct but Statement I is incorrect

13 Let $\rho(r) = \frac{Q}{\pi R^4} r$ be the charge density distribution for a solid sphere of radius R and total charge Q. For a point P inside the sphere at distance r_1 from the centre of the sphere, the magnitude of electric field is **(AIEEE 2009)**

a. zero b. $\frac{Q}{4\pi\varepsilon_0 r_1^2}$

c. $\frac{Q r_1^2}{4\pi\varepsilon_0 R^4}$ d. $\frac{Q r_1^2}{3\pi\varepsilon_0 R^4}$

14 Two points P and Q are maintained at the potentials of 10 V and – 4 V, respectively. The work done in moving 100 electrons from P to Q is **(AIEEE 2009)**

a. -19×10^{-17} J
b. 9.60×10^{-17} J
c. -2.24×10^{-16} J
d. 2.24×10^{-16} J

15 A charge Q is placed at each of the opposite corners of a square. A charge q is placed at each of the other two corners. If the net electrical force on Q is zero, then $\frac{Q}{q}$ equals **(AIEEE 2009)**

a. $-2\sqrt{2}$ b. -1

c. 1 d. $-\frac{1}{\sqrt{2}}$

Passage for Q. Nos. (16-17)

Consider a block of conducting material of resistivity ρ shown in the figure below. Current I enters at A and leaves from D. We apply superposition principle to find voltage ΔV developed between B and C. The calculation is done in the following steps :

I. Take current I entering from A and assume it to spread over a hemispherical surface on the block.

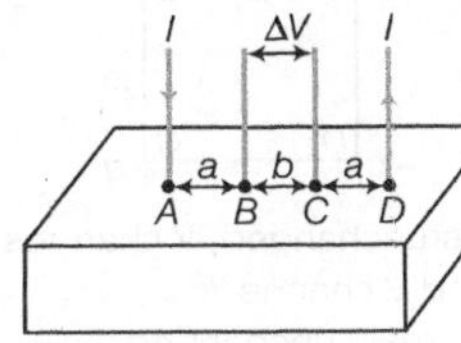

II. Calculate field $E(r)$ at distance r from A by using Ohm's law $= \rho J$, where J is the current per unit area at r.
III. From the r dependence of $E(r)$, obtain the potential $V(r)$ at r.
IV. Repeat steps (I), (II) and (III) for current I leaving D and superpose results for A and D.

16 ΔV measured between points B and C is **(AIEEE 2008)**

a. $\frac{\rho I}{\pi a} - \frac{\rho I}{\pi(a+b)}$ b. $\frac{\rho I}{a} - \frac{\rho I}{(a+b)}$

c. $\frac{\rho I}{2\pi a} - \frac{\rho I}{2\pi(a+b)}$ d. $\frac{\rho I}{2\pi(a-b)}$

17 For current entering at point A, the electric field at a distance r from point A is **(AIEEE 2008)**

a. $\frac{\rho I}{8\pi r^2}$ b. $\frac{\rho I}{r^2}$

c. $\frac{\rho I}{2\pi r^2}$ d. $\frac{\rho I}{4\pi r^2}$

18 A thin spherical shell of radius R has charge Q spread uniformly over its surface. Which of the following graphs most closely represents the electric field $E(r)$ produced by the shell in the range $0 \le r < x$, where r is the distance from the centre of the shell? **(AIEEE 2008)**

a. 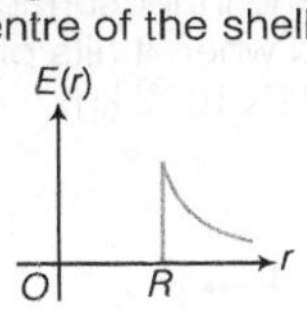

b. 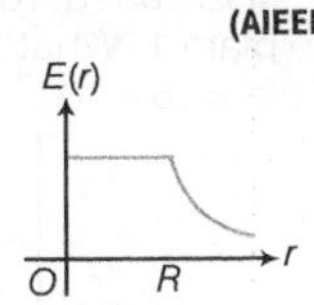

c. 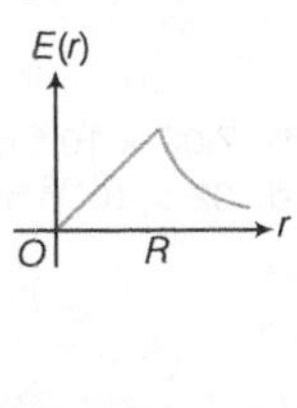

d. 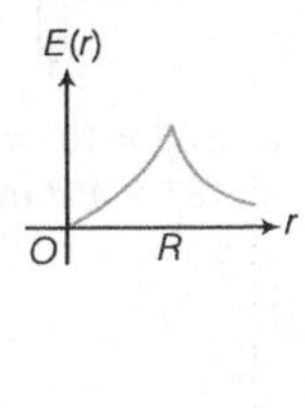

JEE Archive

19 The potential at a point x (measured in μm) due to some charges situated on the X-axis is given by $V(x) = 20/(x^2 - 4)$ V. The electric field E at $x = 4\,\mu m$ is given by **(AIEEE 2007)**

a. $\frac{5}{3}$ V/μm and in the –ve X-direction

b. $\frac{5}{3}$ V/μm and in the +ve X-direction

c. $\frac{10}{9}$ V/μm and in the –ve X-direction

d. $\frac{10}{9}$ V/μm and in the +ve X-direction

20 Charges are placed on the vertices of a square as shown in figure below. Let **E** be the electric field and V be the potential at the centre. If the charges on A and B are interchanged with those on D and C respectively, then **(AIEEE 2007)**

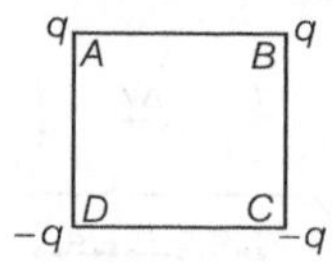

a. **E** remains unchanged, V changes

b. both **E** and V change

c. **E** and V remain unchanged

d. **E** changes, V remains unchanged

21 An electric charge 10^{-3} μC is placed at the origin (0, 0) of xy-coordinate system. Two points A and B are situated at $(\sqrt{2}, \sqrt{2})$ and (2, 0) respectively. The potential difference between the points A and B will be **(AIEEE 2007)**

a. 9 V b. zero
c. 2 V d. 4.5 V

22 Two spherical conductors A and B of radii 1 mm and 2 mm are separated by a distance of 5 cm and are uniformly charged. If the spheres are connected by a conducting wire, then in equilibrium condition, the ratio of the magnitude of the electric fields at the surfaces of spheres A and B is **(AIEEE 2006)**

a. 4 : 1 b. 1 : 2
c. 2 : 1 d. 1 : 4

23 Two insulating plates are uniformly charged in such a way that the potential difference between them is $V_2 - V_1 = 20$ V (i.e. plate 2 is at a higher potential). The plates are separated by $d = 0.1$ m and can be treated as infinitely large. An electron is released from rest on the inner surface of plate 1. What is its speed when it hits plate 2? ($e = 1.6 \times 10^{-19}$ C, $m_0 = 9.11 \times 10^{-31}$ kg) **(AIEEE 2006)**

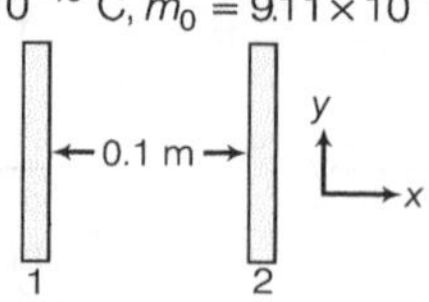

a. 2.65×10^6 ms^{-1} b. 7.02×10^{12} ms^{-1}
c. 1.87×10^6 ms^{-1} d. 32×10^{-19} ms^{-1}

24 An electric dipole is placed at an angle of 30° to a non-uniform electric field. The dipole will experience **(AIEEE 2006)**

a. a translational force only in the direction of the field

b. a translational force only in a direction normal to the direction of the field

c. a torque as well as a translational force

d. a torque only

25 A charged ball B hangs from a silk thread S, which makes an angle θ with a large charged conducting sheet P, as shown in the figure below. The surface charge density σ of the sheet is proportional to **(AIEEE 2005)**

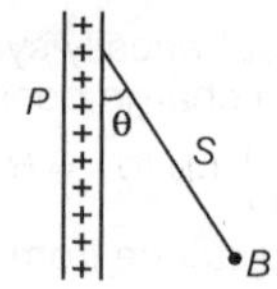

a. $\cos\theta$ b. $\cot\theta$
c. $\sin\theta$ d. $\tan\theta$

26 Two point charges $+8\,q$ and $-2\,q$ are located at $x = 0$ and $x = L$, respectively. The location of a point on the X-axis at which the net electric field due to these two point charges is zero is **(AIEEE 2005)**

a. $2L$

b. $\frac{L}{4}$

c. $8L$

d. $4L$

27 Two thin wire rings each having a radius R are placed at a distance d apart with their axes coinciding. The charges on the two rings are $+q$ and $-q$. The potential difference between the centres of the two rings is **(AIEEE 2005)**

a. $\frac{qR}{4\pi\varepsilon_0 d^2}$

b. $\frac{q}{2\pi\varepsilon_0}\left[\frac{1}{R} - \frac{1}{\sqrt{R^2 + d^2}}\right]$

c. zero

d. $\frac{q}{4\pi\varepsilon_0}\left[\frac{1}{R} - \frac{1}{\sqrt{R^2 + d^2}}\right]$

28 Two spherical conductors B and C having equal radii and carrying equal charges in them repel each other with a force F when kept apart at some distance. A third spherical conductor having same radius as that of B but uncharged, is brought in contact with B, then brought in contact with C and finally removed away from both. The new force of repulsion between B and C is **(AIEEE 2004)**

a. $\frac{F}{4}$ b. $\frac{3F}{4}$
c. $\frac{F}{8}$ d. $\frac{3F}{8}$

29 A charged particle q is shot towards another charged particle Q which is fixed with a speed v. It approaches Q up to a closet distance r and then returns. If q was given a speed $2v$, then closet distance of approach would be

a. r b. $2r$ **(AIEEE 2004)**

c. $\frac{r}{2}$ d. $\frac{r}{4}$

30 Four charges equal to $-Q$ are placed at the four corners of a square and a charge q is at its centre. If the system is in equilibrium, then the value of q is **(AIEEE 2004)**

a. $-\frac{Q}{4}(1+2\sqrt{2})$

b. $\frac{Q}{4}(1+2\sqrt{2})$

c. $-\frac{Q}{2}(1+2\sqrt{2})$

d. $\frac{Q}{2}(1+2\sqrt{2})$

31 If the electric flux entering and leaving an enclosed surface respectively is ϕ_1 and ϕ_2, then electric charge inside the surface will be **(AIEEE 2003)**

a. $(\phi_2 - \phi_1)\varepsilon_0$ b. $\frac{(\phi_1 + \phi_2)}{\varepsilon_0}$

c. $\frac{(\phi_2 - \phi_1)}{\varepsilon_0}$ d. $(\phi_1 + \phi_2)\varepsilon_0$

32 A thin spherical conducting shell of radius R has a charge q. Another charge Q is placed at the centre of the shell. The electrostatic potential at a point P at a distance $R/2$ from the centre of the shell is **(AIEEE 2003)**

a. $\frac{2Q}{4\pi\varepsilon_0 R}$

b. $\frac{2Q}{4\pi\varepsilon_0 R} - \frac{2q}{4\pi\varepsilon_0 R}$

c. $\frac{2Q}{4\pi\varepsilon_0 R} + \frac{q}{4\pi\varepsilon_0 R}$

d. $\frac{(q+Q)}{4\pi\varepsilon_0}\frac{2}{R}$

33 Three charges $-q_1$, $+q_2$ and $-q_3$ are placed as shown in the figure below. The X-component of the force on $-q_1$ is proportional to **(AIEEE 2003)**

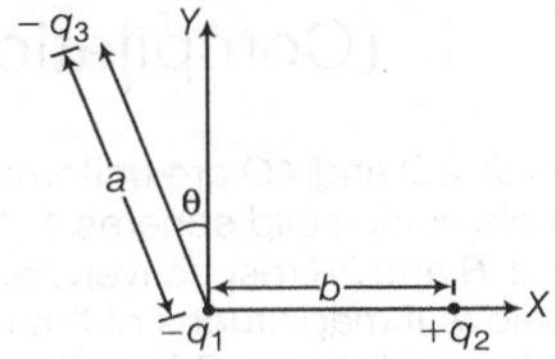

a. $\frac{q_2}{b^2} - \frac{q_3}{a^2}\cos\theta$ b. $\frac{q_2}{b^2} + \frac{q_3}{a^2}\sin\theta$

c. $\frac{q_2}{b^2} + \frac{q_3}{a^2}\cos\theta$ d. $\frac{q_2}{b^2} - \frac{q_3}{a^2}\sin\theta$

34 A charged particle q is placed at the centre O of cube of length $L(ABCDEFGH)$. Another same charge q is placed at a distance L from O. Then, the electric flux through $ABCD$ is **(AIEEE 2002)**

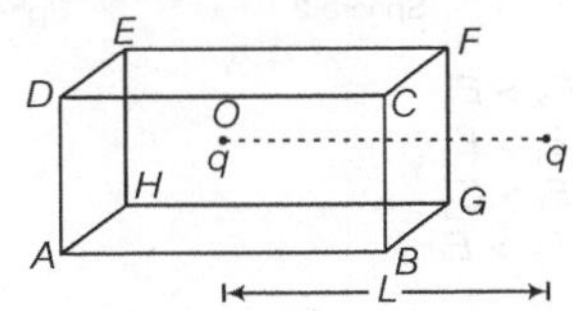

a. $\frac{q}{4\pi\varepsilon_0 L}$ b. $\frac{q}{2\pi\varepsilon_0 L}$

c. $\frac{q}{2\pi\varepsilon_0 L}$ d. None of these

35 If a charge q is placed at the centre of the line joining two equal charges Q such that the system is in equilibrium, then the value of q is **(AIEEE 2002)**

a. $\frac{Q}{2}$ b. $-\frac{Q}{2}$

c. $\frac{Q}{4}$ d. $-\frac{Q}{4}$

36 On moving a charge of 20 C by 2 cm, 2 J of work is done, then the potential difference between the points is **(AIEEE 2002)**

a. 0.1 V   b. 8 V

c. 2 V   d. 0.5 V

JEE Advanced & IIT JEE Archive

(Compilation of Last 10 Years Questions)

1 Charges $Q, 2Q$ and $4Q$ are uniformly distributed in three dielectric solid spheres 1, 2 and 3 of radii $R/2$, R and $2R$ respectively, as shown in figure below. If magnitudes of the electric fields at point P at a distance R from the centre of spheres 1, 2 and 3 are E_1, E_2 and E_3 respectively, then **(2014 Adv., Only One Option Correct Type)**

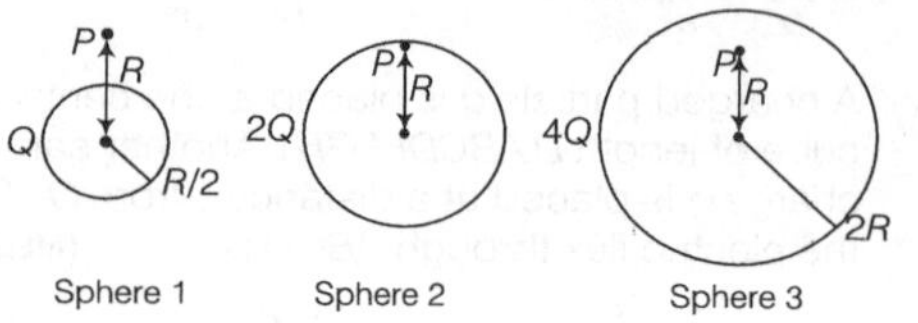

a. $E_1 > E_2 > E_3$
b. $E_3 > E_1 > E_2$
c. $E_2 > E_1 > E_3$
d. $E_3 > E_2 > E_1$

2 Four charges Q_1, Q_2, Q_3 and Q_4 of same magnitude are fixed along the X-axis at $x = -2a, -a, +a$ and $+2a$ respectively. A positive charge q is placed on the positive Y-axis at a distance $b > 0$. Four options of the signs of these charges are given in Column I. The direction of the forces on the charge q is given in Column II. Match Column I with Column II and select the correct answer using the codes given below the columns. **(2014 Adv., Matching Type)**

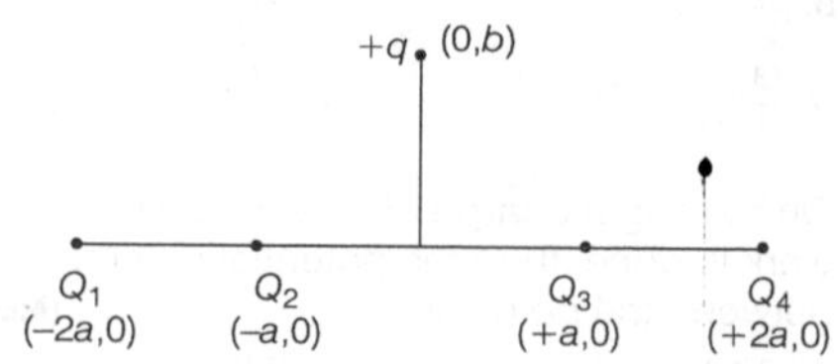

	Column I		Column II
i.	Q_1, Q_2, Q_3, Q_4 all positive	p.	$+X$
ii.	Q_1, Q_2 positive; Q_3, Q_4 negative	q.	$-X$
iii.	Q_1, Q_4 positive; Q_2, Q_3 negative	r.	$+Y$
iv.	Q_1, Q_3 positive; Q_2, Q_4 negative	s.	$-Y$

Codes

	i	ii	iii	iv
a.	r	p	s	q
b.	s	q	r	p
c.	r	p	q	s
d.	s	q	p	r

3 Let $E_1(r)$, $E_2(r)$ and $E_3(r)$ be the respective electric fields at a distance r from a point charge Q, an infinitely long wire with constant linear charge density λ, and an infinite plane with uniform surface charge density σ. If $E_1(r_0) = E_2(r_0) = E_3(r_0)$ at a given distance r_0, then **(2014 Adv., One or More than One Options Correct Type)**

a. $Q = 4\sigma\pi r_0^2$ b. $r_0 = \dfrac{\lambda}{2\pi\sigma}$
c. $E_1(r_0/2) = 2E_2(r_0/2)$ d. $E_2(r_0/2) = 4E_3(r_0/2)$

4 Two non-conducting solid spheres of radii R and $2R$, having uniform volume charge densities ρ_1 and ρ_2 respectively, touch each other. The net electric field at a distance $2R$ from the centre of the smaller sphere, along the line joining the centre of the spheres is zero. The ratio ρ_1/ρ_2 can be **(2013 Adv., One or More than One Options Correct Type)**

a. -4 b. $-\dfrac{32}{25}$
c. $\dfrac{32}{25}$ d. 4

5 Two non-conducting spheres of radii R_1 and R_2 and carrying uniform volume charge densities $+\rho$ and $-\rho$, respectively, are placed such that they partially overlap, as shown in the figure alongside. At all points in the overlapping region **(2013 Adv., One or More than One Options Correct Type)**

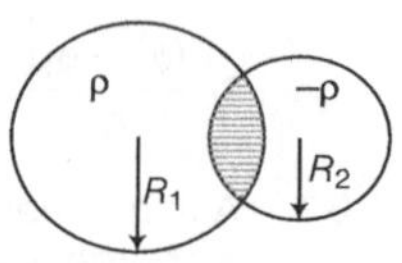

a. the electrostatic field is zero
b. the electrostatic potential is constant
c. the electrostatic field is constant in magnitude
d. the electrostatic field has same direction

6 Six point charges are kept at the vertices of a regular hexagon of side L and centre O as shown in the figure below. Given that $k = \dfrac{1}{4\pi\varepsilon_0}\dfrac{q}{L^2}$, which of the following statements(s) is(are) correct? **(2012, One or More than One Options Correct Type)**

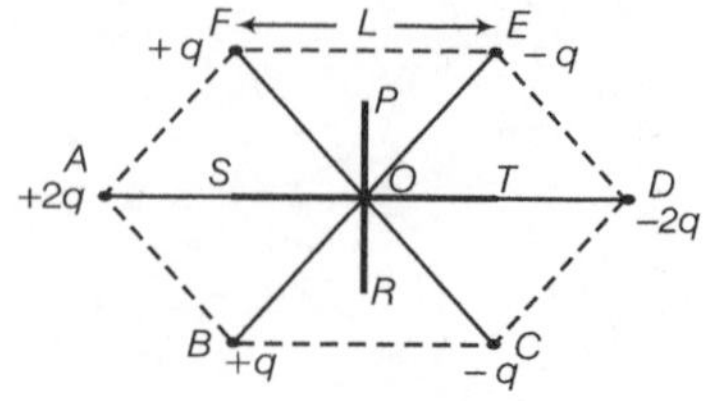

a. The electric field at O is $6k$ along line OD
b. The potential at point O is zero
c. The potential at all points on the line PR is same
d. The potential at all points on the line ST is same

7 A cubical region of side a has its centre at the origin. It encloses three fixed point charges, $-q$ at $(0, -a/4, 0)$, $+3q$ at $(0, 0, 0)$ and $-q$ at $(0, +a/4, 0)$. Choose the correct option(s).

(2012, One or More Than One Options Correct Type)

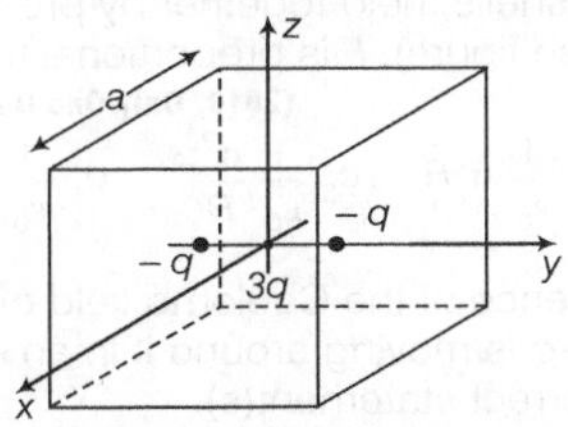

a. The net electric flux crossing the plane $x = +a/2$ is equal to the net electric flux crossing the plane $x = -a/2$

b. The net electric flux crossing the plane $y = +a/2$ is more than the net electric flux crossing the plane $y = -a/2$

c. The net electric flux crossing the entire region is $\frac{q}{\varepsilon_0}$

d. The net electric flux crossing the plane $z = +a/2$ is equal to the net electric flux crossing the plane $x = +a/2$

8 An infinitely long solid cylinder of radius R has a uniform volume charge density ρ. It has a spherical cavity of radius $R/2$ with its centre on the axis of the cylinder, as shown in the figure below. The magnitude of the electric field at the point P, which is at a distance $2R$ from the axis of the cylinder, is given by the expression $\frac{23\rho R}{16k\varepsilon_0}$. Then, find the value of k is **(2012, Integer Type)**

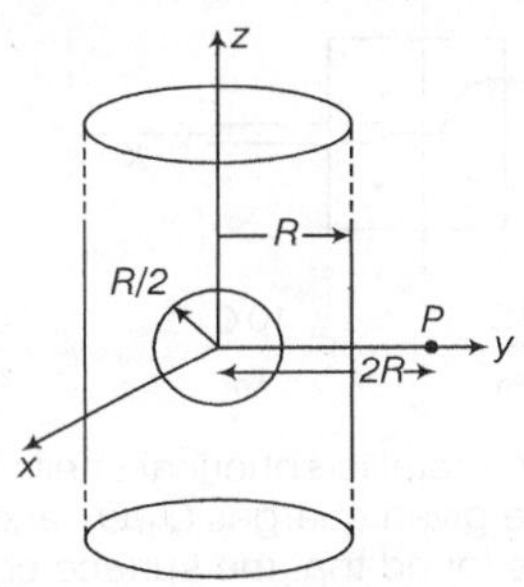

9 Two large vertical and parallel metal plates having a separation of 1 cm are connected to a DC voltage source of potential difference X. A proton is released at rest midway between the two plates. It is found to move at 45° to the vertical just after release. Then, X is nearly

(2012, Only One Option Correct Type)

a. 1×10^{-5} V

b. 1×10^{-7} V

c. 1×10^{-9} V

d. 1×10^{-10} V

10 Consider a thin spherical shell of radius R with its centre at the origin, carrying uniform positive surface charge density. The variation of the magnitude of the electric field $|\mathbf{E}(r)|$ and the electric potential $V(r)$ with the distance r from the centre, is best represented by which graph?

(2012, Only One Option Correct Type)

a.

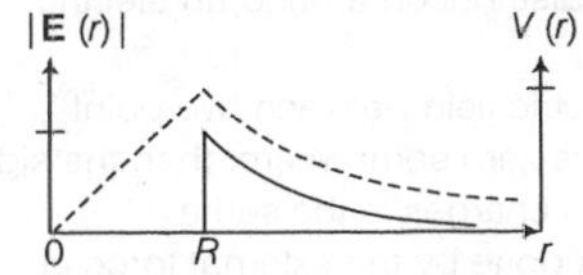

b.

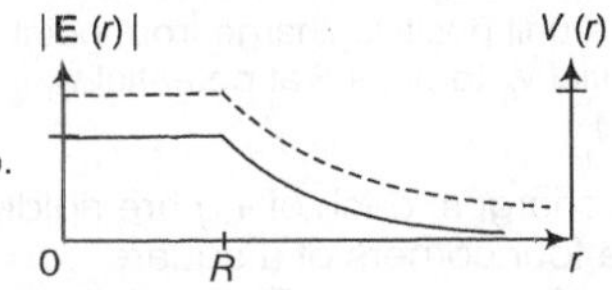

c.

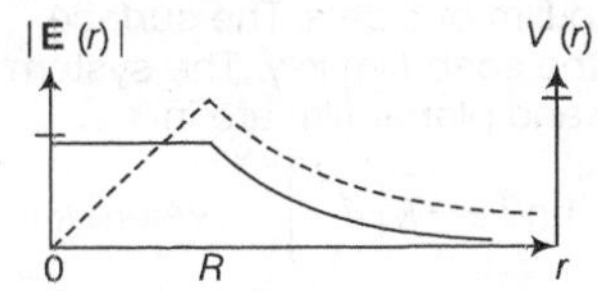

d.

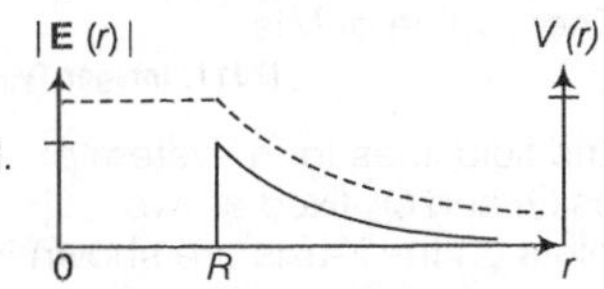

11 Consider an electric field $\mathbf{E} = E_0\,\hat{\mathbf{x}}$, where E_0 is a constant. The flux through the shaded area (as shown in the figure below) due to this field is

(2011, Only One Option Correct Type)

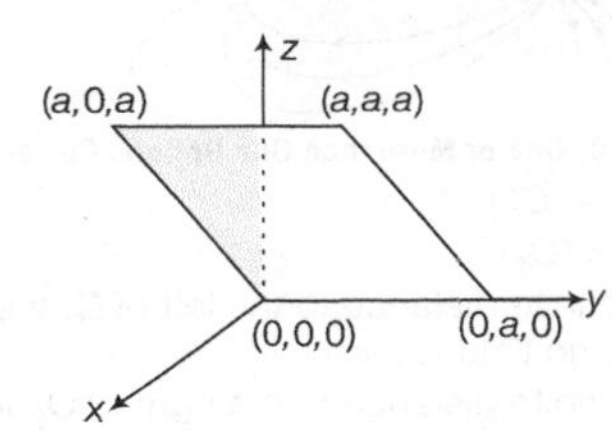

a. $2E_0a^2$

b. $\sqrt{2}E_0a^2$

c. E_0a^2

d. $\frac{E_0a^2}{\sqrt{2}}$

12 A spherical metal shell A of radius R_A and a solid metal sphere B of radius $R_B (< R_A)$ are kept far apart and each is given charge $+Q$. Now, they are connected by a thin metal wire. Then,

(2011, One or More than One Options Correct Type)

a. $E_A^{\text{inside}} = 0$

b. $Q_A > Q_B$

c. $\frac{\sigma_A}{\sigma_B} = \frac{R_B}{R_A}$

d. $E_A^{\text{on surface}} < E_B^{\text{on surface}}$

JEE Archive

13 Which of the following statement(s) is/are correct?

(2011, One or More than One Options Correct Type)

a. If the electric field due to a point charge varies as $r^{-2.5}$ instead of r^{-2}, then the Gauss' law will still be valid

b. The Gauss' law can be used to calculate the field distribution around an electric dipole

c. If the electric field between two point charges is zero somewhere, then the sign of the two charges is the same

d. The work done by the external force in moving a unit positive charge from point A at potential V_A to point B at potential V_B is $(V_B - V_A)$

14 Four point charges, each of $+q$ are rigidly fixed at the four corners of a square planar soap film of side a. The surface tension of the soap film is γ. The system of charges and planar film are in equilibrium and $a = k\left[\frac{q^2}{\gamma}\right]^{1/N}$, where k is a constant. Then, value of N is

(2011, Integer Type)

15 A few electric field lines for a system of two charges Q_1 and Q_2 fixed at two different points on the X-axis are shown in the figure along side. These lines suggest that

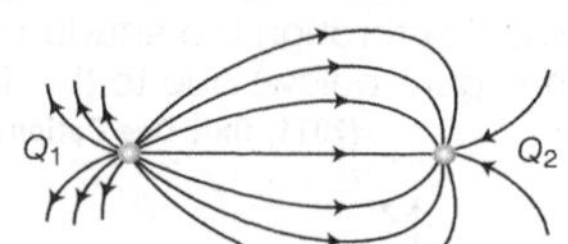

(2010, One or More than One Options Correct Type)

a. $|Q_1| > |Q_2|$

b. $|Q_1| < |Q_2|$

c. at a finite distance to the left of Q_1 the electric field is zero

d. at a finite distance to the right of Q_2 the electric field is zero

16 A tiny spherical oil drop carrying a net charge q is balanced in still air with a vertical uniform electric field of strength $\frac{81\pi}{7} \times 10^5$ Vm^{-1}. When the field is switched off, the drop is observed to fall with terminal velocity 2×10^{-3} ms^{-1}. Given $g = 9.8$ ms^{-2}, viscosity of the air $= 1.8 \times 10^{-5}$ Ns m^{-2} and the density of oil $= 900$ kg m^{-3}, the magnitude of q is

(2010, Only One Option Correct Type)

a. 1.6×10^{-19} C

b. 3.2×10^{-19} C

c. 4.8×10^{-19} C

d. 8.0×10^{-19} C

17 A uniformly charged thin spherical shell of radius R carries uniform surface charge density of σ per unit area. It is made of two hemispherical shells, held together by pressing them with force F (see figure). F is proportional to

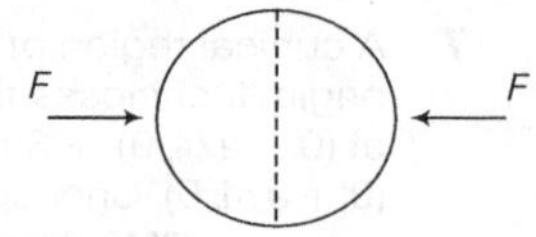

(2010, Only One Option Correct Type)

a. $\frac{1}{\varepsilon_0}\sigma^2 R^2$ b. $\frac{1}{\varepsilon_0}\sigma^2 R$ c. $\frac{1}{\varepsilon_0}\frac{\sigma^2}{R}$ d. $\frac{1}{\varepsilon_0}\frac{\sigma^2}{R^2}$

18 Under the influence of the Coulomb field of charge $+Q$, a charge $-q$ is moving around it in an elliptical orbit. Find out the correct statement(s).

(2009, One or More than One Options Correct Type)

a. The angular momentum of the charge $-q$ is constant

b. The linear momentum of the charge $-q$ is constant

c. The angular velocity of the charge $-q$ is constant

d. The linear speed of the charge $-q$ is constant

19 A disc of radius $\frac{a}{4}$ having a uniformly distributed charge $6C$ is placed in the xy-plane with its centre at $\left(\frac{-a}{2}, 0, 0\right)$. A rod of length a carrying a uniformly distributed charge $8C$ is placed on the X-axis from $x = \frac{a}{4}$ to $x = \frac{5a}{4}$. Two point charges $-7C$ and $3C$ are placed at $\left(\frac{a}{4}, \frac{-a}{4}, 0\right)$ and $\left(\frac{-3a}{4}, \frac{3a}{4}, 0\right)$, respectively. Consider a cubical surface formed by six surfaces $x = \pm\frac{a}{2}$, $y = \pm\frac{a}{2}$, $z = \pm\frac{a}{2}$. The electric flux through this cubical surface is

(2009, Only One Option Correct Type)

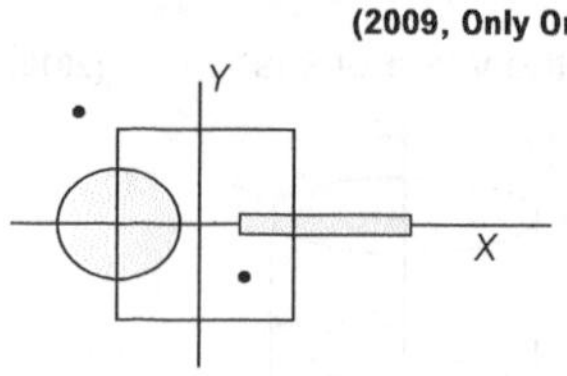

a. $\frac{-2C}{\varepsilon_0}$ b. $\frac{2C}{\varepsilon_0}$ c. $\frac{10C}{\varepsilon_0}$ d. $\frac{12C}{\varepsilon_0}$

20 Three concentric metallic spherical shells of radii R, $2R$ and $3R$ are given charges Q_1, Q_2 and Q_3 respectively. It is found that the surface charge densities on the outer surfaces of the shells are equal. Then, the ratio of the charges given to the shells, $Q_1 : Q_2 : Q_3$ is

(2009, Only One Option Correct Type)

a. 1 : 2 : 3 b. 1 : 3 : 5 c. 1 : 4 : 9 d. 1 : 8 : 18

21 Six point charges, each of the same magnitude q are arranged in different manners as shown in Column II. In each case, a point M and a line PQ passing through M are shown. Let E be the electric field and V be the electric potential at M (potential at infinity is zero) due to the given charge distribution when it is at rest. Now, the whole system is set into rotation with a constant angular velocity about the line PQ. Let B be the magnetic field at M and μ be the magnetic moment of the system in this condition.

Assume each rotating charge to be equivalent to a steady current. **(2009, Matching Type)**

	Column I			Column II
i.	$E = 0$	p.	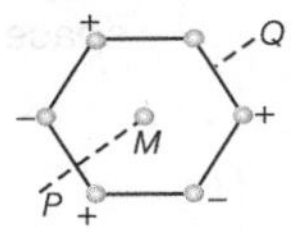	Charges are at the corners of a regular hexagon. M is at the centre of the hexagon. PQ is perpendicular to the plane of the hexagon.
ii.	$V \neq 0$	q.	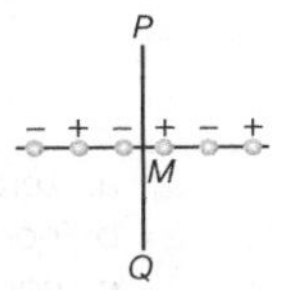	Charges are on a line perpendicular to PQ at equal intervals. M is the mid-point between the two innermost charges.
iii.	$B = 0$	r.	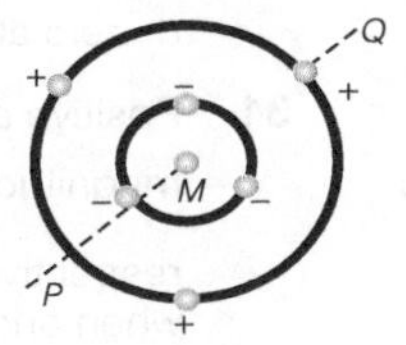	Charges are placed on two coplanar insulating rings at equal intervals. M is the common centre of the rings. PQ is perpendicular to the plane of the rings.
iv.	$\mu \neq 0$	s.	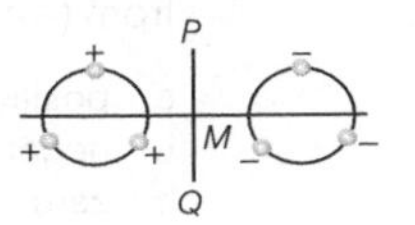	Charges are placed at the corners of a rectangle of sides a and $2a$ and at the mid points of the longer sides. M is at the centre of the rectangle. PQ is parallel to the longer sides.
		t.	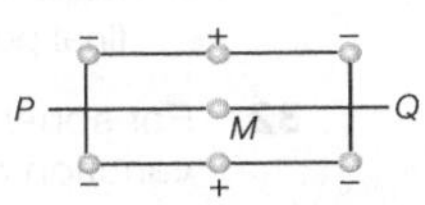	Charges are placed on two coplanar, identical insulating rings at equal intervals. M is the mid-points between the centres of the rings. PQ is perpendicular to the line joining the centres and coplanar to the rings.

Codes

	i	ii	iii	iv
a.	p,r,s	r,s	p,q,t	r,s
b.	p,s	r,s	q,t	p,q
c.	r,s	q,s	p,t	q,s
d.	p,r,s	q,t	r,s,t	q,s

22 A solid sphere of radius R has a charge Q distributed in its volume with a charge density $\rho = kr^a$, where k and a are constants and r is the distance from its centre. If the electric field at $r = \frac{R}{2}$ is $\frac{1}{8}$ times that at $r = R$, then find the value of a. **(2009, Integer Type)**

23 **Statement I** For practical purposes, the earth is used as a reference at zero potential in electrical circuits.

Statement II The electrical potential of a sphere of radius R with charge Q uniformly distributed on the surface is given by $\frac{Q}{4\pi\varepsilon_0 R}$. **(2008, Statement Type)**

a. Both Statement I and Statement II are correct and Statement II is the correct explanation of Statement I
b. Both Statement I and Statement II are correct and Statement II is not the correct explanation of Statement I
c. Statement I is correct but Statement II is incorrect
d. Statement II is correct but Statement I is incorrect

Passage for Q. Nos. (24-26)

The nuclear charge (Ze) is non-uniformly distributed within a nucleus of radius R. The charge density $\rho(r)$ (charge per unit volume) is dependent only on the radial distance r from the centre of the nucleus as shown in figure. The electric field is only along the radial direction. **(2008, Comprehension Type)**

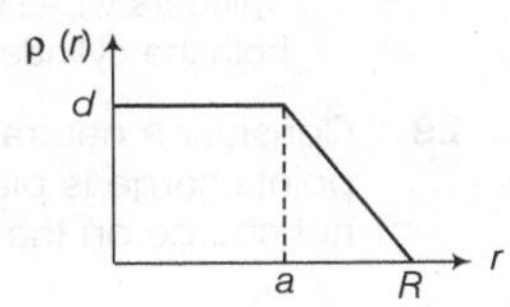

24 The electric field at $r = R$ is

a. independent of a
b. directly proportional to a
c. directly proportional to a^2
d. inversely proportional to a

25 For $a = 0$, the value of d (maximum value of ρ as shown in the figure) is

a. $\frac{3\,Ze}{4\pi R^3}$ b. $\frac{3\,Ze}{\pi R^3}$ c. $\frac{4\,Ze}{3\pi R^3}$ d. $\frac{Ze}{3\pi R^3}$

26 The electric field within the nucleus is generally observed to be linearly dependent on r. This implies

a. $a = 0$ b. $a = \frac{R}{2}$

c. $a = R$ d. $a = \frac{2R}{3}$

27 Consider a system of three charges $\frac{q}{3}, \frac{q}{3}$ and $-\frac{2q}{3}$ placed at points A, B and C respectively, as shown in the figure below. Take O to be the centre of the circle of radius R and angle $CAB = 60°$ **(2008, Only One Option Correct Type)**

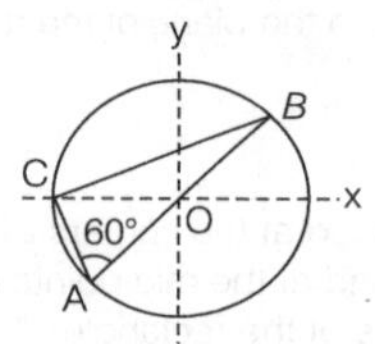

a. The electric field at point O is $\frac{q}{8\pi\varepsilon_0 R^2}$ directed along the negative X-axis
b. The potential energy of the system is zero
c. The magnitude of the force between the charges at points C and B is $\frac{q^2}{54\pi\varepsilon_0 R^2}$
d. The potential at point O is $\frac{q}{12\pi\varepsilon_0 R}$

28 A long, hollow conducting cylinder is kept coaxially inside another long, hollow conducting cylinder of larger radius. Both the cylinders are initially electrically neutral. **(2007, Only One Option Correct Type)**

a. A potential difference appears between the two cylinders when a charge density is given to the inner cylinder
b. A potential difference appears between the two cylinders when a charge density is given to the outer cylinder
c. No potential difference appears between the two cylinders when a uniform line charge is kept along the axis of the cylinders
d. No potential difference appears between the two cylinders when same charge density is given to both the cylinders

29 Consider a neutral conducting sphere. A positive point charge is placed outside the sphere. The net charge on the sphere is, then **(2007, Only One Option Correct Type)**

a. negative and distributed uniformly over the surface of the sphere
b. negative and appears only at the point on the sphere closet to the point charge
c. negative and distributed non-uniformly over the entire surface of the sphere
d. zero

30 A spherical portion has been removed from a solid sphere having a charge distributed uniformly in its volume as shown in the figure below. The electric field inside the emptied space is

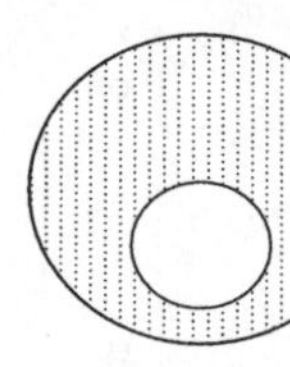

a. zero everywhere **(2007, Only One Option Correct Type)**
b. non-zero and uniform
c. non-uniform
d. zero at its centre

31 Positive and negative point charges of equal magnitude are kept at $\left(0, 0, \frac{a}{2}\right)$ and $\left(0, 0, \frac{-a}{2}\right)$ respectively. The work done by the electric field when another positive point charge is moved from $(-a, 0, 0)$ to $(0, a, 0)$ is **(2007, Only One Option Correct Type)**

a. positive
b. negative
c. zero
d. depends on the path connecting the initial and final positions

32 For spherical symmetrical charge distribution, variation of electric potential with distance from centre is given in figure below. **(2006, One or More than One Options Correct Type)**

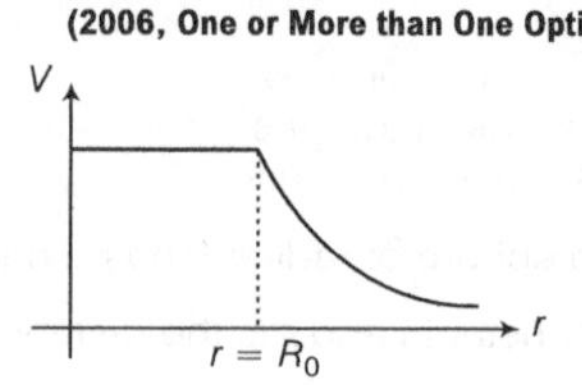

$V = \frac{q}{4\pi\,\varepsilon_0\,R_0}$ for $r \le R_0$ and $V = \frac{q}{4\pi\varepsilon_0\,r}$ for $r \ge R_0$

Then, which option(s) are correct?

a. Total charge within $2R_0$ is q
b. Total electrostatic energy for $r \le R_0$ is zero
c. At $r = R_0$, electric field is discontinuous
d. There will be no charge anywhere except at $r = R_0$

33 Three infinitely long charge sheets are placed as shown in figure below. The electric field at point P is **(2005, Only One Option Correct Type)**

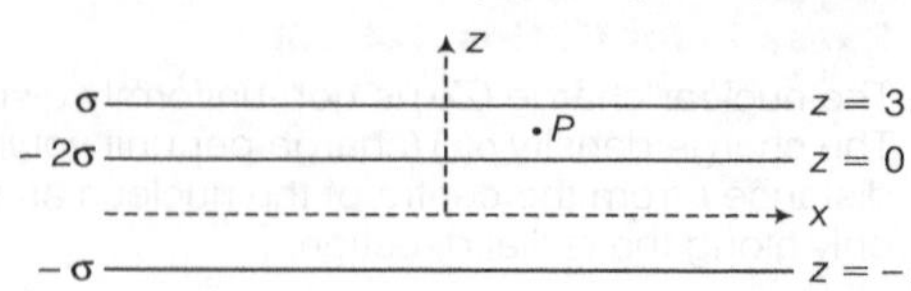

a. $\frac{2\sigma}{\varepsilon_0}\hat{k}$ b. $-\frac{2\sigma}{\varepsilon_0}\hat{k}$ c. $\frac{4\sigma}{\varepsilon_0}\hat{k}$ d. $-\frac{4\sigma}{\varepsilon_0}\hat{k}$

34 A conducting bubble of radius a, thickness t ($t << a$) has potential V. Now, the bubble collapses into a droplet. Find the potential of the droplet. **(2005, Descriptive Type)**

2. Capacitance

DPP

DAILY PRACTICE PROBLEMS

DPP-1 Capacitance

Subjective Questions

Working Space

Direction (Q. Nos. 1- 4) *These questions are subjective in nature, need to be solved completely on notebook.*

1 Two metal objects of arbitrary shapes are shown in figure have charges of $+20$ pC and -20 pC, which result in a potential difference of 20 V between them.

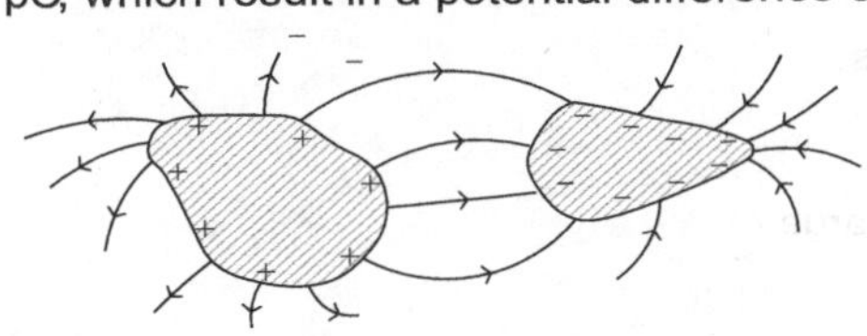

i. What is the capacity of system?

ii. If charges are changed to $+200$ pC and -200 pC, what will be the new capacitance of the system.

iii. If charges are changed to $+200$ pC and -150 pC, then how capacity of the system changes?

2 i. Does the maximum charge given to a metallic sphere depend on whether it is solid or hollow? Find an expression for the maximum charge hold by an isolated sphere? On what factors it depends?

ii. What is dielectric strength of dry air? Can a sphere of radius 1 cm placed in air be given a charge of 1 C?

iii. What maximum charge on a sphere of radius R, can be stored without being leak off in sparks?

3 i. Farad is a large unit. What is the radius of a metal sphere with a capacity of 1 μF How can the size of a sphere be reduced to one third without altering its capacity?

ii. A radioactive source is in the form of a metal sphere of diameter 10^{-3} m emits β-particles at a rate of 6.25×10^{10} per sec. Assuming the rate of decay to be remained constant and source to be insulated, also assume that 80% of emitted β-particles escape from the surface, calculate required time for potential of a sphere to reach 1 V.

iii. What is the difference between dielectric strength and dielectric constant?

iv. A metallic sphere is placed in a laboratory room filled with dry air is being charged continuously. It discharges when field's magnitude is just greater than dielectric strength of the air. Does the maximum value of charge it can hold depend on its distance from other object? If not, then on what factor it depends?

4 i. Draw a graph to show how capacitance C of a capacitor varies with charge Q?

ii. A Q *versus* V graph for two metallic spheres is given. Which sphere has a largest radius?

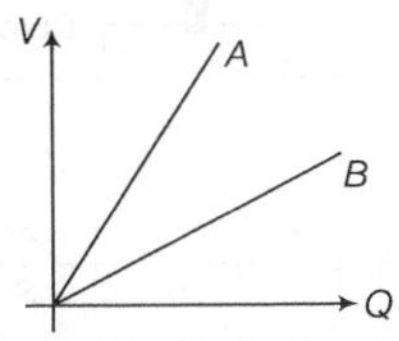

iii. In the graph, V on X-axis and Q on Y-axis, what physical quantity is represented by the slope of graph. What quantity is given by area covered under the curve?

Working Space

iv. In given circuit, a capacitor of capacitance 25 μF is connected with a battery of 120 V potential difference. If switch S is closed at $t = 0$ and capacitor is initially uncharged, then how much charge will flow through the ammeter till the capacitor is fully charged?

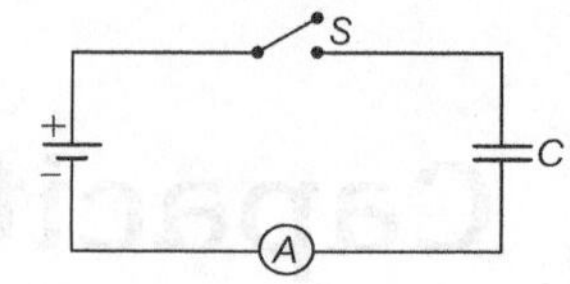

Only One Option Correct Type

Direction *(Q. Nos. 5-14) This section contains 10 multiple choice questions. Each question has four choices (a), (b), (c) and (d), out of which ONLY ONE option is correct.*

5 The capacitance of a charged conductor is 5 μF when it contains 5 μC charge. When it is completely discharged, its capacity is

a. 0 b. 2.5 μF c. 5 μF d. None of these

6 In a charged capacitor, the energy resides

a. with the positive charge
b. with the negative charge
c. equally with the positive and negative charge
d. None of the above

7 Energy stored in a capacitor is in the form of

a. kinetic b. potential c. thermal d. magnetic

8 Consider a finite insulated uncharged object placed near a finite positively charged conductor. The uncharged body must have a potential of

a. less than the charged conductor and more than at infinity
b. more than the charged conductor and less than at infinity
c. zero
d. less than the charged conductor and less than at infinity

9 64 small drops of mercury each with a charge q and radius r combined to form a single drop. The ratio of the surface charge density of each small drop and surface charge density of a large drop will be

a. 1 : 64 b. 1 : 16 c. 1 : 8 d. 1 : 4

10 Eight small charged drops of mercury are combined to form a big drop. The ratio of capacitance of big drop and small drop is

a. 8 : 1 b. 4 : 1 c. 16 : 1 d. 2 : 1

11 Let K and x are dielectric constant and dielectric strength of a dielectric, respectively. A suitable material for using as a dielectric in a capacitor has

a. low k, high x b. low k, low x c. low x, high k d. high k, high x

12 Two capacitors C_1 and C_2 are charged to 120 V and 200 V respectively. It is found that by connecting them together the potential on each one can be made zero. Then, **[JEE Main 2013]**

a. $5C_1 = 3C_2$ b. $3C_1 = 5C_2$ c. $5C_1 + 3C_2 = 0$ d. $9C_1 = 4C_2$

13 Consider the circuit shown as below.

When switch S is closed, what status will ammeter show?

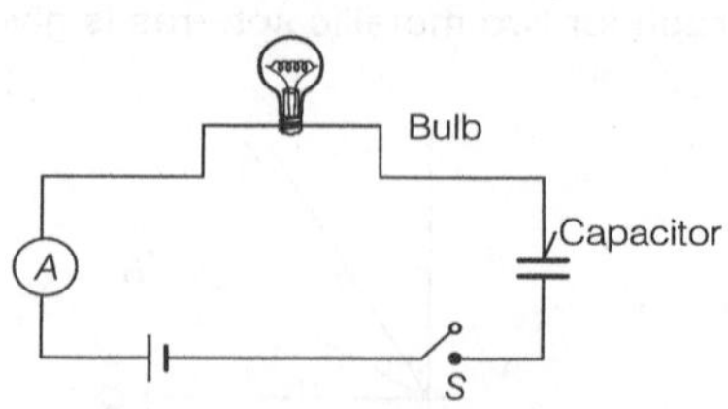

a. ammeter reading remains zero all the times
b. ammeter reading remains same all the times
c. ammeter initially shows some deflection and then its reading becomes zero
d. ammeter reading is initially zero but it shows a small gradually increasing current

14 Capacitance of a capacitor is C farad, when it is connected to a source of V volt, it shows a charge of Q coulomb. The capacitor in its uncharged state is taken and connected to a source of 2 V.

a. Its capacitance becomes $2C$
b. Its capacitance becomes $\frac{C}{2}$
c. Its charge becomes $2Q$
d. Its charge becomes $\frac{Q}{2}$

One or More than One Options Correct Type

Direction *(Q. Nos. 15-21) This section contains 7 multiple choice questions. Each question has four choices (a), (b), (c) and (d), out of which ONE or MORE THAN ONE are correct.*

15 A parallel plate capacitor is charged and the charging battery is then disconnected eventually. If the plates of the capacitor are moved farther apart by means of insulating handles, then mark the correct statements :
a. the charge on the capacitor increases
b. the voltage across the plates increases
c. the capacitance increases
d. the electrostatic energy stored in the capacitor increases

16 Consider a sphere of radius R. What relationships will it satisfy?

a. its capacitance $\propto R$
b. its capacitance $\propto \frac{1}{R}$
c. stored electrostatic PE $\propto \frac{1}{R}$
d. stored electrostatic PE $\propto \frac{1}{R^2}$

17 A spherical metallic ball of radius R is being charged, then
a. its capacitance increases
b. its capacitance remains same
c. its potential increases
d. its potential remains same

18 A rubber balloon is charged and its capacitance is measured. Now, it is inflated and when its radius becomes twice of its original value, then its capacitance is again measured. Correct option(s) are
a. Final capacitance is less than initial capacitance
b. Final capacitance is more than initial capacitance
c. Final potential at a point inside the balloon is less than its surface potential
d. Final surface potential is less than its initial value

19 Consider two hollow identical metallic spheres, first is placed in vacuum and second is placed in water. Mark the correct option(s).

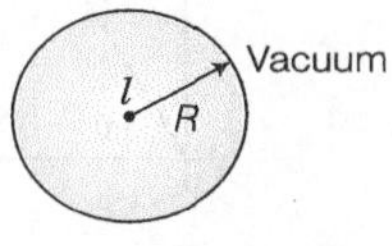

Sphere 1

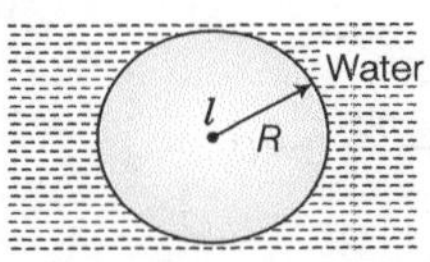

Sphere 2

a. Sphere 1 Can hold more charge than Sphere 2
b. Sphere 2 Can hold more charge than Sphere 1
c. Capacitance of Sphere 1 is greater than that of Sphere 2
d. Capacitance ratio, when they are fully charged is 1

20 Which of the following are true regarding permittivity of free space?
a. Its value is 8.85 pF/m
b. Its dimensions are $[M^{-1}L^{-3}T^4A^2]$
c. Its value increases to $9\,C^2/Nm^2$ with increase in temperature
d. Its dimensions are $[ML^3T^{-4}A^{-2}]$

Working Space

21 A capacitor is an arrangement which is used to store electrical potential energy. Then,

a. It may contain a large sphere
b. It may contain two oppositely large conductors placed side by side
c. It may contain many small spheres kept insulated from each other
d. It may contain two insulator plates with a conducting metal plate in between

Matching List Type

Direction (Q. No. 22) Choices for the correct combination of elements from Column I and Column II are given as options (a), (b), (c) and (d), out of which one is correct.

22 Let n identical drops each of radius r, capacitance c, charge $-q$ and potential $-v$ are combined to form a larger drop. Radius of big drop is R, capacitance $-C$, charge $-Q$ and potential V. Match the following columns.

	Column I		Column II
i.	Ratio $\frac{R}{r}$	p.	$n^{1/3}$
ii.	Ratio $\frac{C}{c}$	q.	$n^{2/3}$
iii.	Ratio $\frac{V}{v}$	r.	$n^{5/3}$

Codes

	i	ii	iii		i	ii	iii
a.	p	p	q	b.	p	q	r
c.	q	q	r	d.	r	p	q

One Integer Value Correct Type

Direction (Q. No. 23 -25) This section contains 3 question. When worked out will result in an integer from 0 to 9 (both inclusive).

23 Two different capacitors with capacity C_1 and C_2 are being charged. Graphs for C_1 and C_2 are as follow. Find the value of $\frac{C_1}{C_2}$.

For capacitor C_1,

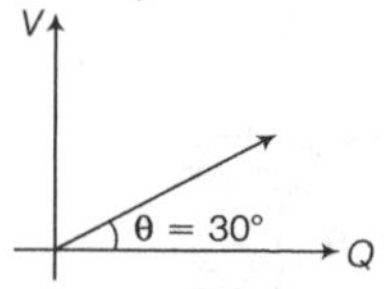

For capacitor C_2,

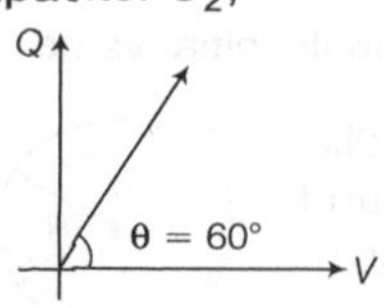

24 In a photoelectric effect study experiment, a small metallie sphere is placed in front of a point source of light having frequency greater than threshold frequency of metal. The following observations are made.

i. Number of photons falling over sphere in 1sec = 10^{11} per sec.
ii. Electron emission efficiency of metal is lone only 1 electron is emitted for every 10^6 photons.
iii. Stopping potential of a metal is 2 V and remains constant throught out the experiment.
iv. After 111 s, no electron is emitted from the sphere. If C is capacity of the metal sphere.

Find the value of $9C \times 10^6$.

Working Space

25 Two metallic spheres are placed in a liquid of dielectric constant a such that their centres are nearly 1 m apart. It is given that radius of each sphere is 2 cm and capacitance of system of spheres is 10^N farads. Find the value of N.

Comprehension Type

Direction *(Q. Nos. 26-28) This section contains a paragraph describing theory, experiments, data, etc. Three questions related to the paragraph have been given. Each question has only one correct answer among the four given options (a), (b), (c) and (d).*

Passage

Consider a metal sphere of radius r over which a light beam of sufficiently high frequency is incident. Due to photo-emission, metal sphere started to lose electrons.

26 Potential of a sphere varies with time as

a. 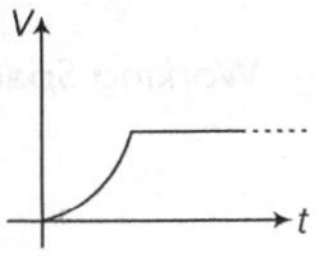

b. 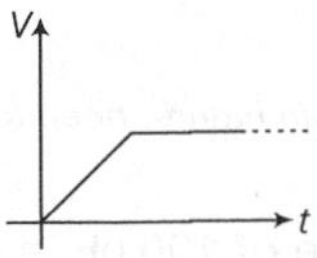

c. 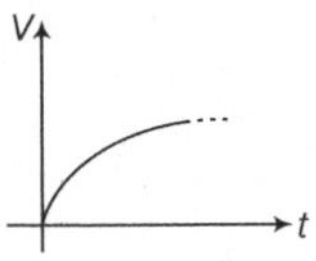

d. 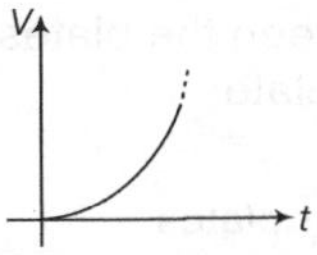

27 Capacity of a sphere varies with time as

a. 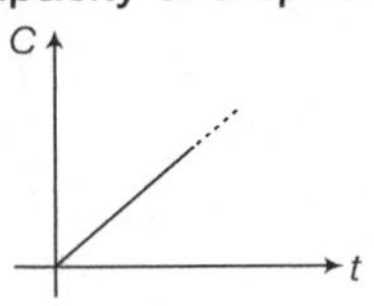

b. 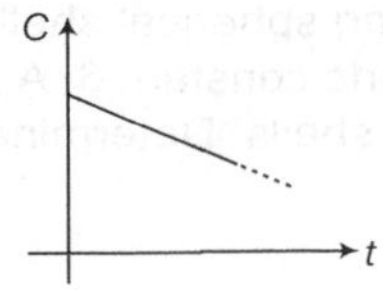

c. 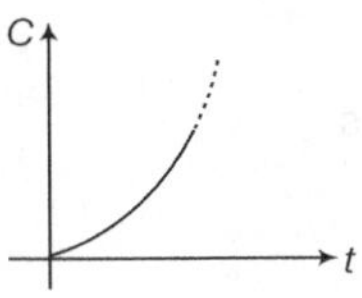

d. 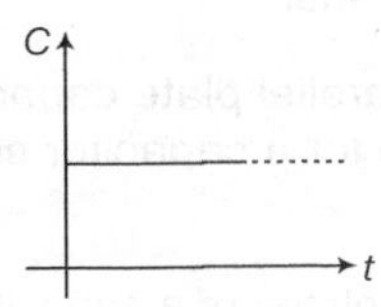

28 Charge on the sphere varies with time as

a. 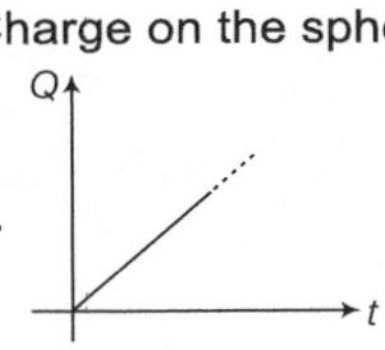

b. 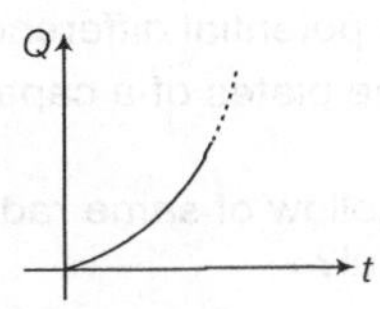

c. 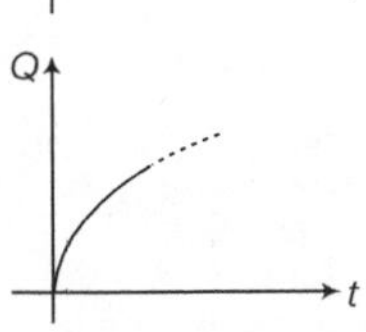

d. 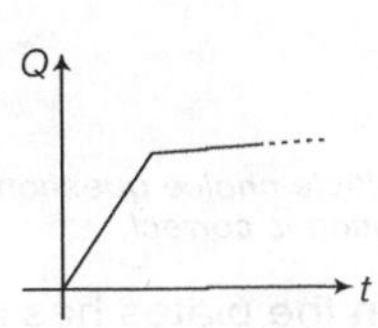

DPP-2 Capacitance of Different Types of Capacitors

Subjective Questions

Working Space

Direction (Q. Nos. 1-4) These questions are subjective in nature, need to be solved completely on notebook.

1 A parallel plate capacitor has a capacitance of 100 pF, a plate area of 100 cm^2 and a mica dielectric ($K = 5.4$) is filled between the plates. When this capacitor is charged to 50 V potential difference. Calculate

i. electric field in the mica
ii. magnitude of free charge on the capacitor plates
iii. magnitude of the induced surface charge on the mica

2 Space between two concentric conducting spherical shells of radii 38 mm and 40 mm is filled with a dielectric of dielectric constant 6. A potential difference of 50 V exists between the inner and outer shells. Determine

i. capacity of device
ii. free charge on inner shell
iii. charge induced on the surface of inner shell

3 When a battery is connected across a parallel plate capacitor, each plate acquires a charge of equal magnitude. Is this true for a capacitor consisting of plates of different sizes or shapes?

4 i. Why a dielectric is used for separating plates of a capacitor?
ii. How the introduction of a dielectric slab between the plates of a capacitor decreases the electric field and electric potential difference between the plates?
iii. If a metal slab is introduced between the plates of a capacitor, its capacitance is infinity. What does this means?
iv. Why two spheres one solid and other hollow of same radius have same charge, when they are charged to same potential?

Only One Option Correct Type

Direction (Q. Nos. 5-15) This section contains 11 multiple choice questions. Each question has four choices (a), (b), (c) and (d), out of which ONLY ONE option is correct.

5 A parallel plate capacitor with air between the plates has a capacitance of 8 pF. When the distance between plates is reduced by half and the space between them is filled with a material of dielectric constant 6, capacitance is increased by

a. 96 pF
b. 48 pF
c. 88 pF
d. 6 pF

Working Space

6 A parallel plate capacitor is made as follows

Capacitance of the arrangement is

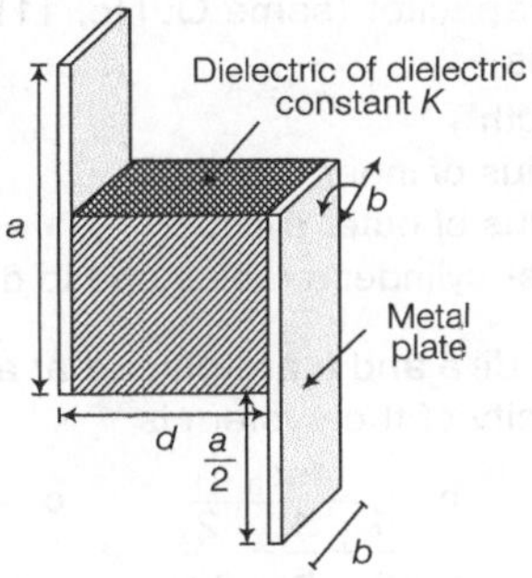

a. $\frac{\varepsilon_0 Kab}{d}$

b. $\frac{\varepsilon_0 Kab}{2d}$

c. $\frac{\varepsilon_0 Kad}{b}$

d. $\frac{\varepsilon_0 Kad}{2b}$

7 Two identical metal plates each of area A are given unequal charges $+Q_1$ and $Q_2 (Q_1 > Q_2)$ and then they are placed d distant apart.

Capacity of formed capacitor is

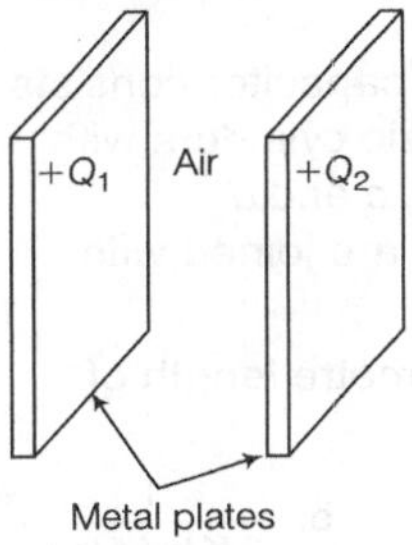

a. $\frac{\varepsilon_0 A}{4d}$

b. $\frac{Q_1 + Q_2}{2V}$

c. $\frac{Q_1 - Q_2}{2V}$

d. $\frac{\varepsilon_0 A}{d}$

8 A parallel plate capacitor C with plates of area 1 square unit and separation d is filled with a liquid of dielectric constant $K = 2$. The level of liquid is $d/3$, initially. If the liquid level decreases at a constant speed v. Then, the capacity of capacitor after t second is **[IIT JEE 2008]**

a. $\frac{6\varepsilon_0}{5d + 3vt}$

b. $\frac{(15d + 9vt)\varepsilon_0}{2d^2 - 3dvt - 9v^2t^2}$

c. $\frac{6\varepsilon_0}{5d - 3vt}$

d. $\frac{(15d - 9vt)\varepsilon_0}{2d^2 + 3dvt - 9v^2t^2}$

9 A variable air capacitor used in radio tuning circuit is made of N semi-circular plates each of radius R and positioned a distance d from its neighbouring plates. A second set of similar plates is enmeshed with its plates half-way between those of first set. Second set of plates can rotate as a complete unit.

Capacitance of arrangement is

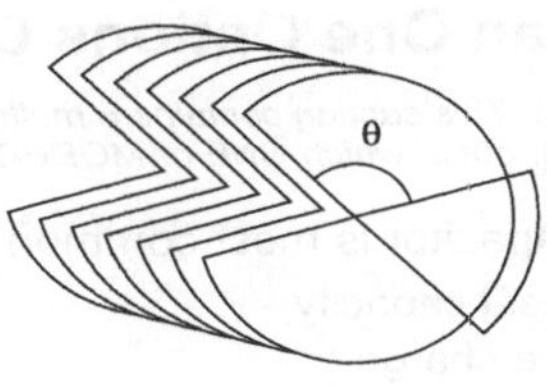

a. $\frac{N\varepsilon_0(\pi - \theta)R^2}{d}$

b. $\frac{(N-1)\varepsilon_0(\pi - \theta)R^2}{d}$

c. $\frac{(N-2)\varepsilon_0(\pi - \theta)R^2}{d}$

d. $\frac{(2N-1)\varepsilon_0(\pi - \theta)R^2}{d}$

10 A spherical capacitor composed of two concentric metal spheres one having a radius twice as large as the other. The region in which the energy is stored has a volume of

a. $\frac{1}{4}\frac{C^3}{\pi^2\varepsilon_0^3}$

b. $\frac{3}{16}\frac{C^3}{\pi^2\varepsilon_0^3}$

c. $\frac{1}{384}\frac{C^3}{\pi^2\varepsilon_0^3}$

d. $\frac{7C^3}{384\pi^2\varepsilon_0^3}$

11 A solid cylindrical conductor of radius a and charge q is placed coaxially with a cylindrical shell of negligible thickness and radius b, $(b > a)$. If outer cylinder is earthed, then capacity per metre length of system is

a. $\frac{1}{2K\ln(b/a)}$

b. $\frac{1}{2K\ln(a/b)}$

c. $\frac{1}{4K\ln(b/a)}$

d. $\frac{1}{4K\ln(a/b)}$

Working Space

12 For a cylindrical capacitor (same Q. No. 11) to increase the capacitance, most effective method is

a. to increase length
b. to increase radius of inner cylinder
c. to increase radius of outer cylinder
d. to increase inner cylinder's radius and to decrease outer cylinder's radius

13 Two spheres of radii *a* and *b* are placed at a distance *d* apart (*d* is large compared to *a* and *b*). Capacity of the system is

a. $\dfrac{4\pi\varepsilon_0}{\dfrac{1}{a}+\dfrac{1}{b}}$ b. $\dfrac{4\pi\varepsilon_0}{\dfrac{1}{a}+\dfrac{1}{b}-\dfrac{2}{ab}}$ c. $\dfrac{4\pi\varepsilon_0}{\dfrac{1}{a}+\dfrac{1}{b}-\dfrac{2}{d}}$ d. $\dfrac{4\pi\varepsilon_0}{\dfrac{1}{a}+\dfrac{1}{b}-\dfrac{1}{d}}$

14 A long cylindrical capacitor consists of 4 long concentric cylinders with effective radii *a*, *b*, *c* and *d*. Cylinders *a* and *b* are joined with metal strips.

Capacitance per metre length of arrangement is

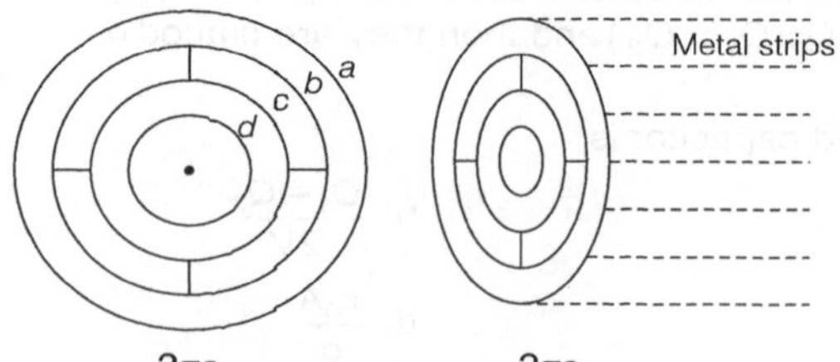

a. $\dfrac{1}{2K\ln(a/d)}$ b. $\dfrac{1}{2K\ln(d/a)}$ c. $\dfrac{2\pi\varepsilon_0}{\ln(ac/bd)}$ d. $\dfrac{2\pi\varepsilon_0}{\ln(ab/cd)}$

15 A parallel plate capacitor is made of two circular plates separated by a distance of 5 mm and with a dielectric of dielectric constant 2.2 between them. When the electric field in the dielectric is 3×10^4 V/m, the charge density of the positive plate will be close to **[JEE Main 2014]**

a. 6×10^{-7} C/m^3 b. 3×10^{-7} C/m^3 c. 3×10^{4} C/m^3 d. 6×10^{4} C/m^3

One or More than One Options Correct Type

Direction *(Q. Nos. 16-20) This section contains 5 multiple choice questions. Each question has four choices (a), (b), (c) and (d), out of which ONE or MORE THAN ONE are correct.*

16 A parallel plate capacitor is most commonly used, because

a. it can have largest capacity
b. it can store more charge
c. it can have very large dielectric breakdown potential
d. it can have a uniform field

17 A spherical capacitor has an inner sphere of radius 12 cm and an outer sphere of radius 13 cm. The outer sphere is earthed and inner sphere is given a charge of $2.5\,\mu$C. Then

a. Arrangement has a capacity of 0.16 nF
b. Arrangement has a capacity of 5.5 pF
c. Potential diffrence of inner sphere is 5.7×10^4 V
d. Potential diffrence of outer sphere is 4.5×10^2 V

18 During charging of a capacitor consisting of two plates of metal which are separated by a certain distance. The correct option(s) are

a. Charge can momentarily flow from one plate to another till both plates are fully charged
b. Induction occurs between the plates
c. Potential difference of region between the plates is increased by using two dissimilar charged plates
d. Potential difference of region between the plates is reduced by using two dissimilar charged plates

Working Space

19 Area of a parallel plate capacitor is 0.2 m^2 and distance between them is 0.01 m. Potential difference between the plates is 3000 V. When a 0.01 m thick plate of an insulating material is put between plates of capacitor, then potential difference decreases to 1000 V.

Correct option(s) are

a. Capacity with air between plates is 1.77×10^{-10} F
b. Dielectric constant of medium is 3
c. Capacity after putting dielectric is 6.31×10^{-10} F
d. Permittivity of dielectric is 9.65×10^{-11} Fm^{-1}

20 A capacitor is practically used as a

a. vital component of all tuning circuits like television transmitters and receivers
b. switching device for ON and OFF operations in a circuit
c. high resistance
d. memory bank of computers

One Integer Value Correct Type

Direction (Q. Nos. 21-23) This section contains 3 questions. When worked out will result in an integer from 0 to 9 (both inclusive).

21 Capacitance of a parallel plate capacitor with plates area A and separation d which is filled with two wedges of dielectric medium of dielectric constants K_1 and K_2 is

$C = N \cdot \frac{\varepsilon_0 A}{d} \log 2$

If $K_2 = 8$ and $K_1 = 4$, then find the value of N.

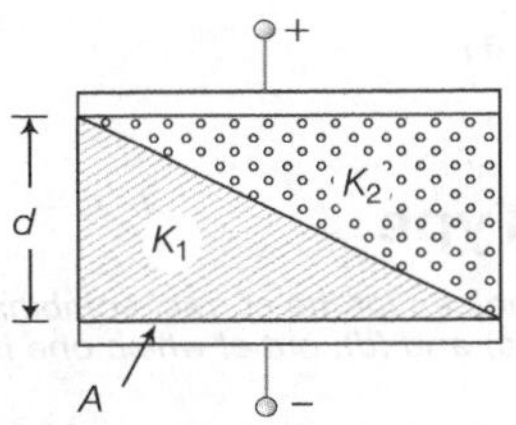

22 Consider a capacitor consisting of a metal ball and an infinite conducting plate. Radius of ball is 2 cm and it is kept at a large distance from the plate. If C is capacitance of a system, then find the value of $9C \times 10^{-7}$.

23 Two straight wires with equal cross-sectional radii are placed parallel to each other in a medium of dielectric constant 6. Distance between wires is b and $b >> a$. Capacitance per metre length of system is $\frac{N\pi\varepsilon_0}{\log_e(b/a)}$. Find the value of N.

Comprehension Type

Direction (Q. Nos. 24-28) This section contains a paragraph describing theory, experiments, data, etc. Five questions related to the paragraph have been given. Each question has only one correct answer among the four given options (a), (b), (c) and (d).

Passage

Three conducting shells A, B and C of radii a, b and c are as shown.

This assembly can be asked as a capacitor by earthing any of the shells A, B or C.

Suppose, a source of emf is connected between shells A and C so that shell A acquires a charge q and shell C has a charge $-q$.

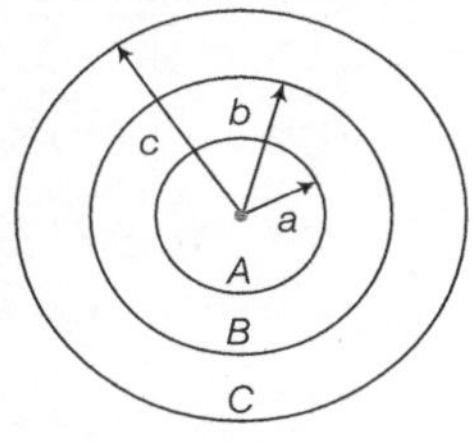

24 Potential of shell A is

a. $\frac{q}{4\pi\varepsilon_0 a}$ b. $\frac{q}{4\pi\varepsilon_0 c}$ c. $\frac{q}{4\pi\varepsilon_0}\left(\frac{1}{c} - \frac{1}{a}\right)$ d. $\frac{q}{4\pi\varepsilon_0}\left(\frac{1}{a} - \frac{1}{c}\right)$

Working Space

25 Potential difference between shells A and C is

a. $\dfrac{q}{4\pi\varepsilon_0 a}$ b. $\dfrac{q}{4\pi\varepsilon_0 c}$ c. $\dfrac{q}{4\pi\varepsilon_0}\left(\dfrac{1}{c}-\dfrac{1}{a}\right)$ d. $\dfrac{q}{4\pi\varepsilon_0}\left(\dfrac{1}{a}-\dfrac{1}{c}\right)$

26 Capacity of the system is

a. $4\pi\varepsilon_0 a$ b. $4\pi\varepsilon_0 b$ c. $4\pi\varepsilon_0\left(\dfrac{ac}{c-a}\right)$ d. $\dfrac{4\pi\varepsilon_0 ac}{(a-c)}$

27 Suppose, a dielectric medium of dielectric constant K is filled between A and B. Then, potential difference between A and C is

a. $\dfrac{q}{4\pi\varepsilon_0 Kabc}[c(b-a)+Ka(c-b)]$ b. $\dfrac{q}{4\pi\varepsilon_0 Kabc}[Kc(b-a)+a(c-b)]$

c. $\dfrac{q}{4\pi\varepsilon_0 abc}K[c(b-a)+a(c-b)]$ d. $\dfrac{q}{4\pi\varepsilon_0 K}(abc)[(a-b)+(b-c)]$

28 Suppose, a medium of dielectric constant K is filled between A and B. Then, capacitance between A and C becomes

a. $4\pi\varepsilon_0 aK$ b. $\dfrac{4\pi\varepsilon_0 abc}{Ka(c-b)+c(b-a)}$

c. $\dfrac{4\pi\varepsilon_0 Kabc}{K(c-b)+(b-a)}$ d. $\dfrac{4\pi\varepsilon_0 Kabc}{Ka(c-b)+c(b-a)}$

Matching List Type

Direction *(Q. No. 29) Choices for the correct combination of elements from Column I and Column II are given as options (a), (b), (c) and (d), out of which one is correct.*

29 In the given arrangement, the area of each plate is A.

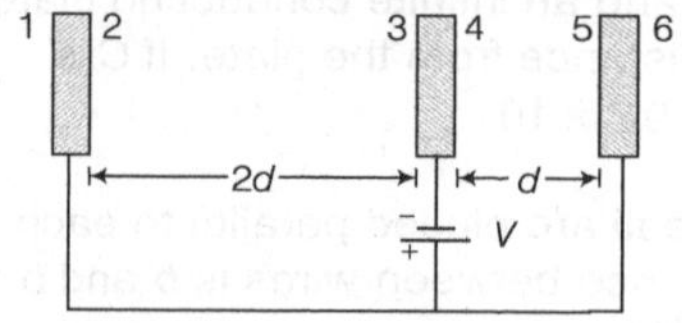

Match the Column I with Column II.

	Column I		Column II
i.	Charge on plate 3	p.	Zero
ii.	Charge on plate 5	q.	V
iii.	Potential difference between plates 2 and 3	r.	$\dfrac{\varepsilon_0 A}{2d}V$
iv.	Potential difference between plates 2 and 5	s.	$\dfrac{\varepsilon_0 A}{d}V$

Codes

	i	ii	iii	iv		i	ii	iii	iv
a.	r	s	q	p	b.	p	r	s	q
c.	p	q	r	s	d.	q	r	p	s

DPP-3 Energy of a Charged Capacitor, Effect of Dielectric and Force between Plates of a Capacitor

Subjective Questions

Working Space

Direction *(Q. Nos. 1-4) These questions are subjective in nature, need to be solved completely on notebook.*

1 A parallel plate capacitor with plate area 40 cm^2 and plate spacing 1 mm is charged using a 600 V DC supply.

i. Find stored energy.
ii. Find electric field between plates.
iii. How much energy is lost in charging of capacitor?
iv. Why there is a loss of energy?
v. What is the magnitude of force between plates of a charged capacitor?

2 A parallel plate capacitor of plate area 40 cm^2 and plate separation 1 mm is charged using a 12 V battery.

i. If plate separation is made 2 mm while the battery remains connected, find the work done by the external source.
ii. Find the charge transferred to battery in above process.
iii. Find work done by battery in the above process.
iv. Is battery being charged or discharged in the above process?

3 A parallel plate capacitor with circular plates of 8.2 cm radii and 1.3 mm separation is connected to a 120 V DC supply.

i. Find the capacitance of a capacitor.
ii. When capacitor is fully charged, what value of charge appears on each plate? If capacitor is disconnected from the battery and its plates are moved apart so that plate separation is doubled.
iii. What will be the new value of potential difference between plates?
iv. What is the work done in the process?
v. What is the force between plates of capacitor now. Is it more or less than the previous value?
vi. By what amount energy of capacitor increases? How do you account for this increase of energy?

4 i. How a dielectric increases the maximum operating voltage without changing the physical size of a capacitor?
ii. Why water molecule is permanently polarised? What type of molecules has no permanent polarisation?
iii. If a dielectric filled in a capacitor's plates is heated, how will its capacitance change?

Only One Option Correct Type

Working Space

Direction *(Q. Nos. 5-16) This section contains 12 multiple choice questions. Each question has four choices (a), (b), (c) and (d), out of which ONLY ONE option is correct.*

5 A parallel plate capacitor with dielectric of dielectric constant K between the plates has a capacity C and is charged to a potential of V volt and battery remains connected. The dielectric slab is slowly removed from the plates and then reinserted. The net work done by the system in the process is

a. $\frac{1}{2}(k-1)CV^2$ b. $CV^2(k-1)K$ c. $(k-1)CV^2$ d. zero

6 A fully discharged capacitor is connected to a battery till the potential difference between plates of a capacitor becomes equal to the terminal potential difference of battery. The ratio of energy stored in the capacitor and the work done by the battery is

a. 1 b. 2 c. $\frac{1}{4}$ d. $\frac{1}{2}$

7 Two capacitors A and B with capacities $3\,\mu F$ and $2\,\mu F$ are charged to a potential difference of 100 V and 180 V respectively. The plates of the capacitors are connected as shown in the figure with one wire of each capacitor free. The upper plate of A is positive and that of B is negative. An uncharged $2\,\mu F$ capacitor C with lead wires falls on the free ends to complete the circuit, then charges on the three capacitors are

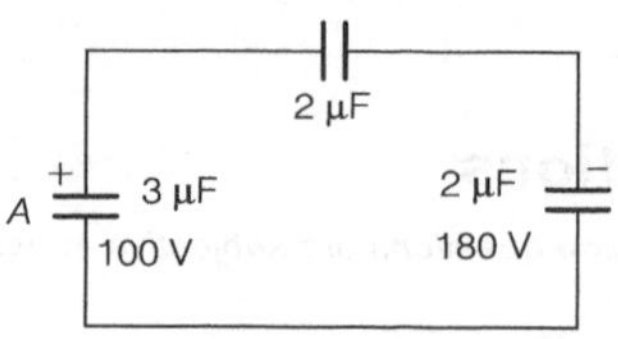

a. $q_1 = 90\,\mu C$, $q_2 = 150\,\mu C$ and $q_3 = 210\,\mu C$
b. $q_1 = 210\,\mu C$, $q_2 = 90\,\mu C$ and $q_3 = 150\,\mu C$
c. $q_1 = 90\,\mu C$, $q_2 = 210\,\mu C$ and $q_3 = 150\,\mu C$
d. $q_1 = 150\,\mu C$, $q_2 = 90\,\mu C$ and $q_3 = 210\,\mu C$

8 Two identical capacitors have same capacity C. One of them is charged to potential V_1 and other to V_2. Likely charged plates are then connected. Then, the decrease in energy of the combined system is **[IIT JEE 2002]**

a. $\frac{1}{4}C(V_1^2 - V_2^2)$ b. $\frac{1}{4}C(V_1^2 + V_2^2)$ c. $\frac{1}{4}C(V_1 - V_2)^2$ d. $\frac{1}{4}C(V_1 + V_2)^2$

9 A dielectric has a dielectric strength of 18 MV/m and a dielectric constant of 2.8. It is used to make a capacitor which withstand a potential difference of 4.0 kV and have a capacity of $7 \times 10^{-2}\,\mu F$. The minimum area of plates required is

a. $63 \times 10^{-12}\,m^2$ b. $81 \times 10^{-12}\,m^2$ c. $27 \times 10^{-12}\,m^2$ d. $5 \times 10^{-12}\,m^2$

10 The ratio of polarisation vector and electric field vector is known as electrical susceptibility. Susceptibility has dimensions of

a. $[M^0L^{-3}T^{-2}]$ b. $[M^{-1}L^{-3}T^4A^2]$ c. $[ML^2T^{-3}A^{-1}]$ d. $[M^0L^0T^0A^0]$

11 A parallel plate capacitor with capacity 100 pF, plate area $100\,cm^2$ and a mica dielectric constant $K = 5.4$ which completely fills the space between plates. The magnitude of induced surface charge on the mica is (σ is the charge on capacitor's plates)

a. $5.4\,\sigma$ b. $\frac{\sigma}{5.4}$ c. $\frac{27}{22}\sigma$ d. $\frac{22}{27}\sigma$

12 A parallel plate capacitor with plate separation d and plate area A is filled with a dielectric slab of thickness t. If it is required that energy stored by the dielectric is half of total energy stored in capacitor, then t equals to (given, $K = 3$)

a. $\frac{d}{4}$ b. $\frac{d}{2}$ c. $\frac{3d}{4}$ d. $0.8\,d$

13 A parallel plate capacitor is isolated with charge $\pm Q$ on its plates. If distance between plates is reduced to $\frac{1}{2}$ and a dielectric slab of dielectric constant K is put between the plates of capacitor, then stored energy is

a. reduced by a factor of $2K$
b. increased by a factor of $2K$
c. reduced by a factor of $\frac{1}{2K}$
d. increased by a factor of $\frac{1}{2K}$

14 A 3500 pF air capacitor is connected to a 32 V battery. If a piece of mica fills the space between plates, then charge that flows from battery to the capacitor is ($K_{mica} = 6$)

a. 5.5×10^{-4} C
b. 6.2×10^{-5} C
c. 6.7×10^{-7} C
d. 3.5×10^{-9} C

15 A 2 μF capacitor is charged as shown in the figure. The percentage of its stored energy dissipated after the switch S is turned to position 2 is **[IIT JEE 2011]**

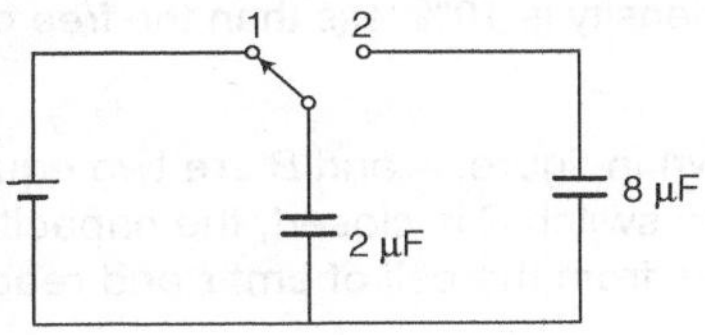

a. 0%
b. 20%
c. 75%
d. 80%

16 A dielectric slab of thickness d is inserted in a parallel plate capacitor whose negative plate is at $x = 0$ and positive plate is at $x = 3d$. The slab is equidistant from the plates. The capacitor is given some charges. As we go from 0 to $3d$, potential difference variation is

a. 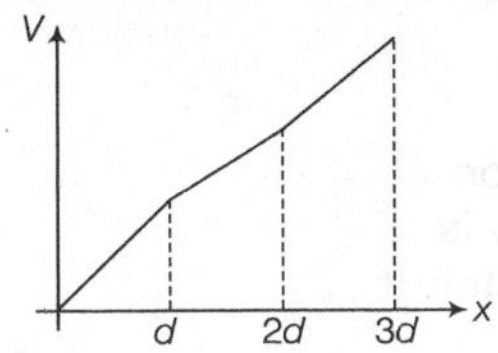
b. 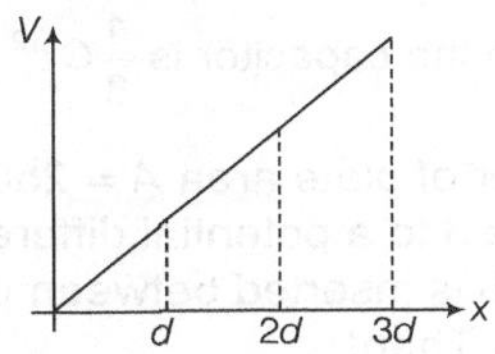
c. 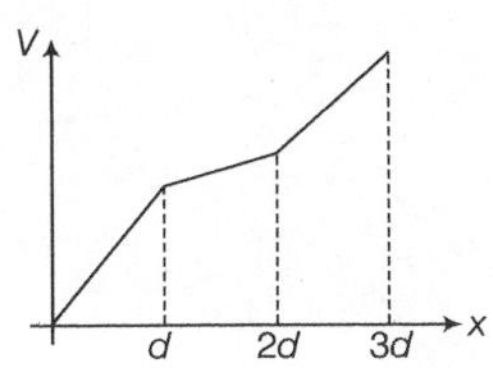
d. 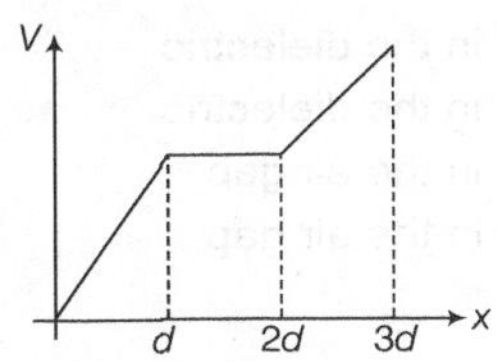

One or More than One Options Correct Type

***Direction** (Q. Nos. 17-21) This section contains 5 multiple choice questions. Each question has four choices (a), (b), (c) and (d), out of which ONE or MORE THAN ONE are correct.*

17 A capacitor is charged by using a battery and then a dielectric slab is released from the position shown in the figure (battery remains connected). Then,

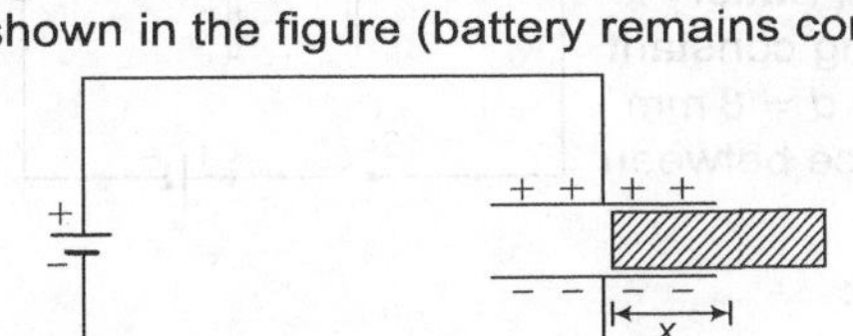

a. Force on dielectric slab does not depend on the dimension x
b. Force on dielectric slab depends on the dimension x
c. Time period of oscillation does not depend on the dimension x
d. Time period of oscillation depends on the dimension x

18 A parallel plate capacitor with plate area 6×10^{-3} m^2 and plate separation 3 mm is charged by a 100 V battery. After being fully charged, a dielectric slab of mica ($K = 6$) is inserted between the plates while the battery remains connected.

a. capacitance increases by 500%
b. capacitance increases by 200%
c. potential increases by 600%
d. charge increases by 500%

19 A parallel plate capacitor is partially filled with a dielectric of dielectric constant $K = 5$. Then,

a. electric field inside the dielectric is 20% less than the electric field due to free charges in air gap
b. electric field inside the dielectric is 10% less than the electric field in air gap
c. bound charge density on the surface of dielectric is 20% less than free charge density on plates of capacitor
d. bound charge density is 10% less than the free charge density on the plates of capacitor

20 In the circuit shown in figure, A and B are two equal resistances. When switch S is closed, the capacitor of capacity C charges from the cell of emf ε and reaches a steady state.

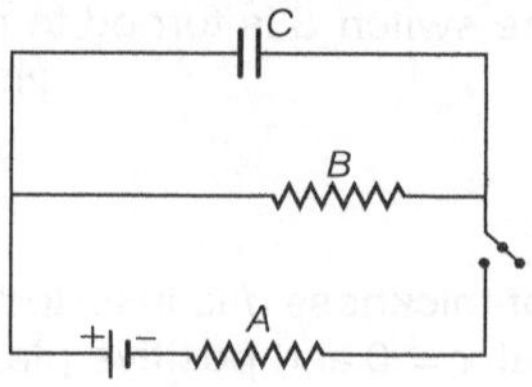

Correct option(s) are

a. During charging, more heat is produced in A than in B
b. In steady state, heat is produced at same rate in A and B
c. In steady state, energy stored in capacitor is $\frac{1}{4}C\varepsilon^2$
d. In steady state, energy stored in the capacitor is $\frac{1}{8}C\varepsilon^2$

21 Consider a parallel plate capacitor of plate area $A = 250$ cm^2 and separation $d = 2$ mm. The capacitor is charged to a potential difference 150 V. Battery is disconnected and a dielectric slab is inserted between plates of the capacitor. If dielectric constant inserted is 3.5. Then,

a. around 22% of energy is stored in the dielectric
b. around 78% of energy is stored in the dielectric
c. around 22% of energy is stored in the air gap
d. around 78% of energy is stored in the air gap

One Integer Value Correct Type

Direction (Q. Nos. 22-24) This section contains 3 questions. When worked out will result in an integer from 0 to 9 (both inclusive).

22 In the given circuit,

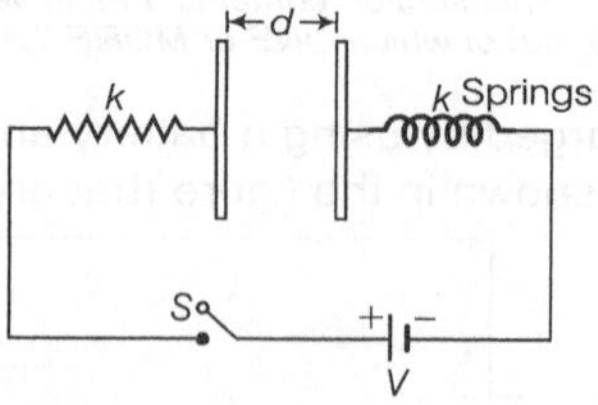

When switch S is opened, the plates are uncharged and are separated by distance $d = 8.00$ mm. Battery is of 100 V and springs are identical with spring constant K. Capacitance of capacitor with separation $d = 8$ mm is 2 μF. When switch S is closed, the distance between plates is reduced to 4 mm.
Find the sum of spring constants of springs.

23 A parallel plate capacitor is charged by a DC supply of 500 V. Plate separation is 1 mm. If capacitor is lowered into water with water filling in the gap between the plates, then change in pressure at any point between the plates is $k\varepsilon_0 \times 10^{13}$ (N/m^2). Find the value of K, given ε_r (for water) = 81.

24 A cylindrical capacitor has radii 4 cm and 9 cm. Half of stored energy lies within a cylindrical volume of radius R. Find the value of R.

Working Space

Comprehension Type

Direction *(Q. Nos. 25-27) This section contains a paragraph describing theory, experiments, data, etc. Three questions related to the paragraph have been given. Each question has only one correct answer among the four given options (a), (b), (c) and (d).*

Passage

Two parallel plate capacitors A and B having same separation $d = 8.85 \times 10^{-4}$ m with plate areas 0.04 m^2 and 0.02 m^2, respectively are taken. A slab of dielectric material ($K = 9$) has dimensions such that it can exactly fill the space in the capacitor B. The dielectric slab is put between the plates of A as shown.

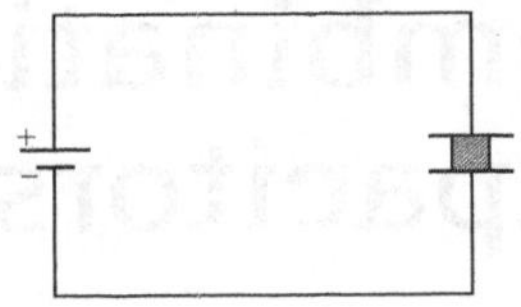

25 Capacity and energy stored in A is around

a. 2nF, 12 μJ
b. 4nF, 12 μJ
c. 2nF, 24 μJ
d. 1nF, 6 μJ

26 Work done in removing slab from capacitor A after disconnecting the battery is

a. 68.5 μJ
b. 24.25 μJ
c. 48.5 μJ
d. 0 μJ

27 If dielectric slab is inserted in A and then B is connected with A as shown.

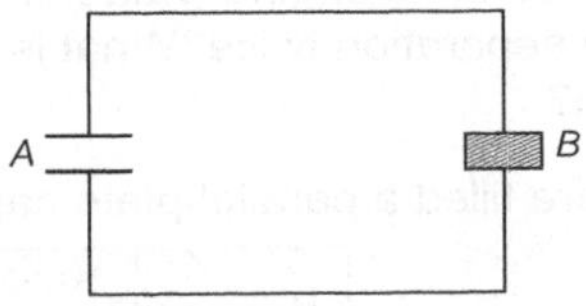

Final energy stored in the system is

a. 33 μJ
b. 22 μJ
c. 44 μJ
d. 11 μJ

Matching List Type

Direction *(Q. No. 28) Choices for the correct combination of elements from Column I and Column II are given as options (a), (b), (c) and (d), out of which one is correct.*

28 The parallel plates of a capacitor have an area 0.2 m^2 and 10^{-2} m apart . The original potential difference between them is 3000 V and it decreases to 1000 V when a dielectric sheet is inserted between the plates filling the full space between plates. (take, $\varepsilon_0 = 9 \times 10^{-12}$ C^2 N^{-1} m^{-2}).

Match the Column I with Column II and select the correct option among the codes given below.

	Column I		Column II
i.	Original capacity in pF.	p.	3×10^5
ii.	Capacity after insertion of dielectric (pF)	q.	54×10^{-8}
iii.	Permittivity of dielectric ($C^2 N^{-1} m^{-2}$)	r.	27×10^{-12}
iv.	Charge on each plate (C)	s.	540
v.	Initial value of field between plates (V/m)	t.	180

Codes

	i	ii	iii	iv	v
a.	t	s	r	q	p
b.	p	q	r	s	t
c.	q	t	p	s	r
d.	r	s	t	q	p

DPP-4 Series and Parallel Combinations of Capacitors

Subjective Questions

Working Space

Direction *(Q. Nos. 1-4) These questions are subjective in nature, need to be solved completely on notebook.*

1 Two identical parallel plate capacitors are first connected in series and then in parallel. In each case, plates of one of the capacitors are brought closer to reduce the separation to half of its original value and plates of other capacitor are moved away to make the separation twice. What is the ratio of capacity of system of capacitors formed?

2 If two dielectrics are filled a parallel plate capacitor as shown.

Find the equivalent dielectric constant of the capacitor.

3 If two dielectrics are placed in a parallel plate capacitor as shown.

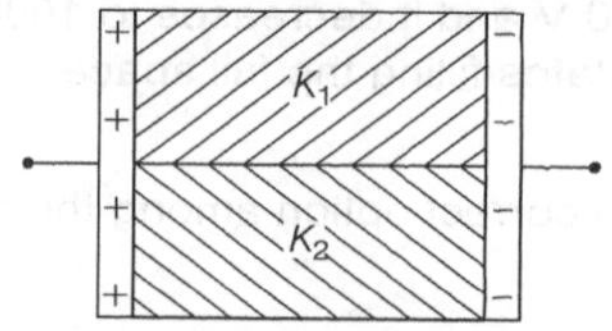

Then, find the equivalent dielectric constant of the system :

4 The graph of V *versus* Q is plotted for two capacitors C_1 and C_2

Capacitor	Plate area	Separation of plates
C_1	A	d
C_2	$2A$	$d/2$

If both capacitors are connected in (i) series and (ii) in parallel and graph of V *versus* Q is plotted again. Then, which of the graph is of

i. C_1
ii. C_2 and
iii. series combination of C_1 and C_2
iv. parallel combination of C_1 and C_2

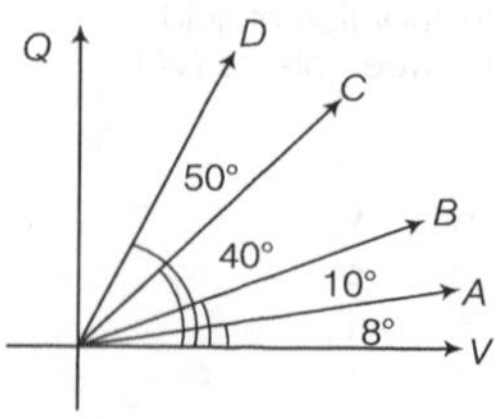

Only One Option Correct Type

Working Space

Direction (Q. Nos. 5-14) *This section contains 10 multiple choice questions. Each question has four choices (a), (b), (c) and (d), out of which ONLY ONE option is correct.*

5 In the network shown in figure, a constant potential difference of 50 V is maintained across a and b. Then, total charge stored in the network is

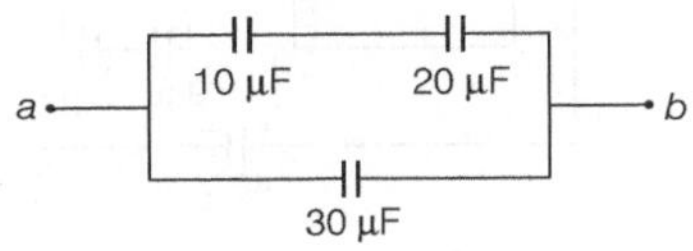

a. $\frac{5500}{3}\,\mu C$

b. $750\,\mu C$

c. $3000\,\mu C$

d. $1280\,\mu C$

6 In the circuit shown in figure,

Initially K_1 is closed and K_2 is open, let charge on capacitor C_2 is Q_2.

Then, K_1 was opened while K_2 kept closed, let charge on capacitor C_2 is Q_2'.

(Take, $C = 1\,\mu F$) Then, ratio $\frac{Q_2}{Q_2'}$ is

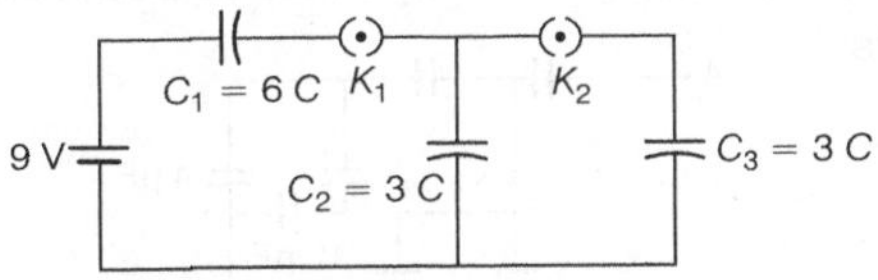

a. 1 : 1 b. 2 : 1

c. 3 : 1 d. 1 : 3

7 In a capacitor bridge, a voltage V_0 is applied and the variable capacitor C_1 has been adjusted till voltmeter shows zero reading. If $C_1 = 8.9\,\mu F$, $C_2 = 18\,\mu F$ and $C_3 = 4.8\,\mu F$ at balance point, then C_x is

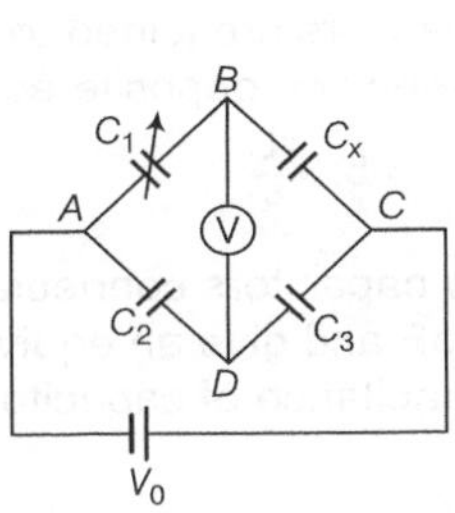

a. $5.4\,\mu F$

b. $7.4\,\mu F$

c. $2.4\,\mu F$

d. $3.8\,\mu F$

8 A parallel plate capacitor is filled with 3 different dielectrics as

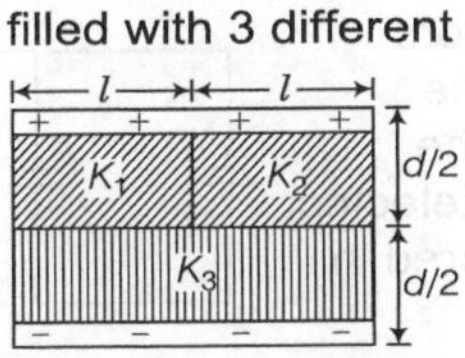

Another capacitor with a single dielectric between its plates is

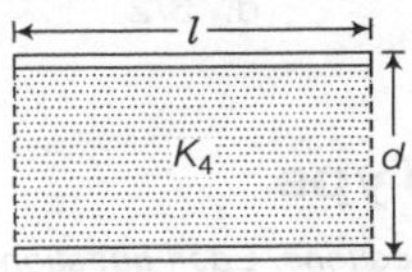

If both capacitors have same capacity and same plate area, then **[IIT JEE 2000]**

a. $\frac{1}{K_4} = \frac{1}{K_1} + \frac{1}{K_2} + \frac{1}{2K_3}$ b. $\frac{1}{K_4} = \frac{1}{K_1 + K_2} + \frac{1}{2K_3}$

c. $K_4 = \frac{K_1 K_2}{K_1 + K_2} + 2K_3$ d. $K_4 = K_1 + K_2 + 2K_3$

9 In the given circuit, key K is closed at time $t = 0$, then

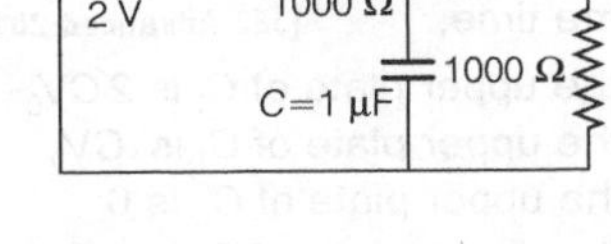

a. $i = 2$ mA at all t

b. i oscillates between 1 mA and 2 mA

c. $i = 1$ mA at all t

d. at $t = 0$, $i = 2$ mA and with time it settles to 1 mA

10 In the circuit given,

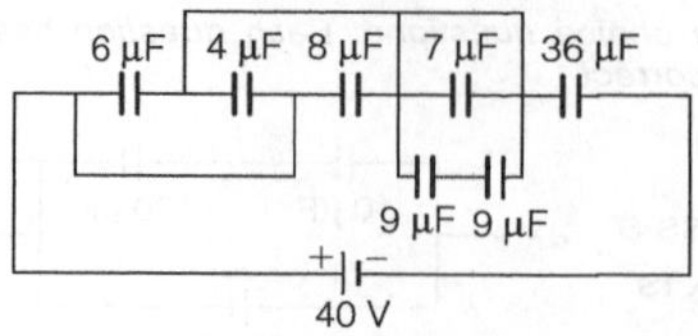

The charge and potential difference across 8 μF capacitor will be respectively

a. 320 μC, 40 V
b. 420 μC, 50 V
c. 214 μC, 27 V
d. 360 μC, 45 V

11 In the following circuit, the resultant capacitance between A and B is 1μF. Then, value of C is

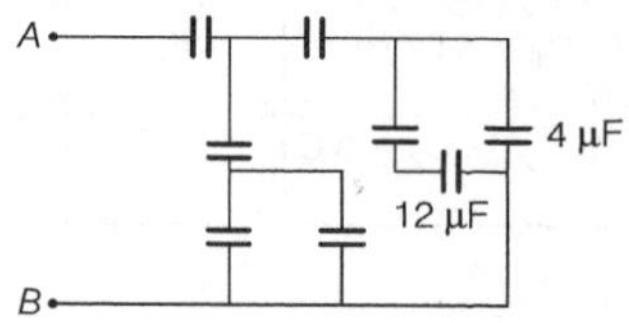

a. $\frac{32}{11}$ μF
b. $\frac{11}{32}$ μF
c. $\frac{23}{32}$ μF
d. $\frac{32}{23}$ μF

12 Twelve equal capacitors are joined to form a cube. Find the equivalent capacitance between two opposite edges on the same plane?

a. $\frac{3C}{4}$
b. $\frac{4C}{3}$
c. $\frac{3C}{2}$
d. $\frac{C}{2}$

13 C_1 and C_2 are two capacitors connected in parallel, give an equivalent capacitance of 9 pF and give an equivalent capacitance of 2 pF when connected in series. The capacitance of capacitor C_2 will be

a. 3 pF
b. 9 pF
c. 6 pF
d. 12 pF

14 The figure shows two identical parallel plate capacitors connected to a battery with the switch S (closed). The switch is now opened and the free space between the plates of the capacitors is filled with a dielectric of dielectric constant 3. The ratio of total electrostatic energy stored in both capacitors before and after the introduction of dielectric is

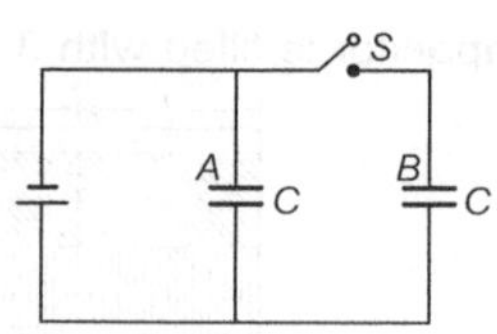

a. 2/3
b. 5/3
c. 3/5
d. 3/2

One or More than One Options Correct Type

Direction (Q. Nos. 15-20) *This section contains 6 multiple choice questions. Each question has four choices (a), (b), (c) and (d), out of which ONE or MORE THAN ONE are correct.*

15 In the circuit shown in the figure, there are two parallel plate capacitors each of capacitance C. The switch S_1 is pressed first fo fully charge the capacitor C_1 and then released. The switch S_2 is then pressed to charge the capacitor C_2. After some time, S_2 is released and then S_3 is pressed. After some time, **[JEE Advanced 2013]**

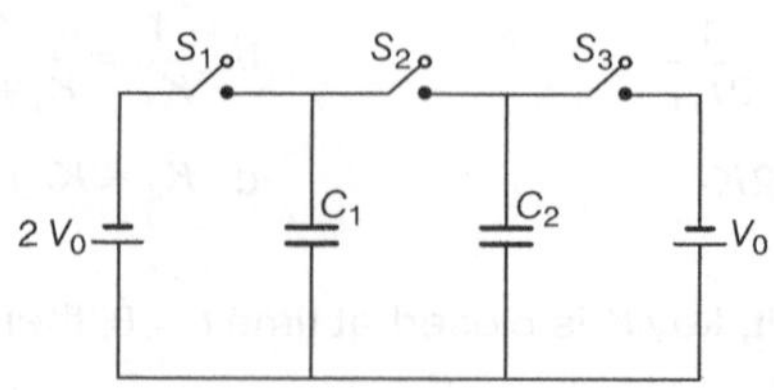

a. the charge on the upper plate of C_1 is $2CV_0$
b. the charge on the upper plate of C_1 is CV_0
c. the charge on the upper plate of C_2 is 0
d. the charge on the upper plate of C_2 is $-CV_0$

Working Space

16 Three capacitors each of 6 μF are available, then

a. minimum capacity obtained is 3 μF
b. minimum capacity obtained is 2 μF
c. maximum capacity obtained is 12 μF
d. maximum capacity obtained is 18 μF

17 Two capacitors are connected in parallel, then

a. the ratio of charge on them is $\frac{C_1}{C_2}$
b. there is no exchange of energy
c. stored energy is maximum
d. they lost some energy when battery across the combination is removed

18 Three conducting plates, each of area A are connected as shown in figure:
Correct options are

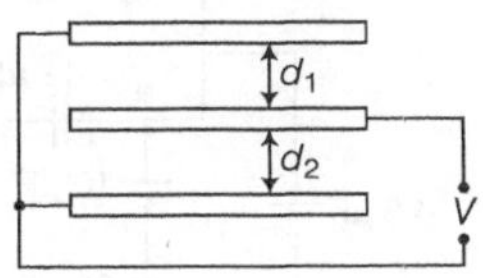

a. If the middle plate can be moved (changing the values of d_1 and d_2), the minimum capacitance will be $\frac{4\varepsilon_0 A}{d_1 + d_2}$
b. The total capacitance will be $\frac{\varepsilon_0(d_1 + d_2)}{(d_1 d_2)}$
c. Both capacitors have their low voltage plates at the same potential (the outer plates) which are connected
d. Both capacitors have their high voltage plates at the same potential (the middle plate). Which are connected

19 Two capacitors C_1 and C_2 are connected in series, potential of point A is V_1 and potential of point B is V_2. Then,

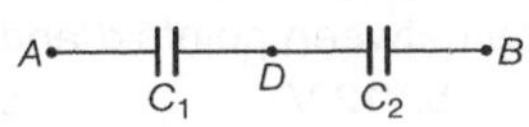

a. potential of D is $\frac{V_1 + V_2}{2}$
b. potential of D is $\frac{C_1V_1 + C_2V_2}{C_1 + C_2}$
c. charge on C_1 = charge on C_2
d. charge on C_1 > charge on C_2

20 In the given circuit,
The relationship between voltages of point A and B are relationship

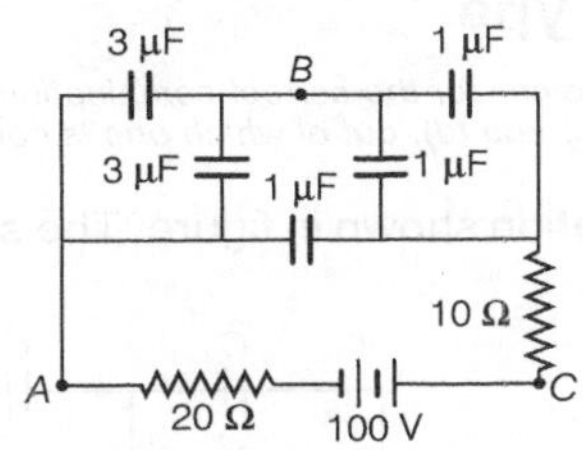

a. $V_{AB} = 25$ V
b. $V_{AB} = 27$ V
c. $V_{BC} = 75$ V
d. $V_{BC} = 50$ V

Comprehension Type

***Direction** (Q. Nos. 21-26) This section contains 2 paragraphs, describing theory, experiments, data, etc. Six questions related to the paragraphs have been given. Each question has only one correct answer among the four given options (a), (b), (c) and (d).*

Passage I

A sensor to find liquid level in a tank is a cylindrical capacitor of length with outer and inner conductor radii R_0 and R_i respectively. When a liquid fills the tank up to height h, $(h \le l)$, from tank's bottom the dielectric in lower part is liquid (K_l) and dielectric in upper part is vapour of liquid (K_v)

21 When liquid fills the tank up to height h, the capacity of capacitor is

a. $\frac{2\pi\varepsilon_0 l}{\ln(R_0/R_i)}\left\{(K_l - K_v)\frac{h}{l} + K_v\right\}$
b. $\frac{2\pi\varepsilon_0 l}{\ln(R_0/R_i)}\left\{(K_v - K_l)\frac{h}{l} + K_l\right\}$
c. $\frac{2\pi\varepsilon_0 l}{\ln(R_i/R_0)}\left\{(K_v + K_l)\frac{h}{l} + K_v\right\}$
d. $\frac{2\pi\varepsilon_0 l}{\ln(R_i/R_0)}\left\{(K_v - K_l)\frac{h}{l} + K_v\right\}$

22 If $l = 2$ m, $R_0 = 5$ mm, $R_i = 4.5$ mm, $K_l = 1.4, K_v = 1.0$. Then, capacitance of capacitor when tank is fully filled is

a. 4.5×10^{-9} F b. 7.5×10^{-9} F c. 3.5×10^{-9} F d. 1.5×10^{-9} F

23 Value of capacitance of capacitor when tank is completely empty is

a. 2.5×10^{-9} F b. 3.5×10^{-9} F c. 1×10^{-9} F d. 1.1×10^{-9} F

Passage II

A combination of capacitors given is charged by a cell of emf E as shown :

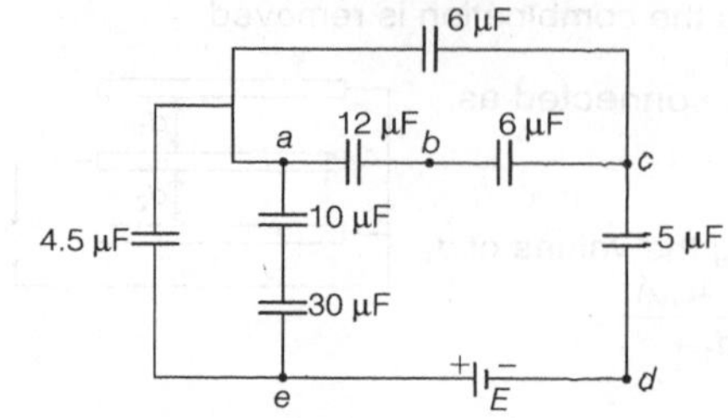

If it is given that V_{ab}, i.e. potential difference between points a and b is 4 V, then answer the given questions.

24 Potential difference between points a and c will be

a. 4 V b. 12 V c. 10 V d. 8 V

25 Potential difference between points d and a will be

a. 4 V b. 12 V c. 10 V d. 8 V

26 EMF E of the charging battery is

a. 46 V b. 12 V c. 20 V d. 18 V

Matching List Type

***Direction** (Q. No. 27) Choices for the correct combination of elements from Column I and Column II are given as options (a), (b), (c) and (d), out of which one is correct.*

27 Consider the situation shown in figure. The switch S is open for a long time and then closed.

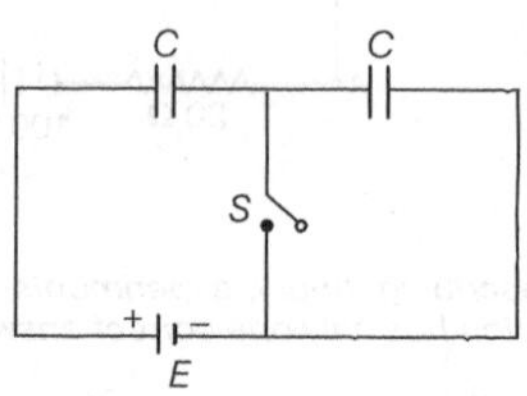

Match the Column I and Column II and mark the correct option codes.

	Column I		Column II
i.	Charge flows through cell when S is closed.	p.	$CE^2/2$
ii.	Work done by battery when S is closed.	q.	$CE/2$
iii.	Change in energy stored when S is closed.	r.	$CE^2/4$
iv.	Heat developed when S is closed.	s.	$CE^2/8$

Codes

	i	ii	iii	iv		i	ii	iii	iv
a.	q	p	r	r	b.	p	p	r	s
c.	p	q	r	s	d.	q	p	s	r

DPP-5 Charging of a Capacitor and *R*-*C* Circuits

Working Space

Subjective Questions

Direction *(Q. Nos. 1-4) These questions are subjective in nature, need to be solved completely on notebook.*

1 Discuss an R - C circuit of following points during charging of a capacitor.

i. Relation of emf of supply and potential drop occuring across resistance and capacitor.

ii. Instantaneous charge on capacitor.

iii. Potential drop across capacitor.

iv. Time constant of the circuit.

v. Current in the circuit.

2 Discuss discharging of a capacitor in R-C circuit.

i. Give instant value of charge on capacitor.

ii. Give instantaneous value of potential drop across the capacitor.

iii. Define time constant giving instantaneous value of current in circuit.

3 In an R-C circuit current exists till the capacitor is fully charged. Is the total energy supplied by battery is equal to total energy stored by the capacitor? If not then, explain the energy loss.

4 i. The voltage across a 1 μF capacitor is 10 V for $t < 0$. At $t = 0$, a 1MΩ resistor is connected across the capacitor terminals. Find the time constant, the voltage $V(t)$, value of voltage at $t = 5$ s, current through the resistor $i(t)$. Also, draw the graphs of $V(t)$ and $i(t)$.

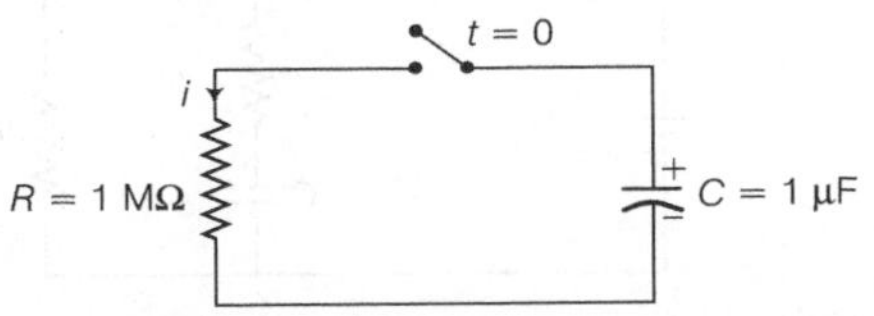

ii. A 4μF capacitor with an initial voltage of $V(t = 0) = 2$ V is connected to a 12 V battery through a resistor $R = 5$ kΩ at $t = 0$. Find the voltage and current through the capacitor for $t > 0$.

Only One Option Correct Type

Direction *(Q. Nos. 5-15) This section contains 11 multiple choice questions. Each question has four choices (a), (b), (c) and (d), out of which ONLY ONE option is correct.*

5 The plates of a capacitor are charged to a potential difference of 320 V and then connected across a resistor.
If potential difference between plates of capacitor becomes 240 V after 1 s, then potential difference after 2 and 3 s will be

a. 200 and 180 V b. 180 and 135 V c. 160 and 180 V d. 140 and 20 V

Working Space

6 A 4 μF capacitor and a resistance of $2.5 \times 10^6\ \Omega$ are connected in series with a battery of 12 V. Time after which potential difference across the capacitor is 3 times of potential difference across the resistor is

a. 13.86 s b. 6.93 s c. 7 s d. 14 s

7 A capacitor is charged steadily from a DC source. Correct variation of potential difference across the plates of capacitor with charge on the plates of capacitor is

a. 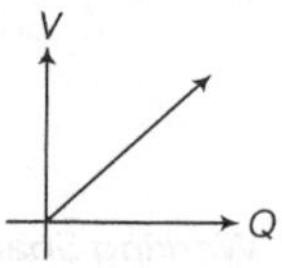
b. 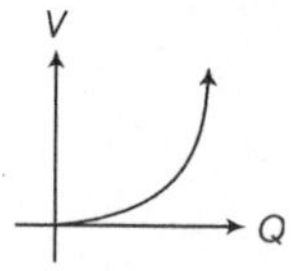
c. 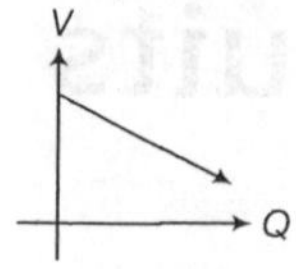
d. 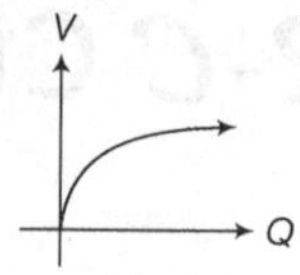

8 A capacitor is made to discharge across a pure resistor. A plot of log I *versus* t (I = instantaneous current and t = instantaneous time) is recorded as shown in graph 1. The same experiment is repeated with a change and graph 2 is observed. Then,

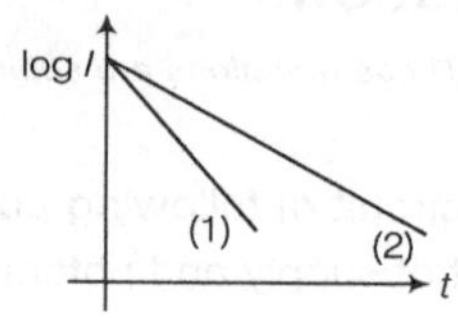

a. R is increased b. R is decreased
c. C is increased d. C is decreased

9 Three identical capacitors are charged by connecting them in parallel across a battery of V volt. They are then allowed to discharge *via* resistor R_1, R_2 and R_3. Here, Q charge *versus* t time graphs are shown. Then,

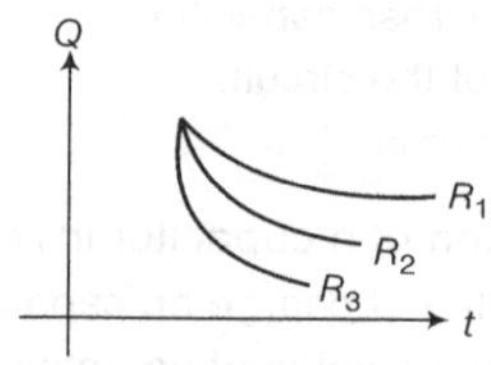

a. R_1 is smallest b. R_2 is smallest
c. R_3 is smallest d. $R_3 > R_2 < R_1$

10 For the circuit shown in figure below,
At $t = 0$, switch is closed, the initial current through resistor and final charge on capacitor are

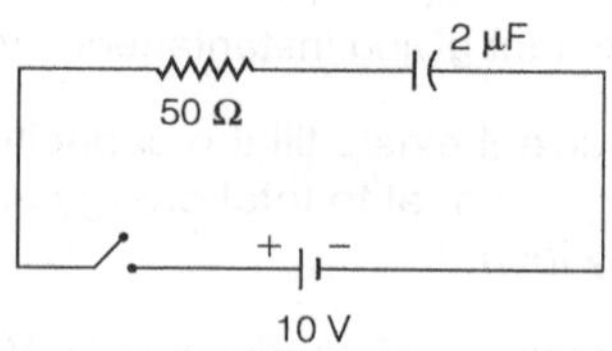

a. 5A, 5 μC
b. 5 A, 20 μC
c. 0.2 A, 5 μC
d. 0.2 A, 20 μC

11 For the given circuit shown in figure below,
Time constant is

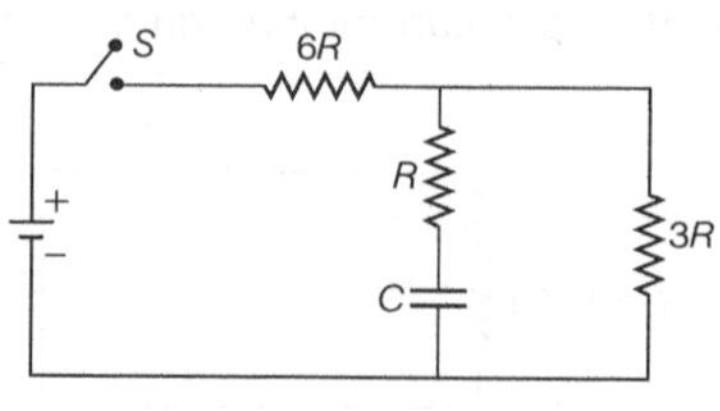

a. $\tau = RC$
b. $\tau = 2.1RC$
c. $\tau = 27/4\,RC$
d. $\tau = 3RC$

12 During charging of capacitor in the circuit shown,

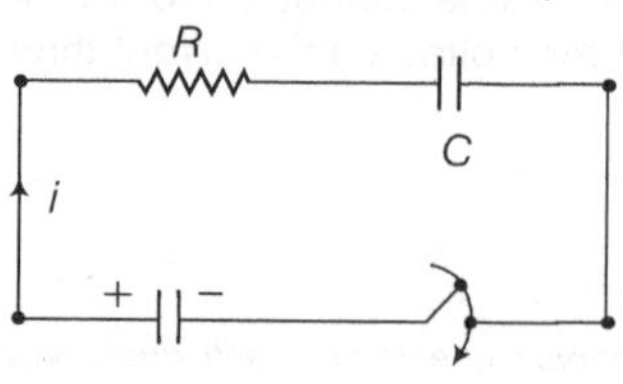

Circuit current *versus* time graph is

a. 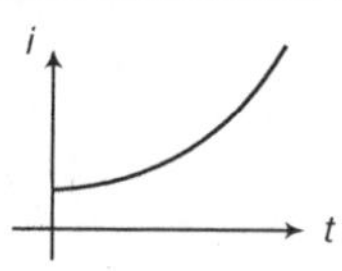
b. 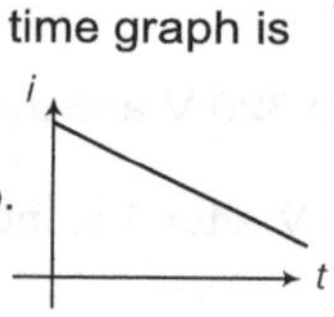
c. 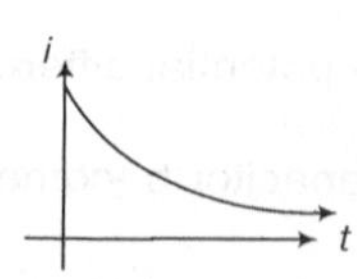
d. 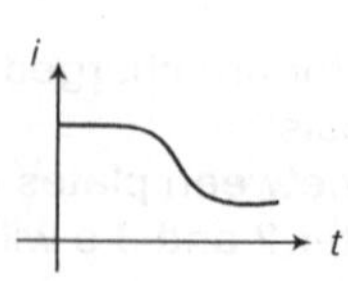

Working Space

13 During discharging of a capacitor *via* a resistor.

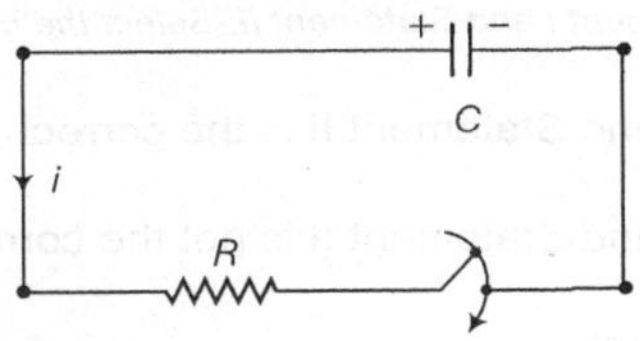

Circuit current varies with time as

a. 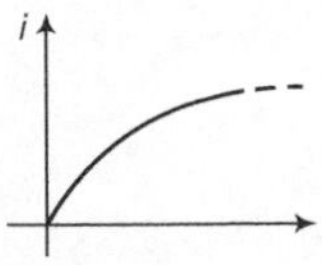

b. 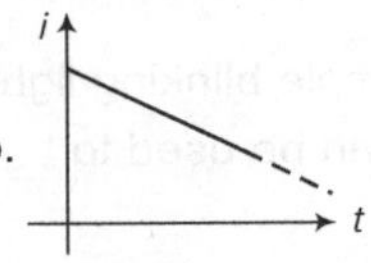

c. 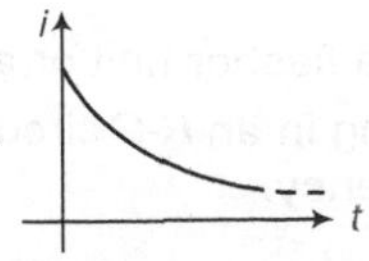

d. 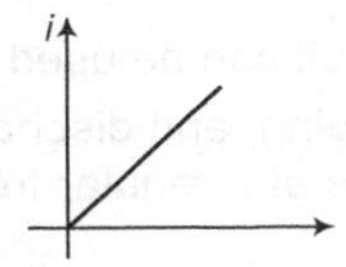

14 If t indicates instantaneous time, then $\frac{t}{RC}$ has dimensions

a. $[M^0 L^0 T^0]$ b. $[M^0 L^0 T^{-1}]$
c. $[M^0 L^0 T^1]$ d. $[M^0 L^0 T^{-2}]$

15 In a capacitor shown in the circuit is charged to 5 V and left in the circuit in 12 s, the charge on the capacitor will becomes

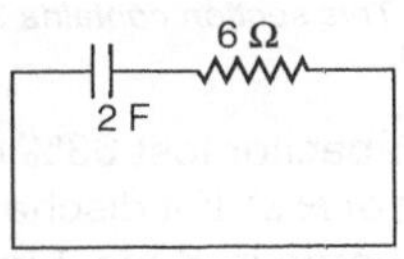

a. $\frac{10}{e}$ C b. $\frac{e}{10}$ C
c. $\frac{10}{e^2}$ C d. $\frac{e^2}{10}$ C $(e = 2.718)$

One or More than One Options Correct Type

Direction *(Q. Nos. 16 and 17) This section contains 2 multiple choice questions. Each question has four choices (a), (b), (c) and (d), out of which ONE or MORE THAN ONE are correct.*

16 Consider the following circuits.

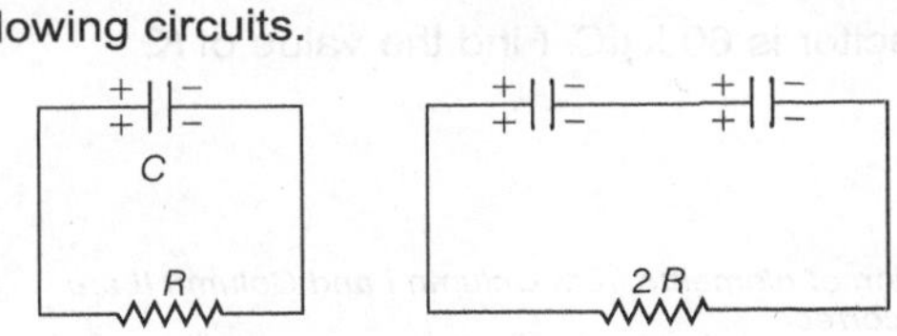

(i) (ii)

Both capacitors are identical and they have same initial charge Q. Mark the correct options.

a. Heat produced in (i) is more than (ii)
b. Heat produced in (ii) is more than (i)
c. Time in which capacitor of (i) discharge is more than that of (ii)
d. Time in which capacitor of (ii) is discharged is same that of (i)

17 In an R-C circuit, $R = 15\ k\Omega$, battery emf = 24 V and time constant $\tau = 24\ \mu s$. Then,

a. total capacitance of circuit is 1.6×10^{-9} F
b. time taken by voltage across resistor to reach 16 V is more than 10 μs
c. total capacitance of circuit is 2.8×10^{-9} F
d. time taken by voltage across resistor to reach 16 V is less than 10 μs

Working Space

Statement Type

Direction (Q. Nos. 18 and 19) *This section is based on Statement I and Statement II. Select the correct answer from the codes given below.*

a. Both Statement I and Statement II are correct and Statement II is the correct explanation of Statement I
b. Both Statement I and Statement II are correct and Statement II is not the correct explanation of Statement I
c. Statement I is correct but Statement II is incorrect
d. Statement II is correct but Statement I is incorrect

18 **Statement I** An *R-C* circuit can be used as a flasher unit or a simple blinking light.

Statement II The charging and discharging in an *R-C* circuit can be used to produce voltage pulses at a regular frequency.

19 **Statement I** When a capacitor is connected to a battery it is charged instantly.

Statement II Current travels in a wire at speed of light.

One Integer Value Correct Type

Direction (Q. No. 20-22) *This section contains 3 questions. When worked out will result in an integer from 0 to 9 (both inclusive).*

20 In an *R-C* circuit capacitor lost 63% of initial charge in 5 ms when it discharges through a resistor of $R\ \Omega$. If it discharges through a resistor of $2R\ \Omega$, then it lost 63% of its initial charge in *K* ms. Find the value of *K*.

21 In the given circuit,

At $t = 0^-$, before the switch closed, V_c was measured as 100 V. If energy liberated in resistor in *t* seconds is $k \times 10^{-1}(1 - e^{-125t})$, then find the value of *k*.

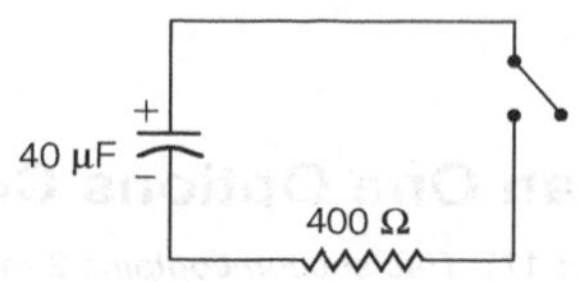

22 An *RC* transient has a power transient given by $P = 360\, e^{-t \times 10^{-5}}$. Value of resistor in circuit is 10 Ω and initial charge on capacitor is 60 k μC. Find the value of *K*.

Matching List Type

Direction (Q. No. 23) *Choices for the correct combination of elements from Column I and Column II are given as options (a), (b), (c) and (d), out of which one is correct.*

23 Consider charging and discharging of a capacitor and select the correct option code, matching the following columns.

	Column I		Column II
i.	A large capacitor	p.	Slower rate of growth of charge
ii.	A small capacitor	q.	Faster rate of growth of charge
iii.	A large resistor	r.	Smaller rate of decay of charge
iv.	A small resistor	s.	Faster rate of decay of charge

Codes

	i	ii	iii	iv		i	ii	iii	iv
a.	q	p	r,s	t	b.	p,r	q,s	p,r	q,s
c.	p	s	r	p,r	d.	q	p,q	r	s

DPP-6 Redistribution of Charges and Van de Graaff Generator

Subjective Questions

Working Space

Direction (Q. Nos. 1-3) *These questions are subjective in nature, need to be solved completely on notebook.*

1 If two charged capacitors with capacitances C_1 and C_2 are charged at potentials V_1 and V_2 respectively, are joined together, show that there is always a loss of energy. Why this loss of energy occurs?

2 Show that force between plates of a charged capacitor is $\frac{1}{2}\frac{Q^2}{A\varepsilon_0}$. Why the factor $\frac{1}{2}$ appears in expression?

3 If two concentric spheres have charges q and Q on their surface, show that inner sphere is always at a higher potential irrespective of charge value.

Only One Option Correct Type

Direction (Q. Nos. 4-13) *This section contains 10 multiple choice questions. Each question has four choices (a), (b), (c) and (d), out of which ONLY ONE option is correct.*

4 If two conducting spheres are separately charged and then bought in contact,

a. total energy of two spheres is conserved
b. total charge on two spheres is conserved
c. both energy and charge of two spheres are conserved
d. final potential is always mean of the original potentials of the two spheres

5 Two insulated spheres of radii 20 cm and 25 cm, respectively and having equal charges Q coulomb each are connected by a copper wire, then they are separated

a. both spheres have same charge Q
b. charge on 20 cm radius sphere is more than that of 25 cm radius sphere
c. charge on 25 cm sphere will be greater than that on the 20 cm sphere
d. charge on each of the sphere is $2Q$

6 Eight drops of mercury of equal radii possessing equal charges combine to form a big drop. Then, ratio of capacity of big drop with one of small drop is

a. 8　　b. 4
c. 2　　d. 32

7 Force acting upon a charged particle kept between plates of a capacitor is F. If one of the plates of capacitor is removed, then the force acting on the same particle will become

a. 0　　b. $F/2$
c. F　　d. $2F$

Working Space

8 Three capacitors C_1, C_2 and C_3 are connected as shown in the figure to a battery of V volt. If the capacitor C_3 breaks down electrically, the change in total charge on the combination of capacitors is

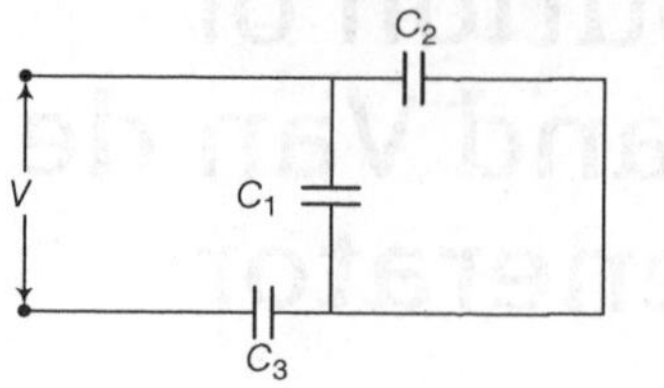

a. $(C_1 + C_2)V\left[1 - \frac{C_3}{C_1 + C_2 + C_3}\right]$

b. $(C_1 + C_2)V\left[1 - \frac{C_1 + C_2}{C_1 + C_2 + C_3}\right]$

c. $(C_1 + C_2)V\left[1 + \frac{C_3}{C_1 + C_2 + C_3}\right]$

d. $(C_1 + C_2)V\left[1 + \frac{C_2}{C_1 + C_2 + C_3}\right]$

9 Two spheres of metal have same surface charge densities (σ). Radius of one of the spheres is R while another one is $2R$. They are bought in contact and separated. The ratio of final charge densities on their surfaces is

a. 1 b. 2

c. $\frac{5}{3}$ d. $\frac{5}{6}$

10 Two identical capacitors are connected as shown.

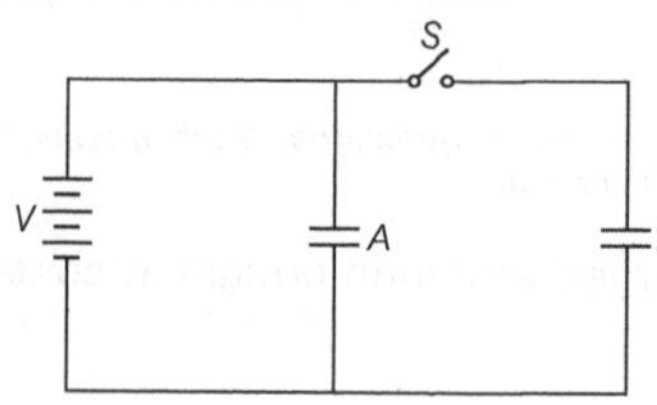

Now consider the following operations.

I. Switch S is initially closed,

II. Switch S is opened and

III. A dielectric medium $K = 3$ is filled between plates of both capacitors.

The ratio of the total electrostatic energy stored in both capacitors before and after the introduction of dielectric slab is

a. $\frac{5}{3}$ b. $\frac{5}{6}$

c. 1 d. $\frac{3}{5}$

11 A solid conducting sphere having a charge Q is surrounded by an uncharged concentric conducting hollow spherical shell. Let the potential difference between the surface of the solid sphere and the outer surface of hollow shell be V.

If shell is given a charge $-3Q$, the new potential difference between the same two surfaces is

a. V b. $2V$

c. $4V$ d. $-2V$

12 Two insulated spheres of radii R_1 and R_2 having charges Q_1 and Q_2, respectively are connected to each other. There is

a. no change in energy of the system
b. an increase in energy of the system
c. always a decrease in energy of the system
d. a decrease in energy of the system unless $Q_1 R_2 = Q_2 R_1$

13 A 4 μF capacitor is charged to 80 V and another capacitor of 6 μF capacity is charged to 30 V. When they are connected the energy lost by 4 μF capacitor is

a. 7.8 mJ
b. 4.6 mJ
c. 3.2 mJ
d. 2.5 mJ

One or More than One Options Correct Type

Direction (Q. Nos. 14-17) This section contains 4 multiple choice questions. Each question has four choices (a), (b), (c) and (d), out of which ONE or MORE THAN ONE are correct.

14 A Van de Graaff generator has a dome of diameter 30 cm placed in dry air with dielectric strength 3×10^6 V/m. Mark the correct options.

a. Maximum potential of dome is 450 kV
b. Maximum charge hold by dome is 7.51/μC
c. Maximum charge hold by dome increases, if it is placed in vacuum
d. Maximum potential at which dome can be raised will be reduced, if a pointed lightning conductor is placed near the dome

15 A Van de Graaff generator can produce potentials as large as 20 million volt. Best method to reduce chances of discharging of dome by dielectric breakdown among the followings are

a. reducing the shell radius and thickness
b. partially evacuating the space surrounding the dome
c. heating the air surrounding the dome
d. placing the dome in a container filled with high pressure gas

16 Two identical capacitors with capacity C are charged to potential V and then are connected in parallel.
Correct options are

a. If plate separation of one of capacitors is doubled then stored energy of both capacitors is reduced
b. If plate separation of one of the capacitors is doubled then stored energy becomes 4/3 times of initial energy
c. Potential of system of capacitors remains constant
d. Potential difference across the plates of capacitors increases

17 A parallel plate capacitor is filled by a dielectric whose relative permittivity varies with the applied voltage V as $\varepsilon_r = \alpha V$ where $\alpha = 2\ V^{-1}$.
A similar capacitor with no dielectric is charged to 78 V. It is then connected to uncharged capacitor with the dielectric.
Mark the correct options.

a. Final voltage of capacitors is 6 V
b. Final voltage of capacitors is 39 V
c. Final charge on capacitor with dielectric is 78 C
d. Final charge on capacitor with dielectric is 6 C

Comprehension Type

Working Space

Direction *(Q. Nos. 18-23) This section contains two paragraphs, describing theory, experiments, data, etc. Six questions related to the paragraphs have been given. Each question has only one correct answer among the four given options (a), (b), (c) and (d).*

Passage I

Two parallel plate capacitors A and B have the same separation ($d = 8.85 \times 10^{-4}$ m) between the plates. The plate area of A and B are $0.04\ m^2$ and $0.02\ m^2$ respectively. A slab of a medium with dielectric constant $K = 9$ has same dimensions such that it can exactly fills the space between the plates of capacitor B.

18 Dielectric slab is put between plates of A and it is charged to a potential of 110 V. Energy stored is

a. 1.21×10^{-5} J
b. 1.21×10^{-3} J
c. 1.21×10^{-2} J
d. 1.21×10^{-6} J

19 Battery across A is disconnected and dielectric slab is removed slowly from A. Work done in removing the slab is

a. 4.84×10^{-6} J
b. 6.05×10^{-5} J
c. 1.21×10^{-5} J
d. 4.84×10^{-5} J

20 The same dielectric is now put in B and then A and B are connected as

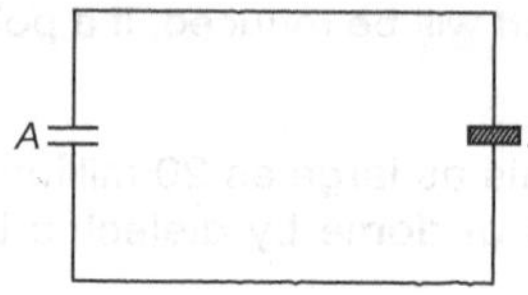

Energy stored in the system is

a. 1.21×10^{-5} J
b. 6.05×10^{-5} J
c. 4.84×10^{-5} J
d. 1.01×10^{-5} J

Passage II

A 600 pF parallel plate capacitor A is charged to 200 V. After sometime, it is disconnected from the supply and connected to another uncharged similar capacitor B of 600 pF.
Plates of capacitor B are pulled apart to make separation twice of its initial value

21 Percentage of energy gained in the process is

a. 11.11%
b 22.22%
c. 33.33%
d. 0%

22 What is the ratio of energy stored in A to energy stored in B?

a. 1 : 1
b 1 : 2
c. 2 : 1
d. 1 : 4

23 What is the ratio of charges on A and B?

a. 1 : 1
b 1 : 2
c. 2 : 1
d. 1 : 4

Statement Type

Direction (Q. Nos. 24-26) *This section is based on Statement I and Statement II. Select the correct answer from the codes given below.*

a. Both Statement I and Statement II are correct and Statement II is the correct explanation of Statement I
b. Both Statement I and Statement II are correct and Statement II is not the correct explanation of Statement I
c. Statement I is correct but Statement II is incorrect
d. Statement II is correct but Statement I is incorrect

24 **Statement I** In a Van de Graaff generator, lower comb sprays positive charge over the insulating belt and upper comb transfers positive charge to the spherical shell of the generator.

Statement II In a Van de Graaff generator, transfer of charge from combs to belt takes place through ionization of air. This process is called 'corona discharge'.

25 **Statement I** If three concentric shells *A*, *B* and *C* with equal charges on their surface are connected with conducting wires then charge always flows to the surface of largest shell.

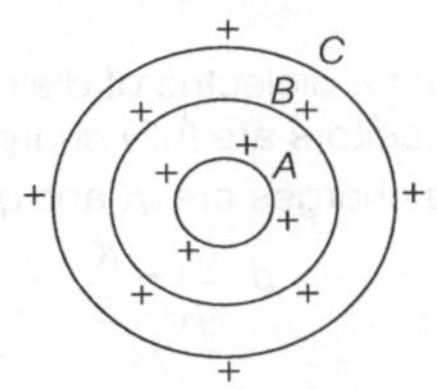

Statement II Capacity of a shell is proportional to its radius.

26 **Statement I** To store more charge over a sphere, it is kept in a vacuum chamber.

Statement II Capacity of sphere changes with the property of medium in which sphere is kept.

Matching List Type

Direction (Q. No. 27) *Choices for the correct combination of elements from Column I and Column II are given as options (a), (b), (c) and (d), out of which one is correct.*

27 A conducting sphere *A* of radius *r* is attached on an insulating hand. Another sphere *B* of radius *R* is mounted on an insulating stand. Initially *A* is given a charge *Q*, it is bought in contact with *B* and removed. *A* is charged to *Q* after each contact. This process is repeated *n* times. Match the Column I with Column II and choose the correct option code.

	Column I		Column II
i.	Electrostatic energy of sphere '*B*' after '*n*' contacts.	p.	$\frac{Q^2R}{8\pi\varepsilon_0 r^2}$
ii.	Charge on '*B*' after '*n*' contacts with '*A*'.	q.	$Q\left(\frac{R}{r}\right)$
iii.	When $n \to \infty$, q_∞ is	r.	$\frac{Q^2\frac{R^2}{r^2}\left[\left(1-\frac{R}{R+r}\right)^n\right]^2}{8\pi\varepsilon_0 R}$
iv.	When $n \to \infty$, U_∞ is	s.	$Q\frac{R}{r}\left[1-\left(\frac{R}{R+r}\right)^n\right]$

Codes

	i	ii	iii	iv		i	ii	iii	iv
a.	r	s	q	p	b.	p	r	s	q
c.	s	r	q	p	d.	p	q	r	s

SET - 1

Revisal Problems for JEE Main

Only One Option Correct Type

Direction *(Q. Nos. 1-23) This section contains 23 multiple choice questions. Each question has four choices (a), (b), (c) and (d), out of which ONLY ONE option is correct.*

Working Space

1 Two identical capacitors C_1 and C_2 are connected as shown :

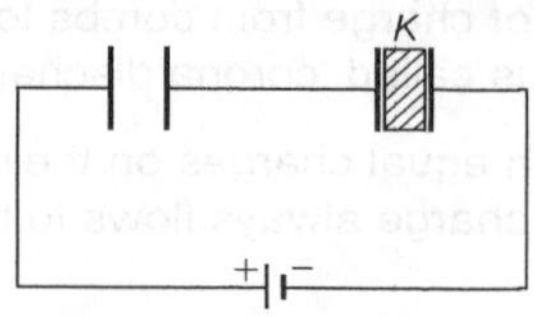

Capacitor 1 has air between plates and capacitor 2 contains a dielectric of dielectric constant K. Let q_1 and q_2 are the charges stored when capacitors are fully charged. Now, the dielectric slab is removed and the corresponding charges are q_1' and q_2'. Then,

a. $\frac{q_1'}{q_1} = \frac{K+1}{K}$ b. $\frac{q_2'}{q_2} = \frac{K+1}{K}$ c. $\frac{q_2'}{q_2} = \frac{K+1}{2K}$ d. $\frac{q_1'}{q_1} = \frac{K}{2}$

2 Two concentric shells of radii r and $2r$ are placed far away from a parallel plate capacitor. The charge on plates is q and $-q$. Then, the keys K_1 and K_2 are closed.

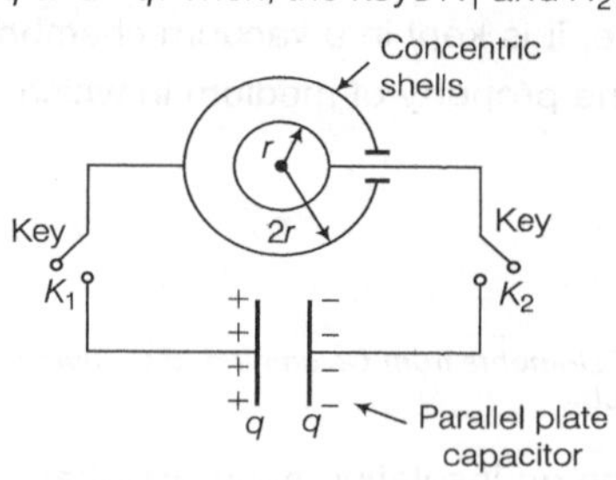

Then, ratio of energy stored in capacitor initially and after closing of switches is

a. $\frac{U_i}{U_f} = 3$ b. $\frac{U_i}{U_f} = 9$ c. $\frac{U_i}{U_f} = 6$ d. $\frac{U_i}{U_f} = 1$

3 A parallel plate capacitor with plates of width 4 cm, length 10 cm and separation between plates in 4 cm, is connected across a 12 V emf battery. A dielectric slab of dielectric constant 7 is slowly introduced between the plates.

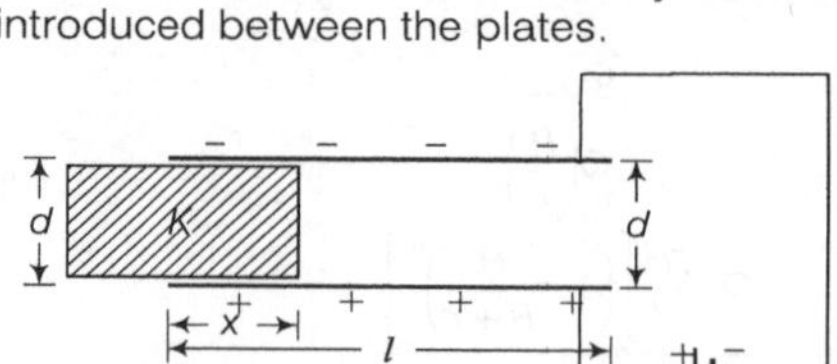

Force exerted on dielectric slab by the field is

a. $114\varepsilon_0$ b. $432\varepsilon_0$ c. $7\varepsilon_0$ d. $326\varepsilon_0$

4 Eight equal charges, 2 μC each are placed on vertices of a cube of side 1 cm. Work done in removing all the charges infinitely away from each other is

a. 0 b. $\frac{8q^2}{4\pi\varepsilon_0 a}$ c. $\frac{q^2}{4\pi\varepsilon_0 a}$ d. $\frac{q^2}{4\pi\varepsilon_0 a}\left(12 - 6\sqrt{2} + \frac{4}{3}\right)$

Working Space

5 A particle of mass 100 g and charge 2 μC is released from a distance of 50 cm from a fixed charge of 5μC. The speed of the particle when its distance from fixed charge is 3 m, is (neglect presence of any other force)

a. $\sqrt{2}$ m/s b. $2\sqrt{3}$ m/s c. $\sqrt{3}$ m/s d. $\sqrt{4}$ m/s

6 Two concentric shells of radii R and $2R$ are arranged as shown.

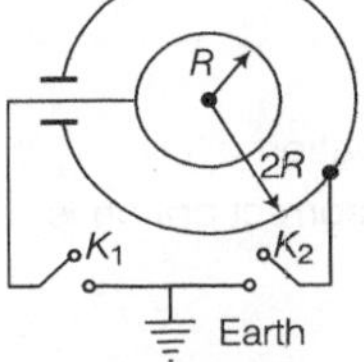

As charge of 1.024×10^3 μC is initially provided to inner sphere. Now keys K_1 and K_2 are alternatively closed for 5 times each.

Potential difference between the shells finally will be

a. $\frac{Kq}{64R}$ b. $\frac{Kq}{32R}$

c. $\frac{Kq}{R}$ d. $\frac{Kq}{28R}$

7 A parallel plate capacitor has a dielectric slab of dielectric constant 7 between the plates. When capacitor is fully charged, plates of capacitor has a charge density of σ_0.

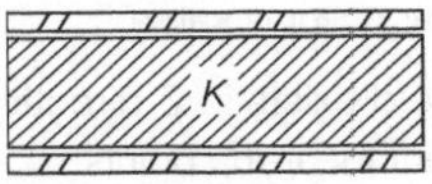

The charge density on the surface of dielectric is

a. $7\sigma_0$ b. $\frac{6}{7}\sigma_0$ c. $\frac{5}{7}\sigma_0$ d. $\frac{1}{7}\sigma_0$

8 A parallel plate capacitor has sphere plates of side 10 cm (each) with a plate separation of 5 mm. A dielectric slab of mica $(K=7)$ and thickness 5 mm is inserted between plates by pushing it with a constant velocity of 1 m/s. Current flowing in the conducting wires in amperes will be (take, emf of source = 12 V)

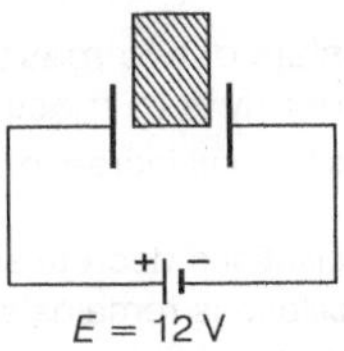

a. $14\varepsilon_0$ b. $144\varepsilon_0$

c. $1440\varepsilon_0$ d. $14400\varepsilon_0$

9 A smooth dielectric slab of mass m and dielectric constant K is placed between plates of a capacitor and is connected with another block of mass m using a light pulley and an inextensible string as shown.

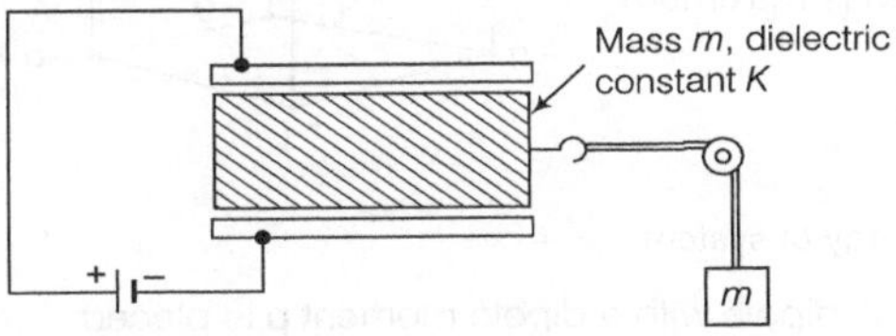

If emf of battery is $E=\sqrt{\frac{mgd}{b(k-1)\varepsilon_o}}$. Plate separation is d, plates have a width b, then acceleration of the dielectric slab will be

a. g b. $\frac{g}{2}$ c. $\frac{g}{4}$ d. $\frac{g}{8}$

10 Three identical capacitors A, B and C are changed to same potential and then are made to discharge through three resistors R_A, R_B and R_C, where $R_A > R_B > R_C$. Their potential difference V *versus* time t graphs are 1, 2 and 3 respectively.

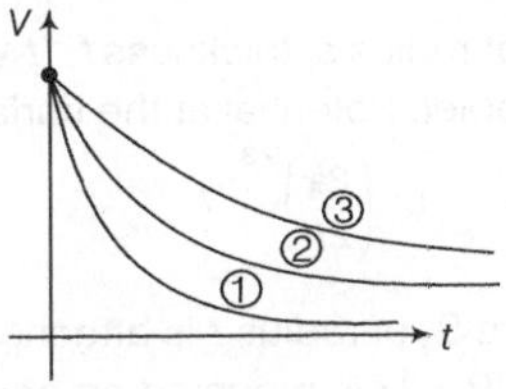

Then, correlation between A, B and C and 1, 2 and 3 is

a. $1\to B, 2\to C, 3\to A$ b. $1\to C, 2\to B, 3\to C$

c. $1\to A, 2\to B, 3\to C$ d. $1\to B, 2\to A, 3\to C$

11 In given combination of capacitors each capacitor has a breakdown voltage of 15 V.

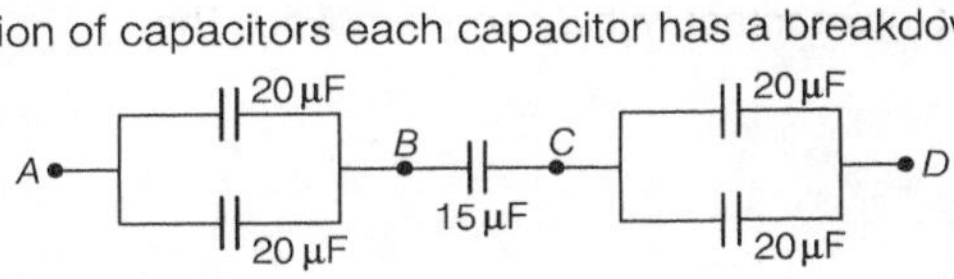

Breakdown voltage of combination will be

a. 15 V b. 3.5 V c. 22.5 V d. 30 V

Working Space

12 A small rigid dumbell carries positive and negative charges of 3.5 nC at each of its end. It is oriented such that its positive charge has coordinates (– 1.2 mm, 1.1 mm) and negative charge is at point (1.4 mm, –1.3 mm). An electric field $\mathbf{E} = (7800\hat{i} - 4900\hat{j})$ also exists in the region.

If orientation of dumbell can be changed, then difference of maximum and minimum potential energy of the system will be

a. 114 nJ
b. – 144 nJ
c. 228 nJ
d. – 228 nJ

13 In the given combination

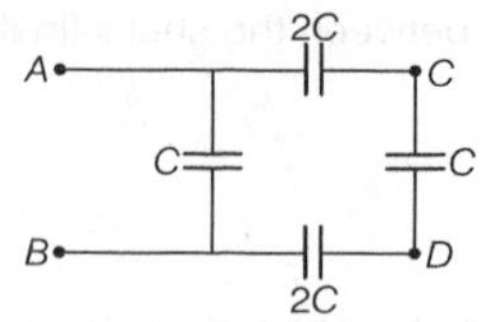

If $V_{AB} = 100$ V, then correct option is

a. $V_{AC} = 50$ V
b. $V_{CD} = 50$ V
c. $V_{BC} = 50$ V
d. charge on each capacitor is same

14 A uniform electric field exists in *Z*-direction, then

a. potential is constant for all the points
b. potential is constant on *YOZ* plane for a given *X*
c. potential is constant on *XOZ* plane for a given *Y*
d. potential is constant on *XOY* plane for a given *Z*

15 For the system consists of two mass particles and an insulated spring of nearly zero mass.

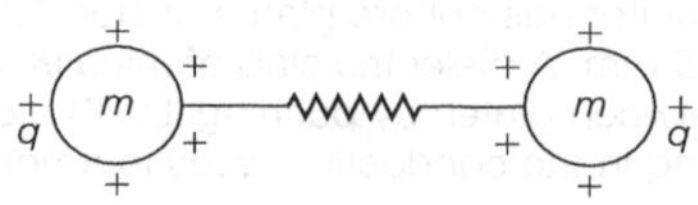

a. Time period of oscillations increases when masses are charged
b. Time period of oscillations decreases when masses are charged
c. Time period of oscillations remains same irrespective of masses are charged or not
d. Oscillations stops when masses are charged

16 Eight point charges are placed at the corners of a cube of side *a* unit as shown.

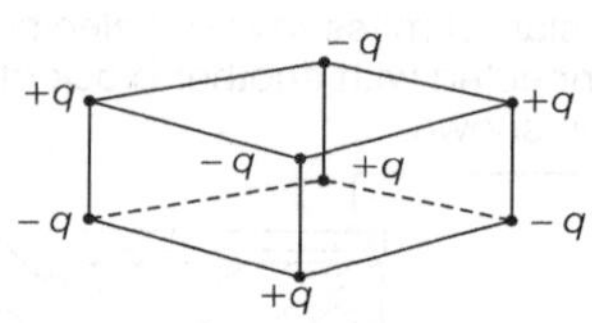

Contribution to total potential energy of system is maximum by

a. 12 pairs of charges at a separation *a*
b. 12 pairs of charges at a separation of $\sqrt{2}a$
c. 4 pairs of charges at a separation of $\sqrt{3}\,a$
d. All charges contributes equally to potential energy of system

17 A positive point charge *q* is fixed at the origin. A dipole with a dipole moment **p** is placed along the *X*-axis far away from the origin with dipole moment is along positive *X*-axis. Kinetic energy of dipole when it reaches a distance *d* from the origin is **[IIT JEE 2009]**

a. $\dfrac{qp}{4\pi\varepsilon_0 d^2}$
b. $\dfrac{2qp}{4\pi\varepsilon_0 d^2}$
c. $\dfrac{qp}{8\pi\varepsilon_0 d^2}$
d. depends on orientation of dipole

18 A mercury bubble of radius *a*, thickness *t*, ($t << a$) has a potential *V*. None the bubble collapses into a droplet. Potential at the surface of droplet will be

a. $\left(\dfrac{a}{2t}\right)^{1/3} V$
b. $\left(\dfrac{3a}{2t}\right)^{1/3} V$
c. $\left(\dfrac{a}{3t}\right)^{1/3} V$
d. $\left(\dfrac{3a}{2t}\right)^{1/3} V^2$

19 A conducting sphere S_1 of radius *r* is attached to an insulating handle. Another conducting sphere S_2 of radius $R = 2r$ is mounted on an insulating stand. S_2 is initially uncharged. S_1 is given a charge *Q*, bought into contact with S_2 and removed S_1 is recharged such that charge on it is again *Q*. It is then again bought into contact with S_2 and removed. The procedure is repeated *n* times.

a. Limiting value of energy of S_2 or $n \to \infty$ is ∞
b. Limiting value of energy of S_2 as $n \to \infty$ is $\dfrac{Q^2}{4\pi\varepsilon_0 r}$
c. Limiting value of charge on S_2 as $n \to \infty$ is ∞
d. Limiting value of charge on S_2 or $n \to \infty$ is *Q*

20 Three capacitors A, B and C are put in circuit as shown.

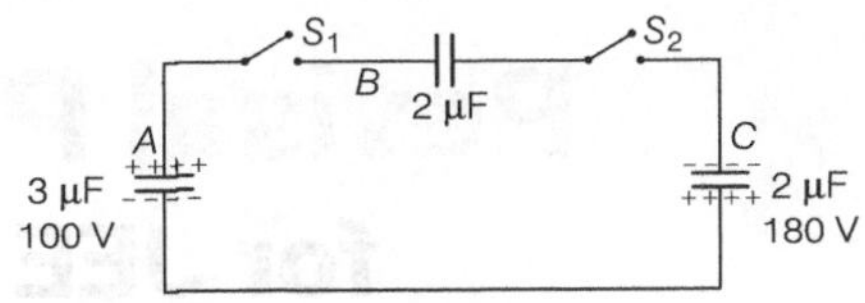

Initially A and C are charged at 100 V and 180 V, respectively. Switches S_1 and S_2 are closed and charges on A, B and C after stabilisation are

a. 90 μC, 90 μC, 90 μC
b. 90 μC, 210 μC, 210 μC
c. 90 μC, 210 μC, 150 μC
d. 90 μC, 150 μC, 150 μC

21 In the given circuit,

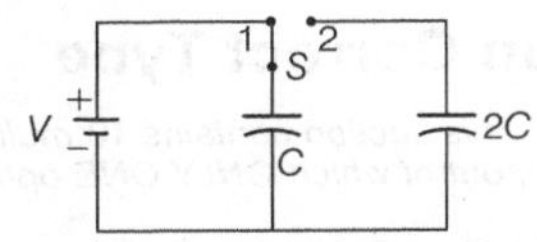

When switch S is shifted from position 1 to position 2, percentage of energy dissipated is

a. 33.33%
b. 22.22%
c. 66.66%
d. 0%

22 A parallel combination of 0.1 MΩ resistor and a 10 μF capacitor is connected across a 1.5 V cell of negligible internal resistance. Time required for the capacitor upto 0.75 V is approximately (in seconds) is

a. infinite
b. $\log_e 2$
c. $\log_{10} 2$
d. zero

23 In the given circuit, capacitor C, charges from cell of emf E and reaches a steady state.

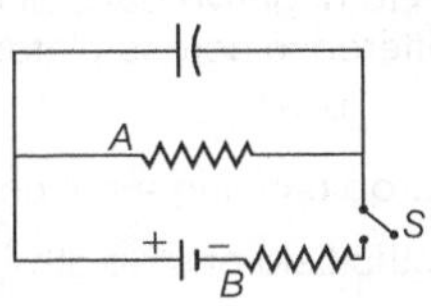

At $t = 0$, switch S is closed and A and B are equal resistances. Now, consider the following statements.

A. In steady state rate of heat production in A and B is different
B. During charging more heat is produced in B than in A
C. In steady state rate of heat production is same in A and B
D. In steady state energy stored in capacitor is $\frac{1}{8}CE^2$

Correct option is

a. ACD
b. AB
c. BCD
d. ABD

Statement Type

Direction *(Q. Nos. 24-27) This section is based on Statement I and Statement II. Select the correct answer from the codes given below.*

a. Both Statement I and Statement II are correct and Statement II is the correct explanation of Statement I
b. Both Statement I and Statement II are correct and Statement II is not the correct explanation of Statement I
c. Statement I is correct but Statement II is incorrect
d. Statement II is correct but Statement I is incorrect

24 **Statement I** For practical purposes, the earth is used as a reference at zero potential in electrical circuits.

Statement II The electrical potential of a sphere of radius R with charge Q on its surface is $\frac{Q}{4\pi\varepsilon_0 R}$.

[IIT JEE 2008]

25 **Statement I** When charges are shared between two bodies, there is no loss of charge but there is loss of energy.

Statement II Electrostatic field is non-conservative in nature.

26 **Statement I** When a capacitor is charged by a cell, half of energy is dissipated.

Statement II Part of electrical energy is dissipated in form of heat due to resistance of connecting wires.

27 **Statement I** If plates of a charged capacitor are allowed to come closer under action of electrostatic force. Then, temperature of plates increases.

Statement II Part of stored energy of capacitor is dissipated in the form of heat.

SET - 2

Revisal Problems
for JEE Advanced

Only One Option Correct Type

Direction (Q. Nos. 1-10) *This section contains 10 multiple choice questions. Each question has four choices (a), (b), (c) and (d), out of which ONLY ONE option is correct.*

1 A charged capacitor of capacitance 2 μF having a charge of 2×10^{-4} C is joined with another capacitor of capacitance C'.
For heat liberation to be maximum, capacity of capacitor C' will be

a. 0 b. 1 μF c. 2 μF d. ∞

2 A capacitor of capacity 5 μF is charged to a potential difference of 10 V. A charge of +10 μC is now given to negative plate of capacitor after disconnecting it from charging battery. Potential difference across plates of capacitor is

a. 10 V b. 9 V c. 15 V d. 0 V

3 A tiny spherical oil drop carrying a net charge of q coulomb is balanced in still air with a vertical uniform electric field of strength $\frac{81}{7}\pi \times 10^5 \frac{\text{V}}{\text{m}}$.

When field is switched OFF the drop is observed to fall with terminal velocity of 2×10^{-3} m/s.

($g = 9.8$ m/s^2, viscosity of air $= 1.8 \times 10^{-5} \frac{\text{Ns}}{\text{m}^2}$, density of oil $= 900$ kg/m^3). The charge q on oil drop is

a. 1.6×10^{-19} C b. 3.2×10^{-19} C c. 4.8×10^{-19} C d. 8×10^{-19} C

4 A parallel plate capacitor has a capacitance C and charge Q. Another plate is inserted between the plates of capacitor at the centre and is connected by a long wire to another far off spherical conductor of radius R.
What is the potential difference between edge of sphere and one of the outer plates?

a. $\frac{Q}{C}$ b. $\frac{Q}{4C}$ c. $\frac{Q}{2C}$ d. $\frac{2Q}{C}$

5 A sphere of radius r_1 carries a uniformly distributed charge of ρ C/m^3. There is a concentric cavity inside a sphere of radius r_2. If $r_1 = 2r_0$ and $r_2 = r_0$, then potential at a point p distant $\frac{3}{2} r_0$ from the centre is

a. $\frac{7}{6}\frac{\rho_0^2}{\varepsilon_0}$ b. $\frac{101}{72}\frac{\rho r_0^2}{\varepsilon_0}$ c. $\frac{17}{72}\frac{\rho r_0^2}{\varepsilon_0}$ d. $\frac{84}{72}\frac{\rho r_0^2}{\varepsilon_0}$

6 A small dipole is placed in region of a non-uniform field $\mathbf{E} = E(x)\,\hat{\mathbf{i}}$, (magnitude of field depends on X-coordinate of point). If θ = angle between dipole moment and X-direction, then

a. dipole feels only a torque
b. dipole feels only a force
c. force on dipole is given by $F = p\left(\frac{dE}{dx}\right)\cos\theta$
d. force on dipole is given by $F = p\left(\frac{dE}{dx}\right)\sin\theta$

7 Consider two long parallel wires (oppositely charged) of radius d with their centres separated by distance D. Capacitance of pair of wires in farad per metre is

a. $\frac{\pi\varepsilon_0}{\log[(D-d)/d]}$ b. $\frac{2\pi\varepsilon_0}{\log(D/d)}$

c. $\frac{4\pi\varepsilon_0}{\log(D/d)}$ d. $\frac{4\pi\varepsilon_0}{\log[(D-d)/d]}$

Working Space

8 A small block of mass m is kept on a smooth inclined plane of angle 30° placed in an elevator going upward with acceleration a. Electric field E exists between the vertical sides of the wall of the elevator. The charge on the block is $+q$. The time taken by the block to come to the lowest point of the inclined plane is (take the surface to be smooth)

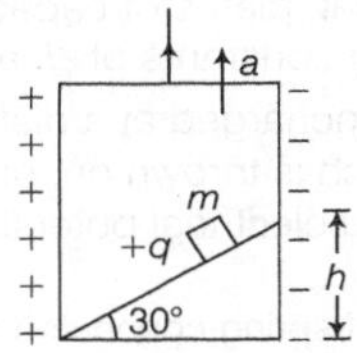

a. $\sqrt{\frac{2h}{g}}$

b. $\sqrt{\frac{2h}{(g-a)+\frac{gE}{m}}}$

c. $2\sqrt{\frac{2h}{(g+a)-\frac{qE}{m}\sqrt{3}}}$

d. $\sqrt{\frac{2h}{(g+a)^2-\left(\frac{qE}{m}\right)h^2}}$

9 If V_0 be the potential at origin in an electric field $\mathbf{E}=E_x\hat{\mathbf{i}}+E_y\hat{\mathbf{j}}$, then the potential at point $P(x, y)$ is

a. $V_0+xE_x+yE_y$ b. $V_0+xE_x-yE_y$ c. $V_0-xE_x-yE_y$ d. $\sqrt{(x^2+y^2)}\sqrt{E_x^2+E_y^2}-V_0$

10 The figure shows an infinite line charge of density σ C/m. The work done by the electrostatic force on a unit charge, when it is moved along the path ABC, is (plane of the curve ABC contains the line charge)

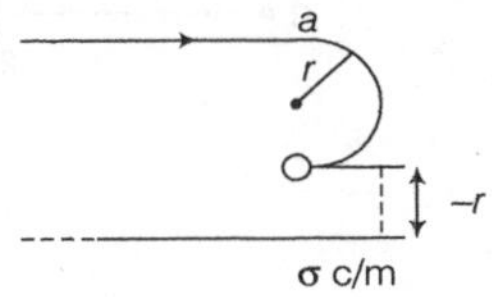

a. zero

b. $(-\sigma/2\pi\varepsilon_0)\ln 2$

c. $(\sigma/2\pi\varepsilon_0)\ln 3$

d. $(-\sigma/2\pi\varepsilon_0)\ln 3$

One or More than One Options Correct Type

Direction (Q. Nos. 11-23) *This section contains 13 multiple choice questions. Each question has four choices (a), (b), (c) and (d), out of which ONE or MORE THAN ONE are correct.*

11 The given figure shows three concentric thin spherical shells A, B and C of radii a, b and c. The shells A and C are given charges q and $-q$, respectively, and shell B is earthed. Then,

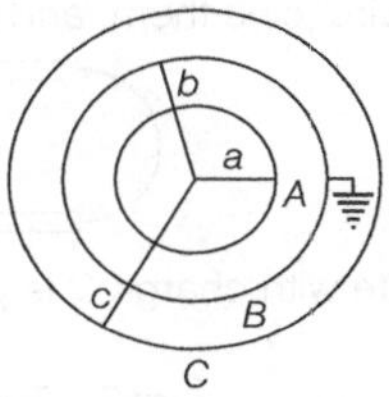

a. the charge q will be there at outer surface A

b. the charge $\frac{-bq}{c}$ will be there at inner surface of C

c. the charge $-q$ will be at inner surface of B

d. the charge $q(1-b/c)$ will be there at outer surface of C

12 Consider a parallel plate capacitor having charge Q. Then,

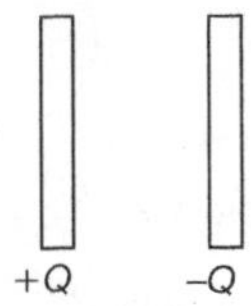

a. there is no force acting on plates of the capacitor

b. the electric field at outer side of plates is zero

c. the electric field between the plates of the capacitor is $\frac{\sigma}{\varepsilon_0}$

d. the force between the plates is given by $\frac{Q^2}{2A\varepsilon_0}$

13 Two conducting plates A and B are placed parallel to each other. A is given a charge Q_1 and B a charge Q_2. Then,

a. the charge on the outer plate A is $\frac{Q_1+Q_2}{2}$

b. the charge on the outer plate B is $\frac{Q_1+Q_2}{2}$

c. the charge on the inner plate A is $\frac{Q_1-Q_2}{2}$

d. the charge on the inner plate B is $-\frac{Q_1-Q_2}{2}$

14 Electric field inside a charged spherical ball depends only on distance from the centre, as $E(r)=ar, a=$ constant. Then,

a. flux through a thin spherical layer of thickness dr and radius r is equals to $\frac{\rho 4\pi r^2 dr}{\varepsilon_0}$

b. space charge distribution inside the ball is $3a\varepsilon_0$

c. field is of constant magnitude at every point inside the ball and $E=\frac{\sigma}{\varepsilon_0}$

d. field is constant outside the ball and $E=\frac{\sigma}{\varepsilon_0}$

Working Space

15 In given arrangement, plates of capacitor are free to move and springs have spring constants of $2k$ and k.

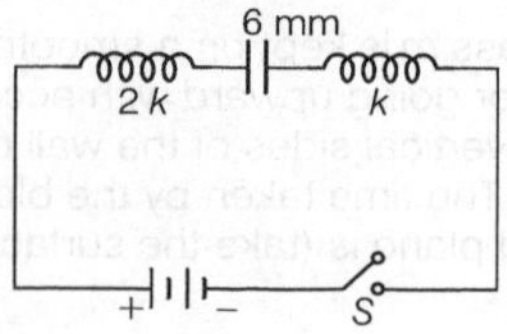

Initially plates are uncharged at a distance of 6 mm and switch S is open. When switch is thrown on, distance between plates becomes 2 mm and electrical potential energy of capacitor is 7.5×10^{-4} J. Then,

a. spring constant of spring connected to positive plate is 281.25 N/m
b. spring constant of spring connected to negative plate is 281.25 N/m
c. total stored energy in system = 1.5×10^{-3} J
d. total stored energy is purely of electrical nature

16 A potential difference of 100 V is applied across points A and B.

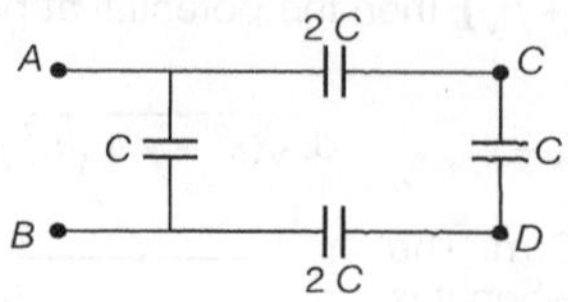

Then,

a. $V_{CD} = 50$ V
b. $V_{AC} = 25$ V
c. $V_{BC} = 25$ V
d. Charge on each capacitor is same

17 An uniform electric field exists in Z-direction. Then,

a. potential is constant for all points
b. potential is constant for any X for a given Z-direction
c. potential is constant for any Y for a given Z-direction
d. potential is constant on X-O-Y plane for a given Z-direction

18 Two large metal plates are originally oriented horizontally and separated by a distance of $3d$. A conducting wire joins them, and initially plats are uncharged.

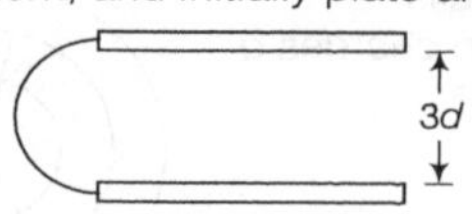

A third identical plate with charge Q is put parallel to them at a distance of d from upper plate.

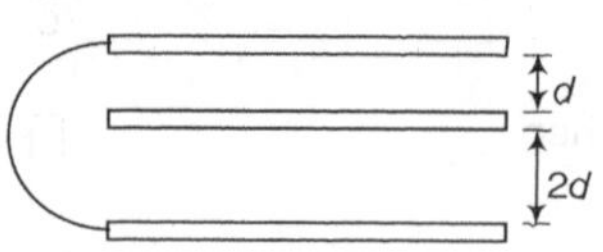

Then,

a. charge on lower plate is $-\frac{Q}{3}$
b. charge on upper plate is $-\frac{2Q}{3}$
c. potential difference between upper and middle plate is $\frac{2Qd}{3\varepsilon_0 A}$
d. potential difference between upper and lower plate is $\frac{Qd}{3\varepsilon_0 A}$

19 A system consists of two mass particles and an insulated spring of nearly zero mass. (Given $m = 10$ g each, $k = 5$ N/m)

Then,

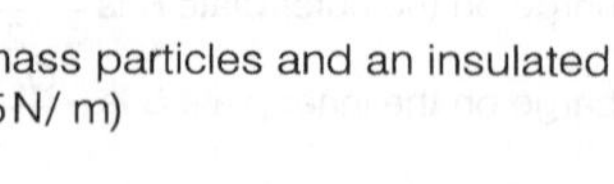

a. time period before charging of mass is $2\pi\sqrt{\frac{m}{2k}}$
b. time period of oscillations before charging and after charging is same
c. time period of oscillations increases if masses are charged
d. time period of oscillations decreases if masses are charged

20 A capacitor of capacitance C is connected to two voltmeters A and B as shown in figure.

Working Space

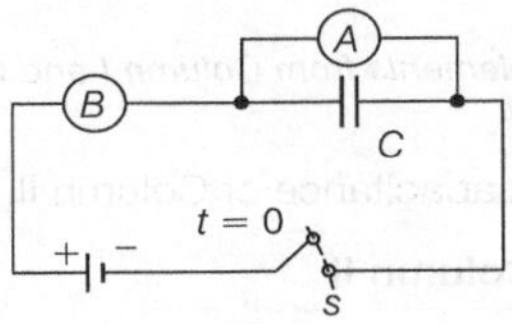

A has infinite resistance and B has R resistance. The capacitor is initially uncharged, at $t = 0$ the switch is closed.

a. Readings of B and A will be ε and zero at $t = 0$
b. During time interval $0 \le t < \infty$ readings of B and A are changing
c. Reading of A and B will be equal at $t = RC \log_e 2$
d. No charge appears on capacitor plates at $t = 1$ s

21 In the circuit shown $t < 0$ the switch S is open and capacitor S is open and capacitor is uncharged. Switch is closed at $t = 0$

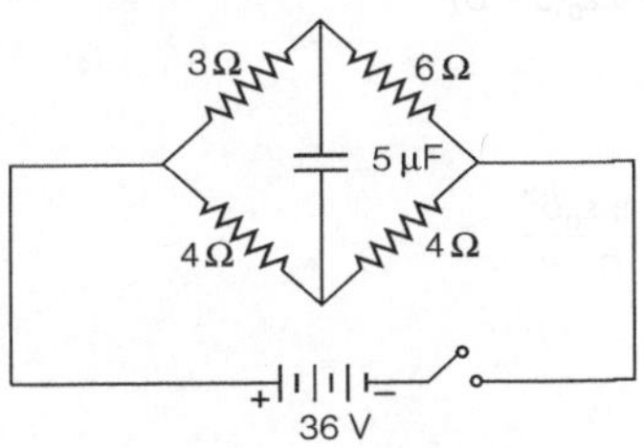

Then,

a. current through battery just after closing the switch is 8.75 A
b. maximum charge on capacitor plates is 30 μC
c. current through the capacitor after a long time is 2 A
d. capacitor plates are fully charged only after 20 μs

22 Three very long plates are given charges as shown in figure.

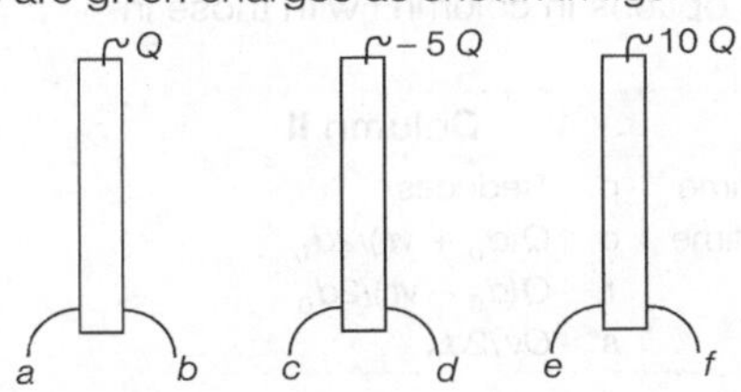

Finally charges are

a. $-7Q$ on surface e and $3Q$ on surface f
b. $-2Q$ on surface b and $3Q$ on surface a
c. $-7Q$ on surface d and $2Q$ on surface e
d. charges on all surfaces a, b, c, d are equal in magnitude

23 In the circuit shown in the figure a capacitor charges from cell of emf x and reaches a steady state.

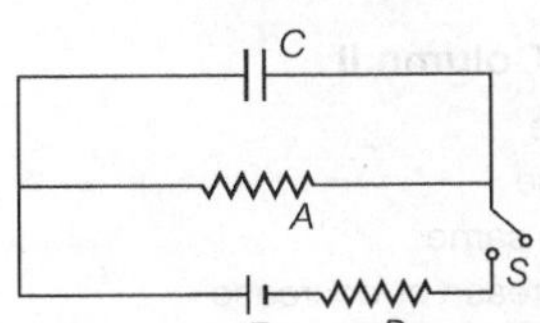

At $t = 0$ switch S is closed, if A and B are equal resistances then incorrect options are

a. During charging more heat is produced in A than in B
b. In the steady state rate of heat production in A and B is different
c. In steady state energy of capacitor is $\frac{1}{4} C \varepsilon^2$
d. In steady state energy stored in capacitor is $\frac{1}{2} Cx^2$

Matching List Type

Working Space

Direction *(Q. No. 24-26) Choices for the correct combination of elements from Column I and Column II are given as options (a), (b), (c) and (d), out of which one is correct.*

24 Match the arrangement of Column I with the equivalent capacitance of Column II.

	Column I		Column II
i.	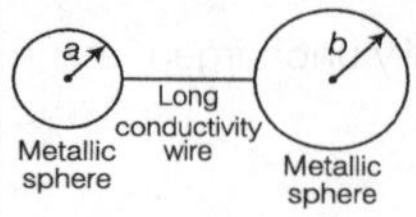	p.	$\dfrac{\varepsilon_0 A[b+2a]}{2ab}$
ii.	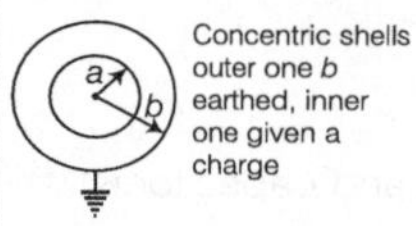	q.	$4\pi\varepsilon_0\left(\dfrac{ab}{b-a}\right)$
iii.	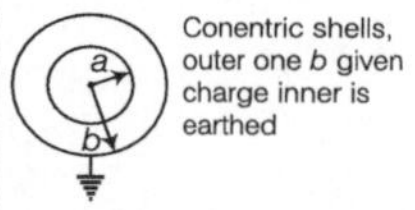	r.	$4\pi\varepsilon_0(a+b)$
iv.	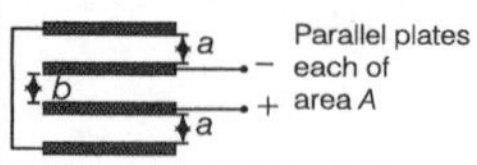	s.	$\dfrac{4\pi\varepsilon_0 b^2}{b-a}$

Codes

	i	ii	iii	iv
a.	p	q	r	s
b.	p	q	s	r
c.	s	r	q	p
d.	r	q	s	p

25 Charge Q is distributed on two identical capacitors in parallel. Separation of plates in each capacitor is d_0. If the first plate of capacitor C_1 and second plate of capacitor C_2 starts moving to left with a constant speed v, match the options in column I with those in column II.

	Column I		Column II
i.	Charge on capacitor C_1 as a function of time	p.	Reduces
ii.	Charge on capacitor C_2 as a function of time	q.	$Q(d_0 + vt)/2d_0$
iii.	Current in circuit	r.	$Q(d_0 - vt)/2d_0$
iv.	Energy of the system	s.	$Qv/2d_0$

Codes

	i	ii	iii	iv		i	ii	iii	iv
a.	p	q	s	r	b.	r	q	s	p
c.	p	s	q	r	d.	q	s	p	r

26 A parallel plate capacitor of capacity C_0 is charged by a battery. After charging battery is disconnected and a dielectric is filled between the plates of capacitor. Column I represents quantity and column II represents the change occurred. Match the columns.

	Column I		Column II
i.	PE of capacitor	p.	increase
ii.	PD between plates	q.	decrease
iii.	Capacity of capacitor	r.	remains same
iv.	Charge on capacitor	s.	may increase or decrease

Codes

	i	ii	iii	iv		i	ii	iii	iv
a.	q	q	s	p	b.	p	q	s	r
c.	p	s	q	r	d.	p	q	r	s

Working Space

Comprehension Type

***Direction** (Q. Nos. 27-38) This section contains 4 paragraphs, describing theory, experiments, data, etc. Eleven questions related to the paragraphs have been given. Each question has only one correct answer among the four given options (a), (b), (c) and (d).*

Passage I

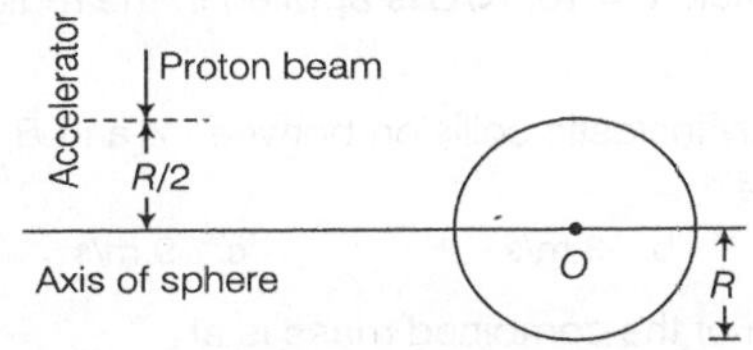

An accelerator produces a narrow beam of protons, each having an initial speed of v_0. The beam is directed towards an initially uncharged distant metal sphere of radius R and centred at point O. The initial path of the beam is parallel to the axis of the sphere at a distance of $(R/2)$ from the axis, as indicated in the diagram.

The protons in the beam that collide with the sphere will cause it to become charged. The subsequent potential field at the accelerator due to the sphere can be neglected. The angular momentum of a particle is defined in a similar way to the moment of force. It is defined as the moment of its linear momentum, linear momentum replacing the force. We may assume the angular momentum of a proton about point O to be conserved. Assume the mass of the proton as m_p and the charge on it as e. Given that the potential of the sphere increases with time and eventually reaches a constant value.

27 The total energy (E) of a proton in the beam, travelling with speed v at a distance of r $(r \geq R)$ from point O, assuming that the sphere has acquired an electrostatic charge Q, is

a. $\frac{eQ}{4\pi\varepsilon_0 r}$ b. less than $\frac{eQ}{4\pi\varepsilon_0 r}$

c. greater than $\frac{eQ}{4\pi\varepsilon_0 r}$ d. zero

28 After a long time, when the potential of the sphere reaches a constant value, the trajectory of proton is correctly sketched as

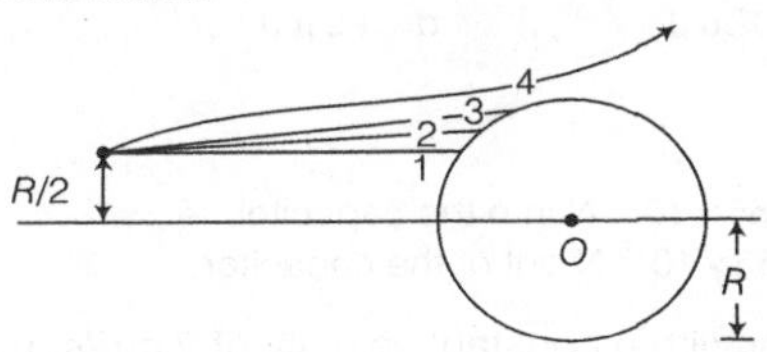

a. 1 b. 2 c. 3 d. 4

29 Once the potential of the sphere has reached its final, constant value, the minimum speed v of a proton along its trajectory path is given by

a. v_0 b. $\frac{v_0}{2}$ c. $2v_0$ d. None of these

Passage II

Electric field in a cubical volume is given by

$$E = E_0\left[1+\frac{2}{a}\right]\hat{i} + E_0\left[\frac{2}{a}\right]\hat{j}$$

One of the corner of cube is at origin, its sides are along coordinate axes and each side length is of d units.

30 Net outward flux through face which is parallel to ZOY plane is

a. $E_0\left(1+\frac{Z}{a}\right)d^2$ b. $-E_0\left(1+\frac{Z}{a}\right)d^2$ c. $E_0\frac{Z}{a}\cdot d^2$ d. $-E_0\frac{Z}{a}d^2$

31 Net outward flux through face parallel to XOZ plane is

a. $E_0\left(1+\frac{Z}{a}\right)d^2$ b. $-E_0\left(1+\frac{Z}{a}\right)d^2$ c. $E_0\frac{Z}{a}\cdot d^2$ d. $-E_0\frac{Z}{a}d^2$

32 Charge enclosed by the cube is of magnitude

a. $4\pi\varepsilon_0 E_0\frac{Z}{a}$ b. $4\pi\varepsilon_0 E_0\left(1+\frac{Z}{a}\right)$ c. $4\pi\varepsilon_0 E_0 d^2$ d. Zero

Working Space

Passage III

In given situation, $m_A = m_B = 1\,\text{kg}$, and $q_A = 0$ but $q_B = -1\,\text{C}$.
Initially, spring ($k = 18\,\text{N/m}$) is relaxed and masses A and B are at rest as shown above.

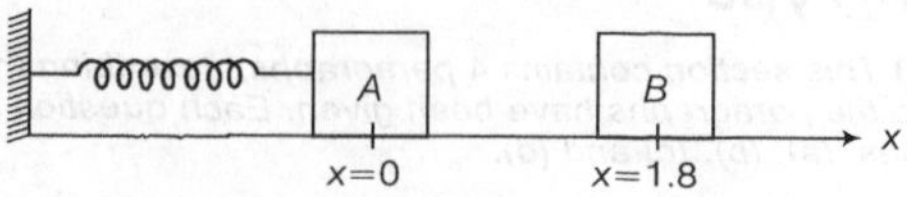

At $t = 0$, an electric field $E = 10\hat{i}$ N/C is applied in the region. Surface on which A and B are placed is smooth.

33 Assuming a perfectly inelastic collision between A and B, velocity of combined mass just after collision will be

a. 6 m/s b. 3 m/s c. 9 m/s d. 12 m/s

34 Equilibrium position of the combined mass is at

a. $x = -\frac{2}{9}$ m b. $x = -\frac{1}{3}$ m c. $x = -\frac{5}{9}$ m d. $x = -\frac{7}{9}$ m

35 Amplitude of oscillation of the combined mass will be

a. $\frac{2}{3}$ m b. $\frac{\sqrt{124}}{3}$ m c. $\frac{\sqrt{72}}{9}$ m d. $\frac{\sqrt{106}}{9}$ m

Passage IV

A capacitor is constructed from two square plates of sides l and separation d. A material of dielectric constant k is inserted a distance x into the capacitor as shown in figure.

Given, $l = 5$ cm, $\Delta V = 2000$ V, $d = 2$ mm, dielectric constant of medium, $k = 4.5$

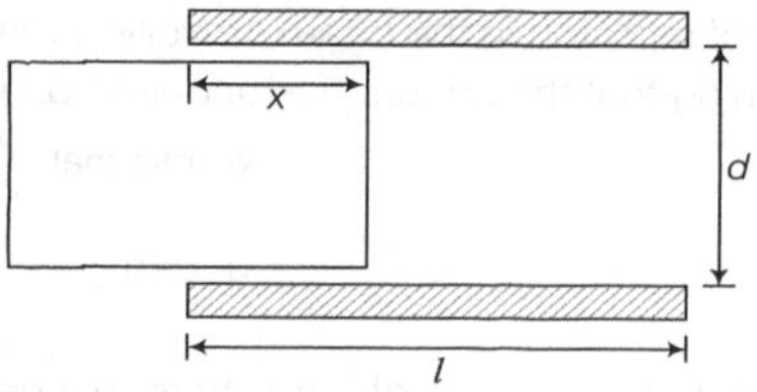

36 Energy stored in the capacitor, when $x = \frac{l}{2}$ is about

a. 61 μJ b. 30 μJ c. 102 μJ d. 89 μJ

37 Force on the dielectric slab, when $x = \frac{l}{2}$ is

a. 1.55×10^{-3} N into the capacitor b. 2.55×10^{-3} N into the capacitor
c. 2.55×10^{-3} N out of the capacitor d. 1.55×10^{-3} N out of the capacitor

38 If dielectric slab is pushed slowly into the capacitor with a constant velocity of 2 cm/s, the rate of change of energy of the capacitor is

a. -3.1×10^{-5} J/s b. $+5.1 \times 10^{-5}$ J/s c. $+3.1 \times 10^{-5}$ J/s d. -5.1×10^{-5} J/s

One Integer Value Correct Type

Direction (Q. No. 39-41) *This section contains 3 questions. When worked out will result in an integer from 0 to 9 (both inclusive).*

39 On planet tehar, the free fall acceleration is same as that of the earth but there is also a strong downward electric field that is nearly uniform closer to planets surface. A 2 kg ball with a charge of 5.00 μC is thrown upwards at a speed of 20.1 m/s and it hits the ground after an interval of 4.1 s. If the potential difference between the starting point and the top point of the trajectory is around 20 k then, find the value of k.

40 Two circular loops of radii 0.05 m and 0.09 m, respectively are placed such that their axis coincides and their centres are 0.12 m apart. Charge of 10^{-4} coulomb is non-uniformly spread on each of loop such that charge on upper half is twice of that on the lower half. If the potential difference between centres of loop is $N \times 71$ kV, then find the value of N.

41 A strip of length l having linear charge density λ is placed near a negatively charged particle of mass m and charge q at a distance l from the end A of the strip.

Let velocity of particle p when it reaches at a distance $\frac{l}{2}$ from end A is v and, $v^2 \alpha \frac{\sigma q}{2\pi\varepsilon_0 m}$. If proportionality constant is $\log N$, then find the value of $2N$.

JEE Main & AIEEE Archive

(Compilation of Last 13 Years Questions)

1 A parallel plate capacitor is made of two circular plates separated by a distance of 5 mm with a dielectric of dielectric constant 2.2 between them. When the electric field in the dielectric is 3×10^4 V/m, the charge density of the positive plate will be close to **(JEE Main 2014)**

a. 6×10^{-7} C/m^2 b. 3×10^{-7} C/m^2
c. 3×10^{4} C/m^2 d. 6×10^{4} C/m^2

2 Two capacitors C_1 and C_2 are charged to 120 V and 200 V, respectively. It is found that by connecting them together, the potential on each one can be made zero. Then, **(JEE Main 2013)**

a. $5C_1 = 3C_2$ b. $3C_1 = 5C_2$
c. $3C_1 + 5C_2 = 0$ d. $9C_1 = 4C_2$

3 Combination of two identical capacitors, a resistor R and a DC voltage source of voltage 6 V is used in an experiment on R - C circuit. It is found that for a parallel combination of the capacitor, the time in which the voltage of the fully charged combination reduces to half its original voltage is 10 s. For series combination, the time needed for reducing the voltage of the fully charged series combination by half is **(AIEEE 2011)**

a. 20 s b. 10 s c. 5 s d. 2.5

4 A resistor R and 2 μF capacitor in series is connected through a switch to 200 V direct supply. Across the capacitor is a neon bulb that lights up at 120 V. Calculate the value of R to make the bulb light up 5 s after the switch has been closed (take, $\log_{10}2.5 = 0.4$) **(AIEEE 2011)**

a. 1.7×10^5 Ω b. 2.7×10^6 Ω
c. 3.3×10^7 Ω d. 1.3×10^4 Ω

5 Let C be the capacitance of a capacitor discharging through a resistor R. Suppose t_1 is the time taken for the energy stored in the capacitor to reduce to half its initial value and t_2 is the time taken for the charge to reduce to one-fourth its initial value. Then, the ratio t_1/t_2 will be **(AIEEE 2010)**

a. 1 b. 1/2 c. 1/4 d. 2

6 A parallel plate capacitor with air between the plates has a capacitance of 9 pF. The separation between its plates is d. The space between the plate is now filled with two dielectrics. One of the dielectrics has dielectric constant $K_1 = 3$ and thickness $\frac{d}{3}$ while the other one has dielectric constant $K_2 = 6$ and thickness $\frac{2d}{3}$. Capacitance of the capacitor is now **(AIEEE 2008)**

a. 1.8 pF b. 45 pF
c. 40.5 pF d. 20.25 pF

7 A parallel plate condenser with a dielectric of dielectric constant K between the plates has a capacity C and is charged to a potential V volts. The dielectric slab is slowly removed from between the plates and then reinserted. The net work done by the system in this process is **(AIEEE 2007)**

a. $\frac{1}{2}(K-1)CV^2$ b. $CV^2(K-1)/K$
c. $(K-1)CV^2$ d. zero

8 A battery is used to charge a parallel plate capacitor till the potential difference between the plates becomes equal to the electromotive force of the battery. The ratio of the energy stored in the capacitor and the work done by the battery will be **(AIEEE 2007)**

a. 1 b. 2 c. $\frac{1}{4}$ d. $\frac{1}{2}$

9 A parallel plate capacitor is made by stacking n equally spaced plates connected alternatively. If the capacitance between any two adjacent plates is C, then the resultant capacitance is **(AIEEE 2005)**

a. $(n-1)C$ b. $(n+1)C$
c. C d. nC

10 A fully charged capacitor has a capacitance C. It is discharged through a small coil of resistance wire embedded in a thermally insulated block of specific heat capacity s and mass m. If the temperature of the block is raised by ΔT, the potential difference V across the capacitance is **(AIEEE 2005)**

a. $\sqrt{\frac{2mC\Delta T}{s}}$ b. $\frac{mC\Delta T}{s}$ c. $\frac{ms\Delta T}{C}$ d. $\sqrt{\frac{2ms\Delta T}{C}}$

11 A sheet of aluminium foil of negligible thickness is introduced between the plates of a capacitor. The capacitance of the capacitor **(AIEEE 2003)**

a. decreases b. remains unchanged
c. becomes infinite d. increases

12 The work done in placing a charge of 8×10^{-18} C on a condenser of capacity 100 μF is **(AIEEE 2003)**

a. 16×10^{-32} J b. 3.1×10^{-26} J
c. 4×10^{-10} J d. 32×10^{-32} J

13 If there are n capacitors in parallel connected to V volt source, then the energy stored is equal to **(AIEEE 2002)**

a. CV b. $\frac{1}{2}nCV^2$ c. CV^2 d. $\frac{1}{2n}CV^2$

14 Capacitance (in F) of a spherical conductor having radius 1 m, is **(AIEEE 2002)**

a. 1.1×10^{-10} b. 10^{-6}
c. 9×10^{-9} d. 10^{-3}

JEE Advanced & IIT JEE Archive

(Compilation of Last 10 Years Questions)

1 A parallel plate capacitor has a dielectric slab of dielectric constant K between its plates that covers 1/3 of the area of its plates, as shown in the figure. The total capacitance of the capacitor is C while that of the portion with dielectric in between is C_1. When the capacitor is charged, the plate area covered by the dielectric gets charge Q_1 and the rest of the area gets charge Q_2. The electric field in the dielectric is E_1 and that in the other portion is E_2. Choose the correct option/options, ignoring edge effects.

(2014 Adv.,One or More Than One Options Correct Type)

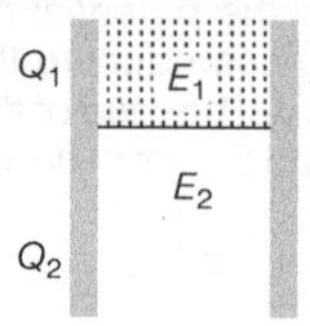

a. $\frac{E_1}{E_2} = 1$ b. $\frac{E_1}{E_2} = \frac{1}{K}$

c. $\frac{Q_1}{Q_2} = \frac{3}{K}$ d. $\frac{C}{C_1} = \frac{2 + K}{K}$

2 In the circuit shown in the figure, there are two parallel plate capacitors each of capacitance C. The switch S_1 is pressed first to fully charge the capacitor C_1 and then released. The switch S_2 is then pressed to charge the capacitor C_2. After some time, S_2 is released and then S_3 is pressed. After some time

(2013 Adv., One or More Than One Options Correct Type)

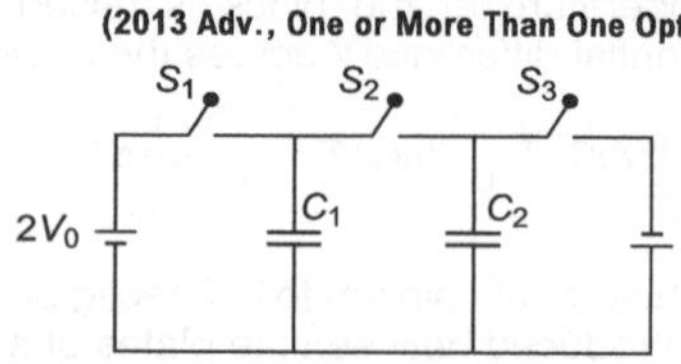

a. the charge on the upper plate of C_1 is $2\ CV_0$

b. the charge on the upper plate of C_1 is CV_0

c. the charge on the upper plate of C_2 is 0

d. the charge on the upper plate of C_2 is $-CV_0$

3 A 2 μF capacitor is charged as shown in the figure. The percentage of its stored energy dissipated after the switch S is turned to position 2 is **(2011, Only One Option Correct Type)**

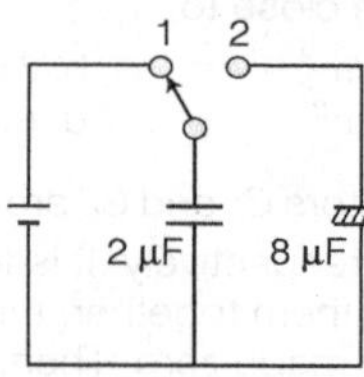

a. 0% b. 20%

c. 75% d. 80%

4 A parallel plate capacitor C with plates of unit area and separation d is filled with a liquid of dielectric constant $K = 2$. The level of liquid is $\frac{d}{3}$ initially. Suppose the liquid level decreases at a constant speed v, the time constant as a function of time t is

(2008, Only One Option Correct Type)

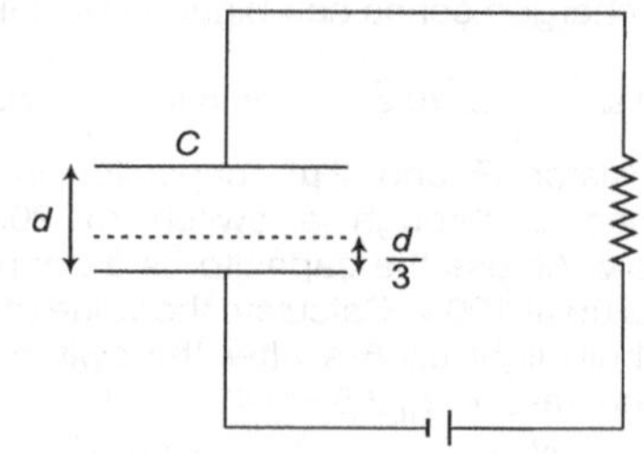

a. $\frac{6\varepsilon_0 R}{5d + 3vt}$

b. $\frac{(15d + 9vt)\varepsilon_0 R}{2d^2 - 3dvt - 9v^2t^2}$

c. $\frac{6\varepsilon_0 R}{5d - 3vt}$

d. $\frac{(15d - 9vt)\varepsilon_0 R}{2d^2 + 3dvt - 9v^2t^2}$

3. Current Electricity

DPP
DAILY PRACTICE PROBLEMS

DPP-1 Electric Current

Subjective Questions

Working Space

Direction (Q. Nos. 1-4) These questions are subjective in nature, need to be solved completely on notebook.

1 An art-piece with a surface area of 700 cm^2 is to be silver plated. It is attached to the negative electrode of an electrolytic cell containing silver nitrate ($Ag^+NO_3^-$). If the cell is powered by a 12 V battery which supplies a current of 5 A continuously, how time does it take for a 0.13 mm layer of silver to build up. Density of silver is 10.5×10^3 kg/m^3 and molar mass of silver is 107.9 g/mol.

2 In a current-carrying conductor,

i. Are all of the electrons moving in the same direction?
ii. Is drift velocity same for all electrons in magnitude and also in the direction?
iii. Why its temperature increases if current flows through it for a time duration?
iv. State the meaning of term mobility, is it a scalar quantity or vector quantity?
v. Although, the current flows from positive end to negative end of a wire but it is not taken as a vector quantity. Why?

3 Electrons in an electric circuit pass through a resistor. The wire on either side of resistor has the same diameter.

i. How does the drift speed of electron before entering the resistor compare with the speed after leaving the resistor? Explain your answer with proper reason.
ii. How does the potential energy of electron before entering the resistor compare with the potential energy of electron leaving the resistor?

4 In a hydrogen discharge tube, the number of hydrogen ions (protons) drifting across a cross-section per second is 1.0×10^{18} while the number of electrons drifting in the opposite direction across another cross-section is 2.7×10^{18}.

i. Find the current flowing in the tube.
ii. If the voltage applied across the tube is 230 V, then find the effective resistance of the tube.

Only One Option Correct Type

Direction (Q. Nos. 5-14) This section contains 10 multiple choice questions. Each question has four choices (a), (b), (c) and (d), out of which ONLY ONE option is correct.

5 A current of 1.30 A flows in a wire. How many electrons are flowing past any point in the wire per second?

a. 8.125×10^{18}
b. 8.125×10^{19}
c. 8.125×10^{12}
d. 8125

Working Space

6 The graph of current *versus* time in a wire is given by

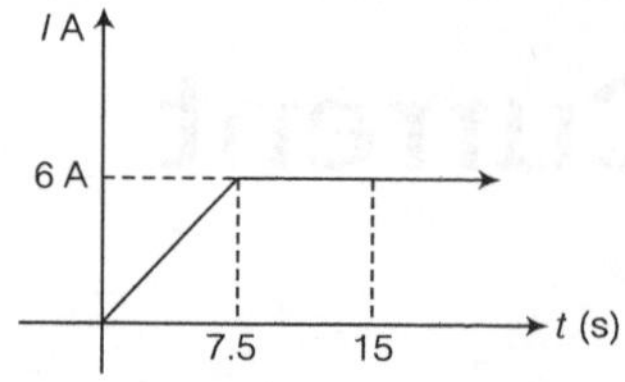

If charge that flows through a wire in 7.5 s is q_1 and that flows through the wire in 15 s is q_2, then ratio $q_1 : q_2$ is

a. 2 : 1　　b. 1 : 2　　c. 4 : 1　　d. 1 : 4

7 The current through a wire is $I = I_0 + \alpha t$, where $I_0 = 15$ A and $\alpha = 6$ A/s, charge that flows through the wire in 12 s, is

a. 252 C　　b. 180 C
c. 612 C　　d. 414 C

8 The charge passing through a conductor is a function of time and it is given by $q = (2t^2 - 4t + 3)$ mC. The current through the conductor at $t = 4$ s, is

a. 7 mA　　b. 19 mA　　c. 15 mA　　d. 12 mA

9 When a wire carries a current, then an electric field

a. exists outside the wire
b. outside the wire but perpendicular to its length
c. inside the wire and perpendicular to its length
d. inside the wire along its length

10 We are able to get fairly a large current in the conductor, because

a. a drift speed of electrons is very high
b. a number density of electrons is very high
c. both number density and drift speed of electrons are very high
d. the charge on an electron is quite high

11 An observer is moving with respect to a stationary electron with constant speed, then he observes

a. only magnetic field　　b. only electric field
c. both electric and magnetic fields　　d. no field at all

12 When a current flows through a conductor, its temperature

a. may increase or decrease　　b. remains constant
c. increases　　d. decreases

13 Consider a conductor of varying cross-section. A cell is connected across ends of the conductor. If ratio of area of section *A* and area of section *B* is 1 : 2, then ratio of currents through these sections is

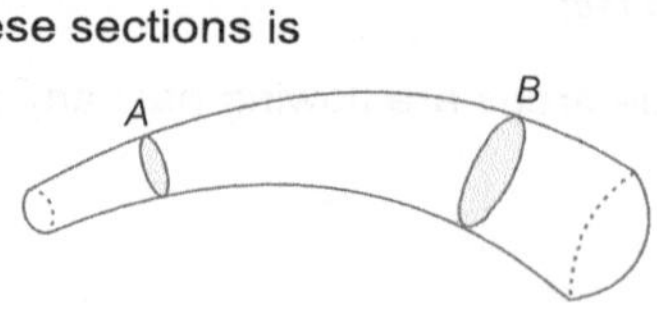

a. 1 : 2　　b. 2 : 1
c. 1 : 1　　d. 4 : 1

14 In a hydrogen discharge tube, it is observed that 3.13×10^{15} electrons are moving from right to left and 3.12×10^{15} protons are moving from left to right through a cross-section per second, then electric current in the discharge tube is

a. 1 mA towards right　　b. 1 mA towards left
c. 2 mA towards left　　d. 2 mA towards right

Working Space

One or More than One Options Correct Type

Direction (Q. Nos. 15 and 16) *This section contains 2 multiple choice questions. Each question has four choices (a), (b), (c) and (d), out of which ONE or MORE THAN ONE are correct.*

15 Let current in a conductor is $I = I_0 e^{-t/\tau}$, where τ is a constant with dimensions of time. Then,

a. current in the wire when switch is just closed is I_0
b. charge passed through the wire in a time interval of $t = \tau$ is $0.632 I_0 \tau$
c. charge passed through wire in a time $t = 10\tau$ sec is $0.632 I_0 \tau$
d. total charge that flows through the wire in duration $t = 0$ to $t = \infty$ is $3 I_0 \tau$

16 A Van de Graaff generator produces a deuteron beam of 2.00 MeV. It is found that beam current is 10 μA.

a. Distance between two nearby (travelling) deuterons is 2.21×10^{-2} m
b. Distance between two nearby (travelling) deuterons is 2.21×10^{-13} m
c. Repulsion between two deuterons causes a significant change in the flow of current
d. Repulsion between two deuterons does not have any effect on the flow of current

Matching List Type

Direction (Q. No. 17) *Choices for the correct combination of elements from Column I and Column II are given as options (a), (b), (c) and (d), out of which one is correct.*

17 Match the particles responsible for conduction with the condition of current flow.

	Column I		Column II
i.	Positive and negative ions	p.	Metal wires
ii.	Electrons	q.	Discharge tubes
iii.	Positive ions and electrons	r.	Cell
iv.	Electrons and holes	s.	Semiconductor

Codes

	i.	ii	iii	iv		i.	ii	iii	iv
a.	q	p	r	s	b.	r	p	q	s
c.	p	r	s	q	d.	s	r	p	q

Comprehension Type

Direction (Q. Nos. 18-21) *This section contains a paragraph, describing theory, experiments, data, etc. Four questions related to the paragraph have been given. Each question has only one correct answer among the four given options (a), (b), (c) and (d).*

Passage

An electron gun is designed to produce pulses of fast moving electrons of 50 MeV kinetic energy.

Each pulse has a duration of 150 ns and time interval between two pulses is 5 ms.

Current constituted by electrons in pulse is 175 mA and current is zero between two consecutive pulses.

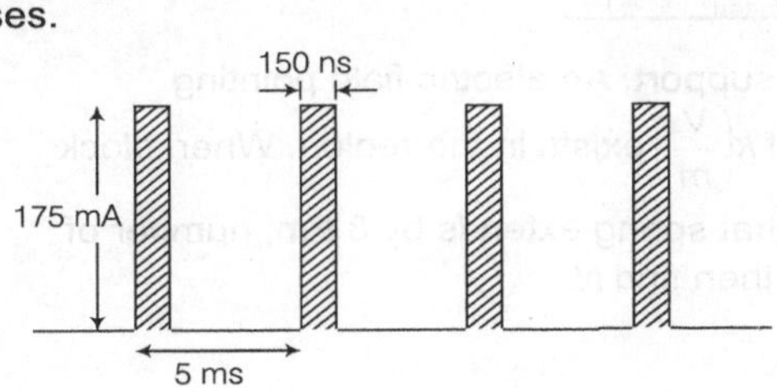

18 Number of electrons per pulse is

a. 2.25×10^{10} b. 1.64×10^{11}

c. 4.2×10^{10} d. 3.8×10^{12}

19 Average current per pulse is

a. 9.5×10^{-3} mA b. 6.75×10^{-3} mA

c. 7.75×10^{-3} mA d. 5.25×10^{-3} mA

20 Total charge delivered per second is

a. $9.75\,\mu C$ b. $7.5\,\mu C$

c. $5.25\,\mu C$ d. $3.25\,\mu C$

21 If electrons are made to pass through an absorber which absorbs 80% of their energy, energy absorbed by the absorber in one second will be nearby

a. 346 J b. 210 J

c. 269 J d. 125 J

One Integer Value Correct Type

Direction *(Q. No. 22-24) This section contains 3 questions. When worked out will result in an integer from 0 to 9 (both inclusive).*

22 In the junction shown below, consider an electron moving with a drift velocity of $v_d = 1.3 \times 10^{-4}$ m/s approaching the juction from *B* to *A*.

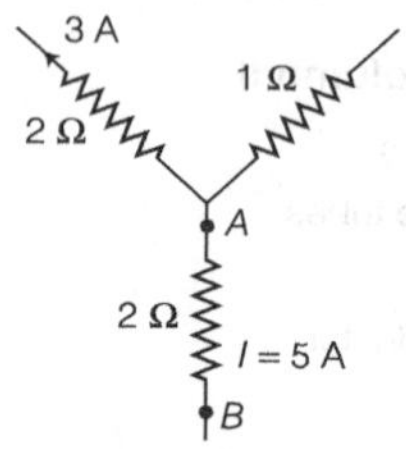

If momentum of electron considered is $p = 1.18 \times 10^{-34}$ kg-m/s (at *B*) and its momentum at *A* is *Np*, then find *N*.

a. 346 J b. 210 J

c. 269 J d. 125 J

23 A positive ion with a kinetic energy of 5.4×10^{-17} J is accelerated through a potential difference of 5 V. It is found that its final kinetic energy is 8.6×10^{-17} J. If ion has 10 N electrons less than protons on it, then find the value of *N*.

24 A mass *m* is connected to a very light (hypothetical) spring having spring constant $k\left(\frac{N_0}{m}\right)$ as shown in below figure.

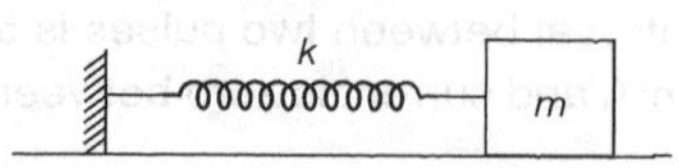

One end of spring is connected to a rigid support. An electric field pointing towards right hand side with magnitude of $k\left(\frac{V}{m}\right)$ exists in the region. When block is given some charge, spring is found to that spring extends by 8 Fm, number of electrons removed from block is $N \times 10^4$, then find *N*.

DPP-2 Relation among I, v_d, j, η, ρ and τ

Subjective Questions

Working Space

Direction *(Q. Nos. 1-4) These questions are subjective in nature, need to be solved completely on notebook.*

1 Explain/Give reasons.

i. A good electrical conductor is also a good conductor of heat.

ii. What would happen to drift velocity of electrons if wire has no resistance?

iii. If a charge flow very slowly through a metal, why does it not require several hours for a light to come on when a switch is on?

iv. An electric field is required inside a conductor so that a current flows through the conductor. In electrostatics, we concluded that electric field inside a conductor is zero. Why is it so?

2 An aluminium wire having a cross-sectional area of 4×10^{-6} m^2 carries a current of 5 A. Find drift speed of electrons in the wire. The density of aluminium is 2.7 g/cm^3. Assuming that one conduction electron is supplied by each atom.

3 Copper has 8.5×10^{28} electrons per cubic metre. If a copper wire of length 25 cm and diameter 2.05 mm carries a current of 1.55 A, then estimate the drift speed of electrons along the wire.

4 i. The relaxation time τ is nearly independent of applied electric field E wherever it changes significantly with temperature T. First fact is responsible for Ohm's law and second fact leads to variation of resistivity ρ with temperature. Explain how?

ii. Three materials have electrical conductivities σ, 2σ and 2σ and their number densities of free electrons are n, $2n$ and n, respectively.
For which material, the average collision time of free electrons is maximum?

Only One Option Correct Type

Direction *(Q. Nos. 5-16) This section contains 12 multiple choice questions. Each question has four choices (a), (b), (c) and (d), out of which ONLY ONE option is correct.*

5 What is the order of drift speed of electrons?

a. 10 cm/s b. 10^{-2} cm/s c. 10^{-4} cm/s d. 10^{-6} cm/s

6 Mean relaxation time of electrons in the conductor varies with temperature as

a. T^{-1} b. T^{-2} c. T d. T^2

7 Two manganin wires whose lengths are in the ratio 1: 2 and whose resistances are in ratio 1: 2 are connected in series with a battery.
The ratio of drift velocities of free electrons in the two wires is

a. 2 : 1
b. 4 : 1
c. 1 : 1
d. 1 : 4

Working Space

8 A conductor of length l is connected to a DC source of emf E. If length of a conductor is tripled by stretching it keeping E constant, then drift speed of electrons

a. is tripled b. is doubled c. becomes half d. becomes one third

9 Current flows through the given conductor. Which of these quantities remains constant throughout the complete length of a conductor?

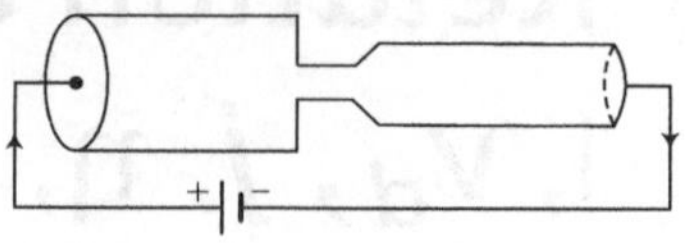

a. Current
b. Current density
c. Electron mobility
d. Electric field

10 For the conductor shown alongside

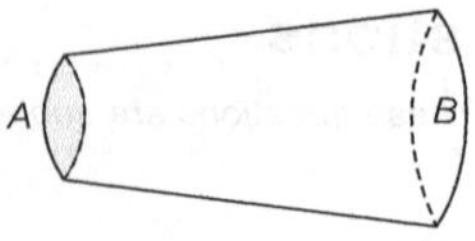

a. $v_{d_A} = v_{d_B}$
b. $v_{d_A} > v_{d_B}$
c. $v_{d_A} < v_{d_B}$
d. the drift speed varies randomly

11 A copper wire of diameter 1 mm carries a current of 1.1 A. The drift speed of electrons is (Given, density of Cu = 9 g/cm^3, atomic weight of Cu = 63 g and one electron is contributed by each Cu atom)

a. 0.1 mm/s b. 0.2 mm/s c. 0.3 mm/s d. 0.2 cm/s

12 In a metallic block, no potential difference is applied across it. Then, mean velocity of free electrons is

a. proportional to T b. proportional to $\sqrt{T}$
c. proportional to $T^{3/2}$ d. independent of temperature

13 In a particle accelerator, a current of 0.5 mA is carried by a 5 MeV proton beam that has a radius of 1.5 mm.
The number of protons crossing any cross-section of a beam in 1 s, is

a. 1.43×10^{13} protons b. 3.10×10^{7} protons
c. 2.48×10^{16} protons d. 2.5×10^{8} protons

14 A 10 mm thick Cu wire and a 14 mm thick Cu wire are joined in series. If wires carry a current of 15 A, then ratio of drift velocities of electrons in the wires is

a. 5 : 7 b. 7 : 5 c. 49 : 25 d. 25 : 49

15 An electron moving through a wire has an average drift velocity that is very small, (order of drift speed is around 10^{-4} m), because

a. instantaneous speed of a electron is also very less
b. electron suffers many collisions
c. electron is repelled by other electrons
d. electrons have random velocities

16 Two bars of radius r and $2r$ are kept constant as shown. An electric current I is passed through the bars. Which one of the following is correct? **[IIT JEE 2006]**

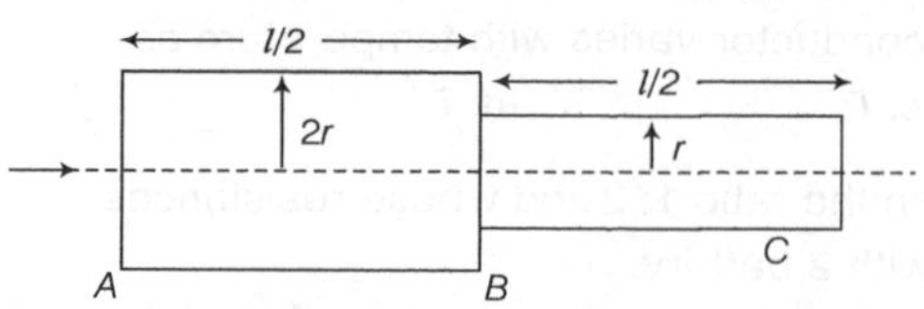

a. Heat produced in bar BC is 4 times the heat produced in bar AB
b. Electric field in both halves is equal
c. Current density across AB is double that of across BC
d. Potential difference across AB is 4 times that of across BC

One or More than One Options Correct Type

Direction (Q. Nos. 17-20) This section contains 4 multiple choice questions. Each question has four choices (a), (b), (c) and (d), out of which ONE or MORE THAN ONE are correct.

17 Temperature dependence of drift velocity depends on the following factors.

a. Number of charge carriers can change with temperature
b. Time interval between two successive collisions can depend on the temperature
c. Length of a material is a function of temperature
d. Mass of a material is a function of temperature

18 In the circuit shown, switch S is closed at $t = 0$. Then,

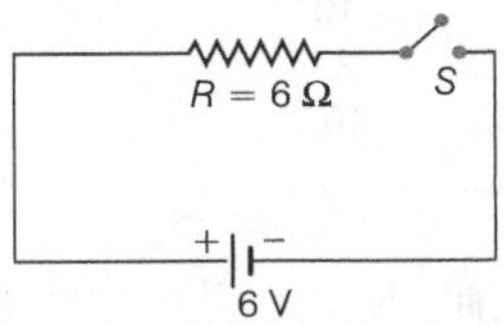

a. no energy is absorbed by electron till current reaches its maximum value
b. from $t = 0$, the instant current in the circuit is maximum, energy absorbed by electrons is 2×10^{-17} J
c. kinetic energy of electrons when current in the circuit is 1 A, is 2×10^{-17}J
d. electrons lost all of their kinetic energy in 10^{-17} s

19 A copper wire of diameter 0.65 mm carries a current of 2.3 μA, then

a. drift speed of electron is 5.1×10^{-10} m/s
b. drift speed of electron is 5.1×10^{-4} m/s
c. the electric field in wire is 1.2×10^{-7} V/m
d. the electric field in wire is 1.2×10^{-8} V/m

20 A wire of 5.8 m long, 2 mm diameter carries 750 mA current when 22 mV potential difference is applied at its ends. If drift speed of electrons is found 1.7×10^{-5} m/s, then

a. current density is 2.4×10^{5} A/m^2
b. current density is 1.6×10^{-4} A/m^2
c. number of free electrons per unit volume is 2.7×10^{16} m^{-3}
d. number of free electrons per unit volume is 8.8×10^{28} m^{-3}

One Integer Value Correct Type

Direction (Q. Nos. 21-23) This section contains 3 questions. When worked out will result in an integer from 0 to 9 (both inclusive).

21 At 27 °C in a copper cube, a number of free electrons is 8.5×10^{28}m^{-3}, resistivity is 1.7×10^{-8} Ω-m. If mean free path of electrons is estimated as 5 NÅ, then N is

22 A cylindrical conductor carries a current such that current density varries with radial distance r or $j = ar^2$, with a = a constant. The current through the outer portion of the wire between radial distances $\frac{R}{2}$ and R is proportional to R^k, where R is radius of wire. Find the value of k.

23 A synthetic metal wire with molar mass $M = 6.02 \times 10^{-1}$ kg/mol has a mass density through the cross-section of wire is uniform and only one conduction electron per 10^6 atoms of metal is available. If number of conduction electrons per unit volume of wire is $N \times 10^{22}$. Find N.

Working Space

Matching List Type

***Direction** (Q. No. 24) Choices for the correct combination of elements from Column I and Column II are given as options (a), (b), (c) and (d) out of which one is correct.*

24 Match the quantities in Column I with their orders in Column II.

	Column I		Column II
i.	Drift velocity of electron in a metallic conductor	p.	10^5
ii.	Speed of electron in an atomic orbit	q.	10^{-3}
iii.	Resistivity of metals	r.	10^{-8}
iv.	Temperature coefficient of resistivity of metals	s.	10^6
v.	Average thermal speed of conduction electrons	t.	10^{-4}

Codes

	i.	ii	iii	iv	v		i.	ii	iii	iv	v
a.	t	s	r	q	p	b.	p	q	r	s	t
c.	s	t	q	r	p	d.	p	q	t	r	s

Comprehension Type

***Direction** (Q. Nos. 25-30) This section contains two paragraphs, describing theory, experiments, data, etc. Six questions related to the paragraphs have been given. Each question has only one correct answer among the four given options (a), (b), (c) and (d).*

Passage I

A straight cylindrical wire lying along the X-axis has a length of 0.5 m and a diameter of 0.2 mm. It has a resistivity of $\rho = 4 \times 10^{-8}$ Ω-m.

Let a potential of 4 V is maintained at $x = 0$ at $x = 0.5$ m.

25 Resistance of the wire is

a. 0.600 Ω
b. 0.318 Ω
c. 0.637 Ω
d. 0.928 Ω

26 Current in the wire is

a. 6 A
b. 7 A
c. 6.28 A
d. 3.14 A

27 Current density in the wire is

a. $-200\hat{i}$ MA/m^2
b. $200\hat{i}$ MA/m^2
c. $100\hat{i}$ MA/m^2
d. $-100\hat{i}$ MA/m^2

Passage II

A steady beam of α-particles travelling with a constant kinetic energy of 20 MeV carries a current of 0.25 μA in space.

28 How many α-particles strike a surface perpendicular to their path in 1.6 s.

a 1.5×10^{12}
b. 1.5×10^{20}
c. 1.5×10^{18}
d. 1.5×10^{19}

29 How many α-particles are there in given 10 cm of beam?

a 2.5×10^6
b. 2.5×10^3
c. 2.5×10^5
d. 2.5×10^{12}

30 What is the value of accelerating potential for the α-particles?

a 10^6 V
b. 10^5 V
c. 10^7 V
d. 10^{10} V

DPP-3 Resistance, Resistivity, Temperature Dependence and Ohm's Law

Subjective Questions

Working Space

Direction *(Q. Nos. 1-4) These questions are subjective in nature, need to be solved completely on notebook.*

1 A 3 V flashlight bulb has a variable resistance as shown in figure.

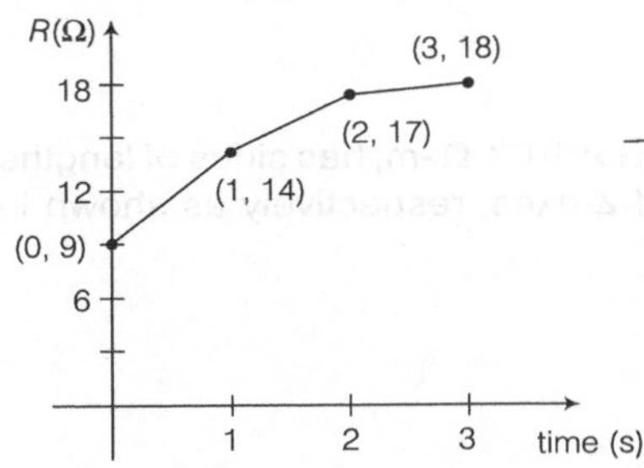

t	R
0	9
1	14
2	17
3	18

Calculate average current through the bulb in first three seconds.

2 In a gas chamber, two electrodes of area 100 cm^2 are placed 6 cm apart. An electric field produces 10^9 pair of monovalent ions/sec/cm^3. Mobility of each ion is 3.5×10^{-4} m^2/ Vs. Calculate the value of saturation current and applied field.

3 Show that $\rho = \rho_0[1 + \alpha (T - T_0)]$, where ρ is the resistivity at temperature T and ρ_0 is the resistivity at temperature T_0.

4 The dielectric material between plates of a capacitor (parallel plate) always have some non-zero conductivity σ. Show that the resistance and capacitance of a capacitor are related as $RC = \frac{k\varepsilon_0}{d}$.

Only One Option Correct Type

Direction *(Q. Nos. 5-15) This section contains 11 multiple choice questions. Each question has four choices (a), (b), (c) and (d), out of which ONLY ONE option is correct.*

5 The resistance of a wire at 20°C is 20 Ω and at 500°C is 60 Ω. At which temperature, resistance will be 25 Ω.

a. 50°C
b. 60°C
c. 70°C
d. 80°C

6 Ohm's law is true for

a. electrolytes when a current is passed through them
b. gases in a discharge tube
c. metallic conductors at constant temperature
d. metallic conductors at varying temperature

7 The length of a metal cylinder is increased by 100%, then its resistance changes by

a. 300% b. 200%
c. 100% d. 50%

8 The alloys constantan and manganin are used to make standard resistances as they have

a. low resistivity
b. low conductivity
c. low elasticity
d. low temperature coefficient of resistance

9 To increase resistance of a copper wire by 15%, then rise in temperature required is of (Given, $\alpha = 0.0068\,^\circ C^{-1}$, initial temperature is 20°C)

a. 22°C b. 28°C
c. 42°C d. 31°C

10 A rectangular block of a material of resistivity $\rho = 3 \times 10^{-5}\ \Omega$-m, has sides of lengths 1.0 cm, 2.0 cm and 4.0 cm lying along *X*, *Y* and *Z*-axes, respectively as shown in figure.

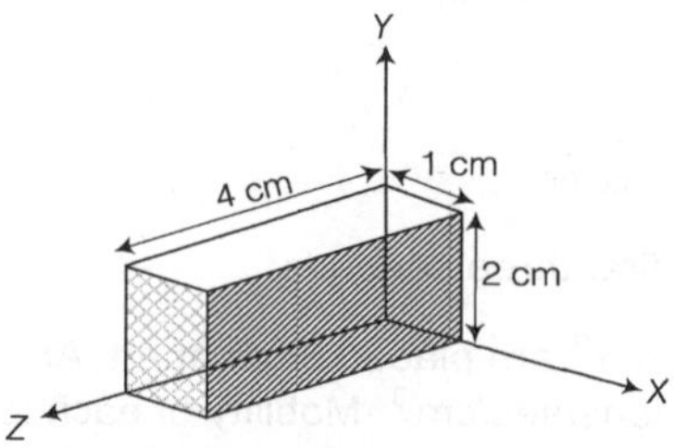

Then,

a. $\frac{R_x}{R_y} = 2$ b. $\frac{R_x}{R_z} = \frac{1}{2}$ c. $\frac{R_z}{R_y} = 4$ d. $\frac{R_z}{R_x} = 2$

11 Three samples of same material and same length with square cross-section are shown in figure.

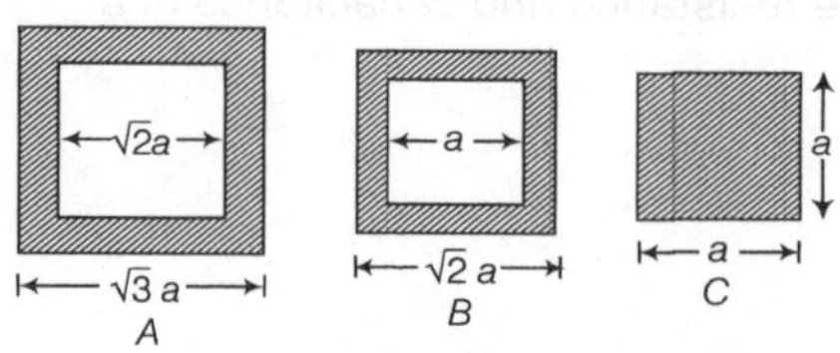

Then,

a. $R_A > R_B > R_C$ b. $R_A < R_B < R_C$
c. $R_A = R_B = R_C$ d. No relation among R_A, R_B and R_C

12 Two rods *A* and *B* are of same material and same length have their resistances in the ratio 1 : 2. When both rods are dipped into water, then

a. *A* has more loss of weight
b. *B* has more loss of weight
c. both *A* and *B* have same loss of weight
d. loss of weight have the ratio 1 : 2

13 *V-I* graph for a conductor is shown below.

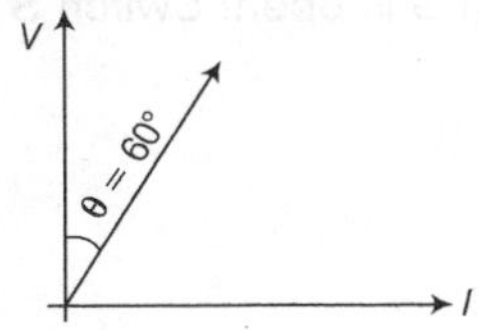

The resistance of the conductor is

a. $\sqrt{3}\ \Omega$ b. $\frac{1}{\sqrt{3}}\ \Omega$ c. 1Ω d. cannot be predicted

14 *V-I* graph for a conductor at temperatures T_1 and T_2 are given by

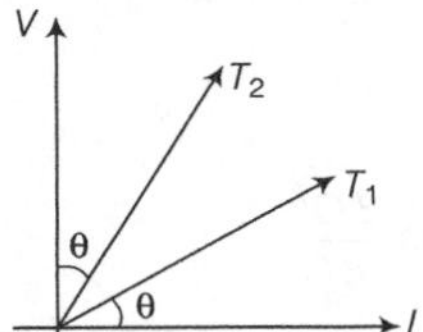

Then, $(T_2 - T_1)$ is proportional to

a. $\cos 2\theta$ b. $\sin\theta$

c. $\cot 2\theta$ d. $\tan\theta$

15 To verify Ohm's law, a student is provided with a test resistor R_T, a high resistance R_1, a small resistance R_2, two identical galvanometers G_1 and G_2 and a variable voltage source V. The correct circuit to carry out the experiment is **[IIT JEE 2010]**

a. 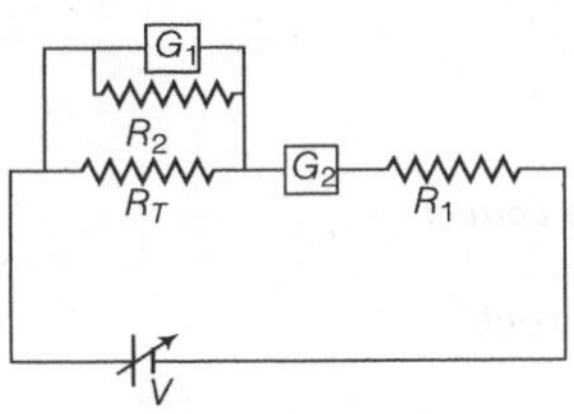

c. 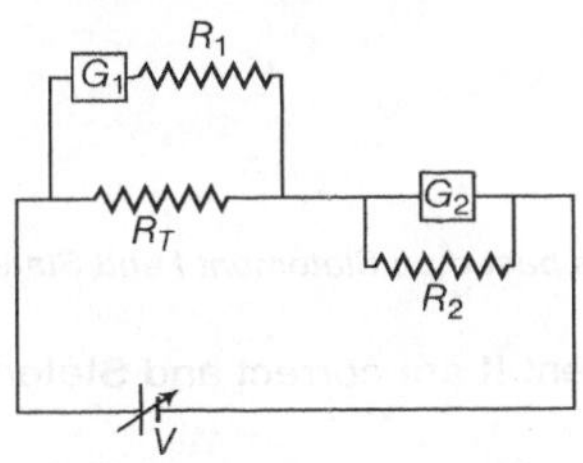

b. 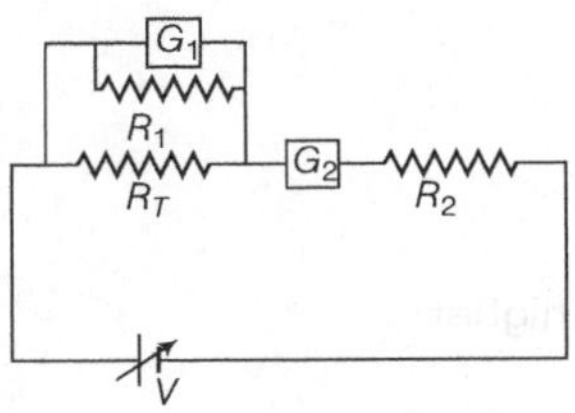

d. 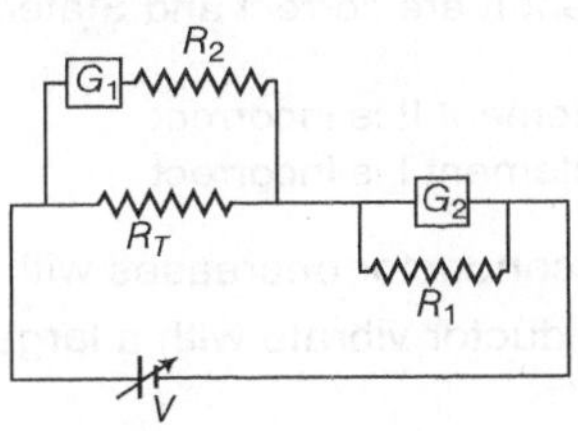

One or More than One Options Correct Type

Direction (Q. Nos. 16 -18) *This section contains 3 multiple choice questions. Each question has four choices (a), (b), (c) and (d), out of which ONE or MORE THAN ONE are correct.*

16 A carbon and a nichrome wire are connected in series. This combination has a resistance of $10\ k\Omega$ at 0°C. If it is required that the resistance of combination does not change with temperature, then

$(\alpha_{\text{nichrome}} = 0.4 \times 10^{-3}{}^\circ C^{-1}$ and $\alpha_{\text{carbon}} = -0.5 \times 10^{-3}{}^\circ C)$

a. resistance of nichrome wire at 0°C is 5.56 kΩ

b. resistance of carbon wire at 0°C is 4.44 kΩ

c. resistance of nichrome wire at 0°C is 8.2 kΩ

d. resistance of carbon wire at 0°C is 1.8 kΩ

Working Space

17. In the circuit given, a charge Q is placed on the capacitor of capacitance C, capacitor of capacitance $3C$ is uncharged and switch S is open. Switch S is closed and the circuit comes to equilibrium.

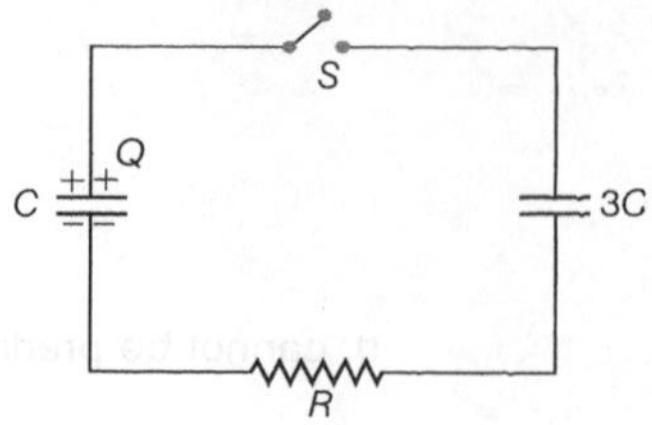

a. Final energy stored in a smaller capacitor is $\dfrac{Q^2}{32C}$

b. Final energy stored in a larger capacitor is $\dfrac{3Q^2}{32C}$

c. Energy dissipated in the form of heat in the resistor is $\dfrac{3Q^2}{8C}$

d. Energy dissipated in the form of heat in the resistor is less than $\dfrac{3Q^2}{8C}$

18. Let σ_1 and σ_2 are conductivities of germanium and sodium, respectively. Then, with the increase in temperature

a. σ_1 increases
b. σ_2 increases
c. σ_1 decreases
d. σ_2 decreases

Statement Type

Direction (*Q. Nos.* 19 -21) *This section is based on Statement I and Statement II. Select the correct answer from the codes given below.*

a. Both Statement I and Statement II are correct and Statement II is the correct explanation of Statement I
b. Both Statement I and Statement II are correct and Statement II is not the correct explanation of Statement I
c. Statement I is correct but Statement II is incorrect
d. Statement II is correct but Statement I is incorrect

19 **Statement I** Resistivity of a semiconductor decreases with temperature.

Statement II Atoms of a semiconductor vibrate with a larger amplitude at higher temperature.

20 **Statement I** Electric appliances with metallic body have three pin connections whereas electric bulb has two pin connections.

Statement II Three pin connections reduce the heating of connecting wires.

21 **Statement I** Bending of a wire does not affect its resistance.

Statement II Resistance of a wire depends on its resistivity.

Matching List Type

***Direction** (Q. No. 22) Choices for the correct combination of elements from Column I and Column II are given as options (a), (b), (c) and (d), out of which one is correct.*

22 Match the column I with column II.

	Column I		Column II
i.	Resistivity of a metal with temperature	p.	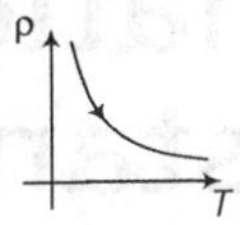
ii.	Resistivity of a semiconductor with temperature	q.	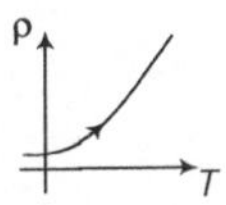
iii.	Resistivity of a metal alloy with temperature	r.	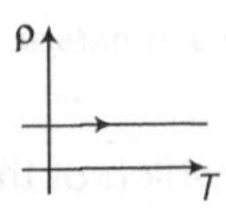
iv.	Resistivity of a diamagnetic substance	s.	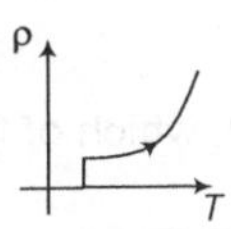

Codes

	i.	ii	iii	iv		i.	ii	iii	iv
a.	q	p	r	s	b.	p	q	r	s
c.	q	r	s	p	d.	q	p	s	r

One Integer Value Correct Type

***Direction** (Q. No. 23-25) This section contains 3 questions. When worked out will result in an integer from 0 to 9 (both inclusive).*

23 For a household wiring, copper alloy is used which has $\rho = 1.7 \times 10^{-8}\ \Omega$-m and $n = 8.49 \times 10^{28}\ \text{m}^{-3}$.

If atomic spacing of copper is about 0.2 nm and an electron on an average travels about $N \times 100$ atomic spacings between two collisions, then find N. Given average speed of free electrons of copper is about 1.6×10^{6} m/s.

24 Two resistance wires A and B of same material are connected in series. It is given that

Lengths of resistance wires are : $l_A = 2l,\ l_B = l$

Areas of resistance wires are : $A_A = \frac{A}{2},\ A_B = A.$

Temperature coefficients of resistance wires are $\alpha_A = \alpha_1$ and $\alpha_B = 2\alpha_1$.

When resistance wires are connected in series, then temperature coefficient of composite conductor is $\frac{6N\alpha}{5}$. Find the value of N.

25 Consider a copper cylinder of volume x, resistivity f, resistance across its length is r. If diameter of cylindrical conductor is $\left(\frac{16fx}{\pi^2 r}\right)^{1/k}$, then find the value of k.

DPP-4 **Series and Parallel Combination of Resistances**

DPP-4 Series and Parallel Combination of Resistances

Subjective Questions

Working Space

Direction (Q. Nos. 1 - 4) These questions are subjective in nature, need to be solved completely on notebook.

1 i. When resistors are connected in series, which of the following is same for each resistor?

Potential difference, current and power

ii. When resistors are connected in parallel, which of the following is same for each resistor?

Potential difference, current and power

2 i. Explain (a) A voltage divider circuit and (b) A current divider circuit.

ii. A voltage divider circuit of two resistors has a total resistance of 50 Ω. If output voltage is10% of the input voltage, obtain the values of two resistors in the circuit.

iii. A current of 30 mA is to be divided into two branch currents 20 mA and 10 mA by a network with an equivalent resistance equal to or greater than 10 Ω. Obtain the branch resistances.

3 i. In an experiment, two bulbs are connected in series. It is found that second light bulb is less bright than first.

A student claims that the second light bulb is less bright, because the first bulb uses up some of the current. What is the correct cause?

ii. Household appliances, bulbs, tubelights, fans, etc., are connected in parallel. What is the advantage of a parallel wiring in our homes?

iii. Two resistors are first connected in series and then in parallel.

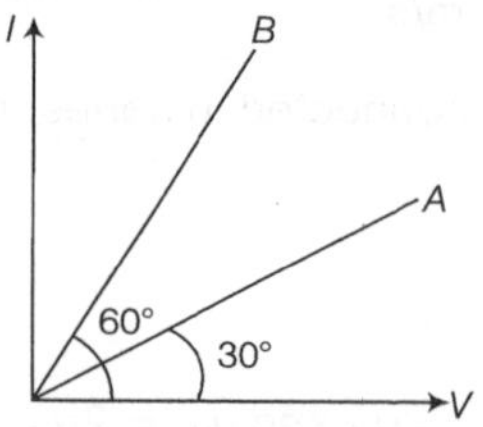

A graph of V and I is prepared for the combination using a suitable circuit. What is the ratio of equivalent resistance in series and equivalent resistance of combination in parallel?

4 An electric kettle has two heating coils. When only first coil is switched on, the boiling begins in 6 min. When only second coil is switched on, the boiling begins in 8 min.

In what time, the boiling begins if both the coils are switched on simultaneously

i. in series and

ii. in parallel?

Working Space

Only One Option Correct Type

Direction (Q. Nos. 5-15) This section contains 11 multiple choice questions. Each question has four choices (a), (b), (c) and (d), out of which ONLY ONE option is correct.

5 Five equal resistances of $R\ \Omega$ each are connected as shown in the figure. A battery of 12 V is connected between A and B. If $R = 2\ \Omega$, then current through the battery will be

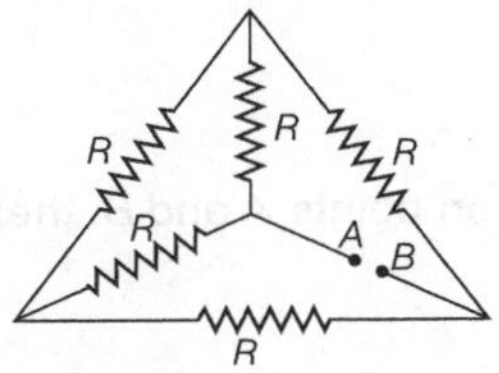

a. 1 A
b. 2 A
c. 4 A
d. 6 A

6 Equivalent resistance between points A and B is

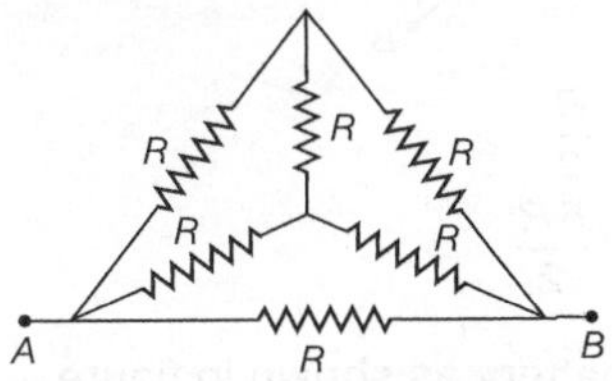

a. $6R\ \Omega$
b. $\frac{3}{4}R\ \Omega$
c. $2R\ \Omega$
d. $\frac{R}{2}\ \Omega$

7 Twelve resistors of equal value R are connected in the form of a cube. Effective resistance between diagonals of the cube is

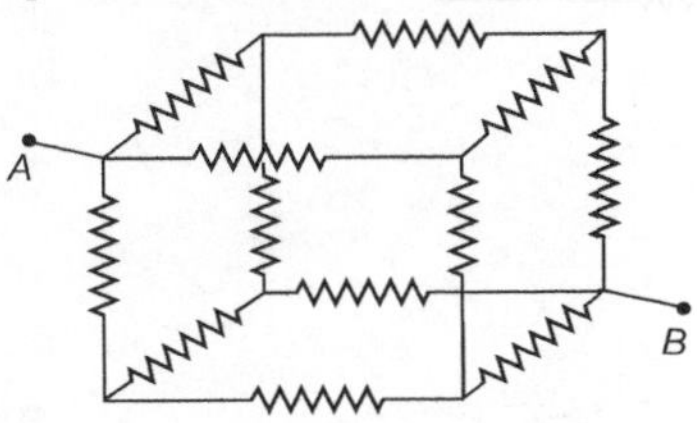

a. $\frac{12}{5}R\ \Omega$
b. $\frac{5}{6}R\ \Omega$
c. $\frac{7}{9}R\ \Omega$
d. $\frac{11}{6}R\ \Omega$

8 The potential difference across $8\ \Omega$ resistance is 48 V as shown in the figure. The value of potential difference across X and Y point will be

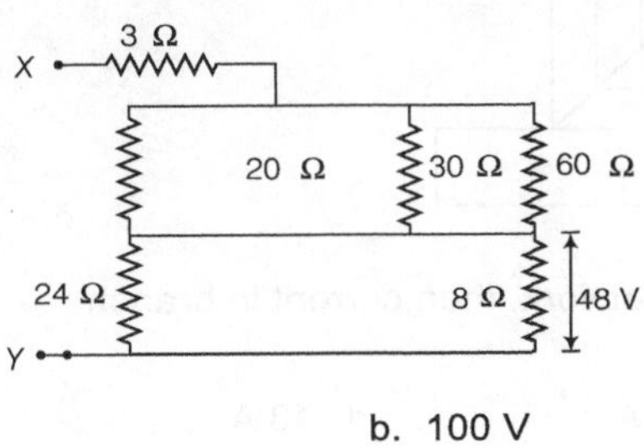

a. 168 V
b. 100 V
c. 152 V
d. 48 V

Working Space

9 Equivalent resistance between points A and B is

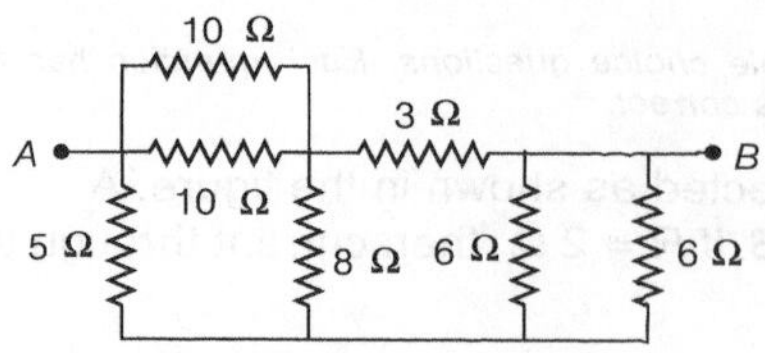

a. 3 Ω
b. 4 Ω
c. 4.5 Ω
d. 5 Ω

10 If twelve resistors each of 5 Ω are connected between points A and B, then

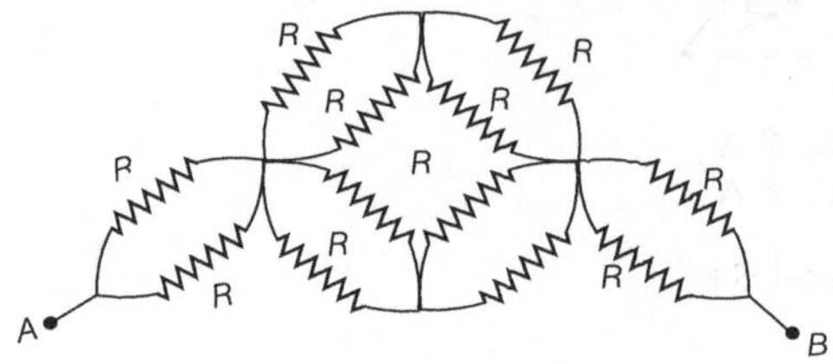

a. $R_{AB} = R$
b. $R_{AB} = \frac{R}{2}$
c. $R_{AB} = \frac{3R}{2}$
d. $R_{AB} = \frac{5R}{2}$

11 A network of resistors is connected with a 12 V battery as shown in figure.

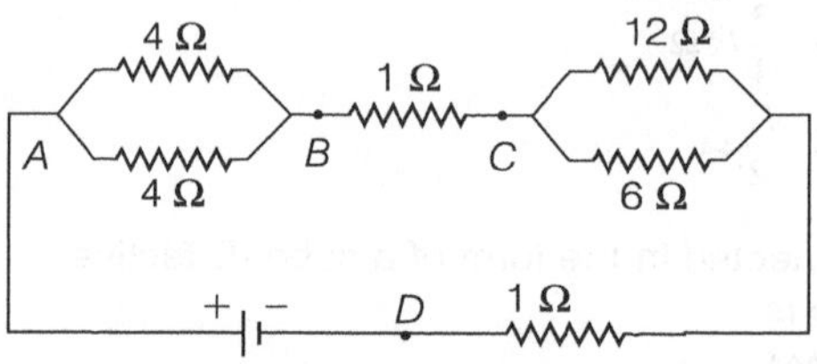

Then, V_{AB}, V_{BC} and V_{CD}, respectively are

a. $V_{AB} = 3\text{ V}, V_{BC} = 1.5\text{ V}, V_{CD} = 7.5\text{ V}$
b. $V_{AB} = 3\text{ V}, V_{BC} = 1\text{ V}, V_{CD} = 8\text{ V}$
c. $V_{AB} = 2\text{ V}, V_{BC} = 3\text{ V}, V_{CD} = 3\text{ V}$
d. $V_{AB} = 2\text{ V}, V_{BC} = 2\text{ V}, V_{CD} = 8\text{ V}$

12 Sixteen resistances each of 5 Ω are connected as shown in the figure.

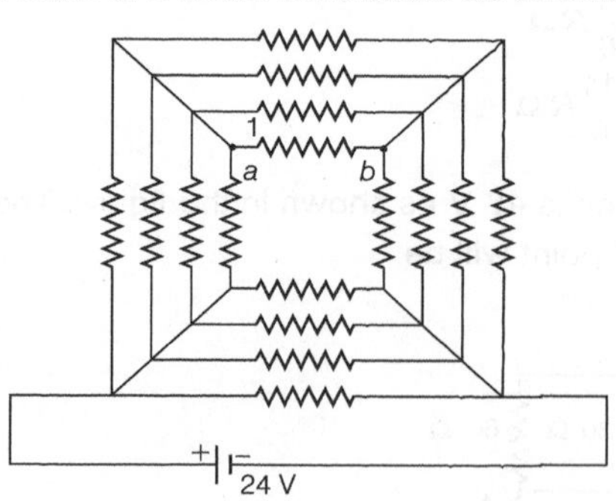

If a battery of 24 V is connected across the network, then current in branch marked ab is

a. $\frac{96}{75}$ A
b. $\frac{48}{75}$ A
c. 5 A
d. 13 A

Working Space

13 In the circuit shown, the ammeter reading is

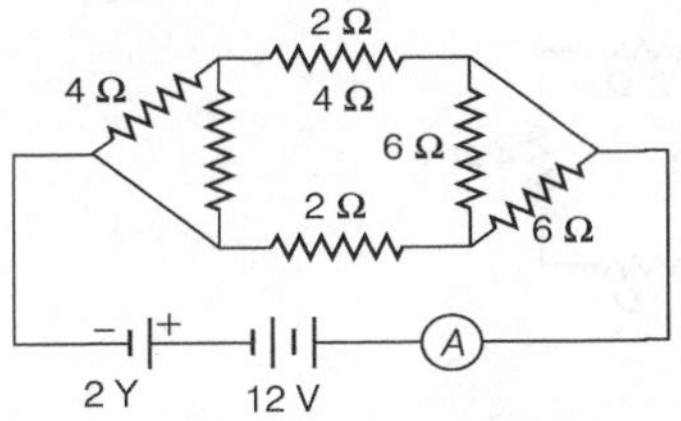

a. $\frac{2}{5}$ A b. $\frac{5}{3}$ A

c. $\frac{3}{7}$ A d. $\frac{4}{5}$ A

14 In the given circuit, it is observed that the current I is independent of the value of the resistance R_6. Then, the resistance values must satisfy **[IIT JEE 2001]**

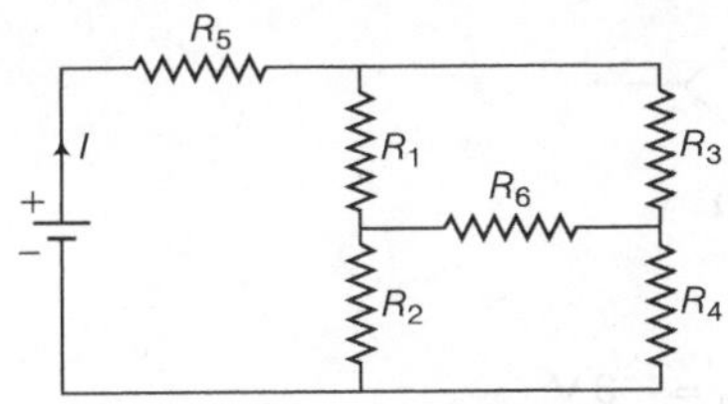

a. $R_1 R_2 R_5 = R_3 R_4 R_6$ b. $\frac{1}{R_5} + \frac{1}{R_6} = \frac{1}{R_1 + R_2} + \frac{1}{R_3 + R_4}$

c. $R_1 R_4 = R_2 R_3$ d. $R_1 R_3 = R_2 R_4$

15 The effective resistance between points P and Q of the electrical circuit in the figure is **[IIT JEE 2002]**

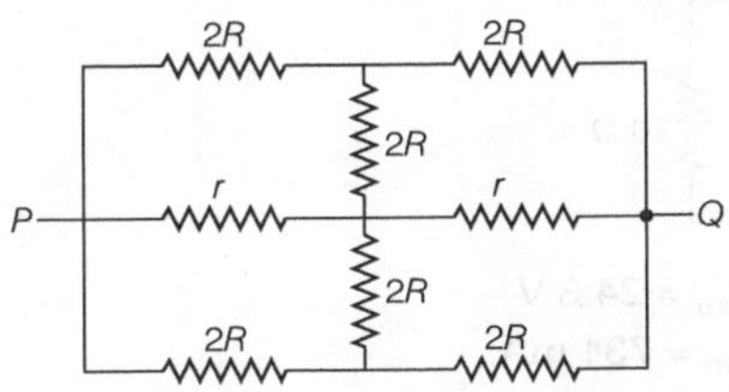

a. $\frac{2Rr}{R+r}$ b. $\frac{8R(R+r)}{3R+r}$

c. $2r + 4R$ d. $\frac{5R}{2} + 2r$

One or More than One Options Correct Type

Direction (Q. Nos. 16-19) *This section contains 4 multiple choice questions. Each question has four choices (a), (b), (c) and (d), out of which ONE or MORE THAN ONE are correct.*

16 In the circuit shown some potential difference is applied between A and B. The equivalent resistance between A and B is R. Then

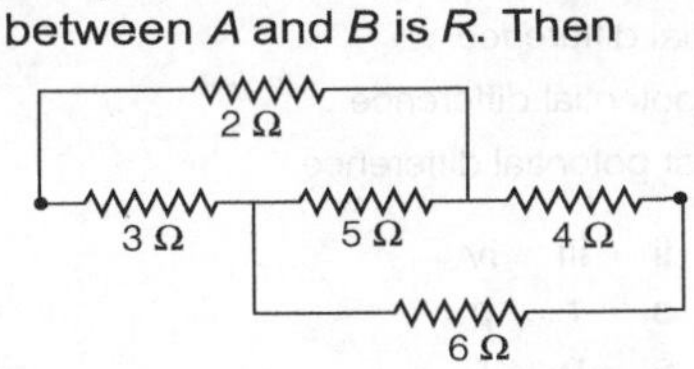

a. no current flows through 5 Ω resistor b. $R = 15\ \Omega$

c. $R = 12.5\ \Omega$ d. $R = \frac{18}{5}\ \Omega$

17 In the circuit shown in the figure, **[IIT JEE 1998]**

Working Space

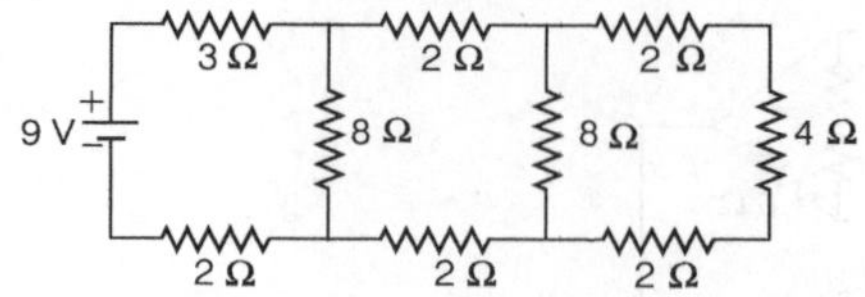

The current through

a. the 3 Ω resistor is 0.5 A
b. the 3 Ω resistor is 1 A
c. the 4 Ω resistor is 0.5 A
d. the 4 Ω resistor is 0.25 A

18 For the network shown in the figure, mark the correct option(s).

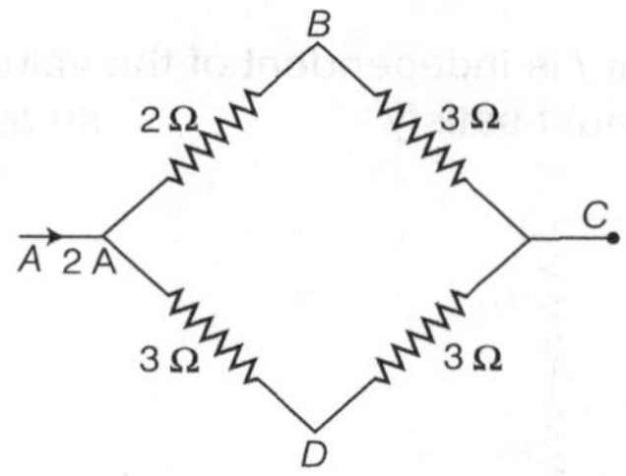

a. $V_{AB} = 2\text{ V}$
b. $V_{DA} = -3\text{ V}$
c. $V_{BD} = 1\text{V}$
d. $V_{CD} = 2\text{ V}$

19 For the circuit shown in the figure, mark the correct option(s).

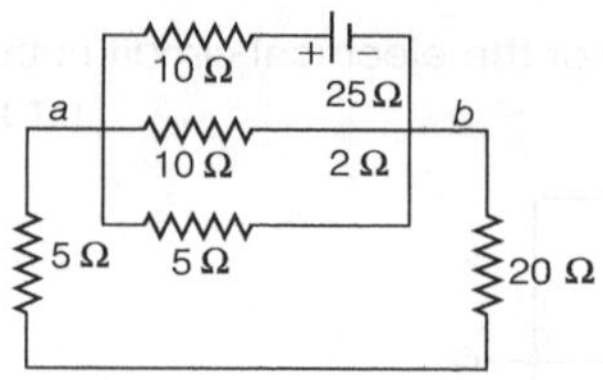

a. $V_{ab} = 5.68\text{ V}$
b. $V_{ab} = 24.3\text{ V}$
c. $I_{ab} = 1.93\text{ A}$
d. $I_{ab} = 731\text{ mA}$

Matching List Type

Direction *(Q. No. 20) Choices for the correct combination of elements from Column I and Column II are given as options (a), (b), (c) and (d), out of which one is correct.*

20 Four bulbs of powers 25 W, 40 W, 60 W and 100 W are connected in series. Match Column I of watts of bulbs with potential difference across them in Column II.

	Column I		Column II
i.	25 W	p.	Highest potential difference
ii.	40 W	q.	Lowest potential difference
iii.	60 W	r.	Third highest potential difference
iv.	100 W	s.	Second highest potential difference

	i	ii	iii	iv
a.	p	q	r	s
b.	q	s	r	p
c.	q	r	s	p
d.	q	s	p	r

Working Space

Comprehension Type

Direction (Q. Nos. 21-26) This section contains 2 paragraphs, describing theory, experiments, data, etc. Six questions related to the paragraphs have been given. Each question has only one correct answer among the four given options (a), (b), (c) and (d).

Passage I

A circuit having few resistances and the capacitors is shown as in the figure.

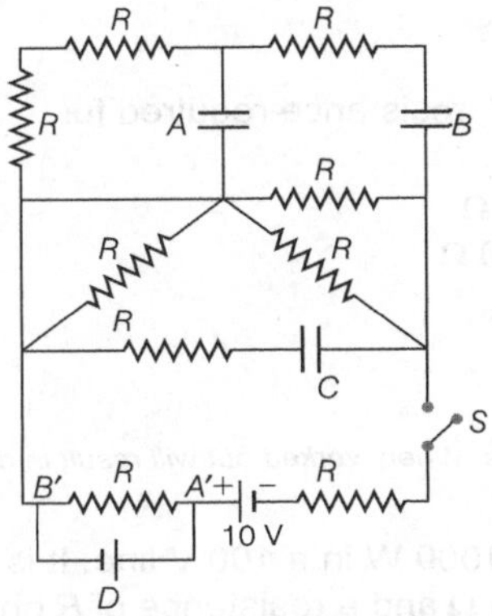

A, *B*, *C* and *D* are four capacitors each of 2 μF. Resistors are of 1 Ω each and battery is of 10 V.

21 Current supplied by battery just after closing of switch *S* is

a. 0 b. 8 A
c. 2 A d. 17 A

22 Current supplied by battery after a long time is

a. zero b. 8 A
c. 4 A d. 2 A

23 Heat dissipated in the circuit after a long time is

a. 56 μJ b. 40 μJ
c. 10 μJ d. 13 μJ

Passage II

A galvanometer is a simple deflection instrument which can be used to measure current or voltage. In both cases, a resistor *R* must be connected to the galvanometer.

To turn the galvanometer into an ammeter, the resistor *R* is connected in parallel.

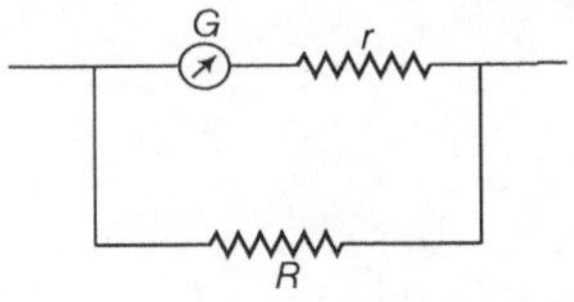

To convert galvanometer into a voltmeter, the resistor *R* is connected in series with the galvanometer. The resistance of galvanometer is *r*.

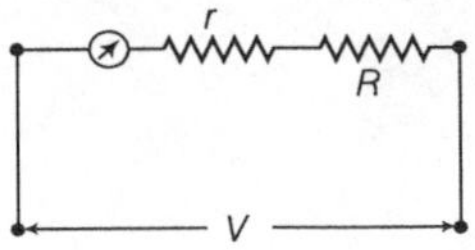

Let current required in galvanometer is 10 mA and galvanometer internal resistance is 100 Ω. Let V_r is the voltage across *r* and V_R is voltage across *R*.

Working Space

24 Which are true for an ammeter and a voltmeter respectively.

a. $V_r > V_R, V_r < V_R$
b. $V_r > V_R, V_r = V_R$
c. $V_r < V_R, V_r > V_R$
d. $V_r = V_R, V_r < V_R$

25 To convert galvanometer into an ammeter capable of reading currents up to 10.01 A, resistance required is

a. $0.1\,\Omega$
b. $1\,\Omega$
c. $10\,\Omega$
d. $100\,\Omega$

26 To make a voltmeter which can read up to 100 V, resistance required for galvanometer is

a. $900\,\Omega$
b. $1000\,\Omega$
c. $9900\,\Omega$
d. $10000\,\Omega$

One Integer Value Correct Type

Direction *(Q. Nos. 27 and 28) This section contains 2 questions. When worked out will result in an integer from 0 to 9 (both inclusive).*

27 A heater is designed to operate with a power of 1000 W in a 100 V line. It is connected in combination with a resistance of $10\,\Omega$ and a resistance of R ohms to a 100 V mains or shown in the figure.

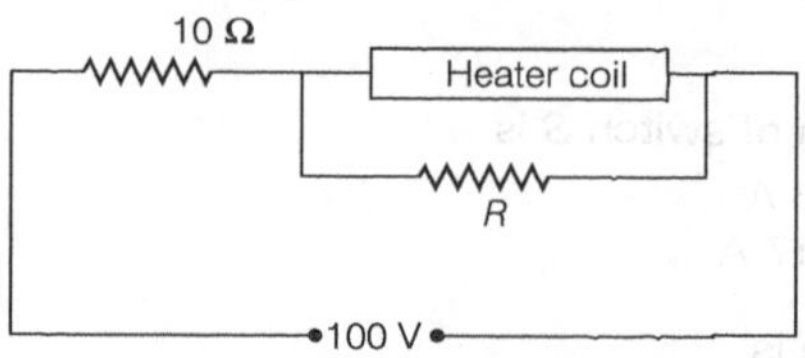

If heater operates with a power of 62.5 W, then find the value of R.

28 In the given circuit, resistance of galvanometer is $25\,\Omega$.

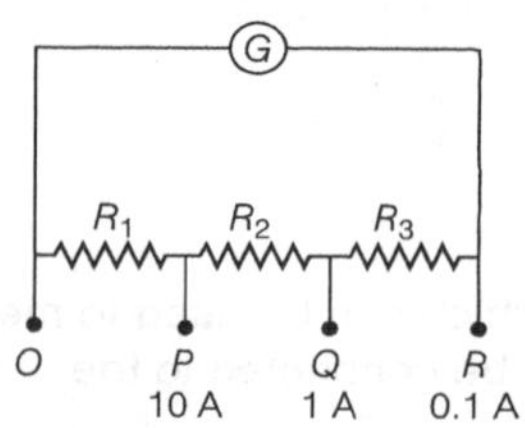

Range of ammeter OP is 0 – 10 A.
Range of ammeter OQ is 0 – 1 A.
Range of ammeter OR is 0 – 0.1 A.
Calculate the value of $2R_3$.

DPP-5 Cells and Combination of Cells

Subjective Questions

Working Space

Direction (Q. Nos. 1-3) *These questions are subjective in nature, need to be solved completely on notebook.*

1 In the given circuit, cells have negligible internal resistance.

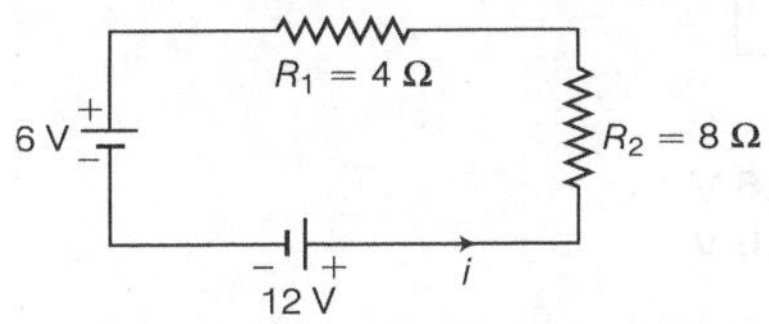

Find the following

i. Current in the circuit

ii. Power dissipated in each resistor

iii. Power of each cell

2 A car battery with 12 V emf and 0.04 Ω internal resistance is being charged with a current of 50 A.

i. What is the potential difference across its terminals?

ii. At what rate energy is dissipated as heat in the battery?

iii. At what rate the electrical energy is converted into chemical energy?

3 A battery of 12 cells is required to send the largest possible current through a resistance of 10 Ω.

Find type of grouping which will give the largest current and estimate its value. Resistance of each cell is 6.3 Ω and emf is 1.4 V.

Only One Option Correct Type

Direction (Q. Nos. 4-14) *This section contains 11 multiple choice questions. Each question has four choices (a), (b), (c) and (d), out of which ONLY ONE option is correct.*

4 n identical cells are connected initially in series and then in parallel. What is the ratio of maximum possible powers dissipated in both cases?

a. n b. n^2

c. $1/n$ d. 1

5 Two identical cells of emf 1.5 V each connected in parallel provide a supply to an external circuit consisting of two resistors of 7 Ω each joined in parallel. A very high resistance voltmeter reads the terminal voltage of the cells to be 1.4 V. What is the internal resistance of each cell?

a. 0.2 Ω b. 0.3 Ω

c. 0.4 Ω d. 0.5 Ω

Working Space

6 Two cells of the same emf E and different internal resistances r_1 and r_2 are connected in series to an external resistance R. The value of R for which the potential difference across the first cell is zero is given by

a. $R = \frac{r_1}{r_2}$ b. $R = r_1 + r_2$

c. $R = r_1 - r_2$ d. $R = r_1 r_2$

7 The circuit given has two cells in opposition to each other. One cell has an emf of 6 V and internal resistance of 2 Ω and the other cell has an emf of 4 V and internal resistance of 8 Ω. The potential difference across the terminals X and Y is

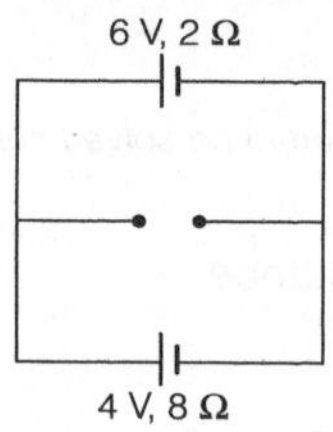

a. 5.4 V b. 5.6 V

c. 5.8 V d. 6.0 V

8 Eight cells marked 1 to 8 each of emf 5 V and internal resistance 0.2 Ω are connected as shown in the figure. What is the reading of the ideal voltmeter V?

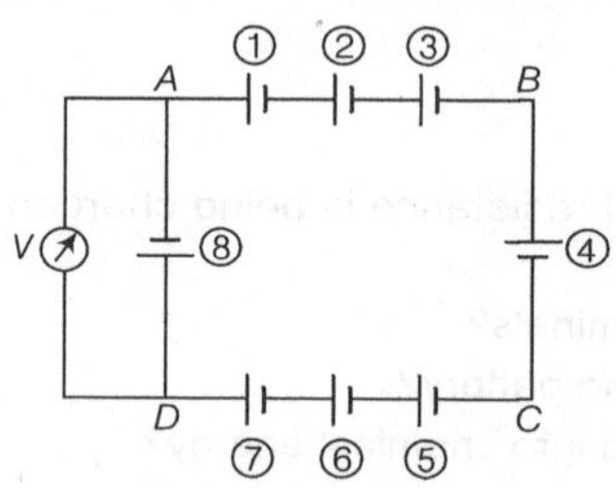

a. 40 V b. 20 V

c. 5 V d. Zero

9 The internal resistances of cells in the circuit shown in the figure are negligible. The current in the circuit is

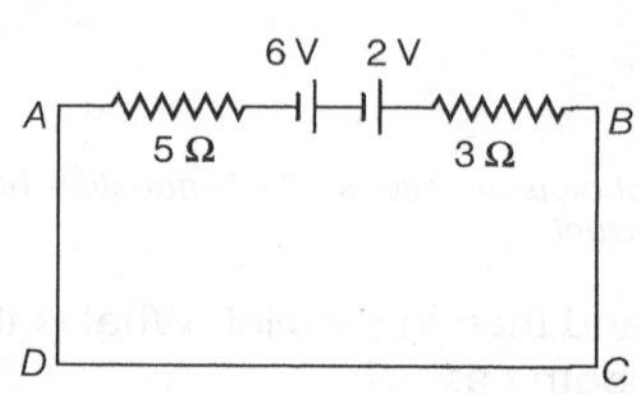

a. 0.5 A flowing from A to D b. 0.5 A flowing from B to C

c. 1.0 A flowing from A to B d. 1.0 A flowing from B to A

10 Twelve cells each having the same emf are connected in series and are kept in a closed box. Some of the cells are wrongly connected. This battery of cells is connected in series with an ammeter and two cells identical with the others. The current is 3 A when the cells and the battery aid each other and is 2 A when the cells and the battery oppose each other. How many cells in the battery are wrongly connected?

a. 1 b. 6

c. 4 d. 8

Working Space

11 A wire of length L and 3 identical cells of negligible internal resistance are connected in series. Due to the current, the temperature of the wire is raised by ΔT in a time t. A number N of similar cells is now connected in series with a wire of the same material and cross-section but of length $2L$. The temperature of the wire is raised by the same amount ΔT in the same time t. The value of N is

a. 4 b. 6
c. 8 d. 9

12 In the given circuit, if potential difference between points A and B is 4 V. Then, resistance X is

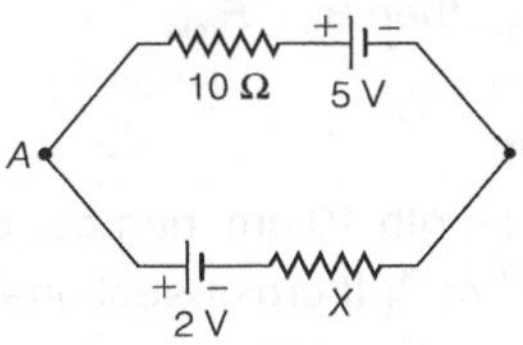

a. 5 Ω b. 10 Ω
c. 15 Ω d. 20 Ω

13 In the given circuit,

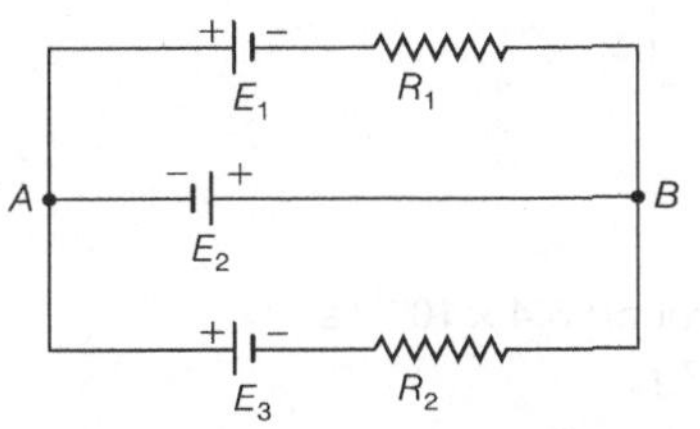

$$E_1 = E_2 = E_3 = 2\text{ V} \Rightarrow R_1 = R_2 = 4\,\Omega$$

The current through branch AB is

a. zero b. 2 A from A to B
c. 2 A from B to A d. 5 A from A to B

14 A storage cell is charged by a 5 A DC current for 18 h. Its strength after charging is

a. 18 Ah b. 5 Ah
c. 90 Ah d. 15 Ah

One or More than One Options Correct Type

Direction (Q. Nos. 15-18) *This section contains 4 multiple choice questions. Each question has four choices (a), (b), (c) and (d), out of which ONE or MORE THAN ONE are correct.*

15 Choose the correct statements from the following.

a. If n identical cells, each of emf E and internal resistance r are connected in series, the emf of combination is nE and the internal resistance of the combination is nr.
b. If n identical cells, each of emf E and internal resistance r are connected in parallel, the emf of the combination is E/n and the internal resistance of the combination is r/n
c. Cells should be connected in series, if the external resistance R is greater than internal resistance r
d. Cells should be connected in parallel of R is smaller than r

Working Space

16 24 cells, each of emf 1.5 V and internal resistance is 2 Ω. Then,

a. in order to send the maximum current through a 12 Ω resistor, connect the two rows of 12 cells in series
b. the current in each row is 0.375 A
c. the current in each cell is 0.75 A
d. potential difference across 12 Ω resistor is 4.5 V

17 There are n resistors $R_1, R_2, R_3, \cdots, R_n$ with $R_{max} = \max\{R_1, R_2, R_3 \cdots, R_n\}$ and $R_{min} = \min\{R_1, R_2, R_3, \cdots, R_n\}$. Then,

a. when they are connected in parallel, then $R_p < R_{min}$
b. when they are connected in series, then $R_s > R_{max}$, then $R_p = R_{min}$
c. when they are connected in parallel, then $R_p = R_{min}$
d. when they are connected in series, then $R_s = R_{max}$

18 In the given circuit, AB is a wire of resistance 6 Ω, length 10 cm, number of electrons per unit volume in material of wire is $10^{29}\ m^{-3}$. If cross-sectional area of wire is $1\ mm^2$, then which of the following options are correct?

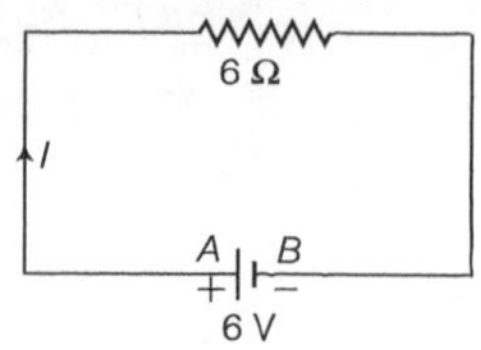

a. Energy absorbed by electrons is 2×10^{-17} J
b. Ohmic loss in wire is 16 J/s
c. All kinetic energy of electrons will be lost in around 3.4×10^{-18} s
d. Energy lost by an electron per sec is 2×10^{-17} J

One Integer Value Correct Type

Direction *(Q. Nos. 19 and 20) This section contains 2 questions. When worked out will result in an integer from 0 to 9 (both inclusive).*

19 A set of n equal resistors of resistance R each are connected in series to a battery of emf E and internal resistance R. A current I is observed to flow in the circuit. Then, n-resistors are connected in parallel to the same battery. It is observed that current is increased 10 times. What is the value of n?

a. 5
b. 8
c. 10
d. 7

20 Two cells of same emf E but internal resistances $r_1 = 3\ \Omega$ and $r_2 = 1\ \Omega$ are connected in series to an external resistance R.

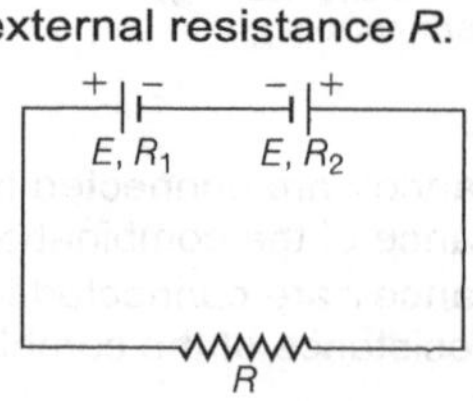

The value of R for which the potential difference across the terminals of the first cell becomes zero, is

a. 4 Ω
b. 3 Ω
c. 2 Ω
d. $\frac{1}{3}$ Ω

Working Space

Comprehension Type

***Direction** (Q. Nos. 22-25) This section contains a paragraph, describing theory, experiments, data, etc. Four questions related to the paragraph have been given. Each question has only one correct answer among the four given options (a), (b), (c) and (d).*

Passage

In the circuit shown in figure,

$E_1 = 3\,V, E_2 = 2\,V, E_3 = 1$ V and $R = r_1 = r_2 = r_3 = 1\ \Omega$.

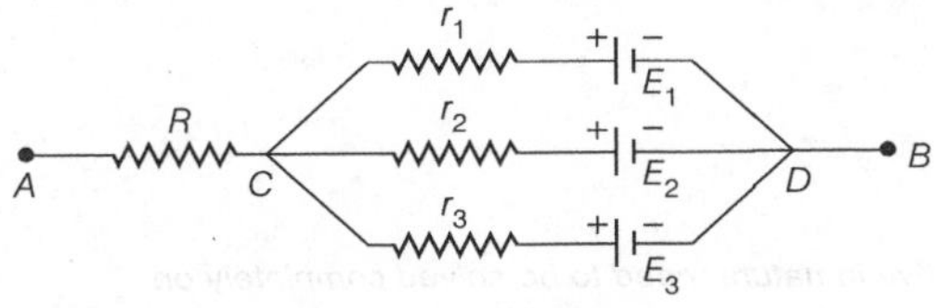

21 The current through branch containing cell E_1, is

a. 1 A b. zero
c. −1 A d. 2 A

22 The current through branch containing cell E_2, is

a. 1 A b. zero
c. −1 A d. 2 A

23 The current through branch containing cell E_3, is

a. 1 A b. zero
c. −1 A d. 2 A

24 If r_2 is short-circuited and points A and B are connected with each other, then current through middle branch is

a. 1 A b. zero
c. −1 A d. 2 A

Matching List Type

***Direction** (Q. No. 25) Choices for the correct combination of elements from Column I and Column II are given as options (a), (b), (c) and (d), out of which one is correct.*

25 Consider two identical cells each of emf E and internal resistance r connected to a load resistance R. Match the following columns.

	Column I		Column II
i.	Maximum power transferred to load if cells are connected in series	p.	$E^2/4r$
ii.	Maximum power transferred to load if cells are connected in parallel	q.	$E^2/2r$
iii.	Series combination of cells	r.	$E, r/2$
iv.	Parallel combination of cells	s.	$2E, 2r$

Codes

	i	ii	iii	iv		i	ii	iii	iv
a.	q	q	r	r	b.	p	q	s	r
c.	p	q	r	s	d.	s	s	p	q

DPP-6 Network Analysis and Kirchhoff's Rules

Subjective Questions

Working Space

Direction (Q. Nos. 1-5) *These questions are subjective in nature, need to be solved completely on notebook.*

1 In the given circuit shown below,

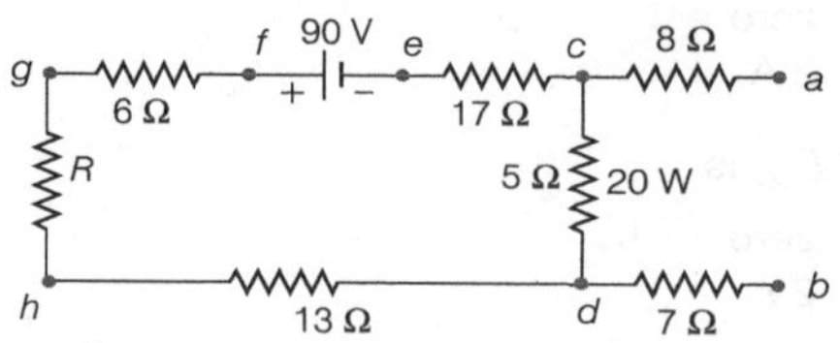

Average power dissipated in 5 Ω resistor is 20 W. Determine the readings of an ideal voltmeter when it is connected between

i. terminals *a* and *b*

ii. terminals *c* and *g*

2 Using Kirchhoff's junction rule and Ohm's law. Find the magnitude and polarity of voltage *V* in the given circuit.

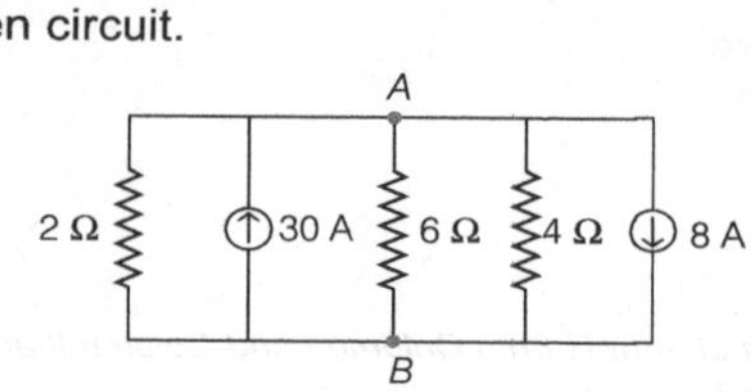

3 Explain sign convention to be used in case of Kirchhoff's rule. Is it necessary to take potential drop across a resistor always with a negative sign?

4 For the circuit shown in the figure,

$$E_1 = 3\text{ V}$$
$$E_2 = 2\text{ V}$$
$$E_3 = 1\text{V}$$

and $$R = r_1 = r_2 = r_3 = 1\Omega.$$

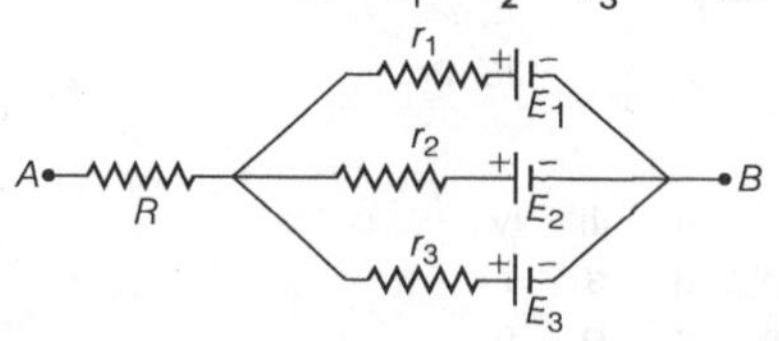

i. Find the potential difference between the points *A* and *B* and the current through each branch.

ii. If r_2 is short-circuited and the point *A* is connected to point *B*, find the currents through E_1, E_2, E_3 and the resistor *R*.

Working Space

5 In the circuit shown in the figure, the battery is an ideal with emf V. The capacitor is initially uncharged. The switch S is closed at time $t = 0$.

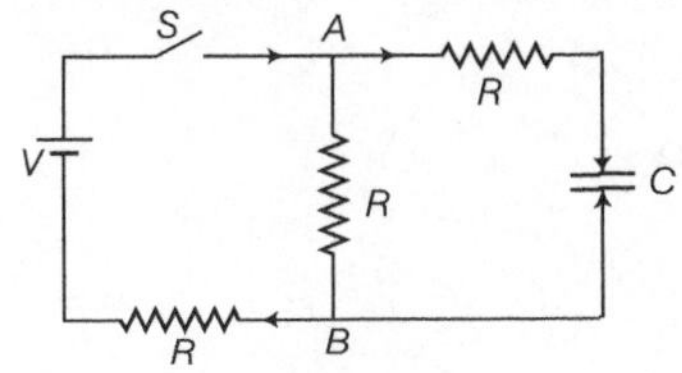

i. Find the charge Q on the capacitor at time t.

ii. Find the current in AB at time t. What is its limiting value as $t \to \infty$?

Only One Option Correct Type

Direction *(Q. Nos. 6-18) This section contains 13 multiple choice questions. Each question has four choices (a), (b), (c) and (d), out of which ONLY ONE option is correct.*

6 The figure shown as below is a part of a circuit.

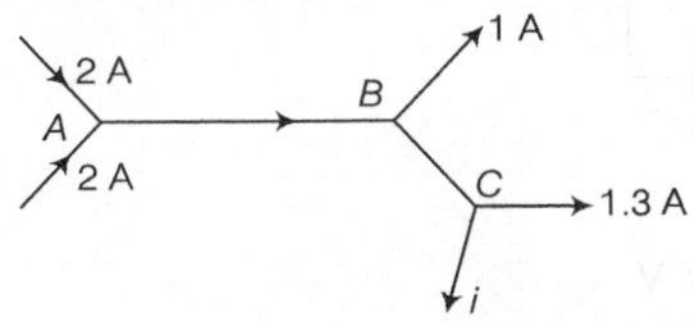

Then, curent i is

a. 1.7 A b. 3.7 A

c. 1.3 A d. 1 A

7 Current in the arm CD of the circuit is

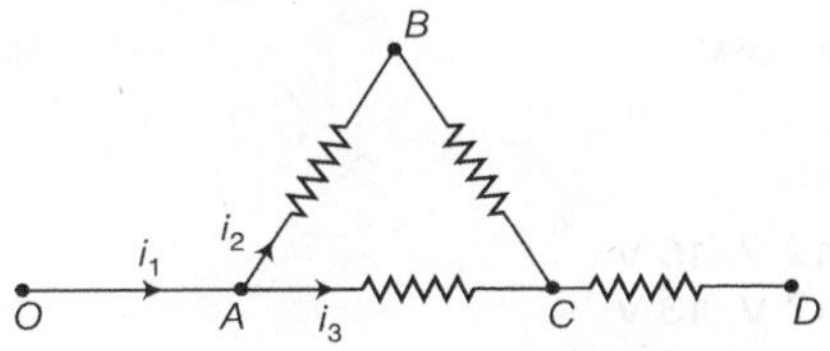

a. $i_1 + i_2$ b. $i_2 + i_3$

c. $i_1 + i_3$ d. $i_1 - i_2 + i_3$

8 For the circuit shown as below,

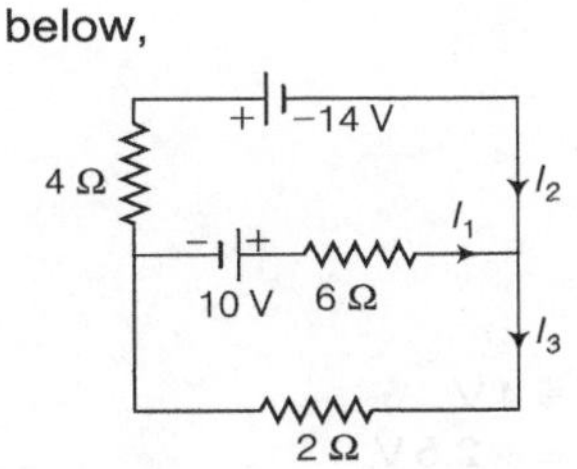

Let I_1, I_2 and I_3 are the currents marked in respective branches, then

a. $I_1 = 2\text{A}, I_2 = -3\text{A}, I_3 = -1\text{A}$

b. $I_1 = -2\text{A}, I_2 = -3\text{A}, I_3 = -1\text{A}$

c. $I_1 = 2\text{A}, I_2 = -3\text{A}, I_3 = -1\text{A}$

d. $I_1 = -2\text{A}, I_2 = 3\text{A}, I_3 = 1\text{A}$

Working Space

9 For the circuit given below, the charge on the capacitor is

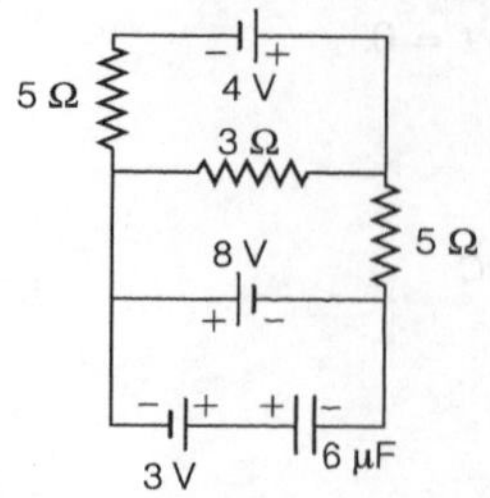

a. 45 μC b. 75 μC c. 66 μC d. 78 μC

10 Two cells with unequal emfs of 2V and 3V are connected as shown in the figure. If each cell has an internal resistance $r = 0.45\ \Omega$ and external resistance is of 4 Ω. Then, potential drop across resistor r is

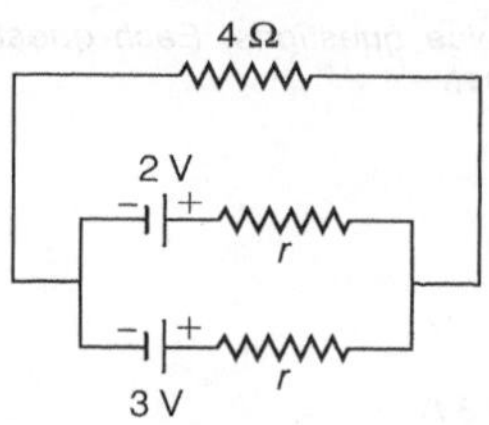

a. 5.42 V
b. 2.37 V
c. 3.28 V
d. 1.5 V

11 For the given circuit, terminal potential differences of cells are around

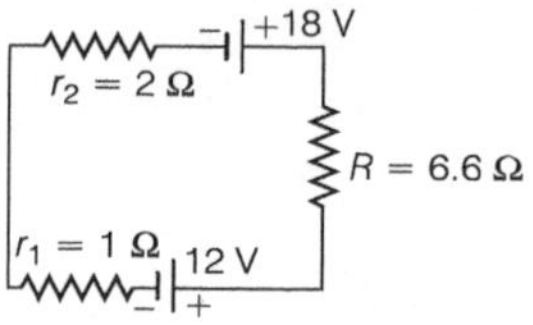

a. 15 V, 10 V
b. 12 V, 15 V
c. 6 V, 12 V
d. 17 V, 13 V

12 For the given circuit,

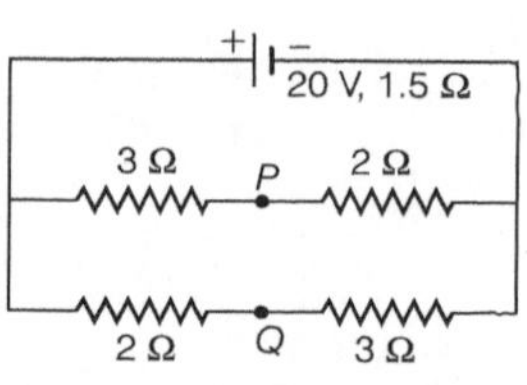

If internal resistance of cell is 1.5 Ω, then

a. $V_P - V_Q = 0$
b. $V_P - V_Q = 4$ V
c. $V_P - V_Q = -4$ V
d. $V_P - V_Q = -2.5$ V

13 A battery of emf 2 V and internal resistance 0.1 Ω and being charged with a current of 0.5 A. Then, $(V_A - V_B)$ is

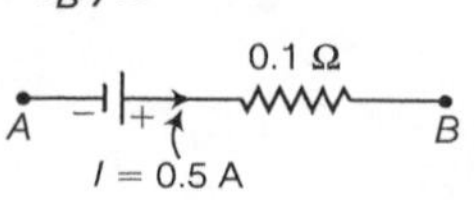

a. 1.95 V
b. −1.95 V
c. 2.05 V
d. −2.05 V

Working Space

14 In the circuit shown in figure, if the ammeter reads 2 A. Then, I_1 and I_2 are

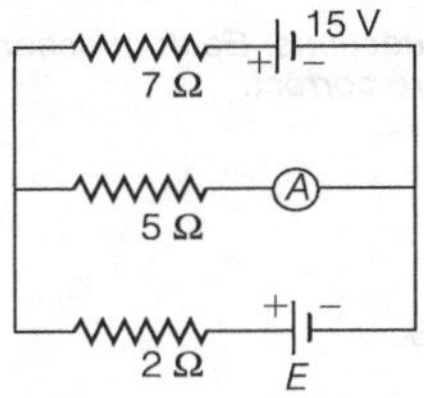

a. $I_1 = 1.8$ A, $I_2 = 0.2$ A
b. $I_1 = 0.8$ A, $I_2 = 1.2$ A
c. $I_1 = 0.75$ A, $I_2 = 1.25$ A
d. $I_1 = 0.714$ A, $I_2 = 1.29$ A

15 For the given network, mark the correct statement.

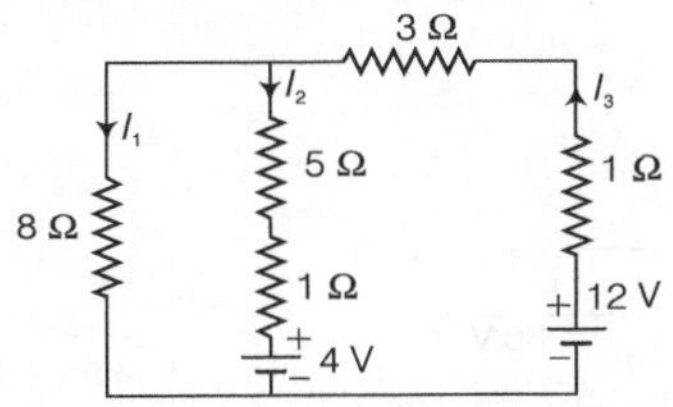

a. 8 Ω resistor is 846 mA
b. 1Ω resistor is 846 mA
c. 1Ω resistor is 400 mA
d. 3 Ω resistor is 2 A

16 For the given circuit, potential difference between points *c* and *f* is

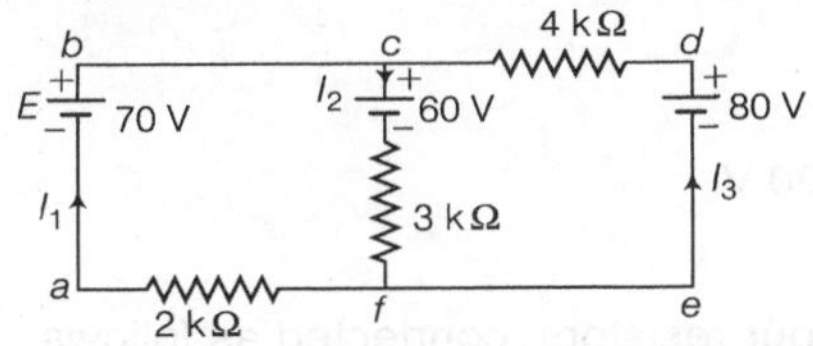

a. $V_c - V_f = 69.2$ V
b. $V_c - V_f = -69.2$ V
c. $V_c - V_f = 48.3$ V
d. $V_c - V_f = -48.3$ V

17 In the circuit shown the current through 2 Ω resistance is **[IIT JEE 2005]**

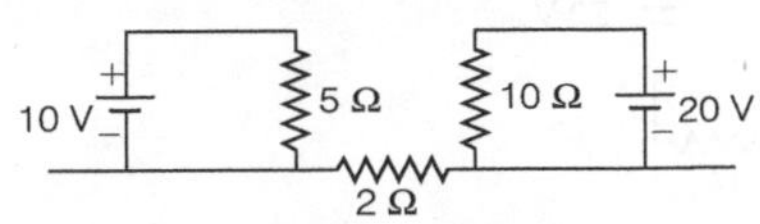

a. 5 A
b. 2 A
c. 0
d. 4 A

18 For the circuit shown, energy stored by the capacitor is

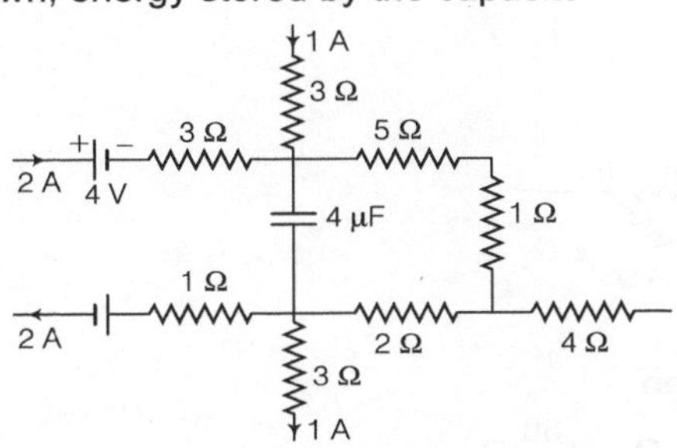

a. around 0.3 mJ
b. around 0.5 mJ
c. around 0.8 mJ
d. around 0.1 mJ

One or More than One Options Correct Type

Working Space

Direction (Q. Nos. 19 -22) *This section contains 4 multiple choice questions. Each question has four choices (a), (b), (c) and (d), out of which ONE or MORE THAN ONE are correct.*

19 Taking $R = 1\,k\Omega$ and $E = 250$ V, then

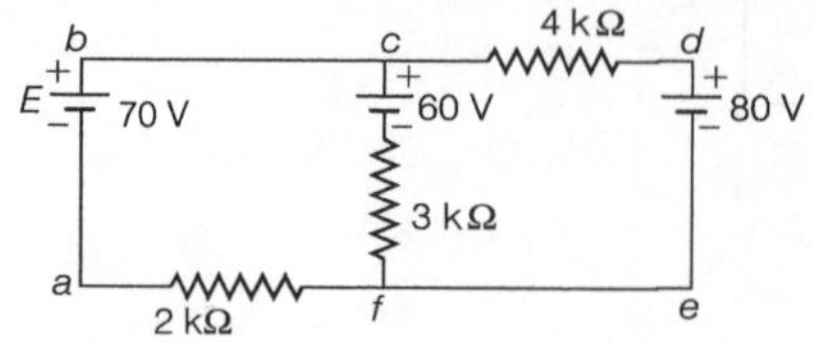

For the above circuit,

a. current in branch ac = 60 mA
b. current in branch ab = 20 mA
c. current in branch ae = 50 mA
d. current in branch ce = 20 mA

20 Consider the given DC circuit,

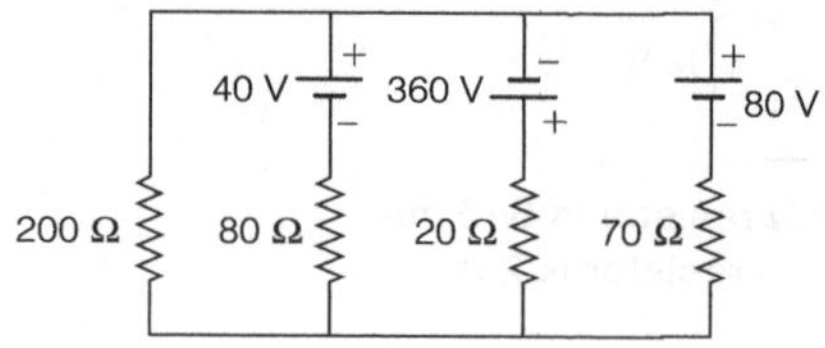

Then,

a. current in 70 Ω resistance is 4 A
b. current in 80 Ω resistance is 5 A
c. potential drop across 200 Ω resistance is 200 V
d. current through 200 Ω resistor is 0.5 A

21 For the circuit containing two batteries and four resistors, connected as follows.

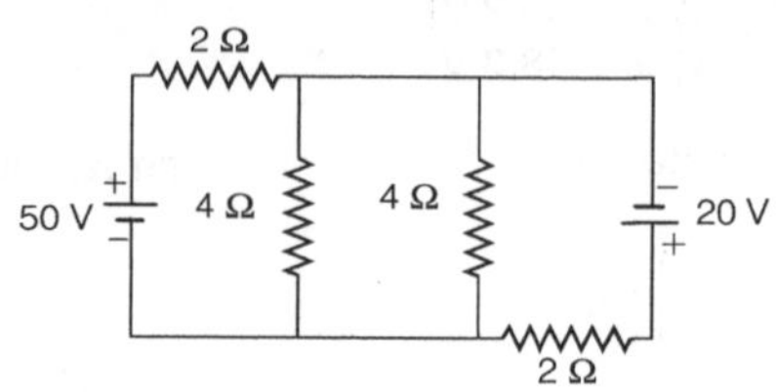

Then,

a. power of 2 Ω resistor is 800 W
b. power of 4 Ω resistor is 25 W
c. power of 2 Ω resistor is 450 W
d. power of 4 Ω resistor is 50 W

22 For the network shown in figure,

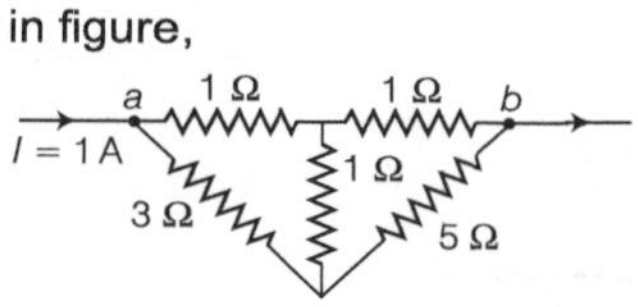

If 1 A current enter the circuit through a, then

a. $R_{ab} = \frac{27}{17}\,\Omega$
b. $R_{ab} = \frac{30}{17}\,\Omega$
c. $V_{ab} = \frac{27}{17}$ V
d. $V_{ab} = \frac{30}{17}$ V

Working Space

Matching List Type

Direction (Q. No. 23) Choices for the correct combination of elements from Column I and Column II are given as options (a), (b), (c) and (d), out of which one is correct.

23 Twelve resistors, each of resistance R, are connected as the edges of a cube.

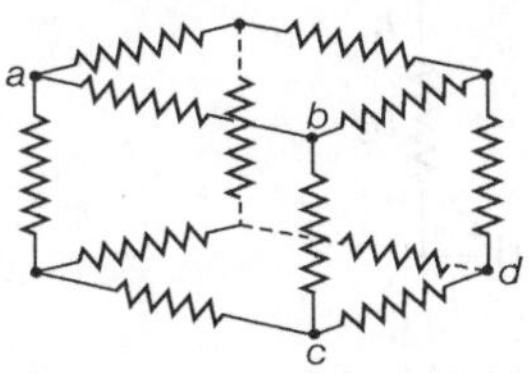

Now, match Column I with the Column II

	Column I		Column II
i.	R_{ab}, resistance between ends of a side	p.	$\frac{5}{6}R$
ii.	R_{ad}, resistance across ends of a diagonal	q.	$\frac{7}{12}R$
iii.	R_{ac}, resistance across a face diagonal	r.	$\frac{3R}{4}$

Codes

	i	ii	iii		i	ii	iii
a.	p	q	r	b.	p	r	q
c.	r	q	p	d.	q	p	r

Comprehension Type

Direction (Q. Nos. 24-27) This section contains a paragraph, describing theory, experiments, data, etc. Four questions related to the paragraph have been given. Each question has only one correct answer among the four given options (a), (b), (c) and (d).

Passage

Kirchhoff's current law (KCL) can be extended in the form of 'NODAL' analysis with the help of 'NODAL' analysis, unknown voltages at different points of a circuit can be found. For solving a circuit with nodal analysis, one can proceed or follows.

Consider an electrical network for example,

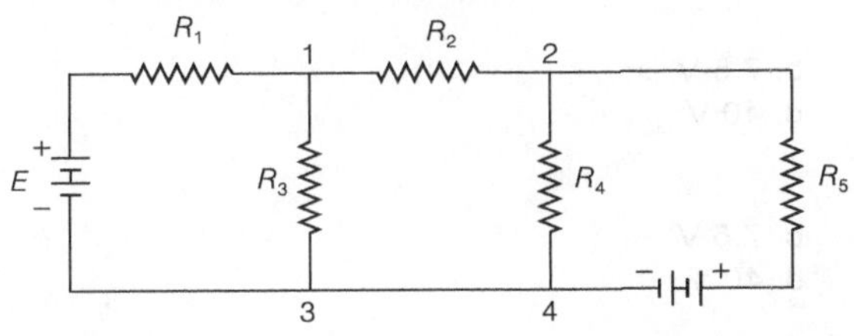

Step 1 Find number of nodes that require sometimes reconstruction of circuit.

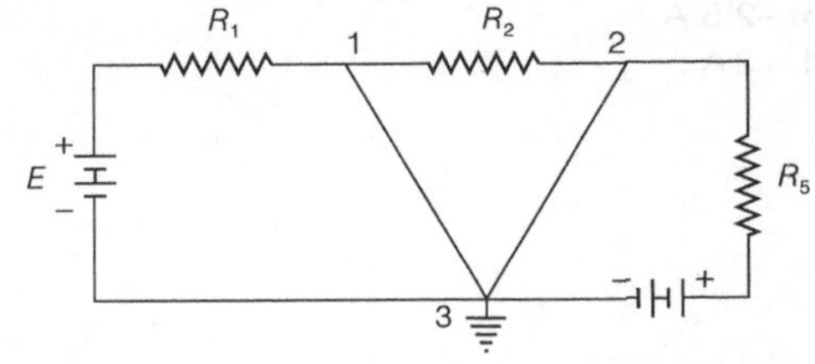

Step 2 Mark any node as reference mode, it is called ground node. Take potential of reference (ground node) as 0. In our example, node 3 is a ground node.

Working Space

Step 3 Mark currents in the all branches and name them as I_1, I_2, I_3, ..., etc. The direction of current is arbitrarily chosen.

Step 4 Apply KCL on each node.

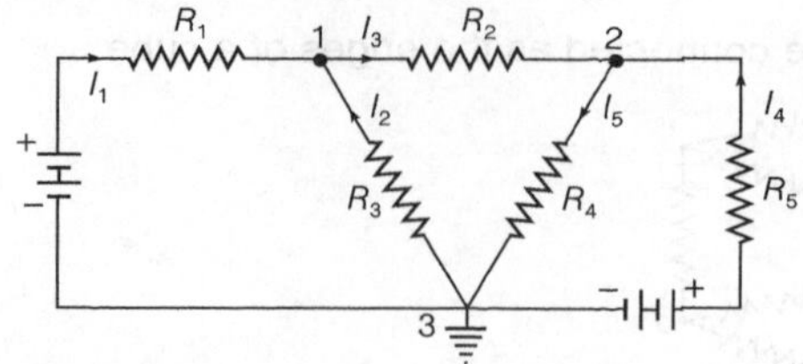

Step 5 Find the value of their branch currents in term of node potentials and branch resistances.

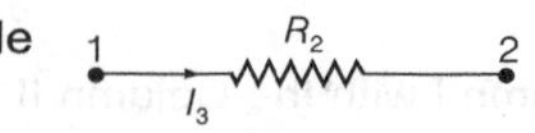

As, $$I_3 = \frac{V_1 - V_2}{R}$$

Ffor the starting point at reference node,

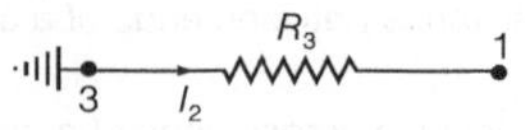

$$I_2 = \frac{V_3 - V_1}{R_3} = \frac{0 - V_1}{R_3} = \frac{-V_1}{R_3}$$

Step 6 Write each branch current and solve equations to get the potential or current which is not known.

Now, consider the circuit shown as below:

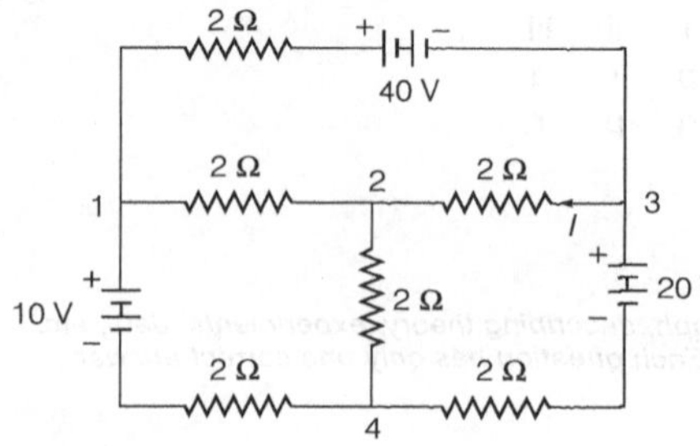

Take point 4 as a ground node.

24 Potential of node 1 is

a. 20 V
b. 7.5 V
c. 2.5 V
d. 40 V

25 Potential of node 2 is

a. 20 V
b. 7.5 V
c. 2.5 V
d. 40 V

26 Potential of node 3 is

a. 20 V
b. 7.5 V
c. 2.5 V
d. 40 V

27 Current I in the 2 Ω resistor is

a. 2.5 A
b. –2.5 A
c. 2 A
d. –2 A

DPP-7 Electrical Energy

Subjective Questions

Working Space

Direction *(Q. Nos. 1-4) These questions are subjective in nature, need to be solved completely on notebook.*

1 An electric motor runs for 5 h a day at 220 V. The supply wiring consists of copper wire of 1 mm radius and length 10 m. Power consumption per day is 10 commercial units. What fraction of it goes in the joule heating in wires? What change occurs when aluminium wire of same dimensions is used?
[For copper, $\rho = 1.7 \times 10^{-8}\ \Omega$-m and for aluminium, $\rho = 2.7 \times 10^{-8}\ \Omega$-m]

2 A resistance of 6 Ω is connected with a cell of 6 V (with negligible internal resistance by using ideal wires)

i. How much energy is absorbed by electrons from initial state of current to the state of drift velocity?

ii. Electrons give up energy at the rate of I^2R in the form of thermal energy. In how much time, electrons are able to give up total energy?

[use : Number of electrons per unit volume = 10^{29}, length of wire = 10 cm, area of cross-section = 1 mm^2]

3 What is rated power and rated voltage of an electrical appliance like a bulb? Why there is no rated current value?

4 Let a load resistance of $R\ \Omega$ is connected across an emf source of E volts and internal resistance of $r\ \Omega$.

Draw graphs of

i. power output *versus* load resistance
ii. efficiency of circuit *versus* load resistance
iii. terminal potential difference *versus* load resistance
iv. current in the circuit *versus* load resistance

Only One Option Correct Type

Direction *(Q. Nos. 5-17) This section contains 13 multiple choice questions. Each question has four choices (a), (b), (c) and (d), out of which ONLY ONE option is correct.*

5 An electric heater of resistance 6 Ω is run for 10 min on a 120 V line. The energy liberated in this period of time is

a. 7.2×10^3 J b. 14.4×10^5 J
c. 43.2×10^4 J d. 28.8×10^4 J

6 A bulb of 220 V and 110 W is operated on 110 V. The power consumed by bulb is

a. 50 W b. 75 W
c. 90 W d. 25 W

7 A heating filament of 500 W, 115 V is being operated at 110 V, percentage drop in output power (with comparison to rated output power) is

a. 10.20%   b. 8.1%
c. 8.6%   d. 7.6%

8 For the given circuit,

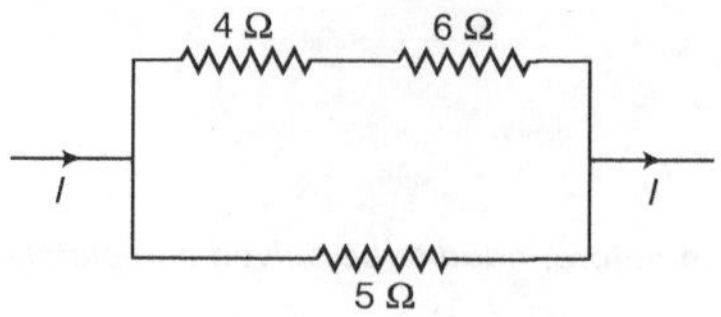

If heat produced in 5 Ω resistor is 10 J/s, then heat produced in 4 Ω resistor is

a. 1 J/s   b. 2 J/s
c. 3 J/s   d. 4 J/s

9 For a bulb of 75 W, resistance and current through the bulb, when it is connected to a 110 V source, are

a. 154 Ω, 0.9 A
b. 153 Ω, 0.7 A
c. 155 Ω, 0.8 A
d. None of the above

10 A 120 V heater has two settings, 850 W and 1250 W. Then,

a. resistance of 850 W is large
b. resistance of 1250 W is large
c. both have same resistances
d. data is insufficient for reaching any conclusion

11 In a large building, there are 15 bulbs of 40 W, 5 bulbs of 100 W, 5 fans of 80 W and 1 heater of 1 kW. The voltage of the electric mains is 220 V. The minimum capacity of the main fuse of the building will be **[JEE Main 2014]**

a. 8 A   b. 10 A
c. 12 A   d. 14 A

12 A power station delivers 750 kW of power at 12000 V, to a factory through a cable of resistance 3 Ω. If electricity is delivered at 50000 V instead of 12000 V, then power wasted will be

a. 1.1×10^5 W   b. 1.1×10^4 W
c. 1.1×10^6 W   d. 1.1×10^3 W

13 For two incandescent bulbs of rated power P_1 and P_2, if $P_1 > P_2$, then

a. filament of bulb 1 is more thicker than filament of bulb 2
b. filament of bulb 1 is thinner than the filament of bulb 2
c. filament of both bulbs is of same thickness
d. rated power of a bulb is independent of filament thickness

14 Cost of 1 unit of electricity in our city is (say) ₹ 5. Then, cost of using a 40 W bulb, eight hours per day for one month is

a. ₹ 48   b. ₹ 50
c. ₹ 108   d. ₹ 92

15 Power of a heater is 500 W at 800°C. What will be its power at 200°C if $\alpha = 4\times 10^{-4}$ per°C?

a. 484 W   b. 672 W
c. 526 W   d. 611 W

Working Space

16 Water boils in an electric kettle for 15 min. If the length of heating wire is decreased to $\frac{2}{3}$ of its initial value, then same amount of water will boil with same supply voltage in

a. 15 min
b. 12 min
c. 10 min
d. 8 min

17 Mark the correct option for the circuit shown in the figure. **[IIT JEE 2009]**

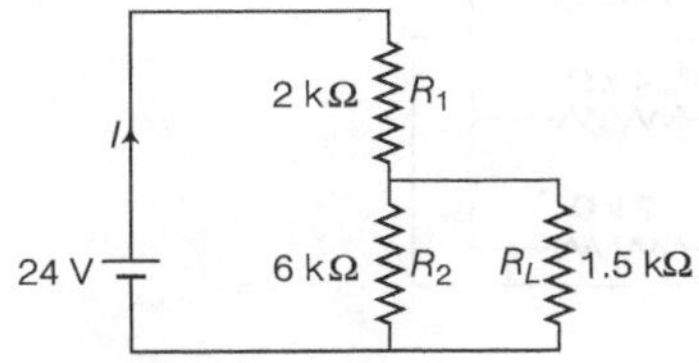

a. The current I through the battery is 7.5 mA
b. The potential difference across R_L is 18 V
c. Ratio of powers dissipated in R_1 and R_2 is 3
d. If R_1 and R_2 are interchanged, magnitude of the power dissipated in R_L will decrease by a factor of 9

One or More than One Options Correct Type

Direction *(Q. Nos. 18-20) This section contains 3 multiple choice questions. Each question has four choices (a), (b), (c) and (d), out of which ONE or MORE THAN ONE are correct.*

18 In an experiment, a student prepares the following circuit. On pressing the key K,

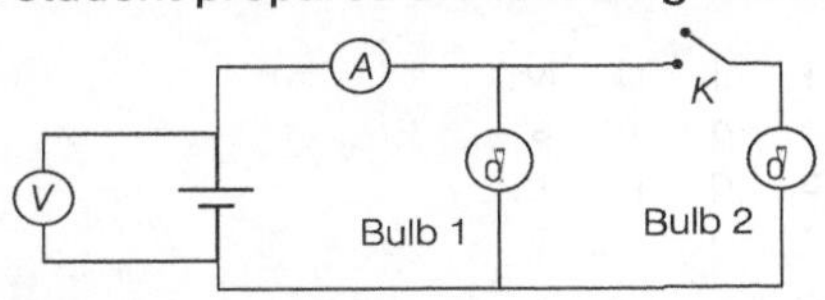

The observations made by the student are (given, bulbs are identical).

a. bulb 1 glows brighter
b. bulb 2 glows brighter
c. reading of ammeter becomes twice
d. reading of voltmeter remains same

19 A flash-light bulb of rated power 2 W and rated voltage of 3 V is to be operated by using a 9 V battery. Then,

a. it can be used directly by connecting the bulb in parallel with battery
b. a resistance of 9 Ω is required in series with the bulb
c. resistance of bulb is 4 .5 Ω
d. a 9 V battery cannot light up a 3 V bulb

20 To produce an equivalent 3.2 kΩ, 3.5 W resistor, which of the given arrangements are appropriate if power in each resistor does not exceed $\frac{1}{2}$ W.

a. 7 resistors of 460 Ω each, rated at $\frac{1}{2}$ W in series
b. 7 resistors of 22.4 kΩ each, rated at $\frac{1}{2}$ W in parallel series
c. 5 resistors of 460 Ω in series and one of 22.4 k Ω in parallel with the combination each, rated at $\frac{1}{2}$ W
d. 3 resistors of 22.4 kΩ in series with two resistors of 460 Ω in parallel

Working Space

Matching List Type

Direction *(Q. No. 21) Choices for the correct combination of elements from Column I and Column II are given as options (a), (b), (c) and (d), out of which one is correct.*

21 The circuit shown is a old version of digital-to-analog converter (DAC).

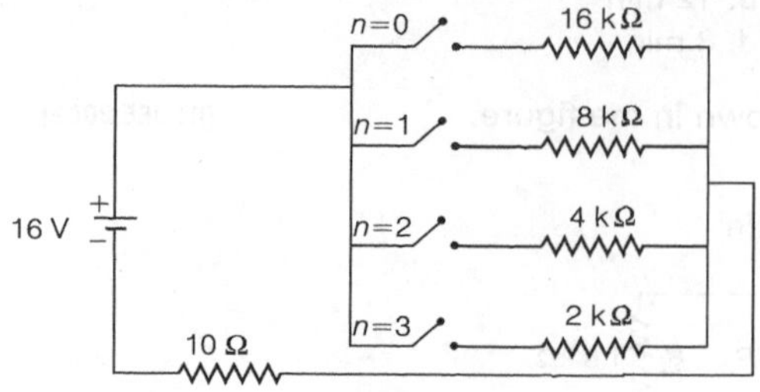

In this circuit to represent each digit (2^n) of a binary number, a 1 has the nth switch closed, whereas 0 has the switch open.

e.g. 0010 is represented by closing a switch $n = 1$ and all other switches are open.

Now, match the Column I of binary digit with Column II of potential drop across resistor of 1 Ω.

	Column I		Column II
i.	0001	p.	10 mV
ii.	0010	q.	4 mV
iii.	0100	r.	2 mV
iv.	1010	s.	1 mV

Codes

	i	ii	iii	iv		i	ii	iii	iv
a.	s	q	r	p	b.	p	q	r	s
c.	s	r	q	p	d.	s	q	p	r

Comprehension Type

Direction *(Q. Nos. 22-24) This section contains a paragraph, describing theory, experiments, data, etc. Three questions related to the paragraphs have been given. Each question has only one correct answer among the four given options (a), (b), (c) and (d).*

Passage

For a series combination of resistors, $R_s = \Sigma R_i$ and current through each resistor is same. In a parallel combination of resistors, $\frac{1}{R_p} = \frac{1}{R_1} + \frac{1}{R_2} + \ldots$ and potential drop across each resistor is same.

22 If we connect n bulbs each with a rated power P in series, then total power consumed by combination at rated current, is

a. $\frac{P}{n}$ b. $\frac{P}{n^2}$ c. nP d. P

23 If we connect n bulbs each with a rated power P in parallel, then total power consumed by combination at rated voltage, is

a. $\frac{P}{n}$ b. $\frac{P}{n^2}$ c. nP d. P

24 The power consumed by n equal resistances in parallel is x times that of power consumed in series if the voltage supply is same. Then, x is equal to

a. n b. n^2 c. $\frac{1}{n}$ d. $\frac{1}{n^2}$

DPP-8 Measuring Instruments

Subjective Questions

Working Space

Direction *(Q. Nos. 1-4) These questions are subjective in nature, need to be solved completely on notebook.*

1 i. For a balanced meter bridge, if cell and galvanometer are interchanged, then its balance point remains unaffected why?

ii. In case of a meter bridge what is end error? How can this be corrected?

2 What is meant by sensitivity of a potentiometer? How can you define potential gradient of a potentiometer wire? How potential gradient and sensitivity are related? How sensitivity of a potentiometer can be increased?

3 Give reasons for

i. area of cross-section of the meter bridge wire be uniform.

ii. potentiometer is preferred over voltmeter for measuring emf of a cell.

iii. emf of driver cell (auxiliary battery) in the potentiometer experiment should be greater than emf of cell to be determined.

iv. wires used in meter bridge or potentiometer are made up of metal alloys.

4 A thin uniform wire *AB* of length 1 m, an unknown resistance *X* and a resistance of 12 Ω are connected by thick conducting strips as shown in the figure. A battery and galvanometer with a sliding jockey connected to it are also available. Connections are made to measure the unknown resistance *X* using the principle of Wheatstone bridge. Answer the following questions. **[IIT JEE 2002]**

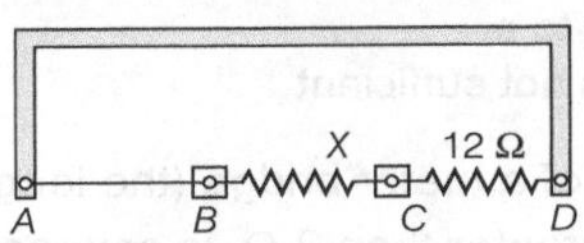

i. Are there positive and negative terminals on the galvanometer?

ii. Copy the figure in your answer book and show the battery and the galvanometer (with jockey) connected at appropriate points.

iii. After appropriate connections are made, it is found that no deflection takes place in the galvanometer when the sliding jockey touches the wire at a distance of 60 cm from *A*. Obtain the value of the resistance *X*.

Only One Option Correct Type

Direction *(Q. Nos. 5-16) This section contains 12 multiple choice questions. Each question has four choices (a), (b), (c) and (d), out of which ONLY ONE option is correct.*

5 In a meter bridge, potential gradient along the resistance wire is 10 V/m. *B* and *C* are two points at distances 30 cm and 60 cm on the wire. Then,

a. $V_B - V_C = 3\text{ V}$ b. $V_B - V_C = 0.4\text{ V}$

c. $V_B - V_C = 7\text{ V}$ d. $V_B - V_C = 4\text{ V}$

6 For the given Wheatstone bridge,

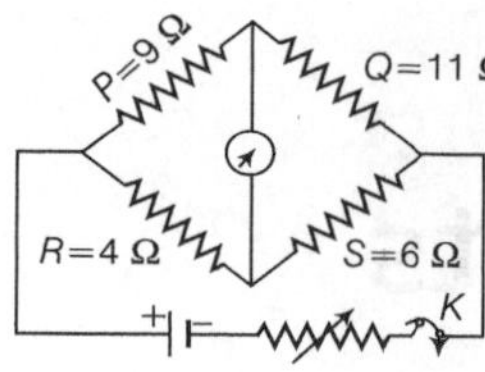

The balance is obtained, if the following resistance is connected in parallel with resistor *S*.

a. 24 Ω b. $\frac{44}{9}$ Ω
c. 26.4 Ω d. 18.7 Ω

7 In a potentiometer, an auxiliary cell of 2 V is connected along with a rheostat at 5 Ω mark. If potentiometer wire is 1000 cm long and its resistance is 15 Ω, then balance length obtained for a cell of 1.5 V is around

a. 1 cm b. 100 cm c. 1000 cm d. 1.5 m

8 While finding internal resistance of a cell, a student observes that the balance length is 2 m when the cell is shunted by 5 Ω resistance and balance length is 3 m when the cell is shunted by a 10 Ω resistance. Internal resistance of cell is

a. 1.5 Ω b. 10 Ω c. 1 Ω d. 15 Ω

9 For the Wheatstone bridge shown, $X > Y$ and $P = Q$. The direction of current in *ab* path is

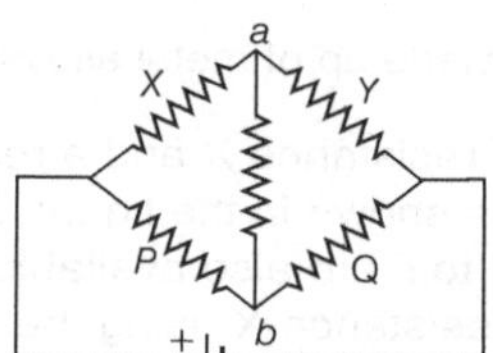

a. from *a* to *b* b. from *b* to *a*
c. no current flows in *ab* d. data is not sufficient

10 A resistance of 2 Ω is connected across one gap of a meter bridge (the length of the wire is 100 cm) and an unknown resistance greater than 2 Ω, is connected across the other gap. When these resistances are interchanged, then balance point shifts by 20 cm. Neglecting any correction, the unknown resistance is

[IIT JEE 2007]

a. 3 Ω b. 4 Ω
c. 5 Ω d. 6 Ω

11 For the given circuit, potentiometer wire is of length 5 m and resistance is 5 Ω. Resistance of 4 Ω is connected across the potentiometer wire with a cell of emf 10 V and internal resistance 1 Ω. If the balance length is found at 300 cm, then emf *E* of each cell is

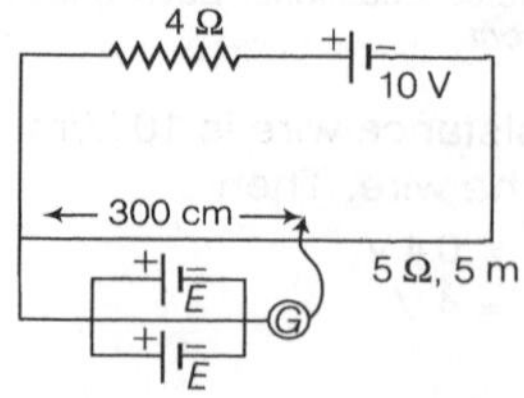

a. 1.5 V b. 3.0 V c. 0.67 V d. 3.33 V

12 A meter bridge is set-up as shown in the figure, to determine an unknown resistance X using a standard $10\,\Omega$ resistor. The galvanometer shows null point, when tapping-key is at 52 cm mark. The end corrections are 1 cm and 2 cm respective for the ends A and B. Then determine the value of X is **[IIT JEE 2011]**

Working Space

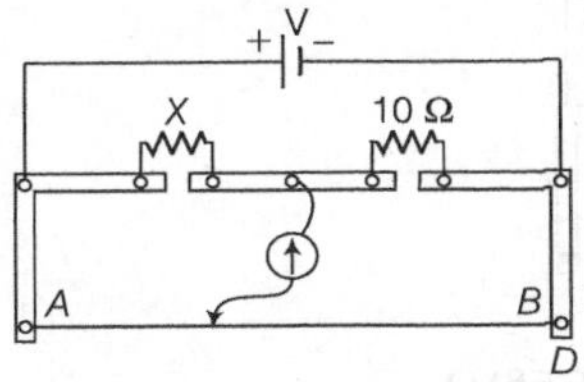

a. $10.2\,\Omega$
b. $10.6\,\Omega$
c. $10.8\,\Omega$
d. $11.1\,\Omega$

13 A potentiometer has a resistance wire of 4 m length and resistance $4\,\Omega$, a cell of emf 12 V is connected across the wire.

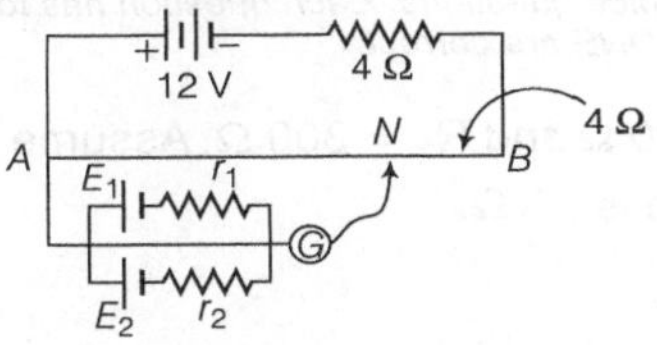

If cells having emfs $E_1 = 2$ V and $E_2 = 4$ V and internal resistance $r_1 = 2\,\Omega$ and $r_2 = 6\,\Omega$ are connected as shown in the figure, then null deflection on length AN, is

a. $\frac{1}{6}$ m
b. $\frac{5}{3}$ m
c. $\frac{3}{2}$ m
d. $\frac{4}{5}$ m

14 In the meter bridge, a network of 10 resistors each of resistance E is balanced against a length of 30 cm as shown, below

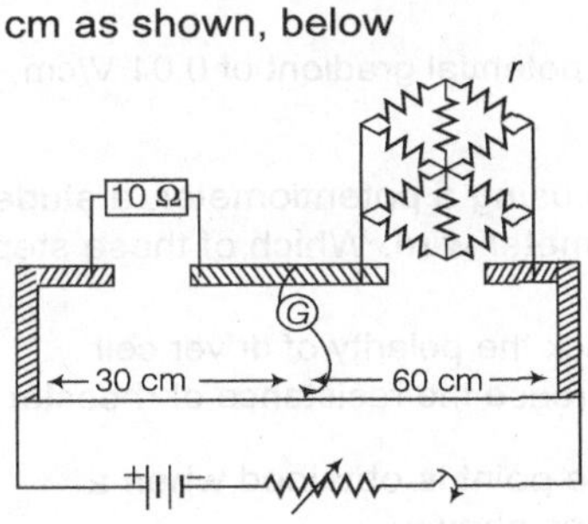

Then, the value of r is

a. $10\,\Omega$
b. $20\,\Omega$
c. $30\,\Omega$
d. $40\,\Omega$

15 R_1, R_2 and R_3 are different values of a resistance R.

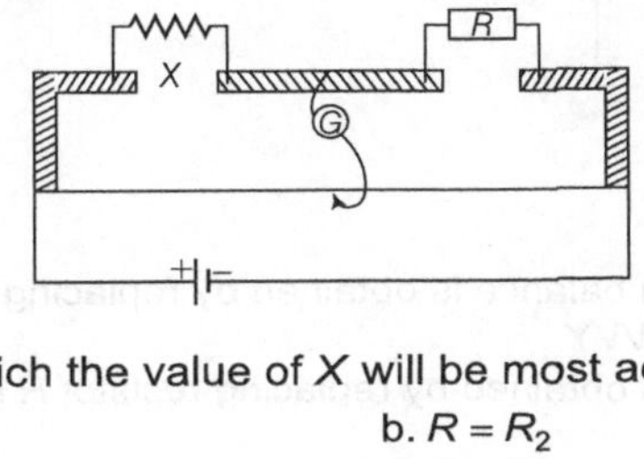

The value of R for which the value of X will be most accurate, is

a. $R = R_1$
b. $R = R_2$
c. $R = R_3$
d. $R = X$

16 During an experiment with a meter bridge, the galvanometer shows a null point when the jockey is pressed at 40.0 cm using a standard resistance of 90Ω as shown in the scale used in the meter bridge is 1mm. The unknown resistance is **[IIT JEE 2014]**

Working Space

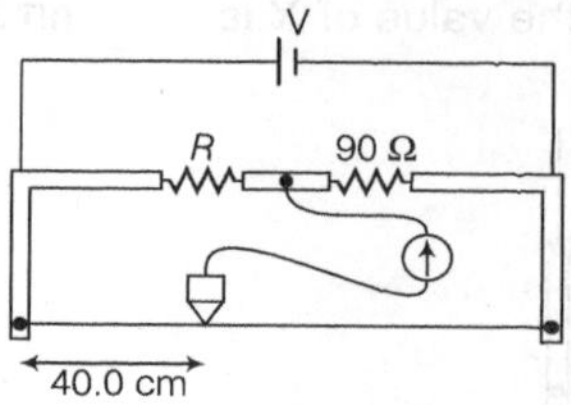

a. $(60 \pm 0.15)\,\Omega$
b. $(135 \pm 0.56)\,\Omega$
c. $(60 \pm 0.25)\,\Omega$
d. $(135 \pm 0.23)\,\Omega$

One or More than One Options Correct Type

Direction (Q. Nos. 17-19) *This section contains 3 multiple choice questions. Each question has four choices (a), (b), (c) and (d), out of which ONE or MORE THAN ONE are correct.*

17 For given circuit, $E_1 = 3\text{V}$, $R_1 = 100\,\Omega$, $R_3 = 250\,\Omega$ and $R_2 = 300\,\Omega$. Assume cell and ammeter as ideal. The voltmeter resistance is $5\,\text{k}\Omega$.

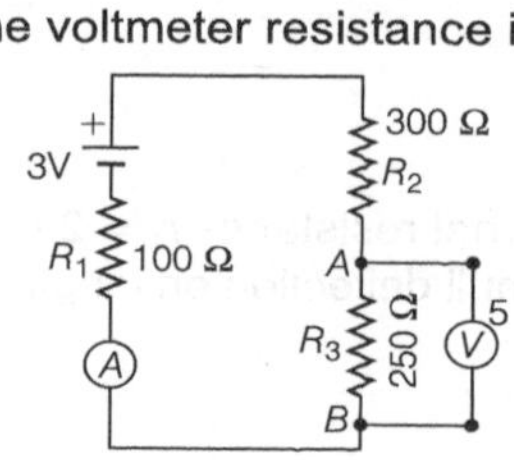

a. Voltmeter reads 3% lower value of potential drop across R_3
b. Voltmeter reads 3% higher value of potential drop across R_3
c. If V_{AB} is balanced against a potentiometer with potential gradient of 0.01 V/cm, then balance length is 115 cm
d. If V_{AB} is balanced against a potentiometer with potential gradient of 0.01 V/cm, then balance length is 112 cm

18 In an experiment of comparison of emfs of cells using a potentiometer, a student does not find any balance point on the potentiometer wire. Which of these steps he must take to make correct observation?

a. Check the polarity of slave cell
b. Check the polarity of driver cell
c. Increase the resistance of rheostat
d. Decrease the resistance of rheostat

19 For the Wheatstone bridge given below, balance point is obtained when a resistance X with colour code *RRO* is attached as shown.

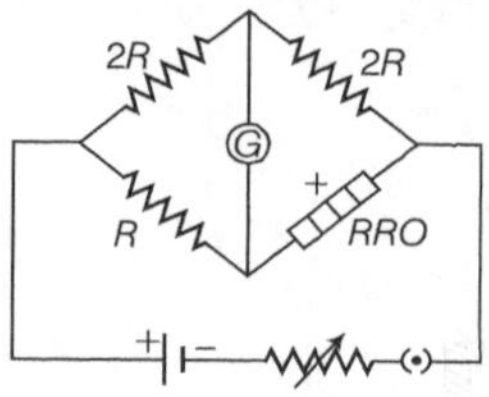

a. Value of $R = 11 \times 10^3\,\Omega$
b. Value of $R = 22 \times 10^3\,\Omega$
c. If X and resistance $2R$ are interchanged, then balance is obtained by replacing resistance R by a resistance with colour code *VVY*
d. If X and $2R$ are interchanged, then balance is obtained by replacing resistor R by a resistance with colour code *GGY*

Statement Type

Direction (Q. Nos. 20 and 21) *This section is based on Statement I and Statement II. Select the correct answer from the codes given below.*

a. Both Statement I and Statement II are correct and Statement II is the correct explanation of Statement I
b. Both Statement I and Statement II are correct and Statement II is not the correct explanation of Statement I
c. Statement I is correct but Statement II is incorrect
d. Statement II is correct but Statement I is incorrect

20 **Statement I** To determine an unknown resistance with a meter bridge, unknown resistance is put in right gap then in left gap and average value of the observations is taken or resistance value of unknown resistance.

Statement II In any experiment, when average value of observations is taken, the net random error is minimised (or eradicated).

21 **Statement I** For accurate measurement a potentiometer of longer length (potentiometer with a wire of longer length) is used.

Statement II Sensitivity of a potentiometer is inversely proportional to its potential gradient.

Matching List Type

Direction (Q. No. 22) *Choices for the correct combination of elements from Column I and Column II are given as options (a), (b), (c) and (d), out of which one is correct.*

22 Consider the given Wheatstone bridge, let current in branch AD is I_2.

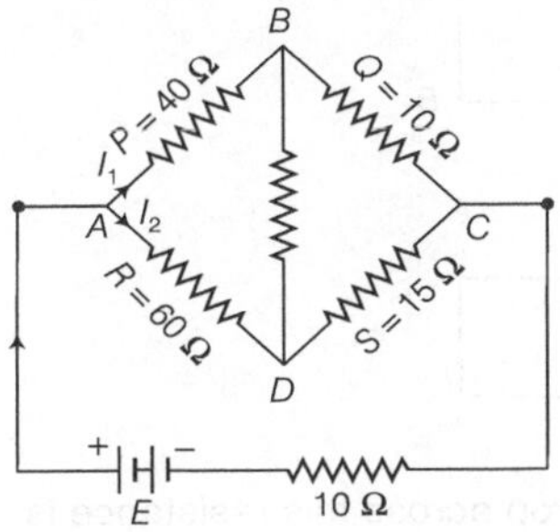

	Column I		Column II
i.	Heat produced in t sec in branch AB	p.	$15\, I_2^2 t$
ii.	Heat produced in t sec in branch BC	q.	$60\, I_2^2 t$
iii.	Heat produced in t sec in branch AD	r.	$22.5\, I_2^2 t$
iv.	Heat produced in t sec in branch DC	s.	$90\, I_2^2 t$

Codes

	i	ii	iii	iv
a.	s	r	q	p
b.	p	q	r	s
c.	s	q	q	r
d.	s	r	p	q

Comprehension Type

Direction *(Q. Nos. 23-25) This section contains a paragraph, describing theory, experiments, data, etc. Three questions related to the paragraphs have been given. Each question has only one correct answer among the four given options (a), (b), (c) and (d).*

A potentiometer can be used to

i. measure internal resistance of a DC source.

ii. compare emfs of 2 cells.

It can also be used to

(i) to determine a small emf like thermo-emf of a thermocouple or output of a photocell. For this purpose, potential gradient of potentiometer is kept very low (around 10^{-4} V/m). Circuit used is

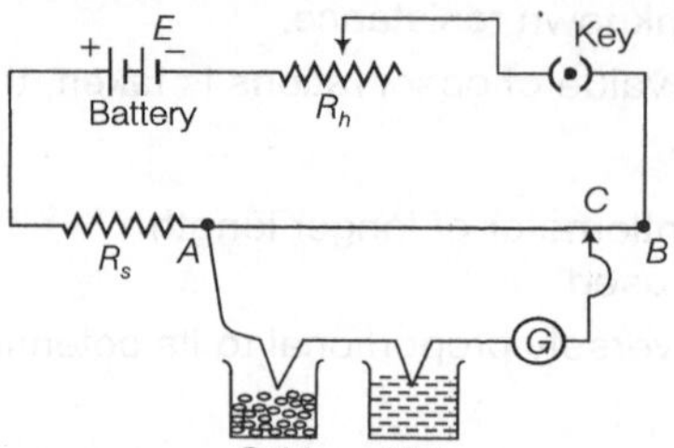

(ii) **To calibrate an ammeter**

Current through a resistor is measured with a potentiometer and also with an ammeter. Then, both values are compared to check the error in ammeter reading. Circuit used for this purpose is

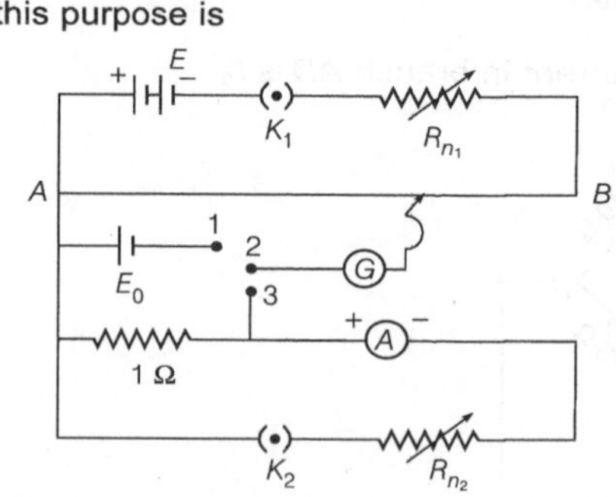

A 1Ω standard resistance is used or potential drop across this resistance is numerically equal to the current flowing through it.

$$V = iR = i \times 1 = i$$

(iii) **Calibration of voltmeter**

Practical voltmeter always measures a lower potential. The error can be found by comparing the voltmeter reading with calculated value of potential difference by using potentiometer.

Circuit used for this process is

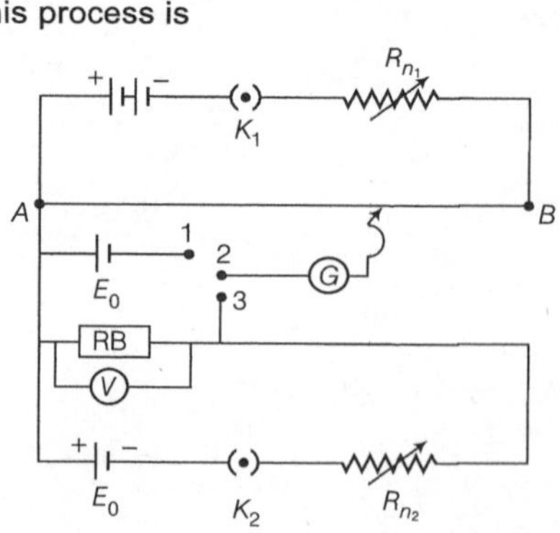

Working Space

23 In a setup for measurement of emf generated in a thermocouple, driver cell of emf $E = 2$ V with negligible internal resistance is used. A high resistance of 990 Ω (R_s) is connected with potentiometer wire of length 1 m and resistance of 10 Ω. The balance point C is found to be 400 mm from A. Emf of thermocouple is

a. 9.1 mV
b. 8 mV
c. 4 mV
d. 3 mV

24 While calibration of an ammeter, balance length for a standard cell of known emf E_0 is found to be l_0. (By connecting points 1 and 2). Let current measured by ammeter is i and potential drop across 1 Ω resistor is balanced for length l_1, then error of ammeter is

a. $i - \frac{E_0 l_1}{l_0}$
b. $i - \frac{E_0 l_0}{l_1}$
c. $i\left(1 - \frac{l_0}{l_1}\right)$
d. $i\left(1 - \frac{l_1}{l_0}\right)$

25 If l_0 is balance length for E_0 and balancing length is l_e for known potential difference V. Then, error in measurement of voltmeter is

a. $V - E_0 \frac{l_0}{l_1}$
b. $V - E_0 \frac{l_1}{l_0}$
c. $V\left(1 - \frac{l_0}{l_1}\right)$
d. $V\left(1 - \frac{l_1}{l_0}\right)$

One Integer Value Correct Type

Direction *(Q. Nos. 26 and 27) This section contains 2 questions. When worked out will result in an integer from 0 to 9 (both inclusive).*

26 In a potentiometer circuit, two wires of same material of resistivity ρ one with radius of cross-section a and other with radius $2a$ are joined in series. Their lengths are l and $2l$, respectively. The combination acts like resistance wire of length $3l$ for the potentiometer. Emf of primary cell in potentiometer is E and internal resistance is $\frac{\rho l}{2\pi a^2}$ (Ω m). If maximum voltage which can be balanced on the potentiometer wire is $\frac{N \cdot E}{4}$ volt. Then, find the value of N.

27 A thin uniform wire AB of length 1 m and an unknown resistance R and a resistance of 12 Ω are connected by thick conducting strips as shown in figure.

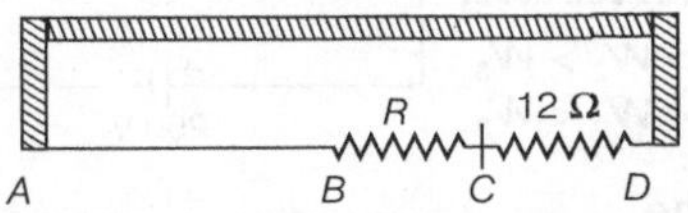

A battery, a galvanometer and a jockey are connected to the above arrangement, and it is found that galvanometer's null deflection point is at a distance of 60 cm from A. Find the value of unknown resistance R.

DPP-9 DC Circuits, Power and Energy Considerations

Subjective Questions

Working Space

Direction *(Q. Nos. 1-3) These questions are subjective in nature, need to be solved completely on notebook.*

1 A battery of emf ε and internal resistance r is connected across a pure resistor of R ohm. Determine the maximum output power and also state the condition for output power to be maximum.

2 When two identical batteries of internal resistance $1\,\Omega$ each, are connected in series across a resistor R, the rate of heat produced in R is J_1. When the same batteries are connected in parallel across R, the rate is J_2. If $J_1 = 2.25\,J_2$, then find the value of R in Ω. **[IIT JEE 2010]**

3 State the conditions for

i. terminal potential difference of a cell and its emf are equal.

ii. terminal potential difference of a cell is less than that of emf of cell.

iii. terminal potential difference of a cell is more than that of emf of cell.

Only One Option Correct Type

Direction *(Q. Nos. 4-16) This section contains 13 multiple choice questions. Each question has four choices (a), (b), (c) and (d), out of which ONLY ONE option is correct.*

4 In a house having 220 V power line, the following appliances are operating

i. 60 W bulb
ii. 1000 W heater
iii. 40 W tubelight

Current through the fuse wire at that instant is

a. 3/11 A b. 2/11 A c. 5 A d. 6 A

5 A 100 W bulb B_1, two 60 W bulbs B_2 and B_3 are connected to a 250 V source as shown. If W_1, W_2 and W_3 are powers of the bulbs, then **[IIT JEE 2002]**

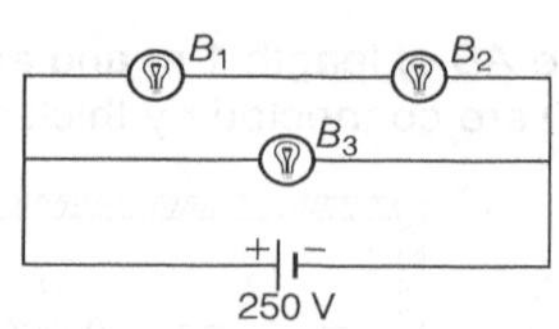

a. $W_1 > W_2 = W_3$
b. $W_1 > W_2 > W_3$
c. $W_1 < W_2 = W_3$
d. $W_1 < W_2 < W_3$

6 A, B and C are three identical bulbs, which of the following change occurs in the brightness of bulb, when switch S is closed

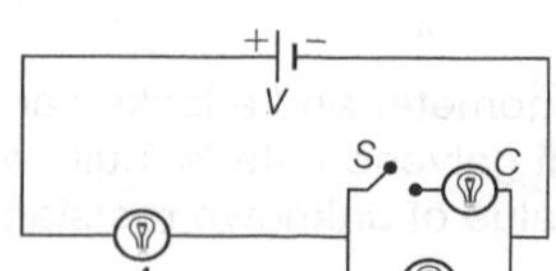

a. A remains same, then B decreases
b. A increases then B decreases
c. A increases then B remains same
d. A decreases then B increases

7 If electric power from 100 MW power station is transmitted to a substation at 20000 V, the power loss during transmission is P_1. If the same power is transmitted at 200 V, the power loss is P_2. The value of P_2/P_1 will be

a. 10^4 b. 10^3 c. 10^2 d. 10

Working Space

8 The walls of a closed cubical box of edge 50 cm are made of a material of thickness 1mm and thermal conductivity $1.68 \times 10^{-1}\ Wm^{-1}K^{-1}$. The interior of the box is maintained at 100 °C above the outside temperature by a heater placed inside the box and connected to a 420 V DC source. What must be the resistance of the heater element?

a. 5 Ω b. 6 Ω c. 7 Ω d. 8 Ω

9 The three resistances of equal value are arranged in the different combinations as shown below. Arrange them in increasing order of power dissipation. **[IIT JEE 2003]**

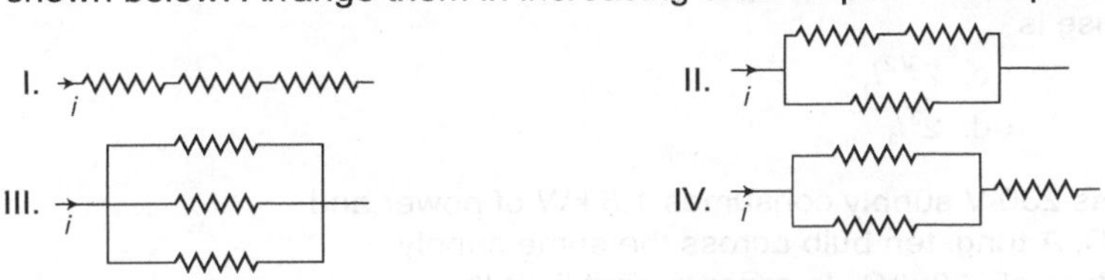

a. III < II < IV < I b. II < III < IV < I

c. I < IV < III < II d. I < III < II < IV

10 Figure shows three configurations resistors of R_1, R_2 and R_3 connected to 3 V battery. If the power dissipated by the configuration of R_1, R_2 and R_3 is P_1, P_2 and P_3 respectively, then **[IIT JEE 2008]**

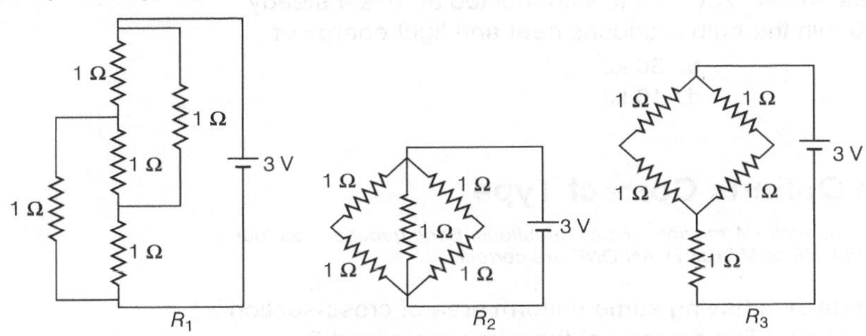

a. $P_1 > P_2 > P_3$ b. $P_1 > P_3 > P_2$

c. $P_2 > P_1 > P_3$ d. $P_3 > P_2 > P_1$

11 The maximum power rating of a 20 Ω resistor is 2 kW. Then, it cannot be used with a 300 V DC source because the

a. current drawn is very large and it can blow the fuse

b. current drawn is very less and appliance does not work

c. resistor may melt

d. None of the above

12 Two ribbon fuses *A* and *B* have following particulars.

A Alloy constantan	*B* Nichrome
Length (m) 8.456	4.235
Width (mm) 1.0	2.0
Thickness (mm) 0.03	0.06
Resistivity ($\times 10^{-7}$ Ω-m) 4.9	11
Temperature coefficient of resistance ~ 0	~ 0

If a more sensitive fuse is required, then

a. *A* is a better option b. *B* is a better option

c. both *A* and *B* are same in use d. *A* and *B* together can be used

Working Space

13 Two wires made of tin and copper having identical cross-section (10^{-6} m^2) and lengths 10 cm and 15 cm are being used as fuses. Then,

a. wire of larger length will melt earlier
b. wire of smaller length will melt earlier
c. both wire melt simultaneously
d. for tin and copper wire melting point is very high to be used as fuse

14 A fuse of cross-section area A_1 melts at current I_1. The current value for a fuse of area $2A_1$ required for melting of fuse is

a. $2^2 I_1$
b. $2^{3/2} I_1$
c. $2^{3/4} I_1$
d. $2^3 I_1$

15 A nichrome heating element across 230 V supply consumes 1.5 kW of power and heats upto a temperature of 750°C. A tungsten bulb across the same supply operates at much higher temperature of 1600°C. In order to emit light the

a. power of bulb is much higher than that of heating element
b. power of heating element is more than that of bulb
c. power of both is nearly same
d. data is insufficient to reach any conclusion

16 An electric bulb is marked as 100 W, 230 V. If it is connected across a steady supply of 115 V. Then, in 20 min the bulb produces heat and light energy of

a. 20 kJ
b. 30 kJ
c. 25 kJ
d. 10 kJ

One or More than One Options Correct Type

Direction *(Q. Nos. 17-20) This section contains 4 multiple choice questions. Each question has four choices (a), (b), (c) and (d), out of which ONE or MORE THAN ONE are correct.*

17 Two wires of the same material and having same uniform area of cross-section are connected in an electric circuit. The masses of the wires are m and $2m$ respectively. When a current I flows through the circuit, the heat produced in them in a given time is in the ratio

a. 2 :1 when they are connected in series
b. 2 :1 when they are connected in parallel
c. 1 : 2 when they are connected in parallel
d. 1 : 2 when they are connected in series

18 If three bulbs of 40 W, 60 W and 100 W are connected in series with a 200 V power supply, then

a. the potential difference will be maximum across the 40 W bulb
b. the current will be the maximum through the 100 W bulb
c. the 40 W bulb has the maximum resistance
d. the current in the circuit is 1.2 A

19 A dry cell of emf 1.5 V and internal resistance 0.10 Ω is connected across a resistor in series with a very low resistance ammeter. When current is switched ON, the ammeter settles to steady value of 2 A. In steady state,

a. the rate of chemical energy consumption of cell is 3.0 W
b. the rate of energy dissipation inside the cell is 0.40 W
c. the rate of energy dissipation inside the resistor is 2.6 W
d. the output power of source is 3.6 W

Working Space

20 A series battery of 6 lead accumulators each of 2.0 V and internal resistance 0.50 Ω is charged by a 100 V DC supply. If charging current is limited to 8 A, then

a. power supplied by source is 800 W
b. power dissipated as heat is 704 W
c. energy stored in battery in 15 min is 86 J
d. series resistor used is of 100 Ω

Comprehension Type

Direction (Q. Nos. 21-25) This section contains 2 paragraphs, describing theory, experiments, data etc. Five questions related to the paragraphs have been given. Each question has only one correct answer among the four given options (a), (b), (c) and (d).

Passage I

Power from a 64 V DC supply goes to charge a battery of 8 lead acid accumulators each of emf 2 V and internal resistance (1/8) Ω. The charging current also runs an electric motor placed in series with the battery. Resistance of windings of motor is 7 Ω and steady supply current is 3.5 A.

21 Mechanical energy yielded by motor in 1h is

a. 2.52×10^5 J
b. 3.6×10^5 J
c. 4.82×10^5 J
d. 0.9×10^5 J

22 Chemical energy stored in the battery during charging in 1h is

a. 201600 J
b. 300000 J
c. 302600 J
d. 542000 J

Passage II

Consider the given circuit.

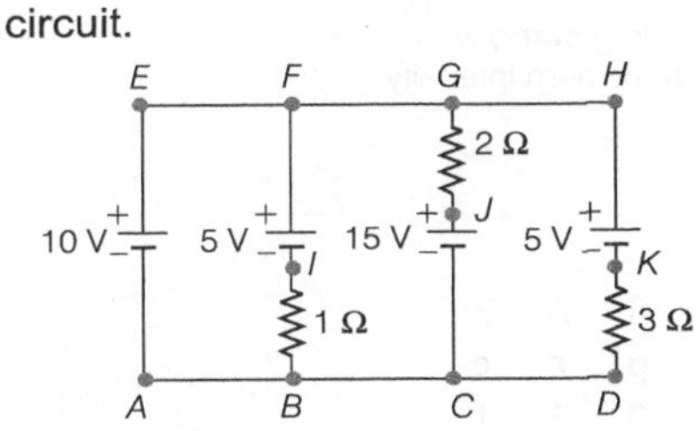

It is known that potential at point A is zero. Now, answer the following questions.

23 In the given circuit,

a. 10 V battery acts like source
b. 15 V battery acts like source
c. 5 V battery acts like source
d. All batteries act like source

24 Which resistor consumes maximum power?

a. 1 Ω resistor
b. 2 Ω resistor
c. 3 Ω resistor
d. None of the above

25 Which battery consumes or gives maximum power?

a. 10 V battery
b. 15 V battery
c. 5 V battery between points F and B
d. 5 V battery between points H and D

Matching List Type

Working Space

Direction *(Q. No. 26) Choices for the correct combination of elements from Column I and Column II are given as options (a), (b), (c) and (d), out of which one is correct.*

26 Three bulbs *A*, *B* and *C* having rated power P_A, P_B and P_C, respectively. Each bulb is designed to operate at rated voltage of *V* volts.
It is given that $P_A > P_B > P_C$. In Column I, different configurations of bulbs are given while in Column II, intensities are mentioned. Match entries of Column I with entries of Column II, neglecting the change of resistances of bulbs due to temperature change.

	Column I		Column II
i.	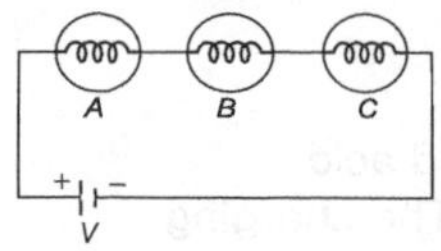	p.	*A* glows with minimum intensity
ii.	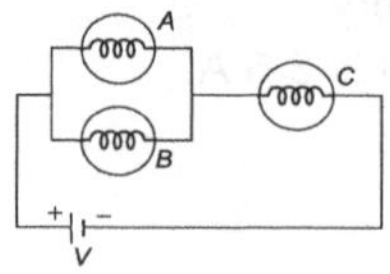	q.	*C* is glowing with maximum intensity
iii.	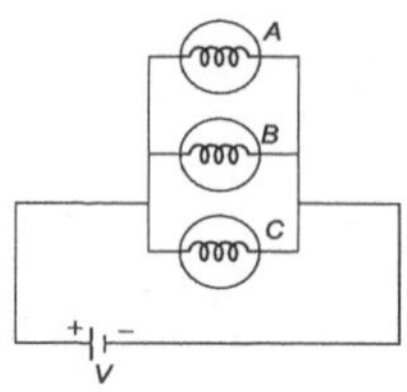	r.	*B* is glowing with minimum intensity
iv.	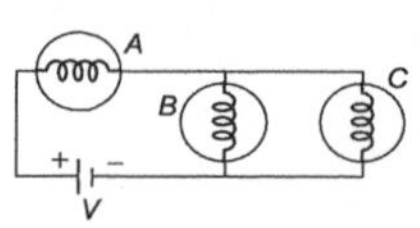	s.	*A* is glowing with maximum intensity

Codes

	i	ii	iii	iv		i	ii	iii	iv
a.	q	r	s	p	b.	s	p	r	q
c.	p	q	r	s	d.	q	q	s	r

One Integer Value Correct Type

Direction *(Q. Nos. 27 and 28) This section contains 2 questions. When worked out will result in an integer from 0 to 9 (both inclusive).*

27 A resistor develops 400 J of thermal energy in 10 s when a current of 2 A passed through it. When the current is increased to 4 A, then energy dissipated in 20 s as heat is found to be 400 *k* J. Find the value of *k*.

28 One kilowatt electric heater is to be used with 220 V DC supply. It converts $\frac{N \times 10}{3}$ g of water at 100°C into steam at 100°C in one minute. Find the value of *N*.

Revisal Problems
for JEE Main

Only One Option Correct Type

Working Space

Direction (Q. Nos. 1-20) *This section contains 20 multiple choice questions. Each question has four choices (a), (b), (c) and (d), out of which ONLY ONE option is correct.*

1 In a copper wire at room temperature 27°C, the number density of charge carriers is $n = 8.5 \times 10^2\ m^{-3}$. Resistivity of copper is $\rho = 1.7 \times 10^{-8}\ \Omega$-m and Boltzmann constant $k_B = 1.38 \times 10^{-23}$ J/K. Mean free path for electrons is

a. 25 Å b. 30 Å
c. 25 nm d. 30 nm

2 A potential difference of 30 V is applied between ends of a conductor of length 100 m and resistance 0.5 Ω. The total drift linear momentum of electrons is

a. 3.4×10^{-10} kg m/s b. 3.4×10^{-8} kg m/s
c. 3.4×10^{-12} kg m/s d. 3.4×10^{-13} kg m/s

3 A cylindrical tube of length l has inner radius a while outer b. The ratio of resistance of tube between its ends and its inner and outer surfaces is

a. $\dfrac{L}{\pi(b^2 - a^2)}$ b. $\dfrac{\log_e b/a}{2\pi L}$
c. $\dfrac{2L^2}{(b^2 - a^2)\log_e \dfrac{b}{a}}$ d. $\dfrac{1}{\pi}\dfrac{L}{ab}$

4 Two identical cells with emf E and internal resistance r are connected across a resistance R in (a) series (b) parallel. If load R is adjusted in both cases so that power transfer is maximum in both cases, then ratio of power output of load will be

a. 1:2 b. 2:1
c. 1:3 d. 1:1

5 In the given circuit, the resistance between points P and Q is

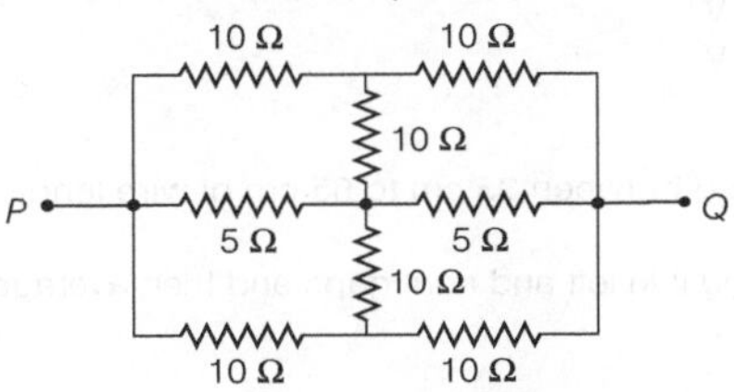

a. 25 Ω b. 10 Ω
c. 30 Ω d. 5 Ω

6 At $t = 0$, switch S is closed. The charge on the capacitor is varying with time is $Q = Q_0(1 - e^{-\alpha t})$.

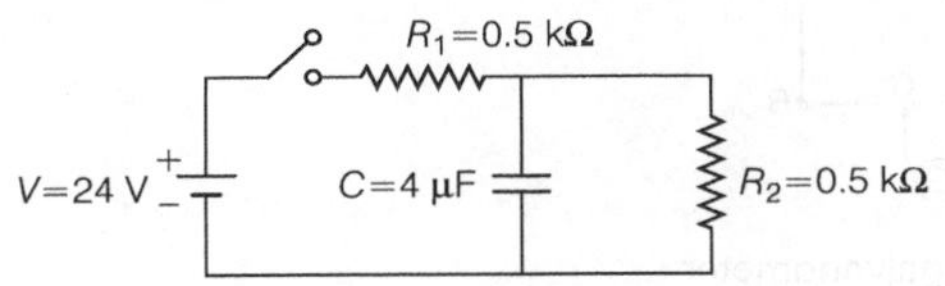

Then, α is

a. 10^{-3} b. 10^6
c. 10^9 d. 10^3

Working Space

7 When two identical batteries of internal resistance 1 Ω each, are connected in series across a resistor R, the rate of heat produced in R is P_1. When the same batteries are connected in parallel across R, the rate of heat produced is $\frac{P_1}{2.25}$. Then, the value of R is

a. 2 Ω b. 3 Ω
c. 4 Ω d. 5 Ω

8 If switch is closed at $t = 0$, then the time at which reading of both voltmeters will be equal

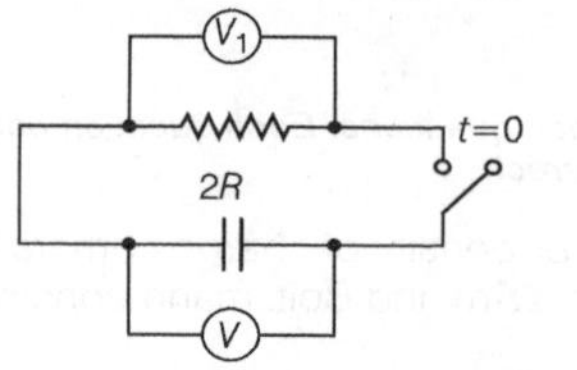

a. at $t = 2RC$ b. at $t = 3RC$
c. at $t = RC \log 2$ d. at all times

9 For the given figure, if current in R_3 is i, then current in R_4 is

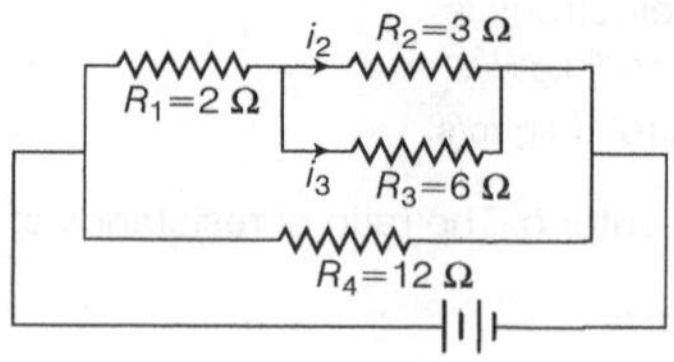

a. $2i$ b. i
c. $i/2$ d. cannot be determined

10 Potential difference between points C and E is

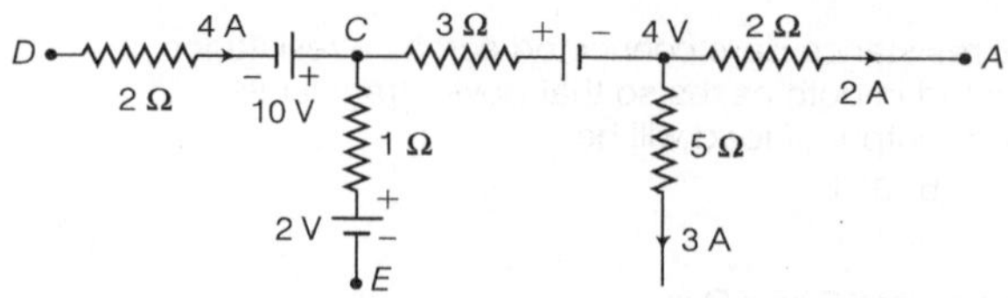

a. 1 V b. 2 V
c. 5 V d. 0 V

11 To avoid end correction in a meter bridge,

a. readings are taken such that null point is obtained between 35 cm to 65 cm of wire length
b. all connections are made properly tight
c. the value of unknown resistance is found by putting it in left and right gaps and then average value is taken
d. ends of resistance wire are clamped properly

12 For the given potentiometer experiment, the deflection of galvanometer increases as we move jockey from A to B and no null point is obtained. Then, to obtain the null point

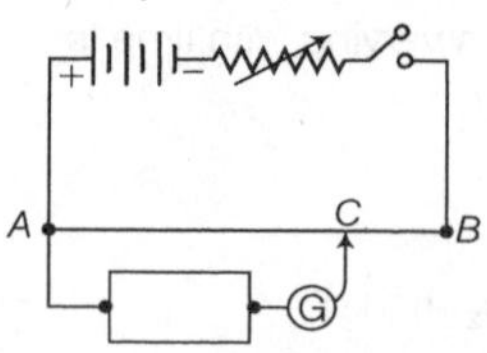

a. we connect a high resistance in series with galvanometer
b. we connect a low resistance in parallel with galvanometer
c. we increase the rheostat's resistance
d. we change (reverse) the polarity of potentiometer's driver cell

13 In a potentiometer for determining emf e_i of a cell of negligible internal resistance, a high resistance R_2 is connected in series with cell as shown in the figure.

Working Space

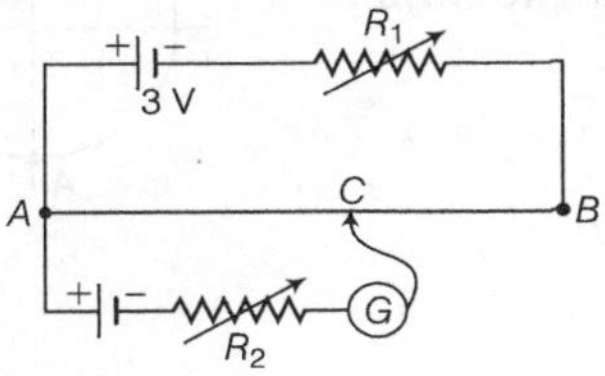

When R_2 increases

a. balance point shifts towards B
b. balance point shifts towards A
c. balance point is obtained at same position C
d. balance point is not obtained at any point on wire AB

14 For the given circuits, galvanometer shows no deflection when switches S_1 and S_2 are closed.

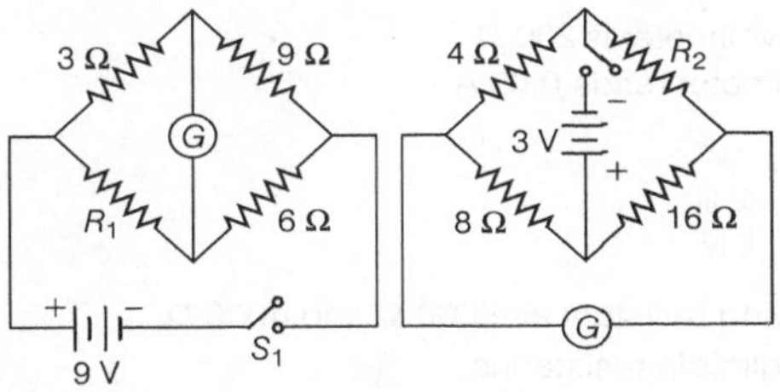

If $R_2 = kR$, then the value of k is

a. 4
b. 6
c. 8
d. 12

15 For the given potentiometer circuit, potential gradient is 0.025 V/m and the ammeter reading is 0.1 A.

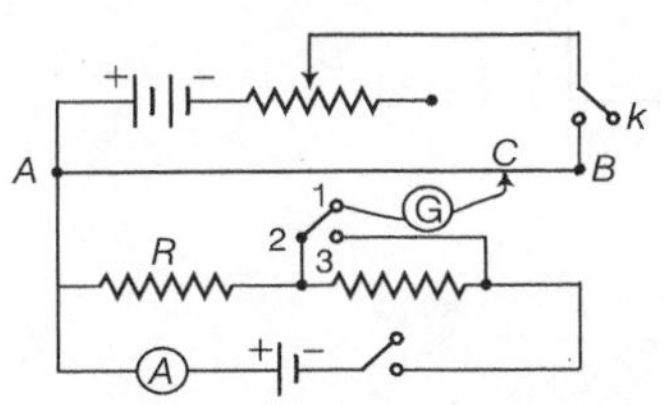

Now, when terminals 1 and 2 are connected, balance point is obtained at 40 cm also when terminals 1 and 3 are connected, balance point is obtained at 100 cm. If $3R = kX$, then the value of k will be

a. 1
b. 2
c. 3
d. 4

16 Consider given network of resistances. If equivalent resistance of circuit is nR then, then the value of n may be

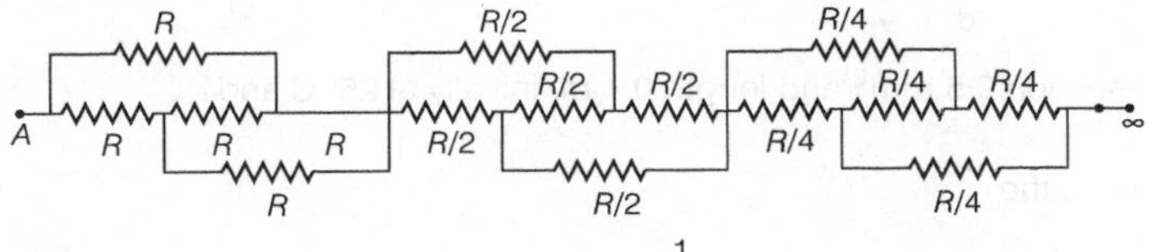

a. 1
b. $\frac{1}{2}$
c. 2
d. 3

17 A uniform wire of resistance R is bent in the form of a ring. Four points A, B, C and D exist on the ring as shown in figure..

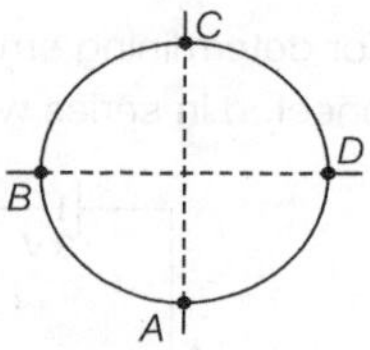

A battery of emf E may be connected across the ring in two ways:

***Case* I** Battery is connected between B and D

***Case* II** Battery is connected between A and B

Then, choose the correct option.

i. In each case same current flows through the battery.
ii. In case I more current is drawn from the battery.
iii. In case II more current is drawn from the battery.
iv. In case I resistance is more than in case II.

a. i, ii b. ii, iii
c. iii, iv d. i, iv

18 A battery of emf 1.4 V and internal resistance 2 Ω is connected to a 100 Ω resistor through an ammeter. The resistance of the ammeter is 4/3 Ω. A voltmeter is also connected to find the potential difference across the resistor.

i. If voltmeter reads 1.10 V, then error in reading is 0.23 V
ii. Voltmeter reads 1.33 V
iii. If ammeter reads 0.02 A, then resistance of voltmeter is 200 Ω
iv. If voltmeter is of in finite resistance, then ammeter reads 0.05 A

Choose the correct option.

a. i, ii b. ii, iii
c. iii, iv d. i, iv

19 For the circuit, V_1 and V_2 are two voltmeters having resistances 6000 Ω and 4000 Ω, respectively emf of battery is 250 V, having negligible resistance.

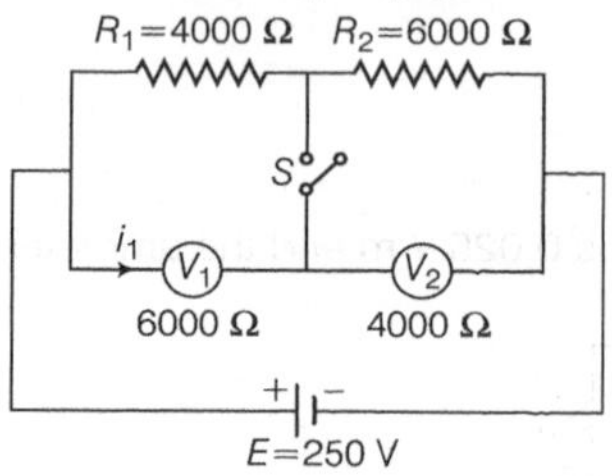

Then,

i. when switch S is open, $\frac{V_1}{V_2} = 1$
ii. when switch S is open, $\frac{V_1}{V_2} = 1.5$
iii. when switch S is closed, $\frac{V_1}{V_2} = 1$
iv. when switch S is closed, $\frac{V_1}{V_2} = 1.5$

Choose the correct option.

a. i, ii b. ii, iii
c. iii, iv d. i, iv

20 A copper wire having area of cross-section 0.5 mm^2 and length 0.1 m, initially at 25° C and is insulated from the surroundings.

If a current of 1.0 A is set up in the wire, the

i. Wire will melt after sometime.
ii. Wire will not melt or heat will be dissipated in atmosphere.
iii. If length of wire is increased, it will require more time to melt.
iv. If length of wire is increased, it melts in same time or before.

Choose the correct option.

a. i, ii b. ii, iiii
c. iii, iv d. i, iv

Working Space

Revisal Problems
for JEE Advanced

Only One Option Correct Type

Working Space

Direction (Q. Nos. 1-8) *This section contains 8 multiple choice questions. Each question has four choices (a), (b), (c) and (d), out of which ONLY ONE option is correct.*

1 For the circuit shown in figure, switch S is closed at $t = 0$. Which statement is not correct?

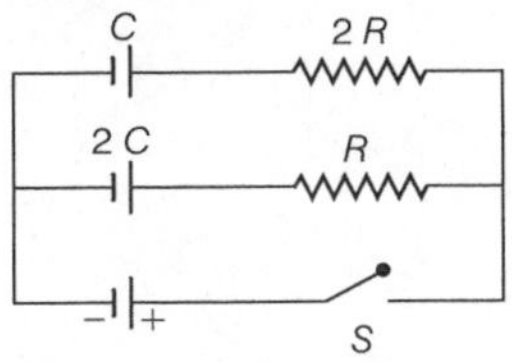

a. Rate of increase of charge is same in both the capacitors
b. Ratio of charges stored in capacitors C and $2C$ at any time t would be 1:2
c. Time constant of both the capacitors are equal
d. Steady state charges in capacitors C and $2C$ are in the ratio of 1:2

2 A resistance R carries a current i. The power lost to the surroundings is $\lambda(\theta - \theta_0)$. Here, λ is a constant, θ is the temperature of the resistance and θ_0 is the temperature of the atmosphere. If the coefficient of linear expansion is α, then strain in the resistance is

a. $\frac{\alpha}{\lambda} i^2 R$
b. $\alpha \lambda i R$
c. $\frac{\alpha i^2 R}{2\lambda}$
d. proportional to the length of the resistance wire

3 A battery of internal resistance $4\,\Omega$ is connected to the network of resistances are shown below :

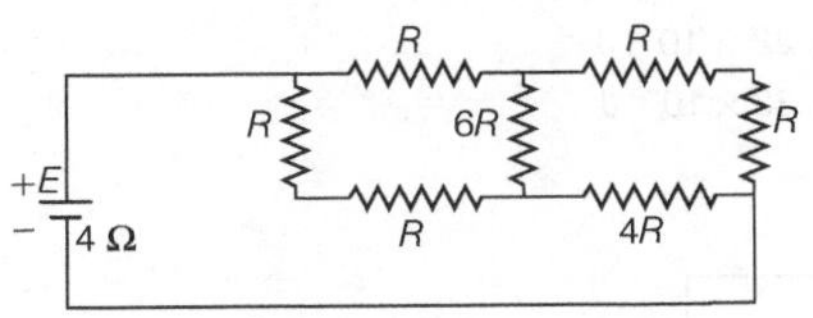

In order to deliver the maximum power to network, the value of R in ohms should be

a. $\frac{4}{9}$ b. 2
c. $\frac{8}{3}$ d. 18

4 For the circuit shown below:

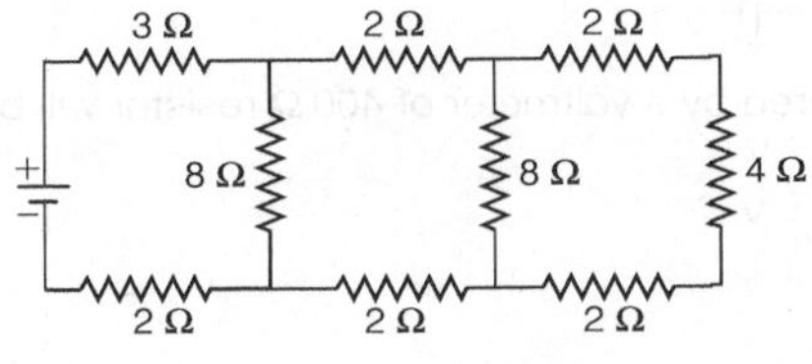

The current through the

a. $3\,\Omega$ resistor is 0.5 A b. $3\,\Omega$ resistor is 0.25 A
c. $4\,\Omega$ resistor is 0.5 A d. $4\,\Omega$ resistor is 0.25 A

Working Space

5 For the given circuit the current through 6 V battery is

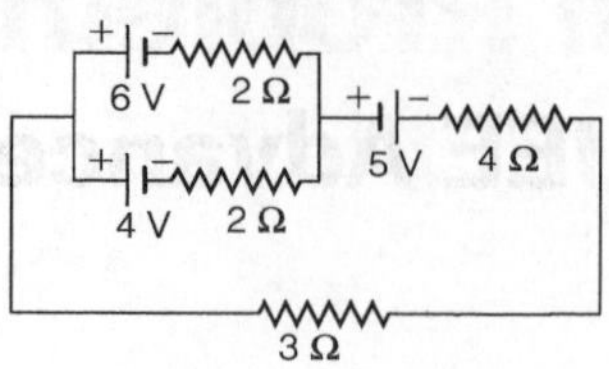

a. 1/4 A
b. 1/8 A
c. 1/2 A
d. 1 A

6 A meter bridge wire *AB* is 40 cm long and having a resistance of 50 Ω/m. An ideal voltmeter is touching the meter bridge wire as shown below:

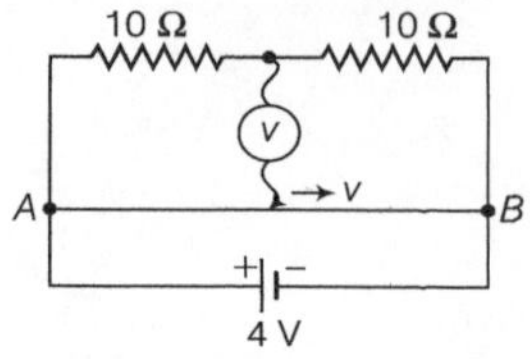

If reading with voltmeter is varying with time as $V = 2\sin \pi t$
Then, velocity *v* of the jockey is
(assume negative terminal of battery is at 0 V)

a. $10\pi\sin\pi t$ cm/s
b. $10\pi\cos\pi t$ cm/s
c. $20\pi\sin\pi t$ cm/s
d. $20\pi\cos\pi t$ cm/s

7 For the given circuit,

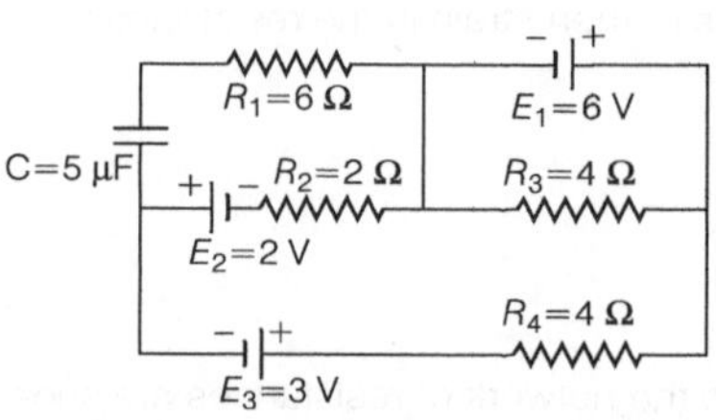

Energy stored in the capacitor in steady state is

a. 1.44×10^{-5} J
b. 1.22×10^{-5} J
c. 1.66×10^{-5} J
d. 1.33×10^{-5} J

8 For the given circuit,

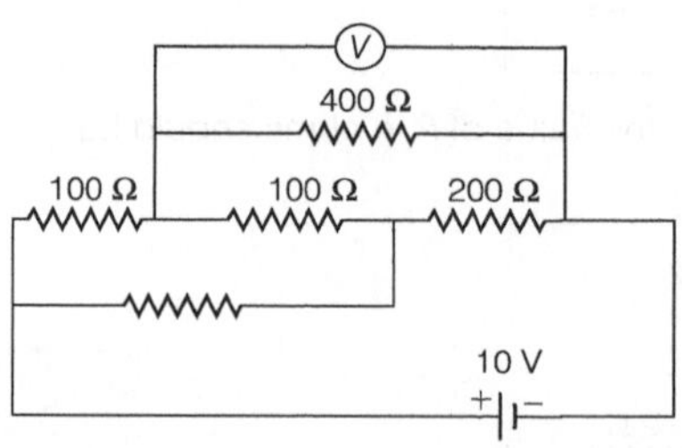

Potential difference across 400 Ω resistor measured by a voltmeter of 400 Ω resistor will be

a. 20 V
b. 17 V
c. $\frac{20}{3}$ V
d. $\frac{17}{3}$ V

One Integer Value Correct Type

Direction *(Q. Nos. 9-10) This section contains 2 questions. When worked out will result in one integer from 0 to 9 (both inclusive).*

9 Equivalent resistance across points *A* and *B* is

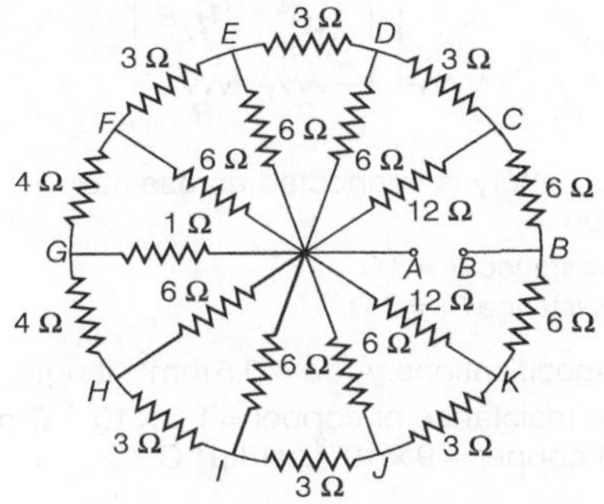

a. 9 Ω b. 7 Ω c. 6 Ω d. 5 Ω

10 In the potentiometer arrangement shown, the driving cell *A* has emf *x* and internal resistance *r*. The emf of cell *B* is *x*/2 and internal resistance is 2*r*. Potentiometer wire is 100 cm long.

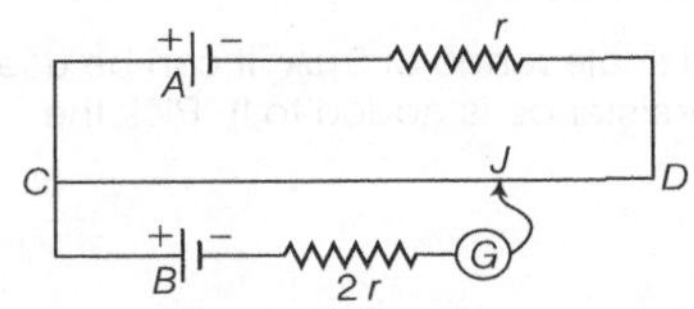

If balance length *Cj* is greater than $k \times 10$ cms, then the value of *k* will be

a. 2 b. 3 c. 4 d. 5

One or More than One Options Correct Type

Direction *(Q. Nos. 11-15) This section contains 5 multiple choice questions. Each question has four choices (a), (b), (c) and (d), out of which ONE or MORE THAN ONE are correct.*

11 An infinite network of equal resistances, 1Ω each is given a supply through 1 V battery as shown below:

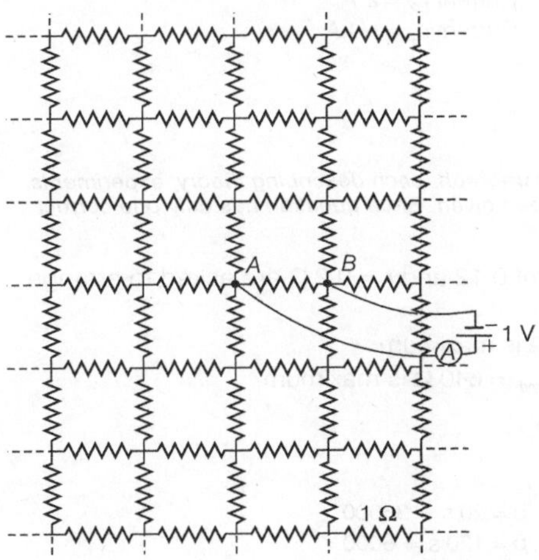

(supply is not connected anywhere except points *A* and *B*) Then,

a. ammeter reading is 0.75 A
b. ammeter reading is 0.5 A
c. equivalent resistance of network is 4 Ω
d. equivalent resistance of network is 6 Ω

12 For the network of resistors shown as below:

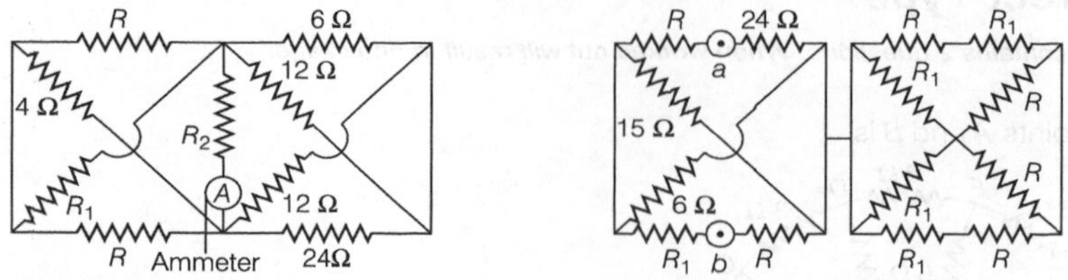

The reading of ammeter is found zero when a 12 V battery is connected across the network. If points *a* and *b* are at same potential, then

a. resistance $R = 6\,\Omega$
b. resistance $R = 8\,\Omega$
c. resistance $R_1 = 1.5\,\Omega$
d. resistance $R_1 = 2\,\Omega$

13 Consider an insulated copper wire with following specifications. Area $= 0.5\,\text{mm}^2$, length $= 0.1\,\text{m}$, melting point of copper $= 1075°\,\text{C}$, specific resistance of copper $= 1.6 \times 10^{-8}\,\Omega\text{-m}$, density of copper $= 9 \times 10^3\,\text{kg/m}^3$, specific heat of copper $= 9 \times 10^{-2}\,\text{cal/kg}\,°\text{C}$.

If current in wire is 1 A and wire is initially at 25° C, then

a. time in which wire starts for melting is 60 s
b. time in which wire starts for melting is 55.5 s
c. if length of wire is doubled, then time required for melting will be 120 s
d. if length of wire is doubled, then also wire melts in same time duration

14 A microammeter has a resistance of 100 Ω and a full scale range of 5 μA. It can be used as a voltmeter or a higher range ammeter providing a resistance is added to it. Pick the correct range and resistance combinations.

a. 50 V range, 10 kΩ resistance in series
b. 10 V range, 200 kΩ resistance in series
c. 5 mA range, 1 Ω resistance in parallel
d. 10 mA range, 1 kΩ resistance in parallel

15 For the given circuit,

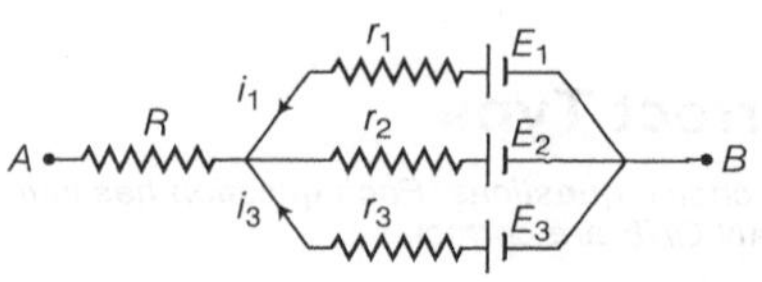

Where, $E_1 = 3\,\text{V}$, $E_2 = 2\,\text{V}$, $E_3 = 1\,\text{V}$, $r_1 = r_2 = r_3 = 1\,\Omega$, $R = 1\,\Omega$, then

a. Current $i_1 = 1\,\text{A}$
b. Current $i_2 = 2\,\text{A}$
c. Current $i_3 = 1\,\text{A}$
d. Current $i_3 = -1\,\text{A}$

Comprehension Type

***Direction** (Q. Nos. 16 and 17) This section contains a paragraph, each describing theory, experiments, data, etc. Two questions related to the paragraph have been given. Each question has only one correct answer among the four given options (a), (b), (c) and (d).*

Consider a number of identical cells each of emf 0.12 and $r = 0.2\,\Omega$ are asked to arrange the cells in a mixed grouping such that

I. current in an external resistance $R_{ext} = 640\,\Omega$ is maximum.
II. power delivered to the external resistance $R_{ext} = 640\,\Omega$ is maximum.

If there are *p* rows each containing *s* cells, then

16 In case I,

a. $p = 15, s = 48000$
b. $p = 20, s = 36000$
c. $p = 18, s = 40000$
d. $p = 120, s = 6000$

17 In case II,

a. $p = 20, s = 36000$
b. $p = 120, s = 6000$
c. $p = 18, s = 40000$
d. $p = 15, s = 48000$

Working Space

Statement Type

Direction ***(Q. Nos. 18 and 19) This section is based on Statement I and Statement II. Select the correct answer from the codes given below.***

a. Both Statement I and Statement II are correct and Statement II is the correct explanation of Statement I
b. Both Statement I and Statement II are correct and Statement II is not the correct explanation of Statement I
c. Statement I is the correct but Statement II is the incorrect
d. Statement II is the correct but Statement I is the incorrect

18 **Statement I** When positive electric charge is transported from positive terminal to the negative terminal through the battery, then energy stored in the battery increases.

Statement II Energy is released when charge is transported through the battery in reverse direction.

a. *A* b. *B* c. *C* d. *D*

19 **Statement I** When an unknown resistance is measured using a Wheatstone bridge, the measured value is most accurate when ratio of resistances in ratio arms is least $\frac{P}{Q}$ is smallest.

Statement II Since, $\frac{P}{Q} = \frac{R}{X} \Rightarrow X = \frac{Q}{P} \times R$

So, δx is least, when ratio $\frac{Q}{P}$ is least.

a. *A* b. *B* c. *C* d. *D*

Comprehension Type

Direction ***(Q. Nos. 20-23) This section contains a paragraphs, each describing theory, experiments, data, etc. Four questions related to the paragraphs have been given. Each question has only one correct answer among the four given options (a), (b), (c) and (d).***

Consider a potentiometer arrangement having a uniform wire *AB* of resistance 99 Ω.

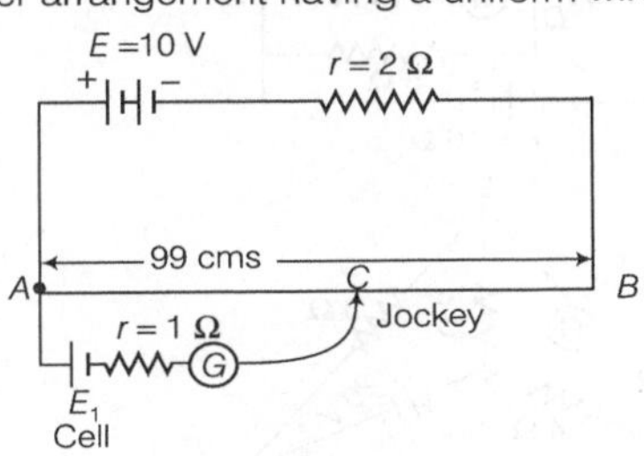

A driving battery of 10 V emf and 1 Ω internal resistance is used in the potentiometer. A cell of unknown emf E_1 and internal resistance of 1 Ω is balanced against *AC* length of potentiometer wire.

20 When jockey is touched with wire *AB* at point *C* such that *AC* = 40 cm, there is no deflection in galvanometer. The emf of cell is

a. 4 V b. 5 V
c. $\frac{5}{4}$ V d. 6 V

21 If jockey is touched at point *B*, then potential difference across wire *AB* is

a. 6.93 V b. 8.93 V
c. 9.93 V d. 0 V

22 If jockey is touched at point *A*, then potential difference across the cell is

a. 3 V b. 2.5 V c. 2 V d. None of these

23 When jockey is touched at point *A*, then potential difference across the terminals of battery of 10 V is

a. 9.9 V b. 9.0 V c. 8 V d. 7.8 V

Matching List Type

Direction *(Q. No. 24) Choices for the correct combination of elements from Column I and Column II are given as options (a), (b), (c) and (d), out of which one is correct.*

24 . Match the circuits given in column II with the respective column I. select the correct code. More than one match is possible.

<table>
<tr><th></th><th>Column I</th><th></th><th>Column II</th></tr>
<tr><td>i.</td><td>Wheatstone bridge</td><td>p.</td><td>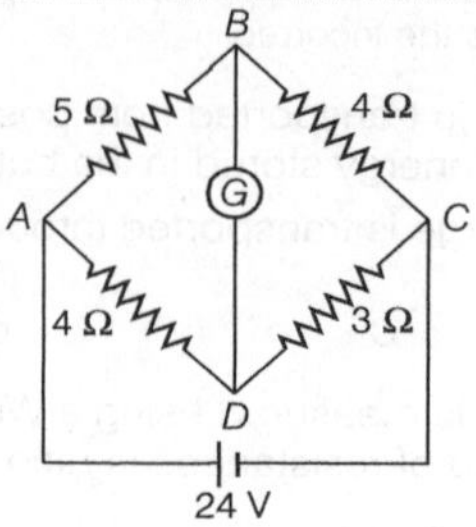
</td></tr>
<tr><td>ii.</td><td>Balanced Wheatstone bridge</td><td>q.</td><td>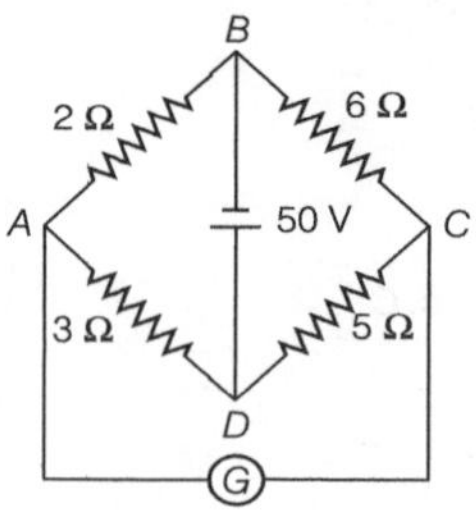
</td></tr>
<tr><td>iii.</td><td>Direction of current from B to D in branch BD, is</td><td>r.</td><td>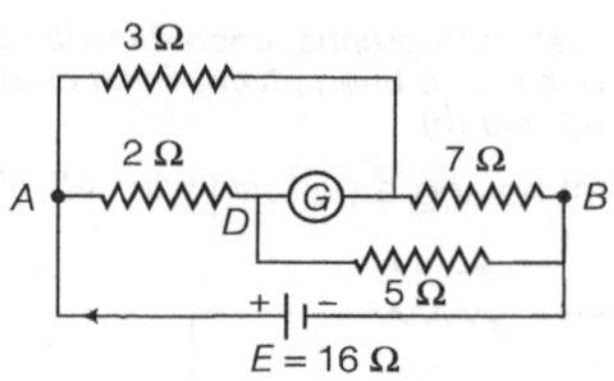
</td></tr>
<tr><td>iv.</td><td>Direction of current from D to B in branch BD ,is</td><td>s.</td><td>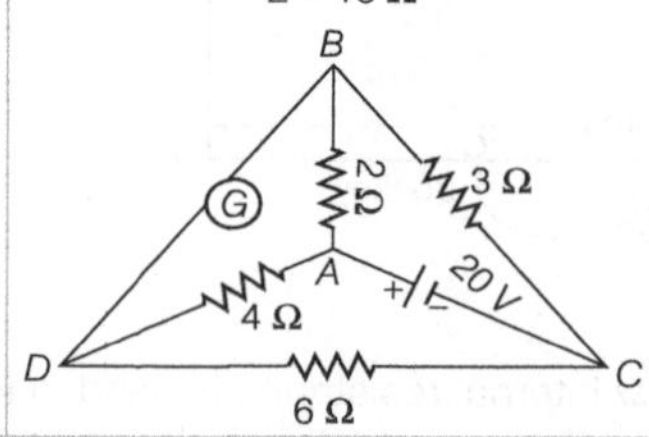
</td></tr>
</table>

Codes

	i	ii	iii	iv
a.	p	q	r,s	q,r
b.	p,q,r,s	s	p,q	r
c.	q,s	p,r	s	p,r
d.	p,r	q,s	p	q

JEE Main & AIEEE Archive

(Compilation of Last 13 Years Questions)

1 In a large building, there are 15 bulbs of 40 W, 5 bulbs of 100 W, 5 fans of 80 W and 1 heater of 1 kW. The voltage of the electric mains is 220 V. Then, the minimum current carrying capacity of fuse should be **(JEE Main 2014)**

a. 8 A b. 10 A
c. 12 A d. 14 A

2 The supply voltage to room is 120 V. The resistance of the lead wires is 6 Ω. A 60 W bulb is already switched on. What is the decrease of voltage across the bulb, when a 240 W heater is switched on in parallel to the bulb? **(JEE Main 2013)**

a. Zero b. 2.9 V
c. 13.3 V d. 10.04 V

3 **Statement I** Higher the range, greater is the resistance of ammeter.

Statement II To increase the range of ammeter, additional shunt needs to be used across it. **(JEE Main 2013)**

a. Both Statement I and Statement II are correct and Statement II is the correct explanation of Statement I
b. Both Statement I and Statement II are correct and Statement II is not the correct explanation of Statement I
c. Statement I is correct but Statement II is incorrect
d. Statement II is correct but Statement I is incorrect

4 Two electric bulbs marked 25 W - 220 V and 100 W - 220 V are connected in series to a 440 V supply. Which of the bulbs will fuse? **(AIEEE 2012)**

a. Both b. 100 W
c. 25 W d. None of these

5 Resistance of a given wire is obtained by measuring the current flowing in it and the voltage difference applied across it. If the percentage errors in the measurement of the current and the voltage difference are 3% each, then error in the value of resistance of the wire is

a. 6% b. zero
c. 1% d. 3%

6 If 400 Ω of resistance is made by adding four 100 Ω resistance of tolerance 5%, then the tolerance of the combination is **(AIEEE 2011)**

a. 20% b. 5%
c. 10% d. 15%

7 The current in the primary circuit of a potentiometer is 0.2 A. The specific resistance and cross-section of the potentiometer wire are 4×10^{-7} Ω-m and 8×10^{-7} m^2, respectively. The potential gradient will be equal to **(AIEEE 2011)**

a. 0.2 V/m b. 1 V/m
c. 0.3 V/m d. 0.1 V/m

8 If a wire is stretched to make, it 0.1% longer, its resistance will **(AIEEE 2011)**

a. increase by 0.2% b. decrease by 0.2%
c. decrease by 0.05% d. increase by 0.05%

9 A rectangular loop has a sliding connector PQ of length l and resistance R Ω and it is moving with a speed v as shown. The set up is placed in a uniform magnetic field going into the plane of the paper. The three currents I_1, I_2 and I are **(AIEEE 2010)**

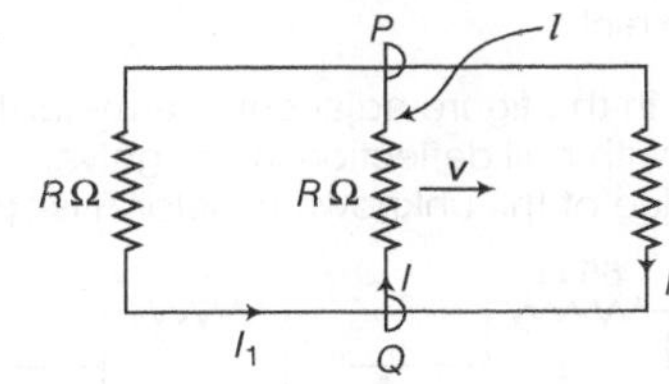

a. $I_1 = -I_2 = \frac{Blv}{R}, I = \frac{2Blv}{R}$

b. $I_1 = I_2 = \frac{Blv}{3R}, I = \frac{2Blv}{3R}$

c. $I_1 = I_2 = I = \frac{Blv}{R}$

d. $I_1 = I_2 = \frac{Blv}{6R}, I = \frac{Blv}{3R}$

10 For the circuit shown below, the key K is closed at $t = 0$. The current through the battery is **(AIEEE 2010)**

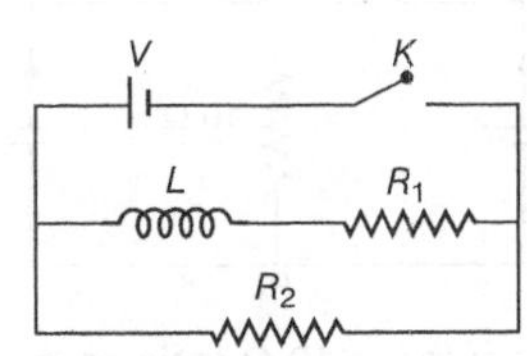

a. $\frac{VR_1R_2}{\sqrt{R_1^2 + R_2^2}}$ at $t = 0$ and $\frac{V}{R_2}$ at $t = \infty$

b. $\frac{V}{R_2}$ at $t = 0$ and $\frac{V(R_1 + R_2)}{R_1R_2}$ at $t = \infty$

c. $\frac{V}{R_2}$ at $t = 0$ and $\frac{VR_1R_2}{\sqrt{R_1^2 + R_2^2}}$ at $t = \infty$

d. $\frac{V(R_1 + R_2)}{R_1R_2}$ at $t = 0$ and $\frac{V}{R_2}$ at $t = \infty$

11 Two conductors have the same resistance at 0°C but their temperature coefficients of resistance are α_1 and α_2. The respective temperature coefficients of their series and parallel combinations are nearly **(AIEEE 2010)**

a. $\frac{\alpha_1 + \alpha_2}{2}, \alpha_1 + \alpha_2$ b. $\alpha_1 + \alpha_2, \frac{\alpha_1 + \alpha_2}{2}$
c. $\alpha_1 + \alpha_2, \frac{\alpha_1\alpha_2}{\alpha_1 + \alpha_2}$ d. $\frac{\alpha_1 + \alpha_2}{2}, \frac{\alpha_1 + \alpha_2}{2}$

12 **Statement I** The temperature dependence of resistance is usually given as $R = R_0(1 + \alpha\Delta t)$. The resistance of a wire changes from 100 Ω to 150 Ω, when its temperature is increased from 27°C to 227°C. This implies that $\alpha = 2.5 \times 10^{-3}/°C$.

Statement II $R = R_i(1 + \alpha\Delta T)$ is valid only when the change in the temperature ΔT is small and $\Delta R = (R - R_0) << R_0$. **(AIEEE 2009)**

a. Both Statement I and Statement II are the correct and Statement II is the correct explanation of Statement I
b. Both Statement I and Statement II are correct and Statement II is not the correct explanation of Statement I
c. Statement I is the correct but Statement II is incorrect
d. Statement II is the correct but Statement I is incorrect

13 Shown in the figure adjacent is a meter-bridge set up with null deflection in the galvanometer. The value of the unknown resistor R is **(AIEEE 2008)**

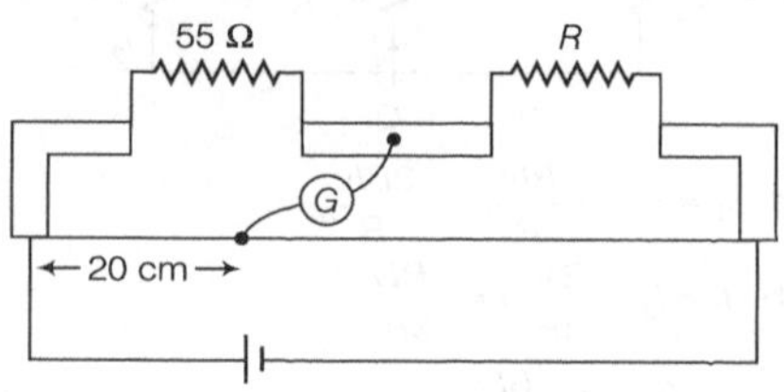

a. 13.75 Ω b. 220 Ω c. 110 Ω d. 55 Ω

14 A 5 V battery with internal resistance 2 Ω and a 2 V battery with internal resistance 1 Ω are connected to a 10 Ω resistor as shown in the figure.

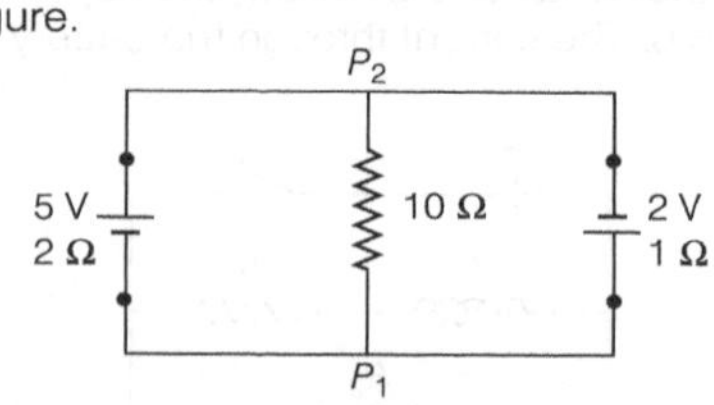

The current in the 10 Ω resistor is **(AIEEE 2008)**

a. 0.27 A, P_2 to P_1 b. 0.03 A, P_1 to P_2
c. 0.03 A, P_2 to P_1 d. 0.27 A, P_1 to P_2.

15 The resistance of a wire is 5 Ω at 50°C and 6 Ω at 100°C. The resistance of the wire at 0°C will be **(AIEEE 2007)**

a. 2 Ω b. 1 Ω c. 4 Ω d. 3 Ω

16 A material B has twice the specific resistance of A. A circular wire made of B has twice the diameter of a wire made of A. Then, for the two wires to have the same resistance, the ratio l_B/l_A of their respective lengths must be **(AIEEE 2006)**

a. 1 b. 1/2 c. 1/4 d. 2

17 An electric bulb is rated 220 V-100 W. The power consumed by it when operated on 110 V will be **(AIEEE 2006)**

a. 75 W b. 40 W c. 25 W d. 50 W

18 The resistance of a bulb filament is 100 Ω at a temperature of 100°C. If its temperature coefficient of resistance is 0.005 per °C, its resistance will become 200 Ω at a temperature of **(AIEEE 2006)**

a. 300°C b. 400°C c. 500°C d. 200°C

19 In a Wheatstone's bridge, three resistances P, Q and R are connected in the three arms and the fourth arm is formed by two resistances S_1 and S_2 connected in parallel. The condition for the bridge to be balanced will be **(AIEEE 2006)**

a. $\frac{P}{Q} = \frac{2R}{S_1 + S_2}$ b. $\frac{P}{Q} = \frac{R(S_1 + S_2)}{S_1S_2}$
c. $\frac{P}{Q} = \frac{R(S_1 + S_2)}{2S_1S_2}$ d. $\frac{P}{Q} = \frac{R}{S_1 + S_2}$

d. flow from Antimony to Bismuth at the cold junction

20 The Kirchhoff's first law $(\Sigma i = 0)$ and second law $(\Sigma iR = \Sigma E)$, where the symbols have their usual meanings, are respectively based on **(AIEEE 2006)**

a. conservation of charge, conservation of momentum
b. conservation of energy, conservation of charge
c. conservation of momentum, conservation of charge
d. conservation of charge, conservation of energy

21 The current I drawn from the 5 V source will be **(AIEEE 2006)**

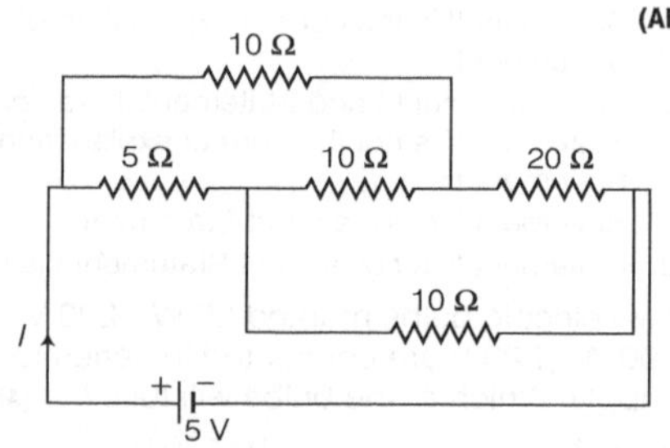

a. 0.33 A b. 0.5 A c. 0.67 A d. 0.17 A

22 In the circuit, the galvanometer G shows zero deflection. If the batteries A and B have negligible internal resistance, the value of the resistor R will be **(AIEEE 2005)**

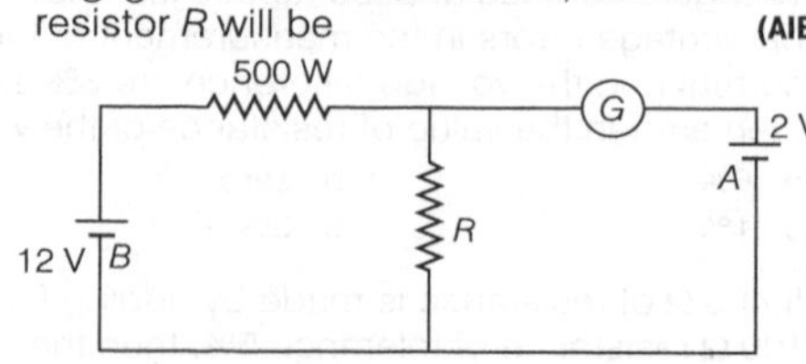

a. 200 Ω b. 100 Ω c. 500 Ω d. 1000 Ω

23 Two sources of equal emf are connected to an external resistance R. The internal resistances of the two sources are R_1 and R_2, $(R_2 > R_1)$. If the potential difference across the source having internal resistance R_2 is zero, then **(AIEEE 2005)**

a. $R = \frac{R_2 \times (R_1 + R_2)}{(R_2 - R_1)}$ b. $R = R_2 - R_1$
c. $R = \frac{R_1R_2}{(R_1 + R_2)}$ d. $R = \frac{R_1R_2}{(R_2 - R_1)}$

24 A heater coil is cut into two equal parts and only one part is now used in the heater. The heat generated will now be **(AIEEE 2005)**

a. doubled b. four times
c. one-fourth d. halved

25 An energy source will supply a constant current into the load, if its internal resistance is **(AIEEE 2005)**

a. equal to the resistance of the load
b. very large as compared to the load resistance
c. zero
d. non-zero but less than the resistance of the load

26 In a potentiometer experiment, the balancing with a cell is at length 240 cm. On shunting the cell with a resistance of 2 Ω, the balancing length becomes 120 cm. The internal resistance of the cell is **(AIEEE 2005)**

a. 1 Ω b. 0.5 Ω c. 4 Ω d. 2 Ω

27 The resistance of hot tungsten filament is about 10 times the cold resistance. What will be the resistance of 100 W and 200 V lamp, when not in use? **(AIEEE 2005)**

a. 40 Ω b. 20 Ω c. 400 Ω d. 200 Ω

28 The total current supplied to the circuit by the battery is **(AIEEE 2004)**

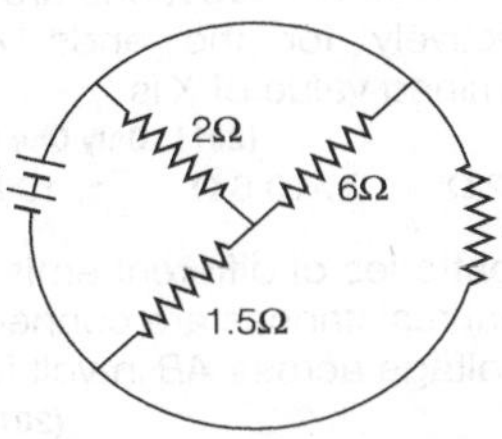

a. 1 A b. 2 A c. 4 A d. 6 A

29 The resistance of the series combination of two resistances is S. When they are joined in parallel, the total resistance is P. If $S = nP$, then the minimum possible value of n is **(AIEEE 2004)**

a. 4 b. 3 c. 2 d. 1

30 In a meter bridge experiment, null point is obtained at 20 cm from one end of the wire, when resistance X is balanced against another resistance Y. If $X < Y$, then where will be the new position of the null point from the same end, if one decides to balance a resistance of $4X$ against Y? **(AIEEE 2004)**

a. 50 cm b. 80 cm c. 40 cm d. 70 cm

31 The thermistors are usually made of **(AIEEE 2004)**

a. metals with low temperature coefficient of resistivity
b. metals with high temperature coefficient of resistivity
c. metal oxides with high temperature coefficient of resistivity
d. semiconducting materials having low temperature coefficient of resistivity

32 Time taken by a 836 W heater to heat 1 L of water from 10°C to 40°C is **(AIEEE 2004)**

a. 50 s b. 100 s c. 150 s d. 200 s

33 The length of a wire of a potentiometer is 100 cm and the emf of its stand and cell is E volt. It is employed to measure the emf of a battery whose internal resistance is 0.5 Ω. If the balance point is obtained at $l = 30$ cm from the positive end, the emf of the battery is **(AIEEE 2003)**

a. $\frac{30E}{100.5}$

b. $\frac{30E}{100 - 0.5}$

c. $\frac{30(E - 0.5I)}{100}$, where I is the current in the potentiometer wire

d. $\frac{30E}{100}$

34 A 3 V battery with negligible internal resistance is connected in a circuit as shown in the figure. The current I in the circuit will be **(AIEEE 2003)**

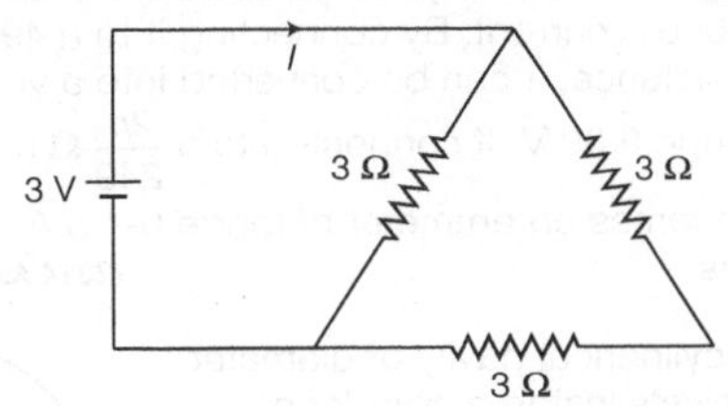

a. 1 A b. 1.5 A c. 2 A d. $\frac{1}{3}$ A

35 The length of a given cylindrical wire is increased by 100%. Due to the consequent decrease in diameter, the change in the resistance of the wire will be **(AIEEE 2003)**

a. 200% b. 100% c. 50% d. 300%

36 A 220 V–1000 W bulb is connected across a 110 V mains supply. The power consumed will be **(AIEEE 2003)**

a. 750 W b. 500 W c. 250 W d. 1000 W

37 A wire when connected to 220 V mains supply has power dissipation P_1. Now, the wire is cut into two equal pieces, which are connected in parallel to the same supply. Power dissipation in this case is P_2. Then, $P_2 : P_1$ is **(AIEEE 2002)**

a. 1 b. 4 c. 2 d. 3

38 If in the circuit, power dissipation is 150 W, then R is **(AIEEE 2002)**

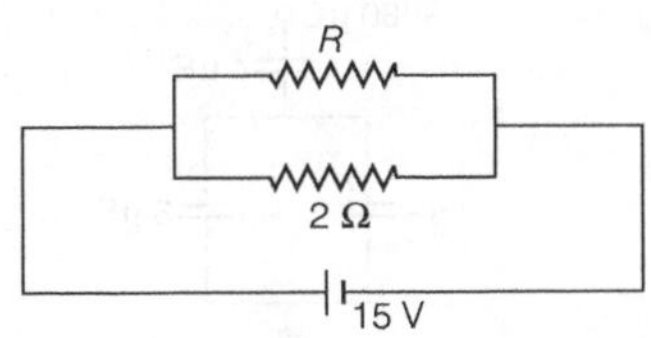

a. 2 Ω b. 6 Ω c. 5 Ω d. 4 Ω

JEE Advanced & IIT JEE Archive

(Compilation of Last 10 Years Questions)

1 During an experiment with a meter bridge, the galvanometer shows a null point, when the jockey is pressed at 40.0 cm using a standard resistance of 90 Ω, as show in the scale used in the meter bridge is 1 mm. The unknown resistance is

(2014 Adv., Only One Option Correct Type)

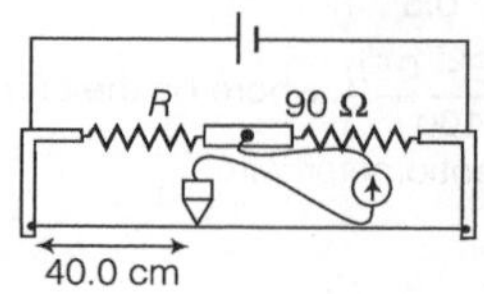

a. 60± 0.15 Ω b. 135± 0.56 Ω
c. 60± 0.25 Ω d. 135± 0.23 Ω

2 A galvanometer gives full scale deflection with 0.006 A current. By connecting it to a 4990 Ω resistance, it can be converted into a voltmeter of range 0-30 V. If connected to a $\frac{2n}{249}$ Ω resistance, it becomes an ammeter of range 0-1.5 A. The value of n is

(2014 Adv., Integer Type)

3 A cylindrical cavity of diameter a exists inside a cylinder of diameter $2a$ as shown in the figure. Both the cylinder and the cavity are infinitely long. A uniform current density J flows along the length. If the magnitude of the magnetic field at the point P is given by $\frac{N}{12}\mu_0 aJ$, then the value of N is

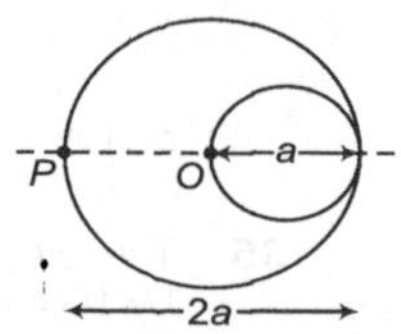

(2012, Integer Type)

4 In the given circuit, the AC source has $\omega = 100$ rad/s. Considering the inductor and capacitor to be ideal, the correct choice(s) is(are)

(2012, One or More than one Options Correct Type)

a. The current through the circuit, I is 0.3 A
b. The current through the circuit, I is $0.3\sqrt{2}$ A
c. The voltage across 100 Ω resistor = $10\sqrt{2}$ V
d. The voltage across 50 Ω resistor = 10 V

5 In the given circuit, a charge of +80 μC is given to the upper plate of the 4 μF capacitor. Then, in the steady state, the charge on the upper plate of the 3 μF capacitor is **(2012, Only One Option Correct Type)**

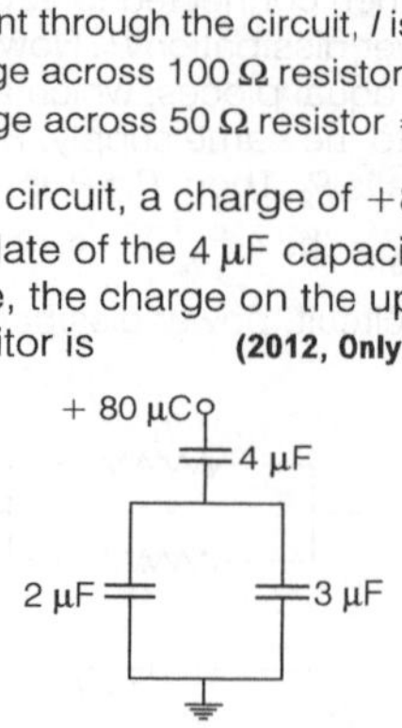

a. + 32 μC b. + 40 μC
c. + 48 μC d. + 80 μC

6 An infinitely long hollow conducting cylinder with inner radius $R/2$ and outer radius R carries a uniform current density along its length. The magnitude of the magnetic field, $|\mathbf{B}|$ as a function of the radial distance r from the axis is best represented by

(2012, Only One Option Correct Type)

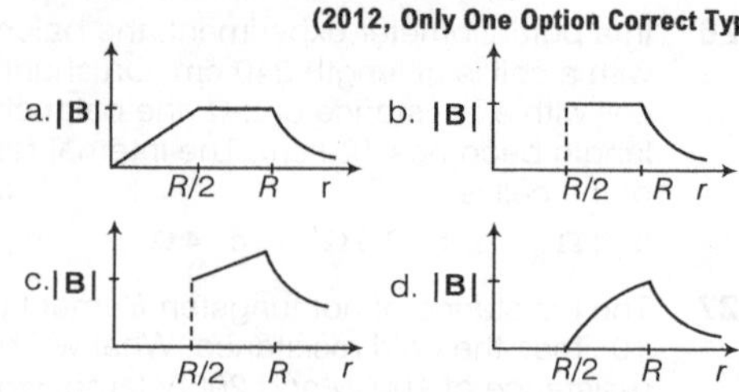

7 A meter bridge is set up as shown in the figure, to determine an unknown resistance X using a standard 10 Ω resistor. The galvanometer shows null point, when tapping-key is at 52 cm mark. The end-corrections are 1 cm and 2 cm respectively for the ends A and B. The determined value of X is

(2011, Only One Option Correct Type)

a. 10.2 Ω b. 10.6 Ω c. 10.8 Ω d. 11.1 Ω

8 Two batteries of different emfs and different internal resistances are connected as shown. The voltage across AB in volt is.

(2011, Descriptive Type)

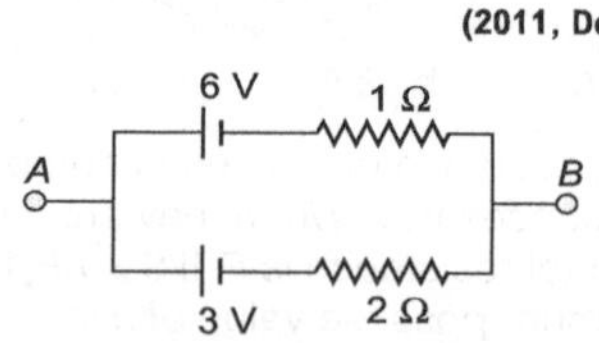

9 To verify Ohm's law, a student is provided with a test resistor R_T, a high resistance R_1, a small resistance R_2, two identical galvanometers G_1 and G_2, and a variable voltage source V. The correct circuit to carry out the experiment is

(2010, Only One Option Correct Type)

a.

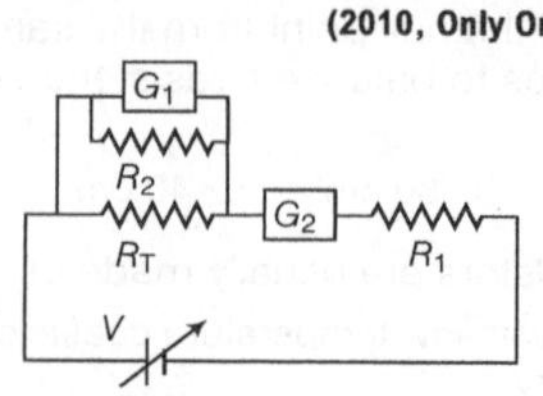

b.

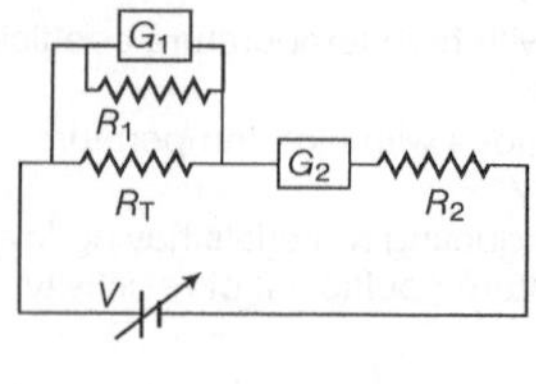

c. 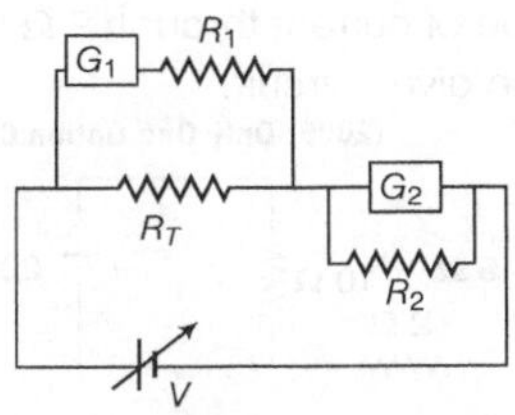

d. 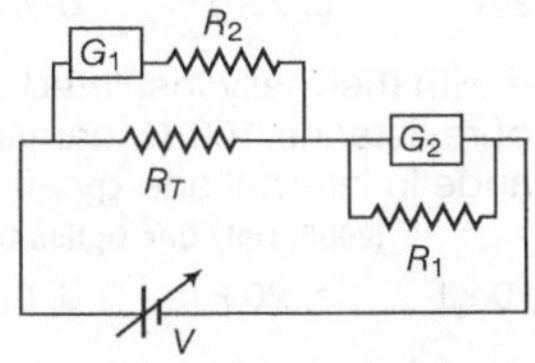

10 Incandescent bulbs are designed by keeping in mind that the resistance of their filament increases with the increase in temperature. If at room temperature 100 W, 60 W and 40 W bulbs have filament resistances R_{100}, R_{60} and R_{40}, respectively, the relation between these resistances is **(2010, Only One Option Correct Type)**

a. $\frac{1}{R_{100}} = \frac{1}{R_{40}} + \frac{1}{R_{60}}$ b. $R_{100} = R_{40} + R_{60}$

c. $R_{100} > R_{60} > R_{40}$ d. $\frac{1}{R_{100}} > \frac{1}{R_{60}} > \frac{1}{R_{40}}$

11 Consider a thin square sheet of side L and thickness t, made of a material of resistivity ρ. The resistance between two opposite faces, shown by the shaded areas in the figure is **(2010, Only One Option Correct Type)**

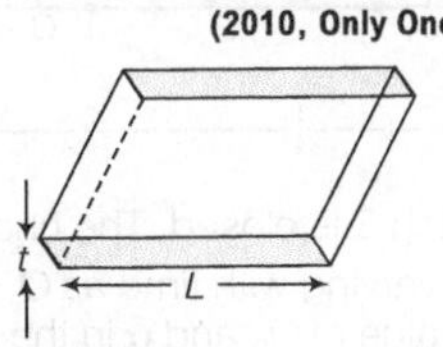

a. directly proportional to L
b. directly proportional to t
c. independent of L
d. independent of t

12 At time $t = 0$, a battery of 10 V is connected across points A and B in the given circuit. If the capacitors have no charge initially, at what time (in second) does the voltage across them become 4 V? Take : ln 5 = 1.6, ln 3 = 1.1 **(2010, Descriptive Type)**

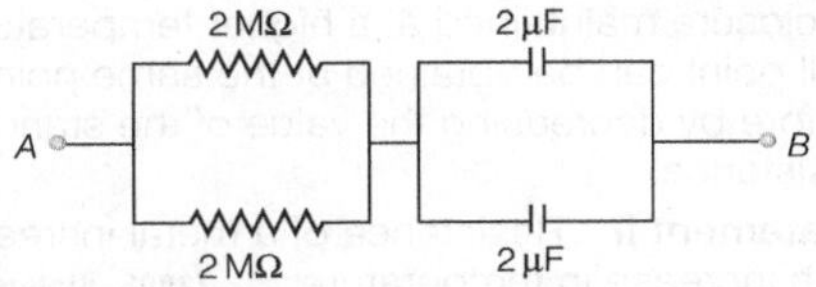

13 When two identical batteries of internal resistance 1 Ω each are connected in series across a resistor R, the rate of heat produced in R is J_1. When the same batteries are connected in parallel across R, the rate is J_2. If $J_1 = 2.25 J_2$, then the value of R in Ω is **(2010, Descriptive Type)**

14 For the resistance network shown in the figure, choose the correct option(s).

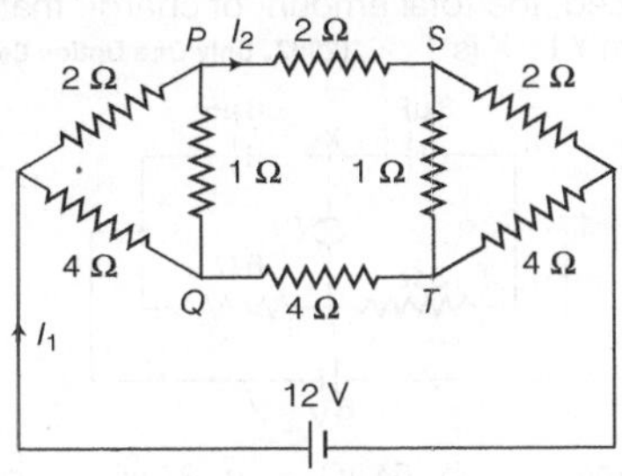

a. The current through PQ is zero
b. $I_1 = 3$ A
c. The potential at S is less than that at Q
d. $I_2 = 2$ A

15 For the circuit shown in the figure **(2009, One or More than One Options Correct Type)**

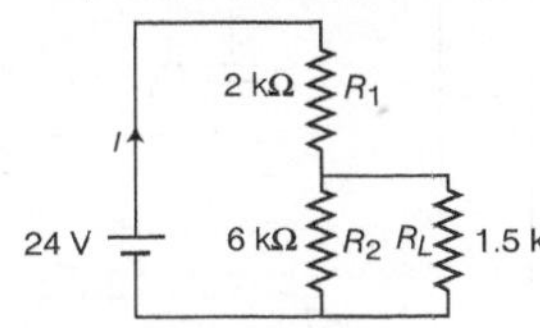

a. the current I through the battery is 7.5 mA
b. the potential difference across R_L is 18 V
c. ratio of powers dissipated in R_1 and R_2 is 3
d. if R_1 and R_2 are interchanged, magnitude of the power dissipated in R_L will decrease by a factor of 9

16 Figure shows three resistor configurations R_1, R_2 and R_3 connected to 3 V battery. If the power dissipated by the configuration R_1, R_2 and R_3 is P_1, P_2 and P_3, respectively, then **(2008, Only One Option Correct Type)**

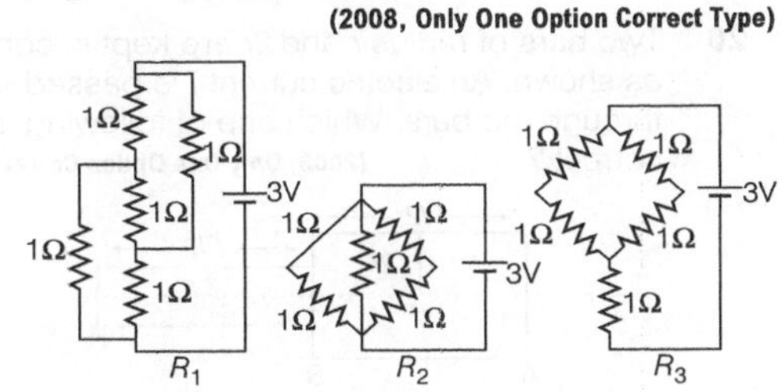

a. $P_1 > P_2 > P_3$
b. $P_1 > P_3 > P_2$
c. $P_2 > P_1 > P_3$
d. $P_3 > P_2 > P_1$

17 A resistance of 2 Ω is connected across one gap of a meter-bridge (the length of the wire is 100 cm) and an unknown resistance, greater than 2 Ω, is connected across the other gap. When these resistances are interchanged, the balance point shifts by 20 cm. Neglecting any corrections, the unknown resistance is **(2007, Only One Option Correct Type)**

a. 3 Ω b. 4 Ω
c. 5 Ω d. 6 Ω

JEE Archive

18 A circuit is connected as shown in the figure with the switch S open. When the switch is closed, the total amount of charge that flows from Y to X is **(2007, Only One Option Correct Type)**

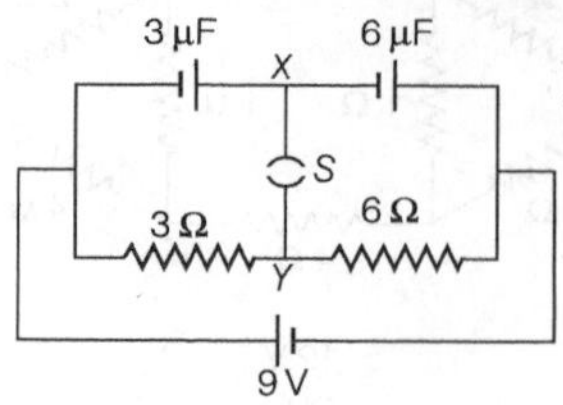

a. zero b. 54 μC c. 27 μC d. 81 μC

19 Find the time constant for the given RC circuits in correct order (in μs). **(2006, Only One Option Correct Type)**

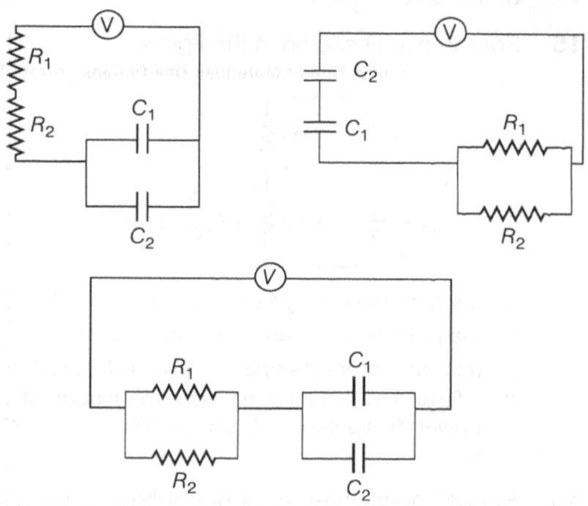

$R_1 = 1\,\Omega, R_2 = 2\,\Omega, C_1 = 4\,\mu F, C_2 = 2\,\mu F$.

a. 18, 4, 8/9 b. 18, 8/9, 4
c. 4, 18, 8/9 d. 4, 8/9, 18

20 Two bars of radius r and $2r$ are kept in contact as shown. An electric current I is passed through the bars. Which one of following is correct ? **(2006, Only One Option Correct Type)**

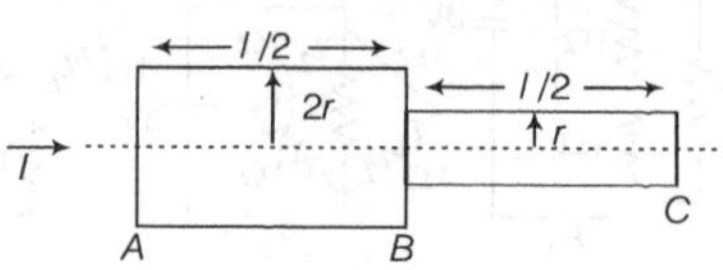

a. Heat produced in bar BC is 4 times the heat produced in bar AB
b. Electric field in both halves is equal
c. Current density across AB is double that of across BC
d. Potential difference across AB is 4 times that of across BC

21 A 4μF capacitor and a resistance of 2.5 M Ω are in series with 12 V battery. Find the time after which the potential difference across the capacitor is 3 times the potential difference across the resistor. (Given, ln (2) = 0.693) **(2005, Only One Option Correct Type)**

a. 13.86 s b. 6.93 s c. 7 s d. 14 s

22 Find out the value of current through 2 Ω resistance for the given circuit. **(2005, Only One Option Correct Type)**

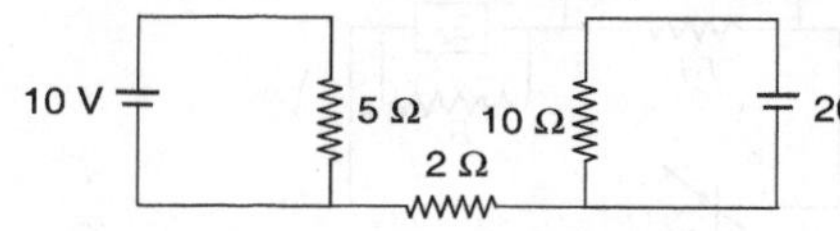

a. 5 A b. 2 A c. zero d. 4 A

23 A rigid container with thermally insulated walls contains a coil of resistance 100 Ω, carrying current 1 A. Change in internal energy after 5 min will be **(2005, Only One Option Correct Type)**

a. zero b. 10 kJ c. 20 kJ d. 30 kJ

24 A moving coil galvanometer of resistance 100 Ω is used as an ammeter using a resistance 0.1 Ω. The maximum deflection current in the galvanometer is 100 μA. Find the current in the circuit, so that the ammeter shows maximum deflection **(2005, Only One Option Correct Type)**

a. 100.1 mA b. 1000.1 mA
c. 10.01 mA d. 1.01 mA

25 R_1, R_2 and R_3 are different values of R. A, B and C are the null points obtained corresponding to R_1, R_2 and R_3, respectively. For which resistor, the value of X will be the most accurate and why? **(2005, Descriptive Type)**

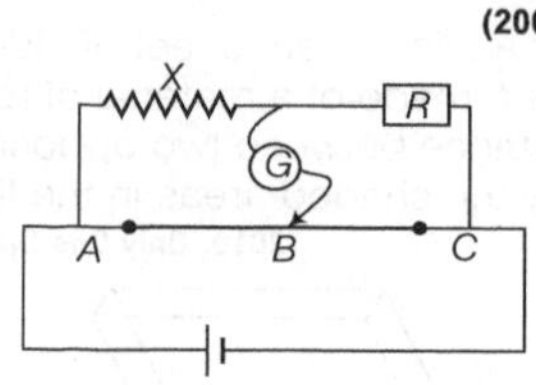

26 At $t = 0$, switch S is closed. The charge on the capacitor is varying with time as $Q = Q_0(1 - e^{-\alpha t})$. Obtain the value of Q_0 and α in the given circuit parameters. **(2005, Subjective Type)**

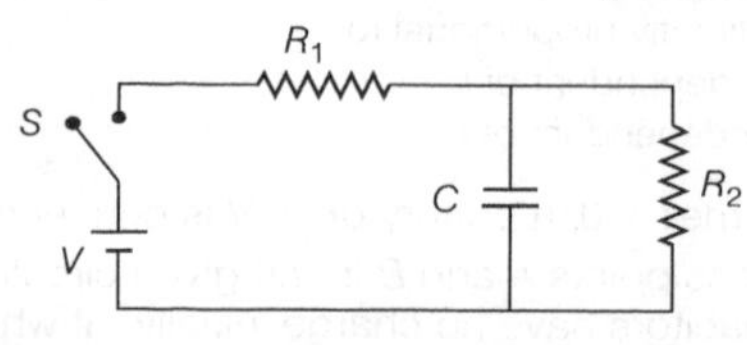

27 **Statement I** In a meter bridge experiment, null point for an unknown resistance is measured. Now, the unknown resistance is put inside an enclosure maintained at a higher temperature. The null point can be obtained at the same point as before by decreasing the value of the standard resistance.

Statement II Resistance of a metal increases with increase in temperature. **(2008, Statement Type)**

a. Both Statement I and Statement II are correct and Statement II is the correct explanation of Statement I
b. Both Statement I and Statement II are correct and Statement II is not the correct explanation of Statement I
c. Statement I is correct but Statement II is incorrect
d. Statement II is correct but Statement I is incorrect

Answers *with* Explanations

Answers

1. Electrostatics

DPP-1 Charge and Properties of Charges

1. (i) 6nC (ii) Electrons are removed (iii) 3.75×10^{10}

4. (i) 2×10^{12} electrons (ii) 2×10^{18} kg **5.** No

6. *d* **7.** *a* **8.** *c* **9.** *c* **10.** *b* **11.** *c* **12.** *c* **13.** *d* **14.** *d*

15. *d* **16.** *d* **17.** *a* **18.** *d* **19.** *b,c,d* **20.** *a,b,d* **21.** *a,b,c,d* **22.** *d*

23. *c* **24.** 4 **25.** 5

DPP-2 Coulomb's Law and Principle of Superposition

3. (i) 12×10^{-12} (ii) 6×10^{-6} C, -2×10^{-6} C

4. 13 m

5. (i) Mass decreases (ii) Force increased (iii) Lesser than the velocity of electron

6. *c* **7.** *c* **8.** *c* **9.** *a* **10.** *b* **11.** *a* **12.** *a* **13.** *c* **14.** *c* **15.** *a*

16. *a,b* **17.** *a,b* **18.** *a,b,c* **19.** *a,c* **20.** *b,d* **21.** *c,d* **22.** *b* **23.** *c* **24.** *a* **25.** *c*

26. 3 **27.** 2

DPP-3 Electric Field, Field Intensity and Superposition of Electric Field

2. $\dfrac{q\lambda}{4\pi\varepsilon_0 R}$ N **5.** *a* **6.** *c* **7.** *a* **8.** *b* **9.** *b* **10.** *c* **11.** *d* **12.** *c*

13. *c* **14.** *b,c* **15.** *b,c* **16.** *b,d* **17.** *b,d* **18.** *b,d* **19.** *c* **20.** *d* **21.** *a* **22.** *a*

23. *a* **24.** *c* **25.** *b* **26.** *d* **27.** *c* **28.** *c* **29.** *a* **30.** 5 **31.** 2

DPP-4 Electric Flux and Gauss' Law

4. (i) Flux remains same (ii) Flux remains same

(iii) Flux remains same (iv) Flux does not change

5. *a* **6.** *b* **7.** *c* **8.** *b* **9.** *c* **10.** *d* **11.** *c* **12.** *b* **13.** *c* **14.** *d*

15. *c,d* **16.** *a,b,d* **17.** *a,c,d* **18.** *b,c* **19.** *c,d* **20.** *a,b,c* **21.** *a* **22.** 1 **23.** 2 **24.** 6

25. 9 **26.** 24 **27.** 3 **28.** d **29.** c **30.** *d* **31.** *c*

DPP-5 Applications of Gauss' Law

1. (i) zero (ii) $\dfrac{Q(r^3 - r_1^3)}{(r_0^3 - r_1^3)4\pi\varepsilon_0 r^2}$ (iii) $\dfrac{Q}{4\pi\varepsilon_0 r^2}$

(a) $\dfrac{q}{4\pi\varepsilon_0 r^2}$ (b) $\dfrac{1}{4\pi\varepsilon_0 r^2}\left[q + Q\dfrac{(r^3 - r_1^3)}{(r_0^3 - r_1^3)}\right]$ (c) $\dfrac{q+Q}{4\pi\varepsilon_0 r^2}$

2. (i) $\dfrac{-\rho_E r_0}{6\varepsilon_0}$ (ii) $\dfrac{17\rho_E r_0}{54\varepsilon_0}$

3. $\varepsilon_0 a l^3$

4. (i) $\dfrac{kQ}{r_0^{\,2}\left(1+\dfrac{t}{T}\right)^2}$ (ii) $\dfrac{kQ}{(3.2r_0)^2}$

5. *b* **6.** *c* **7.** *c* **8.** *a* **9.** *a* **10.** *b* **11.** *c* **12.** *d* **13.** *d* **14.** *b*

15. *b* **16.** *c* **17.** *a,b* **18.** *a,c* **19.** *c,d* **20.** *c,d* **21.** *a,b,c,d* **22.** *a* **23.** *a*

24. *a* **25.** 2 **26.** 3 **27.** 2 **28.** *c* **29.** *c* **30.** *d* **31.** *d* **32.** *c* **33.** *a*

34. *a*

DPP-6 Electric Dipole, Dipole in an External Field

1. (i) 1.05×10^2 cm (ii) No
2. (i) p (ii) $\frac{1}{2}$PE, torque is directed along negative Z- axis.

4. c 5. a 6. b 7. c 8. c 9. c 10. a 11. a 12. c 13. a
14. a,b,c,b 15. b,d 16. a,c 17. b,d 18. b,c 19. d 20. c 21. 4 22. 4
23. a 24. a 25. d 26. c 27. c 28. a 29. a

DPP-7 Electric Potential and Potential Energy

1. (i) 2.25×10^5V, -2.25×10^5V
   (ii) 0.9 J
   (iii) Work done is not path dependent
   (iv) Potential energy increases
2. (i) A negative charge moves from a lower potential point to higher potential point
   (ii) -8.0×10^{-16} J, 4.2×10^7 m / s
3. (i) $\frac{k \times 32 \times 10^{-12}}{5\sqrt{2} \times 10^{-12}}$J (ii) Condition Satisfies

5. a 6. c 7. d 8. a 9. c 10. d 11. c 12. d 13. d 14. b
15. b 16. b 17. c,d 18. b,c,d 19. b,c,d 20. 6 21. 3 22. b 23. c 24. a
25. d 26. d 27. d 28. d 29. a 30. a

DPP-8 Potential Due to Charge Distributions

4. $kq\left(\frac{1}{d-b} - \frac{1}{b} - \frac{1}{b} + \frac{1}{d-b}\right)$

5. zero 6. a 7. c 8. d 9. a 10. a 11. a 12. c 13. d 14. b
15. a 16. a,b 17. b,c 18. a,c 19. a,c,d 20. a 21. d 22. b 23. 2 24. 2
25. c 26. d 27. c 28. b 29. b 30. b

DPP-9 Motion of a Charged Particle inside an Electric Field

2. 1.45×10^7 m/s
3. 5.86 m/s
4. (i) Yes
   (ii) At first plate (at $x = 2$)

5. b 6. c 7. c 8. a 9. d 10. b 11. c 12. d 13. b,c 14. a,d
15. a,c 16. a,c 17. b,d 18. d 19. d 20. 2 21. b 22. d 23. d

DPP-10 Electric Field and Potential Equipotentials

4. (i) $\mathbf{E} = (-5+6xy)\hat{\mathbf{i}} + (3x^2 - 2z^2)\hat{\mathbf{j}} + (-4yz)\hat{\mathbf{k}}$ (ii) $5\sqrt{2}$ N/C

5. a 6. d 7. b 8. b 9. a 10. d 11. d 12. c 13. d 14. d
15. d 16. c 17. d 18. a,c 19. b,c,d 20. b,c 21. b,c 22. a,b,c 23. 4 24. 4
25. d 26. b 27. c 28. a

DPP-11 Electrical Potential Energy

1. (ii) $\frac{1}{4\pi\varepsilon_0}\frac{3q^2}{5R}$

2. $\frac{Qq}{8\pi\varepsilon_0}\left(\frac{1}{r_1}-\frac{1}{r_2}\right)$

3. (i) 0.225 J (ii) $a_1 : a_2 = 2:1$ (iii) $v_1 : v_2 = 2:1$

4. $-27\sqrt{2}$ V

5. (i) $F = \frac{-dU}{dr}$ (ii) $V = \frac{U}{q}$ (iii) $\frac{dV}{dr} = -E$ (iv) zero

6. *d*	7. *d*	8. *d*	9. *d*	10. *b*	11. *c*	12. *b*	13. *a,b*	14. *b,d*	15. *b,d*
16. *b,c*	17. *a,d*	18. *a,d*	19. *b*	20. *d*	21. *c*	22. *a*	23. *b*	24. *b*	25. *c*
26. *c*	27. *a*								

DPP-12 Potential and Potential Energy of Dipole

1. $r = \left(\frac{p}{4\pi\varepsilon_0 E}\right)^{\frac{1}{3}}$

2. (i) zero (ii) $\frac{-\lambda}{2\pi\varepsilon_0 r^2}$ (iii) $\frac{-\rho\lambda}{2\pi\varepsilon_0 r^2}$

3. (i) 2.6×10^{-3} V (ii) 1.8×10^{-3} V

4. $\frac{1}{4\pi\varepsilon_0}\frac{p\cos\theta}{r^2}$

5. *a*	6. *b*	7. *c*	8. *c*	9. *a*	10. *d*	11. *c*	12. *a*	13. *d*	14. *c*
15. *b,d*	16. *a,c*	17. *a,c*	18. *a,c,d*	19. *d*	20. *a*	21. *c*	22. c		

Revisal Problems (JEE Main)

1. *c*	2. *a*	3. *c*	4. *d*	5. *b*	6. *d*	7. *c*	8. *b*	9. *b*	10. *d*
11. *b*	12. *b*	13. *c*	14. *a*	15. *c*	16. *d*	17. *c*	18. *d*	19. *b*	20. *b*

Revisal Problems (JEE Advanced)

1. *a*	2. *b*	3. *c*	4. *d*	5. *c*	6. *a*	7. *c*	8. *d*	9. *b*	10. *c*
11. *d*	12. *b*	13. *d*	14. *b*	15. *a*	16. *c*	17. *c,d*	18. *a,b,c,d*		
19. *a,b,d*	20. *b,c*	21. *b*	22. *a*	23. *b*	24. *d*	25. 4	26. 8		

JEE Main & AIEEE Archive

1. *c*	2. *d*	3. *b*	4. *c*	5. *d*	6. *a*	7. *d*	8. *a*	9. *c*	10. *c*
11. *b*	12. *b*	13. *c*	14. *d*	15. *a*	16. *c*	17. *c*	18. *a*	19. *d*	20. *d*
21. *b*	22. *c*	23. *a*	24. *c*	25. *d*	26. *a*	27. *b*	28. *d*	29. *d*	30. *a*
31. *a*	32. *c*	33. *b*	34. *d*	35. *d*	36. *a*				

JEE Advanced & IIT JEE Archive

1. *c*	2. *a*	3. *c*	4. *b,d*	5. *c,d*	6. *a,b,c*	7. *a,c*	8. 6	9. *c*	10. *d*
11. *c,*	12. *a,b,c,d*		13. *c,d*	14. 3	15. *a,d*	16. *d*	17. *a*	18. *a*	19. *a*
20. *b*	21. *a*	22. 2	23. *b*	24. *a*	25. *b*	26. *c*	27. *c*	28. *a*	29. *d*
30. *b*	31. *c*	32. *a,b,c,d*		33. *b*	34. $V' = V\left(\frac{a}{3t}\right)^{\frac{1}{3}}$				

2. Capacitance

DPP-1 Capacitance

1. (i) 1 pF (ii) $C = 1$ pF (iii) Capacity remains same
2. (i) $Q_{max} = 4\pi\varepsilon_0 kR.V_{max}$ (ii) 3×10^6 V/m, (iii) 3.34×10^{-4} RC
3. (i) 9 m (ii) 6.95 μs 4. (iv) 3×10^{-3} C 5. *c* 6. *d* 7. *b* 8. *a*

9. *d* 10. *d* 11. *d* 12. *b* 13. *c* 14. *c* 15. *b,d* 16. *a,c* 17. *b,c* 18. *b,d*
19. *a,d* 20. *a,b* 21. *a,b* 22. *a* 23. 1 24. 8 25. 7 26. *b* 27. *d* 28. *d*

DPP-2 Capacitance of Different Types of Capacitors

1. (i) 1.0×10^4 V/m (ii) 5 nC (iii) 4.1 nC
2. (i) 507 pF (ii) 2.5 nC (iii) 2.08 nC

5. *a* 6. *b* 7. *d* 8. *a* 9. *d* 10. *d* 11. *a* 12. *d* 13. *c* 14. *c*
15. *a* 16. *a,b* 17. *a,c* 18. *b,d* 19. *a,b* 20. *a,d* 21. 8 22. 2 23. 6 24. *d*
25. *d* 26. *c* 27. *a* 28. *d* 29. *a*

DPP-3 Energy of a Charged Capacitor, Effect of Dielectric and Force between Plates of Capacitor

1. (i) 6.37 μJ (ii) 6×10^5 V/*m* (iii) 6.37 μJ (iv) 1.27×10^{-2} N
2. (i) 1.27 nJ (ii) 2.124×10^{-10} C (iii) −2.54 nJ (iv) Battery is Charged
3. (i) 140 pF (ii) 17 nC (iii) 243 V (iv) 2.06 μJ
   (v) Force remains same (vi) Increase is energy (2.06 μJ)

5. *d* 6. *d* 7. *c* 8. *c* 9. *a* 10. *d* 11. *d* 12. *c* 13. *c* 14. *c*
15. *d* 16. *c* 17. *a,c* 18. *a,d* 19. *a,c* 20. *a,b* 21. *a,d* 22. 5 23. 1 24. 6
25. *a* 26. *c* 27. *d* 28. *a*

DPP-4 Series and Parallel Combination of Capacitors

1. 4 : 25
2. $\frac{1}{k_1}+\frac{1}{k_2}=\frac{2}{k_{eq}}$ 3. $k_{eq}=\frac{k_1+k_2}{2}$ 5. *a* 6. *b* 7. *c* 8. *b* 9. *d* 10. *c*

11. *d* 12. *b* 13. *a* 14. *c* 15. *b,d* 16. *d* 17. *a* 18. *c* 19. *b,c* 20. *a,c*
21. *a* 22. *d* 23. *d* 24. *b* 25. *c* 26. *a* 27. a

DPP-5 Charging of a Capacitor and R-C Circuits

1. (i) $E = IR + \frac{Q}{C}$ (ii) $Q = CE(1-e^{-t/RC})$ (iii) $V = E(1-e^{-t/RC})$ (iv) $t = RC$ (v) $I = \frac{E}{R}e^{-t/RC}$
2. (i) $Q = Q_0 e^{-t/RC}$ (ii) $V_C = V_0 e^{-t/RC}$ (iii) $t = RC$
4. (i) $V(5) = 0.067$ V, $i(t) = 10^{-5}e^{-t}$A (ii) $V(0) = 2$ V, $i(t) = 2\times10^{-13}e^{-50t}$

5. *b* 6. *a* 7. *a* 8. *c* 9. *a* 10. *d* 11. *d* 12. *c* 13. *c* 14. *a*
15. *a* 16. *a,d* 17. *a,d* 18. *a* 19. *d* 20. 10 21. 2 22. 2 23. *b*

DPP-6 Redistribution of Charges and Van de Graaff Generator

4. *b* 5. *c* 6. *c* 7. *b* 8. *a* 9. *b* 10. *d* 11. *a* 12. *d* 13. *a*
14. *a,b,c,d* 15. *a,d* 16. *b,d* 17. *a,d* 18. *a* 19. *b* 20. *c* 21. *c* 22. *c*
23. *c* 24. *d* 25. *b* 26. *b* 27. *a*

Revisal Problems (JEE Main)

1. *c* 2. *b* 3. *b* 4. *d* 5. *c* 6. *a* 7. *b* 8. *c* 9. *c* 10. *b*
11. *c* 12. *c* 13. *b* 14. *d* 15. *b* 16. *b* 17. *a* 18. *c* 19. *b* 20. *c*
21. *c* 22. *d* 23. *c* 24. *b* 25. *c* 26. *c* 27. *a*

Revisal Problems (JEE Advanced)

1. *d* 2. *b* 3. *d* 4. *c* 5. *b* 6. *c* 7. *a* 8. *c* 9. *c* 10. *d*
11. *a,b,c* 12. *b,c,d* 13. *a,b,c,d* 14. *a,b* 15. *a,c* 16. *a,b,c,d* 17. *b,c,d*
18. *a,b,c* 19. *a,b* 20. *b,c* 21. *a,b,d* 22. *a,b,c* 23. *b,c,d* 24. *d* 25. *b* 26. *b* 27. *c*
28. *d* 29. *b* 30. *a* 31. *c* 32. *d* 33. *b* 34. *c* 35. *d* 36. *a* 37. *d*
38. *a* 39. 2 40. 1 41. 3

JEE Main & AIEEE Archive

1. *a* 2. *b* 3. *d* 4. *b* 5. *c* 6. *c* 7. *d* 8. *d* 9. *a* 10. *d*
11. *b* 12. *d* 13. *b* 14. *a*

JEE Advanced & IIT JEE Archive

1. *a,d* 2. *b,d* 3. *d* 4. *a*

3. Current Electricity

DPP-1 Electric Current

1. 4.83 h 4. (i) 5.92 A (ii) 388 Ω
5. *a* 6. *b* 7. *c* 8. *d* 9. *d* 10. *b* 11. *b* 12. *c* 13. *c* 14. *a*
15. *a,b* 16. *a,d* 17. *b* 18. *b* 19. *d* 20. *c* 21. *b* 22. 1 23. 4 24. 5

DPP-2 Relation among I, v_d, j, η, ρ and τ

2. 0.130 mm/s 5. *b* 6. *a* 7. *c* 8. *d* 9. *a* 10. *b* 11. *a* 12. *b*
13. *a* 14. *c* 15. *d* 16. *d* 17. *a,b* 18. *b,c,d* 19. *a,c* 20. *a,d* 21. 5 22. 4
23. 9 24. *a* 25. *c* 26. *c* 27. *b* 28. *a* 29. *b* 30. *c*

DPP-3 Resistance, Resistivity, Temperature Dependence and Ohm's Law

1. 0.2 A 2. 0.192 μA 5. *d* 6. *c* 7. *a* 8. *d* 9. *a* 10. *c* 11. *c*
12. *a* 13. *b* 14. *c* 15. *c* 16. *a,b* 17. *a,c* 18. *a,d* 19. *b* 20. *c* 21. *b*
22. *a* 23. 3 24. $\frac{6}{5}\alpha$ 25. 4

DPP-4 Series and Parallel Combination of Resistances

2. (ii) 45 Ω (iii) 15 Ω, 30 Ω 4. (i) 10.5 min (ii) 3.8 min
5. *d* 6. *d* 7. *b* 8. *c* 9. *b* 10. *c* 11. *a* 12. *a* 13. *b* 14. *c*
15. *a* 16. *a,d* 17. *b,d* 18. *a,b* 19. *a,c* 20. *b* 21. *b* 22. *c* 23. *a* 24. *d*
25. *a* 26. *c* 27. 5 28. 5

DPP-5 Cells and Combination of Cells

1. (i) 0.5 A (ii) 2 W (iii) 6 W
2. (i) 14 V (ii) 100 W (iii) 600 W
4. *d* 5. *d* 6. *c* 7. *b* 8. *d* 9. *b* 10. *a* 11. *b* 12. *d* 13. *b*
14. *c* 15. *a,c,d* 16. *a,b* 17. *a,b* 18. *a,c* 19. 10 20. 2 21. *a* 22. *b* 23. *c*
24. *d* 25. *b*

DPP-6 Network Analysis and Kirchhoff's Rules

1. (i) –10 V (ii) 44 V

2. 24 V, 12 A, –4 A, 6 A

4. (i) 2 V, 1 A, 0, –1 A (ii) 1 A, 2 A, –1 A

5. (i) $\frac{CV}{2}(1-e^{-2t/3RC})$ (ii) $\frac{V}{2R}$

6. *a*	7. *b*	8. *a*	9. *c*	10. *b*	11. *d*	12. *d*	13. *b*	14. *d*	15. *a*
16. *b*	17. *c*	18. *a*	19. *a,c*	20. *a,c*	21. *a,b,c*	22. *a,c*	23. *d*	24. *a*	25. *b*
26. *c*	27. *b*								

DPP-7 Electrical Energy

1. 0.32 %

2. (i) 2×10^{-17} J (ii) 10^{-17} s

5. *b*	6. *d*	7. *c*	8. *b*	9. *d*	10. *a*	11. *c*	12. *b*	13. *a*	14. *a*
15. *d*	16. *c*	17. *d*	18. *c,d*	19. *b,c*	20. *a,b*	21. *c*	22. *a*	23. *c*	24. *b*

DPP-8 Measuring Instruments

4. (iii) 8 Ω	5. *a*	6. *c*	7. *c*	8. *b*	9. *b*	10. *a*	11. *d*	12. *b*	13. *b*
14. *d*	15. *d*	16. *c*	17. *a,c*	18. *a,c*	19. *b,d*	20. *b*	21. *a*	22. *a*	23. *b*
24. *a*	25. *b*	26. 3	27. 8						

DPP-9 DC Circuits, Power and Energy Consideration

1. $\frac{\varepsilon^2}{4r}$	2. 4 Ω	4. *c*	5. *d*	6. *c*	7. *a*	8. *c*	9. *a*	10. *c*	
11. *c*	12. *b*	13. *c*	14. *c*	15. *b*	16. *c*	17. *b,d*	18. *b,c,d*	19. *a,b,c*	20. *a*
21. *a*	22. *a*	23. *d*	24. *a*	25. *a*	26. *a*	27. 8	28. 8		

Revisal Problems (JEE Main)

1. *a*	2. *b*	3. *c*	4. *d*	5. *d*	6. *c*	7. *c*	8. *c*	9. *b*	10. *a*
11. *c*	12. *d*	13. *c*	14. *a*	15. *b*	16. *c*	17. *b*	18. *a*	19. *b*	20. *b*

Revisal Problems (JEE Advanced)

1. *a*	2. *a*	3. *b*	4. *d*	5. *c*	6. *d*	7. *a*	8. *c*	9. *d*	10. *d*
11. *b,c*	12. *a,c*	13. *b,d*	14. *b,c*	15. *a,d*	16. *a*	17. *d*	18. *b*	19. *b*	20. *a*
21. *a*	22. *d*	23. *a*	24. *b*						

JEE Main & AIEEE Archive

1. *c*	2. *d*	3. *d*	4. *c*	5. *a*	6. *b*	7. *d*	8. *d*	9. *b*	10. *b*
11. *d*	12. *d*	13. *b*	14. *c*	15. *c*	16. *d*	17. *c*	18. *b*	19. *b*	20. *d*
21. *b*	22. *b*	23. *b*	24. *a*	25. *c*	26. *d*	27. *a*	28. *c*	29. *a*	30. *a*
31. *c*	32. *c*	33. *d*	34. *b*	35. *d*	36. *c*	37. *b*	38. *b*		

JEE Advanced & IIT JEE Archive

1. *c*	2. 5	3. 5	4. *a,c*	5. *c*	6. *d*	7. *b*	8. 5	9. *c*	10. *d*
11. *c*	12. 2	13. 4	14. *a,b,d*	15. *a,d*	16. *c*	17. *a*	18. *c*	19. *b*	20. *a*
21. *a*	22. *c*	23. *d*	24. *a*	25. *b*	26. $\frac{R_1 + R_2}{CR_1R_2}$		27. *d*		

1. Electrostatics

DPP-1 Charge and Properties of Charges

1. (i) When two identical (same size) metal spheres are in contact, an excess charge will be evenly distributed between the two.

So, charge on sphere C when it is touched with A and removed is

$$q_1 = +6\,\text{nC}$$

(ii) As, C has acquired a positive charge hence, electrons are removed from C.

(iii) Number of electrons removed from C

$$= n_1 = \frac{q_1}{e} = \frac{6 \times 10^{-9}}{1.6 \times 10^{-19}} = 3.75 \times 10^{10} \text{ electrons.}$$

Now, C is touched with B,

Total charge of B and C is $q_2 = (+6\,\text{nC}) + (-6\,\text{nC}) = 0$

So net charge on B and C is zero which is equally shared.

So, charge on C after removal of $B = 0$.

As C had a positive charge and now charge is zero that means electrons are transferred from B to C.

Also, number of electrons transferred

$$= n_2 = \frac{q_2}{e} = \frac{6 \times 10^{-9}\text{C}}{1.6 \times 10^{19}\text{C}} = 3.75 \times 10^{10} \text{ electrons.}$$

So, in the whole process, final charge on C is zero.

Number of electrons removed from C = Number of electrons given to $C = 3.75 \times 10^{10}$ electrons.

2. In all of given cases, body acquires a positive charge.

If body is rubbed with another body, electrons are released from the body with lower work function as they need less energy to get removed.

Induction produces opposite charge. Hence, body is positively charged.

Conduction produces same type of charge, so body is positively charged.

Excessive heating causes few of electrons to leave metal surface (thermionic emission), so body is positively charged.

Photoemission causes ejection of electrons, so body is being positively charged.

A high intensity electric field also causes emission of electrons (field emission).

3. The balloon induces polarisation of the molecules in the wall, so that a layer of positive charge exists near the balloon.

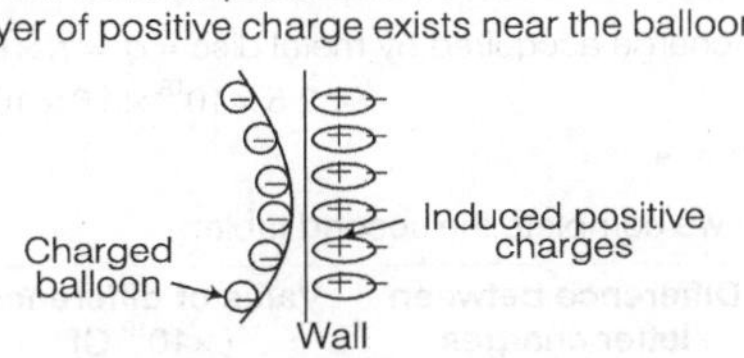

The attraction between their charges and negative charges on the balloon causes attraction between wall and balloon. Ionisation process in the air surrounding the balloon causes transfer of excess electrons and thereby reducing charge on the balloon. Eventually attractive force is insufficient to support the weight of balloon.

4. (i) Using $q = ne$, n = number of electrons removed

$$= \frac{q}{e} = \frac{3 \times 10^{-7}}{1.6 \times 10^{-19}} = 2 \times 10^{12} \text{ electrons}$$

(ii) Decrease in mass of conductor

= mass of 1 electron × number of electrons removed

$= 9.1 \times 10^{-31} \times 2 \times 10^{12}$

$= 2 \times 10^{-18}$ kg

5. No, charge produced due to induction is exactly equal and opposite to the charge transferred from (or to) the body to (or from) the ground.

6. (d) Negative charge on body $q = -2 \times 10^{-6}$ C

So, number of excess electrons on the body

$$n = \frac{q}{e} = \frac{2 \times 10^{-6}}{1.6 \times 10^{-19}} = 1.25 \times 10^{12} \text{ electrons}$$

As body has got 2.25×10^{12} protons. So, in neutral state, it also has 2.25×10^{12} electrons.

Hence, total number of electrons $= 2.25 \times 10^{12} + 1.25 \times 10^{12}$

$= 3.50 \times 10^{12}$ electrons.

7. (a) An uncharged body and a charged body share equal charges only when they are of identical shape and size.

If the bodies are of different shapes (and size), then they share unequal amount of charge. All other options are correct.

8. (c) Few of electrons from plastic scale are transferred to woolen cloth and few of atoms will be ionised.

9. (c) $q = ne$, (quantisation)

$$\Rightarrow \quad n = \frac{q}{e} = \frac{14.4 \times 10^{-19}}{1.6 \times 10^{-19}}$$

As conductor has a positive charge.

Thus, it has lost 9 electrons.

10. (b) Glass rod is positive charged.

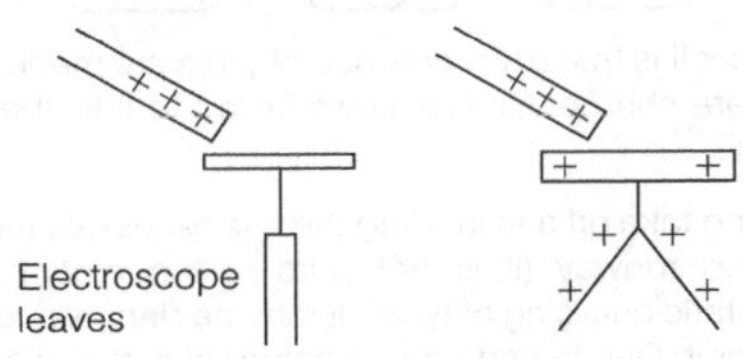

Due to touching, some of positive charge is acquired by leaves of electroscope and so, they show divergence.

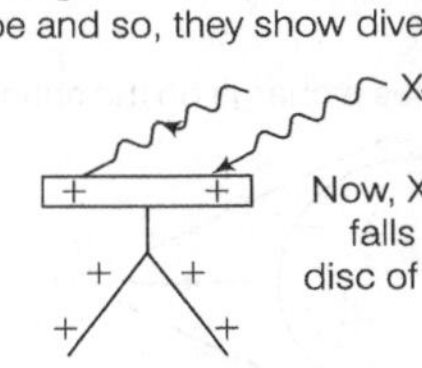

Few of electrons absorbs X-ray photons and are ejected from the metal surface.

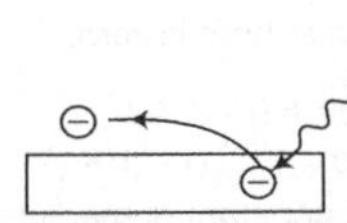

This causes more positive charge on metal disc and on leaves of electroscope.

Thus, leaves show divergence.

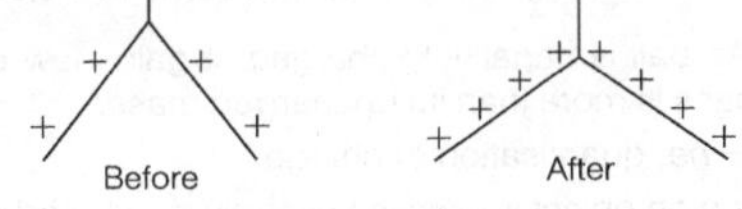

11. (c) As charge is quantised, $q = ne$ and $n = \frac{q}{e}$ = a whole number

In case, $q = 1.6 \times 10^{-20}$C

$$n = \frac{1.6 \times 10^{-20}}{1.6 \times 10^{-19}} = \frac{1}{10} = 0.1$$

12. (c) As ball 1 is not repelling any other ball, it may be uncharged.
Now, ball 2 and ball 4 attract each other.

So, ball 2 and ball 4 both are dissimilarily charged. As ball 2 and ball 4 both show repulsion (with ball 3 and ball 5 respectively).

So, ball 2 and ball 4 both are charged. Remember repulsion is surest test. As ball 1 attracted by both balls 2 and 4. So, this is possible only if ball 1 is uncharged.

13. (d) Frictional electrostatic charging creates equal and opposite charges.

$\therefore \quad q_{net} = 0$

14. (d) Paper bits are initially uncharged and get attracted by comb due to induced polarisation of molecules. But they acquire similar charge when they touch the comb and they are repelled fly away from the comb.

15. (d) During an American winter, cold dry air allows accumulation of significant excess. Charge giving the potential for a shock.

16. (d) Statement I is false, complete charge can be transferred from an object to another, e.g. If we place a conductor inside other and connect both with a wire or in given case.

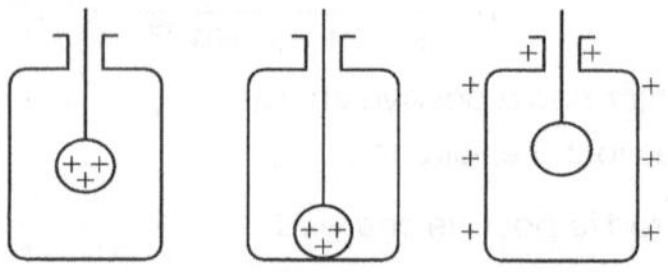

Statement II is true if two charged objects are made to touch, then they share charge. Charge flows from I to II till they reach same potential.

17. (a) During take off and landing there is excessive friction between tyres and runway (it is not pure rolling motion) this causes electrostatic charging of tyres. It may be dangerous if no check is kept over it. Due to conducting nature of material of tyre charged so formed is conducted to ground and so, electrical sparking is avoided.

18. (d) Outer charge induces a charge on the sphere.

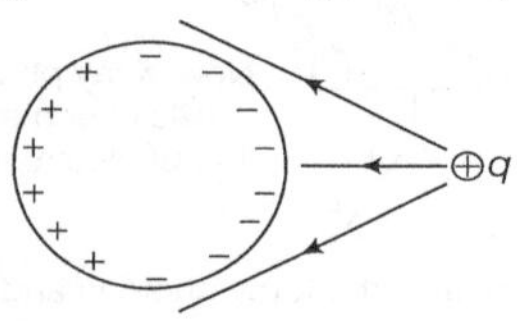

But total charge on the sphere is zero.

19. (b,c,d) Induced charge is given by

$$q_B = q_A(1 - 1/K_A)$$

where, K_A = dielectric constant of medium of body which is being charged by induction.

$q_A = -q_B$ it may be less than that of $q_B \cdot q_A = -q_B$ only when A is metallic ($K_A = \infty$).

But charge on A is always of opposite nature to that of B. Also, induction results only when A has got some free electrons.

20. (a,b,d) As ball is negatively charged, it gains few electrons and so, its mass is more than its uncharged mass.

Also, $q = ne$, quantisation of charge.

Charge on an object is always uniformly distributed due to mutual repulsion.

21. (a,b,c,d) Plastics are positively charged when rubbed with nylon or silk cloth. Rubbing causes a very small amount of heating and so, comb gets warm.

Change of mass = $n \times m_e$ or $m_e = 9.1 \times 10^{-31}$kg

Change of mass although it occurs, is very-very slight. A charged comb attracts hair and so, it is difficult to comb hair by charged comb.

22. (d) When two objects are charged by rubbing them against each other they acquire equal and opposite charge.

$$|q_A| = |q_B|$$

or $\quad q_A = q_B$

Induced charge may be less than inducing charge when bodies are not having sufficient free electrons.

$$q_A \geq q_B$$

If two objects (say two spheres) of same material and same size are taken, 1st is charged ($q = q_A$) is touched with IInd (initially charged), then $q_B = \frac{q_A}{2}$.

23. (c) B is negatively charged, so mass of B increases due to gain of electrons. As A looses electrons its mass decreases. Work function (minimum energy required by an e^- to come onto a surface) of A is less, so it readily looses e^-.

24. (4)

Number of photons emitted by source in 1s which moves in all direction

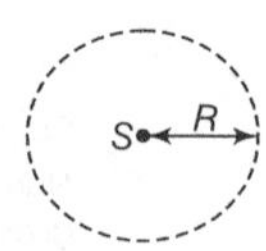

$$= \frac{\text{Power of source}}{\text{Energy of 1 photon}} = \frac{2}{2 \times 10^{-19}} \frac{\text{W}}{\text{J}}$$

$$= 1 \times 10^{19} \text{ per second}$$

This photon flow reaches a sphere of radius, $R = \frac{1}{\sqrt{\pi}}$ m.

So, number of photons falling per unit area over this sphere in/s

$$= \frac{\text{Total number of photons / s}}{\text{Area of sphere}} = \frac{1 \times 10^{19}}{4\pi\left(\frac{1}{\sqrt{\pi}}\right)^2} = \frac{1}{4} \times 10^{19}$$

$$= 2.5 \times 10^{18} \text{ m}^{-2}\text{s}^{-1}$$

As each falling photon over the metal disc can eject an electron, number of electrons ejected in 1s from metal disc

= (Number of photons falling over unit area in unit time) × Area of metal disc

$$= 2.5 \times 10^{18} \text{ m}^{-2} \text{ s}^{-1} \times 1 \text{ m}^2$$

$$= 2.5 \times 10^{18} \text{ s}^{-1}$$

Hence, charge accquired by metal disc = $q = n \times e \times t$

$$= 2.5 \times 10^{18} \times 1.6 \times 10^{-19} \times 1 = 0.4 \text{ C}.$$

So, $10q = 4$

25. (5) First we complete the second table;

Difference between letter charges	Value of difference ($\times 10^{19}$ C)
D-E	1.56
D-C	1.48
A-D	1.64
B-A	1.55

So, difference of average value $= \frac{1.56 + 1.48 + 1.64 + 1.55}{4} = 1.56$

So, our experimental value of charge is $q_1 = 1.56 \times 10^{-19}$ C

And elementary charge is $q_2 = 1.60 \times 10^{-19}$ C

$\therefore$ % Difference $= \frac{(1.60 - 1.56) \times 10^{-19} \times 100}{1.6 \times 10^{-19}} = 2.5\% \Rightarrow x = 2.5$

Hence, $2x = 5$

DPP-2 Coulomb's Law and Principle of Superposition

1. (i) Since, two charged particles attract each other, they must possess opposite charges. On touching each other, there is a redistribution of charge and each particle gets the same type of charge. Hence, the particle repel.

(ii) Since, on sharing charges, the magnitude of charge on each particle is smaller than before. So, the force of repulsion is smaller.

2. For the pair of charges ($1\,\mu C, 2\,\mu C$) placed at a distance $r(m)$ apart,

$$F_1 = \left(9 \times 10^9 \frac{Nm^2}{C^2}\right) \frac{(1\,\mu C)\,(2\,\mu C)}{[r(m)]^2}$$

$$= \frac{18 \times 10^{-3}\,Nm^2}{r^2(m^2)}$$

$$= (18 \times 10^{-3}\,N)\,(1/r^2)$$

Graph between F_1 and $(1/r^2)$ is a straight line with slope $(18 \times 10^{-3}\,N)$

For pair of charges ($2\,\mu C, -3\,\mu C$) placed at the same distance apart.

$$F_2 = \left(9 \times 10^9 \frac{Nm^2}{C}\right)\left(\frac{(2\,\mu C)\,(-3\,\mu C)}{[r(m)]^2}\right)$$

$$= \frac{-54 \times 10^{-3}\,Nm^2}{r^2 m^2}$$

$$= (-54 \times 10^{-3}\,N)\,(1/r^2)$$

Graph between F_2 and $(1/r^2)$ is a straight line with slope $(-54 \times 10^{-3}\,N)$

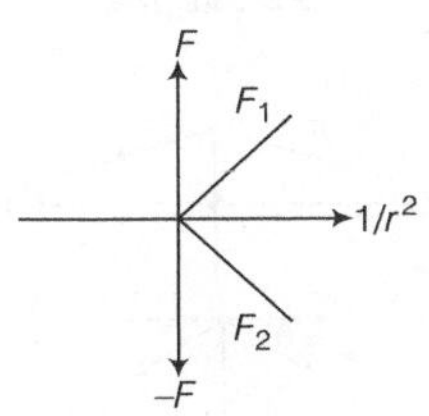

Slope of graph $\propto$ product of charges.

3. (i) Let charges on spheres are q_1 and q_2.

Initially, they attract each other.

Thus, charges are of opposite signs.

Also, $$F = \frac{Kq_1q_2}{r^2}$$

$$\Rightarrow \quad 1.728 = 9 \times 10^9 \times \frac{q_1q_2}{(25 \times 10^{-2})^2}$$

$$\Rightarrow \quad q_1q_2 = 12 \times 10^{-12} \quad \ldots\text{(i)}$$

(ii) When charge is redistributed due to connecting spheres. Then, new value of charges on spheres be q_1' and q_2'

and $$q_1' = q_2' = \frac{q_1 - q_2}{2} \quad \text{(as spheres are identical)}$$

Now, repulsion is with 0.576 N

$$\Rightarrow \quad \frac{Kq_1'q_2'}{r^2} = 0.576$$

$$\Rightarrow \quad \frac{9 \times 10^9 \times \left(\frac{q_1 - q_2}{2}\right)^2}{(25 \times 10^{-2})^2} = 0.576$$

$$\therefore \quad q_1 - q_2 = 4 \times 10^{-6} \quad \ldots\text{(ii)}$$

On solving Eqs. (i) and (ii), we get

$$q_1 = 6 \times 10^{-6}\,C$$

and $$q_2 = -2 \times 10^{-6}\,C$$

4. In equilibrium,

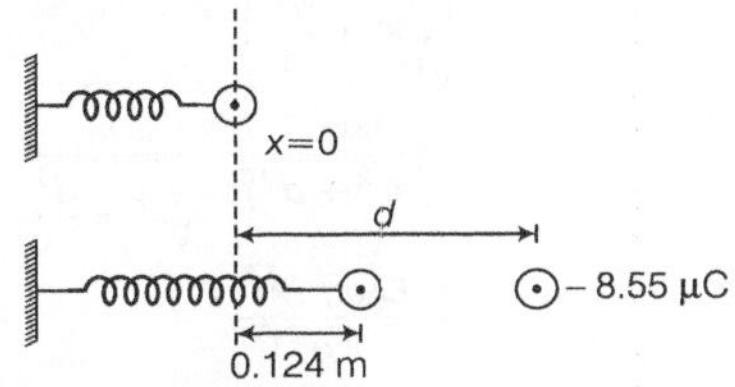

In equilibrium, force of attraction between spheres = spring force

$$\frac{kq_1q_2}{r^2} = Cx$$

where, C = spring constant

$$\frac{9 \times 10^9 \times 2.44 \times 8.55 \times 10^{-12}}{(d - 0.124)^2} = 89.2 \times 0.124$$

$$\Rightarrow \quad (d - 0.124)^2 = \frac{9 \times 10^9 \times 2.44 \times 8.55 \times 10^{-12}}{89.2 \times 0.124}$$

$$\Rightarrow \quad (d - 0.124)^2 = 169 \times 10^{12} \times 10^{-12}$$

$$\Rightarrow \quad d - 0.124 = 13$$

$$\Rightarrow \quad d \approx 13\text{ m}$$

5. (i) To make an object positively charge, only electrons due to be removed, so its mass decreases.

(ii) In case of insulating sphere, charge is uniformly distributed throughout its volume.

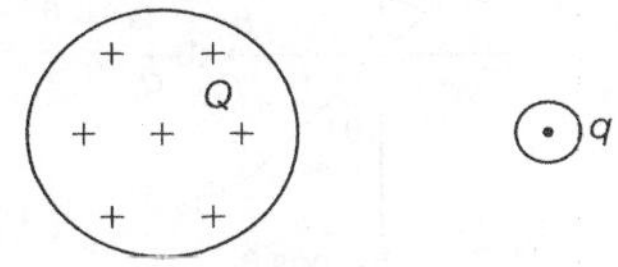

In case of a conducting sphere, charge resides only on the surface.

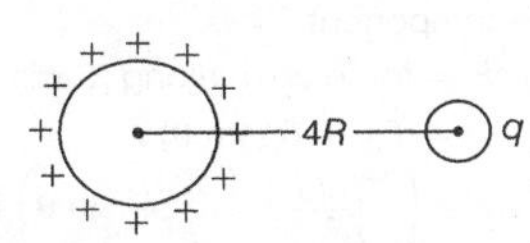

Note that $4R$ is not very large compared to radius R. Mean distance of charge in sphere is less in case I. In case I, force is slightly larger.

(iii) Total potential energy is converted into kinetic energy and kinetic energy is equally shared by electron and proton.

As proton is much heavier than electron, its velocity is much lesser than the velocity of electron.

6. (c)

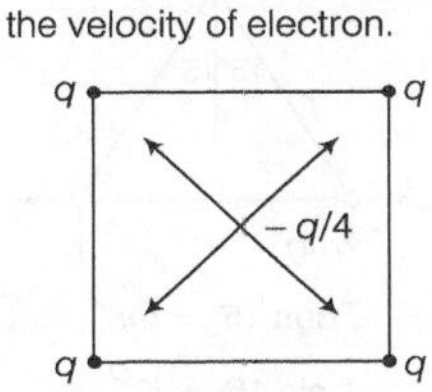

Force due to each q charge on $-q/4$ is balanced by diagonally opposite charge.

$$F_{net} = 0$$

7. (c) Using $F \propto q_1q_2$,

we get,
$$(q_1 + q_2)q_3 \propto 2$$
$$(q_2 + q_3)q_1 \propto 4$$
$$(q_1 + q_3)q_2 \propto 18$$

8. (c) Force on third charge

$$= F_1 \sin\theta + F_2 \sin\theta \text{ (along perpendicular)}$$

$$= \frac{2Kqq'}{\left(\sqrt{x^2 + \frac{d^2}{4}}\right)^2} \sin\theta$$

$$= \frac{4Kqq'}{(\sqrt{4x^2 + d^2})^2} \cdot \frac{x}{\sqrt{x^2 + \frac{d^2}{4}}}$$

or $$F_{net} = \frac{8Kqq'\,x}{(4x^2 + d^2)^{3/2}}$$

Now, F is maximum when $\frac{dF}{dx} = 0$

$$\Rightarrow \quad 8kqq'\left(\frac{(4x^2 + d^2)^{3/2} - \frac{3}{2}x\,(4x^2 + d^2)^{1/2} \cdot 8x}{(4x^2 + d^2)^3}\right) = 0$$

$$(4x^2 + d^2)^{3/2} - \frac{3}{2}x \cdot 8x(4x^2 + d^2)^{1/2} = 0$$

$$(4x^2 + d^2)^{3/2} = 12\,x^2(4x^2 + d^2)^{1/2}$$

$$4x^2 + d^2 = 12x^2$$

On solving, we get

$$d^2 = 8x^2 \text{ or } x = \frac{d}{2\sqrt{2}}$$

9. (a)

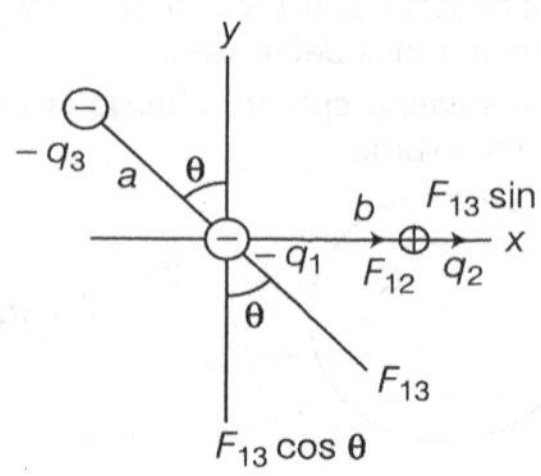

F_{13} can be resolved into components along $-Y$-axis and $+X$-axis. F_{12} has only a x-component.

$\therefore$ F_x = force on q_1 along X-axis

$$= (F_{12} + F_{13}\sin\theta)\,\hat{\mathbf{i}}$$

$$= \left(\frac{kq_1q_2}{b^2} + \frac{kq_3q_1}{a^2}\sin\theta\right)\hat{\mathbf{i}}$$

and F_y = force on q_1 along Y-axis

$$= (F_{13}\cos\theta)\,(-\hat{\mathbf{j}})$$

$$= -\frac{kq_1q_3}{a^2}\cos\theta \cdot \hat{\mathbf{j}}$$

10. (b) In air,

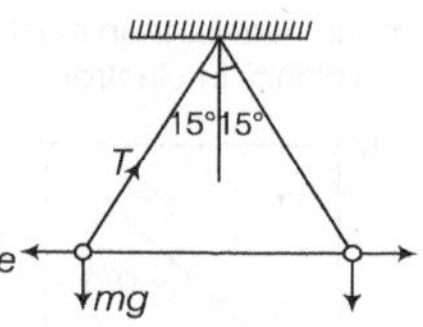

$$T\cos 15° = mg$$

$$T\sin 15° = K\frac{q^2}{r^2}$$

$$\Rightarrow \quad \tan 15° = \frac{1}{4\pi\varepsilon_0} \cdot \frac{q^2}{mgr^2}$$

On immersing whole set up in liquid

$$T'\cos 15° = mg\left(1 - \frac{800}{1600}\right)$$

$$\Rightarrow \quad T'\cos 15° = \frac{mg}{2}$$

Weight of sphere in fluid = Weight – Upthrust

$$= mg - mg\frac{\rho_f}{\rho}$$

and $$T'\sin 15° = \frac{1}{4\pi\varepsilon_0 K}\frac{q^2}{r^2}$$

$$\Rightarrow \quad \tan 15° = \frac{1}{4\pi\varepsilon_0 K}\frac{2q^2}{mgr^2}$$

Equating both, we get

$$\frac{1}{4\pi\varepsilon_0 k}\frac{2q^2}{mgr^2} = \frac{1}{4\pi\varepsilon_0} \cdot \frac{q^2}{mgr^2}$$

$$\therefore \quad K = 2$$

11. (a) ***Case (i)***

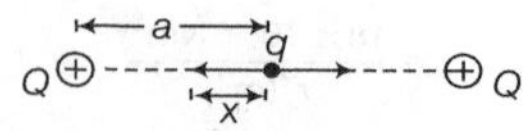

For a displacement x, net force towards equilibrium position is

$$F = \frac{KQq}{(a-x)^2} - \frac{KQq}{(a+x)^2}$$

$$F \approx \frac{4KQqx}{a^3}, \text{ as } x << a$$

So, acceleration is

$$a = -\frac{F}{m} = -\frac{4KQq}{ma^3}x$$

Using $a = -\omega^2 x$

and $$n_a = \frac{1}{T} = \frac{\omega}{2\pi},$$

we have, $$n_a = \frac{1}{2\pi}\sqrt{\frac{Qq}{\pi\varepsilon_0 ma^3}}$$

Case (ii)

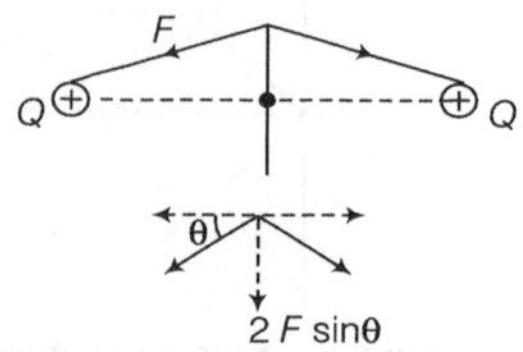

$$\text{Restoring force} = 2F\sin\theta$$

$$= \frac{2KQq}{(a^2 + x^2)} \cdot \frac{x}{\sqrt{a^2 + x^2}}$$

$$= \frac{2KQq}{ma^3} \cdot x$$

$\therefore$ $$\text{Acceleration } a = -\omega^2 x = -\frac{F}{m}$$

and $$\text{frequency } n_b = \frac{1}{T} = \frac{1}{2\pi}\sqrt{\frac{Qq}{2\pi\varepsilon_0 ma^3}}$$

$$\frac{n_a}{n_b} = \frac{T_b}{T_a} = \sqrt{2}$$

$\therefore$ $$\sqrt{2}\,\frac{n_a}{n_b} = 2$$

12. (a) For an element of length dx, at a distance x from point charge Q, force will be

$$dF = \frac{KQdq}{x^2} = \frac{KQq}{Lx^2}dx$$

or $$dq = \frac{q}{L}dx$$

$\therefore$ $$F = \frac{KQq}{L}\int_d^{d+L}\frac{dx}{x^2} = \frac{KqQ}{d(d+L)}$$

13. (c) Initially, $F = \dfrac{Kq_1q_2}{r^2}$

Finally, $F = \dfrac{Kq_1q_2}{(r - t + t\sqrt{k})^2} = \dfrac{Kq_1q_2}{\left(r - \dfrac{r}{2} + \dfrac{r}{2}\sqrt{4}\right)^2}$

$$= \frac{Kq_1q_2}{\left\{\frac{r}{2}(1+2)\right\}^2} = \frac{Kq_1q_2}{\frac{r^2}{4} \times 9} = \frac{Kq_1q_2}{r^2} \times \frac{4}{9}$$

14. (c) In $\Delta\, ABC$, distance $BC = AB \sin 60^\circ$

$$= 2R \times \frac{\sqrt{3}}{2} = \sqrt{3}R$$

Net force acting on BC i.e.

$$|\mathbf{F}_{eBC}| = \frac{1}{4\pi\varepsilon_0} \frac{(q/3)\,(2q/3)}{(\sqrt{3}R)^2} = \frac{q^2}{54\pi\varepsilon_0 R^2}$$

15. (a) Net force acting on vertical component i.e. $F_{net} = 2F \cos \theta$

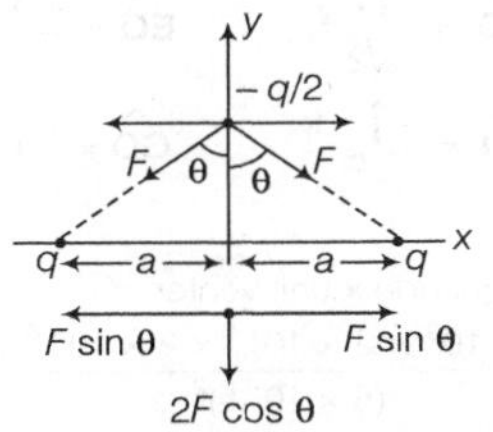

$$F_{net} = \frac{2Kq\left(\frac{q}{2}\right)}{(\sqrt{y^2 + a^2})^2} \cdot \frac{y}{\sqrt{y^2 + a^2}}$$

$$F_{net} = \frac{2Kq\left(\frac{q}{2}\right)y}{(y^2 + a^2)^{3/2}} \Rightarrow \frac{Kq^2y}{a^3} \propto y$$

16. (a,b) If all of the angles to the vertical (in both cases) are assumed to be small, then the sphere only have horizontal displacement and so the electric force of repulsion is always horizontal. Likewise, the small angle condition leads to $\tan\theta \approx \sin\theta \approx \theta$ for all small angles.

In the free-body diagram, each sphere has three forces gravity, tension and the electrostatic force. Taking to the right as the positive horizontal direction and up to the positive vertical direction. Since, the sphere 1 in equilibrium the net force in each direction is zero.

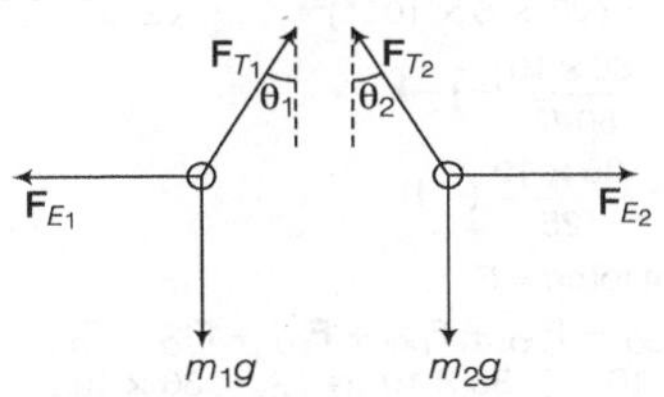

(a) $\Sigma F_{1x} = F_{T_1} \sin\theta_1 - F_{E_1} = 0 \rightarrow F_{E_1} = F_{T_1} \sin\theta_1$

$\Sigma F_{1y} = F_{T_1} \cos\theta_1 - m_1 g \rightarrow F_{T_1}$

$$= \frac{m_1 g}{\cos\theta_1} \rightarrow F_{E1} = \frac{m_1 g}{\cos\theta_1}$$

$$= m_1 g \tan\theta_1 = m_1 g\theta_1$$

A completely parallel analysis would give $F_{E_2} = m_2 g\theta_2$. Since, the electric forces are a Newton's third law pair, they can be set equal to each other in magnitude.

$$F_{E_1} = F_{E_2} \rightarrow m_1 g\theta_1$$

$$= m_2 g\theta_2 \rightarrow \frac{\theta_1}{\theta_2} = \frac{m_2}{m_1} = 1$$

(b) The same analysis can be done for this case

$$F_{E_1} = F_{E_2} = m_1 g\theta_1$$

$$= m_2 g\theta_2 \rightarrow \frac{\theta_1}{\theta_2} = \frac{m_2}{m_1} = 2$$

17. (a,b)

(a) The charges will experience a force that is always pointing towards the origin. In the diagram, there is a greater force of $\dfrac{Qq}{4\pi\varepsilon_0(d - x)^2}$ to the left and a lesser force of $\dfrac{Qq}{4\pi\varepsilon_0(d + x)^2}$ to the right. So, the net force is towards the origin.

The same would be true if the mass were to the left of the origin, the net force is

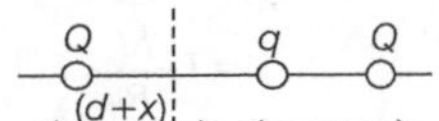

$$F_{net} = \frac{Qq}{4\pi\varepsilon_0(d + x)^2} - \frac{Qq}{4\pi\varepsilon_0(d - x)^2}$$

$$= \frac{Qq}{4\pi\varepsilon_0(d + x)^2(d - x)^2}[(d - x)^2 - (d + x)^2]$$

$$= \frac{-4Qqdx}{4\pi\varepsilon_0(d + x)^2(d - x)^2}$$

$$= \frac{-Qqdx}{\pi\varepsilon_0(d + x)^2(d - x)^2}$$

We assume that $x << d$.

$$F_{net} = \frac{-Qqd}{\pi\varepsilon_0(d + x)^2(d - x)^2} \approx \frac{-Qq}{\pi\varepsilon_0 d^3}x$$

This has the form of simple harmonic motion, where the spring constant is $k_{elastic} = \dfrac{Qq}{\pi\varepsilon_0 d^3}$

(b) $T = 2\pi\sqrt{\dfrac{m}{k_{elastic}}} = 2\pi\sqrt{\dfrac{m}{\dfrac{Qq}{\pi\varepsilon_0 d^3}}} = 2\pi\sqrt{\dfrac{m\pi\varepsilon_0 d^3}{Qq}}$

$\therefore \quad T^2 \propto d^3$

18. (a,b,c) The system is never in stable equilibrium, because the negative charges will repel each other and so the third charge must put an opposite force on each of the original charge.

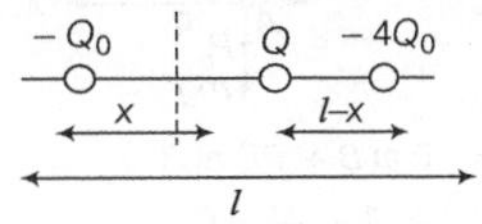

Consideration of the various possible configuration leads to the conclusion that the third charge must be positive and must be between the other two charges.

In diagram the variables for each negative charge, equate the magnitude of the two forces on the charge. Also note that $0 < n < l$.

Left : $\dfrac{Q_0 Q}{x^2} = \dfrac{K4Q_0^2}{l^2}$,

Right: $K\dfrac{4Q_0Q}{(l - x)^2} = K\dfrac{4Q_0^2}{l^2}$

$$\Rightarrow \quad K\frac{Q_0Q}{x^2} = K\frac{4Q_0Q}{(l - x)^2} \rightarrow x = \frac{1}{3}l$$

$$\frac{KQ_0Q}{x^2} = K\frac{4Q_0^2}{l^2} \rightarrow Q = 4Q_0\frac{x^2}{l^2}$$

$$= Q = \frac{4}{(3)^2}Q_0 = \frac{4}{9}Q_0$$

Thus, the charge should be magnitude $\frac{4}{9}Q_0$ and a distance $\frac{1}{3}l$ from $-Q_0$ towards $-4Q_0$.

19. (a,c)

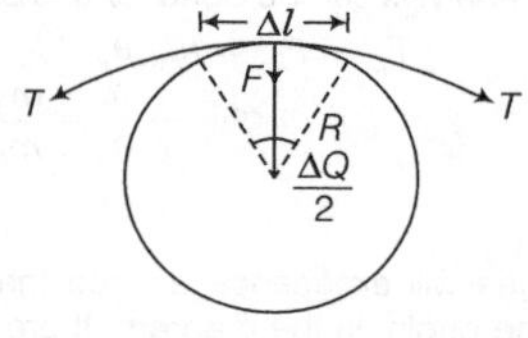

Consider a small segment of length Δl making an angle of $\Delta\theta$ at centre.

Horizontal components of tension cancels out each other and vertical components add upto from these forces **F**.

$$\Rightarrow \quad 2T\sin\frac{\theta}{2} = T\Delta\theta$$

$$= T\cdot\frac{\Delta l}{R}$$

$$\Rightarrow \quad T = \frac{FR}{\Delta l}$$

where, $$F = \frac{1}{4\pi\varepsilon_0}\cdot\frac{5\Delta l \times 10^{-8}}{R^2}$$

$$= 4.5\times 10^4 \times \Delta l$$

So, $$T = \frac{F}{\Delta l}\times R$$

$$= 4.5\times 10^4 \times 0.1$$

$$= 4.5\times 10^3 \text{ N}$$

Also change in radius

$$\Delta R = R \times \frac{T}{AY}$$

$$= 2.25\times 10^{-3} \text{ m}$$

20. (b,d) Let v_A = minimum velocity imparted to ball at A such that it makes a complete loop.

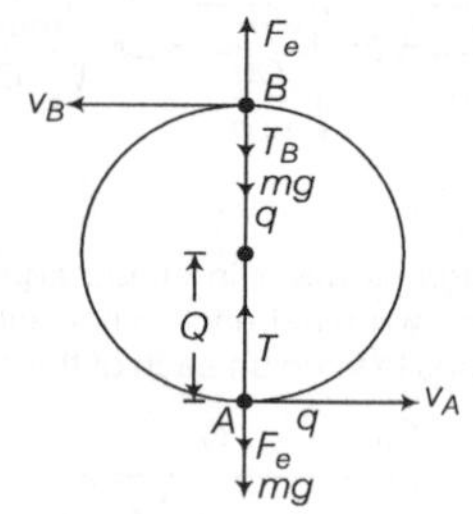

Then, KE at A = KE at B + PE at B

$$\Rightarrow \quad \frac{1}{2}mv_A^2 = \frac{1}{2}mv_B^2 + mg2l$$

$$\Rightarrow \quad v_A^2 = v_B^2 + 4gl \qquad \text{...(i)}$$

To make just one complete revolution i.e. v_A is minimum possible, then

$$T_B = 0$$

$$\therefore \quad \frac{mv_B^2}{l} = mg - F_e$$

Force balance of B,

$$\Rightarrow \quad v_B^2 = gl - \frac{Kq^2}{ml} \qquad \text{...(ii)}$$

From Eqs. (i) and (ii),

$$v_A^2 = gl - \frac{q^2}{4\pi\varepsilon_0 ml} + 4gl$$

$$\Rightarrow \quad v_A = \sqrt{5gl - \frac{q^2}{4\pi\varepsilon_0 ml}}$$

On substituting values,

$$v_A = 5.86 \text{ m/s}$$

21. (c,d) Coordinates of vertices of cube are

O : (0, 0, 0)
A : (0, 0, 5 cm)
B : (5 cm, 0, 5 cm)
C : (5 cm, 0, 0)
D : (5 cm, 5 cm, 0)
E : (5 cm, 5 cm, 5 cm)
F : (0, 5 cm, 5 cm)
G : (0, 5 cm, 0)

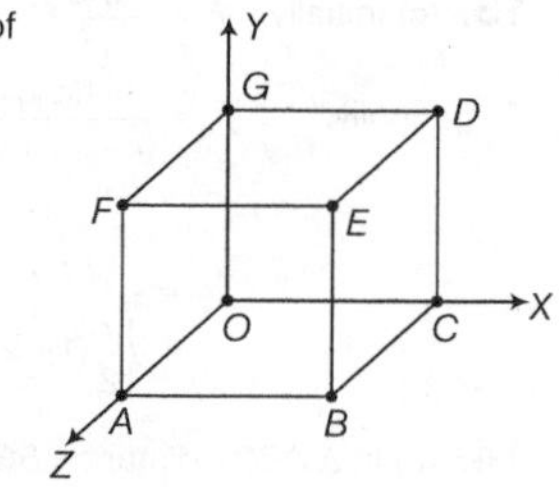

Unit vectors in direction of forces are

$$\widehat{AO} = \frac{\mathbf{AO}}{|\mathbf{AO}|} = \frac{(0\hat{i} + 0\hat{j} + 0\hat{k}) - (0\hat{i} + 0\hat{j} + 5\hat{k})}{5} = -\hat{k}$$

Similarly, $\widehat{BO} = -\dfrac{\hat{i} - \hat{k}}{\sqrt{2}}$ $\qquad \widehat{CO} = -\hat{i}$

$$\widehat{DO} = \frac{-\hat{i} - \hat{j}}{\sqrt{2}} \qquad \widehat{EO} = \frac{-\hat{i} - \hat{j} - \hat{k}}{\sqrt{3}}$$

$$\widehat{FO} = \frac{-\hat{j} - \hat{k}}{\sqrt{2}} \qquad \widehat{GO} = -\hat{j}$$

And forces are

Force = Magnitude × unit vector

$$\therefore \quad \mathbf{F}_{AO} = \frac{9\times 10^9 \times 2\times 10^{-6} \times 2\times 10^{-6}}{(5\times 10^{-2})^2}\cdot(-\hat{k})$$

$$= \frac{36}{25}\times 10\,(-\hat{k})$$

$$\mathbf{F}_{BO} = \frac{9\times 10^9 \times 4\times 10^{-12}}{(\sqrt{2}\times 5\times 10^{-2})^2}\times\left(\frac{-\hat{i} - \hat{k}}{\sqrt{2}}\right)$$

$$= \frac{36\times 10}{50\sqrt{2}}(-\hat{i} - \hat{k})$$

$$\mathbf{F}_{CO} = \frac{9\times 10^9 \times 4\times 10^{-12}}{(5\times 10^{-2})^2}\cdot(-\hat{i}) = \frac{36}{25}\times 10\,(-\hat{i})$$

$$\mathbf{F}_{DO} = \frac{36\times 10}{50\sqrt{2}}(-\hat{i} - \hat{j})$$

$$\mathbf{F}_{EO} = \frac{9\times 10^9 \times 4\times 10^{-12}}{(\sqrt{3}\times 5\times 10^{-2})^2}\times\left(\frac{-\hat{i} - \hat{j} - \hat{k}}{\sqrt{3}}\right)$$

$$= \frac{36\times 10}{75\sqrt{3}}(-\hat{i} - \hat{j} - \hat{k})$$

$$\mathbf{F}_{FO} = \frac{9\times 10^9 \times 4\times 10^{-12}}{(\sqrt{2}\times 5\times 10^{-2})^2}\times\frac{(-\hat{j} - \hat{k})}{\sqrt{2}}$$

$$= \frac{36\times 10}{50\sqrt{2}}(-\hat{j} - \hat{k})$$

$$\mathbf{F}_{GO} = \frac{36\times 10}{25}(-\hat{j})$$

Clearly, net force = $\mathbf{F}_{net}$

$$= \mathbf{F}_{AO} + \mathbf{F}_{BO} + \mathbf{F}_{CO} + \mathbf{F}_{DO} + \mathbf{F}_{EO} + \mathbf{F}_{FO} + \mathbf{F}_{GO}$$

$$= -\left[\frac{36\times 10}{25}\hat{k} + \frac{36\times 10}{50\sqrt{2}}(\hat{i} + \hat{k}) + \frac{36\times 10}{25}\hat{j} + \frac{36\times 10}{50\sqrt{2}}(\hat{i} + \hat{j})\right.$$
$$\left. + \frac{36\times 10}{75\sqrt{3}}(\hat{i} + \hat{j} + \hat{k}) + \frac{36\times 10}{25}\hat{i} + \frac{36\times 10}{25}(\hat{i}) + \frac{36\times 10}{50\sqrt{2}}(\hat{j} + \hat{k})\right]$$

$$= -\left(\frac{36\times 10}{25}\right)\left[\hat{k} + \frac{\hat{i} + \hat{k}}{2\sqrt{2}} + \hat{j} + \frac{\hat{i} + \hat{j}}{2\sqrt{2}} + \frac{\hat{i} + \hat{j} + \hat{k}}{3\sqrt{3}} + \hat{i} + \frac{\hat{j} + \hat{k}}{2\sqrt{2}}\right]$$

$$= -\left(\frac{36\times 10}{25}\right)\left[\hat{i} + \hat{j} + \hat{k} + \frac{\hat{i}}{2\sqrt{2}} + \frac{\hat{i}}{3\sqrt{3}} + \frac{\hat{j}}{2\sqrt{2}} + \frac{\hat{j}}{3\sqrt{3}} + \frac{\hat{k}}{2\sqrt{2}} + \frac{\hat{k}}{3\sqrt{3}}\right]$$

$$= -\left(\frac{36\times 10}{25}\right)\left[\left(1 + \frac{1}{2\sqrt{2}} + \frac{1}{3\sqrt{3}}\right)(\hat{i} + \hat{j} + \hat{k})\right]$$

22. (b) Clearly, force on charge at A is along X-axis or $\hat{i}$ and force on B is OB.

$$\Rightarrow \quad \frac{\hat{i}}{\sqrt{2}} + \frac{\hat{j}}{\sqrt{2}}$$

Force on H is along $OH = \frac{\hat{i}}{\sqrt{2}} - \frac{\hat{i}}{\sqrt{2}}$

and force on D is along $OD = \frac{-\hat{i}}{\sqrt{2}} - \frac{\hat{j}}{\sqrt{2}}$

(i) → (s), (ii) → (r), (iii) → (q), (iv) → (p)

23. (c)

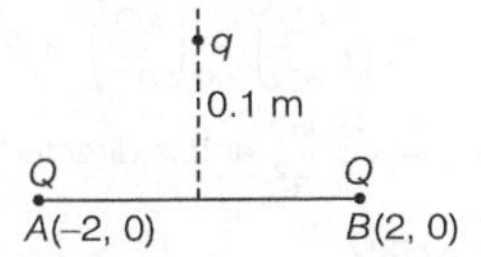

Charge q oscillates, so q is negative. Force on q is

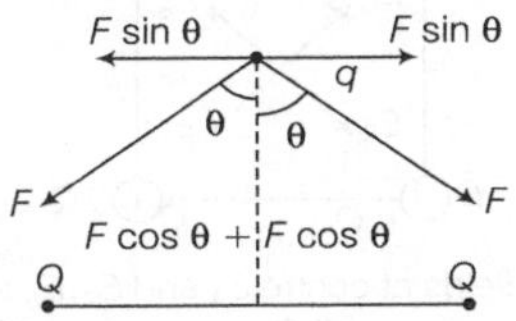

So, net force on q is

$$F_{net} = 2F\cos\theta$$

$$= 2 \times \frac{1}{4\pi\varepsilon_0} \times \frac{Qq}{y^2 + a^2} \times \cos\theta$$

$$= 2 \times \frac{1}{4\pi\varepsilon_0} \times \frac{Qq}{y^2 + a^2} \times \frac{y}{(y^2 + a^2)^{1/2}}$$

$$= \frac{1}{4\pi\varepsilon_0} \cdot \frac{2Qqy}{(y^2 + a^2)^{3/2}}$$

For $y << a$, $F_{net} = \frac{1}{4\pi\varepsilon_0} \cdot \frac{2Qqy}{a^3}$

Now, as $a = 2$ m and $y = 0.1$ m

We can use approximation

$$\therefore \quad F = \frac{1}{4\pi\varepsilon_0} \times \frac{2 \times Qqy}{a^3}$$

$$\Rightarrow \quad 9 \times 10^{-3} = \frac{9 \times 10^9 \times 2 \times 8 \times 10^{-6} \times q \times 0.1}{2^3}$$

$$\Rightarrow \quad q = 5 \times 10^{-6}\ \text{C} = 5\ \mu\text{C}$$

24. (a) At $t = 0$, $y = 0.1$ m assuming an undamped SHM, amplitude of SHM is 10 cm.

25. (c) $T = 2\pi\sqrt{\frac{m}{k}}$

Here, spring constant, $k = \frac{1}{4\pi\varepsilon_0} \cdot \frac{2Qq}{a^3}$

So, frequency, $f = \frac{1}{T}$ (on substituting values, we get)

$$f = 5$$

26. (3)

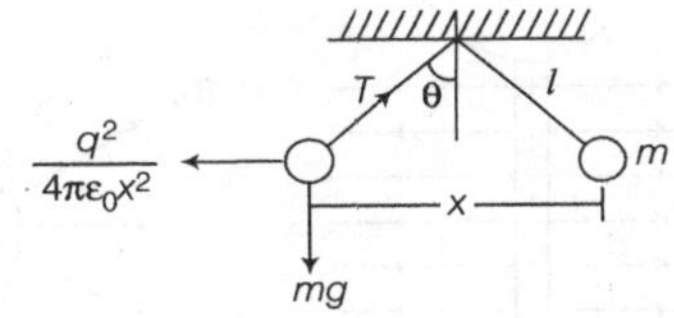

At equilibrium, $\frac{q^2}{4\pi\varepsilon_0 x^2}\cos\theta = mg\sin\theta$

For small θ, $\cos\theta \to 1$ and $\sin\theta \approx \theta = \frac{x/2}{l}$

$$\Rightarrow \quad \frac{q^2}{4\pi\varepsilon_0 x^2} = \frac{mgx}{2l} \quad \Rightarrow q = x^{3/2}\cdot\sqrt{\frac{2\pi\varepsilon_0 mg}{l}}$$

$$\Rightarrow \quad \frac{dq}{dt} = \frac{3}{2}\left(x^{1/2}\cdot\sqrt{\frac{2\pi\varepsilon_0 mg}{l}}\right)\left(\frac{dx}{dt}\right)$$

Put $v = \frac{dx}{dt} \propto x^{1/2} \quad \Rightarrow \quad \frac{dq}{dt} \propto \frac{3}{2}\sqrt{\frac{2\pi\varepsilon_0 mg}{l}}$

hence, $N = 3$

27. (2)

Force on electron causing centripetal acceleration is electrostatic force.

$$\therefore \quad \frac{mv^2}{r} = \frac{Ke^2}{r^2}$$

$$\Rightarrow \quad v = \sqrt{\frac{Ke^2}{mr}} = \sqrt{\frac{9 \times 10^9 \times (1.6 \times 10^{-19})^2}{9.1 \times 10^{-31} \times 0.53 \times 10^{-10}}} = 2.19 \times 10^6\ \text{m/s}$$

Now, $\frac{c}{v} = \frac{2.99 \times 10^8}{2.19 \times 10^6}$

$$\Rightarrow \quad v \approx \frac{c}{138} \quad \Rightarrow \quad v = \frac{c}{2 \times 69}\ \text{m/s}$$

Hence, $N = 2$

DPP-3 Electric Field, Field Intensity and Superposition of Electric Field

1. (i) To investigate the field around a charge (or at a point due to a system of charges), we measure the force on a small positive test charge at rest. Test charge must be small enough so that it does not significantly affect the charges that creates the field.

(ii) The reason for defining electric field $\mathbf{E} = \lim_{q_0 \to 0} \frac{\mathbf{F}}{q_0}$ is so that $\mathbf{E}$ does not depend on test charge q_0. This means field $\mathbf{E}$ describes only the effect of the charges creating the electric field at that point.

We call alteration of space around a charge is its field. (or alteration of space around a mass is called gravitational field or alteration of space around a magnet is called magnetic field). Field model is a very useful tool in visualising and calculating forces for complex arrangement of charges.

Another fundamental reason is clear when we observe effect of field or in case of formation and propagation of electromagnetic waves, etc.

2. Field due to ring at a point x distance from centre on its axis

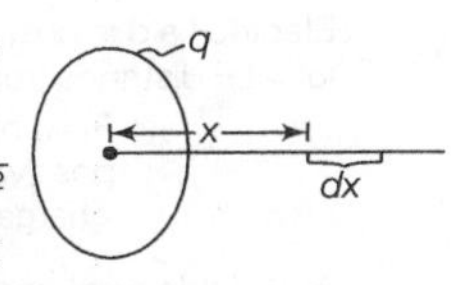

$$= E = \frac{Kqx}{(R^2 + x^2)^{3/2}}$$

Due to this field, force on an elemental length 'dx' of wire will be

$$dF = E.dq = E.\lambda dx$$

$$= \frac{Kqx\lambda dx}{(R^2 + x^2)^{3/2}}$$

Summing up forces; we have force on wire by the ring

= force on ring due to wire

$$= \int_0^\infty dF = \frac{q\lambda}{4\pi\varepsilon_0} \cdot \int_0^\infty \frac{xdx}{(R^2 + x^2)^{3/2}}$$

$$= \frac{q\lambda}{4\pi\varepsilon_0 R}\ \text{N}$$

3.

(i)

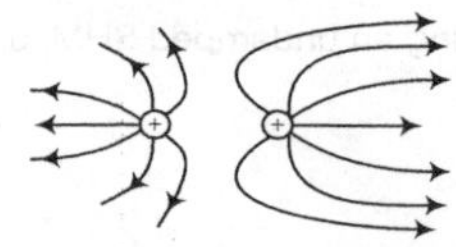

(ii)

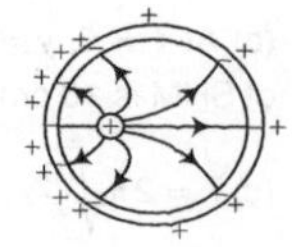

(iii)

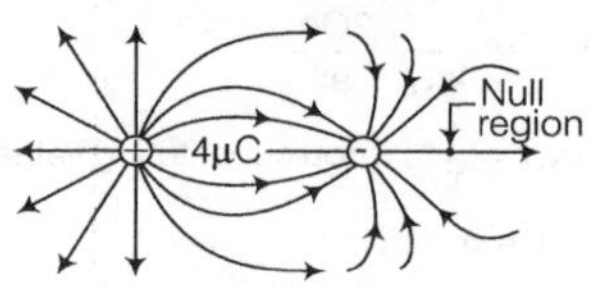

(iv) (a)

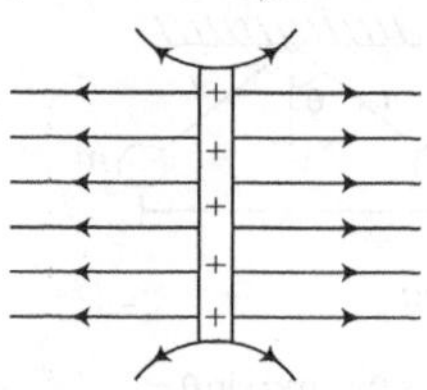

(b)

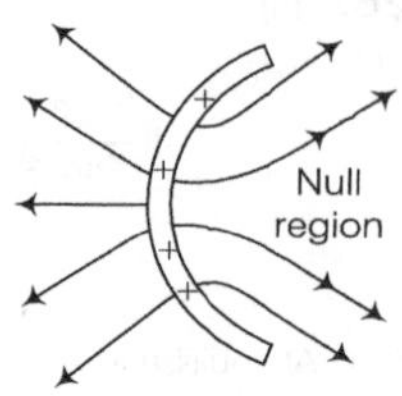

(v) (a)

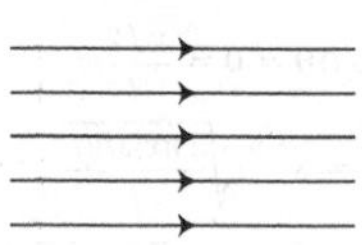

(b)

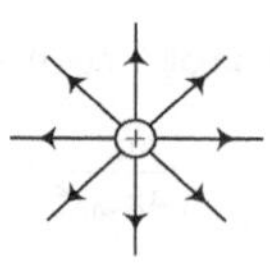

4. (i) If strength of electric field is more, the spacing between field lines is less and *vice-versa*.

So, in given situation, field is more stronger at A than B.

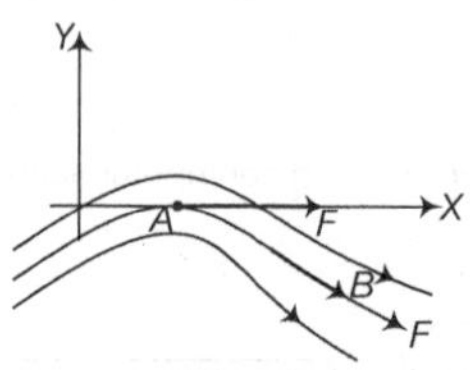

Also, direction of force on a test charge is along a tangent to the field line.

(ii) Hence, at A, force is along positive X-axis and at B, it is along the field line (or shown by tangents drawn at these points).

Also, as $|\mathbf{F}| = q_0|\mathbf{E}|$, so force is more at A and less at B.

i.e. accelerotion of q_0 is higher at point A.

5. (a)

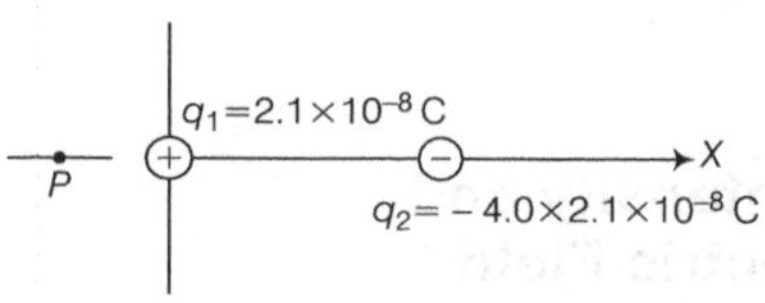

Electric field is possibly zero at some point P on X-axis as shown, let at P distance from q_1.

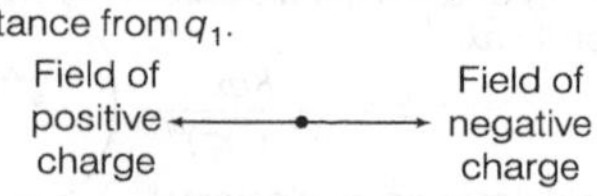

At P, fields exist and are in opposite directions, so fields oppose each other.

If net field is zero, then field of positive charge

= Field of negative charge

$$\Rightarrow \quad \frac{K(q_1)}{x^2 \times 10^{-4}} = \frac{K(q_2)}{(x+50)^2 \times 10^{-4}}$$

$$\Rightarrow \quad q_1(x+50)^2 = q_2(x^2)$$

$$\Rightarrow \quad 2.1 \times 10^{-8} \times (x+50)^2 = 4 \times 2.1 \times 10^{-8} \times x^2$$

$$\Rightarrow \quad x + 50 = 2x \Rightarrow x = 50 \text{ cm}$$

(towards negative side of origin)

6. (c)

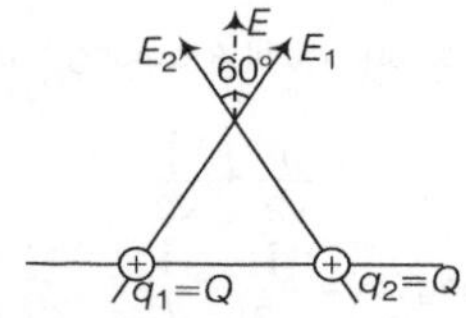

E_{net} due to superposition of fields of both charges.

$$\therefore \quad E_{net} = \sqrt{E_1^2 + E_2^2 + 2E_1E_2\cos\theta}$$

$$\theta = 60°$$

$$= \sqrt{\left(\frac{kQ}{a^2}\right)^2 + \left(\frac{kQ}{a^2}\right)^2 + 2\left(\frac{kQ}{a^2}\right)\left(\frac{kQ}{a^2}\right) \times \frac{1}{2}}$$

$$= \sqrt{3}\,\frac{kQ}{a^2} \text{ and is directed as shown.}$$

7. (a)

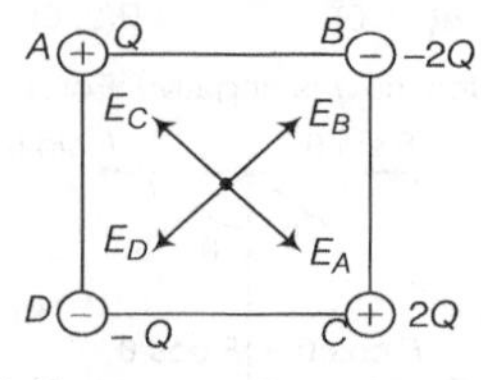

There are four fields at centre E_A and E_C, E_B and E_D

Also,
$$E_A = \frac{k(2Q)}{\left(\frac{5}{\sqrt{2}} \times 10^{-2}\right)^2} = \frac{k2Q}{\left(\frac{5}{\sqrt{2}} \times 10^{-2}\right)^2}$$

$$E_C = \frac{kQ}{\left(\frac{5}{\sqrt{2}} \times 10^{-2}\right)^2}$$

E_A and E_C oppose each other, so their resultant is along E_A (it is stronger) and is

$$E_1 = E_A - E_C = \frac{kQ}{\left(\frac{5}{\sqrt{2}} \times 10^{-2}\right)^2}$$

Similarly, resultant of E_B and E_D is along E_B and is

$$E_2 = E_B - E_D = \frac{kQ}{\left(\frac{5}{\sqrt{2}} \times 10^{-2}\right)^2}$$

They superimpose to form net field.

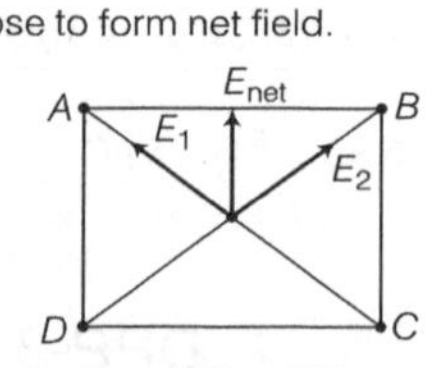

Also, angle between E_1 and E_2 = 90°.

$$\Rightarrow \quad E_{net} = \sqrt{E_1^2 + E_2^2} = 1.02 \times 10^7 \text{ N/C upwards}$$

8. (b)

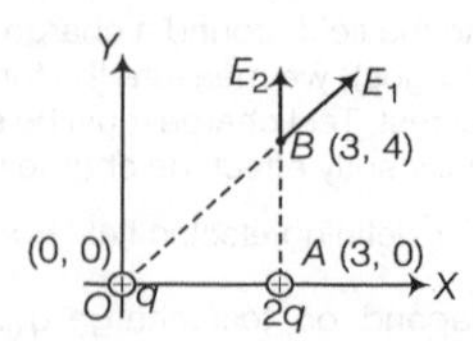

As coordinates are given, we find vector fields and add them to get resultant.

E_1 = (magnitude of unit vector along **OB**)

$$= \left(\frac{kq}{r^2}\right)\left(\frac{\mathbf{OB}}{|\mathbf{OB}|}\right)\{\mathbf{OB} = x\hat{\mathbf{i}} + y\hat{\mathbf{j}}\}$$

$$= \frac{kq\,\mathbf{OB}}{|\mathbf{OB}|^3} = \frac{kq}{(5 \times 10^{-2})^3}(3\hat{\mathbf{i}} + 4\hat{\mathbf{j}}) \times 10^{-2}$$

as dimensions are in m $= \dfrac{9 \times 10^9 \times q \times 10^{-2}}{125 \times 10^{-6}} (3\hat{i} + 4\hat{j})$

$= \dfrac{4}{3} kq (3\hat{i} + 4\hat{j}) \times 10^2 \text{N/C}$

E_2 = Magnitude × unit vector along **AB**

$= \dfrac{kq}{|\mathbf{AB}|^3} \mathbf{AB} = \dfrac{kq}{(4 \times 10^{-2})^3} \times 4\hat{j} \times 10^{-2}$

$= 25kq \times 10^2 \hat{j}$ N/C

So, net field $E_{net} = kq \times 10^2 \left(4\hat{i} + \dfrac{16}{3}\hat{j} + 25\hat{j}\right)$

$= kq \left(4\hat{i} + \dfrac{141}{3}\hat{j}\right) \times 10^2$ N/C

9. (b) When spheres are equally charged (say each have a charge q), they have different surface potentials.

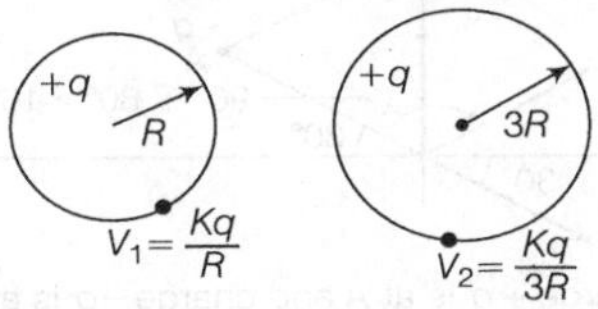

When they are joined by a wire, charge flows from small sphere to large sphere till they have same potentials.

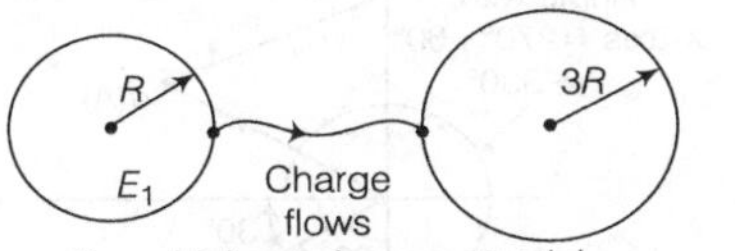

From higher to lower potential

The common potential is given by

$$V = \frac{C_1V_1 + C_2V_2}{C_1 + C_2} = \frac{q_1 + q_2}{C_1 + C_2} = \frac{2q}{4\pi\varepsilon_0(R + 3R)}$$

∴ Capacity of an isolated sphere is $C = 4\pi\varepsilon_0 R = \dfrac{kq}{2R}$

And they finally have are the charges q_1' and q_2', respectively,

$$q_1' = 4\pi\varepsilon_0 R \cdot \frac{1}{4\pi\varepsilon_0} \frac{q}{2R} = CV$$

∴ $C = \dfrac{q}{V} \Rightarrow q = CV = \dfrac{q}{2}$

and $q_2' = 2q - \dfrac{q}{2} = \dfrac{3q}{2}$

So, now fields on their surfaces are

$$E_1 = \frac{Kq_1'}{R^2} = \frac{K\left(\frac{q}{2}\right)}{R^2} \text{ and } E_2 = \frac{Kq'_2}{(3R)^2} = \frac{K \cdot \frac{3q}{2}}{9R^2}$$

So, ratio of electric field, $\dfrac{E_1}{E_2} = \dfrac{3}{1} \Rightarrow 3:1$

10. (c) $E = \dfrac{Kq}{r^2} \cdot \hat{r} = \dfrac{Kq}{r^3} \cdot \mathbf{r}$

Here, $r = \sqrt{(x_2 - x_1)^2 + (y_2 - y_1)^2 + (z_2 - z_1)^2}$

$= \sqrt{(30)^2 + (30)^2 + 0^2} = 30\sqrt{2}$ cm $= 30\sqrt{2} \times 10^{-2}$ m

and $q = 8 \times 10^{-3} \times 10^{-6}$C

Also, $\mathbf{r} = (30\hat{i} + 30\hat{j}) \times 10^{-2}$m

So, $\mathbf{E} = \dfrac{9 \times 10^9 \times 8 \times 10^{-3} \times 10^{-6}}{(30\sqrt{2} \times 10^{-2})^3} \times (30\hat{i} + 30\hat{j}) \times 10^{-2}$

$\Rightarrow \mathbf{E} = \dfrac{9 \times 8 \times 10^9 \times 10^{-11}}{27 \times 2\sqrt{2} \times 10^{-6} \times 10^3} \times 30(\hat{i} + \hat{j})$

$= \dfrac{9 \times 8 \times 10^2 \times 3}{27 \times 2\sqrt{2}} (\hat{i} + \hat{j}) = 200\sqrt{2} (\hat{i} + \hat{j}) \text{NC}^{-1}$

11. (d) When sphere of radius $3R$ is touched with sphere of radius R, the ratio of new values of charges on them,

$$\frac{q_1'}{q_3} = \frac{R}{3R} = \frac{1}{3}$$

Also, $q_1' + q_3 = q$ (initially only Ist is charged)

$q_3 = 3q_1'$

$\Rightarrow q_1' + 3q_1' = q$

or $q_1' = \dfrac{q}{4}$ ...(i)

∴ Sphere of radius $3R$ will have a charge $q_3 = \dfrac{3}{4}q$ when it is again touched with second sphere they share this charge of $\dfrac{3q}{4}$.

So, now let charges (after sharing) is q_2' and q_3'.

Then, $\dfrac{q_2'}{q_3'} = \dfrac{2R}{3R} = \dfrac{2}{3}$ and $q_2' + q_3' = \dfrac{3q}{4}$

Solving these, we get

$$q_2' + \frac{3}{2}q_2' = \frac{3q}{4}$$

$$\Rightarrow \frac{5q_2'}{2} = \frac{3q}{4}$$

q_2' = charge on sphere of radius $2R = \dfrac{3}{10}q$

So, field intensities on surfaces of spheres I and II after removing III are

$$E_1 = \frac{k\left(\frac{q}{4}\right)}{R^2}, \quad E_2 = \frac{k\left(\frac{3}{10}q\right)}{(2R)^2}$$

So, ratio $\dfrac{E_1}{E_2} = 10:3$

12. (c) Electric field is zero everywhere inside metal (conductor) i.e. field lines do not enter a metal. Simultaneously these are perpendicular to metal surface (equipotential surface).

13. (c) As electric field at a sphere 1, we get

$E_1 = \dfrac{kQ}{R^2}$, where $k = \dfrac{1}{4\pi\varepsilon_0}$

$E_2 = \dfrac{K(2Q)}{R^2}$

$\Rightarrow E_2 = \dfrac{2KQ}{R^2}$

$E_3 = \dfrac{K(4Q)R}{(2R)^3}$

$\Rightarrow E_3 = \dfrac{KQ}{2R^2}$

∴ $E_2 < E_1 < E_3$

14. (b,c) $E_{net} = E_1 + E_2$ and is pointing towards negative charge,

$E_1 = \dfrac{kq_1}{r_1^2}$, $E_2 = \dfrac{Kq_2}{r_2^2}$

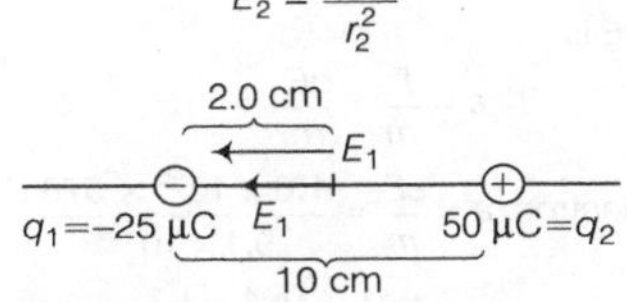

Take, $r_1 = 2.0$ cm and $r_2 = 8$cm

Also, force on electron $F = eE$

and acceleration of electron

$$= a = \frac{F}{m} = \frac{eE}{m} = \frac{1.6 \times 10^{-19} \times 6.3 \times 10^8}{9.1 \times 10^{-31}}$$

$= 1.1 \times 10^{20}$ m/s²

15. (b,c) Field at A is vector sum of fields $\mathbf{E}_{A_1}$ due to Q_1 and $\mathbf{E}_{A_2}$ due to Q_2.

$$E_{A_1} = \frac{Kq}{r^2} = \frac{9 \times 10^9 \times 50 \times 10^{-6}}{(0.60)^2}$$
$$= 1.25 \times 10^6 \text{ N/C}$$
$$\mathbf{E}_{A_2} = \frac{Kq}{\mu^2} = 5 \times 10^6 \text{ N/C}$$

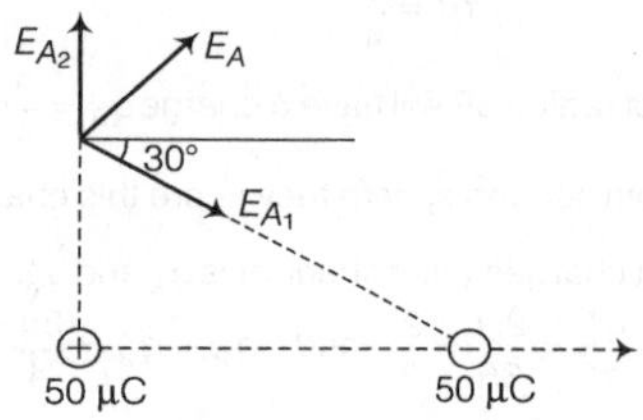

$\mathbf{E}_A$ has components $E_{A_x} = E_{A_1} \cos 30°$

$\Rightarrow$ $E_{A_x} = 1.1 \times 10^6$ N/C

and $E_{A_y} = E_{A_2} - E_{A_1} \sin 30°$
$= 4.4 \times 10^6$ N/C

So, E_A = net field at point A
$$= \sqrt{E_{A_x}^2 + E_{A_y}^2} = 4.5 \times 10^6 \text{ N/C}$$

Similarly, at B,

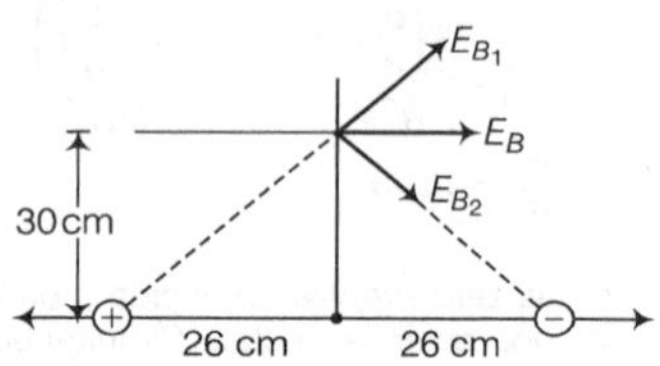

By symmetry,
$$E_{B_1} = E_{B_2} = \frac{KQ}{r^2}$$
$$= \frac{9 \times 10^9 \times 50 \times 10^{-6}}{(0.40)^2}$$
$$= 2.8 \times 10^6 \text{ N/C}$$

Also, $E_B = E_{B_1} \cos\theta + E_{B_2} \cos\theta$
$= 2 \times 2.8 \times 10^6 \times 0.65$
$$= 3.6 \times 10^6 \text{ N/C} \quad \left\{\cos\theta = \frac{26 \text{ cm}}{40 \text{ cm}}\right\}$$

So, ratio $\frac{E_A}{E_B} = \frac{4.5 \times 10^6}{3.6 \times 10^6} = \frac{45}{36} = \frac{5}{4}$

16. (b,d) Acceleration of a charge q, when released from rest in a field of intensity E is
$$a = \frac{F}{m} = \frac{qE}{m}$$
$\therefore$ For electron, $a = \frac{eE}{m} = \frac{1.6 \times 10^{-6} \times 576}{9.1 \times 10^{-31}}$
$$= 1.01 \times 10^{14} \text{ m/s}^2$$

Also, electron is accelerated opposite to field.

17. (b,d) Clearly, field pattern reveals a null point on left hand side of charge q_1.

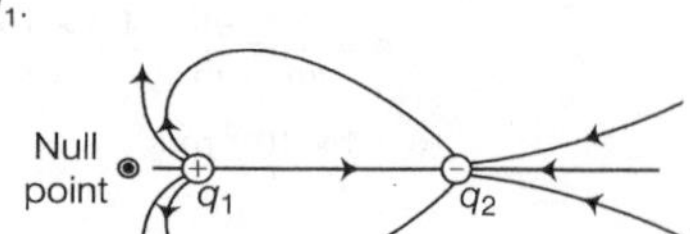

A null point field of both charges opposes each other.

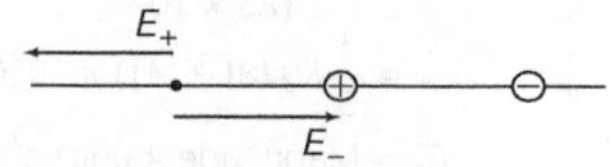

As they cancel out each other, $E_+ = E_-$ and this is possible only when negative charge has higher magnitude.

$\Rightarrow$ $|q_2| > |q_1|$

And if a test charge q_0 is placed at null point, it experiences a force
$$F = q_0 E_{net} = 0$$

18. (b,d)

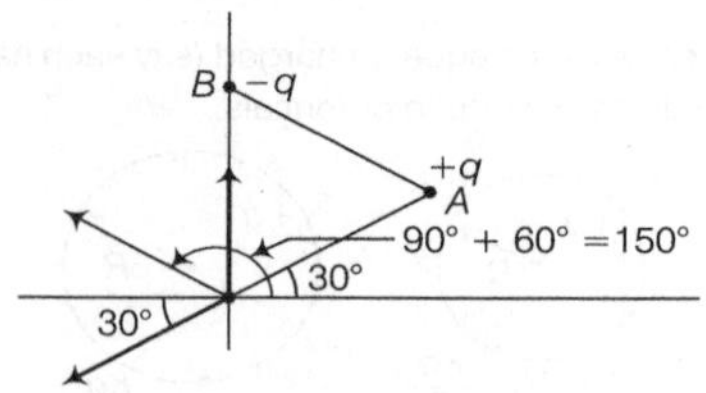

When a charge $+q$ is at A and charge $-q$ is at B field at 150° with X-axis.

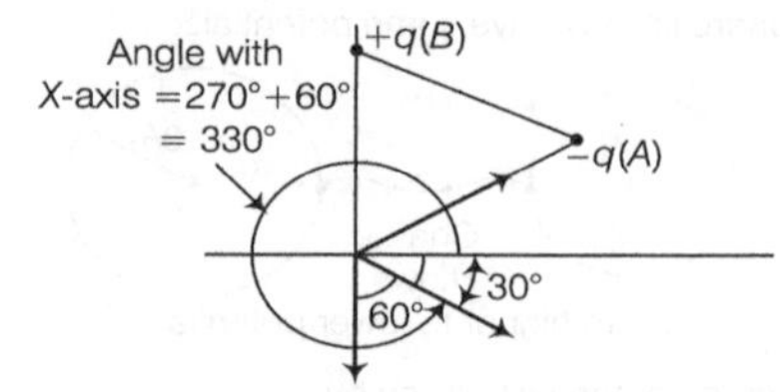

With a charge $-q$ at A and $+q$ at B, field is at 330° with positive X-axis.

19. (c) Electric field does not exist inside a hollow shell, so electrostatic shielding is possible.

But field lines are not reflected from metal surface.

20. (d) Accelerations of proton and electron in field of same magnitude are
$$a_p = \frac{eE}{m_p}$$
and
$$a_e = \frac{eE}{m_e}$$
Clearly, $a_e > a_p$

Also, electric force $\propto q$ and it is mass independent.

21. (a) As net field is zero, the forces must form a cyclic triangle.

22. (a) Field between two positive charges decreases and at some point in between it is zero, so curve Q.

(i) $\rightarrow q$

Field between a positive and a negative charge is more than field due to a single charge.

$\therefore$ (ii) $\rightarrow r$ and (iii) $\rightarrow p$

Explanations for Q. Nos. (23-25)

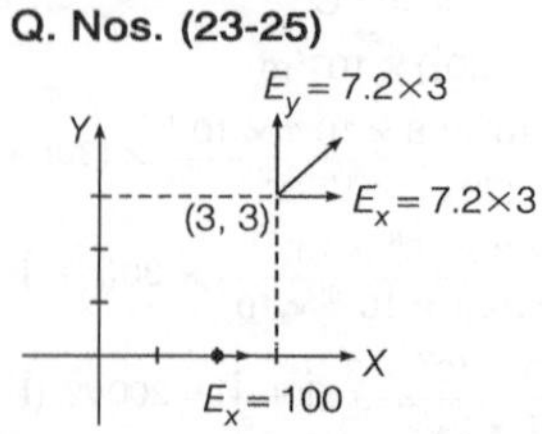

As on X-axis, there is no component of field.

$\therefore$ Point charge must lies on X-axis.

If θ = Angle made by field line at P with X-axis.

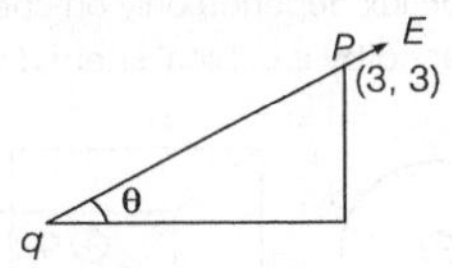

Then, $\tan\theta = \dfrac{\text{perpendicular}}{\text{base}}$

$$= \frac{3}{\text{base}} = \frac{E_y}{E_x} = \frac{3}{4}$$

Hence, base of triangle is of 4 unit.

Clearly, x-coordinate of charge is -1.0 unit.

As charge is on X-axis, y-coordinate of charge is zero.

Given field at (2, 0) is 100 N/C

Using, $\quad E = \dfrac{Kq}{r^2}, r = 3\,\text{cm}$

we have, $\quad 100 = \dfrac{9\times10^9 \times q}{9\times10^{-4}} \Rightarrow q = \dfrac{100\times9\times10^{-4}}{9\times10^9}$

$$= 1\times10^{-11}$$
$$= 10\times10^{-12}\,\text{C} = 10\,\text{pC}$$

23 (a), **24.** (c), **25.** (b)

Explanations for Q. Nos. (26 - 29)

Charge density $\lambda = \lambda_0\cos\theta$, $\cos\theta$ is positive when $\theta \in \left(-\dfrac{\pi}{2}, \dfrac{\pi}{2}\right)$

and $\cos\theta$ is negative when $\theta \in \left(\dfrac{\pi}{2}, \dfrac{3\pi}{2}\right)$.

So, charge on ring is as shown

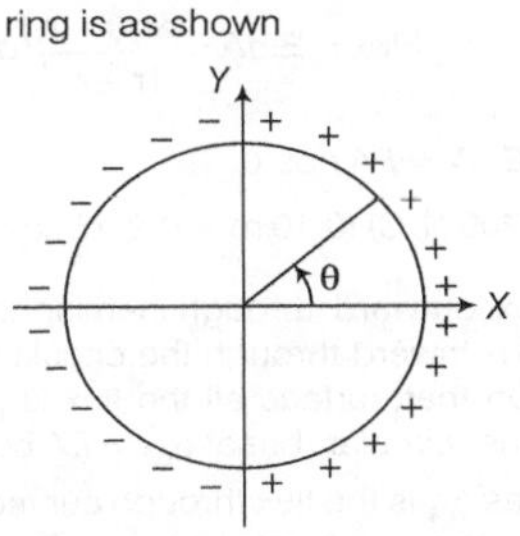

Now, for an elemental part subtending angle $d\theta$ at centre of ring, field produced at centre will be dE, given by

$$dE = \frac{kdq}{R^2}, dq = (Rd\theta)\lambda = \lambda_0\cos\theta\cdot Rd\theta$$

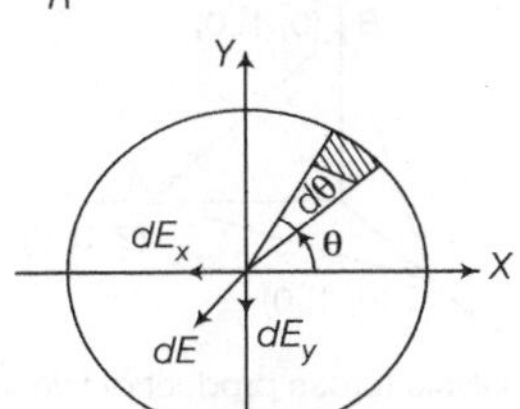

Electric field dE can be resolved into components dE_x and dE_y.

$$dE_x = dE\cos\theta,$$
$$dE_y = dE\sin\theta$$

and $\quad E_x = \int_0^{2\pi} dE_x = \dfrac{\lambda_0}{4\pi\varepsilon_0 R}\int_0^{2\pi}\cos^2\theta d\theta = \dfrac{\lambda_0}{4\varepsilon_0 R}$

Also, $\quad E_y = \int_0^{2\pi} dE_y = 0$

$\Rightarrow \quad E_{\text{net}}$ = Net field at the centre of the ring

$$= \sqrt{E_x^2 + E_y^2} = \frac{\lambda_0}{4\varepsilon_0 R}\cdot\left(\frac{\text{N}}{\text{C}}\right)$$

Also, field on axis is

$$dE = \frac{dq}{4\pi\varepsilon_0(x^2+R^2)} = \frac{\lambda_0\cos\theta Rd\theta}{4\pi\varepsilon_0(x^2+R^2)}$$

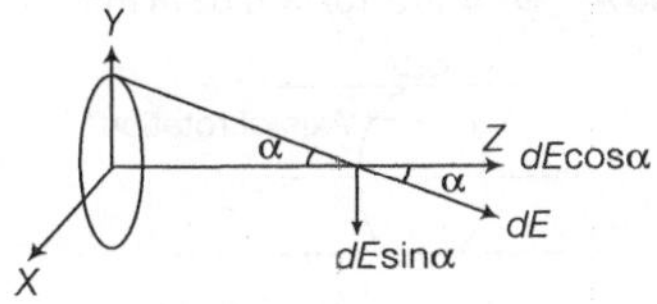

where, $\quad \cos\alpha = \dfrac{x}{\sqrt{x^2+R^2}}$

Also, $\quad E_x = E_z = 0$

and $\quad E_{\text{net}} = \int_0^{2\pi} dE_x = \int_0^{2\pi}\dfrac{\lambda_0 R\cdot\sin\alpha\cos^2\theta}{4\pi\varepsilon_0(x^2+R^2)}d\theta$

$$= \frac{\lambda_0 R\sin\alpha}{4\pi\varepsilon_0(x^2+R^2)}\int_0^{2\pi}\cos^2\theta d\theta$$

$$= \frac{\lambda_0 R\sin\alpha}{4\varepsilon_0(x^2+R^2)}$$

and when $x >> R$, $E_{\text{net}} = \dfrac{\lambda_0 R^2}{4\varepsilon_0 x^3}$ along X-axis.

26. (d) **27.** (c) **28.** (c) **29.** (a)

30. (5) As particle is moving horizontally, only forces acting on it are Stoke's force and electric force.

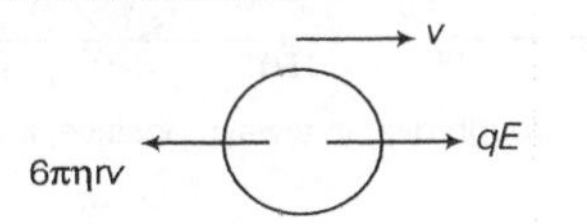

If v = constant, then

$$6\pi\eta rv = qE$$
$$\Rightarrow \quad 6\pi\eta rv = neE$$
$$\Rightarrow \quad n = \frac{6\pi\eta rv}{Ee}$$
$$= \frac{6\times3.14\times1.6\times10^{-5}\times5\times10^{-7}\times0.02}{6.28\times10^5\times1.6\times10^{-19}}$$
$$= 30 = 6k$$
$$\Rightarrow \quad k = 5$$

31. (2) Particle will not hit the upper plate, if the velocity component normal to plate is zero before reaching the plate.

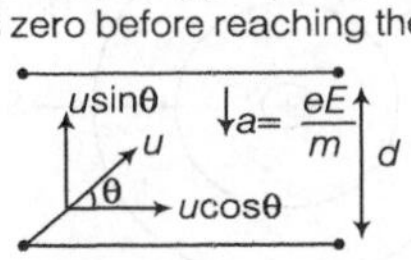

Acceleration due to field (downwards).

$$a_y = \frac{eE}{m}$$
$$v_y^2 = 0 = u^2 - 2(a_y)y$$

where, $y \le d$, d = distance between plates.

So, $\quad u_{\max} = \sqrt{2a_y} = \sqrt{\dfrac{2\times1.6\times10^{-19}\times10^3\times2\times10^{-2}}{1.6\times10^{-30}}}$

$\Rightarrow \quad u_{\max} = 2\times10^6$ m/s

DPP-4 Electric Flux and Gauss' Law

1. Electric field in expression for Gauss' laws refers to the total electric field not just the electric field due to any enclosed charge. If field is due to a charge outside the Gauss' surface, then flux through the surface due to this charge is zero.

2. Area of disc $A = \pi r^2 = \pi(0.10)^2 = 0.0314 \text{ m}^2$

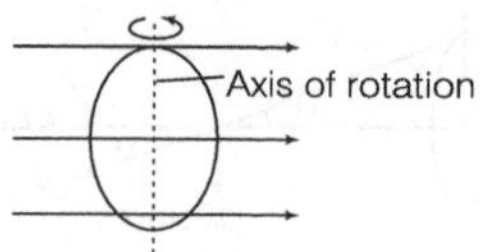

As disc rotates, orientation of area vector and field changes and so flux linked with the disc is also change.

Flux linked with disc when area vector makes an angle θ with field is

$$\phi_E = \mathbf{E} \cdot \mathbf{A} = EA \cos \theta$$

When plane of disc is parallel to field lines,

$$\theta = 90°, \cos \theta = 0$$

$$\Rightarrow \qquad \phi_E = EA\cos 90° = 0$$

When plane of disc is perpendicular to field lines, $\theta = 0°, \cos \theta = 1$

and so $\phi_E = EA\cos 0° = 2 \times 10^3 \times 0.0314 \times 1 = 63 \dfrac{\text{Nm}^2}{\text{C}}$

The graph between ϕ_E *versus* θ is (It starts with $\theta = 0°$)

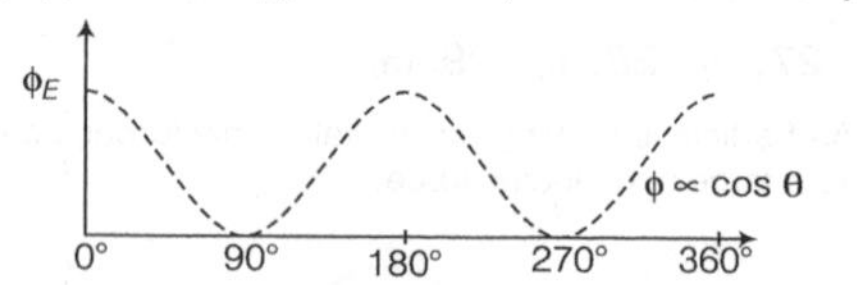

(as the flux through disc is always positive, it is plotted in positive side).

3. Electric flux $\phi_E = \mathbf{E} \cdot \mathbf{A}$. By analogy, the gravitational flux through a surface would be the product of the gravitational field (or the force per unit mass) and the area.

or $\qquad \phi_g = gA \cos \theta = \mathbf{g} \cdot \mathbf{A}$

Gravitational flux is always negative or there are only sinks and no sources (anti-gravity). Gauss' law for gravitational flux would be

$\phi_g \propto$ net mass enclosed by the surface.

or $\phi_g = \oint \mathbf{g} \cdot d\mathbf{A} = -(4\pi G) \cdot m_{\text{enclosed}}$

4.

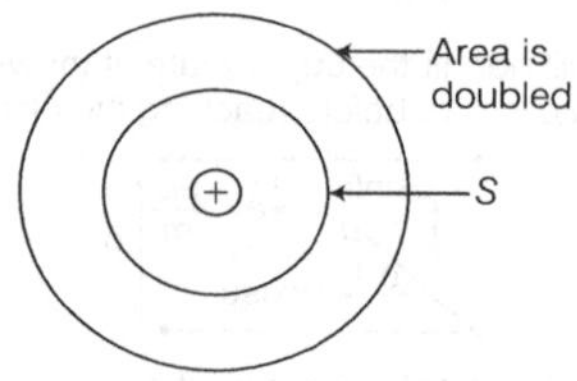

(i) When area is doubled, flux remains same as the charge enclosed is still the same.

(ii) When charge is shifted to one side, field intensity at all points on the surface is not same, but flux linked with surface is still the same.

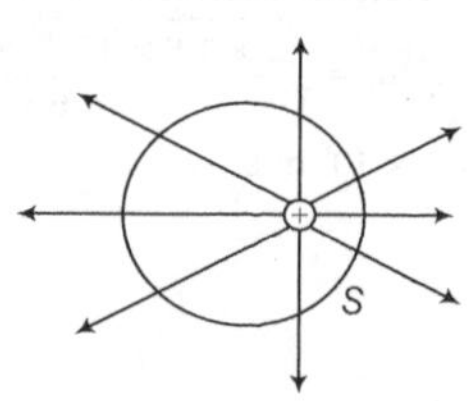

(iii) As quantity of flux depends only on charge enclosed.

$\therefore$ If S is replaced by a cubical shaped surface, flux remains same.

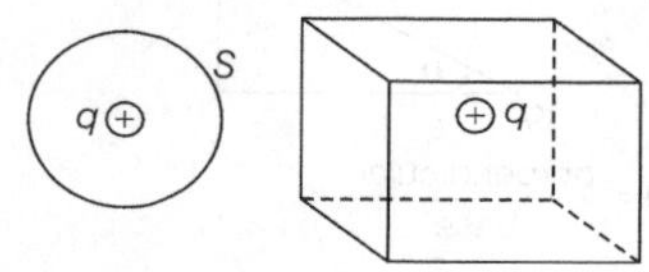

(iv) When another charge is placed just outside the surface S, field lines of outer charge enters and leaves the surface, so net flux is still depending on the charge enclosed.

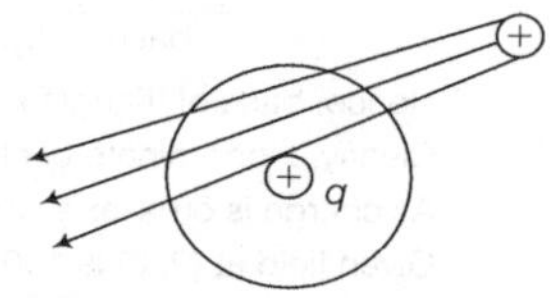

Hence, flux does not change due to presence of any charge outside the surface.

5. (a) Flux through an elemental area is given by

$$\phi_E = \mathbf{E} \cdot d\mathbf{A}$$

Also, $\qquad \mathbf{E} = \dfrac{kq}{r^2} \cdot \hat{\mathbf{r}}^2 = \dfrac{kq\,\mathbf{r}}{r^3}$

where, $\mathbf{r}$ is position vector of point and $\mathbf{E}$ is determined w.r.t. charge q taken as origin. In given figure, charge q is not origin. So, vector from charge to point of observation of field is (at location of ds)

$$\mathbf{r} - \mathbf{r}_1$$

So, $\qquad \mathbf{E} = \dfrac{kq(\mathbf{r} - \mathbf{r}_1)}{|\mathbf{r} - \mathbf{r}_1|^3} \cdot d\mathbf{s}$

$$\text{Flux} = \mathbf{E}.d\mathbf{A} = \frac{kq(\mathbf{r} - \mathbf{r}_1)}{|\mathbf{r} - \mathbf{r}_1|^3} d\mathbf{s}$$

6. (b) Flux, $\phi_E = \mathbf{E} \cdot \mathbf{A} = EA \cos \theta$

$$= (200 \text{ N/C})(0.10 \text{ m} \times 0.2 \text{ m}) \cos 30° = 3.5 \text{ Nm}^2/\text{C}$$

7. (c) We see flux outward through hemispherical surface is the same as the flux inward through the circular surface base of the hemisphere. On that surface all the flux is perpendicular to the surface. On the circular base $\phi_{ps} = EA \cos 180° = -\pi r^2 E$ i.e. plane surface as ϕ_{cs} is the flux through curved surface. As net flux through the hemisphere $\phi = \phi_{ps} + \phi_{cs} = 0$, so $\phi_{cs} = -\phi_{ps} = \pi r^2 E$

8. (b) Flux through a plane area in uniform field is $\phi_E = \mathbf{E} \cdot \mathbf{A}$

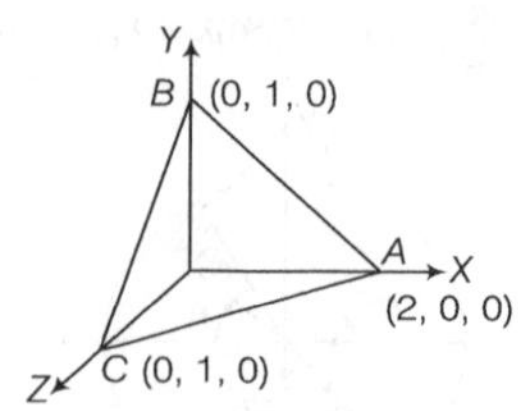

Using vector analysis (cross product of two vectors)

Area of $\Delta ABC = \dfrac{1}{2} |\mathbf{AB} \times \mathbf{AC}|$

$\therefore$ $\mathbf{A}$ = Area vector of ΔABC = Area $\times \hat{\mathbf{n}}$

$$= \frac{1}{2} |\mathbf{AB} \times \mathbf{AC}| = \frac{1}{2} \begin{vmatrix} \hat{\mathbf{i}} & \hat{\mathbf{j}} & \hat{\mathbf{k}} \\ -2 & 1 & 0 \\ -2 & 0 & 1 \end{vmatrix} = \frac{1}{2}(\hat{\mathbf{i}} + 2\hat{\mathbf{j}} + 2\hat{\mathbf{k}}) \text{ m}^2$$

So, flux $x = \phi_E = \mathbf{E} \cdot \mathbf{A} = (3 \times 10^3 \hat{\mathbf{i}}) \cdot \dfrac{1}{2}(\hat{\mathbf{i}} + 2\hat{\mathbf{j}} + 2\hat{\mathbf{k}})$

$$= \frac{3}{2} \times 10^3 \text{ Nm}^2/\text{C} = 1.5 \times 10^3 \text{ Nm}^2/\text{C}$$

9. (c)

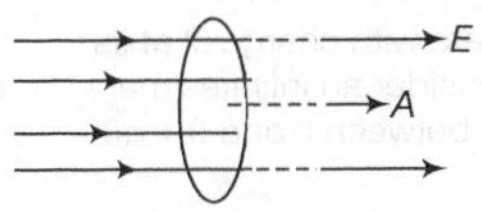

When face of disc is perpendicular to field, flux through the disc

$$\phi_1 = \mathbf{E} \cdot \mathbf{A} = EA \cos 0^\circ \qquad (\text{as } A \parallel E)$$

$$= EA = 5.8 \times 10^2 \times \pi r^2 \qquad (r = \text{radius of disc})$$

When face of disc is 45° to field lines

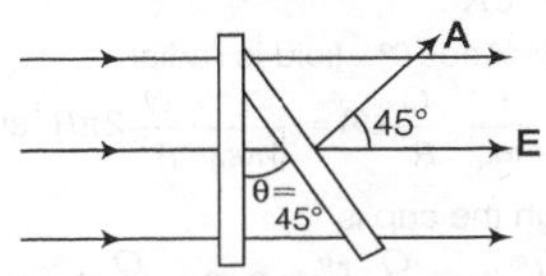

then, angle between **A** and **E** is also 45°.

So, $\phi_2 = \text{flux} = \mathbf{E} \cdot \mathbf{A} = EA \cos \theta = EA \cos 45^\circ$

$\Rightarrow$ Ratio $\dfrac{\phi_1}{\phi_2} = \dfrac{EA \cos 0^\circ}{EA \cos 45^\circ} = \sqrt{2}$

10. (d) The only contributions to the flux from the faces perpendicular to the electric field. Over each of these two surfaces, the magnitude of the field is constant, so the flux is just **E** · **A** on each of these two surfaces.

$$\phi_E = (\mathbf{E} \cdot \mathbf{A})_{\text{right}} + (\mathbf{E} \cdot \mathbf{A})_{\text{left}}$$

$$E_{\text{right}} \cdot l^2 - E_{\text{left}} \cdot l^2 = \frac{Q_{\text{encl}}}{\varepsilon_0}$$

$$Q_{\text{encl}} = (E_{\text{right}} - E_{\text{left}}) l^2 \varepsilon_0$$

$$= (410 \text{ N/C} - 560 \text{ N/C})(25 \text{ m})^2 (8.85 \times 10^{-12} \text{ C}^2/\text{Nm}^2)$$

$$= -8.3 \times 10^{-7} \text{ C}$$

11. (c) The quantity of **E** on the left side of equation $\oint \mathbf{E} \cdot d\mathbf{A}$, however, is the electric field resulting from all changes, both those inside and those outside the Gaussian surface.

12. (b) The quantity of **E** on the left side of equation $\oint \mathbf{E} \cdot d\mathbf{A}$, however is the electric field resulting from all changes, both those inside and those outside the Gaussian surface.

The right side of the equation from the Gauss' law $\phi = \dfrac{q_{\text{enclosed}}}{\varepsilon_0}$ through the surface depends on the net charge q_{encl} enclosed surface.

13. (c) Flux $= \oint \mathbf{E} \cdot d\mathbf{A} \longrightarrow$ for hemisphere $= \oint |\mathbf{E}|\, |d\mathbf{A}| \cos \theta$

Now, for any elemental area of hemisphere, $d\mathbf{A} \parallel \mathbf{E}$

$\therefore \quad \cos \theta = 1$

$\Rightarrow \quad \theta = 0^\circ$

So, $\phi_E = \oint E\, dA$, also, E is uniform or all points on surface and hemisphere are equidistant from centre.

$\Rightarrow \quad \phi_E = E \oint dA = E\, 2\pi r^2 = 2\pi r^2 E$

14. (d) From the Gauss' law, $\phi = \dfrac{q_{\text{enclosed}}}{\varepsilon_0}$

in which q_{enclosed} is the net charge inside an imaginary closed surface (a Gaussian surface) flux does not depend on the radius of imaginary enclosed surface.

15. (c,d) Flux through cube is maximum when wire passes diagonally through cube.

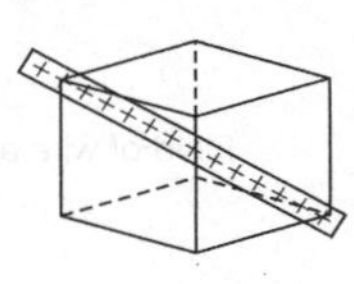

$$\therefore \quad \phi_{\max} = \frac{q_{\text{enclosed}}}{\varepsilon_0}$$

$$= \frac{\lambda \sqrt{3} \times \text{side}}{\varepsilon_0} = \sqrt{3}\, \lambda / \varepsilon_0$$

And also, $\phi_{\min} = \dfrac{q_{\text{enclosed}}}{\varepsilon_0} = \dfrac{\lambda \times \text{side}}{\varepsilon_0} = \dfrac{\lambda}{\varepsilon_0}$

16. (a,b,d) The field **E** is directed into faces 1 and 3 so the fluxes through them are negative **E** is directed out of faces 2 and 4, so the fluxes through them are positive.

We find

$$\phi_{E_1} = \mathbf{E}_1 \cdot \mathbf{n}_1 A = Ea^2 \cos \theta\, (180^\circ - \theta) = -Ea^2 \cos \theta$$

$$\phi_{E_2} = \mathbf{E} \cdot \hat{\mathbf{n}}_2 A = +Ea^2 \cos \theta$$

$$\phi_{E_3} = \mathbf{E} \cdot \hat{\mathbf{n}}_3 A = Ea^2 \cos (90^\circ + \theta) = -Ea^2 \sin \theta$$

$$\phi_{E_4} = \mathbf{E} \cdot \hat{\mathbf{n}}_4 A = Ea^2 \cos (90^\circ - \theta) = Ea^2 \sin \theta$$

$$\phi_{E_5} = \phi_{E_6} = Ea^2 \cos 90^\circ = 0$$

The total flux $\phi_E = \phi_{E_1} + \phi_{E_2} + \phi_{E_3} + \phi_{E_4} + \phi_{E_5} + \phi_{E_6}$ through the surface of the cube is again zero.

17. (a,c,d) For face *CDEF*, $\mathbf{A} = 4\hat{\mathbf{i}}\ \text{m}^2$ and $\mathbf{E} = 15\hat{\mathbf{i}} + 3\hat{\mathbf{j}}$,

As, $\quad x = 3\text{ m}$

$$\phi = \text{flux linked with face} = \mathbf{E} \cdot \mathbf{A} = (15\hat{\mathbf{i}} + 3\hat{\mathbf{j}}) \cdot 4\hat{\mathbf{i}} = 60 \frac{\text{Nm}^2}{\text{C}}$$

For face *ABGH*, flux linked with face

$$= \phi = \mathbf{E} \cdot \mathbf{A} = (5\hat{\mathbf{i}} + 3\hat{\mathbf{j}}) \cdot 4\hat{\mathbf{i}} = 20 \frac{\text{Nm}^2}{\text{C}}$$

For face *EFGH*, x varies from $x = 1$m to $x = 3$ m.

$\therefore \quad \phi$ = Flux linked with face

$$= \int_{x=1}^{x=3} \mathbf{E} \cdot \mathbf{A} = \int_{x=1}^{3} (5x\, \hat{\mathbf{i}} + 3\hat{\mathbf{j}}) \cdot (2dx\, \hat{\mathbf{j}}) = \int_{x=1}^{x=3} 6\, dx = 6x \Big|_{x=1}^{x=3}$$

$$= 6 \times 3 - 6 \times 1 = 18 - 6 = 12\ \frac{\text{Nm}^2}{\text{C}}$$

18. (b,c) If the charge is placed at the corner of a cube, seven more identical cubes are required to enclose the charge completely.

Thus, the total flux through all the faces of cubes is $\phi_{\text{total}} = \dfrac{q}{\varepsilon_0}$. The contribution of this flux through each cube $\phi = \dfrac{q}{8\varepsilon_0}$. This flux emerging equally through three front faces of the cube other three faces of cube junction of which charge is placed have no flux.

Therefore, the flux through the face $= \dfrac{q/8\varepsilon_0}{3} = \dfrac{q}{24\varepsilon_0}$.

Also, only $\dfrac{1}{4}$ th part of sphere is inside the cube, so

$$\phi_{\text{cube}} = \frac{q/4}{\varepsilon_0} = \frac{q}{4\varepsilon_0}$$

19. (c,d) If charges are of opposite signs, then the two fields are along the same direction. So, they cannot be zero. Hence, the charges should be of same sign.

Therefore, option (c) is correct.

Further, work done by external force

= change in potential energy

$\therefore \quad W_{A \to B} = q(\Delta V) = (+1)(V_B - V_A)$

or $\quad W_{A \to B} = V_B - V_A$

Therefore, option (d) is also correct.

20. (a,b,c) Option (a) is correct due to symmetry.

Option (b) is correct again due to symmetry.

Option (c) is correct, because as per Gauss' theorem, net electric flux passing through any closed surface $= \dfrac{q_{\text{in}}}{\varepsilon_0}$

Here, $q_{\text{in}} = 3q - q - q = q$

$\therefore$ Net electric flux $= \dfrac{q}{\varepsilon_0}$

Option (d) is incorrect, because there is no symmetry in two given planes.

21. (a) The net flux through each closed surface is determined by the net charge inside

$\phi_{S_1} = (+Q - 3Q)/\varepsilon_0 = -2Q/\varepsilon_0$, $\phi_{S_2} = (+Q + 2Q - 3Q)/\varepsilon_0 = 0$

$\phi_{S_3} = (+2Q - 3Q)/\varepsilon_0 = -\dfrac{Q}{\varepsilon_0}$, $\quad \phi_{S_4} = +Q/\varepsilon_0$

22. (1) $\phi_{\text{cube}} = \frac{q_{\text{equal}}}{\varepsilon_0} = \frac{\sqrt{3}\,a\lambda}{\varepsilon_0}$, $\phi_{\text{sphere}} = \frac{q_{\text{enclosed}}}{\varepsilon_0} = \frac{2a\lambda}{\varepsilon_0}$

$$\text{Ratio } r = \frac{\sqrt{3}\,a\lambda}{\varepsilon_0} \times \frac{\varepsilon_0}{2a\lambda} = \frac{\sqrt{3}}{2}$$

Then, $\frac{2}{\sqrt{3}} r = \frac{\sqrt{3}}{2} \times \frac{2}{\sqrt{3}} = 1$

23. (2) Flux linked with a Gaussian sphere of radius r

$$= \text{charge enclosed} \times \frac{1}{\varepsilon_0}$$
$$= \text{volume enclosed} \times \text{charge density} \times \frac{1}{\varepsilon_0}$$
$$= \frac{4}{3}\pi r^3 \times \rho \times \frac{1}{\varepsilon_0}$$

So, rate of change of flux $= \frac{d\phi_E}{dt} = \frac{d}{dt}\frac{4}{3}\pi\frac{\rho}{\varepsilon_0} \times r^3$

$$= \frac{4}{3}\pi\frac{\rho}{\varepsilon_0} \times 3r^2\frac{dr}{dt} = 4\pi r^2\frac{\rho}{\varepsilon_0}\left(\frac{dr}{dt}\right) = 8\pi r^2\frac{\rho}{\varepsilon_0}$$

$\therefore \quad \frac{d\phi_E}{dt} \propto r^2, k = 2$

24. (6) When charge is at centre, flux through both faces 1 and 2 will be $\frac{q}{6\varepsilon_0}$ when charge is removed near face 1, flux through face 1 increases and through face 2 decreases.

$\therefore \quad k = 6$

25. (9) With air in the space, between S_1 and S_2,

$$\phi_{S_1} = \frac{q}{\varepsilon_0} \quad \text{and} \quad \phi_{S_2} = \frac{2q}{\varepsilon_0}$$

When a medium of $K = 5$ is filled in space

$$\phi'_{S_1} = \frac{Q}{K\varepsilon_0} = \frac{\phi_{S_1}}{5} = \frac{q}{5\varepsilon_0}$$

And in this case, flux through S_2

$$\phi'_{S_2} = \text{flux of } S_1 + \text{flux of charge } 2q$$
$$= \frac{q}{5\varepsilon_0} + \frac{2q}{\varepsilon_0} = \frac{q + 10q}{5\varepsilon_0} = \frac{11q}{5\varepsilon_0}$$

So, ratio $\frac{\phi'_{S_2}}{\phi'_{S_1}} = 11$

So, $\frac{9K}{11} = 9$

26. (24) Cover the charge such that it is located at centre in a cube whose side is $2a$.

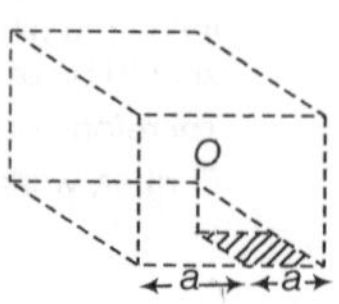

Clearly, flux $= \frac{1}{6} \times \frac{1}{4} \times \frac{q}{\varepsilon_0}$

$\therefore \quad K = 24$

27. (3) Just invert the cone and cover it with another similar cone.

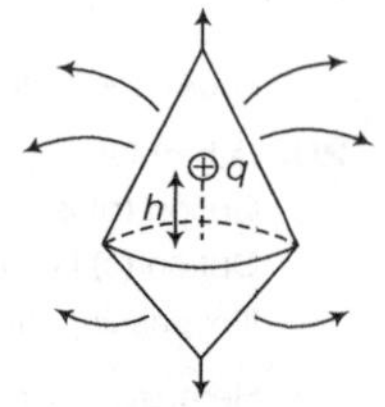

Clearly, flux through upper cone is $\frac{2q}{3\varepsilon_0}$.

So, flux through lower cone is total flux – flux through upper cone.

$\Rightarrow$ Flux through lower cone

$$= \frac{q}{\varepsilon_0} - \frac{2}{3}\frac{q}{\varepsilon_0} = \frac{q}{3\varepsilon_0}$$

Hence, $n = 3$.

28. (d)

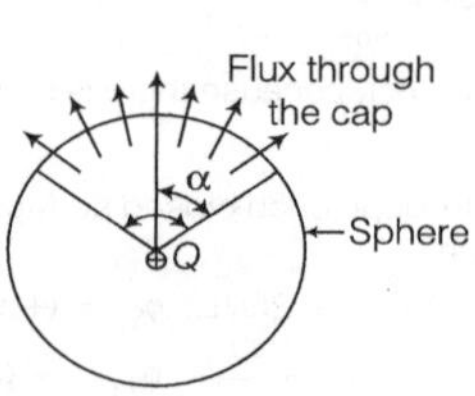

Consider a sphere with charge Q at its centre. Also, consider an infinitesimal area dA located between θ and $\theta + d\theta$ angles.

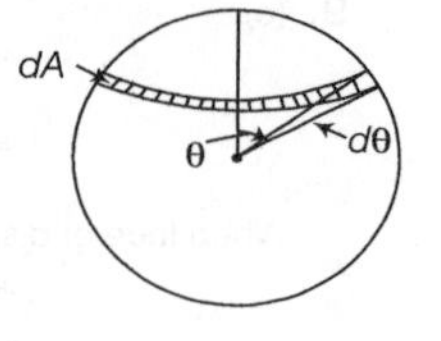

dA = circumfrence of strip $\times$ width of strip
$= (2\pi R\sin\theta) \times R\,d\theta$
$= 2\pi R^2 \sin\theta\,d\theta$

Electric flux of charge Q from this area dA is $d\phi$,

$$d\phi = \mathbf{E}\cdot d\mathbf{A}$$
$$Q = E\,dA\cos 0° \text{– field is radial}$$
$$\Rightarrow \quad d\phi = \frac{1}{4\pi\varepsilon_0}\cdot\frac{Q}{R^2}dA = \frac{1}{4\pi\varepsilon_0}\cdot\frac{Q}{R^2}2\pi R^2\sin\theta\,d\theta = \frac{Q}{2\varepsilon_0}\cdot\sin\theta\,d\theta$$

So, flux through the cap is

$$\phi_E = \int_0^\alpha d\phi = \frac{Q}{2\varepsilon_0}\int_0^\alpha \sin\theta\,d\theta = \frac{Q}{2\varepsilon_0}(-\cos\theta)_0^\alpha$$
$$= \frac{Q}{2\varepsilon_0}(1 - \cos\alpha)$$

29. (c) A disc subtending a semi-vertex angle α is shown.

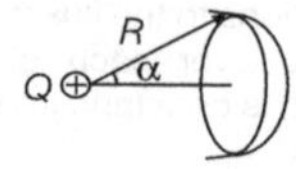

This can be viewed as base of a spherical cap.

So, flux through the disc

$$= \text{flux through the cap} = \frac{Q}{2\varepsilon_0}(1 - \cos\theta)$$

Here, $\theta = \alpha$

$$\therefore \quad \phi_E = \frac{Q}{2\varepsilon_0}(1 - \cos\alpha) = \frac{2Q\sin^2(\alpha/2)}{2\varepsilon_0} = \frac{Q}{\varepsilon_0}\sin^2\left(\frac{\alpha}{2}\right).$$

30. (d) Arrangement of charges is

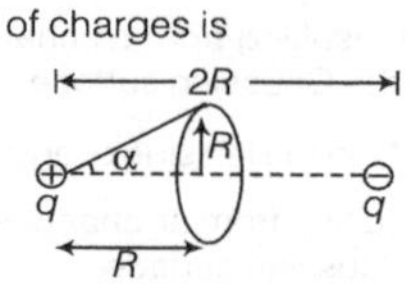

Clearly, angle made at the charge is of 45°. So, using result of sol. 28.

We have,

ϕ_E = Flux associated with circle due to a single charge q on its axis is

$$= \frac{q}{2\varepsilon_0}(1 - \cos\alpha)$$

As, there are two charges q and $-q$. Hence, total flux due to both charges will be twice of above value.

$$\therefore \quad \phi_E = \frac{2q}{2\varepsilon_0}(1 - \cos\alpha), \ \alpha = 45°$$
$$= \frac{q}{\varepsilon_0}\left(1 - \frac{1}{\sqrt{2}}\right) = \frac{q}{\varepsilon_0}\left(\frac{\sqrt{2} - 1}{\sqrt{2}}\right).$$

31. (c) Given situation is

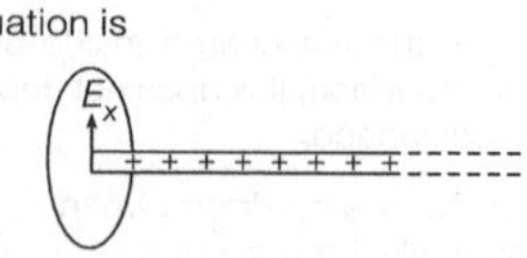

Field of wire at its end at a distance R is

$$E_x = \frac{1}{2}\left(\frac{2k\lambda}{R}\right) = \frac{1}{4\pi\varepsilon_0}\cdot\frac{\lambda}{R}$$

Flux linked with circle is

$$\phi_E = \mathbf{E}\cdot\mathbf{A} = EA\cos 90° = 0$$

Hence, $\mathbf{E} \perp \mathbf{A}$.

DPP-5 Applications of Gauss' Law

1. Due to spherical symmetry of the problem, Gauss' law using a sphere of radius r leads to the following

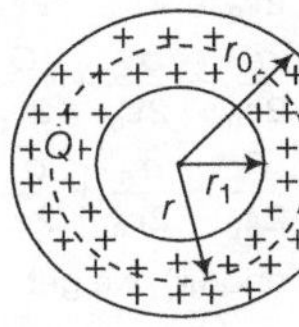

$$\oint \mathbf{E} \cdot d\mathbf{A} = E(4\pi r^2) = \frac{Q_{\text{enclosed}}}{\varepsilon_0}$$

$$E = \frac{Q_{\text{enclosed}}}{4\pi\varepsilon_0 r^2}$$

(i) For the region $0 < r < r_1$, the enclosed charge is 0 (zero)

$$E = \frac{Q_{\text{enclosed}}}{4\pi\varepsilon_0 r^2} = 0$$

(ii) For the region $r_1 < r < r_0$, the enclosed charge is the product of volume and charged density. Charged density is 3 times the volume of charged matherial enclosed.

$$\rho = \frac{Q}{\frac{4}{3}\pi r_0^3 - \frac{4}{3}\pi r_1^3} = \frac{3\,Q}{4\pi(r_0^3 - r_1^3)}$$

$$E = \frac{Q_{\text{enclosed}}}{4\pi\varepsilon_0 r^2} = \frac{\rho V_{\text{enclosed}}}{4\pi\varepsilon_0 r^2}$$

$$= \frac{\rho\left[\frac{4}{3}\pi r^3 - \frac{4}{3}\pi r_1^3\right]}{4\pi\varepsilon_0 r^2}$$

$$= \frac{\frac{3\,Q}{4\pi(r_0^3 - r_1^3)}\left[\frac{4}{3}\pi r^3 - \frac{4}{3}\pi r_1^3\right]}{4\pi\varepsilon_0 r^2}$$

$$\therefore \quad E = \frac{Q(r^3 - r_1^3)}{(r_0^3 - r_1^3)\,4\pi\varepsilon_0 r^2}$$

(iii) For the region $r > r_0$, the enclosed charge is the total charge Q.

$$\therefore \quad E = \frac{Q}{4\pi\varepsilon_0 r^2}$$

When a charge q is at centre, then by the superposition principle for electric field, we find the field for this problem by adding the field due to the point charge at the centre to the field found above. At any location $r > 0$, the field due to the point charge is $E = \dfrac{q}{4\pi\varepsilon_0 r^2}$.

(a) $E = E_q + E_Q = \dfrac{q}{4\pi\varepsilon_0 r^2} + 0 = \dfrac{q}{4\pi\varepsilon_0 r^2}$

(b) $E = E_q + E_Q = \dfrac{q}{4\pi\varepsilon_0 r^2} + \dfrac{Q(r^3 - r_1^3)}{4\pi\varepsilon_0 r^2 (r_0^3 - r_1^3)}$

$$= \frac{1}{4\pi\varepsilon_0 r^2}\left[q + Q\frac{(r^3 - r_1^3)}{(r_0^3 - r_1^3)}\right]$$

(c) $E = E_q + E_Q$

$$= \frac{q}{4\pi\varepsilon_0 r^2} + \frac{Q}{4\pi\varepsilon_0 r^2} = \left(\frac{q + Q}{4\pi\varepsilon_0 r^2}\right)$$

2. Consider given sphere is a combination of two spheres. Sphere 1 is a solid sphere of radius r_0 and charge density ρ_E centred at A and sphere 2 is a second sphere of radius $r_0/2$ and density $-\rho_E$ centred at C.

(i) The electric field at point A will have zero contribution from sphere 1 due to its symmetry about point A.

The electric field is then calculated by creating a Gaussian surface centred at point C with radius $r_0/2$.

$$\oint \mathbf{E} \cdot d\mathbf{A} = \frac{q_{\text{enclosed}}}{\varepsilon_0}$$

$$\Rightarrow \quad E \cdot 4\pi\left(\frac{1}{2}r_0\right)^2 = \frac{\frac{4}{3}\pi\left(\frac{r_0}{2}\right)^3 \cdot (-\rho_E)}{\varepsilon_0}$$

$$E = \frac{-\rho_E r_0}{6\varepsilon_0}$$

Since, the electric field points into the Gaussian surface (negative) the electric field at point A to the right.

(ii) At point B, the electric field will be the sum of the electric fields from each sphere. The electric field from sphere 1 is calculated using a Gaussian surface of radius r_0 centred at A.

$$\oint \mathbf{E}_1 \cdot d\mathbf{A} = \frac{q_{\text{enclosed}}}{\varepsilon_0} \qquad \text{(for point } A\text{)}$$

$$\Rightarrow \quad E_1 \cdot 4\pi r_0^2 = \frac{\frac{4}{3}\pi r_0^3(\rho_E)}{\varepsilon_0}$$

$$\Rightarrow \quad E_1 = \frac{\rho_E r_0}{3\varepsilon_0}$$

At point B, the field from sphere 1 points towards the left. The electric field from sphere 2 is calculated using a Gaussian surface centred at C of radius $\dfrac{3\,r_0}{2}$.

$$\oint \mathbf{E}_2 \cdot d\mathbf{A} = \frac{q_{\text{enclosed}}}{\varepsilon_0} \qquad \text{(for point } B\text{)}$$

$$\Rightarrow \quad E_2 \cdot 4\pi\left(\frac{3}{2}r_0\right)^2 = \frac{(-\rho_E)\frac{4}{3}\pi\left(\frac{1}{2}r_0\right)^3}{\varepsilon_0}$$

$$\Rightarrow \quad E_2 = -\frac{\rho_E r_0}{54\,\varepsilon_0}$$

At point B, the electric field from sphere 2 points towards the right. The net electric field is the sum of these two fields. The net field points to the left.

$$E = E_1 + E_2 = \frac{\rho_E r_0}{3\varepsilon_0} + \frac{-\rho_E r_0}{54\varepsilon_0} = \frac{17\rho_E r_0}{54\varepsilon_0}$$

3. As field is in y-direction only, so flux passes through faces perpendicular to Y-axis (parallel to XOZ-plane.)

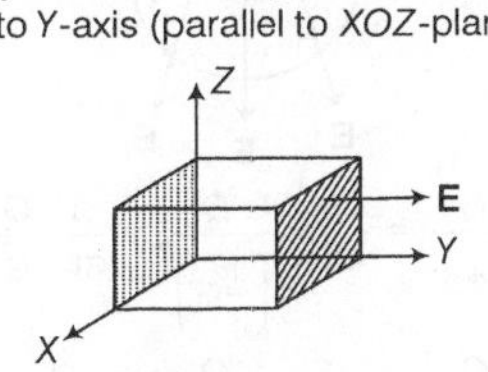

Now, $\quad \phi_E = \displaystyle\int_{\text{face1}} \mathbf{E} \cdot d\mathbf{A} + \int_{\text{face 2}} \mathbf{E} \cdot d\mathbf{A}$

$$= \int b\hat{\mathbf{j}} \cdot (-\hat{\mathbf{j}}dA) + \int (al + b)\hat{\mathbf{j}} \cdot (\hat{\mathbf{j}}\,dA)$$

$$= -bl^2 + al^3 + bl^2$$

$$= al^3$$

By Gauss' law,

$$\phi_E = \frac{q_{\text{enclosed}}}{\varepsilon_0}$$

$$\Rightarrow \quad q_{\text{enclosed}} = \varepsilon_0\, al^3$$

4. Radius increases from r_0 to $2r_0$ in time T_{sec}.

Rate of increase $= \frac{r_0}{T}$.

Radius as a function of time, $r = r_0 + \frac{r_0}{T} \cdot t = r_0\left(1 + \frac{t}{T}\right)$

(i) Field just outside the surface

$$E = \frac{kQ}{r^2} = \frac{kQ}{r_0^2\left(1 + \frac{t}{T}\right)^2}$$

(ii) As radius of balloon always remains smaller than $3.3\,r_0$, so total charge inside the Gaussian surface does not change with time.

$$\Rightarrow \qquad E = \frac{kQ}{r^2} = \frac{kQ}{(3.3r_0)^2}$$

5. (b) When charged metallic ball touches the bottom of can, all of its charge goes to the outer surface of the can. (Net charge on a conductor lies only on its surface).

When ball is withdrawn from the can, it is uncharged.

When ball is placed on the disc of electroscope, it is shared the charge of electroscope. Final charge on the leaves is reduced so leaves of electroscope converges.

6. (c) In region, $r_1 < r < r_2$

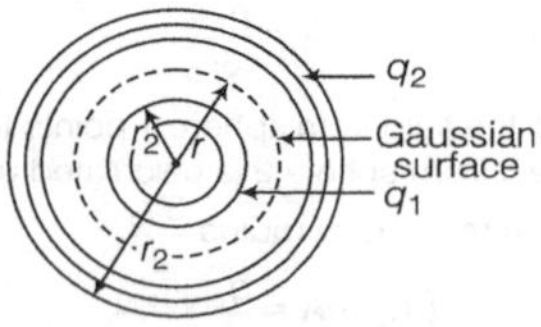

So, net flux will be zero if enclosed charge is zero. As, $q_{net} = \sigma_1 \times 4\pi r_1^2$.

$\therefore$ A charge of $-4\pi\sigma_1 r_1^2$ is to be placed at centre to make flux zero.

7. (c) The maximum electric field will be at the point on the sphere closest to Q which is the top of the sphere. The minimum electric field will be at the point on the sphere farthest from Q, which is the bottom of the sphere.

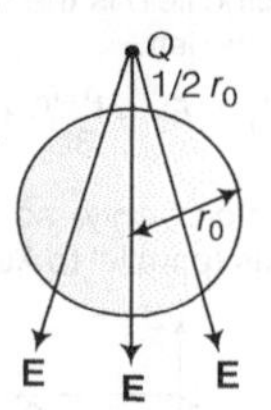

$$E_{max} = \frac{1}{4\pi\varepsilon_0}\frac{Q}{r_{closest}^2} = \frac{1}{4\pi\varepsilon_0}\frac{Q}{\left(\frac{1}{2}r_0\right)^2} = \frac{1}{\pi\varepsilon_0}\frac{Q}{r_0^2}$$

$$E_{min} = \frac{1}{4\pi\varepsilon_0}\frac{Q}{r_{farthest}^2} = \frac{1}{4\pi\varepsilon_0}\frac{Q}{\left(\frac{5}{2}r_0\right)^2} = \frac{1}{25\pi\varepsilon_0}\frac{Q}{r_0^2}$$

Thus, the range of value is $\frac{1}{\pi\varepsilon_0}\frac{Q}{r_0^2} \le E_{\text{sphere surface}} \le \frac{1}{25\pi\varepsilon_0}\frac{Q}{r_0^2}$.

8. (a)

(i) Inside the inner spherical shell (A), i.e. for $x < r_A$, $E = 0$.

(ii) At $x = r_A$, E is constant.

(iii) As Q_B is negative and $|Q_B| > |Q_A|$, in the space between A and $B (x = r_A$ to $x = r_B)$, E decreases with x.

(iv) E is constant at $x = r_B$.

(v) For $x > r_B$ (outside the shell B), E changes from negative value to zero.

Hence, Fig. (a) represents the correct graph.

9. (a) $q_1 + q_2 = 3Q, q_3 + q_4 = Q$

$q_5 + q_6 = Q$

The field inside metal sheets must be zero

$$\therefore \qquad \frac{-q_1}{2\varepsilon_0} + \frac{q_2}{2\varepsilon_0} + \frac{Q}{2\varepsilon_0} + \frac{Q}{2\varepsilon_0} = 0 \qquad \text{...(i)}$$

$$\frac{-3Q}{2\varepsilon_0} - \frac{q_3}{2\varepsilon_0} + \frac{q_4}{2\varepsilon_0} + \frac{Q}{2\varepsilon_0} = 0 \qquad \text{...(ii)}$$

$$\frac{-3Q}{2\varepsilon_0} - \frac{Q}{2\varepsilon_0} - \frac{q_5}{2\varepsilon_0} + \frac{q_6}{2\varepsilon_0} = 0 \qquad \text{...(iii)}$$

From above Eqs. (i) (ii) and (iii), we get

$$q_3 = -\frac{Q}{2}, q_5 = -\frac{3Q}{2}$$

10. (b) For a uniformly charged, conducting solid cylinder of infinite length electric field is

$$E = \frac{\lambda}{2\pi\varepsilon_0 r}$$

$\therefore E \propto \frac{1}{r}$, so the option (b) is correct. (Also, refer the solution of Q. 18)

11. (c) According to question $x < R$, inside the sphere, we choose for our Gaussian surface a concentric sphere of radius $x (x < R)$ in figure. From symmetry, the magnitude of **E** is the same at all points A_1 and **E** is perpendicular to the surface, so

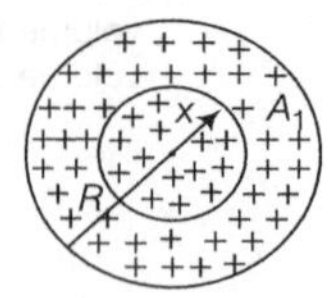

$$\oint \mathbf{E} \cdot d\mathbf{A} = E(4\pi x^2)$$

We must equate this to $Q_{enclosure}/\varepsilon_0$, where $Q_{enclosure}$ is the charge enclosed by A_1. $Q_{enclosure}$ is not total charge Q but only portion of it. We define the charge density ρ_E, as the charge per unit volume ($\rho_E = dQ/dV$) and here we are given that ρ_E = constant. So, the charge enclosed by the Gaussian surface A_1, a sphere of radius x, is

$$Q_{enclosure} = \left(\frac{\frac{4}{3}\pi x^3 \rho_E}{\frac{4}{3}\pi R^3 \rho_E}\right)Q = \frac{x^3}{R^3}Q$$

Hence, from Gauss' law

$$E(4\pi x^2) = \frac{Q_{enclosure}}{\varepsilon_0} = \frac{x^3}{R^3}\frac{Q}{\varepsilon_0}$$

or $\qquad E = \frac{1}{4\pi\varepsilon_0}\frac{Q}{R^3}x[x < R]$

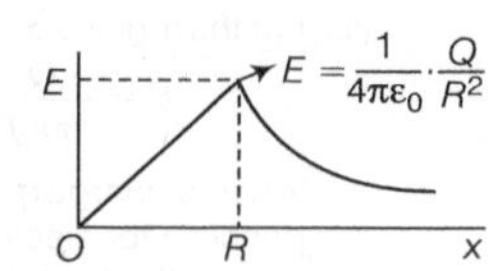

Thus, the field linearly with x, until $x = R$. It then decreases as $1/x^2$, as plotted in figure.

12. (d) Charge dq is enclosed in a shell of thickness dr.

$$dq = 4\pi r^2 dr \rho$$

$$= 4\pi r^2 \rho_0\left(1 - \frac{r}{R}\right)dr$$

$$= 4\pi\rho_0\left\{r^2 - \frac{r^3}{R}\right\}dr$$

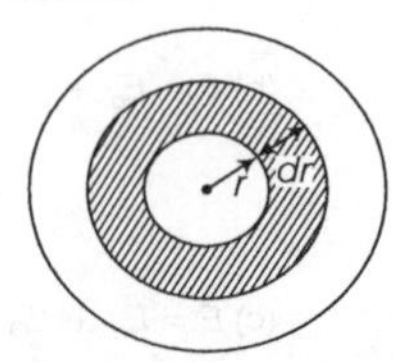

Total charge q enclosed inside radius r.

$$\Rightarrow \qquad q = \int dq = \int_0^r 4\pi\rho_0\left(r^2 - \frac{r^3}{R}\right)dr$$

$$q = 4\pi\rho_0\left[\frac{r^3}{3} - \frac{r^4}{4R}\right]$$

Electric field $E(r)$ is

$$E = \frac{kq}{r^2} = \frac{1}{4\pi\varepsilon_0}\cdot\frac{1}{r^2}\cdot 4\pi\rho_0\left(\frac{r^3}{3} - \frac{r^4}{4R}\right)$$

$$= \frac{\rho_0}{4\varepsilon_0}\left(\frac{r}{3} - \frac{r^2}{4R}\right) = \frac{\rho_0 r}{4\varepsilon_0}\left(\frac{1}{3} - \frac{r}{4R}\right)$$

13. (d) By Gauss' law,

$$\oint \mathbf{E} \cdot d\mathbf{A} = \frac{q_{\text{net}}}{\varepsilon_0} = \frac{1}{\varepsilon_0} q\, dV$$

$$\Rightarrow \quad E\, 4\pi r_1^2 = \frac{Q r_1}{\varepsilon_0 \pi R^4} \cdot \frac{4}{3} \pi r_1^3 \quad \Rightarrow \quad E = \frac{Q r_1^2}{3\pi \varepsilon_0 R^4}$$

14. (b) Flux $(x = a) = \int \mathbf{E} \cdot d\mathbf{A} = \int_0^a \left(E_0\left(1 + \frac{z}{a}\right)\hat{\mathbf{i}} + E_0\left(\frac{z}{a}\right)\hat{\mathbf{j}}\right)(a\, dz\, \hat{\mathbf{i}})$

$$= E_0 a \int_0^a \left(1 + \frac{z}{a}\right) dz = E_0\, a \left[z + \frac{z^2}{2a} \right]_0^a$$

$$= \frac{3}{2} E_0\, a^2 = \frac{3}{2} \times 0.2 \times 1^2 = \frac{3}{10} \frac{\text{N}}{\text{C}}$$

15. (b) All the three plates will produce electric field at point P along negative Z-axis.

Hence, $\quad \mathbf{E}_P = \left[\frac{\sigma}{2\varepsilon_0} + \frac{2\sigma}{2\varepsilon_0} + \frac{\sigma}{2\varepsilon_0} \right] (-\hat{\mathbf{k}}) = \frac{-2\sigma}{\varepsilon_0} \hat{\mathbf{k}}$

16. (c) At any point over the spherical Gaussian surface, net electric field is the vector sum of electric fields due to $+q_1$, $-q_1$ and $+q_2$. Do not confuse with the electric flux which is zero (net) passing over the Gaussian surface as the net charge enclosing the surface is zero.

17. (a,b) Using, $\oint \mathbf{E} \cdot d\mathbf{A} = \frac{q_{\text{enclosure}}}{\varepsilon_0}$

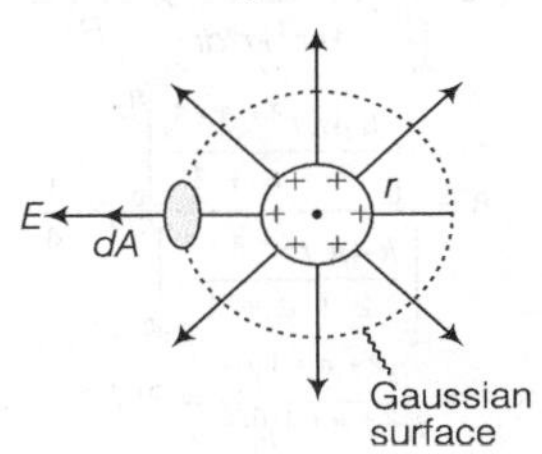

For an outside point $\mathbf{E} \parallel d\mathbf{A}$

$\therefore \quad \mathbf{E} \cdot d\mathbf{A} = E\, dA \quad$ as $\cos\theta = 1$

and $q_{\text{enclosure}}$ = charge density × volume $= \rho \times \frac{4}{3}\pi r_0^3$

So, $\quad E \oint dA = \dfrac{\frac{4}{3}\pi r_0^3}{\varepsilon_0} \quad \Rightarrow E \cdot 4\pi r^2 = \frac{4}{3} \frac{\pi \rho r_0^3}{\varepsilon_0}$

$$E = \frac{\rho r_0^3}{3\varepsilon_0 r^2} \quad (E \text{ at } r > r_0)$$

For a point inside the sphere, $r < r_0$

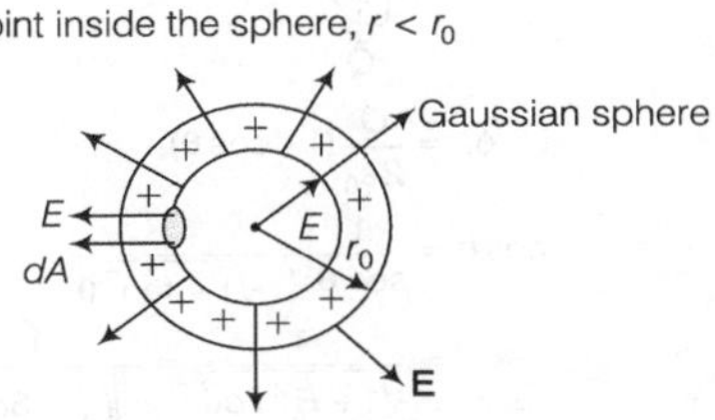

For any elemental patch dA of Gaussian sphere, $\mathbf{E} \parallel d\mathbf{A} \Rightarrow \theta = 0°$ $\Rightarrow \mathbf{E} \cdot d\mathbf{A} = E\, dA$. Also, by symmetry E is constant.

$$\therefore \quad \oint \mathbf{E} \cdot d\mathbf{A} = \frac{q_{\text{enclosure}}}{\varepsilon_0} \quad \Rightarrow \quad E \oint dA = \frac{\rho \cdot \frac{4}{3}\pi r^3}{\varepsilon_0}$$

$$\Rightarrow \quad E \cdot 4\pi r^2 = \frac{\rho \frac{4}{3}\pi r^3}{\varepsilon_0}$$

$$\Rightarrow \quad E_{\text{inside}} = \frac{\rho r}{3\varepsilon_0} \text{ and at } r = r_0$$

$$E_{\text{surface}} = \frac{\rho r_0}{3\varepsilon_0}$$

18. (a,c) A charge of $(Q - q)$ will be on the outer surface of the conductor. The total charge Q is placed on the conductor but since, $+q$ will reside on the inner surface, the left over, $(Q - q)$, will reside on the outer surface.

A charge of $+q$ will reside on the inner surface of the conductor since, that amount is attracted by the charge $-q$ in the cavity. E must be zero inside the conductor.

19. (c,d) The uniform fields from each of the three sheets are indicated in the diagram.

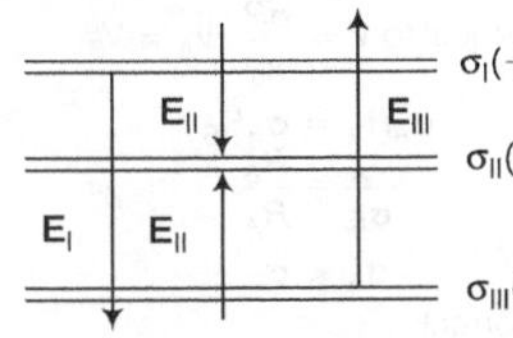

$$F_{\text{on sheet}} = q_{\text{on sheet}}\, \mathbf{E}_{\text{other sheet}} = q_{\text{on sheet}}$$

$$E_{\text{other sheets}} \longrightarrow \left(\frac{F}{A}\right)_{\text{on sheet}} = \left(\frac{q}{A}\right)_{\text{on sheet}}$$

$$\mathbf{E}_{\text{other sheet}} = \sigma_{\text{on sheets}}\, \mathbf{E}_{\text{other sheets}}$$

Take the positive direction as upward. We take the direction from the diagram and use the absolute value of each charge density. The electric field magnitude due to each sheet is given by $E = \frac{\sigma}{2\varepsilon_0}$.

$$\left(\frac{F}{A}\right)_{\text{I}} = \sigma_1(E_{\text{III}} - E_{\text{II}}) = \frac{\sigma_1}{2\,\varepsilon_0}(|\sigma_{\text{III}}| - |\sigma_{\text{II}}|)$$

$$= \frac{6.5 \times 10^{-9}\ \text{C/m}^2}{2\,(8.55 \times 10^{-12}\ \text{C}^2/\text{Nm}^2)} \times [(5.0 - 2.0) \times 10^{-9}\ \text{C/m}^2]$$

$$= 1.1 \times 10^{-6}\ \text{N/m}^2\ \text{(up)}$$

$$\left(\frac{F}{A}\right)_{\text{II}} = \sigma_{\text{II}}(E_{\text{III}} - E_{\text{I}}) = \frac{\sigma_{\text{II}}}{2\,\varepsilon_0}(|\sigma_{\text{III}}| - |\sigma_{\text{I}}|)$$

$$= \frac{-2.0 \times 10^{-9}\ \text{C/m}^2}{2\,(8.85 \times 10^{-12}\ \text{C}^2/\text{Nm}^2)} \times [(5.0 - 6.5) \times 10^{-9}\ \text{C/m}^2]$$

$$= 1.7 \times 10^{-7}\ \text{N/m}^2\ \text{(up)}$$

$$\left(\frac{F}{A}\right)_{\text{III}} = \sigma_{\text{III}}\,(E_{\text{III}} - E_{\text{I}}) = \frac{\sigma_{\text{III}}}{2\varepsilon_0}\,(|\sigma_{\text{II}}| - |\sigma_{\text{I}}|)$$

$$= \frac{5.0 \times 10^{-9}\ \text{C/m}^2}{2\,(8.55 \times 10^{-12}\ \text{C}^2/\text{Nm}^2)} [(2.0 - 6.5) \times 10^{-9}\ \text{C/m}^2]$$

$$= -1.3 \times 10^{-6}\ \text{N/m}^2\ \text{(down)}$$

20. (c,d) The conducting sphere is uncharged and the electric field is zero everywhere within its interior, except in the cavities. When charge Q_1 is placed in the first cavity, a charge $-Q_1$ will be drawn from the conducting material to the inner surface of the cavity and the electric field will remain zero in the conductor.

The charge $-Q_1$ will not be distributed symmetrically on the cavity surface. Since, the conductor is neutral, a compensating charge Q will appear on the outer surface of the conductor (charge can only be on the surface of conductor in electrostatics).

Likewise when charge Q_2 is placed in the second cavity, a charge $-Q_2$ will be drawn from the conducting material and a compensating charge Q_2 will appear on the outer surface.

Since, there is no electric field in the conducting material, there is no way for the changes in the cavities to influence the charge distribution on the outer surface. So, the distribution of charge on the outer surface is uniform just it would be, if there were no inner charges and instead a charge $Q_1 + Q_2$ were placed on the conductor.

Thus, the field outside the conductor is due to a spherically symmetric distribution of $Q_1 + Q_2$. Application of Gauss' law leads to $E = \frac{1}{4\pi\varepsilon_0} \frac{Q_1 + Q_2}{r^2}$. If $Q_1 + Q_2 > 0$, the field will point radially outward. If $Q_1 + Q_2 < 0$, the field will point radially inward.

21. (a,b,c,d) Inside a conducting shell electric field is always zero. Therefore, option (a) is correct. When the two are connected, their potentials become the same.

$\therefore \quad V_A = V_B$

or $\quad \dfrac{Q_A}{R_A} = \dfrac{Q_B}{R_B} \quad \left(V = \dfrac{1}{4\pi\varepsilon_0}\dfrac{Q}{R}\right)$

Since, $\quad R_A > R_B \quad \therefore Q_A > Q_B$

$\therefore$ Option (b) is correct.

Potential is also equal to $V = \dfrac{\sigma R}{\varepsilon_0}$, $V_A = V_B$

$\therefore \quad \sigma_A R_A = \sigma_B R_B$

or $\quad \dfrac{\sigma_A}{\sigma_B} = \dfrac{R_B}{R_A}$

or $\quad \sigma_A < \sigma_B$

$\therefore$ Option (c) is correct.

Electric field on surface, $E = \dfrac{\sigma}{\varepsilon_0}$ or $E \propto \sigma$

Since, $\sigma_A < \sigma_B \quad \therefore E_A < E_B$

$\therefore$ Option (d) is correct.

22. (a)

(i) In the symmetry, we except the field to be directed radially outward to depend on the perpendicular distance r from the wire. In the cylindrical symmetry, the field will be same at all points on a Gaussian surface that is a cylinder with the wire along its axis.

In figure, **E** is perpendicular to this surface at all points. For Gauss's law, we need a closed surface, so we include the flat ends of cylinder. Since, **E** is parallel to the ends, there is no flux through the ends (the cosine of the angle between **E** and d**A** on the end is ($\cos 90° = 0$).

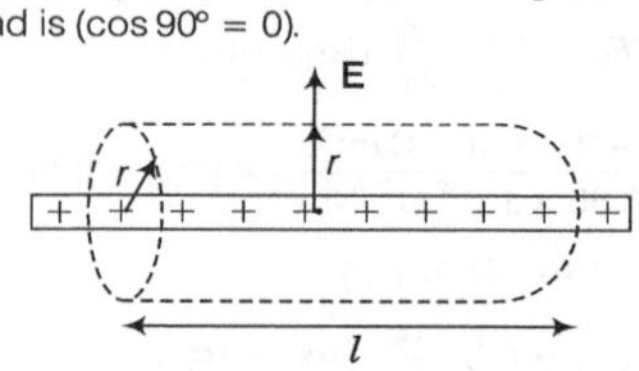

For Gaussian surface, Gauss' law gives

$$\oint \mathbf{E} \cdot d\mathbf{A} = E(2\pi r l) = \frac{Q_{\text{enclosed}}}{\varepsilon_0} = \frac{\lambda l}{\varepsilon_0}$$

where l is the length of our chosen Gaussian surface ($l <<$ length of wire) and $2\pi r$ is its circumference. Hence,

$$E = \frac{1}{2\pi\varepsilon_0} \cdot \frac{\lambda}{r}.$$

(ii) Since, the charge is distributed symmetrically in the sphere, the electric field at all points must again be symmetric. **E** depends only r and is directed radially outward (or inward if $\theta < 0$). (Refer the solution of Q. 19)

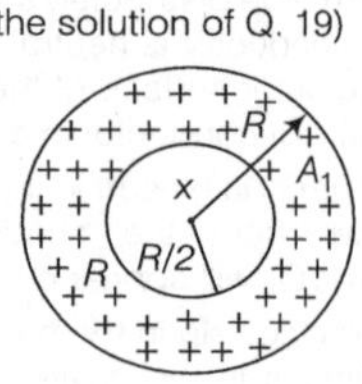

(iii) For the region $r > R$, $E = \dfrac{Q}{4\pi\varepsilon_0 r^2}$

(iv) A Gaussian surface of radius $r < R/2$ only encloses the centre charge q. The electric field will therefore, be the field of the single charge.

$$E(r < R/2) = \frac{q}{4\pi\varepsilon_0 r^2}$$

(i) $\rightarrow$ (q), (ii) $\rightarrow$ (s), (iii) $\rightarrow$ (r), (iv) $\rightarrow$ (p)

23. (a) Gauss' law is useful in determining electric fields when the charge distribution is characterised by a high degree of symmetry. When choosing the surface, we should always take advantage of the symmetry of the charge distribution so that we can remove E from the integral and solve for it.

The goal of this type of calculation is to determine a surface that satisfies one or more of the following conditions

1. The value of the electric field can be argued by symmetry to be constant over the surface.
2. The dot product in equation $\phi_E = \oint E \cdot dA$ can be expressed as a simple algebraic EdA, because product E and dA are parallel.
3. The dot product in equation $\phi_E = \oint E \cdot dA$ is zero, because E and dA are perpendicular.
4. The field can be argued to be zero over the surface.

24. (a) By drawing a symmetrical surface,

$$\mathbf{E} \parallel d\,\mathbf{A}$$

and $\quad \mathbf{E} \cdot d\,\mathbf{A} = EdA$

This leads to $E\oint dA = \dfrac{q_{\text{enclosed}}}{\varepsilon_0}$

which is easier to evaluate.

25. (2) By Gauss' theorem, $E \propto \dfrac{q_{\text{net}}}{r^2}$

$$\Rightarrow \quad 8 = \frac{\int_0^R 4\pi r^2 k r^a dr}{\int_0^{R/2} 4\pi r^2 k r^a dr} \times \frac{(R/2)^2}{R^2}$$

$$\Rightarrow \quad 8 = \frac{\left[\dfrac{k\,4\pi r^{2+a+1}}{2+a+1}\right]_0^R}{\left[\dfrac{k\,4\pi\, r^{2+a+1}}{2+a+1}\right]_0^{R/2}} \times \frac{1}{4}$$

$$\Rightarrow \quad 8 \times 4 = \frac{(r^{2+a+1})_0^R}{(r^{2+a+1})_0^{R/2}} = 2^{2+a+1}$$

$\Rightarrow \quad 2 + a + 1 = 5$

$\Rightarrow \quad a = 2$

26. (3)

Flux through an area with half angle θ as shown is

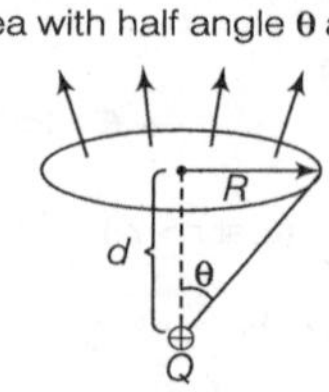

$$\phi_E = \frac{Q}{2\varepsilon_0}(1 - \cos\theta)$$

Here, $\quad \cos\theta = \dfrac{1}{\sec\theta} = \dfrac{1}{\sqrt{1+\tan^2\theta}}$

$$= \frac{1}{\sqrt{1+R^2/d^2}} = \frac{1}{\sqrt{\left(1+\dfrac{3d^2}{d^2}\right)}} = \frac{1}{2}$$

$\therefore$ Flux through disc, $\phi_1 = \dfrac{Q}{4\varepsilon_0}$

So, flux which is not passing through the disc

$$\phi_2 = \frac{Q}{\varepsilon_0} - \frac{Q}{4\varepsilon_0} = \frac{3\,Q}{4\,\varepsilon_0}$$

Hence, ratio is $\quad \dfrac{\phi_2}{\phi_1} = \dfrac{\dfrac{3\,Q}{4\,\varepsilon_0}}{\dfrac{1\,Q}{4\,\varepsilon_0}} = 3$

27. (2)

Flux through an elemental strip of end be integrated to get the flux through an end of cylinder.

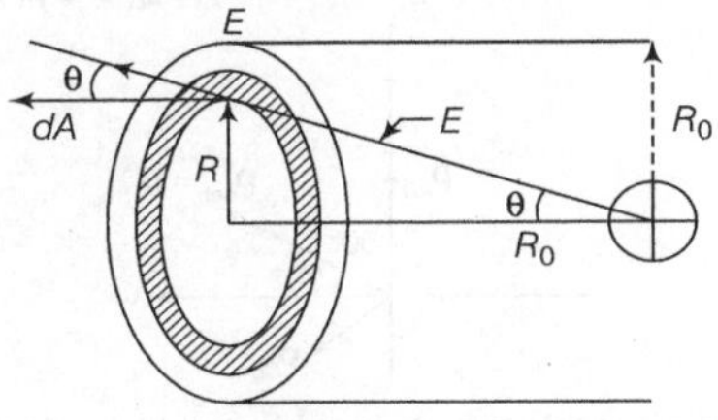

$\therefore \qquad \phi_{\text{left end}} = \int_{\text{end}} \mathbf{E}.d\mathbf{A}$

$$= \int_0^{R_0} \frac{1}{4\pi\varepsilon_0} \frac{Q}{r^2}.\cos\theta\,(2\pi R).dR$$

where, $r = \sqrt{R^2 + R_0^2}, \cos\theta = \frac{R_0}{r}$

Substituting, we get

$$\phi_{\text{left end}} = \int_0^{R_0} \frac{2\pi Q R_0}{4\pi\varepsilon_0} \cdot \frac{RdR}{(\sqrt{R^2 + R_0^2})^3}$$

$$\Rightarrow \qquad \phi_{\text{left end}} = \int_0^{R_0} \frac{2\pi Q R_0}{4\pi\varepsilon_0} \cdot \frac{RdR}{(\sqrt{R^2 + R_0^2})^3}$$

$$\Rightarrow \qquad \phi_{\text{left end}} = \frac{QR_0}{2\varepsilon_0}\left\{\frac{-1}{\sqrt{R^2 + R_0^2}}\right\}_0^{R_0}$$

$$= \frac{Q}{2\varepsilon_0}\left(1 - \frac{1}{\sqrt{2}}\right)$$

$\therefore$ Flux through right end is also

$$\phi_{\text{right end}} = \frac{Q}{2\varepsilon_0}\left(1 - \frac{1}{\sqrt{2}}\right)$$

Hence, $\phi_{\text{curved surface}}$

$$= \phi_{\text{total}} - \phi_{\text{left end}} - \phi_{\text{right end}}$$

$$= \frac{Q}{\varepsilon_0} - 2\left\{\frac{Q}{2\varepsilon_0}\left(1 - \frac{1}{\sqrt{2}}\right)\right\}$$

$$= \frac{Q}{\sqrt{2}\varepsilon_0}$$

$$= \frac{1.77 \times 10^{-11}}{\sqrt{2} \times 8.85 \times 10^{-12}}$$

$$= \frac{2}{\sqrt{2}}$$

So, $\sqrt{2}\,\phi_{\text{curved surface}} = 2$

Explanations for Q. Nos. (28-31)

Let charge on sphere is q

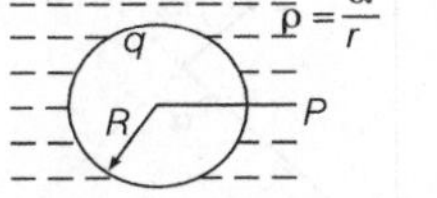

The charge inside and outside the sphere is in spherical symmetry. So, field is in radial direction only.

Using Gauss' law, field at point P is given as

$$E \cdot 4\pi r^2 = \frac{q_{\text{enclosed}}}{\varepsilon_0}$$

$$= q_1 + q_2$$

$$= q + \int_R^r \rho \cdot 4\pi r^2 dr$$

$$= q + \int_R^r \frac{\alpha}{r} \cdot 4\pi r^2 dr$$

$$= q + \frac{4\pi r}{2}(r^2 - R^2)$$

$$E = \frac{q}{4\pi\varepsilon_0 r^2} + \frac{\alpha}{2\varepsilon_0}\left(1 - \frac{R^2}{r^2}\right)$$

$$= \left(\frac{q}{4\pi\varepsilon_0} - \frac{\alpha R^2}{2\varepsilon_0}\right)\frac{1}{r^2} + \frac{\alpha}{2\varepsilon_0}$$

For field independent of r

$$\frac{q}{4\pi\varepsilon_0} - \frac{\alpha R^2}{2\varepsilon_0} = 0$$

$$\Rightarrow \qquad q = 2\pi\alpha R^2$$

and electric field is $E = \frac{\alpha}{2\varepsilon_0}$.

28. (c) **29.** (c) **30.** (d) **31.** (d)

Explanations for Q. Nos. (32-34)

Consider the given situation,

$$\rho = \rho_0 e^{-\alpha r^3}$$

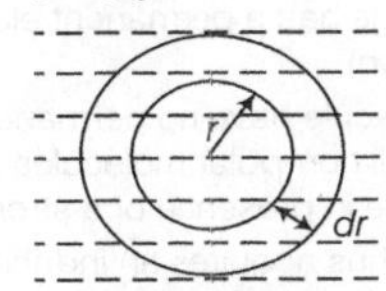

Using Gauss' law, for elemental volume 'dr' of shell of radius r,

$$E \cdot 4\pi r^2 = \frac{q_{\text{enclosed}}}{\varepsilon_0}$$

$$\Rightarrow \qquad E \cdot 4\pi r^2 = \int_0^r \rho \cdot 4\pi r^2 dr$$

$$= 4\pi\rho_0 \int_0^r e^{-\alpha r^3} \cdot r^2 dr$$

$$= \frac{4\pi\rho_0}{3x}(1 - e^{-\alpha r^3})$$

$$\Rightarrow \qquad E = \frac{\rho_0}{3\varepsilon_0 \alpha \varepsilon^2}(1 - e^{-\alpha r^3})$$

If $\alpha r^3 << 1$, $\qquad e^{-\alpha r^3} = 1 - \frac{\alpha r^3}{1!} + \frac{(\alpha r^3)^2}{2!} + \ldots$

$$\Rightarrow \qquad e^{-\alpha r^3} \approx 1 - \alpha r^3$$

$$\Rightarrow \qquad E = \frac{\rho_0 r}{3\varepsilon_0}$$

If $\alpha r^3 >> 1$, $\qquad e^{-\alpha r^3} \to 0$

$$\therefore \qquad E = \frac{\rho_0}{3\varepsilon_0 \alpha r^2}$$

32. (c) **33.** (a) **34.** (a)

DPP-6 Electric Dipole, Dipole in an External Field

1. (i) n = number of atoms per unit volume of bar

$$= \frac{\rho N_A}{M} \quad \text{where } \rho = \text{density}$$

$$= \frac{5 \times 6.02 \times 10^{23}}{3.01}$$

$$= 10 \times 10^{23} = 1 \times 10^{24} \frac{\text{atoms}}{\text{cm}^3}$$

Now, volume of bar = Area of cross-section × length

$$= 5 \times 1 = 5 \text{ cm}^3$$

So, total number of atoms $= 5 \times 1 \times 10^{24}$

$$= 5 \times 10^{24} \text{ atoms}$$

∴ Total dipole moment of bar

$$= \text{Dipole moment of 1 atom} \times \text{number of atoms}$$

$$= 2.1 \times 10^{-23} \times 5 \times 10^{24} = 10.5 \times 10$$

$$= 1.05 \times 10^2 \text{ cm.}$$

(ii) No, all molecules are not polar in nature. Some molecules have a permanent electric dipole moment due to a non-uniform distribution of charge within the molecule. Such molecules are called polar molecules. e.g. A water molecule.

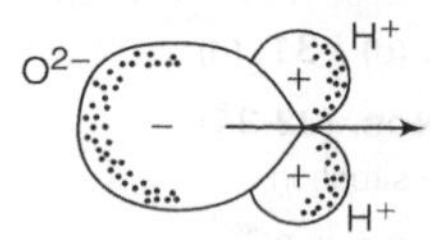

An H_2O molecule has a permanent electric dipole moment (direction shown).

Non-polar molecule have no permanent electric dipole moment. But all non-polar molecules shows a separation of charge in space in presence of a strong field.

The molecule thus acquires an induced dipole moment and said to be polarised.

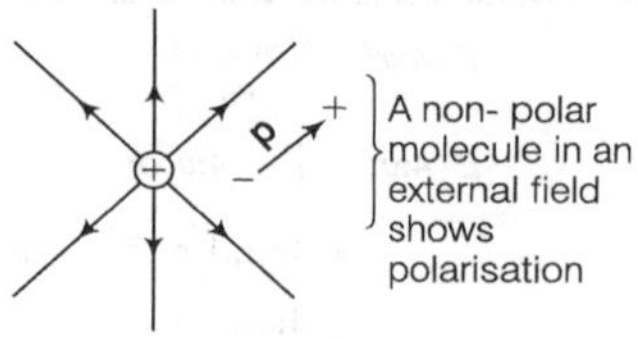

2. (i)

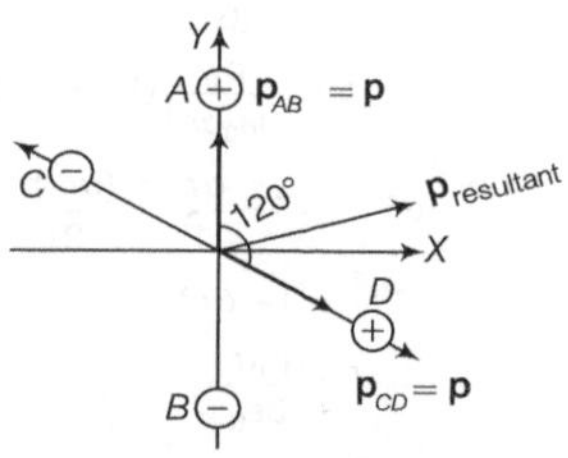

Dipole moment vector points from negative to positive charges. As both dipoles AB and CD have same dipole moments.

$$\mathbf{p}_{AB} = \mathbf{p}_{CD} = \mathbf{p}_i \,|\, \mathbf{p}_{\text{net}}|$$

$$= \sqrt{\mathbf{p}_{AB}^2 + \mathbf{p}_{CD}^2 + 2\mathbf{p}_{AB} \cdot \mathbf{p}_{CD} \cos\theta}$$

In given case,

$$\theta = 120°$$

$$\cos 120° = \cos(180° - 60°) = -\cos 60° = -\frac{1}{2}$$

$$\therefore \quad |\,\mathbf{p}_{\text{net}}\,| = \sqrt{p^2 + p^2 + 2p^2 \times \frac{-1}{2}} = p$$

So, magnitude of resultant dipole moment is p and it is directed along the angle bisector of $\mathbf{p}_{AB}$ and $\mathbf{p}_{CD}$.

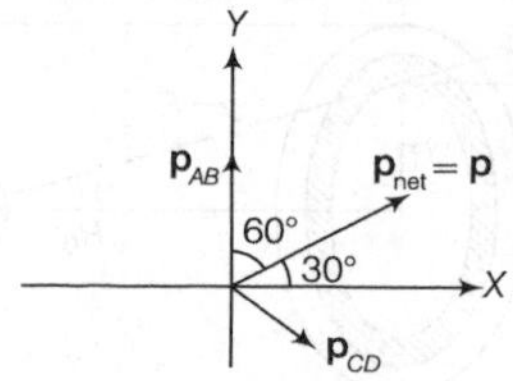

(ii) Now, when this system is subjected to an electric field along the X-direction angle between $\mathbf{p}$ and $\mathbf{E} = 30°$

So, torque has a magnitude, which is given by

$$|\,\tau\,| = pE \sin\theta - pE \sin 30°$$

$$= \frac{1}{2} pE$$

Also, dipole system shows a clockwise rotation about the origin so torque is directed along negative Z-axis (outwards and perpendicular to plane of paper.)

3. To find field of a quadrapole, we add up field of each charge individually,

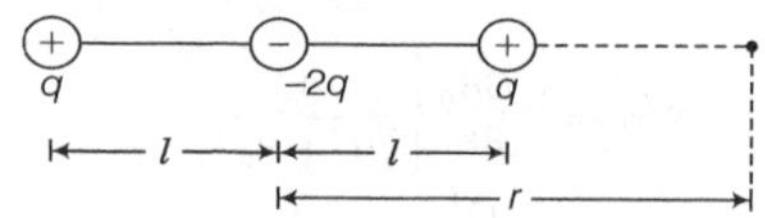

$$E_{\text{net}} = \frac{kq}{(r-l)^2} + \frac{kq}{(r+l)^2} - \frac{2kq}{r^2}$$

$$= kq\left\{\frac{2r^2 + 2l^2}{r^4(1 - l^2/r^2)} - \frac{2}{r^2}\right\}$$

For $r >> l$, $l/r << 1$

$$\therefore \quad E_{\text{net}} \approx kq\left(\frac{2r^2 + 2l^2}{r^4} - \frac{2}{r^2}\right) = \frac{2kql^2}{r^4}$$

$$\text{So,} \quad E_{\text{net}} \propto \frac{1}{r^4}.$$

4. (c) Force on charge q is $F = qE$ and as $E \propto \frac{1}{r^3}$ (field of dipole)

Thus, when distance is doubled force becomes $\frac{1}{8}$ th.

5. (a) In non-uniform field, force as dipole is non-zero and torque is also non-zero.

A dipole shows rotation and translation both.

6. (b)

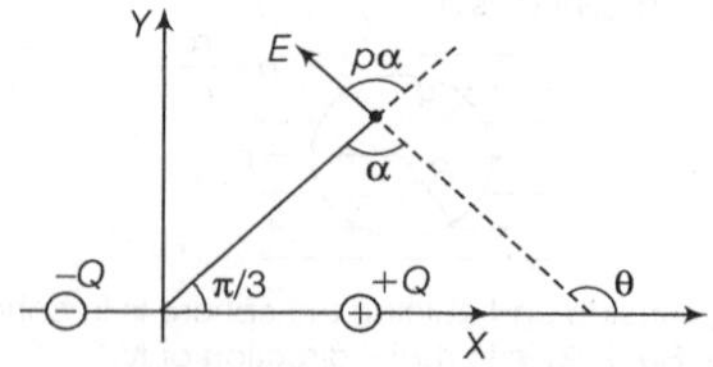

Analysing the figure, we get

$$\theta = \frac{\pi}{3} + \alpha$$

$$\text{where,} \quad \tan\alpha = \frac{1}{2}\tan\frac{\pi}{3}$$

$$\Rightarrow \quad \alpha = \tan^{-1}\frac{\sqrt{3}}{2}$$

$$\therefore \quad \theta = \frac{\pi}{3} + \tan^{-1}\frac{\sqrt{3}}{2}$$

7. (c)

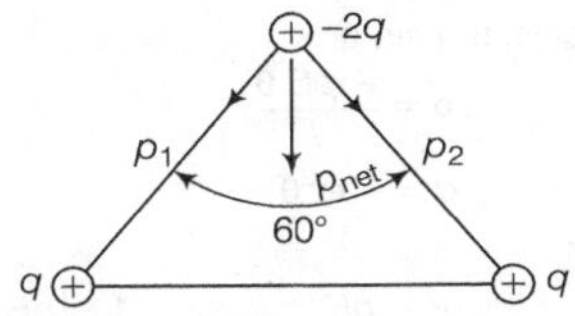

From above figure, we get

System got two dipole moments $|\mathbf{p}_1| = |\mathbf{p}_2| = p$ (say) and are inclined to each other at an angle of 60°.

$\therefore$ p_{net} (is along angle bisector of an angle between $\mathbf{p}_1$ and $\mathbf{p}_2$)

$$= \sqrt{p_1^2 + p_2^2 + 2p_1p_2 \cos 60°} = \sqrt{p^2 + p^2 + 2p^2 \times \frac{1}{2}}$$

$$\sqrt{3}p = \sqrt{3}ql$$

8. (c) Dipole moment of system is equals to product of magnitude of any charge and distance between charges.

$$\therefore \ |\mathbf{p}| = q(d) = (3.2 \times 10^{-19} \times 2.4 \times 10^{-10} \text{ m})$$
$$= 7.68 \times 10^{-29} \text{ C-m}$$

9. (c)

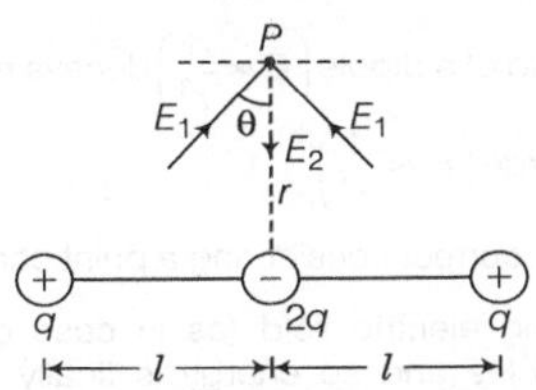

At P, vertical components of positive charges q are in same direction and horizontal components cancels out each other.

So, $\quad E_{net} = 2E_1 \cos\theta - E_2$

where, $\quad E_1$ = field of positive charge

and $\quad E_2$ = field of negative charge.

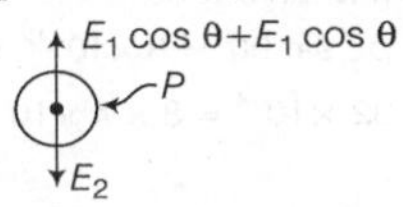

Now, $\quad E_1 = \dfrac{kq}{(\sqrt{r^2 + l^2})^2}, \cos \dfrac{r}{\sqrt{r^2 + l^2}}$

and $\quad E_2 = \dfrac{k(2q)}{r^2}$

$$\therefore \quad E_{net} = \frac{2kqr}{(r^2 + l^2)^{3/2}} - \frac{2kq}{r^2}$$

$$\Rightarrow \quad E_{net} = 2kq\left\{\frac{r}{(r^2 + l^2)^{3/2}} - \frac{1}{r^2}\right\}$$

For $r >> l$, $(r^2 + l^2)^{3/2} \approx r^3$

$$\therefore \ E_{net} = 2kq\left\{\frac{r^3 - (r^2 + l^2)^{3/2}}{r^2(r^2 + l^2)^{3/2}}\right\} = 2kqr^3\left\{\frac{1 - (1 + l^2/r^2)^{3/2}}{r^5(1 + l^2/r^2)^{3/2}}\right\}$$

$$= \frac{2kq\,r^3}{r^5} \cdot \left\{\frac{1 - \left(1 + \frac{3}{2}\cdot\frac{l^2}{r^2}\right)}{1 + \frac{3}{2}\cdot l^2/r^2}\right\}$$

$$= \frac{2kq\,r^3\left(-\frac{3}{2}\frac{l^2}{r^2}\right)}{r^5} = \left(-\frac{3}{2}\right)\frac{2kq\,l^2}{r^4}$$

$$= -\frac{3}{4}\frac{k(q(2l)^2)}{r^4} = -\frac{3}{4}\frac{k\,\mathbf{p}}{r^4}$$

$[(\mathbf{p} = q(2l)^2$ = quadrapole moment$)]$

$$\therefore \quad E \propto \frac{1}{r^4}$$

and it points towards negative charge.

10. (a) Given situation is

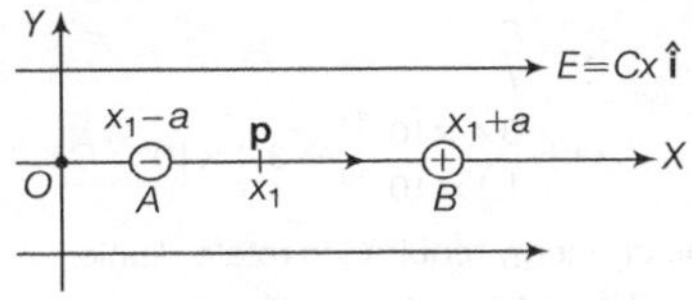

Force on negative charge

$$F_A = qE_A = qC\,(x_1 - a) \qquad \text{(Towards origin)}$$

And force on positive charge is

$$F_B = qC(x_1 + a)$$

(Along X-axis away from origin)

$$\therefore \quad F_{net} = F_B - F_A$$
$$= qC\,(x_1 + a) - qC(x_1 - a) = qC\,2a$$
$$= 2qa \cdot C = pC\,\hat{\mathbf{i}}$$

(As net force is along positive X-axis)

Also, $\quad \dfrac{d}{dx}E_x\ p\hat{\mathbf{i}} = p\hat{\mathbf{i}}\left(\dfrac{d}{dx}E_x\right) = p\hat{\mathbf{i}}\left(\dfrac{d}{dx}Cx\right)$

$$= Cp\hat{\mathbf{i}}\left(\frac{d}{dx}x\right) = Cp\,\hat{\mathbf{i}}$$

11. (a) Restoring couple acting on the dipole is

$$\tau = -\,pE\sin\theta \approx -\,pE\,\theta$$

or $\quad I\dfrac{d^2\theta}{dt^2} = -\,pE\theta$

$\Rightarrow \quad \dfrac{d^2\theta}{dt^2} + \dfrac{pE}{I}\theta = 0 \qquad$ (Angular SHM)

Compare $\quad \dfrac{d^2\theta}{dt^2} + \omega^2\theta = 0 \qquad \left(\text{as } \omega = \dfrac{2\pi}{T}\right)$

$$\therefore \quad T = 2\pi\sqrt{\frac{I}{pE}}$$

12. (c) Polar molecules of water attracts each other and when the surface is deformed, then net force acts in the direction perpendicularly upwards.

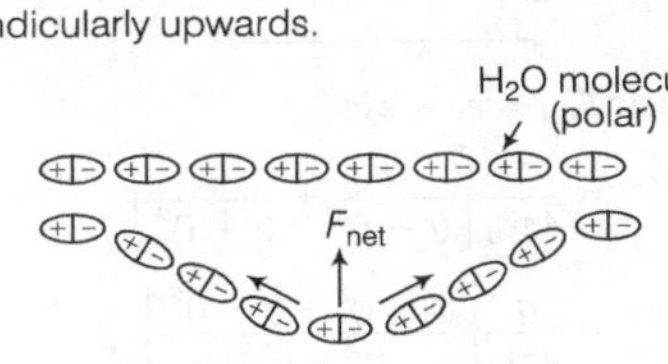

13. (a) A dipole in stable position is shown below.

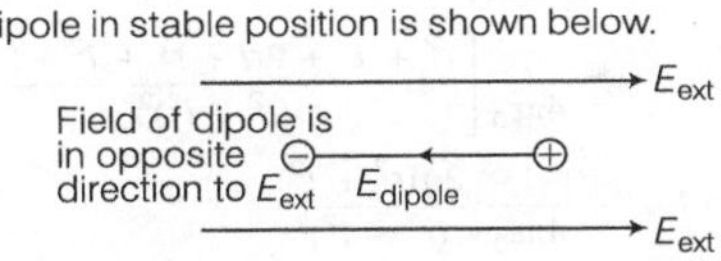

14. (a,b,c,d) $p = q \times d = 1.6 \times 10^{-19} \times 0.68 \times 10^{-9}$

$$\tau = \mathbf{p} \times \mathbf{E} = pE\sin\theta$$

Torque when $\theta = 90°$

$$\tau = (1.1 \times 10^{-28})\,(2.2 \times 10^4)\sin 90°$$
$$= 2.4 \times 10^{-24} \text{ Nm}$$

Torque when $\theta = 45°$ is $\tau = pE\sin\theta$

$$= (1.1 \times 10^{-28})(2.2 \times 10^4)\sin 45°$$
$$= 1.7 \times 10^{-24} \text{ Nm}$$

Work done by an external torque to rotate the dipole

$$\Delta W = \text{change in PE of dipole}$$
$$= \Delta U = (-\,pE\cos\theta_{final}) - (-\,pE\cos\theta_{initial})$$
$$= (1.1 \times 10^{-28})\,(2.2 \times 10^4) \times [1 - (-1)]$$
$$= 4.8 \times 10^{-28} \text{J}$$

15. (b,d) $p = Ql$

$$\Rightarrow \quad Q = \frac{p}{l}$$

$$\Rightarrow \quad Q = \frac{3.4 \times 10^{-30}}{1.1 \times 10^{-10}} = 3.4 \times 10^{-20}\ \text{C}$$

Work done or energy required to rotate dipole

$$\Delta W = \Delta U = pE(\cos\theta_1 - \cos\theta_2)$$

Here, $\quad Q_1 = 0°, Q_2 = 45°$

$\therefore \quad \Delta W = (3.4 \times 10^{-30})$

$(2.5 \times 10^4)(1 - \cos 45°) = 2.5 \times 10^{-26}$ J

The charge on a dipole is not free charge and it may not be quantised as in given case.

$$\frac{Q}{e} = \frac{3.4 \times 10^{-20}}{1.6 \times 10^{-19}} = 0.12$$

Also, electrons spend more time real Cl-atom than the H-atom, giving rise to a net dipole moment.

The electrons are not distributed symmetrically about two nuclei.

16. (a,c) For equatorial position,

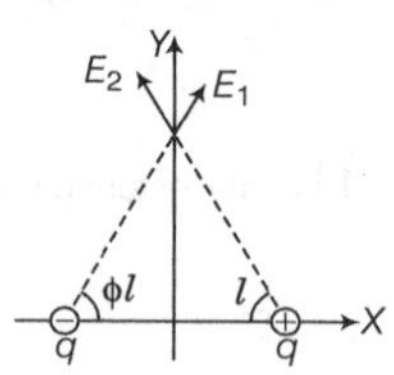

By symmetry, net field is in Y-direction.

Also, $\quad E_{net} = 2E\sin\phi$

$$\Rightarrow \quad E_{net} = 2\frac{q}{4\pi\varepsilon_0(r^2 + l^2)} \cdot \frac{r}{(r^2 + l^2)^{1/2}}$$

$$= \frac{2qr}{4\pi\varepsilon_0(r^2 + l^2)^{3/2}}$$

For large r, $\quad E_{net} = \frac{2qr}{4\pi\varepsilon_0 r^3} = \frac{2q}{4\pi\varepsilon_0 r^2}$

$\therefore E \propto \frac{1}{r^2}$ and from large distance, it appears like a field of $+2q$ charge.

On axial position,

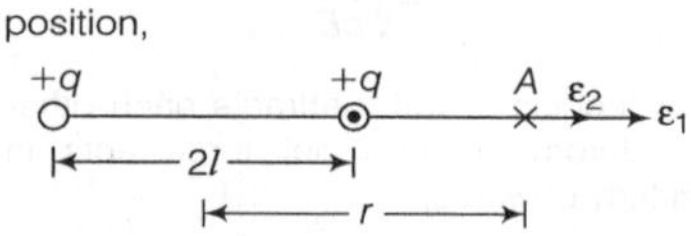

Net electric field $E_{net} = \varepsilon_1 + \varepsilon_2$

$$= \frac{1}{4\pi\varepsilon_0}\left[\frac{q}{(r-l)^2} + \frac{q}{(r+l)^2}\right]$$

$$= \frac{q}{4\pi\varepsilon_0}\left[\frac{(r+l)^2 + (r-l)^2}{(r^2 - l^2)^2}\right]$$

$$= \frac{q}{4\pi\varepsilon_0}\left[\frac{r^2 + l^2 + 2rl + r^2 + l^2 - 2rl}{(r^2 - l^2)^2}\right]$$

$$E_{net} = \frac{1}{4\pi\varepsilon_0} \cdot \frac{2q(r^2 + l^2)}{(r^2 - l^2)^2}$$

If $r >> l$,

Then, $\quad r^2 + l^2 \to r^2$

$$r^2 - l^2 \to r^2$$

$$E_{net} = \frac{1}{4\pi\varepsilon_0}\frac{2qr^2}{r^4}$$

$$E_{net} = \frac{1}{4\pi\varepsilon_0}\frac{2q}{r^2}$$

$$E \propto \frac{1}{r^2}$$

17. (b,d) There is a torque on dipole which gives an angular acceleration to dipole (α).

Then, $\quad \tau = -pE\sin\theta = I\alpha$

$$\Rightarrow \quad \alpha = \frac{d^2\theta}{dt^2} = \frac{-pE}{I}\sin\theta$$

For small angles, $\theta \approx \sin\theta$

$$\Rightarrow \quad \alpha = \frac{-pE\theta}{I} \quad \ldots(i)$$

As, $\quad \alpha = -\omega^2\theta \quad \ldots(ii)$

which is SHM,

$$\omega^2 = \frac{pE}{I} \Rightarrow f = \frac{1}{2\pi}\sqrt{\frac{pE}{I}}$$

Dipole releases oscillation energy in form of heat to achieve stable equilibrium position.

18. (b,c) The potential energy of the parallel dipole is a function of its position.

$$\Rightarrow \quad U = -\mathbf{p}\cdot\mathbf{E}$$

$$\Rightarrow \quad F_x = -\frac{dU}{dx} = \frac{-d}{dx}(-\mathbf{p}\cdot\mathbf{E}) = \mathbf{p}\cdot\frac{d\mathbf{E}}{dx}$$

As the field does not depend on Y and Z-coordinates, other components of force are zero.

$$\Rightarrow \quad \mathbf{F} = F_x\hat{\mathbf{i}} = \left(\mathbf{p}\cdot\frac{d\mathbf{E}}{dx}\right)\hat{\mathbf{i}}$$

19. (d) Electric field of a dipole $\left(E \propto \frac{1}{r^3}\right)$ decays much faster than that of a point charge $\left(E \propto \frac{1}{r^2}\right)$.

Statement II is correct considering a point charge.

20. (c) A changing electric field (as in case of microwave oven) oscillates dipoles and so energy is finally dissipated as heat. Oscillating dipole does not radiate energy in form of a electromagnetic wave.

21. (4) Dipole moment of system

$$\mathbf{p} = q \times d = 4 \times 10^{-8} \times 2 \times 10^{-4}\ \text{C-m} = 8 \times 10^{-12}\ \text{C-m}$$

Torque is maximum when $\theta = 90°$

$$\tau_{max} = pE\sin 90° = 8 \times 10^{-12} \times 4 \times 10^8$$

$$= 32 \times 10^{-4} = 8 \times 4 \times 10^{-4}\ \text{Nm}$$

$\therefore \quad k = 4$

22. (4) Field of dipole varies as $\frac{1}{r^4}$.

As, force $= qE$

$\therefore$ Force $\propto \frac{1}{r^4}$

Explanations for Q. Nos. (23-26)

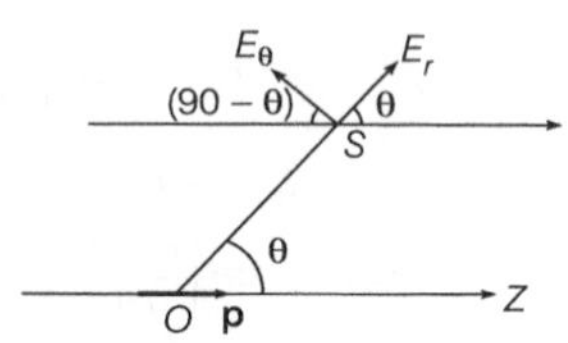

Given condition is radial and perpendicular components of field at S are

$$E_\theta = E_r = \frac{2p\cos\theta}{4\pi\varepsilon_0 r^3}$$

$$E_\theta = E_\theta = \frac{p\sin\theta}{4\pi\varepsilon_0 r^3}$$

$$E_z = E_r\cos\theta - E_\theta\sin\theta = \frac{p}{4\pi\varepsilon_0 r^3}(3\cos^2\theta - 1)$$

$$E_\perp = E_r\sin\theta + E_\theta\cos\theta = \frac{3p\sin\theta\cos\theta}{4\pi\varepsilon_0 r^3}$$

Now, E_{net} at S is perpendicular to $\mathbf{p}$ when $E_z = 0$

$$\Rightarrow \quad 3\cos^2\theta - 1 = 0$$

$$\Rightarrow \quad \cos\theta = \frac{1}{\sqrt{3}} \text{ and}$$

$$\therefore E_{net} = E_\perp = \frac{3p}{4\pi\varepsilon_0 r^3} \times \frac{\sqrt{2}}{\sqrt{3}} \times \frac{1}{\sqrt{3}} = \frac{p\sqrt{2}}{4\pi\varepsilon_0 r^3}.$$

23. (a), **24.** (a)

25. (d), **25.** (c)

Explanations for Q. Nos. (27-29)

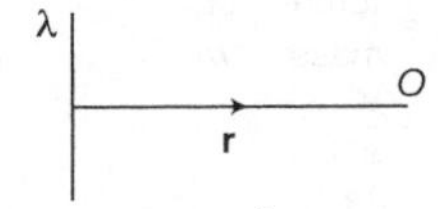

Field of wire at O is $\quad E = \dfrac{\lambda}{2\pi\varepsilon_0 r}\hat{\mathbf{r}}$

Force on dipole is $\quad F = |\mathbf{p}|\left(\dfrac{dE}{dr}\right)$ along $\mathbf{p} = |\mathbf{p}|\cdot 0$

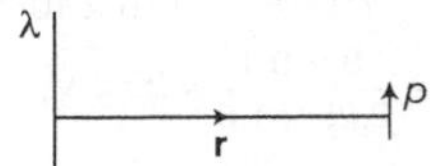

As, field of wire is constant when we move parallel to wire.

$\therefore \quad \dfrac{dE}{dr} = 0.$

Along the radius vector, force $= |\mathbf{p}|\left(\dfrac{dE}{dr}\right)$

$$\frac{dE}{dr} = -\frac{\lambda}{2\pi\varepsilon_0 r^2} \text{ along } \mathbf{r}$$

$\therefore \quad F = \dfrac{\lambda p}{2\pi\varepsilon_0 r^2}$ and force is towards the wire.

$$\text{Force, } F = p\left(\frac{dE_1}{dy}\right) = \frac{p\lambda}{2\pi\varepsilon_0 r^2}$$

27. (c), **28.** (a), **29.** (a)

DPP-7 Electric Potential and Potential Energy

1. (i) Potential at A = potential of q_1 + potential of q_2

$$= \frac{kq_1}{\varepsilon_1} + \frac{kq_2}{\varepsilon_2}$$

$$= \frac{1}{4\pi\varepsilon_0}\left(\frac{5\times10^{-6}}{10\times10^{-2}} + \frac{-5\times10^{-6}}{20\times10^{-2}}\right)$$

$$= \frac{9\times10^{9}\times10^{-6}}{10^{-2}}\left(\frac{5}{10} - \frac{5}{20}\right)$$

$$= 2.25\times10^{5}\text{ V}$$

Similarly, potential at $B = \dfrac{9\times10^{9}\times10^{-6}}{10^{-2}}\left(\dfrac{5}{20} - \dfrac{5}{10}\right)$

$$= -2.25\times10^{5}\text{ V}$$

(ii) Work done by external agency in moving q_3 from A to B = (change in potential energy)

$$= U_A - U_B$$
$$= q_3(V_A - V_B)$$
$$= 2\times10^{-6}\,[2.25\times10^{5} - (-2.25\times10^{5})]$$
$$= 2\times10^{-6}\times4.5\times10^{5}\text{ J}$$
$$= 9\times10^{-1}\text{ J} = 0.9\text{ J}$$

(iii) Work done is not path dependent as electrostatic field is conservative in nature.

(iv) The work done by external agent is positive so the potential energy of 3-charge system increases.

2 (i) A charge moves from point A to B in a field when its potential energy is higher at A than B.

So, a positive test charge moves from point A to B when

$$V_A > V_B$$

But due to its negative sign, if a test charge moves from point A to B, then

U_A = potential energy at $A = (-q)\,V_A$

and U_B = potential energy at $B = (-q)\,V_B$

Also, $\quad U_A > U_B$

$\Rightarrow \quad -q\,V_A > -q\,V_B$

$\Rightarrow \quad V_A < V_B$

So, a negative charge moves from a lower potential point to a higher potential point.

(ii) The charge on an electron is $q = -e$

$$= -1.6\times10^{-19}\text{C}.$$

Therefore, its change in potential energy is

$$\Delta U = qV$$
$$= (-1.6\times10^{-19}\text{C})\,(5000\text{ V})$$
$$= -8.0\times10^{-16}\text{ J}$$

The minus sign indicates that the potential energy decreases. The potential difference has a positive sign, since the final potential is higher than the initial potential. Negative electrons are attracted toward a positive electrode and repelled away from a negative electrode.

The potential energy lost by electron becomes kinetic energy K. From conservation of energy, $\Delta K + \Delta U = 0$, so $\Delta K = -\Delta U$.

$$\frac{1}{2}mv^2 - 0 = -qV$$

where, the initial kinetic energy is zero, since we are given that the electron started from rest.

Now, we solve for $\quad v = \sqrt{-\dfrac{2qV}{m}}$

$$= \sqrt{-\frac{2\,(-1.6\times10^{-19}\text{C})\,(5000\text{ V})}{9.1\times10^{-31}\text{kg}}}$$

$$= 4.2\times10^{7}\text{ m/s}$$

3. (i) Potential at a point distant r from a charge q is given by $V = \dfrac{kq}{r}$.

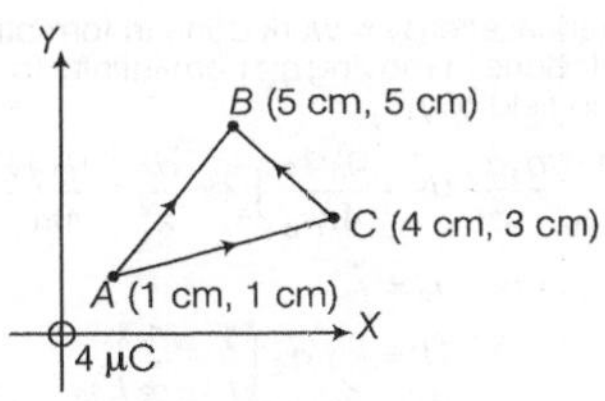

So, $\quad V_A = \dfrac{kq}{r_A} = \dfrac{k(4\times10^{-6})}{(\sqrt{2}\times10^{-2})}\text{ V}$

$$V_B = \frac{k(4\times10^{-6})}{5\sqrt{2}\times10^{-2}}\text{ V}$$

$$V_C = \frac{k(4\times10^{-6})}{5\times10^{-2}}\text{ V}$$

Work done in moving $2\,\mu$C charge from A to B directly is

$$W_{A\to B} = q'\,(V_A - V_B)$$

$$= 2\times10^{-6}\left(\frac{k\times4\times10^{-6}}{\sqrt{2}\times10^{-2}} - \frac{k\times4\times10^{-6}}{5\sqrt{2}\times10^{-2}}\right)$$

$$= \frac{k\times8\times10^{-12}}{\sqrt{2}\times10^{-2}}\left(1 - \frac{1}{5}\right) = \frac{k\times4\times8\times10^{-12}}{5\sqrt{2}\times10^{-2}}$$

$$= \frac{k\times32\times10^{-12}}{5\sqrt{2}\times10^{-2}}\text{ J},\ k = 9\times10^{9}$$

(ii) (a) Also, $W_{A \to C} = q'(V_A - V_C)$

$$= 2 \times 10^{-6}\left(\frac{k \times 4 \times 10^{-6}}{\sqrt{2} \times 10^{-2}} - \frac{k \times 4 \times 10^{-6}}{5 \times 10^{-2}}\right)$$

$$= \frac{k \times 8 \times 10^{-12}}{5\sqrt{2} \times 10^{-2}}(5 - \sqrt{2})\text{ J}$$

$$W_{C \to B} = q'(V_C - V_B)$$

$$= 2 \times 10^{-6}\left(\frac{k \times 4 \times 10^{-6}}{5 \times 10^{-2}} - \frac{k \times 4 \times 10^{-6}}{5\sqrt{2} \times 10^{-2}}\right)$$

$$= \frac{k \times 8 \times 10^{-12}}{5\sqrt{2} \times 10^{-2}}(\sqrt{2} - 1)\text{ J}$$

(b) Work done in taking charge of 2 μC from A to C, then C to B

$$W_{A \to C \to B} = W_{A \to C} + W_{C \to B}$$

$$= \frac{k \times 8 \times 10^{-12}}{5\sqrt{2} \times 10^{-2}}(5 - \sqrt{2}) + \frac{k \times 8 \times 10^{-12}}{5\sqrt{2} \times 10^{-2}}(\sqrt{2} - 1)$$

$$= \frac{k \times 8 \times 10^{-12}}{5\sqrt{2} \times 10^{-2}}(5 - \sqrt{2} + \sqrt{2} - 1)$$

$$= \frac{k \times 4 \times 8 \times 10^{-12}}{5\sqrt{2} \times 10^{-2}}\text{ J}$$

Clearly, $W_{A \to B} = W_{A \to C \to B}$

4. (i,ii) Electric potential energy of two charge system is

$$U = \frac{k q_1 q_2}{r_{12}}$$

So, potential energy of two unlike charges is negative. Potential energy of two like charges is positive.

(iii) In case of unlike charges, work has to be done by external means to separate the charges ($U = -$ ve). So, the system is stable.

In case of like charges, work must be done to bring the charges together. The system is unstable ($U = +$ ve).

(iv) Potential energy of a charge system (two or more) is the work done in formation of the system.

For a two charge system separated by a distance r.

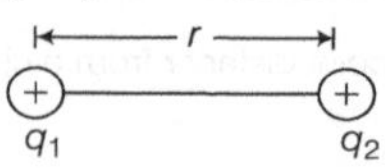

U = potential energy = work done in formation
= work done in moving q_2 from infinity to a point distance r from q_1 in field of q_1.

$$\int_{r_1}^{r_2} -\frac{1}{4\pi\varepsilon_0}\frac{q_1 q_2}{x^2}dx = \frac{q_1 q_2}{4\pi\varepsilon_0}\int_{r_1}^{r_2} -\frac{dx}{x^2} = \frac{q_1 q_2}{4\pi\varepsilon_0}\left[\frac{1}{x}\right]_{r_1}^{r_2}$$

Here, $r_1 = \infty$, $r_2 = r$

$$\therefore \quad U = kq_1q_2\left(\frac{1}{r} - \frac{1}{\infty}\right)$$

$$= \frac{kq_1q_2}{r}$$

$$= \frac{1}{4\pi\varepsilon_0} \cdot \frac{q_1q_2}{r}\text{ J}$$

(assuming, $V = 0$ at $r = \infty$)

(v) When potential at distance $r_1 = 0$, then

$$U = kq_1\int_{r_1}^{r} -\frac{q_2}{x^2}dx$$

$$= kq_1q_2\left(\frac{1}{r} - \frac{1}{r_1}\right)\text{J}$$

5. (a) Let a = acceleration of electron,

as a = constant

$\therefore$ E = uniform

$\Rightarrow$ $s = ut + \frac{1}{2}at^2$ $(\because u = 0)$

$\Rightarrow$ $a = \frac{2s}{t^2}$

Also, $a = \frac{\text{force}}{\text{mass}} = \frac{eE}{m}$

and $E = \frac{V}{s}$

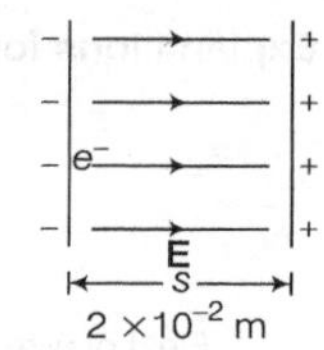

So, $\frac{eV}{sm} = \frac{2s}{t^2}$

$\Rightarrow$ V = potential difference between plates

$$= \frac{2s^2}{t^2} \cdot \frac{m}{e} = \frac{2 \times (2 \times 10^{-2})^2}{(1.5 \times 10^{-8})^2} \times \frac{9.1 \times 10^{-31}}{1.6 \times 10^{-19}}$$

$$= \frac{8 \times 9.1}{2.25 \times 1.6} = 20.2\text{ V}$$

6. (c) Charges will flow from charged sphere to uncharged sphere until their potentials become equal. If charge q flows from charged sphere to uncharged sphere, then

$$\frac{Q - q}{4\pi\varepsilon_0 r_1} = \frac{q}{4\pi\varepsilon_0 r_2} \Rightarrow Q - q = \frac{r_1}{r_2}q$$

$$q = Q\left(\frac{r_2}{r_1 + r_2}\right)$$

7. (d) By the work-energy theorem, the total work done by the external force and the electric field together is the change in the kinetic energy. The work done by the electric field is given by

$$\Rightarrow \quad W_{\text{ext}} + W_{\text{electric}} = \text{KE}_{\text{final}} - \text{KE}_{\text{initial}}$$

$$\Rightarrow \quad W_{\text{ext}} - q(V_B - V_A) = \text{KE}_{\text{final}}$$

$$\Rightarrow \quad (V_B - V_A) = \frac{W_{\text{ext}} - \text{KE}_{\text{initial}}}{q}$$

$$= \frac{7.00 \times 10^{-4}\text{ J} - 2.10 \times 10^{-4}\text{ J}}{-q}$$

$$= \frac{-4.90 \times 10^{-4}}{q}\text{ V}$$

Since, the potential difference is negative, we see that $V_A > V_B$.

8. (a) For $U = k\left[\frac{qq}{l} + \frac{Qq}{l} + \frac{Qq}{l}\right] = 0$

$$\Rightarrow \quad qq + Qq + Qq = 0$$

$$\Rightarrow \quad Q[q + q] = -qq$$

$$\Rightarrow \quad Q \times 2q = -qq$$

$$\Rightarrow \quad Q = -\frac{q}{2}$$

9. (c) Using, $V_f - V_i = -\int_i^f \mathbf{E} \cdot d\mathbf{s}$

$$\Rightarrow \quad V_f - V_i = -Ed \Rightarrow V_C - V_A = -5E$$

$$\Rightarrow \quad V_A - V_C = 5E$$

10. (d) Electric potential at $P = k\sum_{i=1}^{n}\frac{q_i}{r_i} = k\left(\frac{5q}{d} + \frac{5q}{d} - \frac{5q}{d} - \frac{5q}{2d}\right)$

$$= \frac{kq}{d}\left(5 + 5 - 5 - \frac{5}{2}\right)$$

$$= \frac{kq}{d} \times 2.5 = 2.5\frac{kq}{d}$$

11. (c) As, volume constant $\frac{4}{3}\pi R^3 = n\left(\frac{4}{3}\pi r^3\right)$

$$R = n^{1/3}r$$

$$V = \frac{kq}{r} = 400\text{ V} \quad \text{(given)}$$

$$V' = \frac{kq}{R} = k \cdot \frac{nq}{n^{1/3}r}$$

$$V' = n^{2/3}(kq/r)$$

$$V' = n^{2/3}V$$

$$V' = (8)^{2/3} \times 400\text{V} = (2^3)^{2/3} \times 400\text{V}$$

$$= 4 \times 400\text{ V} = 1600\text{ V}$$

12. (d) The potential energy of system of charges

$$U = \frac{1}{4\pi\varepsilon_0}\left[\frac{q \times 4q}{d} + \frac{(-4q)\times q}{d} + \frac{2q \times q}{d}\right]$$

$$U = \frac{1}{4\pi\varepsilon_0}\left[\frac{-4q^2}{d} - \frac{8q^2}{d} + \frac{2q^2}{d}\right]$$

$$U = \frac{1}{4\pi\varepsilon_0}\left[\frac{-10q^2}{d}\right] = \frac{-10q^2}{4\pi\varepsilon_0 d}$$

13. (d) For small displacement $d\mathbf{s}$, work done by the electric field

$(\Delta W)_{\text{electric force}} = F_e \cdot d\mathbf{r}$

The change in potential ΔV is given

$$\Delta V = \frac{-[\Delta W]_{\text{electric force}}}{q'} = \frac{\Delta U}{q'}$$

14. (b) $V_{C_1} = V_{Q_1} + V_{Q_2}$

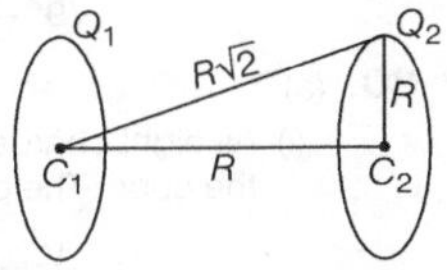

$$= \frac{Q_1}{4\pi\varepsilon_0 R} + \frac{1}{4\pi\varepsilon_0}\frac{Q_2}{R\sqrt{2}}$$

$$= \frac{1}{4\pi\varepsilon_0 R}\left(Q_1 + \frac{Q_2}{\sqrt{2}}\right)$$

Similarly, $V_{C_2} = \frac{1}{4\pi\varepsilon_0 R}\left(Q_2 + \frac{Q_1}{\sqrt{2}}\right)$

$\therefore$ $$\Delta V = V_{C_1} - V_{C_2} = \frac{1}{4\pi\varepsilon_0 R}\left[(Q_1 - Q_2) - \frac{1}{\sqrt{2}}(Q_1 - Q_2)\right]$$

$$= \frac{Q_1 - Q_2}{\sqrt{2}\,(4\pi\varepsilon_0 R)}(\sqrt{2} - 1)$$

$$W = q\Delta V = q\,(Q_1 - Q_2)\,(\sqrt{2} - 1)/\sqrt{2}(4\pi\,\varepsilon_0 R)$$

15. (b) Net electrostatic energy of the configuration will be

$$U = k\left[\frac{q \cdot q}{q} + \frac{Q \cdot q}{\sqrt{2}a} + \frac{Q \cdot q}{a}\right]$$

Here, $k = \frac{1}{4\pi\varepsilon_0}$

Putting $U = 0$, we get $Q = \frac{-2q}{2 + \sqrt{2}}$

16. (b)

q	Q	q		q	Q	q
$x=-a$	$x=0$	$x=+a$		$x=-a$	$x=x$	$x=a$

Initial position   Final position

$$U_i = \frac{2kQq}{a} + \frac{k.q.q}{2a}$$

and $$U_f = kQq\left[\frac{1}{a + x} + \frac{1}{a - x}\right] + \frac{k.q.q}{2a}$$

Here, $k = \frac{1}{4\pi\varepsilon_0}$

$$\Delta U = U_f - U_i \quad \text{or } |\Delta U| = \frac{2kQqx^2}{a^3}$$

For $x << a$ $\quad\therefore\quad$ $\Delta U \propto x^2$

17. (c,d) Charge on electron $= -1.6 \times 10^{-19}$C

Charge on proton $= +1.6 \times 10^{-19}$C

Separation of electron and proton $= r = 0.53\text{Å} = 0.533 \times 10^{-10}$m

Change in energy when atom (electron and proton system) is formed

$$= \Delta U = \frac{kq_1q_2}{r} = \frac{9 \times 10^9\,(1.6 \times 10^{-19})(-1.6 \times 10^{-19})}{0.53 \times 10^{-10}}$$

$$= -4.35 \times 10^{-18}\,\text{J} = \frac{-4.35 \times 10^{-18}}{1.6 \times 10^{-19}}\,\text{eV} = -27.2\,\text{eV}$$

At infinity separation, $U = 0$

Thus, the energy when system is formed is

$$U = 0 + \Delta U = -27.2\,\text{eV}.$$

As, kinetic energy is positive and $K = \left|\frac{U}{2}\right| = \frac{27.2}{2} = 13.6$ eV

So, total energy = KE + PE = 13.6 + −27.2 = −13.6 eV

Potential energy of system when separation = 1.06 Å

$$= U' = \frac{-27.2}{2} = -13.6 \text{ eV} = 0$$

So, total energy = −13.6 + 13.6 = 0

18. (b,c,d) Force, $\mathbf{F} = q\mathbf{E} = 10^{-6} \times 200$

$= 2 \times 10^{-4}$ N along AB.

Potential difference, $\Delta V = V_A - V_B = Ed = 200 \times 0.02 = 4$V

Work done, $\Delta W = \mathbf{F} \cdot \mathbf{d} = 2 \times 10^{-4} \times 0.02 = 4 \times 10^{-6}$ J

19. (b,c,d) Energy for proton field and electron field system is conserved.

For proton, change in potential energy $= \Delta U = qV$

$= 1.6 \times 10^{-19} \times 120$ J

Also, $\Delta U = \Delta K$

$\Rightarrow$ $$qV = \frac{1}{2}mv^2$$

$\Rightarrow$ $$v = \frac{1.6 \times 10^{-19} \times 120}{\frac{1}{2} \times 1.67 \times 10^{-27}}$$

$$= 1.52 \times 10^5 \text{ m/s}$$

For electron, change in potential energy

$$\Delta U = qV = -1.6 \times 10^{-19} \times 120 \text{ J}$$

And using $$qV = \frac{1}{2}mv^2$$

$$v_e = 6.49 \times 10^6 \text{ m/s}$$

20. (6) $\Delta U = -$[work from origin to (20 cm, 0 cm) − work from (20 cm, 0 cm) to (20 cm, 50 cm)].

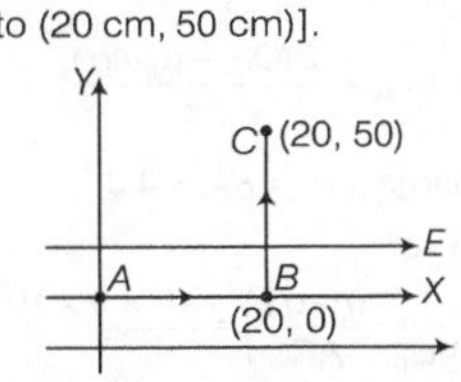

As displacement from B to C is perpendicular to field.

Work done is zero for IInd part,

$\therefore$ $$\Delta U = -(qE_x)\Delta x$$

$$= -(12 \times 10^{-6})\,(250)\,(0.2)$$

$$= -6 \times 10^{-4} \text{ J}$$

21. (3) For PE to be minimum, the bigger charges must be as far as possible.

$\therefore$ $2q$ and $8q$ charge must be 9 cm away.

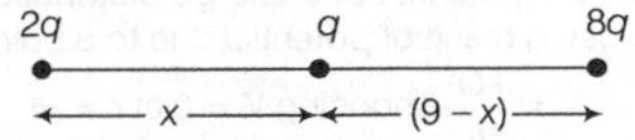

Let x = distance between q and $2q$, then

$$U = k\left(\frac{(2q)(q)}{x} + \frac{(8q)(q)}{9 - x} + \frac{(2q)(8q)}{9}\right)$$

$\Rightarrow$ For U to be minimum, $\frac{2}{x} + \frac{8}{9 - x}$ should be minimum.

$\Rightarrow$ $$\frac{d}{dx}\left(\frac{2}{x} + \frac{8}{9 - x}\right) = 0$$

$\Rightarrow$ $$\frac{-2}{x^2} + \frac{8}{(9 - x)^2} = 0$$

$\therefore$ $$x = 3 \text{ cm}$$

22. (b) Using energy of conservation,

$$(KE + U_s + U_E)_{initial} = (k + U_s + U_E)_{final}$$

$$\Rightarrow \quad 0 + 0 + 0 = 0 + \frac{1}{2}kx_{max}^2 - qEx_{max}$$

$$\Rightarrow \quad x_{max} = \frac{2QE}{k}$$

23. (c) At equilibrium position,

$$\Sigma F_x = -F_s + F_e = 0 \quad \text{or} \quad kx = QE$$

$$\therefore \quad x = \frac{QE}{k} \quad \text{(equilibrium)}$$

24. (a) Equation of motion of block is

$$\Sigma F_x = -kx + QE = \frac{md^2x}{dt^2}$$

Let $x' = x - \frac{QE}{k}$ or $x = x' + \frac{QE}{k}$

So, equation of motion becomes

$$-k\left(x' + \frac{QE}{k}\right) + QE = \frac{md^2(x + QE/k)}{dt^2}$$

or $$\frac{d^2x'}{dt^2} = -\frac{k}{m}x' \quad \ldots(i)$$

This is an equation of SHM,

$$ax' = -\omega^2 x' \quad \ldots(ii)$$

$$\Rightarrow \text{as} \quad T = \frac{2\pi}{\omega} \text{ and } \omega = \sqrt{\frac{k}{m}}$$

From Eqs. (i) and (ii), we have $T = 2\pi\sqrt{\frac{m}{k}}$

25. (d) When friction is present,

$$(k + U_s + U_e)_{initial} + \Delta E_{mechanical} = (k + U_s + U_e)_{final}$$

$$0 + 0 + 0 - \mu_k mgx_{max} = 0 + \frac{1}{2}kx_{max}^2 - QEx_{max}$$

$$\Rightarrow \quad x_{max} = \frac{2(QE - \mu_k mg)}{k}$$

26. (d) At C, KE of charge $-q = K_C = 4$ J

and PE of charge $-q$

$$U_C = 2\left(\frac{1}{4\pi\varepsilon_0}\cdot\frac{q(-q)}{AC}\right) = \frac{-2\times 9\times 10^9 \times (5\times 10^{-5})^2}{5} = -9\text{ J}$$

$\therefore$ Total energy at $C = K_C + U_C = 4\text{J} + (-9\text{J}) = -5\text{J}$

27. (d) Total energy of charge $-q$ is conserved.

$\therefore$ Total energy at C = total energy at D

$\therefore E_D = -5$J

28. (d) At D, energy of charges is only potential, it is on the charge of turning back.

$\therefore$ KE = 0, so PE at $D = U_D = -5$ J

29. (a) Also, $$K_C + U_C = K_B + U_D$$

$$\Rightarrow \quad -5 = 0 + 2\left(\frac{1}{4\pi\varepsilon_0}\frac{q(-q)}{AD}\right)$$

$$\Rightarrow \quad -5 = \frac{2\times 8.99\times 10^9 \times (5\times 10^{-5})^2}{AD}$$

$$\Rightarrow \quad -5 \approx \frac{-45}{AD}$$

$$\Rightarrow \quad AD \approx 9\text{ m}$$

and $$OD = \sqrt{AD^2 - OA^2}$$

$$\sqrt{9^2 - 3^2} = \sqrt{81 - 9} = 8.48\text{ m}$$

30. (a)

(i) All eight charges are at the same distance from the centre of the cube. The potential of a point charge

$$V_{centre} = 8\frac{1}{4\pi\varepsilon_0}\frac{Q}{\frac{\sqrt{3}}{2}l}$$

$$= \frac{16}{\sqrt{3}}\frac{1}{4\pi\varepsilon_0}\frac{Q}{l} \approx 9.24\frac{1}{4\pi\varepsilon_0}\frac{Q}{l}$$

(ii) For the seven charges that produce the potential at a corner, there are a distance l away from that corner, there are a distance $\sqrt{2}l$ away from that corner and one is a distance $\sqrt{3}l$ away from that corner.

$$V_{corner} = 3\frac{1}{4\pi\varepsilon_0}\frac{Q}{l} + 3\frac{1}{4\pi\varepsilon_0}\frac{Q}{\sqrt{2}l} + \frac{1}{4\pi\varepsilon_0}\frac{Q}{\sqrt{3}l}$$

$$= \left[\left(3 + \frac{3}{\sqrt{2}} + \frac{1}{\sqrt{3}}\right)\frac{1}{4\pi\varepsilon_0}\frac{Q}{l}\right]$$

$$= 5.70\frac{1}{4\pi\varepsilon_0}\frac{Q}{l}$$

(iii) The total potential energy of the system is half the energy found by multiplying each charge times the potential at a corner. The factor of half comes from the fact that if you took each charge times the potential at a corner, you would be counting each pair of charges twice.

$$U = \frac{1}{2}8(QV_{corner}) = 4\left(3 + \frac{3}{\sqrt{2}} + \frac{1}{\sqrt{3}}\right)$$

$$\frac{1}{4\pi\varepsilon_0}\frac{Q^2}{l} = 22.8\frac{1}{4\pi\varepsilon_0}\frac{Q^2}{l}$$

(i) $\rightarrow$ (r), (ii) $\rightarrow$ (p), (iii) $\rightarrow$ (q)

DPP-8 Potential Due to Charge Distributions

1. We calculate the potential of a charge distribution or a system of charges by using result of potential due to a point charge,

$$V = \frac{kQ}{r}, \text{ choosing } V = 0 \text{ at } r = \infty$$

Then, we add overall the charges.

If we have n individual point charges, then potential at the point A is

$$V_A = \sum_{i=1}^{n} V_i = k\sum_{i=1}^{n}\frac{Q_1}{r_{ia}}$$

Taking, $V = 0$ at $r = \infty$

r_{ia} = distance of ith charge from point A.

If a charge distribution is continuous, then $V = \frac{1}{4\pi\varepsilon_0}\int\frac{dq}{r}$

where, r is the distance of point of elemental charge dq from the point where potential is being determined.

2. (i) When two spheres are put in contact, they effectively become one conductor and potentials of both will be same, because potential is same everywhere over a conducting surface.

(ii) As spheres are identical a charge $Q/2$ will flow from the initially charge sphere to initially neutral sphere so that they will have equal charges.

(iii) Even if spheres have different radii, the charge from Ist is transferred to IInd till they reach equal potential. However, the shared charge is not equal, the larger sphere gets an larger charge.

(iv) Consider 2 spheres with radii R_1 and R_2 let only Ist one is charged having a charge q. $R_1 > R_2$.

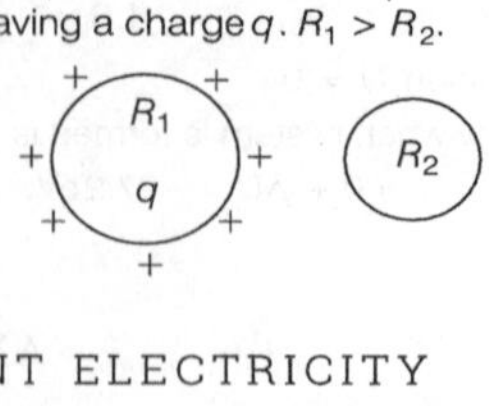

Both spheres are touching now,

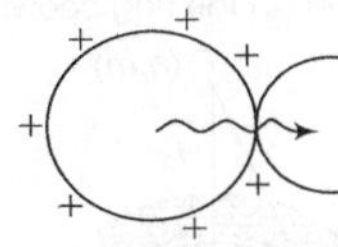

Charge from sphere Ist is transferred to sphere IInd.

Transfer of charge from sphere Ist to sphere IInd occurs till they have same potential.

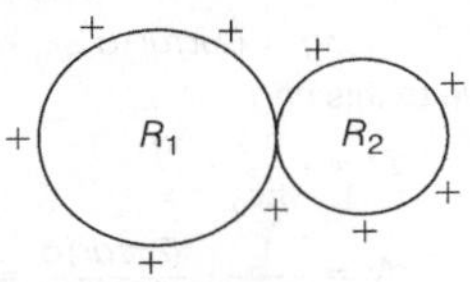

When they are at same potentials, no transfer of charge occurs. They have a common potential

$$V = \frac{k(Q_1 + Q_2)}{(R_1 + R_2)} \begin{matrix}\leftarrow \text{effective charge} \\ \leftarrow \text{effective radius}\end{matrix}$$

but $Q_1 = q$ and $Q_2 = 0$

$$\Rightarrow \qquad V = k\left(\frac{q}{R_1 + R_2}\right), k = \frac{1}{4\pi\varepsilon_0}$$

Now, both are removed far from each other, then charge on each will be calculated as

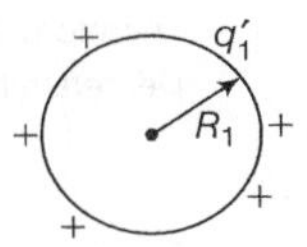

$$V = \frac{kQ}{R}$$

On surface of a metal sphere, let sphere Ist has charge q_1, then

$$V = \frac{kq_1'}{R_1}, \text{ but } V = \frac{kq}{R_1 + R_2}$$

Equating, $q_1' = q\left(\frac{R_1}{R_1 + R_2}\right)$ and similarly

$$q_2' = \text{charge on sphere IInd} = q\left(\frac{R_2}{R_1 + R_2}\right)$$

Now, charge density on surface of Ist and IInd can be obtained as,

σ = final charge density of sphere Ist

$$= \frac{q_1'}{4\pi R_1^2} = \frac{q\left(\frac{R_1}{R_1 + R_2}\right)}{4\pi R_1^2} = \left(\frac{q}{R_1 + R_2}\right) \cdot \frac{1}{R_1}$$

and σ_2 = charge density of sphere IInd $= \left(\frac{q}{R_1 + R_2}\right) \cdot \frac{1}{R_2}$

So, charge density $\sigma \propto \frac{1}{R}$.

When we have a single conductor of non-uniform shape,

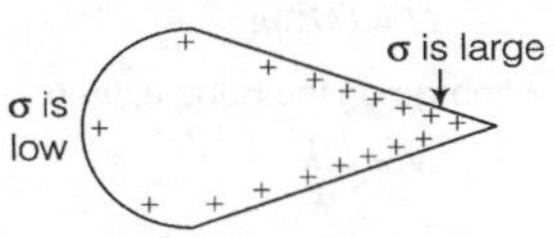

Charge density is higher at pointed ends.

3. (i) Electric field is zero near weaker charge.

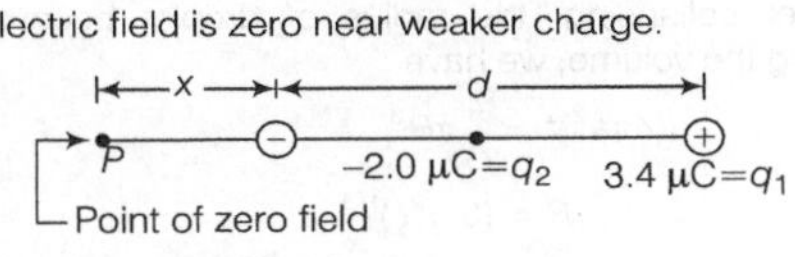

Let field is zero at P, then

$$E = 0 = E_+ + E_-$$

$$= \frac{1}{4\pi\varepsilon_0} \frac{q_2}{x^2} - \frac{1}{4\pi\varepsilon_0} \cdot \frac{q_1}{(d + x)^2}$$

$$\Rightarrow \qquad q_2(d + x)^2 = q_1 x^2$$

Substituting data from problem, we get

$$x = 16 \text{ cm}$$

Also, there is only one point where field intensity is zero.

(ii) There are two possible locations where potential can be zero. Now, as potential due to positive charge is positive and due to negative charge is negative and we can use

$$V_{net} = V_1 + V_2 \text{ and } V = kq/r$$

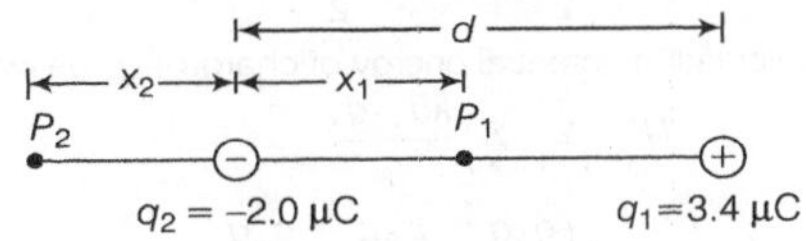

$$V_{P_1} = k\left(\frac{q_1}{d - x_1} + \frac{q_2}{x_2}\right) = 0$$

and $$V_{P_2} = k\left(\frac{q_1}{d + x_2} + \frac{q_2}{x_2}\right) = 0$$

Solving these equations, we get

$x_1 = 1.852$ cm and $x_2 = 7.143$ cm

So, there are 2 locations where potential is zero.

(iii) When $E = 0, V \neq 0$ and when $V = 0, E \neq 0$ in general.

4. Using, $V_{BA} = V_B - V_A$ and $V = \frac{kq}{r}$

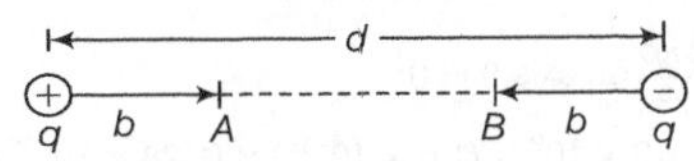

we get,

$$V_{BA} = V_B - V_A$$

$$= k\left(\frac{q}{a - b} + \frac{(-q)}{b}\right) - k\left(\frac{a}{b} + \frac{-q}{d - b}\right)$$

$$= kq\left(\frac{1}{d - b} - \frac{1}{b} - \frac{1}{b} + \frac{1}{d - b}\right)$$

5. For every element with charge $+dq$ on top half of ring there is a element diametrically opposite on cover half of ring with charge $-dq$. The potential due to these two small elements cancel out each other, so the potential due to entire ring is zero.

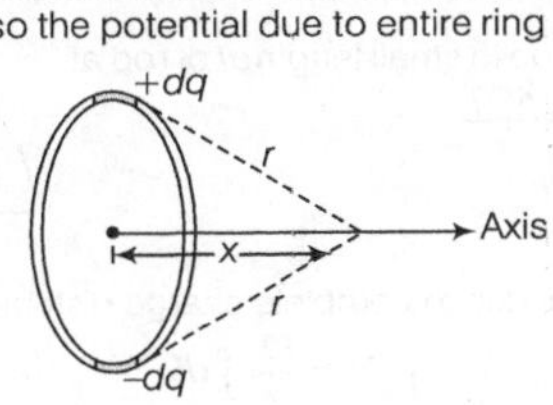

6. (a) Work done for moving a charge q_0 from B to A,

$$W_{B \to A} = q_0(V_A - V_B)$$

Also, V_A = potential at point A

$$= \frac{kq_1}{r_1} + \frac{kq_2}{r_2} = k\left(\frac{q_1}{r_1} + \frac{q_2}{r_2}\right)$$

$$= 10^{10}\left\{\frac{-5 \times 10^{-6}}{15 \times 10^{-2}} + \frac{2 \times 10^{-6}}{5 \times 10^{-2}}\right\} = \frac{10^6}{15} \text{ V}$$

Similarly,

$$V_B = 10^{10}\left\{\frac{2 \times 10^{-6}}{15 \times 10^{-2}} - \frac{5 \times 10^{-6}}{5 \times 10^{-2}}\right\} = \frac{-13}{15} \times 10^6 \text{ V}$$

$$\Rightarrow \qquad W_{B \to A} = 3 \times 10^{-6}\left\{\frac{1}{15} \times 10^6 - \left(\frac{-13}{15} \times 10^6\right)\right\}$$

$$= 3 \times 10^{-6} \times \frac{10^6}{15}[1 - (-13)]$$

$$= \frac{1}{5} \times 14 = 2.8 \text{ J}$$

7. (c) $V = \frac{kq}{r}$, on increasing radius charge remains same.

$$\frac{V_1}{V_2} = \frac{r_2}{r_1} \quad \text{or} \quad V_2 = V_1\frac{r_1}{r_2} = 8\text{V}$$

8. (d) Charge $+q$ is at a distance of $\frac{1}{2}$ diagonal length of cube

$$= \frac{\sqrt{3}}{2} \times \text{side} = \frac{\sqrt{3}}{2} b$$

Now, calculating potential energy of charge $+q$, pairwise, we get

$$U = \sum_{j=1}^{n} \sum_{k \neq j} \frac{kq_j \cdot q_k}{r_{jk}}$$

$$= k\left(\frac{q_1 q}{r_1} + \frac{q_2 q}{r_2} + \frac{q_3 q}{r_3} + \cdots\right)$$

There are eight such pairs,

$$= \frac{8 \times k \times (-q)(q)}{\left(\frac{\sqrt{3}}{2} \times b\right)} = \frac{-4q^2}{\sqrt{3}\,\pi\varepsilon_0 b}\ \text{J}$$

9. (a) Potential at a point inside is same as that of surface (in case of a hollow sphere)

$$\therefore \quad V_{r = 4\text{cm}} = V_{\text{surface}} = \frac{kq}{r} = \frac{9 \times 10^9 \times 3.2 \times 10^{-19}}{10 \times 10^{-2}}$$

$$= 28.8 \times 10^{-9}\ \text{V}$$

10. (a) $V = \frac{2kp}{r^3}$ on axis, $\theta = 0°$

$$\Rightarrow V = \frac{2 \times [9 \times 10^9 \times (1.6 \times 10^{-19}) \times (1.28 \times 10^{-10})]}{(12 \times 10^{-10})^2}$$

$$= 2 \times \frac{18432}{144} = 0.13 \times 2 = 0.26\ \text{V}$$

11. (a) $\Delta V = V_{2R} - V_0$

$$= \frac{kQ}{\sqrt{R^2 + (2R)^2}} - \frac{kQ}{R}$$

$$= \frac{kQ}{R}\left(\frac{1}{\sqrt{5}} - 1\right) = \frac{\sqrt{5} - 1}{\sqrt{5}}\frac{kQ}{R}$$

12. (c) Potential due to small length dl of rod at centre $= dV = \frac{kdq}{r}$

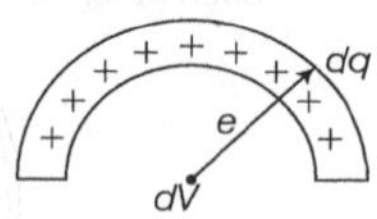

$$= \frac{k\lambda dl}{r}$$

Thus, potential due to complete charge distribution is

$$V = \int dV = \frac{k\lambda}{r}\int dl$$

$$= \frac{k\lambda}{r} \cdot \pi r = k\lambda\pi = k\frac{Q}{l}\pi\ \text{V}$$

13. (d) Using result of a charged ring, potential due to a ring of area dA at P is

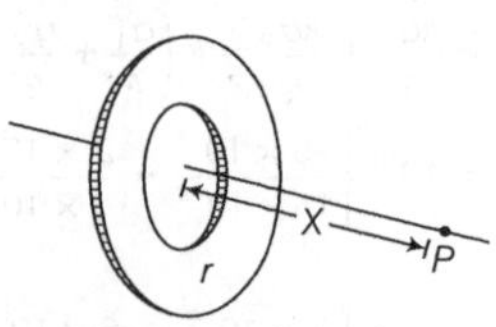

$$dV = \frac{kdq}{\sqrt{r^2 + x^2}}$$

$$dq = \sigma dA = \sigma 2\pi r dr$$

$$\therefore \quad V = \int_a^b 2\pi\sigma k\left(\frac{rdr}{\sqrt{r^2 + x^2}}\right)$$

$$= 2\pi\sigma k\sqrt{x^2 + b^2} - \sqrt{x^2 + a^2}$$

14. (b) Potential at a height H on the axis of the disc $V(P)$.
The charge dq contained in the ring shown in figure,

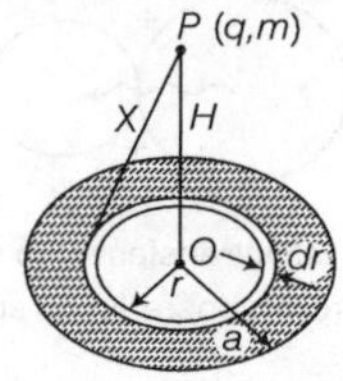

$$dq = (2\pi r dr)\sigma$$

Potential at P due to this ring

$$dV = \frac{1}{4\pi\varepsilon_0} \cdot \frac{dq}{x}, \quad \text{where, } x = \sqrt{H^2 + r^2}$$

$$dV = \frac{1}{4\pi\varepsilon_0} \cdot \frac{(2\pi r dr)\sigma}{\sqrt{H^2 + r^2}} = \frac{\sigma}{2\varepsilon_0}\frac{rdr}{\sqrt{H^2 + r^2}}$$

$\therefore$ Potential due to the complete disc

$$V_P = \int_{r=0}^{r=a} dV = \frac{\sigma}{2\varepsilon_0}\int_{r=0}^{r=a}\frac{rdr}{\sqrt{H^2 + r^2}}$$

$$V_P = \frac{\sigma}{2\varepsilon_0}[\sqrt{a^2 + H^2} - H] \text{ and } V_0 = \frac{\sigma_a}{2\varepsilon_0}$$

Particle is released from point P and it just reaches point O. Therefore, from conservation of mechanical energy.

Decrease in gravitational potential energy

= Increase in electrostatic potential energy.

(ΔKE = 0, because $K_i = K_f = 0$)

$$\therefore \quad mgH = q[V_O - V_P]$$

$$\text{or} \quad gH = \left(\frac{q}{m}\right)\left(\frac{\sigma}{2\varepsilon_0}\right)[a - \sqrt{a^2 + H^2} + H] \quad \ldots\text{(i)}$$

$$\text{As,} \quad \frac{q}{m} = \frac{4\varepsilon_0 g}{\sigma}$$

$$\therefore \quad \frac{q\sigma}{2\varepsilon_0 m} = 2g$$

Substituting in Eq. (i), we get

$$gH = 2g\ [a + H - \sqrt{a^2 + H^2}]$$

$$\text{or} \quad \frac{H}{2} = (a + H) - \sqrt{a^2 + H^2}$$

$$\text{or} \quad \sqrt{a^2 + H^2} = a + \frac{H}{2}$$

$$\text{or} \quad a^2 + H^2 = a^2 + \frac{H^2}{4} + aH$$

$$\text{or} \quad \frac{3}{4}H^2 = aH \quad \text{or} \quad H = \frac{4}{3}a$$

$$\text{and} \quad H = 0$$

$$\therefore \quad H = (4/3)a$$

15 (a) Let q be the charge on the bubble, then

$$V = \frac{kq}{a} \quad \left(\text{Here, } k = \frac{1}{4\pi\varepsilon_0}\right)$$

$$\therefore \quad q = \frac{Va}{k}$$

Let after collapsing, the radius of droplet becomes R, then equating the volume, we have

$$(4\pi a^2)t = \frac{4}{3}\pi R^3$$

$$\therefore \quad R = (3\,a^2 t)^{1/3}$$

Now, potential of droplet will be $V' = \frac{kq}{R}$

Substituting the values, we have

$$V' = \frac{(k)\left(\frac{Va}{k}\right)}{(3\,a^2 t)^{1/3}} \quad \text{or} \quad V' = V\left(\frac{a}{3t}\right)^{1/3}$$

16. (a,b)

(a) Work done by the field in taking a test charge q_0 from point A to point B.

$$W_{A\to B} = -(U_B - U_A)$$
$$= -q_0(V_B - V_A)$$
$$\Rightarrow \quad U_B - U_A = q_0(V_B - V_A)$$

Also, work done by external force

$$= -\{-q_0(V_B - V_A)\}$$
$$= q_0\left(\left\{\frac{kq_1}{r_{B_1}} + \frac{kq_2}{r_{B_2}}\right\} - \left\{\frac{kq_1}{r_{1A}} + \frac{kq_2}{r_{2A}}\right\}\right)$$

(b) Here, $\quad q_0 = -1.5 \times 10^{-6}\text{C}$

$$r_{B_1} = \sqrt{12^2 + 14^2} = 18.44\text{cm}$$
$$r_{B_2} = \sqrt{14^2 + 24^2} = 27.8\text{ cm}$$
$$r_{1A} = 0.12\text{m},\ r_{2A} = 0.24\text{ m}$$

Substituting data, we get

$W_{B\to A}$(external force) $= +1.5$ J

17. (b,c) All eight charges are at a distance of $\frac{\sqrt{3}}{2}l$ from centre of cube.

$$\therefore \quad V_{\text{centre}} = 8\frac{kq}{r} = 8 \times \frac{1}{4\pi\varepsilon_0} \times \frac{Q}{(\sqrt{3}/2\,l)}$$
$$= \frac{16}{\sqrt{3}} \times \frac{1}{4\pi\varepsilon_0} \times \frac{Q}{l}\text{ V}$$

For seven charges producing a potential at a corner point, three at a distance l, three at a distance $\sqrt{2}\,l$ and one at a distance $\sqrt{3}\,l$ from the corner.

$$\therefore \quad V_{\text{corner}} = 3 \times \frac{1}{4\pi\varepsilon_0} \times \frac{Q}{l} + 3 \times \frac{1}{4\pi\varepsilon_0} \times \frac{Q}{\sqrt{2}\,l} + \frac{1}{4\pi\varepsilon_0} \times \frac{Q}{\sqrt{3}\,l}$$
$$= \left(3 + \frac{3}{\sqrt{2}} + \frac{1}{\sqrt{3}}\right) \times \frac{1}{4\pi\varepsilon_0} \times \frac{Q}{l}$$

18. (a,c) Using, $V = k\Sigma\frac{q_i}{r_i}$

we have,

$$V_A = k\left(\frac{-Q}{l/2}\right) + k\left(\frac{-3Q}{l/2}\right) + \frac{kQ}{\left(\frac{\sqrt{3}}{2}l\right)}$$
$$\Rightarrow \quad V_A = k\frac{2Q}{l}\left(-4 + \frac{1}{\sqrt{3}}\right)$$
$$= \frac{Q}{\pi\varepsilon_0 l}\left(\frac{\sqrt{3}}{6} - 2\right)\text{V}$$
$$V_C = k\frac{Q}{l/2} + k\left(\frac{-3Q}{l/2}\right) + \frac{k(-Q)}{\sqrt{3}\cdot l/2}$$
$$= \frac{-Q}{\pi\varepsilon_0 l}\left(1 + \frac{\sqrt{3}}{6}\right)\text{V}$$

19. (a,c,d) We are not calculating it, but **E** is greatest at $r = r_2$.

$\therefore$ Potential at $r = r_2$ surface is greatest.

And inside the cavity, $E = 0$

V = constant and it has same value at it has on the boundary (at $r = r_1$)

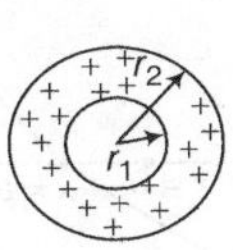

Also, for all exterior points charge distribution acts like a point charge, so

$$V \propto \frac{1}{r}$$

20. (a) Both Statement I and Statement II are true and Statement II explains Statement I.

U = total electrostatic potential energy of three charge system

$$= \frac{kq_1q_2}{r_{12}} + \frac{kq_1q_3}{r_{13}} + \frac{kq_2q_3}{r_{23}} = k\left(\frac{Qq}{\sqrt{2}\,a} + \frac{Qq}{a} + \frac{q^2}{a}\right) = 0$$

(given in problem)

$$\Rightarrow \quad Q = \frac{-2q}{2 + \sqrt{2}}$$

21. (d) Surface potential value also depends on type of charge distribution pattern on the conductor, not only on volume and total charge of the conductor. (except for spherical conductors), so Statement I is incorrect and Statement II is correct.

22. (b) In given problem,

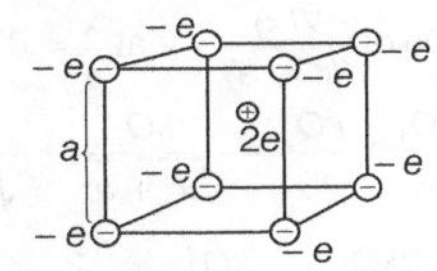

(i) Cube has got 12 pairs separated by distance a,

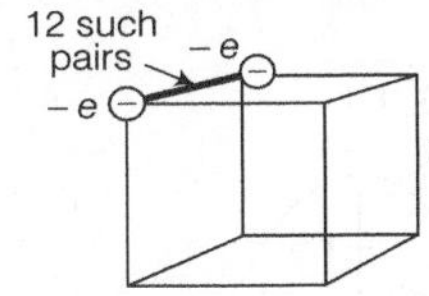

$$U = \frac{k12e^2}{a}$$

(ii) Cube has 8 pairs separated by distance $\sqrt{3}/2\,a$.

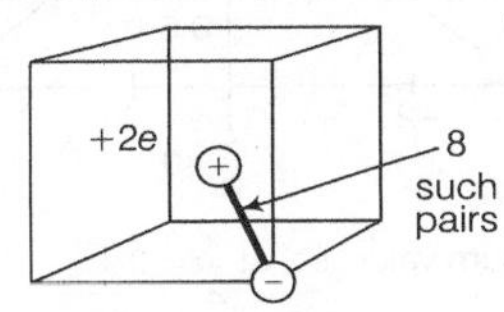

$$U = \frac{k8(-2e^2)}{(\sqrt{3}/2\ a)}$$
$$U = -18.475\frac{ke^2}{a}$$

(iii) Cube has 12 pairs at distance $\sqrt{2}\,a$.

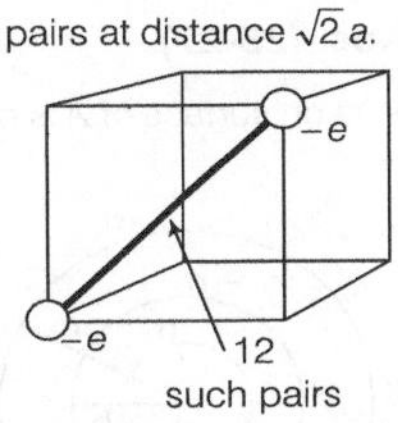

$$U = \frac{12k(e^2)}{\sqrt{2}a} = 8.485\frac{ke^2}{a}$$

(iv) Cube has 4 pairs separated by distance $\sqrt{3}\,a$.

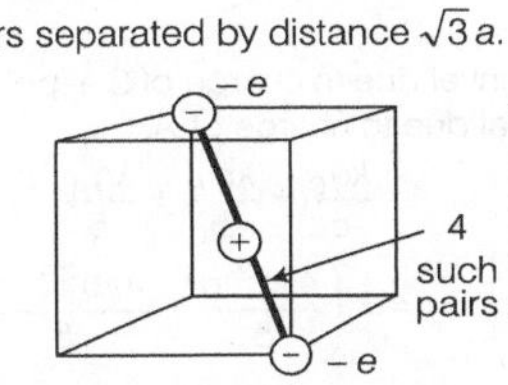

$$U = \frac{4ke^2}{(\sqrt{3}\,a)}$$
$$U = 2.309\frac{ke^2}{a}$$

(i) → (q), (ii) → (p), (iii) → (r), (iv) → (s)

23. (2) Total charge enclosed for any radius r is

$$Q_{\text{enclosed}} = \int \rho_E\, dV$$

(dV = elemental volume)

$$\int_0^r \rho_0\left(1 - \frac{r^2}{r_0^2}\right)4\pi r^2 dr = 4\pi\rho_0\int_0^r\left(r^2 - \frac{r^4}{r_0^2}\right)dr$$
$$= 4\pi\rho_0\left(\frac{r^3}{3} - \frac{r^5}{5r_0^2}\right)$$

when $r = r_0$

Q_{total} (inside the sphere) $= 4\pi\rho_0\left(\frac{r_0^3}{3} - \frac{r_0^5}{5r_0^2}\right) = \frac{8\pi\rho_0 r_0^3}{15}$

So, $\quad V(r = r_0) = \frac{kQ_{total}}{r_0}$

(As, a spherical charge distributing may be treated as a point charge concentrated at the centre)

So, $\quad V(r = r_0) = \frac{k\left(\frac{8\pi\rho_0 r_0^3}{15}\right)}{r_0}$

$$= \frac{1}{4\pi\varepsilon_0} \times \frac{8\pi\rho_0 r_0^3}{15 r_0} = 2\left(\frac{\rho_0 r_0^2}{15\varepsilon_0}\right)$$

24. (2) Maximum value of $\frac{V(x)}{(kQ/a)} = 2$ at $\frac{x}{a} = 0$

$$V(x) = \frac{kQ_1}{\varepsilon_1} + \frac{kQ_2}{\varepsilon_2} = \frac{kQ}{\sqrt{x^2 + a^2}} + \frac{kQ}{\sqrt{x^2 + (-a)^2}}$$

$$= \frac{2kQ}{\sqrt{x^2 + a^2}} = \frac{kQ}{a}\left(\frac{2}{\sqrt{(x/a)^2 + 1}}\right)$$

So, $\frac{V(x)}{(kQ/a)} = \frac{2}{\sqrt{(x/a)^2 + 1}}$

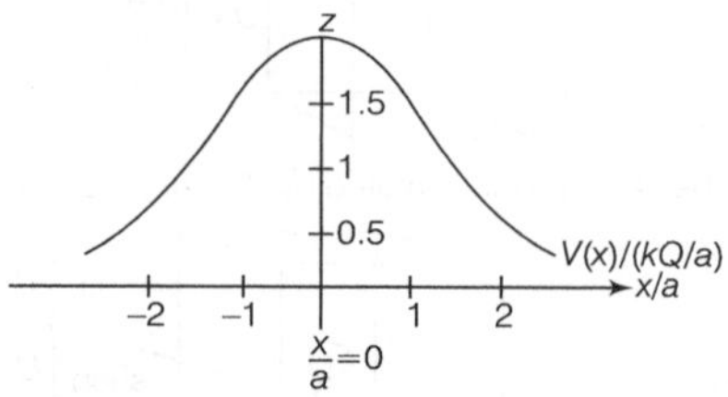

So, maximum value is 2 at $\frac{x}{2} = 0$.

Explanations for Q. Nos. (25-27)

25. (c) Potential at a point on surface of A is due to charge of A, B and C.

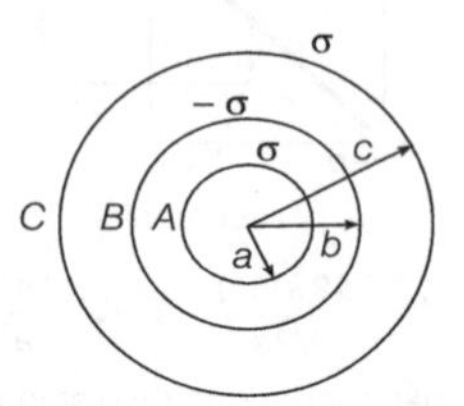

V_A = potential due to charge of C + potential due to charge of B + potential due to charge of A.

$$= \frac{kq_C}{c} + \frac{kq_B}{b} + \frac{kq_A}{a}$$

$$= k\left(\frac{4\pi c^2.\sigma}{c} + \frac{4\pi b^2.(-\sigma)}{b} + \frac{4\pi a^2.\sigma}{a}\right)$$

$$= \frac{1}{4\pi\varepsilon_0}\cdot 4\pi(\sigma c + \sigma a - \sigma b) = \frac{\sigma}{\varepsilon_0}(a - b + c)$$

26. (d) Potential on surface of B is

$$V_B = \frac{1}{4\pi\varepsilon_0}\left(\frac{q_A}{b} + \frac{q_B}{b} + \frac{q_C}{c}\right)$$

$$= \frac{1}{4\pi\varepsilon_0}\left(\frac{4\pi a^2\sigma}{b} - \frac{4\pi b^2\sigma}{b} + \frac{4\pi c^2\sigma}{c}\right)$$

$$= \frac{\sigma}{\varepsilon_0}\left(\frac{a^2}{b} - b + c\right)$$

27. (c) If $V_C = V_A$, then

$$\Rightarrow \quad \frac{1}{4\pi\varepsilon_0}\left(\frac{q_A}{c} + \frac{q_B}{c} + \frac{q_C}{c}\right) = \frac{1}{4\pi\varepsilon_0}\left(\frac{q_C}{c} + \frac{q_B}{b} + \frac{q_A}{a}\right)$$

$$\Rightarrow \quad \frac{q_A}{c} + \frac{q_B}{c} = \frac{q_B}{b} + \frac{q_A}{a}$$

$$\Rightarrow \quad a - b = \frac{a^2}{c} - \frac{b^2}{c}$$

$$\therefore \quad a + b = c$$

28. (b) For any point $P(x, y)$ on XY-plane, electric potential at P is

V = potential due to Q + potential due to $-2Q$

$$\Rightarrow \quad V = k\left(\frac{Q}{\sqrt{(3a - x)^2 + y^2}}\right) + k\left(\frac{-2Q}{\sqrt{(3a + x)^2 + y^2}}\right) \qquad \ldots\text{(i)}$$

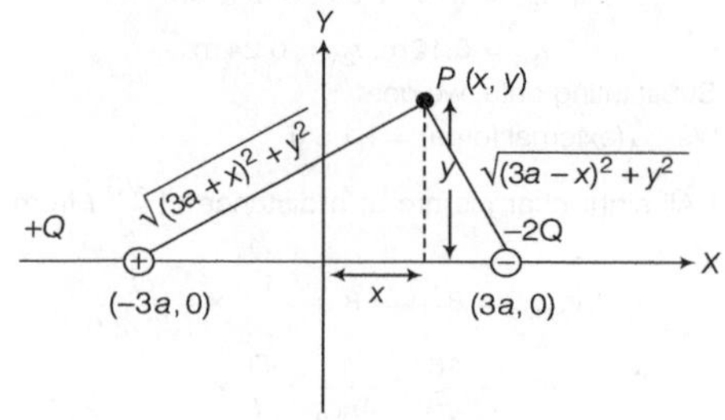

If $V_p = 0$, then

$$4(3a - x)^2 + y^2 = (3a + x)^2 + y^2$$

$$\Rightarrow \quad (x - 5a)^2 + y^2 = 16a^2$$

$\therefore$ This is a circle with centre at $(5a, 0)$ and radius 4.

29. (b) Potential at point $(5a, 0) = \frac{Q}{4\pi\varepsilon_0}\left(\frac{1}{2a} - \frac{2}{8a}\right) = \frac{Q}{16\pi\varepsilon_0 a}$

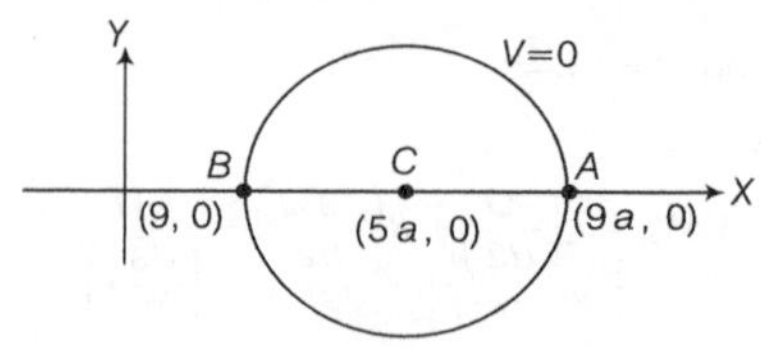

and potential at point $(9a, 0) = 0$

So, when charge is released from centre C, it moves to B i.e. from higher potential to lower potential.

30. (b) On X- axis for $-3a < x < 3a$, we can find potential function $V(x)$ by putting $y = 0$ in Eq. (i) given in sol. 28.

$$\therefore \quad V(x) = \frac{Q}{4\pi\varepsilon_0}\left(\frac{1}{3a - x} - \frac{2}{3a + x}\right)$$

$\Rightarrow \quad V(x) = 0$ at $x = a$

Also, $\quad V \to -\infty$ at $x \to -3a$

and $\quad V \to +\infty$ at $x \to 3a$

Hence, there is no maxima or minima at $x = a$ and $x = a$ is a point of inflexion.

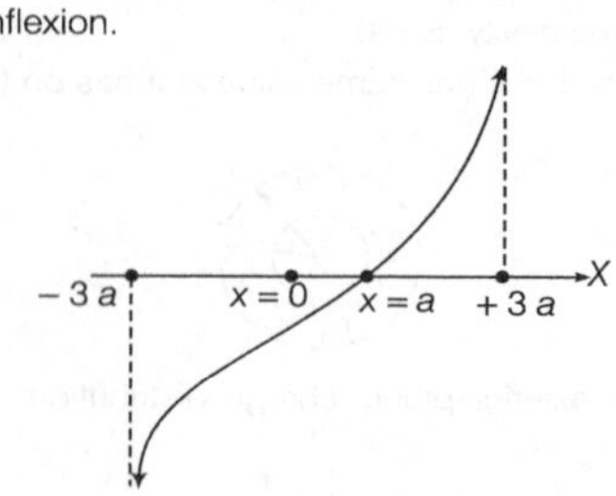

DPP-9 Motion of a Charged Particle inside an Electric Field

1. (i) A charged particle will move along a field line if its initial velocity is zero. Then, acceleration is directed along the tangent at that point on the field line.

But if particle has a non-zero initial velocity, then it moves under the action of two velocities, so it does not follow a field line.

(ii) When a charged particle moves from some point A to B in a field,

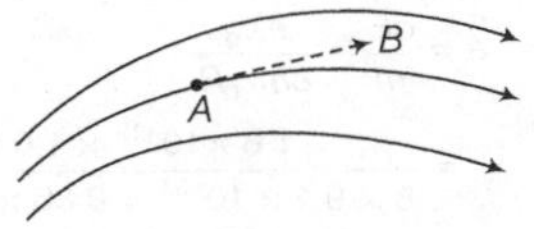

Then, energy balance equation is

$$K_i + U_i + \Delta E_{mech} = K_f + U_f$$

Where, K and U are kinetic and potential energies

and ΔE_{mech} = External mechanical work done on the particle.

2. Potential due to an elemental length dl having charge dq on point P on axis is

$$dV = \frac{kdq}{r} = \frac{k\lambda\, dl}{\sqrt{a^2 + x^2}}$$

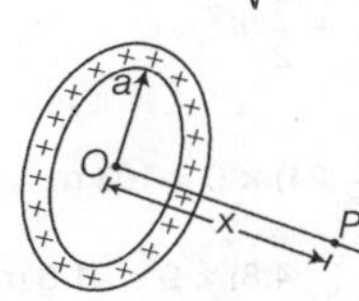

$\therefore V_1$ = potential due to complete ring of charge

$$= \int dV = \int \frac{k\lambda dl}{\sqrt{a^2 + x^2}} = \frac{k\,\lambda(2\pi a)}{\sqrt{a^2 + x^2}} \qquad (\because \int dl = 2\pi a)$$

At centre of ring $x = 0$

$\therefore \qquad V_2$ = potential at centre $= k \cdot 2\pi\lambda$

When electron reaches from P to centre O of ring, its total energy will be conserved.

$$\therefore \qquad K_i + (-eV_i) = K_f + (-eV_f)$$

$$0 - eV_1 = \frac{1}{2}m_e v_f^2 - eV_2$$

$$\Rightarrow \quad v_f^2 = \frac{2e}{m_e}(V_2 - V_1)$$

$$\Rightarrow \quad v_f^2 = \frac{4\pi ek\lambda}{m_e}\left(1 - \frac{a}{\sqrt{x^2 + a^2}}\right)$$

$$= \frac{4\pi(1.6 \times 10^{-19})(9 \times 10^9)(1 \times 10^{-7})}{9.1 \times 10^{-31}}\left(1 - \frac{0.2}{\sqrt{0.1^2 + 0.2^2}}\right)$$

$$= 1.45 \times 10^7 \text{ m/s}$$

3. $q = 1\,\mu\text{C} = 10^{-6}$ C

$$m = 2 \times 10^{-3} \text{ kg}$$

$$l = 0.8 \text{ m}$$

Let u be the speed of particle at lowest position and v is the speed at highest position.

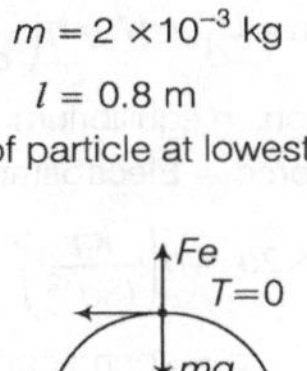

At highest position, 3 forces are acting on the particle.

$$F_e = \frac{1}{4\pi\varepsilon_0}\frac{q^2}{l^2} \qquad \text{(outward)}$$

$$\text{Weight } w = mg \qquad \text{(inward)}$$

and $\qquad$ tension $T = mv^2/l \qquad$ (inward)

For minimum velocity at highest point $T = 0$

$$F_e - mg = \frac{mv^2}{l}$$

$$v^2 = \frac{l}{m}\left(mg - \frac{1}{4\pi\varepsilon_0}\frac{q^2}{l^2}\right)$$

$$v^2 = \frac{0.8}{2 \times 10^{-3}} \times \frac{(2 \times 10^{-3} - 9 \times 10^9 \times (10)^{-6})^2}{(0.8)^2}$$

$$v = \sqrt{2.4} \text{ m/s}$$

Now, the electrostatic potential energy at lowest and highest points are equal. Hence, from conservation of energy,

Increase in gravitational potential energy

= Decrease in kinetic energy

$$mg(2l) = \frac{1}{2}m(u^2 - v^2)$$

$$u^2 = v^2 + 4gl$$

$$u^2 = (\sqrt{2.4})^2 + 4 \times 10 \times 0.8 = 34.4$$

$$u = 5.86 \text{ m/s}$$

So, minimum horizontal velocity to the lower ball, so that it can complete revolution is 5.86 m/s.

4. (i) Lower plate is given positively charged, so force on electron is downwards.

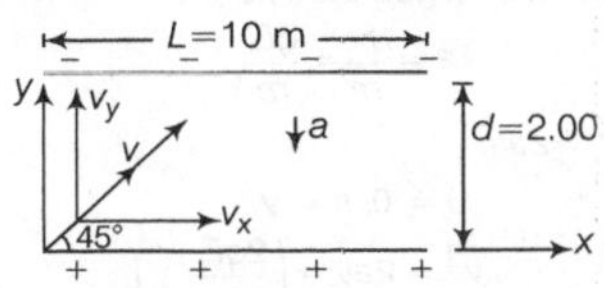

Velocity $v = 6 \times 10^6$ m/s, $\theta = 45°$ velocity of electron can be resolved into parallel (v_x) and perpendicular (v_y) components where v_x is constant and v_y is accelerated due to the field. Acceleration of electron is

$$a_y = \frac{F_y}{m} = \frac{eE}{m}$$

After t second of projection of electron, taking projection point as origin, let electron is at (x, y) point, then

$$x = v_x t$$

and $$y = v_y t - \frac{1}{2}a_y t^2$$

Using $v_x = v\cos\theta$ and $v_y = v\sin\theta$,

we get

$$x = \left(6 \times 10^6 \times \frac{1}{\sqrt{2}}\right)t = 3\sqrt{2} \times 10^6 t \qquad \ldots\text{(i)}$$

and $$y = 3\sqrt{2} \times 10^6 t - \frac{1}{2}\frac{1.6 \times 10^{-19} \times 2 \times 10^3}{9.1 \times 10^{-31}}t^2 \qquad \ldots\text{(ii)}$$

Eliminating t, we get

$$y = 3\sqrt{2} \times 10^6\left(\frac{x}{3\sqrt{2} \times 10^6}\right) - \frac{1}{2}\frac{1.6 \times 10^{-19} \times 2 \times 10^3}{9.1 \times 10^{-31}} \times \frac{x^2}{9 \times 2 \times 10^{12}}$$

or $\qquad y = x - 50x^2$

(A slight approximation taken to keep calculation simple).

(ii) Now, equation of upper plate is $y = 2 \times 10^{-2}$.

Substituting in equation of path, we get

$2 \times 10^{-2} = x - 50x^2$ or $50x^2 - x + 2 \times 10^{-2} = 0$

This equation has a negative discriminant, so no real root means electron does not strike upper plate.

For lower plate, $y = 0$

$\Rightarrow \quad x - 50x^2 = 0 \Rightarrow x(1 - 50x) = 0$

$\Rightarrow \quad x = 0$ (which is point of projection)

and $\quad x = \frac{1}{50}\text{ m} = \frac{100}{50} = 2 \text{ cm}$

So, electron strikes first plate at $x = 2$ cm point from the point of projection.

5. (b) Acceleration of electron $a_e = \frac{F}{m_e} = \frac{eE}{m_e}$

Acceleration of proton $a_p = \frac{F}{m_p} = \frac{eE}{m_p}$

Using, $s = ut + \frac{1}{2}at^2$ and $u = 0$

$$\Rightarrow \quad s = \frac{1}{2}at^2$$

$$s_e = \frac{1}{2}a_e t_1^2 = \frac{1}{2}\frac{eE}{m_e}\cdot t_1^2$$

$$s_p = \frac{1}{2}a_p t_2^2 = \frac{1}{2}\frac{eE}{m_p}\cdot t_2^2$$

$$s_e = s_p$$

$$\therefore \quad \frac{eE}{m_e}t_1^2 = \frac{eE}{m_p}t_2^2$$

$$\Rightarrow \quad \frac{t_1^2}{t_2^2} = \frac{m_e}{m_p} \Rightarrow \frac{t_1}{t_2} = \sqrt{\frac{m_e}{m_p}}$$

6. (c) Acceleration of charged particle

$$a = \frac{F}{m} = \frac{qE}{m}$$

Using $v^2 - u^2 = 2as$,

Here, $\quad u = 0, s = y$

$$v^2 = 2ay = \left(\frac{2qE}{m}\cdot y\right)$$

KE of particle after covering a distance y

$$= \frac{1}{2}mv^2 = \frac{1}{2}m\cdot\frac{2qE}{m}\cdot y = qEy$$

7. (c) Potential energy of particle $= eV$

This PE is fully converted into KE

$$\frac{1}{2}mv^2 = eV \Rightarrow v^2 = \frac{2eV}{m}$$

$$v = \sqrt{\frac{2eV}{m}}$$

8. (a) Consider an element of charge dQ at same angle θ with X-axis.

Magnitude of field element $dE = k\frac{\lambda R d\theta}{R^2}$

As charge distribution is $\lambda = \lambda_0 \sin\theta$ and $\sin(-\theta) = -\sin\theta$, lower half or ring is negatively charged.

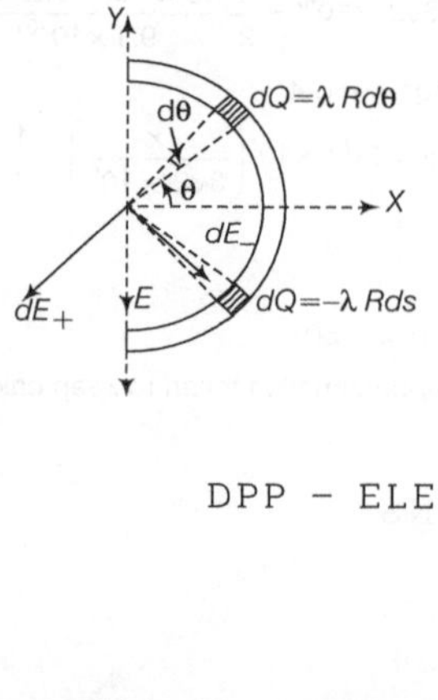

So, **E** has only y-component,

Now, $\quad dE = \frac{k\lambda R d\theta}{R^2}\sin\theta$

$$= \frac{k\lambda_0 \sin^2\theta d\theta}{R}$$

$$E = \int_{-\pi/2}^{\pi/2}\frac{k\cdot\lambda_0}{R}\sin^2\theta\, d\theta$$

$$= \frac{k\lambda_0}{R}\left(\frac{1}{2}\theta - \frac{1}{4}\sin 2\theta\right)_{-\pi/2}^{\pi/2} = \frac{\lambda_0}{8\varepsilon_0 R}$$

For an electron, $F = ma$

$$\Rightarrow \quad qE = ma$$

$$\Rightarrow \quad a = \frac{qE}{m} = \frac{e\lambda_0}{8m\varepsilon_0 R}$$

$$= \frac{1.6 \times 10^{-19} \times 1.0 \times 10^{-6}}{8 \times 9.1 \times 10^{-31} \times 8.85 \times 10^{-12} \times 0.01}$$

$$= 2.5 \times 10^{17} \text{ m/s}^2$$

Solving these, we get $q = q_e$

9. (d) Acceleration of block

$$a = \frac{qE}{m} = \frac{8 \times 10^{-5}(3 \times 10^3\hat{i} - 600\hat{j})}{10 \times 10^{-3}}$$

$$= 8 \times 10^{-3}\{3 \times 10^3\,\hat{i} - 600\,\hat{j}\}$$

$$= (24\,\hat{i} - 4.8\,\hat{j})\text{ m/s}^2$$

After 3 s, using $s = ut + \frac{1}{2}at^2$

$u = 0$, we get

$$x = \frac{1}{2}(24) \times 9 = 108 \text{ m}$$

$$y = \frac{1}{2}(-4.8) \times 9 = 21.6 \text{ m}$$

10. (b) Using energy conservation,

$$\frac{keQ}{r_1} = \frac{keQ}{r_2} + \frac{1}{2}mv^2$$

$$\Rightarrow \quad v = \sqrt{\frac{2keQ}{m}\left(\frac{1}{r_1} - \frac{1}{r_2}\right)}$$

$$= \sqrt{\frac{(2)(9 \times 10^9)(-1.6 \times 10^{-19})(10^{-9})}{(9.11 \times 10^{-3})}} \times \sqrt{\left(\frac{1}{0.03} - \frac{1}{0.02}\right)}$$

$$= 7.26 \times 10^6 \text{ m/s}$$

11. (c) Each charge moves off on its diagonal line. All charges have equal speeds.

$$\Sigma(K + U)_{\text{initially}} = \Sigma(K + U)_{\text{finally}}$$

$$0 + \frac{4kq^2}{L_1} + \frac{2kq^2}{\sqrt{2}L_1} = 4\left(\frac{1}{2}mv^2\right) + \frac{4kq^2}{2L_1} + \frac{2kq^2}{2\sqrt{2}L_1}$$

$$\Rightarrow \quad \left(2 + \frac{1}{\sqrt{2}}\right)\frac{kq^2}{L_1} = 2mv^2$$

$$\Rightarrow \quad v = \sqrt{\left(1 + \frac{1}{\sqrt{8}}\right)\frac{kq^2}{mL_1}}$$

12. (d) Initial PE of system $= U_1 = qV = q\left(\frac{kq}{d}\right)$

In the final configuration, in equilibrium

Spring force = Electrostatic force

$$c \times 2d = q\left(\frac{kq}{(3d)^2}\right)$$

c = spring constant

$$\Rightarrow \quad c = \frac{kq}{18d^3}$$

In final configuration, the total PE is

$$\frac{1}{2}cx^2 + qV = \frac{1}{2}\left(\frac{kq^2}{18d^3}\right)(2d)^2 + \frac{qkq}{3d} = \frac{4}{9}\frac{kq^2}{d}$$

As final PE is less than initial PE and table is frictionless, energy cannot be lost in form of heat. Also, internal kinetic friction of spring causes increase in internal energy.

As system is isolated,

$$\frac{kq^2}{d} = \frac{4\,kq^2}{9d} + \Delta E_{int}$$

$$\Rightarrow \quad \Delta E_{int} = \frac{5}{9}\left(\frac{kq^2}{d}\right)$$

13. (b,c) If charge is positive, force is along the tangent to field line.

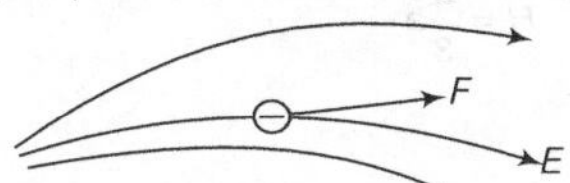

$\therefore$ Acceleration of charge is parallel to force.

Also, charge will move along the field line, if $u = 0$

14. (a,d) Acceleration of electron is directed opposite to the velocity.

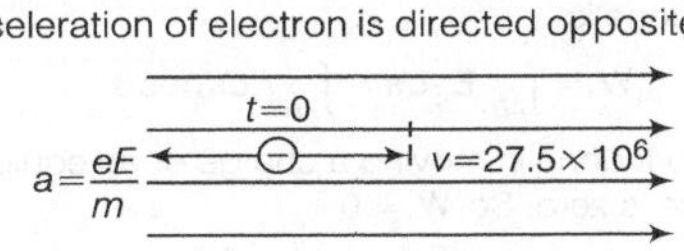

$\therefore \quad u = 27.5 \times 10^6$ m/s

$v = 0$

Using $\quad v^2 - u^2 = 2as$

we get,

$$s = \frac{x^2}{2a} = 0.18 \text{ m}$$

To calculate t, we use $v = u + at$

t = time in which electron comes to rest

where, $\quad v = 0$

$u = 27.5 \times 10^6$ m/s

$$a = \frac{eE}{m}$$

We get total time to return to starting point

$= 2t = 2.89 \times 10^{-8}$ s

15. (a,c) As field is in XOY-plane.

$\therefore$ Force vector is in XOY- plane and so electrons acceleration is

$$a = \frac{eE}{m}$$

$$= \frac{e}{m}(2\hat{i} + 8\hat{j}) \times 10^4 \text{ m/s}$$

which is in XOY - plane.

Also, final velocity is $v = v_0 + at$, clearly it is also in XOY - plane.

16. (a,c) Find acceleration of electron $a = \frac{eE}{m}$

and then use, $s = ut + \frac{1}{2}at^2$ by separating motion along X-axis and Y-axis to find

$x = 3.2$ m after $t = 2.00$ s

and $\quad y = -4.2$ m after $t = 2.00$ s

17. (b,d) For a small displacement x of positive charge, $x << R$.

$$F = qE = \frac{q}{4\pi\varepsilon_0}\frac{-Qx}{(x^2 + R^2)^{3/2}}$$

$$\Rightarrow \quad F = \frac{-qQx}{4\pi\varepsilon_0 R^3}$$

$$\therefore k \text{ (spring constant)} = \frac{Qq}{4\pi\varepsilon_0 R^3}$$

and $\quad T = 2\pi\sqrt{\frac{m}{k}} = 4\pi\sqrt{\frac{m\pi\varepsilon_0 R^3}{Qq}}$

18. (d) Force on an α-particle in field = $2eE$

Force on a deuteron in same field = $e\,E$

So, Statement I is incorrect and Statement II is correct.

19. (d)

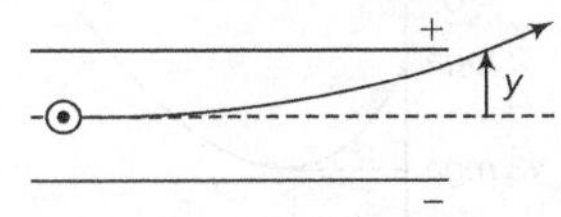

Deflection y is given by $s = \frac{1}{2}\left(\frac{qE}{m}\right)t^2 = \frac{1}{2}\frac{qE}{m}\left(\frac{L}{v^2}\right)$

$\Rightarrow y \propto \frac{q}{m}$, so Statement I is incorrect as deflection is proportional to specific charge of the particle.

20. (2) Field between the plates $= \frac{\sigma}{\varepsilon_0}$

$$= \frac{36 \times 10^{-9}}{2 \times 8.85 \times 10^{-12}}$$

$$= 4.07 \text{ kN/C}$$

Using integral form of $\frac{dV}{dr} = -E$ and taking $V = 0$ at the negative plates,

$$V - 0 = -\int_0^{12.0 \text{ cm}} -4.07\,dx$$

$$= 4.07 \times 10^3 \times 0.12$$

$$= 488 \text{ v}$$

Using energy conservation for charged particle

$$\left(\frac{1}{2}mv^2 + qV\right)_{\text{initial}} = \left(\frac{1}{2}mv^2 + qV\right)_{\text{final}}$$

$$\Rightarrow \quad 1.6 \times 10^{-19} \times 488 = \frac{1}{2}\text{m}\,v_f^2$$

$$\Rightarrow \quad v_f = 306 \text{ km/s} = 2 \times 102 \text{ km/s},$$

$\therefore \quad N = 2$

21. (b) Potential due to an elemental ring of thickness 'dr' on its axis is

$$dV = \frac{1}{4\pi\varepsilon_0}\cdot\frac{dq}{x}$$

$$x = \sqrt{H^2 + r^2}$$

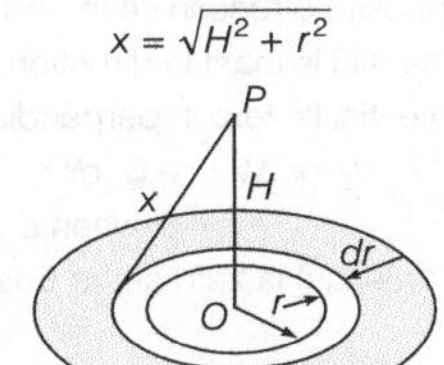

And $\quad dq = 2\pi r dr \sigma$

$$\therefore \quad dV = \frac{\sigma}{2\varepsilon_0}\cdot\frac{rdr}{\sqrt{H^2 + r^2}}$$

Potential due to complete disc

$$V = \int_{r=0}^{a} dV = \frac{\sigma}{2\varepsilon_0}\int_{r=0}^{a}\frac{rdr}{\sqrt{H^2 + r^2}}$$

$$\Rightarrow \quad V_{\text{at }P} = \frac{\sigma}{2\varepsilon_0}\{\sqrt{a^2 + H^2} - H^2\}$$

and potential at centre $V_0 = \frac{\sigma a}{2\varepsilon_0}$

$\therefore \quad H = 0$

22. (d) U-H equation is

$$U = mg(2\sqrt{a^2 + H^2} - H)$$

which is parabolic and $U = 2mga$

at $\quad H = 0$

Also, at $H = \frac{a}{\sqrt{3}}, U = U_{\min}$

$\therefore$ Graph will be

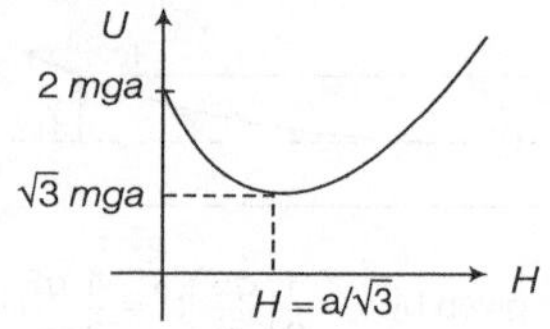

23. (d) From P to O,

Decrease in GPE = Increase in EPE

$\therefore \quad mgH = q(V_0 - V_p)$

$\Rightarrow \quad gH = \left(\frac{q}{m}\right)\left(\frac{\sigma}{2\varepsilon_0}\right)(a - \sqrt{a^2 + H^2} + H)$

As, $\quad \frac{q}{m} = \frac{4\varepsilon_0 g}{\sigma}$

We get,

$\Rightarrow \quad \frac{H}{2} = (a + H) - \sqrt{a^2 + H^2}$

or $\quad \sqrt{(a^2 + H^2)} = a + \frac{H}{2}$

$\therefore \quad H = \frac{4}{3}a$

DPP-10 Electric Field, Potential and Equipotentials

1. (i) No, electric field cannot be calculated by knowing only potential and potential cannot be calculated by knowing only field e.g. if electric potential between two parallel dissimilarily charged plates is known, then to calculate field distance between plates must be known

$$V = Ed$$

Generally to find V, **E** must be known and it must be integrated. To find **E**, V must be known and it must have a derivative. Thus, we need **E** and V in the region around the point and not just at the point in order to be able to find the other variable.

If electric field is uniform in the region, then electric potential is increasing or decreasing uniformly in the region

$$\frac{d\mathbf{V}}{dr} = -\mathbf{E}$$

(ii) If the electric potential of a region of space is uniform, then field must be zero in that region.

$$E = -\frac{d\mathbf{V}}{dr} = 0$$

when $\quad V = \text{constant}$

(iii) From $\frac{d\mathbf{V}}{dr} = -\mathbf{E}$

If field is pointing towards North, then change in potential is greatest in opposite direction i.e in South.

Change in potential is least in direction of field, i.e. in North.

Change in potential is zero in perpendicular direction to field.

$$dV \text{ or } \Delta V = -\mathbf{E}\cdot d\mathbf{r}$$
$$= 0 \text{ when } E \perp dr$$

$\therefore$ Change in potential is zero along East-West direction.

2. (i)

(a) An equipotential is locus of all points in space with same potential. If they cross, intersection point having two different potentials simultaneously, it will become a condition which is not possible.

(b) In region of space where equipotential lines are closely packed, the field is stronger than the region of space where equipotentials are far apart.

(c) Consider an equipotential and let field is inclined at some angle θ to it.

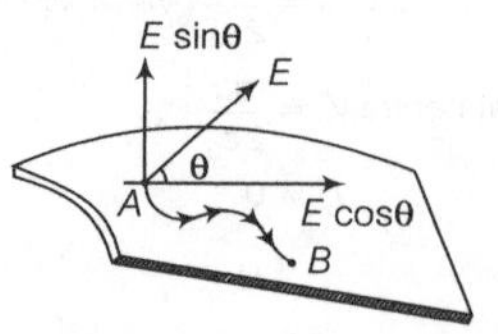

The **E** has a component $E\cos\theta$ parallel to the surface, so when a charge q at A is displaced to some another point B, work done is

$$W = \int_{AB} \mathbf{E}\cdot d\mathbf{s} = \int_{AB} Eds\cos\theta$$

But work done in moving a charge or an equipotential surface is zero. So, $W = 0$

$\Rightarrow \quad Eds\cos\theta = 0$

$\Rightarrow \quad \cos\theta = 0$

$\Rightarrow \quad \theta = 90°$

Hence, field lines are always perpendicular to an equipotential surface.

(ii)

(a) For a negatively charged hollow sphere, no equipotential exists inside and field points towards the sphere, so equipotentials are more closely packed near the surface.

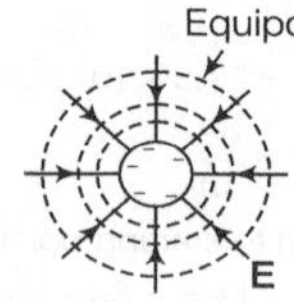

(b) Field near plate is $\frac{\sigma}{2\varepsilon_0}$.

$\therefore$ Using $\Delta V = -\mathbf{E}\cdot d\mathbf{r}$ and integrating between limits x_0 and x_1,

$$\Delta V = -\int_{x_0}^{x_1} \frac{\sigma}{2\varepsilon_0}dx = \frac{\sigma(x_0 - x_1)}{2\varepsilon_0}$$

Here, $\quad \Delta V = 100\text{V}, \sigma = 0.75\,\frac{\mu\text{C}}{\text{m}^2}$

$\Rightarrow \quad x_0 - x_1 = \frac{2\varepsilon_0 \Delta V}{\sigma}$

$$= \frac{2 \times 8.85 \times 10^{-12} \times 100}{0.75 \times 10^{-6}}$$

$$= 2.36 \times 10^{-3}$$

$$= 2 \text{ mm}$$

So, equipotentials are 2 mm apart.

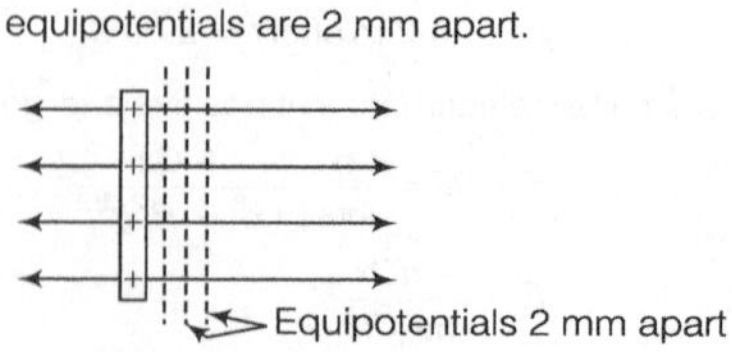

(c)

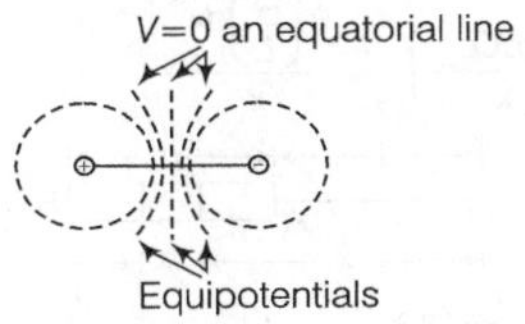

(d)

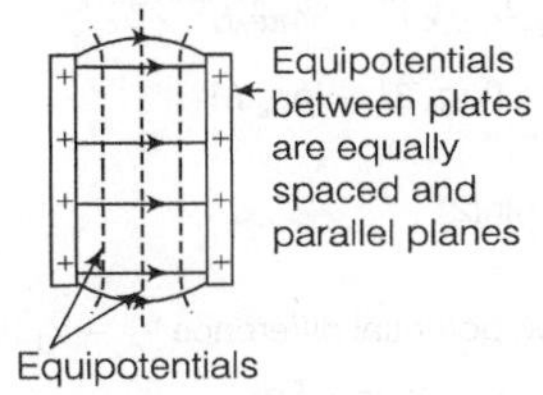

3. (i)

(a) $v = a(x^2 - y^2)$

$$\mathbf{E} = E_x\hat{\mathbf{i}} + E_y\hat{\mathbf{j}} + E_z\hat{\mathbf{k}} = \frac{-\partial V}{\partial x}\hat{\mathbf{i}} + \frac{-\partial V}{\partial y}\hat{\mathbf{j}} + \frac{-\partial V}{\partial z}\hat{\mathbf{k}}$$

$$= \frac{-\partial}{\partial x}a\,(x^2\text{-}y^2)\hat{\mathbf{i}} - \frac{\partial}{\partial y}a\,(x^2\text{-}y^2)\hat{\mathbf{j}} - \frac{\partial}{\partial z}a\,(x^2\text{-}y^2)\,\hat{\mathbf{k}}$$

$$= -2ax\hat{\mathbf{i}} + 2ay\hat{\mathbf{j}} = 2a(-x\hat{\mathbf{i}} + y\hat{\mathbf{j}})$$

(b) $V = axy$

$$E_x = \frac{-\partial V}{\partial x} = -ay$$

$$E_y = \frac{-\partial V}{\partial y} = -ax$$

$$E_z = \frac{-\partial V}{\partial z} = 0$$

$$\therefore \quad \mathbf{E} = -a(y\hat{\mathbf{i}} + x\hat{\mathbf{j}})$$

(c) $V = a(x^2 + y^2) + bz^2$

$$E_x = \frac{-\partial V}{\partial x} = -2ax$$

$$E_y = \frac{-\partial V}{\partial y} = -2ay$$

$$E_z = \frac{-\partial V}{\partial z} = -2bz$$

$$\therefore \quad \mathbf{E} = E_x\hat{\mathbf{i}} + E_y\hat{\mathbf{j}} + E_x\hat{\mathbf{k}}$$

$$= -2(ax\hat{\mathbf{i}} + ay\hat{\mathbf{j}} + bz\hat{\mathbf{k}})$$

(ii)

(a) $\mathbf{E} = a(y\hat{\mathbf{i}} + x\hat{\mathbf{j}})$

$$E_x = \frac{-\partial V}{\partial x} = ay$$

$$E_y = \frac{-\partial V}{\partial y} = ax$$

$$E_z = \frac{-\partial V}{\partial z} = 0$$

$\Rightarrow \quad dV = -E_x dx$

$$\int dV = -\int ay dx$$

$\Rightarrow \quad V = -axy + f(y) + C_1 \quad \ldots(i)$

And $\quad dV = -E_y dy$

$$\int dV = -\int ax dy$$

$\Rightarrow \quad V = -axy + f(x) + C_2 \quad \ldots(ii)$

Also, $\quad dV = 0.dz$

$\Rightarrow \quad V = f(xy) + f(x) + f(y) + f(y) + C_3 \quad \ldots(iii)$

From Eqs. (i), (ii) and (iii), we have

$$C_1 = C_2 = C_3 = C$$

$$f(x) = 0 \;\; f(y) = 0,\; f(xy) = -axy$$

$\Rightarrow \quad V = -axy + C$

(b) $\mathbf{E} = 2axy\hat{\mathbf{i}} + a(x^2 - y^2)\hat{\mathbf{j}}$

$\Rightarrow \quad \int dV = \int -2axy dx$

$$V = \frac{-2ax^2y}{2} + f(y) + C_1 + f(y, z) + f(z) \quad \ldots(i)$$

Also, $\int dV = -a\int (x^2 - y^2)dy$

$$V = -ax^2y + \frac{ay^3}{3} + f(x) + C_2 + f(x, z) + f(z) \quad \ldots(ii)$$

And, $\int dV = \int 0.dz$

$\Rightarrow \quad V = f(x, y) + f(y) + f(x) + C_3 \quad \ldots(iii)$

As the potential function of Eqs. (i), (ii) and (iii) is same, equality we have

$$f(y) = \frac{ay^3}{3},\; f(y, z) = 0$$

$$f(z) = 0,\; f(x) = 0,\; f(x, z) = 0$$

Also, $\quad C_1 = C_2 = C_3 = C$

Hence, $\quad V = -ax^2y + \frac{ay^2}{3} + C$

(c) $\mathbf{E} = ay\hat{\mathbf{i}} + (ax + bz)\hat{\mathbf{j}} + by\hat{\mathbf{k}}$

$$\int dV = -\int ay dx$$

$$\int dV = -\int (ax + by)dy$$

$$\int dV = -\int by dz$$

Integrating we have

$$V = -axy + f(y) + f(y, z) + f(z) + C_1 \quad \ldots(i)$$

$$V = -axy - bzy + f(x) + f(z) + f(x, z) + C_2 \quad \ldots(ii)$$

and $\quad V = -byz + f(x) + f(y) + f(x, z) + C_3 \quad \ldots(iii)$

Equating and solving Eqs. (i), (ii) and (iii), we have

$$C_1 = C_2 = C_3 = C,\; f(y) = 0,\; f(y, z) = -bzy$$

$$f(z) = 0,\; f(x) = 0,\; f(x, z) = 0\; f(x, y) = -axy$$

So, $\quad V = -axy - bzy + C$

4. (i) Given potential function is

$$V = 5x - 3x^2y + 2yz^2$$

$$E_x = \frac{-\partial V}{\partial x} = -5 + 6xy$$

Differentiating w.r.t. x treating y and z as constant.

$$E_y = -\frac{\partial V}{\partial y} = 3x^2 - 2z^2$$

x, z = constant

$$E_z = -\frac{\partial V}{\partial z} = -4yz$$

x, y = constant

So, $\quad E$ = field in the region

$$\mathbf{E} = E_x\hat{\mathbf{i}} + E_y\hat{\mathbf{j}} + E_z\hat{\mathbf{k}}$$

$$\mathbf{E} = (-5 + 6xy)\hat{\mathbf{i}} + (3x^2 - 2z^2)\hat{\mathbf{j}} + (-4yz)\,\hat{\mathbf{k}}$$

(ii) Field at (1, 0, − 2) can be found by substituting $x = 1$, $y = 0$ and $z = -2$ in the expression of field.

$$\mathbf{E}\,(1, 0, -2) = -5\hat{\mathbf{i}} - 5\hat{\mathbf{j}} + 0\,\hat{\mathbf{k}} \text{ or } \mathbf{E} = -5(\hat{\mathbf{i}} + \hat{\mathbf{j}})$$

and $\quad |\mathbf{E}|$ = magnitude of field

$$= \sqrt{(-5)^2 + (-5)^2 + 0^2} = 5\sqrt{2} \text{ N/C}$$

5. (a) Using, $\quad \mathbf{E} = -\left(\frac{dV}{dx}\hat{\mathbf{i}} + \frac{dV}{dy}\hat{\mathbf{j}} + \frac{dV}{dz}\hat{\mathbf{k}}\right)$

$$\Rightarrow \quad \mathbf{E} = -\left(\frac{d}{dx}4x^2\right)\hat{\mathbf{i}} = -8x\hat{\mathbf{i}}$$

When $x = 1$, $\quad \mathbf{E} = -8\hat{\mathbf{i}}$ V/m

So, electron moves along positive X-axis.

6. (d) $V_P = V_Q$ as potential remains same along a surface perpendicular to field lines.

7. (b) $\frac{d\mathbf{V}}{dt} = -\mathbf{E}$

Potential decreases as we move along the field.

$$V_A = V_B > V_C$$

8. (b) Potential gradient is the negative of electric field

$$\therefore \quad \frac{dV}{dr} = -E = -\frac{1}{4\pi\varepsilon_0}\frac{Q}{r^2} = -\frac{(9\times10^9)\times 92\times1.6\times10^{-19}}{(7.5\times10^{-15})^2}$$

$$= -2.4\times10^{21}\ \text{V/m}$$

9. (a) $E = -\left.\frac{dV}{dx}\right\}$ field between plates

Also, $E = \left.\frac{\sigma}{\varepsilon_0}\right\}$ from Gauss' law

$\Rightarrow \quad \sigma = \varepsilon_0 E = \varepsilon_0 \cdot \left.\frac{dV}{dx}\right\}$ dropped negative sign to calculate magnitude

$$\varepsilon_0 \times \frac{d}{dx}(8x+5) = 8\times8.85\times10^{-12} = 7.1\times10^{-11}\ \text{C/m}^2$$

10. (d) Components of field are $E_x = -\frac{\partial V}{\partial x}$

(partially differentiate by treating y as z as constants)

$$= -2.5y + 3.5yz$$

$$E_y = -\frac{\partial V}{\partial y} = -2y - 2.5x + 3.5xz$$

$$E_z = -\frac{\partial V}{\partial z} = 3.5xy$$

So, $\mathbf{E} = (-2.5y + 3.5yz)\hat{\mathbf{i}} + (-2y - 2.5x + 3.5xz)\hat{\mathbf{j}} + 3.5xy\hat{\mathbf{k}}$

$\mathbf{E}(1, 1, 1) = \hat{\mathbf{i}} - \hat{\mathbf{j}} + 3.5\hat{\mathbf{k}}$

{obtained by sub $x = 1$, $y = 1$, $z = 1$}

11. (d) In free space, field cannot be zero abruptly and as

$$E = -\frac{dV}{dr}$$

As, $\frac{dV}{dr} \neq 0$ at any point in space, so potential function can neither have a maxima nor a minima in free space.

12. (c) By work-energy theorem,

net work done = change in KE

$\Rightarrow$ External work done + work done by field

= Final KE − Initial KE

$\Rightarrow$ Work done by field $= W_{ext} - KE_{final}$

$$\Rightarrow V_B - V_A = \frac{W_{ext} - KE_{final}}{q} = \frac{7\times10^{-4} - 2.10\times10^{-4}}{-q\cdot\text{(negative charge)}} = -\text{ve}$$

13. (d) Work done in moving a charge particle between two points A and B is an electric field is

$$W_{A\to B} = q(V_{BA}) = q(V_B - V_A)$$

As, $V_B - V_A = 10$ V in case I

whereas $V_B - V_A = 20$ V in case II and III

Hence, work done in case I is least.

14. (d) For a point charge, potential of V volts exists at a distance r when,

$$V = \frac{1}{4\pi\varepsilon_0}\frac{q}{r} \quad \Rightarrow \quad r = \frac{4\pi\varepsilon_0 \times V}{q}$$

In this case, $V = 30$ V

$q = 4\times10^{-9}$ C

$\Rightarrow$ r = radius of equipotential surface

$$= \frac{30}{9\times10^9\times4\times10^{-9}} = \frac{30}{36} = 0.6\ \text{m}$$

15. (d) $V = \int_L^{2L}\frac{k\,dQ}{x} = \int_L^{2L}\frac{k\left(\frac{Q}{L}\right)dx}{x}$

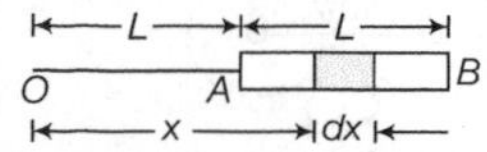

$$= \frac{Q}{4\pi\varepsilon_0 L}\int_L^{2L}\left(\frac{1}{x}\right)dx = \frac{Q}{4\pi\varepsilon_0 L}[\log_e x]_L^{2L}$$

$$= \frac{Q}{4\pi\varepsilon_0 L}[\log_e 2L - \log_e L]$$

$$= \frac{Q}{4\pi\varepsilon_0 L}\ln(2)$$

16. (c) As we know, potential difference $V_A - V_O$ is

$$dV = -E\,dx$$

$$\int_{V_O}^{V_A} dV = -\int_0^2 30x^2\,dx$$

$$V_A - V_O = -30\times\left[\frac{x^3}{3}\right]_0^2$$

$$= -10\times[2^3 - (0)^3]$$

$$= -10\times8 = -80\ \text{J}$$

17. (d) For inside points $(r \le R)$

$$E = 0$$

$\Rightarrow \quad V = \text{constant} = \frac{1}{4\pi\varepsilon_0}\frac{q}{R}$

For outside points $(r \ge R)$

$$E = \frac{1}{4\pi\varepsilon_0}\cdot\frac{q}{r^2} \quad \text{or} \quad E \propto \frac{1}{r^2}$$

and $$V = \frac{1}{4\pi\varepsilon_0}\frac{q}{r} \quad \text{or} \quad V \propto \frac{1}{r}$$

On the surface $(r = R)$

$$V = \frac{1}{4\pi\varepsilon_0}\frac{q}{R}$$

$$E = -\frac{1}{4\pi\varepsilon_0}\frac{q}{R^2} = \frac{\sigma}{\varepsilon_0}$$

where, $\sigma = \frac{q}{4\pi R^2}$ = surface charge density corresponding to above equations the correct graphs are shown in option (d).

18. (a,c) Potential of surface $V_0 = \frac{kQ}{r}$ and

Potential outside the sphere at a distance r is

$$V = \frac{kQ}{r} = \frac{kQ}{r_0}\cdot\frac{r_0}{r} = V_0\cdot\frac{r_0}{r}$$

It decreases as we move away from surface. Difference between potential at a given location (at an equipotential) and the surface is to be a multiple of 100 V.

$$\therefore \quad V_0 - V = V_0 - V_0\frac{r_0}{r} = (100)n$$

$$\Rightarrow \quad r = \frac{V_0}{(V_0 - 100n)}\cdot r_0$$

where, $n = 1, 2, 3\cdots$ etc., indicates number of an equipotential

So, $r(100) = \left(\frac{V_0}{V_0 - 100\times100}\right)r_0$

$$= \left(\frac{10216}{10216 - 10000}\right)\times0.44\left[\because V_0 = \frac{kQ}{r_0} = 10216\ \text{V}\right]$$

$$= \frac{10216}{216}\times0.44 \approx 2\ \text{m}$$

Also, as field is decreasing space between equipotential increases.

19. (b,c,d) Potential is constant for all points in a direction perpendicular to field lines.

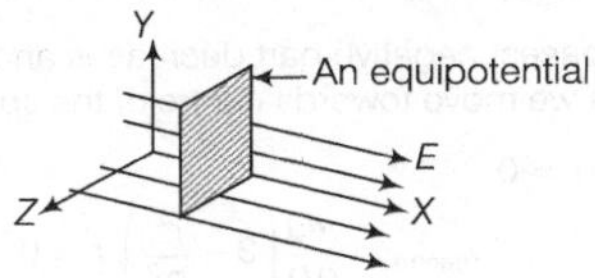

20. (b,c) Equipotentials are not always equally spaced.
They are drawn same potential apart. So, they are closer near pointed ends and where the charge density is larger.

21. (b,c) $\frac{dV}{dr} = -E$, if V = constant, then

$$\frac{dV}{dr} = 0$$

$\Rightarrow$ $E = 0$ also $E = 0$ is possible when $Q_{encl.} = 0$

22. (a,b,c)

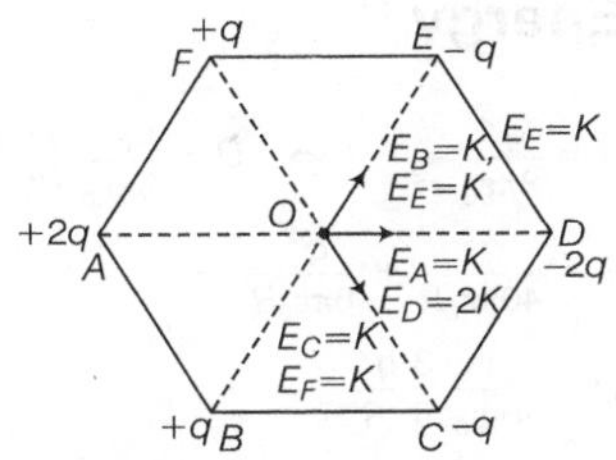

Resultant of $2K$ and $2K$ (at 120°) is also $2K$ towards $4K$.
Therefore, net electric field is $6K$.

Also, $V_0 = \frac{1}{4\pi\varepsilon_0}\left[\frac{q_A}{L} + \frac{q_B}{L} + \frac{q_C}{L} + \frac{q_D}{L} + \frac{q_E}{L} + \frac{q_F}{L}\right]$

$$= \frac{1}{4\pi\varepsilon_0}(q_A + \cdots + q_F) = 0$$

Because $q_A + q_B + q_C + q_D + q_E + q_F = 0$
And, only line PR, potential is same (= 0)

23. (4) Taking $V = 0$ at initial position of rod, then potential after 'd' distance along the field = $V = -Ed$

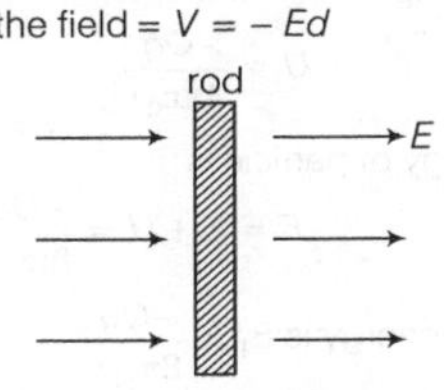

Charge on rod = λL
Potential energy of rod-field system initially = $QV_{initial} = 0$
Potential energy of rod- field system finally

$$= -\lambda LEd \quad \text{(Using, } U = qV\text{)}$$

Now using energy conservation,

$$(KE + PE)_{inital} = (KE + PE)_{final}$$

we get,

$$0 + 0 = \frac{1}{2}mv^2 - \lambda LEd$$

$\Rightarrow$ $v^2 = \frac{2\lambda L\,Ed}{m} = \frac{2\lambda LEd}{\mu L} = \frac{2\lambda Ed}{\mu}$

$\Rightarrow$ $v = \sqrt{\frac{2\lambda Ed}{\mu}} = \sqrt{\left(\frac{2 \times 40 \times 10^{-6} \times 100 \times 2}{0.1}\right)}$

$$= 0.400 \text{ m/s} = \frac{4}{10} \text{ m/s}$$

so, $v = 4$

24. (4) Field between plates = $\frac{2V_0}{d}$

If ball swings a small distance x to right. Voltage is lower by $-Ex = \frac{-2V_0}{d}x$.

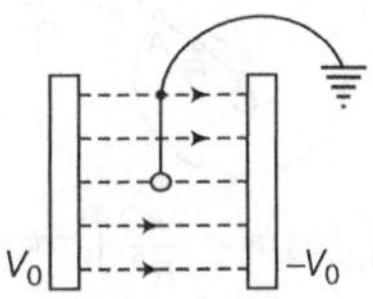

As ground connection keeps ball at zero potential by allowing charge 'q' to flow to ball from the ground.

So, $-\frac{2V_0}{d}x + \frac{kq}{R} = 0 \Big\}\, k = \frac{1}{4\pi\varepsilon_0}$

$\Rightarrow$ $q = \frac{2V_0 xR}{kd}$

The ball
$\therefore$ Feels the force $F = qE = \frac{4V_0^2 xR}{kd^2}$

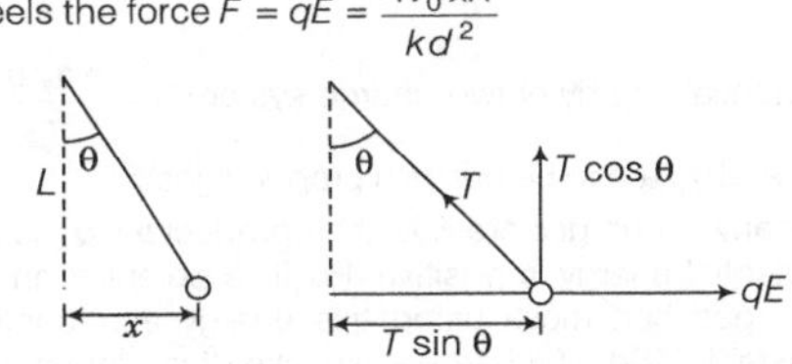

For equilibrium, this force must be balanced by the horizontal component of the string tension T.

From FBD, $T\cos\theta = mg$
and $T\sin\theta = qE$

$\Rightarrow$ $\tan\theta = \frac{qE}{mg} = \frac{4V_0^2 xR}{kd^2 mg} = \frac{x}{L}$

For small displacements x.

So, $V_0 = \left(\frac{kd^2 mg}{4RL}\right)^{1/2}$

If V_0 is less than this, the only equilibrium position of ball is hanging down between the plates. If V_0 exceeds this value the ball will swing over to one plate or the other.

$\therefore$ $N = 2$

25. (d) V_A = Potential due to charge q_A, charge q_B and charge q_C.

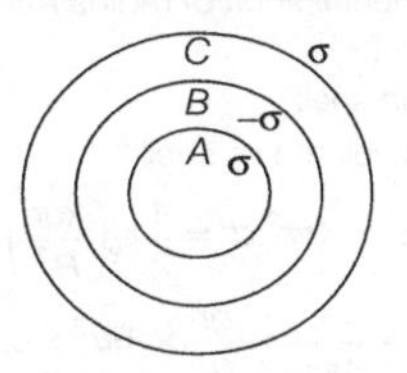

$$= k\left(\frac{q_A}{a} + \frac{q_B}{b} + \frac{q_C}{c}\right)$$

$$= k\left(\frac{4\pi a^2 \cdot \sigma}{a} + \frac{4\pi b^2 \cdot (-\sigma)}{b} + \frac{4\pi c^2 \cdot (\sigma)}{c}\right)$$

$$= \frac{1}{4\pi\varepsilon_0} \cdot 4\pi(\sigma a - \sigma b + \sigma c)$$

$$= \frac{\sigma}{\varepsilon_0}(a - b + c) = \frac{\sigma}{\varepsilon_0}\left(\frac{3}{2} - \frac{5}{2} + \frac{7}{2}\right) = \frac{\sigma}{\varepsilon_0}\left(\frac{5}{2}\right)$$

Also, $V_B = \frac{\sigma}{\varepsilon_0}\left(\frac{a^2}{b} - b + c\right) = \frac{\sigma}{\varepsilon_0}\left(\frac{19}{10}\right)$

and $V_c = \frac{\sigma}{\varepsilon_0}\left(\frac{a^2}{c} - \frac{b^2}{c} + c\right) = \frac{\sigma}{\varepsilon_0}\left(\frac{33}{14}\right)$

(i) → (r), (ii) → (q), (iii) → (p)

26. (b) Electric field inside an insulating sphere is

$$E(r) = \frac{kQ}{R^3}.r, \text{ for } r < R$$

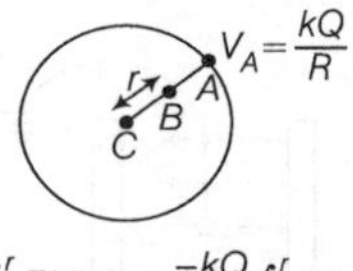

Then, $V_B - V_A = -\int_R^r E(r)\,dr = \frac{-kQ}{R^3}\int_R^r r\,dr = \frac{kQ}{2R^3}(R^2 - r^2)$

As, $V_A = \frac{kQ}{R}$, we get

$$V_B = \frac{kQ}{2R^3}(R^2 - r^2) + \frac{kQ}{R} = \frac{kQ}{2R}(3 - r^2 / R^2)$$

27. (c) As, $V(r) = \frac{kQ}{2R}\left(3 - \frac{r^2}{R^2}\right)$

when r decreases, negative part decreases and potential increases as we move towards centre of the sphere.

28. (a) At centre $r = 0$

$$\therefore \quad V_{\text{centre}} = \frac{kQ}{2R}\left(3 - \frac{r^2}{R^2}\right), r = 0$$

$$\Rightarrow \quad V_{\text{centre}} = \frac{3kQ}{2R} = \frac{3}{2}\left(\frac{kQ}{R}\right)$$

$$V_{\text{centre}} = \frac{3}{2} \times V_{\text{surface}}$$

$$\therefore \quad V_{\text{surface}} = \frac{2}{3}V_0$$

DPP-11 Electrical Potential Energy

1. (i) Potential energy of two charge system $U = \frac{kq_1 q_2}{r_{12}}$; (charges q_1 and q_2 are to be put with proper signs).

Clearly, if charges are like, then product $q_1 \cdot q_2$ is positive, so potential energy is positive. Positive potential energy means charges can move using this energy and configuration is unstable. So, if charges are like they move apart when released. Similarly if q_1 and q_2 are of opposite signs, then potential energy is negative. Means charges attract each other.

i.e. $$F = \frac{-dU}{dr}$$

The configuration is stable.

(ii) As electrostatic potential energy per unit volume $u = \frac{1}{2}\varepsilon_0 E^2$

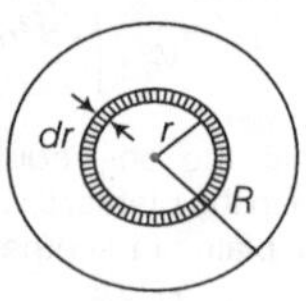

Energy stored with in the sphere

Consider an elemental shell of radius r and thickness dr. Volume of shell $= 4\pi r^2 dr$

$\therefore$ Energy stored in shell

$$dU = u_E \times \text{volume of shell}$$

$$= \frac{1}{2}\varepsilon_0 E^2 \times 4\pi r^2 .dr = \frac{1}{2}\varepsilon_0\left(\frac{kqr}{R^3}\right) \times 4\pi r^2\, dr$$

$$= \frac{1}{2}\varepsilon_0 \times \frac{1}{4\pi\varepsilon_0} \times \frac{qr}{R^3} \times 4\pi r^2 \times dr$$

$$dU = 4\pi r^2\, dr \frac{\varepsilon_0}{2}\left(\frac{1}{4\pi\varepsilon_0}\frac{q}{R^3}r\right)^2$$

$$dU = \frac{1}{8\pi\varepsilon_0}\frac{q^2}{R^6}r^4\, dr$$

$$U = \int_0^R \frac{1}{8\pi\varepsilon_0}\frac{q^2}{R^6}r^4\, dr = \frac{1}{8\pi\varepsilon_0}\frac{q^2}{5}\cdot\frac{R^5}{R^6}$$

$$U = \frac{1}{40\pi\varepsilon_0}\frac{q^2}{R}$$

Energy stored outside the sphere

$$E_{\text{out}} = \frac{1}{4\pi\varepsilon_0}\frac{q}{r^2}$$

$$dU = 4\pi r^2 dr\left[\frac{\varepsilon_0}{2}\left(\frac{1}{4\pi\varepsilon_0}\frac{q}{r^2}\right)^2\right]$$

$$dU = \frac{q^2}{8\pi\varepsilon_0}\frac{dr}{r^2} \quad \Rightarrow \quad U = \frac{q^2}{8\pi\varepsilon_0}\int_R^\infty \frac{dr}{r^2} = \frac{q^2}{8\pi\varepsilon_0 R}$$

$$\text{Total energy} = \frac{q^2}{40\pi\varepsilon_0 R} + \frac{q^2}{8\pi\varepsilon_0 R}$$

$$\text{Total energy} = \frac{1}{4\pi\varepsilon_0}\frac{3}{5}\frac{q^2}{R}$$

2. Rotating particle (mass m, charge $-q$) has both PE and KE.

Also, force on $-q$ charge acts like centripetal force.

$$\frac{Qq}{4\pi\varepsilon_0 r^2} = \frac{mv^2}{r}$$

$$\Rightarrow \quad mv^2 = \frac{Qq}{4\pi\varepsilon_0 r}$$

$\therefore$ KE of particle is

$$K = \frac{1}{2}mv^2 = \frac{Qq}{8\pi\varepsilon_0 r}$$

and PE of particle is

$$U = \frac{-Qq}{4\pi\varepsilon_0 r}$$

Hence, total energy of particle is

$$E = K + U = \frac{-Qq}{8\pi\varepsilon_0 r}$$

when radius is r_1, energy is $E_1 = \frac{-Qq}{8\pi\varepsilon_0 r_1}$

when radius is r_2, energy is $E_2 = \frac{-Qq}{8\pi\varepsilon_0 r_2}$

So, work done by external agent to change the radius

$$W = E_2 - E_1 = \frac{-Qq}{8\pi\varepsilon_0}\left(\frac{1}{r_2} - \frac{1}{r_1}\right) = \frac{Qq}{8\pi\varepsilon_0}\left(\frac{1}{r_1} - \frac{1}{r_2}\right)\text{ J}$$

3. (i) Potential energy of system of spheres is

$$U = \frac{q^2}{4\pi\varepsilon_0 d} = \frac{9 \times 10^9 \times (5 \times 10^{-6})^2}{1} = 0.225\text{ J}$$

With $v = 0$ at ∞

(ii) Force of repulsion on each sphere is

$$F = \frac{q^2}{4\pi\varepsilon_0 d^2} = \frac{9 \times 10^9 \times (5 \times 10^{-6})^2}{(1)^2} = 0.225\text{ N}$$

So, acceleration of spheres are

$$a_1 = \frac{F}{m_1} = \frac{0.255}{5 \times 10^{-3}} = 51\text{ m/s}^2$$

and $$a_2 = \frac{F}{m_2} = \frac{0.255}{10 \times 10^{-3}} = 22.5\text{ m/s}^2$$

(iii) By applying energy conservation,

$$U = \frac{1}{2}m_1v_1^2 + \frac{1}{2}m_2v_2^2$$

And momentum conservation

$$0 = m_1v_1 + m_2v_2$$

Solving these equations for v_1 and v_2, we get

$$v_1 = \sqrt{\left(\frac{2Um_2}{m_1(m_1 + m_2)}\right)} = 7.75 \text{ m/s}$$

and $\quad v_2 = \frac{-m_1}{m_2}.v_1 = -3.87 \text{ m/s}$

4. Potential due to an elemental length of rod at point P is

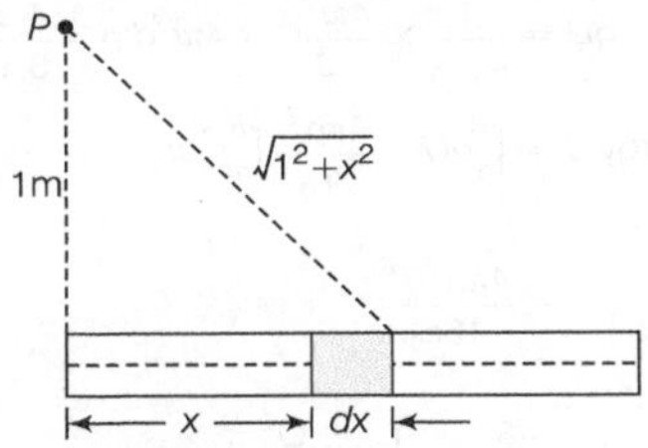

$$dV = \frac{kdq}{r} = \frac{k\lambda\, dx}{r} = \frac{Cx\,dx}{4\pi\varepsilon_0(1 + x^2)^{1/2}}$$

$$V = \int_0^1 dV = \frac{C}{4\pi\varepsilon_0}\int_0^1 \frac{x\,dx}{(1 + x^2)^{1/2}}$$

Let, $\quad 1 + x^2 = t$

$\Rightarrow \quad x\,dx = \frac{dt}{2}, \; \begin{matrix} x = 0 \Rightarrow t = 1 \\ x = 1 \Rightarrow t = 2 \end{matrix}$

$$\Rightarrow \quad V = \frac{C}{4\pi\varepsilon_0}\int_1^2 \frac{dt}{2t^{1/2}} = \frac{C}{8\pi\varepsilon_0}[2t^{1/2}]_0^2$$

$$= \frac{2C}{8\pi\varepsilon_0}(\sqrt{2} - 0) = \frac{\sqrt{2}C}{4\pi\varepsilon_0}$$

$$= \sqrt{2} \times -3 \times 10^{-9} \times 9 \times 10^9$$

$$= -27\sqrt{2} \text{ V}$$

5. (i) $F = -\frac{dU}{dr}$

(ii) $V = \frac{U}{q}$

(iii) $\frac{d\mathbf{V}}{dr} = -\mathbf{E}$

(iv) Potential at infinity is taken zero as field is zero, where infinity is distance at which field is zero.

6. (d) At an equipotential surface, V = constant.

Hence, potential energy of a point charge which is qV is also constant at every point of surface.

7. (d) Potential at P = Potential due to positive charge + Potential due to negative charge.

Clearly, V at $P = 0$

$\therefore$ Potential energy of a point charge at

$$P = qV = 0.$$

8. (d) Work done = change in potential energy

$$W = U_f - U_i$$

$$W = \frac{1}{4\pi\varepsilon_0}\left[\left\{\frac{(-2)(-2)}{2} + \frac{1(-2)}{2} + \frac{(1)(-2)}{2}\right\} - \left\{\frac{(-2)(-2)}{1} + \frac{1(-2)}{1} + \frac{1(-2)}{1}\right\}\right]$$

$$= \frac{1}{4\pi\varepsilon_0}[\{0\} - \{0\}] = 0$$

9. (d) Force on electron $= eE$

Acceleration of electron $a = \frac{F}{m} = \frac{eE}{m}$.

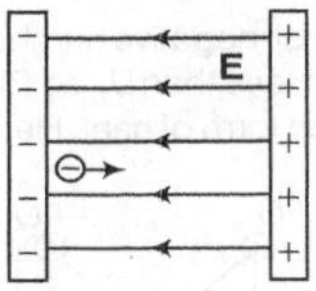

If d = distance between plates, then

$$v^2 - u^2 = 2ad, u = 0$$

$$\Rightarrow \quad v = \sqrt{2ad}$$

If distance reduced to half.

Let velocity of electron on reaching positive plate is v_1.

Then, $\quad v_1^2 - u^2 = \frac{2ad}{2}$

$$\Rightarrow \quad v_1 = \frac{\sqrt{2ad}}{\sqrt{2}} \Rightarrow v_1 = \frac{v}{\sqrt{2}}$$

10. (b) The masses of protons are same, so they have the same velocities.

Total KE at infinite separation of the two protons is the electrostatic potential energy of the system $= \frac{1}{4\pi\varepsilon_0}\frac{e^2}{r}$

$$= \frac{9 \times 10^9 (1.6 \times 10^{-19})^2}{10^{-10}}$$

By conservation law of energy,

Loss in potential energy = Gain in kinetic energy

$$U - 0 = 2 \times \frac{1}{2}mv^2$$

$$\frac{9 \times 10^9 (1.6 \times 10^{-19})^2}{10^{-10}} = 2 \times \frac{1}{2}mv^2$$

So, KE of one proton $= \frac{1}{2}mv^2$

$$= \frac{9 \times 10^9 (1.6 \times 10^{-19})^2}{2 \times 10^{-10}} = 11.5 \times 10^{-19} \text{ J}$$

11. (c) Work done in bringing charges to corners of equilateral triangle is stored in the form of PE of system.

$\therefore$ Work done = PE $= \frac{kq_1q_2}{r_{12}} + \frac{kq_1q_3}{r_{13}} + \frac{kq_2q_3}{r_{23}} = \frac{k}{r}\{3q^2\}$

$\therefore \quad r_{12} = r_{13} = r_{23} = 1 \times 10^{-10}$

and $q_1 = q_2 = q_3 = -1.6 \times 10^{-19}$ C

$$= \frac{9 \times 10^9}{1 \times 10^{-10}} \times 3 \times 1.6 \times 10^{-19}$$

$$= 9 \times 3 \times 1.6 = 43.2 \text{ eV} \approx 43 \text{ eV}$$

12. (b) As, gain of KE = loss of PE

$$\text{KE} = \frac{1}{2}mv^2 = qV$$

$$\Rightarrow \quad V = \frac{\text{KE}}{q} = \frac{125 \times 10^3 \times 1.6 \times 10^{-19}}{2 \times 1.6 \times 10^{-19}} = 62.5 \text{ kV}$$

13. (a,b) $\quad \frac{1}{2}mv^2 = 1.5 \times 10^3 \times 1.6 \times 10^{-19}$

$$\Rightarrow \quad v = \sqrt{\frac{2 \times 1.5 \times 10^3 \times 1.6 \times 10^{-19}}{9.1 \times 10^{-31}}}$$

$$= 2.3 \times 10^7 \text{ m/s}$$

Also, when $\frac{1}{2}mv^2 = 1.5 \times 10^3 \times 1.6 \times 10^{-19}$

and $\quad m = 1.67 \times 10^{-27}$ kg

we have, $\quad v = 5.4 \times 10^5$ m/s

14. (b,d) U_i = initial energy $= \frac{k(e)(-e)}{r_1}$ and

U_f = final energy $= \frac{k(e)(-e)}{r_2}$

As $r_2 < r_1 \Rightarrow U_i$ is a small negative number and U_f is a large negative number. Hence, U_f is less than U_i, so PE is lost by the system. This lost energy is liberated form of heat. Hence, reaction is exothermic.

15. (b,d)

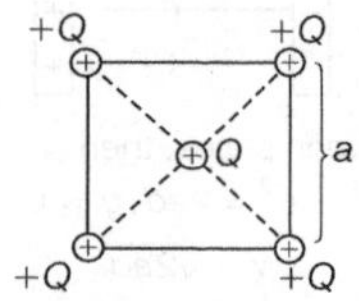

Potential energy of 5th charge is placed at centre $= q\ (V_{centre})$.

Now, V_{centre} (due to 4 charges at vertices) $= q' \Sigma \frac{kq}{r}$

where, $q = +Q$ and $r = \frac{a}{\sqrt{2}}$ and $q' = +Q$

$$\therefore \quad PE = \frac{4kQ^2}{q/\sqrt{2}} = \frac{kQ^2}{a}(4\sqrt{2})$$

Also, consider the motion of charged particle at centre along perpendicular direction to the plane containing square. It moves away

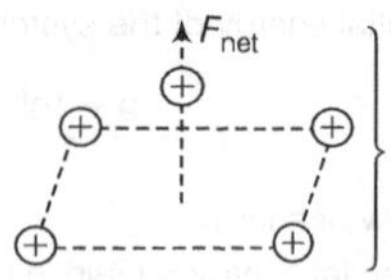

Form the configuration, system is in unstable equilibrium.

16. (b,c) Along path II, charge moves near to positive plate. So, work has to be done by external force. Also, this work is stored in the form of PE.

17. (a,d) Path I and path V are perpendicular to field lines. So, they are equipotentials and hence, work done in moving a test charge along these equipotential lines must be zero.

18. (a,d) Kinetic energy is produced on the expense of PE of system and PE is completely converted into KE only when the force between charges is zero, i.e. at infinity separation.

19. (b) Line along East-West is an equipotential line.

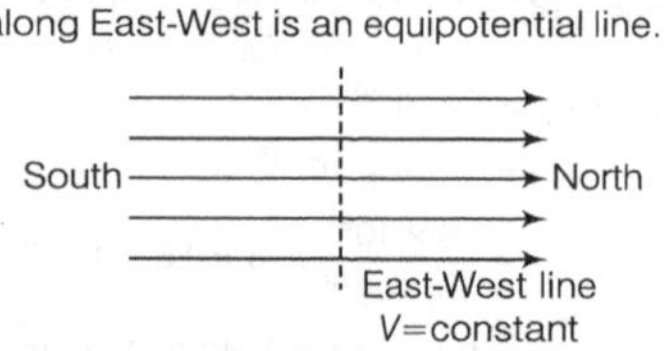

As V = constant along East-West line

qV or PE of charge remains constant along this line.

Also, field of a point charge is non-uniform.

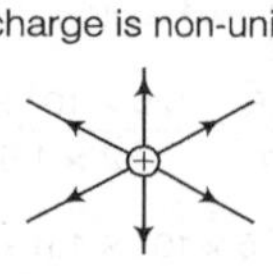

20. (d) PE of 2 positive charges, $U = \frac{kq_1q_2}{r_{12}}$ = positive

So, $\quad U \propto \frac{1}{r}$

When distance between charges decreases, the PE of system of charges increases.

Also, work done in bringing two charges in field of each other is exactly equal to the PE of the system.

21. (c) Let the spherical charge distribution of radius R has been assembled by bringing one by one infinite small thin spherical charge layers.

Work done in bringing a charge layer of radius r, thickness dr and charge dq

$$= dU = \frac{1}{4\pi\varepsilon_0}\frac{q}{r} \times dq$$

Using, $\quad U = \frac{kq_1q_2}{r} = \frac{1}{4\pi\varepsilon_0} \cdot \frac{q}{r}dq$

If ρ = volume charge density, then $q = \frac{4}{3}\pi r^3\rho$

and $\quad dq = 4\pi r^2 dr\rho$

$$\therefore \quad dU = \frac{1}{4\pi\varepsilon_0} \times \frac{4\pi r^3\rho}{3} \times 4\pi r^2 dr\rho = \frac{4}{3}\frac{\pi}{\varepsilon_0}\rho^2 r^4 dr$$

$$\text{Total energy } U = \int_0^R dU = \frac{4\pi\rho^2}{3\varepsilon_0}\int_0^R r^4 dr$$

$$= \frac{4\pi\rho^2 R^5}{15\varepsilon_0} \qquad \left(\because \rho = \frac{Q}{\frac{4}{3}\pi R^3}\right)$$

$$= \frac{3}{5} \times \frac{1}{4\pi\varepsilon_0} \times \frac{Q^2}{R}\ \text{J}$$

22. (a) Using, $U = \frac{3}{5}G\left(\frac{M^2}{R}\right)$,

Also, $\quad g = \frac{GM}{R^2},\ GM = gR^2$

$$\therefore \quad U = \frac{3}{5}MRg = \frac{3}{5} \times 2.5 \times 10^{31} \times 10$$

$$= 1.5 \times 10^{32}\ \text{J}$$

23. (b) For a spherical conductor,

$$C = 4\pi\varepsilon_0 R$$

and $\quad V = \frac{1}{4\pi\varepsilon_0}\frac{Q}{R}$

$$\text{Energy} = \frac{1}{2}CV^2$$

Hence, $\quad U = \frac{1}{2} \times 4\pi\varepsilon_0 R \times \frac{1}{(4\pi\varepsilon_0)^2} \times \frac{Q^2}{R^2}$

$$= \frac{1}{4\pi\varepsilon_0} \times \frac{Q^2}{2R}\ \text{J}$$

24. (b)

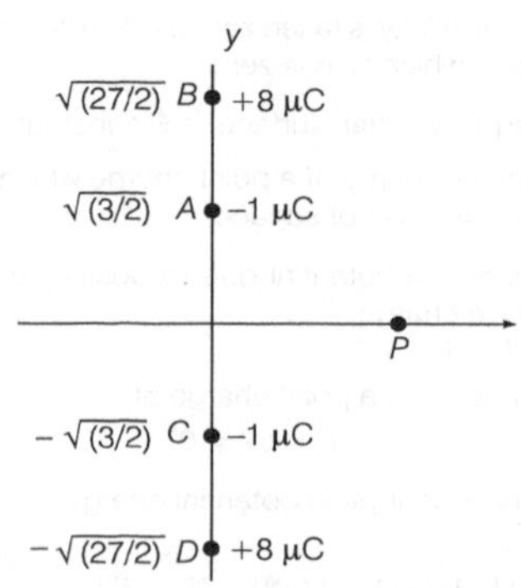

In above figure,

$$AP = CP = \sqrt{\left(\frac{3}{2} + x^2\right)}$$

$$BP = DP = \sqrt{\left(\frac{27}{2} + x^2\right)}$$

So, potential at $P = \Sigma V_i = \sum_{i=1}^{i=4} \frac{kq_i}{r_i}$

$$= 2 \times 9 \times 10^9 \left(\frac{8 \times 10^{-6}}{\sqrt{\left(\frac{27}{2} + x^2\right)}} - \frac{10^{-6}}{\sqrt{\left(\frac{3}{2} + x^2\right)}} \right)$$

$$= 1.8 \times 10^4 \left(\frac{8}{\sqrt{\left(\frac{27}{2} + x^2\right)}} - \frac{1}{\sqrt{\left(\frac{3}{2} + x^2\right)}} \right)$$

25. (c) Applying energy conservation at P, let particles speed at infinity is v_0.

$\therefore \quad \frac{1}{2} m v_0^2 = qV$

$\therefore v_0 = \sqrt{\frac{2qV}{m}} = \sqrt{\frac{2 \times 10^{-7} \times 2.7 \times 10^4}{6 \times 10^{-4}}} = 3 \text{ m/s}$

$$E = -\frac{dV}{dx}$$

$$= 1.8 \times 10^4 \left[8\left(\frac{-1}{2}\right)\left(\frac{27}{2} + x^2\right)^{-3/2} \cdot 2x - \left(-\frac{1}{2}\right)\left(\frac{3}{2} + x^2\right)^{-3/2} \cdot 2x \right]$$

Now, field is zero when, $x = 0$ or when,

$$\Rightarrow \quad \frac{8}{\left(\frac{27}{2} + x^2\right)^{3/2}} = \frac{1}{\left(\frac{3}{2} + x^2\right)^{3/2}}$$

$$\Rightarrow \quad \left(\frac{27}{2} + x^2\right) = 4\left(\frac{3}{2} + x^2\right)$$

$$\Rightarrow \quad x = \pm\sqrt{\frac{5}{2}}$$

$\therefore E = 0$ at $x = 0$, $x = \pm\sqrt{\frac{5}{2}}$ m

Also,

For $x > \sqrt{\frac{5}{2}}$ m, E is towards positive X-axis

For $x < \sqrt{\frac{5}{2}}$ m, E is towards negative X-axis.

26. (c) Let KE of particle at origin be K, then

$$K + qV_0 = \frac{1}{2} m v_0^2$$

But $\quad \frac{1}{2} m v_0^2 = qV$

$\therefore \quad K = q(V - V_0)$

$$= 10^{-7} \times (2.7 \times 10^4 - 2.4 \times 10^4)$$

$$= 3 \times 10^{-4} \text{ J}$$

27. (a) $E = 0$ at $x = 0$ and $x = \pm\sqrt{\frac{5}{2}}$ m but total energy of particle is least at $x = 0$.

So, $x = 0$ is stable equilibrium position and $x = \pm\sqrt{\frac{5}{2}}$ m is unstable equilibrium position.

DPP- 12 Potential and Potential Energy of a Dipole

1. For a dipole, at any point with position vector **r**, field is

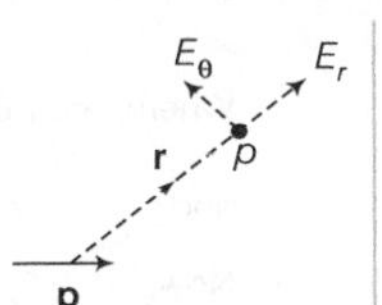

E_r = radial component of field

$$= \frac{2p\cos\theta}{4\pi\varepsilon_0 r^3}$$

and E_θ = perpendicular component of field

$$= \frac{p\sin\theta}{4\pi\varepsilon_0 r^3}$$

When the dipole is placed in an external field of strength E, such that **E** and **p** are parallel to each other then,

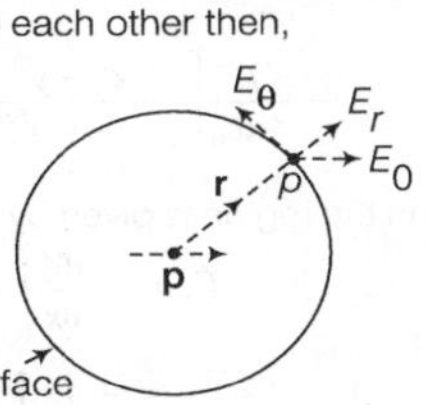

Now, if an equipotential surface exists at P which is spherical then, field at P must be radial only (i.e. perpendicular to equipotential surface).

$\therefore$ Component tangent to this circle must be zero.

$$\Rightarrow \quad E_{\text{net}} = E_\theta - E\sin\theta = 0$$

$$\Rightarrow \quad \frac{p\sin\theta}{4\pi\varepsilon_0 r^3} - E\sin\theta = 0$$

$$r^3 = \frac{p}{4\pi\varepsilon_0 E}$$

So, radius of equipotential surface is

$$r = \left(\frac{p}{4\pi\varepsilon_0 E}\right)^{1/3}$$

2. (i) Electric field due to linear charge, $E = \frac{2k\lambda}{r}$

Force on dipole is $F = |\mathbf{p}| \left|\frac{dE}{dr}\right|$ along **p**

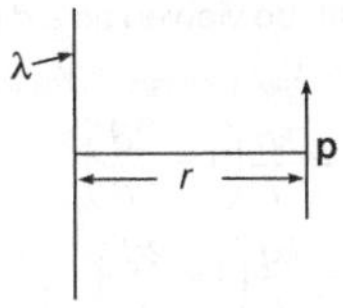

But along direction of **p**,

$$\frac{dE}{dr} = 0$$

So, $\quad$ Force = 0

(ii) $\quad \frac{dE}{dr} = \frac{-\lambda}{2\pi\varepsilon_0 r^2}$

$\therefore \quad \text{Force} = \frac{-p\lambda}{2\pi\varepsilon_0 r^2}$

(iii) $\quad \frac{dE}{dr} = \frac{-\lambda}{2\pi\varepsilon_0 r^2}$

$\therefore \quad \text{Force} = |\mathbf{p}|\left(\frac{dE}{dr}\right) = \frac{-p\lambda}{2\pi\varepsilon_0 r^2}$

3. (i) $V_p = \frac{1}{4\pi\varepsilon_0} \frac{p\cos\theta}{r^2} = \frac{9 \times 10^9 \times 4.8 \times 10^{-30} \times \cos 0°}{(4.1 \times 10^{-9})^2}$

$$= 2.6 \times 10^{-3} \text{ V}$$

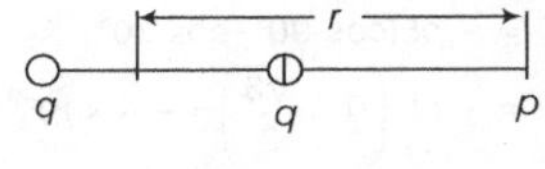

(ii) $V = \dfrac{kp\cos\theta}{r^2}$

$V = \dfrac{9\times 10^9 \times 4.8\times 10^{-30} \times \cos 45^\circ}{(4.1\times 10^{-9})^2}$

$= 1.8\times 10^{-3}$ V

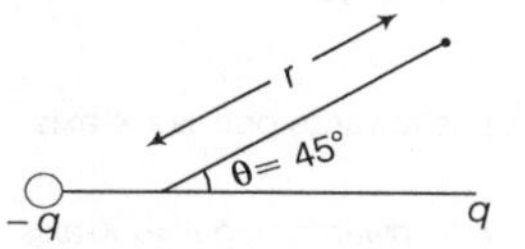

4. $|\mathbf{p}_1| = |\mathbf{p}_2| = \dfrac{p}{2\cos 52^\circ}$

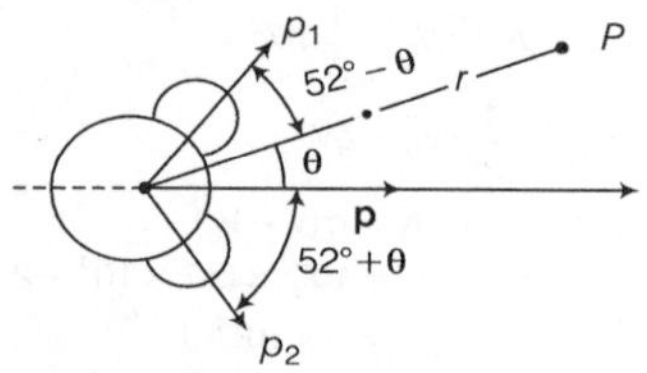

$V = V_{p_1} + V_{p_2}$

$= k\left[\dfrac{p_1\cos(52^\circ - \theta)}{r^2}\right] + k\left[\dfrac{p_2\cos(52^\circ + \theta)}{r^2}\right]$

$= \dfrac{kp}{2\cos 52^\circ}\left[\dfrac{\cos(52^\circ - \theta) + \cos(52^\circ + \theta)}{r^2}\right]$

$= \dfrac{kp}{(2\cos 52^\circ)(r^2)}(\cos 52^\circ\cos\theta + \sin 52^\circ\cos\theta + \cos 52^\circ\cos\theta - \sin 52^\circ\cos\theta)$

$= \dfrac{1}{4\pi\varepsilon_0 r^2}\cdot\dfrac{p}{2\cos 52^\circ}\cdot 2\cos 52^\circ\cos\theta$

$= \dfrac{1}{4\pi\varepsilon_0}\dfrac{p\cos\theta}{r^2}$

5. (a) 1 debye unit = 3.34×10^{-30} Cm

6. (b) Given system can be viewed as a dipole and a point charge.

$\therefore \quad V = V_{\text{point-charge}} + V_{\text{dipole on axis}}$

$\therefore \quad \dfrac{kq}{r} + \dfrac{k|\mathbf{p}|}{r^2} = \dfrac{kq}{r}\left(1 + \dfrac{2d}{r}\right)$

$\dfrac{q}{4\pi\varepsilon_0 r}\left(1 + \dfrac{2d}{r}\right) = \dfrac{kq}{r}\left(1 + \dfrac{2d}{r}\right)$

7. (c) As, $W = pE(\cos 0^\circ - \cos 60^\circ) = \dfrac{1}{2}pE$

$W' = pE(\cos 0^\circ - \cos 180^\circ) = 2pE = 4W$

8. (c)

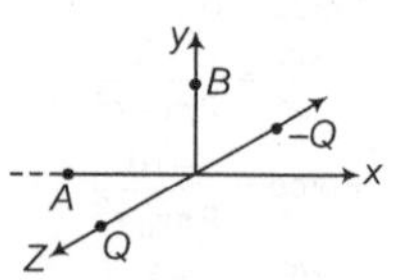

Given situation is shown in figure

$A \equiv (-a, 0, 0),\ B \equiv (0, a, 0)$

Point charge is moved from A to B

$V_A = V_B = 0$

Hence, wore done by electric field is

$W_{A\to B} = q(V_B - V_A) = 0$

9. (a) Work done $= \int_{30^\circ}^{90^\circ} pE\sin\theta\, d\theta = pE(-\cos\theta)_{30^\circ}^{90^\circ}$

$= -pE(\cos 90^\circ - \cos 30^\circ)$

$= -pE\left(0 - \dfrac{\sqrt{3}}{2}\right) = -4\times 10^{-9}\times 5\times 10^4 \times -\dfrac{\sqrt{3}}{2}$

$= \sqrt{3}\times 10^{-4}$ J

10. (d) Work done $= \int_{90^\circ}^{180^\circ} pE\sin\theta d\theta = [-pE\cos\theta]_{90}^{180} = 0$

11. (c) Electric field at the point P_2 due to a dipole P_1 is $E_1 = \dfrac{Kp}{L^3}$

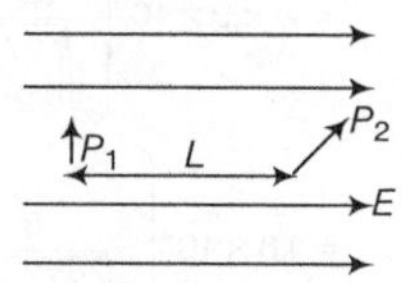

Potential energy due to a dipole in an electric field is given by

$U = -\mathbf{pE}$

$U = -pE\cos 90^\circ - pE\cos 60^\circ - \left(\dfrac{pKp}{L^3}\right)\cdot\cos 60^\circ$

$U = -\left[\dfrac{Kp^2}{2L^3} + \dfrac{pE}{2}\right]$

12. (a) For a given dipole, equipotential surface is y-z plane.

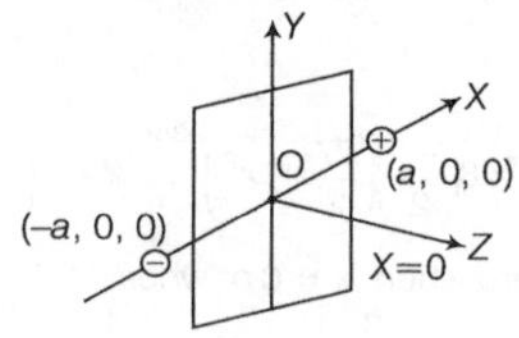

Equation of YOZ plane is, $x = 0$

or $\quad \mathbf{r}\cdot\hat{\mathbf{i}} = 0$

13. (d) Potential of dipole at point P is

$V = \dfrac{p\cos\theta}{r^2} = \dfrac{p}{4\pi\varepsilon_0}\left\{\dfrac{\dfrac{x}{(x^2+y^2)^{1/2}}}{x^2+y^2}\right\}$

Where, $\cos\theta = \dfrac{x}{(x^2+y^2)^{1/2}}$

and $\quad r = (x^2+y^2)^{1/2}$

Now, $\quad E_x = -\dfrac{\partial V}{\partial x}$ {Partial derivative of V w.r.t. x.}

$= \dfrac{-p}{4\pi\varepsilon_0}\left[\dfrac{(x^2+y^2)^{3/2} - x\cdot\frac{3}{2}(x^2+y^2)^{\frac{1}{2}}\cdot 2x}{(x^2+y^2)^3}\right]$

$= \dfrac{p}{4\pi\varepsilon_0}\left[\dfrac{2x^2 - y^2}{(x^2+y^2)^{5/2}}\right] = \dfrac{p}{4\pi\varepsilon_0}\left[\dfrac{2\cos^2\theta - \sin^2\theta}{r^3}\right]$

14. (c) Field in the region is given by

$E_x = -\dfrac{\partial V}{\partial x}\hat{\mathbf{i}} = \left(-\dfrac{\partial}{\partial x}2x^2\right)\hat{\mathbf{i}}$

$= -4x\,\hat{\mathbf{i}}\left(\dfrac{\text{N}}{\text{C}}\right)$

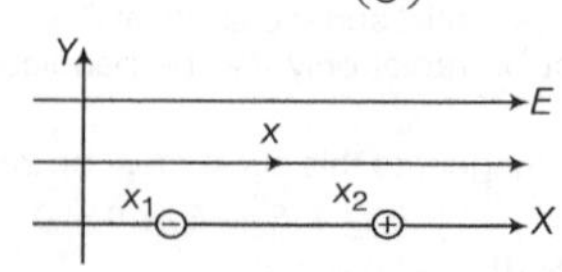

When field is uniform in direction, there is no component of field in any other direction except along X-axis.

So, force on dipole

F_{net} = Force on positive charge – Force on negative charge

$= qE_{(x_2)} - qE_{(x_1)} = q(-4x_2 - (-4x_1))$

$= 4q(x_1 - x_2) = 4\times 2\times 10^{-6}\times 3\times 10^{-10}$

$= 24\times 10^{-16}$ N

15. (b,d) The work done in deflecting a dipole through an angle θ,

$$W = \int_0^\theta pE \sin\theta \, d\theta = pE\,(1 - \cos\theta)$$

The potential energy U of the dipole,

$$U = pE(1 - \cos\theta)$$

16. (a,c) The potential due to the dipole is

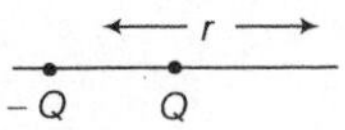

$$V = \frac{1}{4\pi\varepsilon_0}\frac{p\cos\theta}{r^2}$$

$$= \frac{(8.99\times10^9\ \text{Nm}^2/\text{C}^2)(4.8\times10^{-30}\ \text{Cm})\cos 0^\circ}{(4.1\times10^{-9}\ \text{m})^2}$$

$$= 2.6\times10^{-3}\ \text{V}$$

$$V = \frac{1}{4\pi\varepsilon_0}\frac{p\cos\theta}{r^2}$$

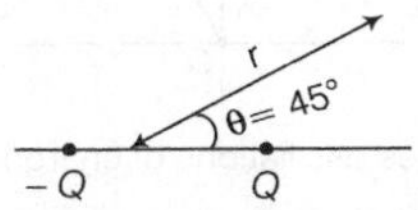

$$= \frac{(8.99\times10^9\ \text{Nm}^2/\text{C}^2)(4.8\times10^{-30}\ \text{Cm})\cos 45^\circ}{(4.1\times10^{9}\ \text{m})^2}$$

$$= 1.8\times10^{-3}\ \text{V}$$

17. (a,c) If the point is located at the right of positive charge, then

$$V = ke\sum\frac{q_i}{r_i} = ke\left[\frac{q}{x-a} - \frac{q}{x+a}\right]$$

$$= \frac{2k_{eq}\,a}{x^2 - a^2}$$

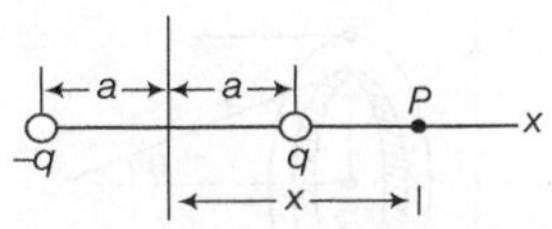

If a point far from the dipole such $x >> a$, then a^2 can be neglected in the term $x^2 - a^2$ and V becomes,

$$V \approx \frac{2k_{eq}a}{x^2}\ (x >> a)$$

18. (a,c,d) When the point P lies on the axial line of the dipole on the side of the positive charge $\theta = 0, \cos\theta = 1$

$$V = \frac{kp}{r^2}$$

When the point P lies on the equatorial line of the dipole, $\theta = 90^\circ$ or $\cos\theta = 0$

$$V = 0$$

At large distance

$V \propto \frac{1}{r^2}$ and not as $V \propto \frac{1}{r}$ (which is the characteristic of the potential due to a single pole).

19. (d) Torque on the arrangement is

$$\tau = -\mathbf{p}\times\mathbf{E} = qL\,E\sin\theta$$

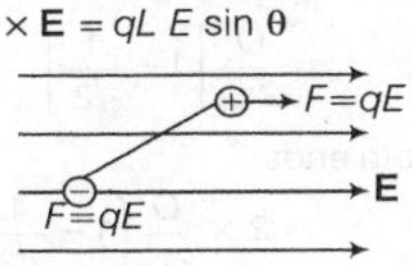

If I = moment of intertia of the arrangement, then

$$I = 2\times m\times\left(\frac{L}{2}\right)^2 = \frac{mL^2}{2}$$

Also torque $= I\alpha$

$$qLE\sin\theta = \frac{ML^2}{2}\alpha$$

$$\Rightarrow \quad \alpha = -\frac{2qE}{mL}\sin\theta$$

For small angles, $\quad \sin\theta \approx \theta$

$$\Rightarrow \quad \alpha = \frac{-2qE}{mL}\theta$$

So, $|\alpha|$ = angular acceleration $= \frac{2qE}{mL}\theta$

20. (a) $\quad \alpha = -\frac{2qE}{mL}\theta$

$$\Rightarrow \quad \frac{d^2\theta}{dt^2} + \frac{2qE}{mL}\cdot\theta = 0 \qquad \text{(SHM)}$$

So, time period is $\quad T = 2\pi\sqrt{\frac{mL}{2qE}}$

So, time required by rod to get aligned with the field is

$$\frac{T}{4} = \frac{\pi}{2}\sqrt{\frac{mL}{2qE}}$$

21. (c) Change in PE of rod = work done

$$= -\,pE(1 - \cos\theta)$$

$$= pE\,(\cos\theta - 1)$$

22. (c)

(i)

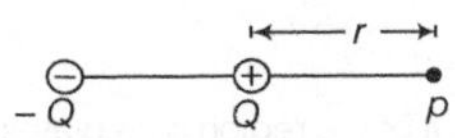

$$V = \frac{kp\cos\theta}{r^2}$$

$$= \frac{9\times10^9\times4.8\times10^{-30}\times\cos 0^\circ}{(4.1\times10^{-9})^2}$$

$$= 2.6\times10^{-3}\ \text{V}$$

(ii) $V = \frac{kp\cos\theta}{r^2} = 1.8\times10^{-3}\ \text{V}$

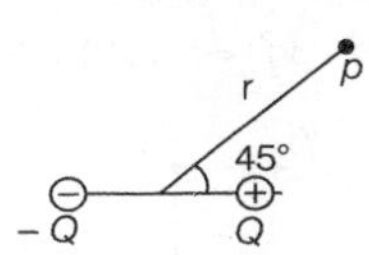

(iii) $V = \frac{1}{4\pi\varepsilon_0}\cdot\frac{p\cos\theta}{r^2}$

$$= \frac{9\times10^9\times4.8\times10^{-30}\times\cos 135^\circ}{(4.1\times10^{-9})^2}$$

$$= -1.8\times10^{-3}\ \text{V}$$

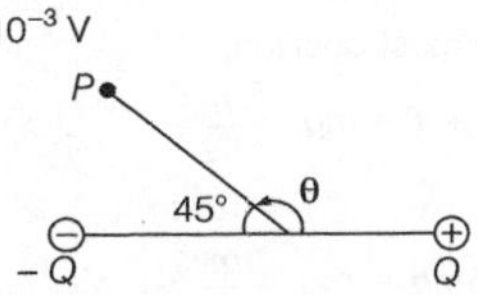

(i) → (r), (ii) → (q), (iii) → (p)

Revisal Problems (JEE Main)

1. (c) As, $E = \lim_{q_0 \to 0} \frac{F}{q_0}$

i.e. test charge q_0 is very small, so that it does not modify the field of charge q.

Hence, when we measure a field by putting a test charge q_0 at a point, measured value of field is smaller than true value of field.

2. (a) Particle will come out of plates, without striking nagative plate if time taken by the particle to cover 1 cm vertically is just equal to (less than) time taken by particle to cover 10 cm horizontally.

Now, if v_x = velocity imparted in X-direction, then time taken by particle to cover 10 cm horizontally is

$$t_1 = \frac{10 \times 10^{-2}}{v_x} = \frac{10^{-1}}{v_x}$$

Acceleration in vertical direction

$$a_y = \frac{F}{m} = \frac{qE}{m} = \frac{qV}{md}$$

$$= \frac{1.6 \times 10^{-19} \times 300}{12 \times 10^{-24} \times 2 \times 10^{-2}}$$

$$= \frac{1.6 \times 3}{12 \times 2} \times 10^{9}$$

$$= 2 \times 10^{8} \text{ m/s}^2$$

Initial velocity in Y-direction is zero

$\therefore \quad u_y = 0$

So, time taken by particle to cover a distance of 1 cm vertically

$$s = ut + \frac{1}{2}at^2 = \frac{1}{2} \times 2 \times 10^{8} \times t_2^2$$

or $\quad t_2 = 10^{-5}$s

As, particle go out of the region blue plates, when

$$t_2 \geq t_1$$

$$\Rightarrow \quad 10^{-5} \geq \frac{10^{-1}}{v_x}$$

$$\Rightarrow \quad v_x \geq 10^{4} \text{ m/s}$$

3. (c) Gain in KE = Work done by external forces

$$= W_{\text{gravity}} + W_{\text{electrical}}$$

$$\Rightarrow \quad \frac{1}{2}m(v_2^2 - v_1^2) = mg\,(2l) + (-q\,E2l) \quad \text{...(i)}$$

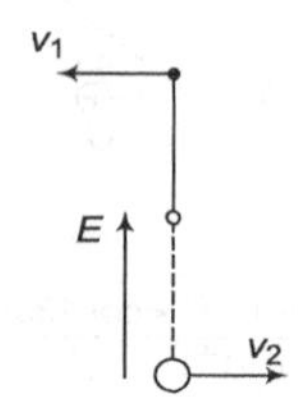

For ball at lower most position,

$$qE + T - mg = \frac{mv_2^2}{l}$$

Given, $\quad T = 15mg$

$$\therefore \quad qE + 15mg - mg = \frac{mv_2^2}{l} \quad \text{...(ii)}$$

From Eqs. (i) and (ii), we have

$$v_1 = \sqrt{\frac{l}{m}(10mg + 5qE)}$$

4. (d) PE first decreases as particle falls towards the ring but when it reaches a certain distance, it is again repelled so the PE increases.

5. (b) Ring is positively charged.

$\therefore$ Its field points away from centre.

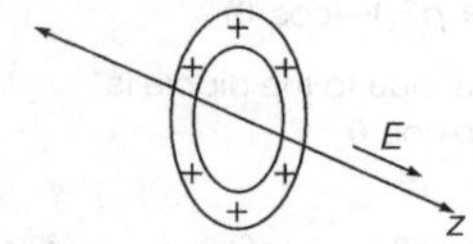

And field exerts a force $F = qE$ on the particle which is directed towards origin. Now, consider the field of the ring.

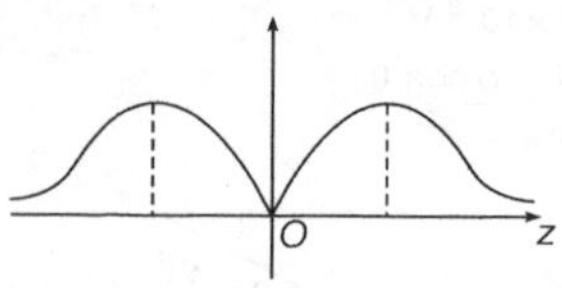

Clearly, field causes oscillations of charged particle.

6. (d) Charge is located at centre of cylinder and total flux through the cylinder is Q/ε_0.

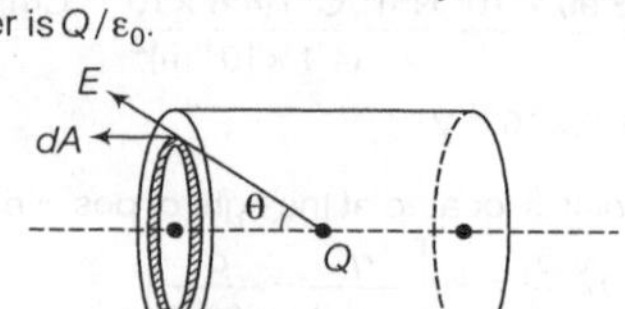

To find flux through ends, we divide the ends into concentric rings of radius R and thickness dR.

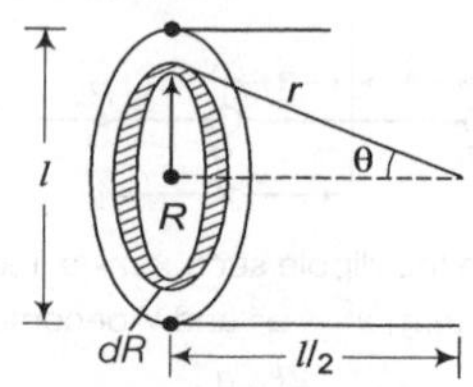

Area of each ring = $2\pi R\,dR$ and angle between **E** and **A** is θ where $\tan\theta = \frac{R}{l/2}$

$$\therefore \quad \phi_{\text{left end}} = \int \mathbf{E} \cdot d\mathbf{A} = \int \frac{kQ}{r^2} \cdot 2\pi R\,dR \cdot \cos\theta$$

where, $\quad r = \sqrt{R^2 + (l/2)^2}$

and $\quad \cos\theta = \frac{l/2}{r} = \frac{l/2}{\sqrt{R^2 + (l/2)^2}}$

Hence, $\quad \phi_{\text{left hand}} = \int_0^{l/2} \frac{2\pi Q\,l/2}{4\pi\varepsilon_0} \cdot \frac{R\,dR}{(R^2 + (l/2)^2)^{3/2}}$

$$= \frac{Q\,l/2}{2\varepsilon_0}\left[\frac{-1}{\sqrt{R^2 + (l/2)^2}}\right]_0^{l/2}$$

$$= \frac{Q}{2\varepsilon_0}\left[1 - \frac{1}{\sqrt{2}}\right]$$

So, flux through both ends

$$= 2 \times \frac{Q}{2\varepsilon_0}\left(1 - \frac{1}{\sqrt{2}}\right)$$

$\therefore$ Flux through curved part

$$= \frac{Q}{\varepsilon_0} - \frac{Q}{\varepsilon_0}\left(1 - \frac{1}{\sqrt{2}}\right)$$

$$= \frac{Q}{\sqrt{2}\,\varepsilon_0}$$

7. (c) Consider a Gaussian cylinder of cross-sectional area A.

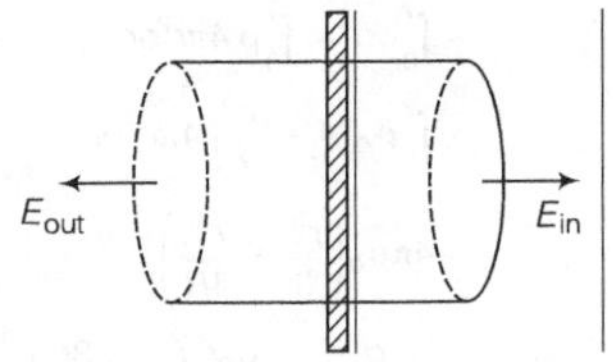

As no flux pass through curved part of cylinder,

$$\oint \mathbf{E}\cdot d\mathbf{A} = \oint_{LHS} \mathbf{E}\cdot d\mathbf{A} + \int_{RHS} \mathbf{E}\cdot d\mathbf{A} = \frac{Q_{encl}}{\varepsilon_0} \quad \ldots(i)$$

As, $\quad Q_{encl} = \sigma A + \rho_E xA$

$\therefore$ Substituting Q_{encl} in Eq. (i), we have

$$\left(\frac{\sigma + \rho_E d}{2\varepsilon_0}\right) A + E_{in} A = \frac{\sigma A + \rho_E xA}{\varepsilon_0}$$

$$\Rightarrow \quad E_{in} = \frac{\sigma + \rho_E(2x - d)}{2\varepsilon_0}$$

where, $x \in (0, d)$

8. (b) This is a situation similar to projectile motion. We use constant acceleration equations to solve for angle.

$$a_y = -\frac{eE}{m}, \quad l = v_0 \cos\theta_0 \cdot t$$

$$t = \frac{l}{v_0 \cos\theta_0}$$

$$v_y(\text{top}) = v_0 y + a_y t$$

$$0 = v_0 \sin\theta_0 - \frac{eE}{m}\left(\frac{1}{2}\frac{l}{v_0\cos\theta_0}\right)$$

$$v_0^2 = \frac{eE}{2m}\left(\frac{l}{\sin\theta_0\cos\theta_0}\right)$$

By using equation of trajectory,

$$h = l\tan\theta_0 - \frac{eEl^2}{4m\cos^2\theta_0 v_0^2}$$

$$\Rightarrow \quad h = l\tan\theta_0 - \frac{1}{2}l\tan\theta_0$$

$$\Rightarrow \quad h = \frac{1}{2}l\tan\theta_0$$

$$\Rightarrow \quad \theta_0 = \tan^{-1}\left(\frac{2h}{l}\right) = \tan^{-1}\left(\frac{2\times 1}{3.464}\right)$$

$$\approx \tan^{-1}\left(\frac{1}{\sqrt{3}}\right) \Rightarrow \theta_0 = 30^\circ$$

9. (b) As, $\frac{3}{4}$th part of charge lies outside the cube.

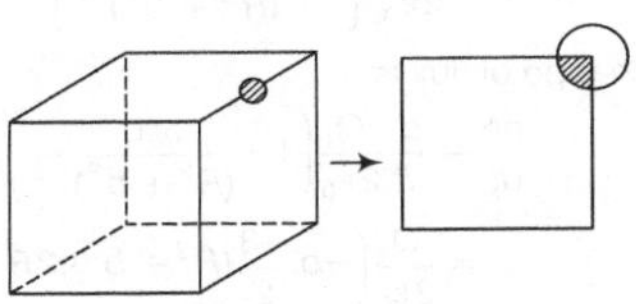

$\therefore \quad$ Charge enclosed $= \frac{1}{4}q$

Hence, $\quad$ flux $\phi = \dfrac{\frac{1}{4}q}{\varepsilon_0}$

or $\quad \phi_{cube} = \dfrac{q}{4\varepsilon_0}$

10. (d) If there is only one type of charge and it exists inside any closed surface, Gauss law will be valid

$$\therefore \quad \oint \mathbf{E}\cdot d\mathbf{s} = \frac{q_{encl}}{\varepsilon_0}$$

11. (b) Let a dipole is placed in a field of strength E and p is dipole moment then force on dipole is

$$F = p\left(\frac{dE}{dx}\right).\cos\theta$$

In given case, situation is as shown

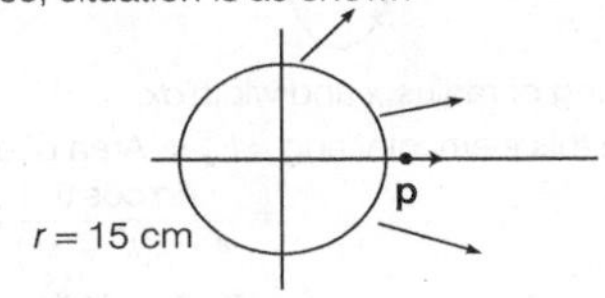

$$E(x) = \frac{kq}{x^2}.\hat{\mathbf{i}}, \quad \frac{dE}{dx} = -\frac{2kq}{x^3}$$

So, force on dipole $= p\left(\dfrac{2kq}{x^3}\right)\cos\theta$

12. (b) Locus of moving particle is a circle of radius r

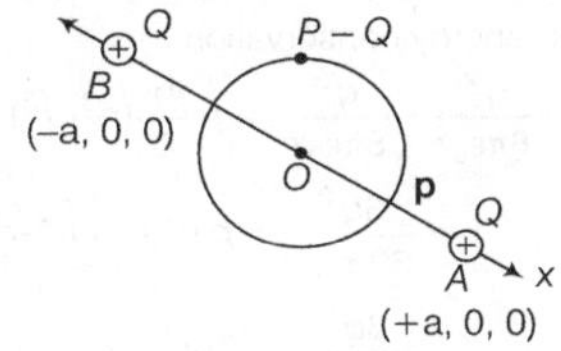

Force due to charge $+Q$ at A on revolving charge $-Q = \dfrac{Q^2}{4\pi\varepsilon_0 AP^2}$

And force due to charge at

$$B = \frac{Q^2}{4\pi\varepsilon_0 BP^2}$$

$$AP = BP = (r^2 + a^2)^{1/2}$$

Radial components of these forces are add up while parallel to X-axis cancels out .

So, resultant radial component of force is

$$F_{radial} = 2\left(\frac{Q^2}{4\pi\varepsilon_0 AP^2}\right).\cos\angle APO = \frac{2Q^2}{4\pi\varepsilon_0(r^2 + a^2)^{3/2}}$$

As $\quad \cos\angle APO = \dfrac{r}{(r^2 + a^2)^{1/2}}$

Radial force produces centripetal acceleration

$$\therefore \quad \frac{mv^2}{r} = \frac{2Q^2 r}{4\pi\varepsilon_0(r^2 + a^2)^{3/2}}$$

$$\frac{1}{2}mv^2 = \frac{Q^2 r^2}{4\pi\varepsilon_0(r^2 + a^2)^{3/2}}$$

$$= \text{KE of revolving particle}$$

13. (c) Pressure inside the soap bubble is

$$p = p_0 + \frac{4T}{r}$$

Due to charging, electrostatic pressure is

$$p_\sigma = \frac{\sigma^2}{2\varepsilon_0}$$

So, net pressure inside the charged bubble is

$$p_0 + \frac{4T}{r} - \frac{\sigma^2}{2\varepsilon_0}$$

By given condition,

$$p_0 + \frac{4T}{r} - \frac{\sigma^2}{2\varepsilon_0} = p_0$$

$$\Rightarrow \quad \frac{4T}{r} = \frac{\sigma^2}{2\varepsilon_0} = \frac{\varepsilon_0 V^2}{2r^2}$$

$$\Rightarrow \quad V = \sqrt{\frac{8Tr}{\varepsilon_0}}$$

14. (a)

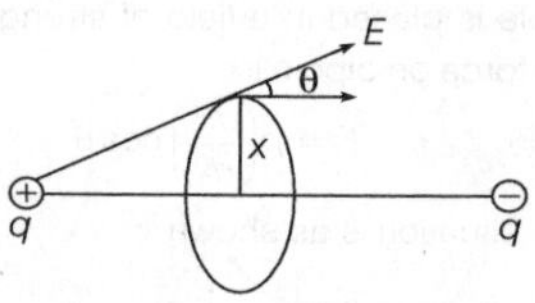

Consider a ring of radius x and width dx.

Flux through this elemental ring = $E_x \times$ Area of element

$$= \frac{2q\cos\theta}{4\pi\varepsilon_0(l^2+x^2)} 2\pi x dx$$

Net flux through complete ring $= \frac{q}{\varepsilon_0}\int_0^r \frac{lx.dx}{(l^2+x^2)^{3/2}}$

$$= \frac{q}{\varepsilon_0}\left(1 - \frac{1}{\sqrt{(1+r^2/l^2)}}\right)$$

15. (c) Let r be the new radius due to expansion of shell.

Then, applying energy conservation

$$\frac{q^2}{8\pi\varepsilon_0 r} - \frac{q^2}{8\pi\varepsilon_0 r'} = p.\frac{4\pi}{3}(r'^3 - r^3)$$

$$\Rightarrow \quad \frac{3q^2}{32\pi^2\varepsilon_0} = p.rr'(r'^2 + rr' - r^2)$$

$$\Rightarrow \quad \frac{3q^2}{32\pi^2\varepsilon_0} = 3pr^4$$

$$\Rightarrow \quad p = \frac{q^2}{32\pi^2\varepsilon_0 r^4}$$

16. (d) Excess of pressure inside a soap bubble is $\frac{4T}{R}$.

When electrostatic pressure exceeds this, soap bubble begins to swell.

i.e. $$\frac{q^2}{32\pi^2\varepsilon_0 R^4} > \frac{4T}{R}$$

$$\Rightarrow \quad q > \sqrt{32\pi.4\pi\varepsilon_0.T.R^3}$$

$$\Rightarrow \quad q > \left(\frac{32\times 3.14\times 3\times 10^{-2}\times 10^{-6}}{9\times 10^9}\right)^{1/2}$$

$$\Rightarrow \quad q > 1.83\times 10^{-8}\text{C}$$

17. (c) Initially, there are three forces in equilibrium.

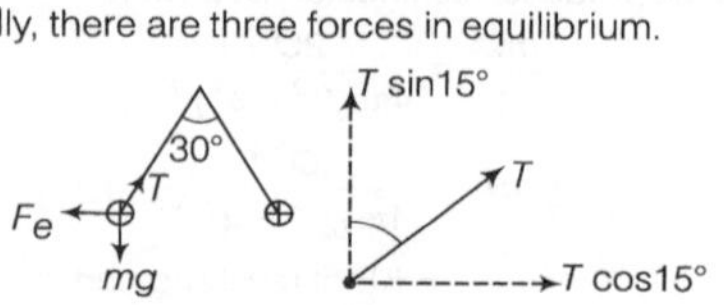

When they are suspended in a liquid, net downward force decreases.

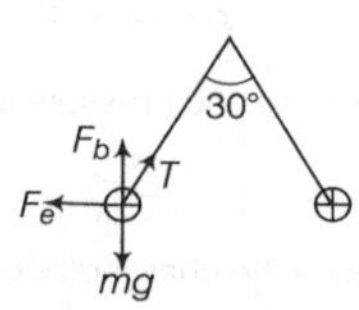

$\therefore$ F_e also decreases.

By applying force balance, we get $K = 2$ and also if dielectric constant is less than 2, angle increases as F_e is more.

18. (d) Volume of elemental shell of thickness dr and radius r is

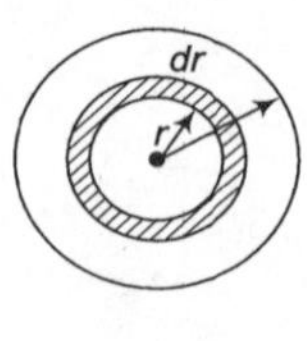

$$dV = 4\pi r^2 dr$$

Charge in shell $dq = \rho dV$

$\therefore$ Net charge in a sphere of radius r

$$= \int_0^r dq = \int_0^r \rho\, 4\pi r^2 dr$$

$$= \int_0^r \rho_0\left(1 - \frac{r}{R}\right) 4\pi r^2 dr$$

$$= 4\pi\rho_0\left(\frac{r^3}{3} - \frac{r^4}{4R}\right)$$

Hence, $$E(r) = \frac{q}{4\pi\varepsilon_0 r^2} = \frac{\rho_0 r}{3\varepsilon_0}\left(1 - \frac{3r}{4R}\right)$$

$\Rightarrow$ When, $r = 1$

$$E = \frac{5}{12}\cdot\frac{\rho_0}{\varepsilon_0}$$

Also, for maximum field

$$\frac{dE}{dr} = 0$$

$$\Rightarrow \quad 1 - \frac{6r}{4R} = 0, r = \frac{2R}{3}$$

or $$r = \frac{4}{3}\text{ m}$$

19. (b) As field is zero at a point nearer to and q_1 and $q_1 < q_2$. Also, both q_1 and q_2 must be positive (or both may be negative).

20. (b) Total flux through a surface closing a charge Q is Q/ε_0.

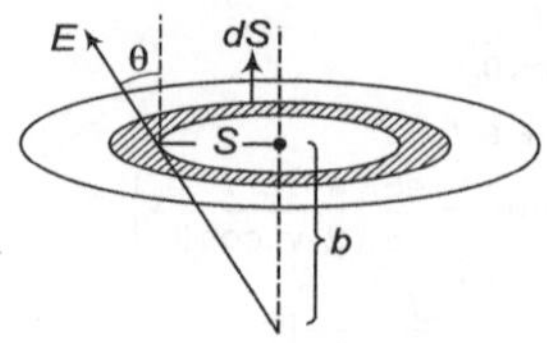

The flux through disc $= \phi = \int \mathbf{E}\cdot d\mathbf{A}$

$$\mathbf{E}\cdot d\mathbf{A} = EdA\cos\theta = E2\pi S\,dS\cos\theta$$

$$= \frac{1}{4\pi\varepsilon_0}\frac{Q}{r^2}\cdot 2\pi SdS\cos\theta$$

$$= \frac{1}{4\pi\varepsilon_0}\frac{Q}{(S^2+b^2)}\cdot 2\pi S\,dS\cdot\frac{b}{r}$$

So, flux through disc is

$$\phi = \int \mathbf{E}\cdot d\mathbf{A} = \int_0^R \frac{Qb}{2\varepsilon_0}\frac{SdS}{(S^2+b^2)^{3/2}}$$

$$= \frac{Qb}{2\varepsilon_0}\cdot[-(S^2+b^2)^{1/2}]_0^R$$

$$= \frac{Q}{2\varepsilon_0}\left\{1 - \frac{b}{(R^2+b^2)^{1/2}}\right\}$$

$\therefore$ Rate of change of flux is

$$\frac{d\phi}{dt} = \frac{d}{dt}\frac{Q}{2\varepsilon_0}\left(1 - \frac{b}{(R^2+b^2)^{3/2}}\right)$$

$$= \frac{Q}{2\varepsilon_0}\left(-b.\frac{-3}{2}(R^2+b^2).2R\frac{dR}{dt}\right)$$

$$= \frac{Q}{2\varepsilon_0}\left(\frac{3Rb}{(R^2+b^2)^{1/2}}\right)\frac{dR}{dt}$$

$$\Rightarrow \quad \frac{d\phi}{dt} = \frac{8.85\times 10^{-9}}{2\times 8.85\times 10^{-12}}\times\frac{3b^2}{\sqrt{2}b}\cdot\frac{dR}{dt}$$

$$\Rightarrow \quad \frac{d\phi}{dt} = \frac{10^3\times 3}{2\times\sqrt{2}}\times b\times 5\times 10^{-2}$$

$$= \frac{150b}{2\sqrt{2}} = \frac{75b}{\sqrt{2}} = \frac{75}{2}\sqrt{2}\,b\,\text{Vms}^{-1}.$$

Revisal Problems (JEE Advanced)

1. (a) For equilibrium of mass m

$$T \sin \theta = \frac{kQ_1Q_2}{r^2} \quad \text{and} \quad T \cos \theta = mg$$

$$\therefore \qquad \tan \theta = \frac{kQ_1Q_2}{mgr^2}$$

For $\theta_1 = \theta_2$, $\quad \tan \theta_1 = \tan \theta_2 \quad \Rightarrow \quad \frac{kQ_1Q_2}{m_1gr^2} = \frac{kQ_1Q_2}{m_2gr^2}$

$$\Rightarrow \qquad m_1 = m_2$$

2. (b) There are five forces on charge at origin each of magnitude $\frac{kq_1q_2}{r^2}$ and is directed along line joining origin with the point charge.

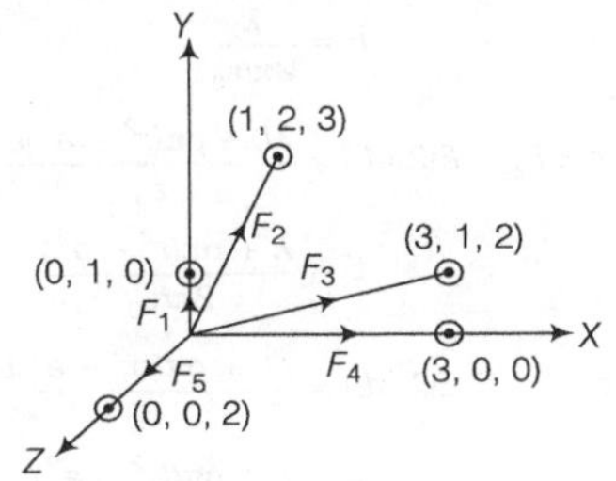

So, $\quad \mathbf{F}_1 = \frac{kq^2(\hat{\mathbf{j}})}{1^3}, \qquad \mathbf{F}_2 = \frac{kq^2(\hat{\mathbf{i}} + 2\hat{\mathbf{j}} + 3\hat{\mathbf{k}})}{(1^2 + 2^2 + 3^2)^{3/2}}$

$$\mathbf{F}_3 = \frac{kq^2(3\hat{\mathbf{i}} + \hat{\mathbf{j}} + 2\hat{\mathbf{k}})}{(1^2 + 2^2 + 3^2)^{3/2}} \qquad \mathbf{F}_4 = \frac{kq^2(3\hat{\mathbf{i}})}{3^3}$$

$$\mathbf{F}_5 = \frac{kq^2 2\hat{\mathbf{k}}}{2^3}$$

So, resultant force is $\quad \mathbf{F} = \Sigma \mathbf{F}$

$$= kq^2\left(\hat{\mathbf{i}}\left(\frac{1}{3^2} + \frac{4}{(14)^{3/2}}\right) + \hat{\mathbf{j}}\left(1 + \frac{3}{(14)^{3/2}}\right) + \hat{\mathbf{k}}\left(\frac{1}{2^2} + \frac{5}{(14)^{3/2}}\right)\right)$$

3. (c) Without any field, there are only two forces on oil drop.

As size of drop is not very large, buoyant force is very small to observe.

$\therefore$ If terminal velocity is v_1 then

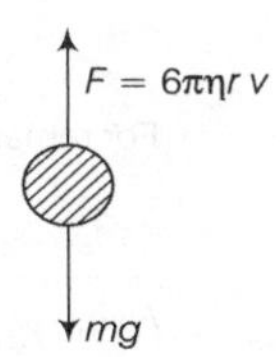

$$6\pi\eta r v_1 = mg \quad \text{or} \quad v_1 = \frac{mg}{6\pi\eta r}$$

With field, resultant force in upward direction is

$$F = qE - mg \qquad \text{or} \qquad v_2 = \frac{qE - mg}{6\pi\eta\varepsilon}$$

$\therefore$ Charge on the drop is

$$q = \frac{6\pi\eta\varepsilon\,(v_1 + v_2)}{E}$$

4. (d) Field at P due to each element dx is

$$dE = \frac{kdq}{x^2 + y^2}$$

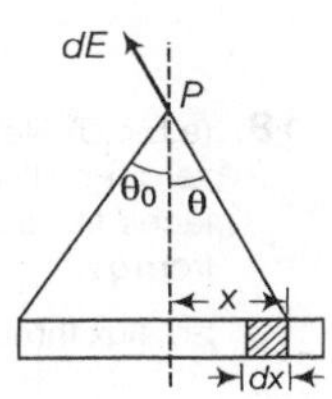

and this field is directed along the line joining the element to the point P.

By symmetry,

$$E_x = \int dE_x = 0 \quad \text{and} \quad E_y = \int dE_y$$

$$\Rightarrow \qquad E_y = \int dE \cos \theta$$

where, $\quad dE = \frac{k\,dq}{x^2 + y^2}, \qquad dq = \lambda dx$

$$\cos \theta = \frac{y}{(x^2 + y^2)^{1/2}}$$

$$\therefore \qquad E_y = \int_0^{l/2} \frac{2k\lambda y\,dx}{(x^2 + y^2)^{3/2}} = \frac{2k\lambda \sin \theta_0}{y}$$

5. (c) Each ion moves in a quarter circle, the electric force causes the centripetal acceleration.

$$\Sigma F = ma \qquad \Rightarrow \qquad qE = \frac{mv^2}{R}$$

$$\Rightarrow \qquad E = \frac{mv^2}{qR}$$

For the x-motion, $\quad v^2 = u^2 + 2as$

$$\Rightarrow \qquad 0 = v^2 + 2a_xR$$

$$\Rightarrow \qquad a_x = \frac{-v^2}{2R} = \frac{F_x}{m} = \frac{qE_x}{m}$$

$$\Rightarrow \qquad E_x = \frac{-mv^2}{2qR}$$

For the y-motion, $\quad v^2 = 0 + 2ayR$

$$a_y = \frac{+v^2}{2R} = \frac{qE_y}{m}$$

$$E_y = \frac{mv^2}{2qR}$$

Magnitude of field is $\quad E = \sqrt{E_x^{\,2} + E_y^{\,2}} = \frac{mv^2}{\sqrt{2}\,qR}$

6. (a)

$$\Sigma F_x = -T \sin \theta + qA = 0 \qquad \text{...(i)}$$

$$\Sigma F_y = T \cos \theta + qB - mg = 0 \qquad \text{...(ii)}$$

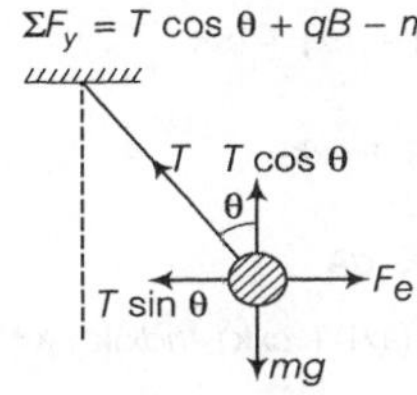

On solving Eqs. (i) and (ii), we get

$$\frac{qA \cos \theta}{\sin \theta} + qB = mg \quad \Rightarrow \quad q = \frac{mg}{(A \cot \theta + B)}$$

Substituting the value of q in Eq. (i),

$$T = \frac{mg\ A}{(A \cos \theta + B \sin \theta)}$$

7. (c)

$$a = \frac{F}{m} = \frac{qE}{m} = \frac{q}{m}(\alpha - \beta x) \qquad \text{...(i)}$$

$$a = 0 \text{ at } x = \frac{\alpha}{\beta}$$

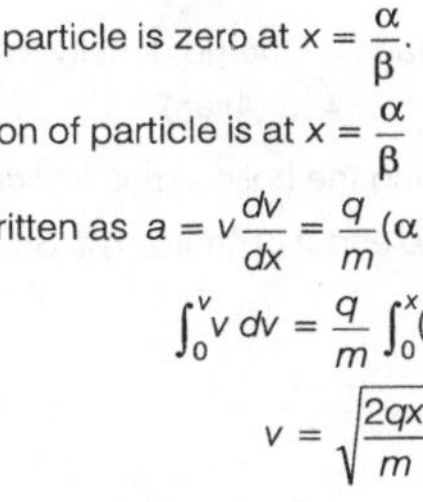

i.e. force on the particle is zero at $x = \frac{\alpha}{\beta}$.

So, mean position of particle is at $x = \frac{\alpha}{\beta}$

Eq. (i) can be written as $a = v\frac{dv}{dx} = \frac{q}{m}(\alpha - \beta x)$

$$\int_0^v v\,dv = \frac{q}{m}\int_0^x (\alpha - \beta x)dx$$

$$v = \sqrt{\frac{2qx}{m}\left(\alpha - \frac{\beta}{2}x\right)}$$

$v = 0$ at $x = 0$ and $x = \frac{2\alpha}{\beta}$ with its mean position at $x = \frac{\alpha}{\beta}$, therefore amplitude of the particle is $\frac{\alpha}{\beta}$. Maximum acceleration of particle is at extreme positions (at $x = 0$ or $x = \frac{2\alpha}{\beta}$ and $a_{max} = \frac{q\alpha}{m}$ from Eq. (i).

8. (d) $\therefore$ **E** is a uniform pointing radially outward, so $\phi_E = EA$

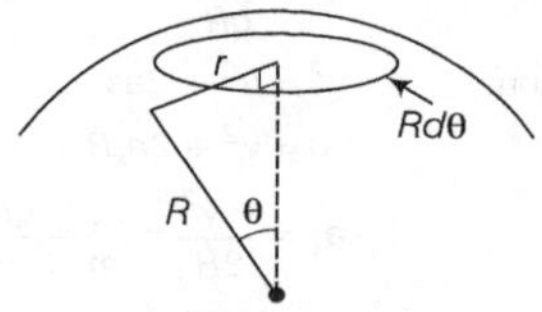

The arc length is $ds = Rd\theta$ and circumference is

$$2\pi r = 2\pi R \sin\theta$$

$$A = \int 2\pi r ds = \int_0^\theta (2\pi R \sin\theta) R d\theta$$

$$= 2\pi R^2 \int_0^\theta \sin\theta \, d\theta = 2\pi R^2 [-\cos\theta]|_0^\theta$$

$$= 2\pi R^2 (1 - \cos\theta)$$

So, flux $\phi_E = \frac{1}{4\pi\varepsilon_0} \frac{Q}{R^2} \cdot 2\pi R^2 (1 - \cos\theta)$

$$= \frac{Q}{2\varepsilon_0}(1 - \cos\theta)$$

9. (b) Given, $\mathbf{E} = ay\hat{\mathbf{i}} + bz\hat{\mathbf{j}} + cx\hat{\mathbf{k}}$

In XY-plane $z = 0$,

$\therefore$ $E = ay\hat{\mathbf{i}} + cx\hat{\mathbf{k}}$

$dA = hdx$

and flux, $\phi_E = \int \mathbf{E} \cdot d\mathbf{A}$

$$= \int (ay\hat{\mathbf{i}} + cx\hat{\mathbf{k}}) \cdot (hdx)\hat{\mathbf{k}}$$

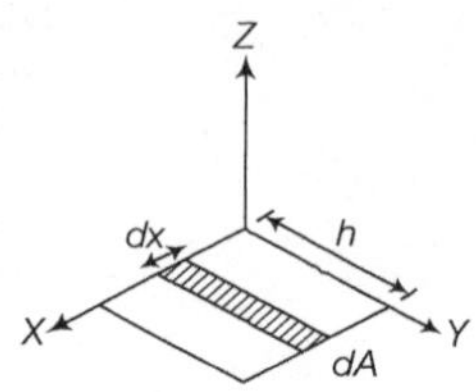

$\therefore$ $\phi_E = \int_{x=0}^{x=w} ch \, x dx = \frac{chw^2}{2}$

10. (c) Bowl exerts a normal force on each bead.

Clearly,

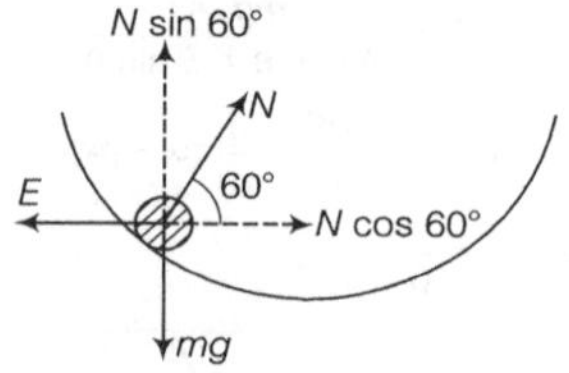

$$N \sin 60° = mg$$

and $N \cos 60° = kq^2/R^2$

So, $\tan 60° = \frac{mg}{\frac{kq^2}{R^2}}$,

or $q = R\left(\frac{mg}{k\sqrt{3}}\right)^{1/2}$

11. (d) Charge on surface = Charge density × Area

$$= \sigma(4\pi a^2)$$

12. (b) Net field reaching the hollow metallic inside surface is zero. Metal sphere shield effect of charge q_B. So, net force on A is zero.

13. (d) Charge distribution on outer surface remains symmetrical even if q_A is not put at centre.

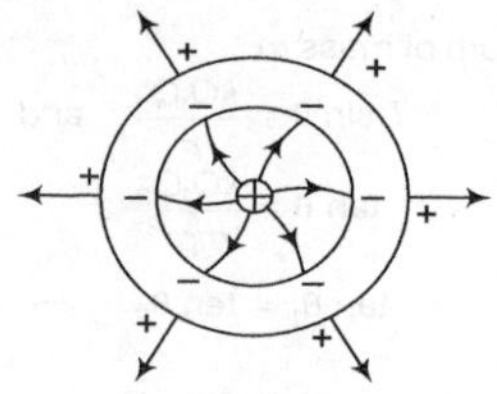

14. (b) Flux, $\phi_E = \frac{q}{\varepsilon_0}$

For $r < a$, $E(2\pi rL) = \frac{\lambda L}{\varepsilon_0}$

$\Rightarrow$ $E = \frac{\lambda}{2\pi r\varepsilon_0}$

15. (a) For $a < r < b$, $E(2\pi rL) = \frac{\lambda L + \rho\pi(r^2 - a^2)L}{\varepsilon_0}$

$\Rightarrow$ $E = \frac{\lambda + \rho\pi(r^2 - a^2)}{2\pi r\varepsilon_0}$

16. (c) For $r > b$, $E(2\pi rL) = \frac{\lambda L + \rho\pi(b^2 - a^2)L}{\varepsilon_0}$

$\Rightarrow$ $E = \frac{\lambda + \rho\pi(b^2 - a^2)}{2\pi r\varepsilon_0}$

17. (c,d) Contribution to net field from charges on same face is zero.

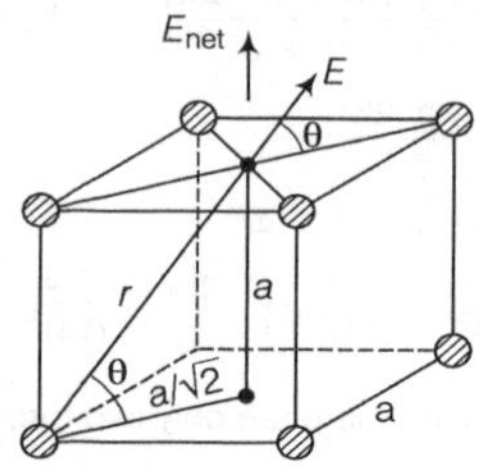

For remaining four charges, distance

$$r = \sqrt{\frac{a^2}{2} + a^2} = \sqrt{\frac{3}{2}}a$$

Also, $\sin\theta = \frac{a}{r}$

$$E_{net} = \frac{4kq}{r^2}\sin\theta = \frac{4kq}{r^2} \cdot \frac{a}{r}$$

$$= \frac{4kqa}{r^3} = \frac{4kq\,a}{\left(\sqrt{\frac{3}{2}}a\right)^3}$$

$$= \frac{2^{7/2}}{3^{3/2}} \cdot \frac{kq}{a^2} \text{ N/C}$$

18. (a,b,c,d) Net charge enclosed = 0. So, net flux through cube is zero. Positive flux comes out of faces meeting at g. These three faces fills solid angle equals to one eighth of a sphere as seen from q.

So, flux through three faces = $\frac{q}{8\varepsilon_0}$

As $\phi_{net} = 0 = 3\phi_{abcd} + \frac{q}{8\varepsilon_0}$

$\Rightarrow$ $\phi_{abcd} = \frac{-q}{24\,\varepsilon_0}$

19. (a,b,d) For $b \le r \le c$, field is zero as field exists inside a conductor.

For $a < r < b$,

$$E(4\pi r^2) = \frac{Q}{\varepsilon_0} \Rightarrow E = \frac{Q}{4\pi\varepsilon_0 r^2}$$

Also, induced charge on outer surface of shell is $+Q$.

20. (b,c) For moving charge,

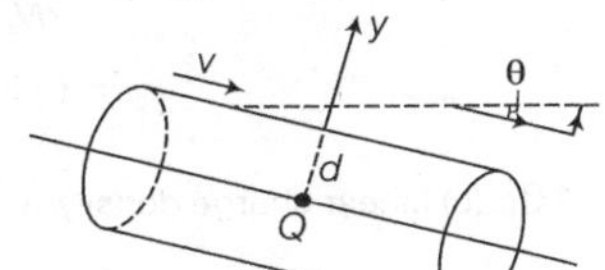

$$m\frac{dv_y}{dt} = F_y$$

$$m\frac{dv_y}{dx}\cdot\frac{dx}{dt} = qE_y$$

$$m\frac{dv_y}{dx}\cdot v_x = qE_y$$

$$\Rightarrow \quad dv_y = \frac{q}{mv}\cdot E_y dx$$

$$v_y = \int_0^{v_y} dv_y = \int_{-\infty}^{\infty} \frac{q}{mv}\cdot E_y dy$$

$$= \frac{q}{mv}\cdot\frac{Q}{2\pi d\varepsilon_0} \qquad \left(\because \int_{-\infty}^{\infty} E_y dy = \frac{Q}{2\pi d\varepsilon_0}\right)$$

$$\Rightarrow \quad v_y = \frac{qQ}{mv\, 2\pi d\varepsilon_0}$$

Also, $\tan\theta = \frac{v_y}{v} = \frac{qQ}{2\pi\varepsilon_0 dmv^2}$

21. (b) For $r < R_1$, $E = 0$ as no charge is inside region $r < R_1$

$\therefore E = 0$ at all points

For $r < r_1$, $\quad \frac{dV}{dr} = -E = 0$

$\Rightarrow \quad V =$ constant for $r < R_1$

For $R_1 < r < R_2$, is constant for q_2 as point lies inside the sphere of radius R_2.

Also, for $R_1 < r < R_2$, charge q_2 does not contribute to the field as it lies outside the surface.

22. (a) Water $K = 81$, mica $K = 7$. This difference is due to polarised molecules of water.

23. (b) $F = qE$ but it does not explain the force experienced by a moving charge.

24. (d) Field of a dipole falls much rapidly than the point charge. Statement I is false.

25. (4) Minimum distance of charge Q from Gaussian surface is $r/2$ and maximum distance $\left(2r + \frac{r}{2}\right)$ is $\frac{5r}{2}$.

So, $\quad E_{max} = \frac{kQ}{d^2} = \frac{kQ}{(r/2)^2} = \frac{4kQ}{r^2}$

and $\quad E_{min} = \frac{kQ}{d^2} = \frac{kQ}{\left(\frac{5}{2}r\right)^2} = \frac{4kQ}{25r^2}$

$$\therefore \quad E_{max} - E_{min} = \frac{4kQ}{r^2} - \frac{4kQ}{25r^2}$$

$$= \frac{24 \times 4kQ}{25r^2} = \frac{24Q}{25\pi\varepsilon_0 r^2}$$

$$= 4 \times \frac{6Q}{25\pi\varepsilon_0 r^2}$$

Hence, $N = 4$

26. (8) Energy density of field between the plates (which is uniform)

$$u_E = \frac{1}{2}\varepsilon_0 E^2 = \frac{1}{2}\varepsilon_0\left(\frac{\sigma}{\varepsilon_0}\right)^2 = \frac{1}{2}\frac{\sigma^2}{\varepsilon_0}$$

Work done in separating plates = Change in energy

= Energy density × Change in volume

$$= \frac{1}{2}\frac{\sigma^2}{\varepsilon_0} \times A \times d$$

$$= \frac{1}{2} \times \frac{(1.5 \times 10^{-6})^2}{8.85 \times 10^{-12}} \times (1.0) \times (5 \times 10^{-3})$$

$$= \frac{1.5 \times 1.5 \times 5 \times 10^{-3}}{2 \times 8.85}$$

$$= 0.64 \times 10^{-3}\ \text{J}$$

$$= 64 \times 10^{-5}\ \text{J}$$

$$= 8 \times 8 \times 10^{-5}\ \text{J}$$

Hence, $N = 8$

JEE Main & AIEEE Archive

1. (c) As we know, potential difference $V_A - V_O$ is $dV = -Edx$

$$\int_{V_O}^{V_A} dV = -\int_0^2 30x^2 dx$$

$$V_A - V_O = -30 \times \left[\frac{x^3}{3}\right]_0^2$$

$$= -10 \times [2^3 - (0)^3]$$

$$= -10 \times 8 = -80\ \text{J}$$

2. (d)

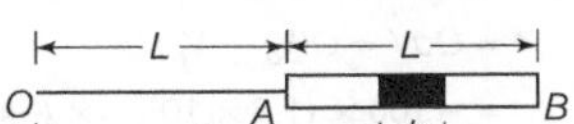

Let an element of length dx, charge dq is at distance x from point O.

$dV = k\frac{dq}{x}$ $\quad$ where, $dq = \frac{Q}{L}dx$

$$\therefore \quad V = \int_L^{2L}\frac{kdq}{x} = \int_L^{2L}\frac{\left(\frac{Q}{L}\right)dx}{x} = \frac{Q}{4\pi\varepsilon_0 L}\int_L^{2L}\left(\frac{1}{x}\right)dx$$

$$= \frac{Q}{4\pi\varepsilon_0 L}[\log_e x]_L^{2L} = \frac{Q}{4\pi\varepsilon_0 L}[\log_e 2L - \log_e L]$$

$$= \frac{Q}{4\pi\varepsilon_0 L}\left[\log_e^2 \frac{2L}{L}\right] = \frac{Q}{4\pi\varepsilon_0 L}\ln(2)$$

3. (b) $F_{net} = 2F\cos\theta$

$$\therefore \quad F_{net} = -\frac{2kq\left(\frac{q}{2}\right)}{(\sqrt{y^2 + a^2})^2}\cdot\frac{y}{\sqrt{y^2 + a^2}}$$

[negative sign indicate the net force is towards the mean position]

$$F_{net} = -\frac{2kq\left(\frac{q}{2}\right)y}{(y^2 + a^2)^{3/2}} \Rightarrow \frac{kq^2 y}{a^3} \propto -y$$

4. (c) Electric field inside the uniformly charged sphere varies linearly, $E = \frac{kQ}{R^3}\cdot r, (r \le R)$, while outside the sphere, it varies as inverse square of distance, $E = \frac{kQ}{r^2}$, $(r \ge R)$ which is correctly represented in option (c).

5. (d) For a solid insulating sphere, potential at centre is

$$V_{centre} = \frac{3}{2}\frac{kq}{R}$$

and potential at surface is $V_{surface} = \frac{kq}{R}$

So, change in potential energy when charge q is taken from centre to surface is,

$$\Delta U = U_f - U_i = q'(V_f - V_i)$$
$$= q'\left(\frac{kq}{R} - \frac{3}{2}\frac{kq}{R}\right)$$
$$= \frac{-q'}{2}\cdot\frac{1}{4\pi\varepsilon_0}\cdot\frac{\frac{4}{3}\pi R^3\rho}{R}$$
$$= \frac{-q'R^2\rho}{6\varepsilon_0}$$

So, Statement I is incorrect, while Statement II is correct.

$\therefore \quad E(n) = \rho r / 3\varepsilon_0$

6. (a) Initial potential energy of the system,

$$U_i = \frac{2kqQ}{a} + \frac{2k(-q)Q}{\sqrt{5}a}$$
$$= \frac{1}{4\pi\varepsilon_0}\cdot\frac{2qQ}{a}\left[1 - \frac{1}{\sqrt{5}}\right]$$

Final potential energy of the system, $U_f = 0$

By conservation law of energy,

Gain in KE = Loss in PE

$$K = \frac{1}{4\pi\varepsilon_0}\cdot\frac{2qQ}{a}\left[1 - \frac{1}{\sqrt{5}}\right]$$

7. (d) At any instant,

$$T\cos\theta = mg \quad \ldots\text{(i)}$$
$$T\sin\theta = F_e = \frac{kq^2}{x^2} \quad \ldots\text{(ii)}$$

From Eqs. (i) and (ii), we get

$$\frac{kq^2}{x^2} = mg\tan\theta$$
$$\Rightarrow \quad q^2 = \frac{mg}{k}\cdot\frac{x}{2l}\cdot x^2 \qquad \left[\because \tan\theta \approx \frac{x}{2l}\right]$$
$$\Rightarrow \quad q^2 = \frac{mg}{2kl}x^3 \quad \ldots\text{(iii)}$$
$$\Rightarrow \quad 2q\frac{dq}{dt} = \frac{3mg}{2kl}x^2\frac{dx}{dt}$$
$$\Rightarrow \quad 2\left(\frac{mg}{2kl}x^3\right)^{1/2}\frac{dq}{dt} = \frac{3mg}{2kl}x^2 v \qquad \left[\because q = \left(\frac{mg}{2kl}x^3\right)^{1/2}\right]$$
$$\Rightarrow \quad vx^{1/2} = \text{constant}$$
$$\Rightarrow \quad v \propto x^{-1/2}$$

8. (a) As initial charge is maximum.

$$\therefore \quad q = q_0\cos\omega t$$
$$\Rightarrow \quad i = \frac{dq}{dt} = -\omega q_0\sin\omega t$$

Given, energy stored in inductor $\left(U = \frac{1}{2}Li^2\right)$ is equal to energy stored in capacitor $\left(U = \frac{q^2}{2C}\right)$.

$$\therefore \quad \frac{1}{2}Li^2 = \frac{q^2}{2C}$$
$$\Rightarrow \quad \frac{1}{2}L(\omega q_0\sin\omega t)^2 = \frac{(q_0\cos\omega t)^2}{2C}$$

But $\quad \omega = \frac{1}{\sqrt{LC}}$

$$\Rightarrow \quad \tan\omega t = 1$$
$$\Rightarrow \quad \omega t = \frac{\pi}{4}$$
$$\Rightarrow \quad t = \frac{\pi}{4\omega} = \frac{\pi}{4}\sqrt{LC}$$

9. (c) Electric field, $E = -\frac{d\phi}{dr} = -2ar$

By Gauss' theorem, $\quad E(4\pi r^2) = \frac{q}{\varepsilon_0}$

$$\Rightarrow \quad q = -8\pi\varepsilon_0 ar^3$$
$$\rho = \frac{dq}{dV} = \frac{dq}{dr}\times\frac{dr}{dV}$$
$$= (-24\pi\varepsilon_0 ar^2)\left(\frac{1}{4\pi r^2}\right) = -6\varepsilon_0 a$$

10. (c) Linear charge density, $\lambda = \left(\frac{q}{\pi r}\right)$

$$E = \int dE\sin\theta\,(-\hat{j})$$
$$= \int_0^{\pi}\int\frac{k\cdot dq}{r^2}\sin\theta\,(-\hat{j})$$
$$E = \frac{k}{r^2}\int_0^{\pi}\frac{qr}{\pi r}d\theta\sin\theta\,(-\hat{j})$$
$$= \frac{k}{r^2}\frac{q}{\pi}\int_0^{\pi}\sin\theta\,d\theta\,(-\hat{j})$$
$$= \frac{1}{4\pi\varepsilon_0}\frac{q}{\pi r^2}[-\cos\theta]_0^{\pi}(-\hat{j})$$
$$= \frac{q}{2\pi^2\varepsilon_0 r^2}(-\hat{j})$$

11. (b) Apply Shell theorem, the total charge upto distance r can be calculated as followed

$$dq = 4\pi r^2\cdot dr\cdot\rho = 4\pi r^2\cdot dr\cdot\rho_0\left[\frac{5}{4} - \frac{r}{R}\right]$$
$$= 4\pi\rho_0\left[\frac{5}{4}r^2\,dr - \frac{r^3}{R}dr\right]$$
$$\int dq = q = 4\pi\rho_0\int_0^r\left(\frac{5}{4}r^2\,dr - \frac{r^3}{R}dr\right)$$
$$q = 4\pi\rho_0\left[\frac{5}{4}\frac{r^3}{3} - \frac{1}{R}\frac{r^4}{4}\right]$$

As electric field intensity,

$$E = \frac{kq}{r^2} = \frac{1}{4\pi\varepsilon_0}\frac{1}{r^2}\cdot 4\pi\rho_0\left[\frac{5}{4}\left(\frac{r^3}{3}\right) - \frac{r^4}{4R}\right]$$
$$= \frac{\rho_0 r}{4\varepsilon_0}\left[\frac{5}{3} - \frac{r}{R}\right]$$

12. (b) Work done by conservative force does not depend on the path. Electrostatic force is a conservative force.

13. (c)

$$E4\pi r_1^2 = \frac{\int_0^{r_1}\frac{Q}{\pi R^4}r4\pi r^2\,dr}{\varepsilon_0}$$
$$\Rightarrow \quad E = \frac{Qr_1^2}{4\pi\varepsilon_0 R^4}$$

14. (d)

$$W = QdV = Q(V_q - V_p)$$
$$= -100\times(1.6\times10^{-19})\times(-4-10)$$
$$= +100\times1.6\times10^{-19}\times14 = +2.24\times10^{-16}\text{ J}$$

15. (a) Resultant on Q becomes zero only when q charges are of negative nature.

$$\therefore \quad F = \frac{1}{4\pi\varepsilon_0}\frac{Q\times Q}{(\sqrt{2}d)^2}$$
$$\Rightarrow \quad \sqrt{2}\frac{qQ}{d^2} = \frac{Q\times Q}{2d^2}$$
$$\Rightarrow \quad \sqrt{2}\times q = \frac{Q}{2}$$
$$\therefore \quad q = \frac{Q}{2\sqrt{2}} \text{ or } \frac{Q}{q} = -2\sqrt{2}$$

16. (c) Electric field at a distance r from A is

$$E = \rho \times \frac{I}{2\pi r^2}$$

$$\Rightarrow \qquad \int dV = -\int E\, dr$$

$$\Rightarrow \qquad V_C - V_B = \Delta V = -\int_a^{a+b} \frac{\rho I}{2\pi} \times \frac{dr}{r^2}$$

or
$$\Delta V = \frac{\rho I}{2\pi}\left[\frac{1}{a} - \frac{1}{a+b}\right]$$

17. (c) As, $E = \frac{V}{l} = \frac{IR}{l} = \frac{I\rho l}{t A} = \rho J$ [$\because$ current density, $J = I/A$]

From $E = \rho J = \frac{\rho I}{2\pi r^2}$

18. (a) For uniformly charged spherical shell,

$$E = 0, \quad r < R$$
$$= \frac{Q}{4\pi\varepsilon_0 r^2}, \quad r \ge R$$

19. (d) $E = -\frac{\partial V}{\partial x}\hat{\mathbf{i}} - \frac{\partial V}{\partial y}\hat{\mathbf{j}} - \frac{\partial V}{\partial z}\hat{\mathbf{k}}$

$$\Rightarrow \qquad E_x = -\frac{\partial V}{\partial x} = -\frac{d}{dx}\left[\frac{20}{x^2 - 4}\right] = \frac{40x}{(x^2-4)^2}$$

$$\Rightarrow \qquad E_x \text{ at } x = 4\,\mu\text{m} = \frac{10}{9}\text{ V }\mu\text{m}$$

and is along positive X-direction.

20. (d) Direction of **E** reverses while magnitude remains same and V remains unchanged.

21. (b) Potential at A due to charge at O,

$$V_A = \frac{1}{4\pi\varepsilon_0}\frac{(10^{-3})}{OA} = \frac{1}{4\pi\varepsilon_0}\cdot\frac{(10^{-3})}{\sqrt{(\sqrt{2})^2 + (\sqrt{2})^2}} = \frac{1}{4\pi\varepsilon_0}\cdot\frac{(10^{-3})}{2}$$

Potential at B due to charge at O,

$$V_B = \frac{1}{4\pi\varepsilon_0}\cdot\frac{(10^{-3})}{OB} = \frac{1}{4\pi\varepsilon_0}\cdot\frac{(10^{-3})}{2}$$

So, $V_A - V_B = 0$

22. (c) When the two conducting spheres are connected by a conducting wire, charge will flow from one sphere (having higher potential) to other (having lower potential) till both acquire the same potential.

Therefore, $E = \frac{V}{r}$

$$\Rightarrow \qquad \frac{E_1}{E_2} = \frac{r_2}{r_1} = \frac{2}{1} = 2:1$$

23. (a) Since, $V_2 > V_1$, so electric field will point from plate 2 to plate 1.

The electron will experience an electric force, opposite to the direction of electric field and hence move towards the plate 2.

Use work-energy theorem to find speed of electron when it strikes the plate 2.

$$\therefore \qquad \frac{m_e v^2}{2} - 0 = e(V_2 - V_1)$$

where, v is the required speed.

$$\Rightarrow \qquad \frac{9.11 \times 10^{-31}}{2} v^2 = 1.6 \times 10^{-19} \times 20$$

or
$$v = \sqrt{\frac{1.6 \times 10^{-19} \times 40}{9.11 \times 10^{-31}}}$$
$$= 2.65 \times 10^6 \text{ ms}^{-1}$$

24. (c) In a non-uniform electric field, the dipole may experience both non-zero torque as well as translational force.

25. (d) Electric field due to a charged conducting sheet of surface charge density σ is given by $E = \frac{\sigma}{\varepsilon_0\varepsilon_r}$.

where, ε_0 is the permittivity in vacuum and ε_r is the relative permittivity of medium.

Here, electrostatic force on B, $QE = \frac{Q\sigma}{\varepsilon_0\varepsilon_r}$

In equilibrium, $T\cos\theta = mg$ and $T\sin\theta = \frac{Q\sigma}{\varepsilon_0\varepsilon_r}$

Thus, $\tan\theta = \frac{Q\sigma}{\varepsilon_0\varepsilon_r mg} \Rightarrow \tan\theta \propto \sigma$

26. (a) Suppose that a point B, where net electric field is zero due to charges $8q$ and $-2q$.

$$\mathbf{E}_{BO} = \frac{-1}{4\pi\varepsilon_0}\cdot\frac{8q}{a^2}\hat{\mathbf{i}}, \quad \mathbf{E}_{BA} = \frac{1}{4\pi\varepsilon_0}\cdot\frac{+2q}{(a+L)^2}\hat{\mathbf{i}}$$

According to condition,

$$\mathbf{E}_{BO} + \mathbf{E}_{BA} = 0$$

$$\Rightarrow \qquad \frac{1}{4\pi\varepsilon_0}\frac{8q}{a^2} = \frac{1}{4\pi\varepsilon_0}\frac{2q}{(a+L)^2} \quad \text{or} \quad \frac{2}{a} = \frac{1}{a+L}$$

or $\qquad 2a + 2L = a$

or $\qquad 2L = -a$

Thus, at distance $2L$ from origin, net electric field will be zero.

27. (b) V_A = Potential due to charge $+q$ on ring A
+ Potential due to charge $-q$ on ring B

$$= \frac{1}{4\pi\varepsilon_0}\left(\frac{q}{R} - \frac{q}{d_1}\right)$$

$$= \frac{1}{4\pi\varepsilon_0}\left(\frac{q}{R} - \frac{q}{\sqrt{R^2 + d^2}}\right) \qquad [\because d_1 = \sqrt{R^2 + d^2}] \ldots \text{(i)}$$

Similarly, $V_B = \frac{1}{4\pi\varepsilon_0}\left(-\frac{q}{R} + \frac{q}{\sqrt{R^2 + d^2}}\right)$

Potential difference, $V_A - V_B$

$$= \frac{1}{4\pi\varepsilon_0}\left(\frac{q}{R} - \frac{q}{\sqrt{R^2+d^2}}\right) - \frac{1}{4\pi\varepsilon_0}\left(\frac{-q}{R} + \frac{q}{\sqrt{R^2+d^2}}\right)$$

$$= \frac{1}{4\pi\varepsilon_0}\frac{q}{R} + \frac{1}{4\pi\varepsilon_0}\frac{q}{R} - \frac{1}{4\pi\varepsilon_0}\frac{q}{\sqrt{R^2+d^2}} - \frac{1}{4\pi\varepsilon_0}\frac{q}{\sqrt{R^2+d^2}}$$

$$= \frac{q}{2\pi\varepsilon_0}\left(\frac{1}{R} - \frac{1}{\sqrt{R^2+d^2}}\right)$$

28. (d) Initially

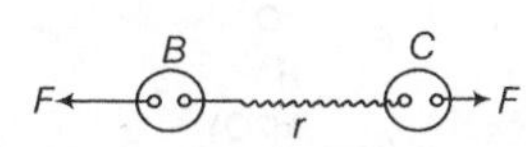

$$F = \frac{kq_1q_2}{r^2} = \frac{kq^2}{r^2} \text{ (repulsive)}$$

When a third conductor is contacted with B, they share equal charges (As conductors are similar in size).

So, charge on B is $\frac{q}{2}$, and on IIIrd conductor $\frac{q}{2}$.

When IIIrd conductor touches C, charge is equally shared,

$\therefore$ C has a charge of $\frac{q + \frac{q}{2}}{2} = \frac{3q}{4}$ finally.

$\therefore$ Force of repulsion is

$$F' = \frac{k\left(\frac{q}{2}\right)\left(\frac{3}{4}q\right)}{r^2} = \frac{kq^2}{r^2} \times \frac{3}{8} = \frac{3}{8}F$$

29. (d) At closest distance, all of KE becomes PE of system.

$$\therefore \qquad \frac{1}{2}mv^2 = \frac{kQq}{r} \qquad \text{...(i)}$$

When speed is $2v$,

$$\frac{1}{2}m(2v)^2 = \frac{kqQ}{r'} \qquad \text{...(ii)}$$

Dividing Eq.(i) by Eq (ii), we have,

$$\frac{1}{4} = \frac{r'}{r} \Rightarrow r' = \frac{r}{4}$$

30. (a)

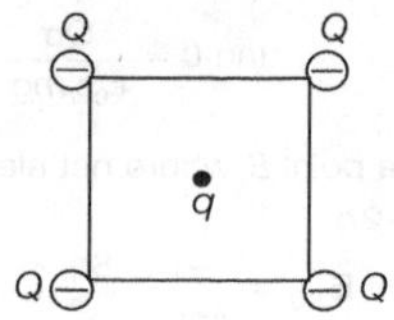

The system is in equilibrium net force on any of charge will be zero.

Consider any corner's charge,

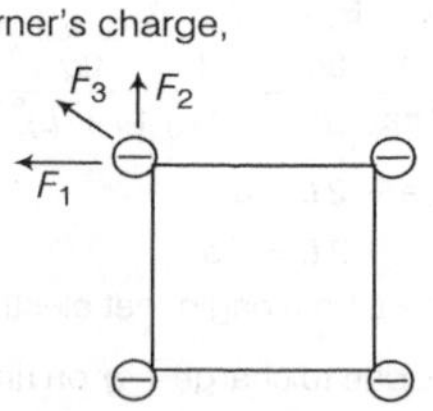

Resultant of all three forces is $\frac{\sqrt{2}kQ^2}{a^2} + \frac{kQ^2}{(\sqrt{2}a)^2}$

This must be balanced by attraction of charge at centre.

$$\frac{kQq}{\left(\frac{1}{2}\sqrt{2}a\right)^2} = \frac{kQq}{a^2/2}$$

On equating, we get

$$\frac{2kQq}{a^2} = \frac{\sqrt{2}kQ^2}{a^2} + \frac{kQ^2}{2a^2}$$

$$2q = \sqrt{2}Q + \frac{Q}{2}$$

$$q = \frac{Q}{\sqrt{2}} + \frac{Q}{4} = Q\left(\frac{1}{\sqrt{2}} + \frac{1}{4}\right) = \frac{Q}{4}(1 + 2\sqrt{2})$$

31. (a) Net flux $= \frac{q}{\varepsilon_0} \Rightarrow q = \varepsilon_0(\text{net flux}) = \varepsilon_0(\phi_2 - \phi_1)$

32. (c) Potential at P = Potential due to charge on surface of shell + potential due to charge Q at a distance of $R/2$.

$$V_p = \frac{kq}{R} + \frac{kQ}{R/2} = \frac{kq}{R} + \frac{2kQ}{R} = \frac{1}{4\pi\varepsilon_0}\left(\frac{q}{R} + \frac{2Q}{R}\right)$$

33. (b) Forces on q_1 are

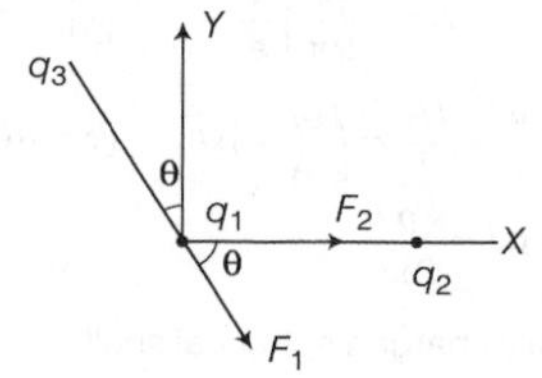

F_1 has a component along X-axis $= F_1 \cos\theta = \frac{kq_1q_3}{a^2}\cos\theta$

Also, F_2 is along X-axis, $F_2 = \frac{kq_1q_2}{b^2}$

$\therefore$ Total force on q_2 along X-axis is

$$k\left(\frac{q_1q_3}{a^2}\cos\theta + \frac{q_1q_2}{b^2}\right) = kq_1\left(\frac{q_3}{a^2}\cos\theta + \frac{q_2}{b^2}\right)$$

34. (d) Flux through $ABCD$ is only due to charge inside the cube. As, charge is centrally placed, flux $= \frac{1}{6} \cdot \frac{q}{\varepsilon_0}$.

35. (d) Clearly, charges Q and q must be opposite in nature.

Also, force balance is

$$\frac{kQ^2}{d^2} = \frac{kQq}{(d/2)^2}$$

$$\frac{kQ^2}{d^2} = \frac{4kQq}{d^2}$$

$$\therefore \qquad q = \frac{Q}{4} \quad \text{i.e.} \quad q = \frac{-Q}{4}$$

36. (a) Work done $= q(\Delta V)$

$$2 = 20(\Delta V)$$

$$\therefore \qquad \Delta V = \frac{2}{20} = \frac{1}{10} = 0.1\,\text{V}$$

JEE Advanced & IIT JEE Archive

1. (c)

$$E_1 = \frac{kQ}{R^2}, \qquad \text{where } k = \frac{1}{4\pi\varepsilon_0}$$

$$E_2 = \frac{k(2Q)}{R^2} \Rightarrow E_2 = \frac{2kQ}{R^2}$$

$$E_3 = \frac{k(4Q)R}{(2R)^3} \Rightarrow E_3 = \frac{kQ}{2R^2}$$

$$\therefore \qquad E_2 > E_1 > E_3$$

2. (a) (i) Component of forces along X-axis will vanish. Net force along positive Y-axis

$$E_3 < E_1 < E_2$$

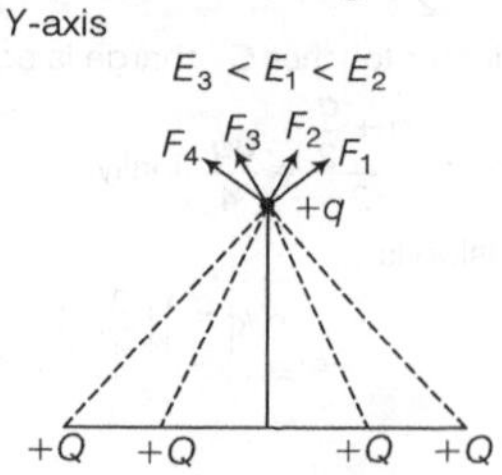

(ii) Component of forces along Y-axis will vanish. Net force along positive X-axis

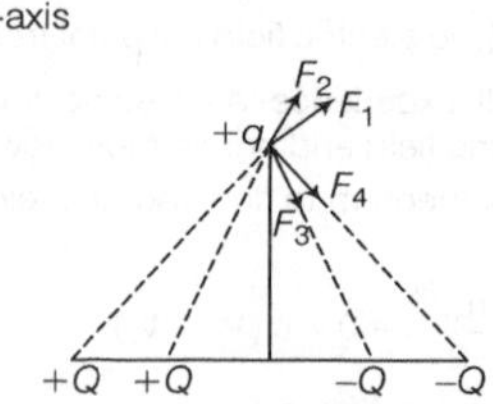

(iii) Component of forces along X-axis will vanish. Net force along negative Y-axis

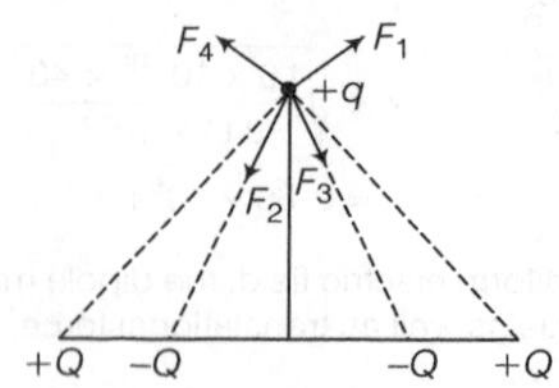

(iv) Component of forces along Y-axis will vanish. Net force along negative X-axis

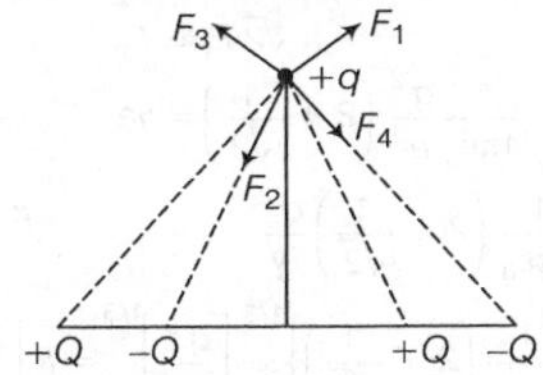

(i) → (r), (ii) → (p), (iii) → (s), (iv) → (q)

3. (c) $\frac{Q}{4\pi\varepsilon_0 r_0^2} = \frac{\lambda}{2\pi\varepsilon_0 r_0} = \frac{\sigma}{2\varepsilon_0} \Rightarrow Q = 2\pi\sigma r_0^2$

(a) is incorrect, $r_0 = \frac{\lambda}{\pi\sigma}$

(b) is incorrect, $E_1\left(\frac{r_0}{2}\right) = 4E_1(r_0)$

As,

$$E_1 \propto \frac{1}{r^2}$$

$\Rightarrow$ $E_2\left(\frac{r_0}{2}\right) = 2E_2(r_0)$ as $E_2 \propto \frac{1}{r}$

(c) is correct

$$E_1\left(\frac{r_0}{2}\right) = E_2(r_0)$$

As, $E_3 \propto r_0 \Rightarrow$ (d) is incorrect

4. (b,d) **At point P**

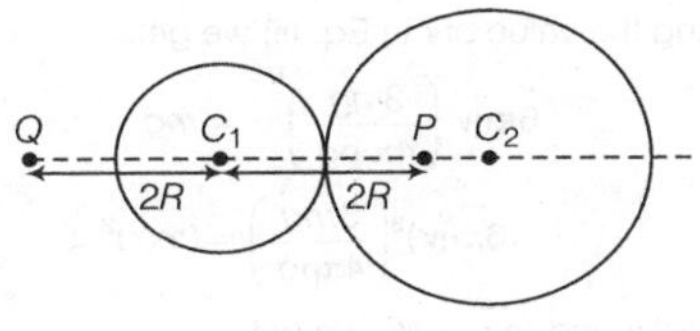

If resultant electric field is zero, then

$$\frac{KQ_1}{4R^2} = \frac{KQ_2}{8R^3}R$$

$$\frac{\rho_1}{\rho_2} = 4$$

At point Q

If resultant electric field is zero, then

$$\frac{KQ_1}{4R^2} + \frac{KQ_2}{25R^2} = 0$$

$$\frac{\rho_1}{\rho_2} = -\frac{32}{25} \quad (\rho_1 \text{ must be negative})$$

5. (c,d)

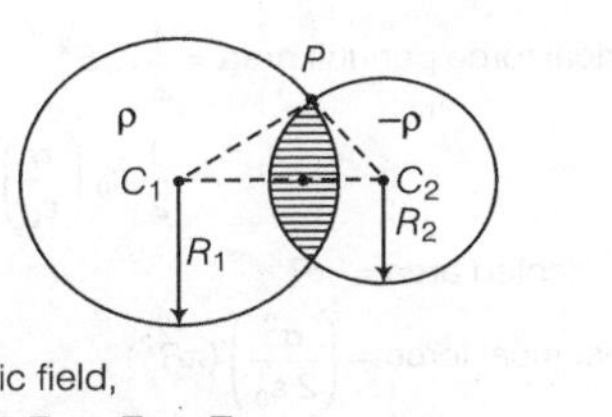

For electrostatic field,

$$\mathbf{E}_P = \mathbf{E}_1 + \mathbf{E}_2$$

$$= \frac{\rho}{3\varepsilon_0}\mathbf{C_1P} + \frac{(-\rho)}{3\varepsilon_0}\mathbf{C_1P}$$

$$= \frac{\rho}{3\varepsilon_0}(\mathbf{C_1P} + \mathbf{PC_2})$$

$$= \frac{\rho}{3\varepsilon_0}\mathbf{C_1C_2}$$

6. (a,b,c)

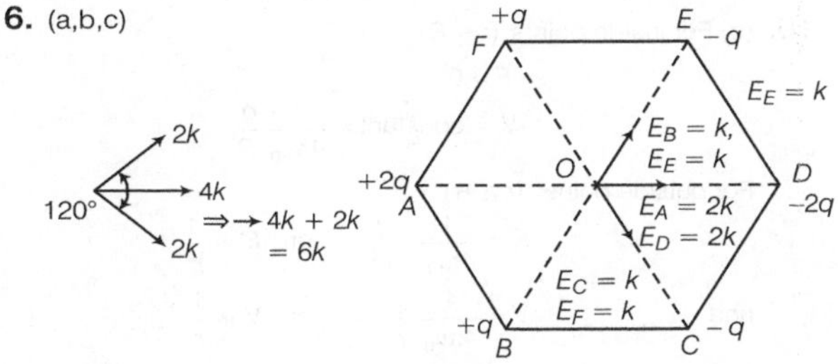

Resultant of $2k$ and $2k$ (at 120°) is also $2k$ towards $4k$. Therefore, net electric field is $6K$.

(b) $V_0 = \frac{1}{4\pi\varepsilon_0}\left[\frac{q_A}{L} + \frac{q_B}{L} + \frac{q_C}{L} + \frac{q_D}{L} + \frac{q_E}{L} + \frac{q_F}{L}\right]$

$$= \frac{1}{4\pi\varepsilon_0 L}(q_A + \ldots + q_F) = 0$$

Because $q_A + q_B + q_C + q_D + q_E + q_F = 0$

(c) Only line PR, potential is same (= 0).

7. (a,c) Option (a) is correct due to symmetry.

Option (b) is incorrect due to symmetry.

Option (c) is correct because as per Gauss' theorem, net electric flux passing through any closed surface = $\frac{q_{in}}{\varepsilon_0}$

Here, $q_{in} = 3q - q - q = q$

$\therefore$ Net electric flux = $\frac{q}{\varepsilon_0}$

Hence, option (d) is incorrect because there is no symmetry in two given planes.

8. (6) Volume of cylinder per unit length (l = 1) is

$$V = \pi R^2 l = (\pi R^2)$$

$\therefore$ Charge per unit length,

$$\lambda = (\text{volume per unit length}) \times (\text{volume charge density}) = (\pi R^2\rho)$$

Now at P, $E_R = E_T - E_C$

R = Remaining portion

T = Total portion and

C = Cavity

$\therefore$ $E_R = \frac{\lambda}{2\pi\varepsilon_0(2R)} - \frac{1}{4\pi\varepsilon_0}\frac{Q}{(2R)^2}$

Q = charge on sphere

$$= \frac{4}{3}\pi\left(\frac{R}{2}\right)^3\rho = \frac{\pi R^3\rho}{6}$$

Substituting the values, we have

$$E_R = \frac{(\pi R^2\rho)}{4\pi\varepsilon_0 R} - \frac{1}{4\pi\varepsilon_0}\cdot\frac{(\pi R^3\rho/6)}{4R^2} = \frac{23\rho R}{96\varepsilon_0} = \frac{23\rho R}{(16)(6)\varepsilon_0}$$

$\therefore k = 6$

9. (c)

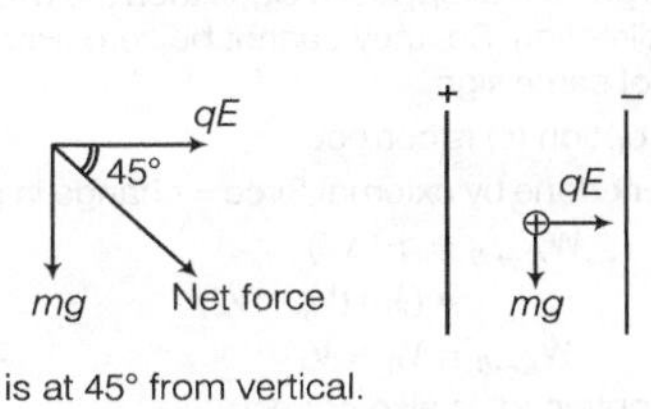

Net force is at 45° from vertical.

$\therefore$ $qE = mg$

or $\frac{qX}{d} = mg$ $\left(\because E = \frac{X}{d}\right)$

or $X = \frac{mgd}{q} = \frac{(1.67 \times 10^{-27})(9.8)(10^{-2})}{(1.6 \times 10^{-19})}$

$$\approx 1 \times 10^{-9}\text{ V}$$

10. (d) For inside points, $(r \le R)$

$$E = 0$$

$$V = \text{constant} = \frac{1}{4\pi\varepsilon_0}\frac{q}{R}$$

For outside points, $(r \ge R)$

$$E = \frac{1}{4\pi\varepsilon_0}\cdot\frac{q}{r^2} \quad \text{or} \quad E \propto \frac{1}{r^2}$$

and $$V = \frac{1}{4\pi\varepsilon_0}\frac{q}{r} \quad \text{or} \quad V \propto \frac{1}{r}$$

On the surface, $(r = R)$

$$V = \frac{1}{4\pi\varepsilon_0}\frac{q}{R}$$

$$\Rightarrow \quad E = \frac{1}{4\pi\varepsilon_0}\cdot\frac{q}{R^2} = \frac{\sigma}{\varepsilon_0}$$

where, $\sigma = \frac{q}{4\pi R^2}$ = surface charge density corresponding to above equations the correct graphs are shown in option (d).

11. (c) Electric flux, $\phi = \mathbf{E}\cdot\mathbf{S}$

or $$\phi = ES\cos\theta$$

Here, θ is the angle between **E** and **S**.

In this question $\theta = 45°$, because **S** is perpendicular to surface.

$$E = E_0$$

$$S = (\sqrt{2}a)(a) = \sqrt{2}a^2$$

$$\therefore \quad \phi = (E_0)(\sqrt{2}a^2)\cos 45° = E_0 a^2$$

Hence, option (c) is correct.

12. (a,b,c,d) Inside a conducting shell electric field is always zero. Therefore, option (a) is correct. When the two are connected, their potentials become the same.

$$\therefore \quad V_A = V_B$$

or $$\frac{Q_A}{R_A} = \frac{Q_B}{R_B} \qquad \left(V = \frac{1}{4\pi\varepsilon_0}\frac{Q}{R}\right)$$

Since, $$R_A > R_B$$

$$\therefore \quad Q_A > Q_B$$

Potential is also equal to, $V = \frac{\sigma R}{\varepsilon_0}$

$$V_A = V_B$$

$$\therefore \quad \sigma_A R_A = \sigma_B R_B \quad \text{or} \quad \frac{\sigma_A}{\sigma_B} = \frac{R_B}{R_A}$$

or $$\sigma_A < \sigma_B$$

Electric field on surface, $E = \frac{\sigma}{\varepsilon_0}$ or $E \propto \sigma$

Since, $$\sigma_A < \sigma_B$$

$$\therefore \quad E_A < E_B$$

Hence, options (a), (b), (c) and (d) are correct.

13. (c,d) If charges are of opposite signs then the two fields are along the same direction. So, they cannot be zero. Hence, the charges should be of same sign.

Therefore, option (c) is correct.

Further, work done by external force = change in potential energy

$$\therefore \quad W_{A\to B} = q\,(\Delta V)$$

$$= (+1)(V_B - V_A)$$

or $$W_{A\to B} = V_B - V_A$$

Therefore, option (d) is also correct.

14. (3) F_1 = Net electrostatic force on any one charge due to rest of three charges

$$= \frac{1}{4\pi\varepsilon_0}\frac{q^2}{a^2}\left(\sqrt{2} + \frac{1}{2}\right)$$

F_2 = Surface tension force = γa

If we see the equilibrium of line BC, then

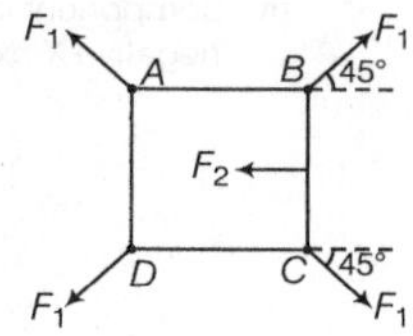

$$2F_1\cos 45° = F_2$$

or $$\sqrt{2}F_1 = F_2$$

or $$\frac{1}{4\pi\varepsilon_0}\frac{q^2}{a^2}\left(2 + \frac{1}{\sqrt{2}}\right) = \gamma a$$

$$\therefore \quad a^3 = \frac{1}{4\pi\varepsilon_0}\left(2 + \frac{1}{\sqrt{2}}\right)\frac{q^2}{\gamma}$$

or $$a = \left\{\frac{1}{4\pi\varepsilon_0}\left(2 + \frac{1}{\sqrt{2}}\right)\right\}^{1/3}\left[\frac{q^2}{\gamma}\right]^{1/3} = k\left[\frac{q^2}{\gamma}\right]^{1/3}$$

where, $k = \left\{\frac{1}{4\pi\varepsilon_0}\left(2 + \frac{1}{\sqrt{2}}\right)\right\}^{1/3}$

Therefore, $N = 3$

15. (a,d) From the behaviour of electric lines, we can say that Q_1 is positive and Q_2 is negative. Further, $|Q_1| > |Q_2|$

At some finite distance to the right of Q_2, electric field will be zero. Because electric field due to Q_1 is towards right (away from Q_1) and due to Q_2 is towards left (towards Q_2). But since magnitude of Q_1 is more, the two fields may cancel each other because distance of that point from Q_1 will also be more.

16. (d) $$qE = mg \qquad \ldots\text{(i)}$$

$$6\pi\eta rv = mg$$

$$\frac{4}{3}\pi r^3\rho g = mg \qquad \ldots\text{(ii)}$$

$$\therefore \quad r = \left(\frac{3mg}{4\pi\rho g}\right)^{1/3} \qquad \ldots\text{(iii)}$$

Substituting the value of r in Eq. (ii) we get,

$$6\pi\eta v\left(\frac{3mg}{4\pi\rho g}\right)^{1/3} = mg$$

or $$(6\pi\eta v)^3\left(\frac{3mg}{4\pi\rho g}\right) = (mg)^3$$

Again substituting $mg = qE$, we get,

$$(qE)^2 = \left(\frac{3}{4\pi\rho g}\right)(6\pi\eta v)^3$$

or $$qE = \left(\frac{3}{4\pi\rho g}\right)^{1/2}(6\pi\eta v)^{3/2}$$

$$\therefore \quad q = \frac{1}{E}\left(\frac{3}{4\pi\rho g}\right)^{1/2}(6\pi\eta v)^{3/2}$$

Substituting the values we get ,

$$q = \frac{7}{81\pi\times 10^5}\sqrt{\frac{3}{4\pi\times 900\times 9.8}}\times 216\pi^3 \times\sqrt{(1.8\times 10^{-5}\times 2\times 10^{-3})^3}$$

$$= 8.0\times 10^{-19}\ \text{C}$$

17. (a) Electrical force per unit area $= \frac{1}{2}\varepsilon_0 E^2$

$$= \frac{1}{2}\varepsilon_0\left(\frac{\sigma}{\varepsilon_0}\right)^2 = \frac{\sigma^2}{2\varepsilon_0}$$

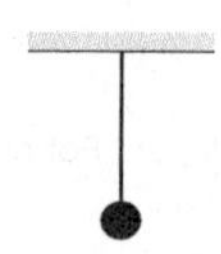

Projected area $= \pi R^2$

$$\therefore \quad \text{Net electrical force} = \left(\frac{\sigma^2}{2\varepsilon_0}\right)(\pi R^2)$$

In equilibrium, this force should be equal to the applied force.

$$\therefore \quad F = \frac{\pi\sigma^2R^2}{2\varepsilon_0}$$

or $$F \propto \frac{\sigma^2R^2}{\varepsilon_0}$$

18. (a) Net torque on $(-q)$ about a point (say P) lying over $+Q$ is zero. Therefore, angular momentum of $(-q)$ about point P should remain constant.

19. (a) Total enclosed charge as already shown is

$$q_{\text{net}} = \frac{6C}{2} + \frac{8C}{4} - 7C = -2C$$

From Gauss' theorem, net flux, $\phi_{\text{net}} = \dfrac{q_{\text{net}}}{\varepsilon_0} = \dfrac{-2C}{\varepsilon_0}$

20. (b) $Q_1 = \sigma(4\pi R^2) = 4\pi\sigma R^2$

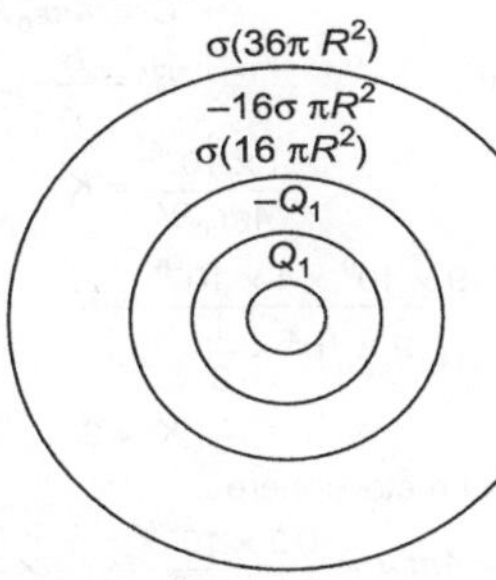

$$Q_2 = 16\pi\sigma R^2 - Q_1 = 12\pi\sigma R^2$$

$$Q_3 = 36\pi\sigma R^2 - 16\pi\sigma R^2 = 20\pi\sigma R^2$$

$$Q_1 : Q_2 : Q_3 = 1 : 3 : 5$$

21. (a) (i) → (p,r,s), (ii) → (r,s), (iii) → (p,q,t). (iv) → (r,s)

22. (2) From Gauss theorem,

$$E \propto \frac{q}{r^2} \qquad (q = \text{charge enclosed})$$

$$\therefore \quad \frac{E_2}{E_1} = \frac{q_2}{q_1} = \frac{r_1^2}{r_2^2}$$

or
$$8 = \frac{\int_0^R (4\pi r^2)kr^a dr}{\int_0^{R/2} (4\pi r^2)kr^a dr} \times \frac{(R/2)^2}{(R)^2}$$

Solving this equation we get, $a = 2$.

23. (b) No solution is required.

24. (a) At $r = R$, from Gauss' law

$$E\,(4\pi R^2) = \frac{q_{\text{net}}}{\varepsilon_0} = \frac{Ze}{\varepsilon_0}$$

or
$$E = \frac{1}{4\pi\varepsilon_0} \cdot \frac{Ze}{R^2}$$

E is independent of a.

Hence, option (a) is correct.

25. (b) For $a = 0 \Rightarrow \rho(r) = \left(-\dfrac{d}{R} \cdot r + d\right)$

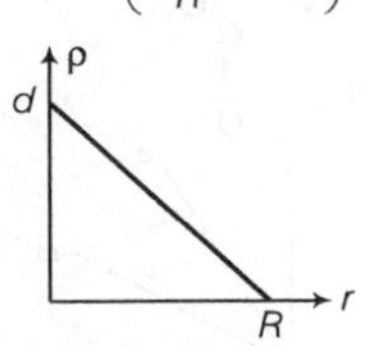

Now, $\int_0^R (4\pi r^2)\left(d - \dfrac{d}{R}r\right)dr = \text{net charge} = Ze$

Solving this equation, we get $d = \dfrac{3Ze}{\pi R^3}$

26. (c) In case of solid sphere of charge of uniform volume density

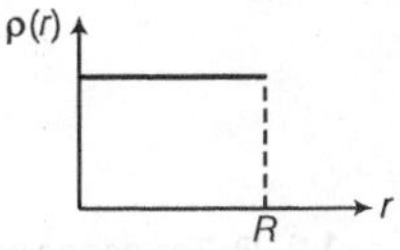

$$E = \frac{1}{4\pi\varepsilon_0} \cdot \frac{q}{R^3} \cdot r$$

or $\quad E \propto r$

Thus, for E to be linearly dependent on r, volume charge density should be constant.

or $\quad a = R$

27. (c) Distance $BC = AB\sin 60°$

$$= (2R)\frac{\sqrt{3}}{2} = \sqrt{3}R$$

$$\therefore \quad |Fe_{BC}| = \frac{1}{4\pi\varepsilon_0} \frac{(q/3)(2q/3)}{(\sqrt{3}R)^2}$$

$$= \frac{q^2}{54\pi\varepsilon_0 R^2}$$

28. (a) There will be an electric field between two cylinders (using Gauss theorem). This electric field will produce a potential difference.

29. (d) Charge will be induced in the conducting sphere, but net charge on it will be zero.

30. (b) Inside the cavity, field at any point is uniform and non-zero.

31. (c) $A \equiv (-a, 0, 0)$, $B \equiv (0, a, 0)$

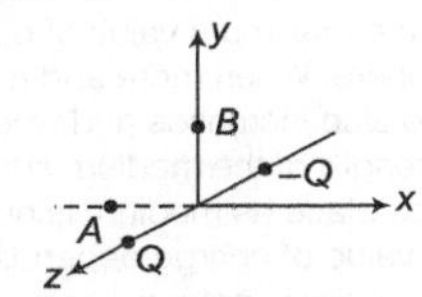

Point charge is moved from A to B

$$V_A = V_B = 0$$

$$\therefore \quad W = 0$$

32. (a,b,c,d) The given graph is of charged conducting sphere of radius R_0. The whole charge q distributes on the surface of the sphere.

33. (b) All the three plates will produce electric field at P along negative Z-axis. Hence,

$$\mathbf{E}_p = \left[\frac{\sigma}{2\varepsilon_0} + \frac{2\sigma}{2\varepsilon_0} + \frac{\sigma}{2\varepsilon_0}\right](-\hat{\mathbf{k}}) = -\frac{2\sigma}{\varepsilon_0}\hat{\mathbf{k}}$$

34. Let q be the charge on the bubble, then

$$V = \frac{kq}{a} \qquad \left(\text{Here } k = \frac{1}{4\pi\varepsilon_0}\right)$$

$$\therefore \quad q = \frac{Va}{k}$$

Let after collapsing, the radius of droplet becomes R, then equating the volume, we have

$$(4\pi a^2)t = \frac{4}{3}\pi R^3$$

$$\therefore \quad R = (3a^2t)^{1/3}$$

Now, potential of droplet will be

$$V' = \frac{kq}{R}$$

Substituting the values, we have

$$V' = \frac{(k)\left(\dfrac{Va}{k}\right)}{(3a^2t)^{1/3}}$$

or
$$V' = V\left(\frac{a}{3t}\right)^{1/3}$$

2. Capacitance

DPP-1 Capacitance

1. (i) The capacity of a system consisting of two isolated conductors (plates) with equal and opposite charges $+q$ and $-q$ is defined as $C = \frac{q}{V}$

So, capacity of given system

$$C = \frac{20 \times 10^{-12}}{20} = 1 \times 10^{-12}\,F = 1\,pF$$

(ii) The capacity of a system of plates (or conductors) depends only on the geometry of the plates and not on their charge or potential difference.

So, $C = 1$ pF when charges are changed.

(iii) The capacity still remains the same $C = 1$ pF

2. (i) The capacitance of an isolated sphere is $C = 4\pi\varepsilon_0 R$, when it is placed in vacuum $C = 4\pi\varepsilon_0 KR$ when it is placed in a medium of dielectric constant K.

As, $C = \frac{Q}{V}$, we get

$$Q = CV$$

$$\Rightarrow \quad Q_{max} = CV_{max} = 4\pi\varepsilon_0 KR \cdot V_{max}$$

where, V_{max} is the maximum value of potential which can be given to the sphere. When more and more charge is given to sphere, its field also increases and when this reaches breakdown strength of the medium surrounding the sphere, discharge takes place as medium is no longer insulating. So, maximum value of charge depends on the radius of sphere and breakdown potential or breakdown strength of the medium surrounding the sphere.

(ii) Dielectric strength of a medium is the maximum value of field, the medium can withstand without being ionised.

When applied field exceeds the dielectric strength of medium, it becomes conductive and sparks flow across the medium.

Dielectric strength or breakdown strength of dry air in room is around 3×10^6 V/m.

With $q = 1$ C, $r = 1$ cm $= 10^{-2}$ m

Field on the surface of sphere is $E = \frac{Kq}{r^2}$

$$\Rightarrow \quad E = \frac{9 \times 10^9 \times 1}{(10^{-2})^2} = 9 \times 10^{13}\left(\frac{V}{m}\right)$$

This value is very high compared to dielectric strength of dry air which is $3 \times 10^6 \frac{V}{m}$. So, air in surroundings is ionised well before sphere gets 1C charge and charge leaks off the surface of sphere. Hence, it is not possible to give 1C charge to a sphere of 1 cm radius when it is placed in air.

(iii) Let Q is the maximum charge on the surface, then with this charge, the field at the surface of sphere of radius R is

$$E = K\frac{Q}{R} = \frac{1}{4\pi\varepsilon_0}\frac{Q}{R}$$

As, the field must be below the breakdown strength of the air surrounding the sphere.

$$\therefore \quad E \le 3 \times 10^6 \text{ V/m} = \frac{1}{4\pi\varepsilon_0} \times \frac{Q}{R}$$

$$\Rightarrow \quad Q_{max} = 4\pi\varepsilon_0 R \times 3 \times 10^6 = R \times \frac{3 \times 10^6}{9 \times 10^9}$$

$$\text{or} \quad Q_{max} = \frac{R}{3 \times 10^3} \approx 3.34 \times 10^{-4} RC$$

3. (i) $C = 4\pi\varepsilon_0 R = 1\,\mu F$

$$R = \frac{10^{-6}}{4\pi\varepsilon_0 R} = 9 \times 10^9 \times 10^{-6} = 9\,m$$

To reduce the radius, we surround the sphere with a dielectric medium of dielectric strength k.

Then, $C = 4\pi\varepsilon_0 KR'$

Here, $C = 1\,\mu F$, $R' = \frac{R}{3}$

$$\Rightarrow \quad \frac{1 \times 10^{-6}}{4\pi\varepsilon_0 R'} = K$$

$$\Rightarrow \quad \frac{9 \times 10^9 \times 1 \times 10^{-6}}{9 \times 10^3 \times \frac{1}{3}} = K$$

$$\Rightarrow \quad K = 3$$

(ii) Capacitance of metal sphere

$$C = 4\pi\varepsilon_0 r = \frac{0.5 \times 10^{-3}}{9 \times 10^9} = \frac{1}{18} \times 10^{-12}\,F$$

Rate of escaping of charge from surface $= \frac{80}{100} \times ne$

$$= \frac{80}{100} \times 6.25 \times 10^{10} \times 1.6 \times 10^{-19} = 8 \times 10^{-9}\,\frac{C}{r}$$

If t = time required, then $q = 8 \times 10^{-9} \times t$

and $C = \frac{q}{V} \Rightarrow q = CV$

$$\Rightarrow \quad 8 \times 10^{-9} \times t = \frac{1}{18} \times 10^{-12} \times 1.0$$

$$\Rightarrow \quad t = \frac{10^{-12}}{8 \times 10^{-9} \times 18} = 6.95\,\mu s$$

(iii) Dielectric strength is maximum field value upto which a medium stays non-conducting. Dielectric constant of a medium is a measure of extent of polarisation of the medium.

(iv) Maximum value of charge depends inversely on distance and also on the shape of neighbouring objects. Pointed objects cause polarisation of medium much earlier.

4. (i) Capacitance C of a capacitor is a ratio of charge to potential of capacitor. It is a constant.

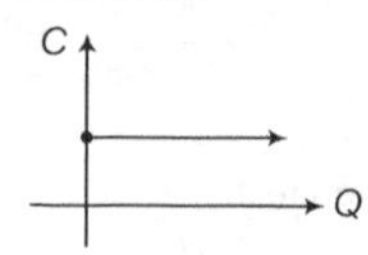

(ii) $C = \frac{Q}{V} \quad \Rightarrow \quad V = \frac{1}{C}Q$

So, slope of graph $= \frac{1}{C}$

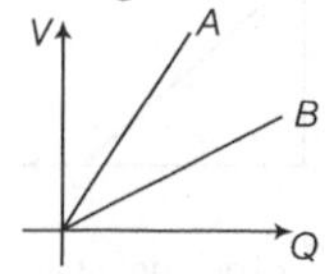

Slope of A > Slope of B

$$\Rightarrow \quad \frac{1}{C_A} > \frac{1}{C_B}$$

$$\Rightarrow \quad C_B > C_A$$

$\therefore$ Capacitance of B is more than that of A.

(iii) When we take charge Q on Y-axis, then slope ≡ capacity

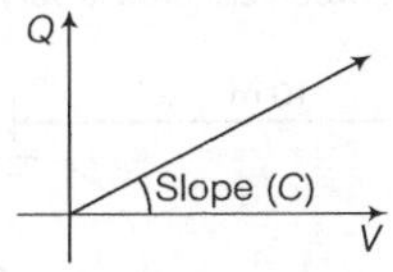

Also, area of graph $= \frac{1}{2}QV$ = energy of the capacitor

(iv) $\because C = \frac{Q}{V} \Rightarrow Q = CV$

$\therefore$ Charge flowing through ammeter $= 25 \times 10^{-6} \times 120$

$$= 3000 \times 10^{-6}$$
$$= 3 \times 10^{-3} \text{ C}$$
$$= 3 \text{ mC}$$

5. (c) Capacitance of a capacitor is a constant depending on the shape and size of the capacitor.

$\therefore$ The capacity in discharged state is 5 μF.

6. (d) In a charged capacitor, energy resides in the form of potential energy of the electric field established between dissimilar charged plates.

7. (b) Energy is stored in the form of potential energy of the electric field.

8. (a) Consider two different paths, one is directly to the conductor from infinity and another is from infinity to insulating object to the conductor.

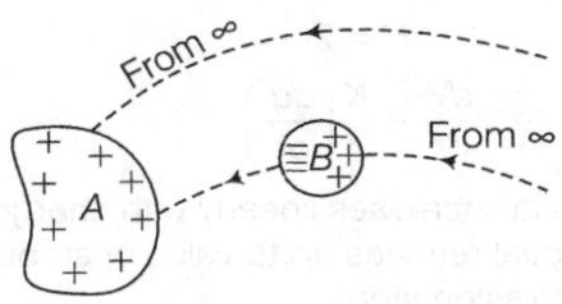

Work done in taking a charge from ∞ to conductor is equal to work done in taking a charge from ∞ to insulator and then to conductor.

$\therefore W_{\infty \to \text{conductor}} = W_{\infty \to \text{insulator}} + W_{\text{insulator} \to \text{conductor}}$

$$\Rightarrow \quad \frac{W_{\infty \to C}}{q} = \frac{W_{\infty \to \text{ins}}}{q} + \frac{W_{\text{ins} \to C}}{q}$$

$$\Rightarrow \quad V_A = V_B + V_{AB}$$

Hence, option (a) is true.

9. (d) Let radius and charge of a small drop is r and q, respectively.

Then, σ_{small} = surface charge density of small drop

$$= \frac{\text{Charge}}{\text{Surface area}} = \frac{q}{4\pi r^2}$$

When 64 such drops are combined to form a large drop (radius = R), Then

Volume of large drop = 64 × Volume of 1 small drop

$$\frac{4}{3}\pi R^3 = 64 \times \frac{4}{3}\pi r^3$$
$$R^3 = 64 \times r^3$$
$$R = 4r$$

So, surface charge density of large drop

$$\sigma_{\text{large}} = \frac{\text{Total charge}}{\text{Surface area}} = \frac{64\,q}{4\pi R^2}$$
$$= \frac{64q}{4\pi(4r)^2} = \frac{q}{\pi r^2}$$

$$\Rightarrow \quad \sigma_{\text{large}} = 4\left(\frac{q}{4\pi r^2}\right) = 4(\sigma_{\text{small}})$$

$$\Rightarrow \quad \frac{\sigma_{\text{small}}}{\sigma_{\text{large}}} = \frac{1}{4}$$

10. (d) Volume of large drop = 8 × Volume of each small drop

$$\Rightarrow \quad \frac{4}{3}\pi R^3 = 8 \times \frac{4}{3}\pi r^3$$

$$\Rightarrow \quad R^3 = 8r^3 \Rightarrow R = 2r$$

Ratio of large drop and small drop

$$\frac{C_{(\text{large drop})}}{C_{(\text{small drop})}} = \frac{4\pi\varepsilon_0 R}{4\pi\varepsilon_0 r} = \frac{R}{r} = 2:1$$

11. (d) Dielectric in a capacitor reduces the occurrence of a spark between plates of a capacitor (by creating an induced opposing field).

Also, dielectric constant K and dielectric strength x must be high for a suitable material for a capacitor.

12. (b) Initially, potential is zero on each capacitor. Then, net charge value $(Q_1 + Q_2)$ must be zero, i.e. magnitude of total charge must be zero.

Now, $\quad Q_1 = C_1V_1,\ Q_2 = C_2V_2$

$\Rightarrow \quad Q_1 = 120C_1$ and $Q_2 = 200C_2$

As $\quad |Q_1| = |Q_2|$

$\therefore \quad 120C_1 = 200C_2 \Rightarrow 3C_1 = 5C_2$

13. (c) During charging of a capacitor, current flows through the capacitor and it gradually decreases as capacitor is being charged. The current stops or capacitor is fully charged.

14. (c) $\because C = \frac{Q}{V}$ = constant

$$\therefore \quad C = \frac{2Q}{2V} = \frac{Q}{V}$$

Hence, charge becomes $2Q$ when V becomes 2 V.

15. (b,d) Charging battery is removed. Therefore, q = constant.

Distance between the plates is increased. Therefore, C decreases.

Now, $V = q/C$, q is constant and C is decreasing.

Therefore, V should increase.

16. (a,c) $\quad C = 4\pi\varepsilon_0 R \Rightarrow C \propto R$ Energy stored

$$U = \frac{1}{2}CV^2 = \frac{1}{2}4\pi\varepsilon_0 R \times \frac{q^2}{(4\pi\varepsilon_0)^2R^2}$$

$$\Rightarrow \quad U \propto \frac{1}{R}$$

17. (b,c) Capacity of a capacitor is a constant, but potential increases with the amount of charge stored.

18. (b,d) Capacitance $C = 4\pi\varepsilon_0 R$

Capacitance increases with increase in radius.

As, $V = \frac{Kq}{r}$ and as no further charge is given to balloon (while inflating), potential reduces as its size increases.

19. (a,d) Dielectric breakdown strength of water is less than that of vacuum.

So, more charge can be given to sphere in vacuum. As, capacitance is directly proportional to R, capacitance ratio is always 1.

20. (a,b) $\quad C = \varepsilon_0 \frac{A}{d}$

$$\Rightarrow \quad \varepsilon_0 = \frac{Cd}{A}$$

$$\Rightarrow \quad \varepsilon_0 = 8.884 \times 10^{-12} \text{ Fm}^{-1}$$
$$= 8.884 \text{ pFm}^{-1}$$

Also, $\quad [\varepsilon_0] = [\text{M}^{-1}\text{L}^{-3}\text{T}^4\text{A}^2]$

21. (a,b) A large sphere can be used as a capacitor or two differently charged plates (conductors) may be placed side by side so that potential of region decreases.

$$C = Q/V$$

22. (a) Volume of big drop = $n \times$ Volume of small drop

$\therefore \quad \frac{4}{3}\pi R^3 = n \times \frac{4}{3}\pi r^3 \Rightarrow R = n^{1/3} r$

Capacitance of big drop, $C = 4\pi\varepsilon_0 R = 4\pi\varepsilon_0 n^{1/3} r = n^{1/3} C$

Potential of big drop $V = \frac{1}{4\pi\varepsilon_0} \cdot \frac{Q}{R} = \frac{1}{4\pi\varepsilon_0} \cdot \frac{nq}{n^{1/3} r}$

$= \frac{n}{n^{1/3}}\left(\frac{1}{4\pi\varepsilon_0} \cdot \frac{q}{r}\right) = n^{2/3} v$

(i) $\rightarrow$ (p), (ii) $\rightarrow$ (p), (iii) $\rightarrow$ (q)

23. (1) As, $C = \frac{Q}{V}$

For first graph, $V = \frac{Q}{C_1}$

$\therefore$ Slope of graph $= \frac{1}{C_1}$

$\Rightarrow \quad \frac{1}{C_1} = \tan 30° = \frac{1}{\sqrt{3}}$

$\Rightarrow \quad C_1 = \sqrt{3}$ units

For second graph, $Q = CV$

$\Rightarrow$ Slope of graph $= C_2 = \tan 60° = \sqrt{3}$ units

Hence, ratio $\frac{C_1}{C_2} = 1$

24. (8) Photons incident per sec over metal surface $= 10^{11}\,\text{s}^{-1}$

Electrons emitted per second

$n = \frac{10^{11}}{10^6} = 10^5\,\text{s}^{-1}$

$\therefore$ Charge acquired by metal sphere in 111 s is

$q = n \times e \times t = 10^5 \times 1.6 \times 10^{-19} \times 111$

$= 1.78 \times 10^{-12}$ C

As, stopping potential or potential on the surface is $V = \frac{kq}{r}$

$r = \frac{9 \times 10^9 \times 1.78 \times 10^{-12}}{2} \quad [r = 8.0 \times 10^3 \text{ m}]$

So, capacitance of sphere is

$C = 4\pi\varepsilon_0 R = \frac{8 \times 10^{-3}}{9 \times 10^9}$

$\Rightarrow \quad 9 \times 10^6 \times C = 8$

25. (7) As, distance between spheres is very large than the radius of sphere.

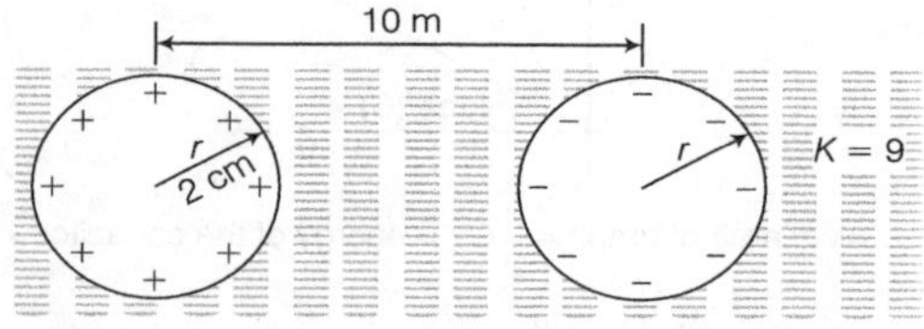

System will be a capacitor, if we place a charge $+q$ over Ist and $-q$ over IInd sphere.

Potential on the surface of Ist sphere is

$V_1 = \frac{Kq}{r_1} = \frac{q}{4\pi\varepsilon_0\varepsilon_r r}$

and potential on the surface of IInd sphere is

$V_2 = \frac{kq}{r_2} = \frac{-q}{4\pi\varepsilon_0\varepsilon_r r}$

So, potential difference between the spheres is given by

$\Delta V = V_1 - V_2 = \frac{2q}{4\pi\varepsilon_0\varepsilon_r r} = \frac{q}{C}$

where, C = capacitance of system

$\Rightarrow \quad C = 2\pi\varepsilon_0\varepsilon_r r$

$\Rightarrow \quad C = 4\pi\varepsilon_0 \cdot \frac{\varepsilon_r}{2} r$

$= \frac{9 \times 2 \times 10^{-2}}{9 \times 10^9 \times 2} = 10^7$ F

$\therefore \quad N = 7$

26. (b) $V = \frac{kq}{r} \Rightarrow \frac{dV}{dt} = \frac{K}{r}\left(\frac{dq}{dt}\right)$

Hence, potential increases linearly with charge and time but soon value of potential reaches up to value of stopping potential and then photo-emission stops.

27. (d) $C = 4\pi\varepsilon_0 R$, as R remains constant. So, C is a constant.

28. (d) Sphere gains charge or more and more electrons are leaving its surface but as $V \rightarrow V_0$, photo-emission stops and charge is constant.

DPP-2 Capacitance of Different Types of Capacitors

1. (i) $E = \frac{V}{d}$, $C = \frac{K\varepsilon_0 A}{d} \Rightarrow d = \frac{K\varepsilon_0 A}{c}$

$\Rightarrow \quad E = \frac{VC}{K\varepsilon_0 A}$

$= \frac{50 \times 100 \times 10^{-12}}{5.4 \times 8.85 \times 10^{-12} \times 100 \times 10^{-4}}$

$= 1.0 \times 10^4$ V/m

(ii) Free charge on the plates of capacitor is given by

$q_{\text{free}} = CV$

$= 100 \times 10^{-12} \times 50$

$= 5 \times 10^{-9}$ C

$= 5$ nC

(iii) Induced charge on the dielectric is given by

$q_{\text{induced}} = q_{\text{free}} - \varepsilon_0 AE$

$= (5 \times 10^{-9}) - (8.85 \times 10^{-12} \times 100 \times 10^{-4} \times 1 \times 10^4)$

$= 4.1 \times 10^{-9}$ C

$= 4.1$ nC

2. (i) Capacity of a spherical capacitor is given by

$C = 4\pi\varepsilon_0 \frac{Kab}{b-a} = \frac{4\pi\varepsilon_0 \times 40 \times 38 \times 10^{-6} \times 6}{2 \times 10^{-3}}$

$= \frac{40 \times 38 \times 10^{-6} \times 6}{9 \times 10^9 \times 2 \times 10^{-3}}$

$= 84.44 \times 10^{-12} \times 6$

≈ 507 pF

(ii) Free charge on inner shell is

$q = CV = 4\pi\varepsilon_0 K \cdot \left(\frac{ab}{b-a}\right) \cdot V$

$= \frac{6 \times 38 \times 40 \times 10^{-6}}{9 \times 10^9 \times 2 \times 10^{-3}} \times 50$

$= 2534 \times 10^{-12}$ C

$= 2.5$ nC

(iii) Induced charge is given by

$q' = \frac{K-1}{K} \cdot q = \frac{5}{6} \times 2.5 = 2.08$ nC

3. Charge appears due to induction between the plates of a capacitor. As, plates are metallic.

So, equal and opposite charges appear. Also, magnitude of charge is same over each plate.

4. (i) A dielectric weak ends the field between plates of a capacitor by producing an opposing field within its volume. This results in an increase in maximum allowable value of charge on plates without breakdown.

(ii) A dielectric polarises and its molecules (or atomic dipoles) are oriented in the direction of field.

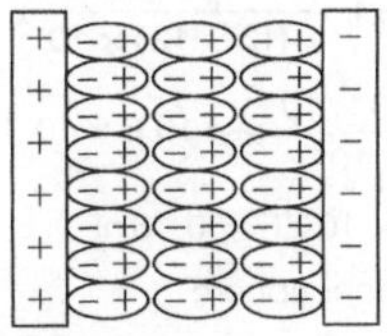

Net effect of polarisation is an appearence of a negative charge layer in front of positive plate and a positive charged layer in front of negative plate.

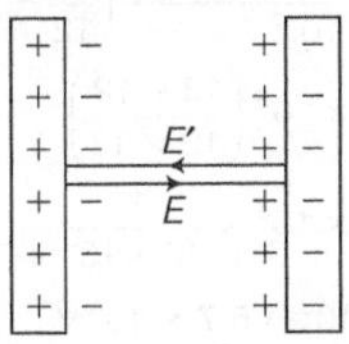

Polarised molecules produce their own field which opposes the applied field and thus, net field is reduced.

(iii) $C = \dfrac{\varepsilon_0 KA}{d}$

For a metal plate, $K = \infty$

$\therefore$ $C = \infty$

This means that capacitor accepts any amount of charge given to it. In fact, it is conducted across the capacitor.

(iv) For a sphere (solid or hollow)

$$C = 4\pi\varepsilon_0 R$$

So, charge on sphere $Q = CV = 4\pi\varepsilon_0 RV$

Thus, Q is a constant for same radius R and same potential V.

5. (a) $C = \dfrac{\varepsilon_0 A}{d} = 8\text{ pF}$

and $C' = \dfrac{\varepsilon_0 KA}{d'} = \dfrac{6\varepsilon_0 A}{d/2} = \dfrac{12\,\varepsilon_0 A}{d} = 12 \times 8 = 96\text{ pF}$

So, capacity increased by 96 pF.

6. (b) $C = \dfrac{\varepsilon_0 KA}{d} = \dfrac{\varepsilon_0 Kb \times \frac{a}{2}}{d} = \dfrac{\varepsilon_0 Kab}{2d}$

7. (d) Capacity of a capacitor depends only on its geometrical factors.

$\therefore$ $C = \dfrac{\varepsilon_0 A}{d}$

8. (a) After time t, thickness of liquid will remain $\left(\dfrac{d}{3} - vt\right)$

So, capacity after t second is

$$C = \frac{\varepsilon_0(1)}{\left(d - \dfrac{d}{3} + vt\right) + \left(\dfrac{d/3 - vt}{2}\right)} \qquad \left(\text{Using } C = \frac{\varepsilon_0 A}{d - t + \dfrac{t}{K}}\right)$$

$\Rightarrow$ $C = \dfrac{6\varepsilon_0}{5d + 3vt}\text{ F}$

9. (d) When $\theta = \pi$, then plates are out of mesh and overlap area is zero and when $\theta = 0°$, the overlap area is $\dfrac{\pi R^2}{2}$.

By proportion effective overlap area of a single sheet is $\dfrac{(\pi - \theta)R^2}{2}$ for angle θ.

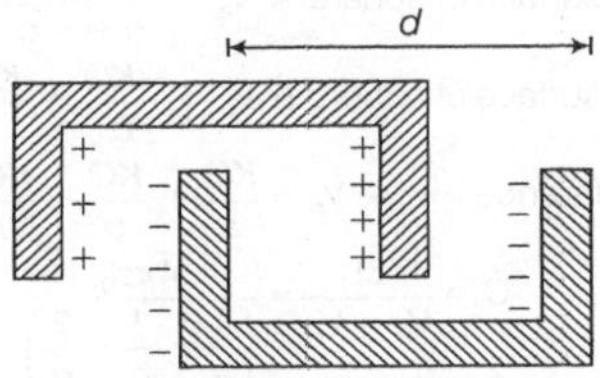

When there are two plates in combination, number of capacitors formed is 3. So, with N plates, number of capacitors is $(2N - 1)$. So, total capacity is

$$C = \frac{(2N-1)\varepsilon_0 A_{\text{effective}}}{\text{Distance}} = \frac{(2N-1)\,\varepsilon_0(\pi - \theta)R^2}{d}$$

10. (d) Let a and b are radii of inner and outer spheres with charges Q and $-Q$ (on outer). Potential of inner sphere is $V_a = \dfrac{KQ}{a}$ and of outer sphere is $V_b = \dfrac{KQ}{b}$

Potential difference, $\Delta V = V_a - V_b = \dfrac{Q}{4\pi\varepsilon_0}\left(\dfrac{b-a}{ab}\right)$

and $C = \dfrac{Q}{V_a - V_b} = \dfrac{4\pi\varepsilon_0 ab}{b - a}$

$\Rightarrow$ $C = \dfrac{4\pi\varepsilon_0 2a^2}{a}$ (As, $b = 2a$)

$\Rightarrow$ $C = 8\pi\varepsilon_0 a$

$\Rightarrow$ $a = \dfrac{C}{8\pi\varepsilon_0}$

Volume in which field exists $= \dfrac{4}{3}\pi b^3 - \dfrac{4}{3}\pi a^3 = 7\left(\dfrac{4}{3}\pi a^3\right)$

$$= \frac{7C^3}{384\pi^2\varepsilon_0^2}$$

11. (a) Using $V_b - V_a = -\int_a^b E\,dr$

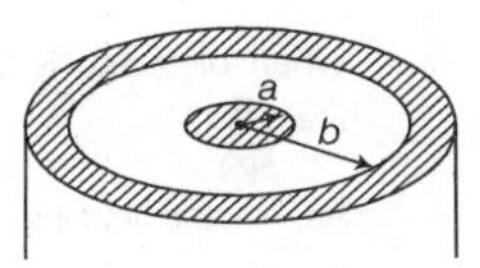

$\Rightarrow$ $V_b - V_a = -\int_a^b 2K\lambda \dfrac{dr}{r}$

$\Rightarrow$ $V_b - V_a = -2K\lambda \ln\left(\dfrac{b}{a}\right)$

As, $C = \dfrac{Q}{\Delta V}$

$C = \dfrac{Q}{2K(Q/l)\ln\left(\dfrac{b}{a}\right)}$

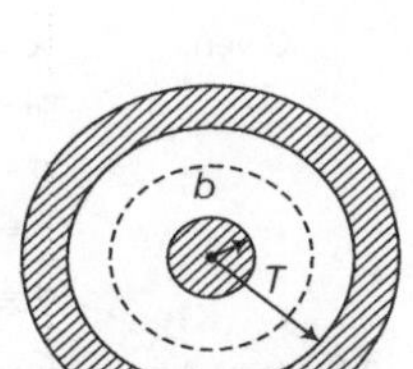

$\Rightarrow$ $C = \dfrac{l}{2K\ln\left(\dfrac{b}{a}\right)}$

$\Rightarrow$ $\dfrac{C}{l}$ = Capacity per metre length

$= \dfrac{1}{2K\ln\left(\dfrac{b}{a}\right)}$

12. (d) Increasing length increases the capacitance, but best method is to reduce the radius of outer cylinder and to increase the radius of inner cylinder. This brings the plates nearer increasing the field and increasing the capacitance.

13. (c)

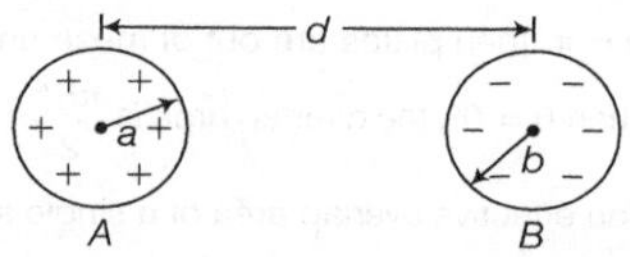

Potential at surface of sphere A, $V_a = \dfrac{KQ}{a} - \dfrac{KQ}{d}$

Potential at surface of sphere B, $V_b = \dfrac{-KQ}{b} + \dfrac{KQ}{d}$

Potential difference $= V_a - V_b = \dfrac{KQ}{a} + \dfrac{KQ}{b} - \dfrac{KQ}{d} - \dfrac{KQ}{d}$

$$\therefore \quad C = \frac{Q}{V_a - V_b} = \frac{4\pi\varepsilon_0}{\left(\frac{1}{a} + \frac{1}{b} - \frac{2}{d}\right)}$$

14. (c) Metal striped joins b and c.

So, b and c are at same potentials. Polarity of charges is as shown :

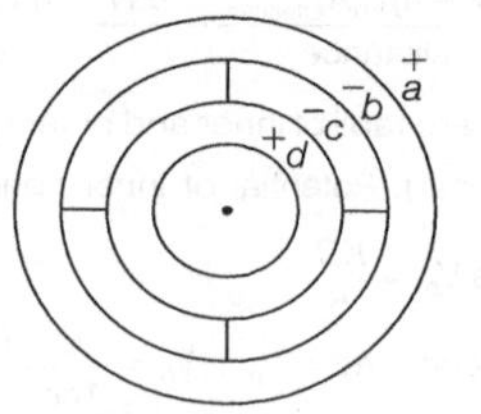

This arrangement is then viewed as two capacitors,

$$C_1 = \frac{2\pi\varepsilon_0 l}{\ln\left(\frac{a}{b}\right)} \quad \text{and} \quad C_2 = \frac{2\pi\varepsilon_0 l}{\ln\left(\frac{c}{d}\right)}$$

Using formula for series combination,

$$\frac{1}{C_{eq}} = \frac{1}{C_1} + \frac{1}{C_2}, \text{ we get}$$

$$C_{net} = \frac{C_1C_2}{C_1 + C_2} = \frac{2\pi\varepsilon_0}{\ln\left(\frac{ac}{bd}\right)}$$

15. (a) When free space between parallel plate capacitor, $\varepsilon = \dfrac{\sigma}{\varepsilon_0 K}$

When dielectric is introduced between parallel plate capacitor $\varepsilon' = \dfrac{\sigma}{\varepsilon_0 K}$.

As, electric field inside dielectric

$$\frac{\sigma}{\varepsilon_0 k} = 3 \times 10^4.$$

Given, $\quad k = 2.2$

$\varepsilon_0 = 8.85 \times 10^{-12}$

$\sigma = 2.2 \times 8.85 \times 10^{-12} \times 3 \times 10^4$

$= 6.6 \times 8.85 \times 10^{-8} = 5.841 \times 10^{-7}$

$\sigma = 6 \times 10^{-7}\ \text{C/m}^3$

16. (a,b) A parallel plate capacitor can be made very large as plates are kept apart using a dielectric and then they are rolled to form a small cylinder.

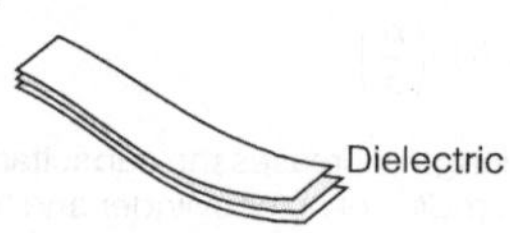

As, plate area is very large, it can store more charge.

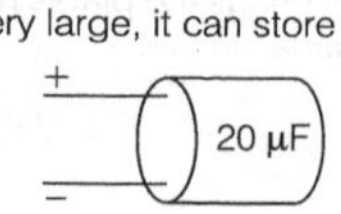

17. (a,c) Capacitance of capacitor

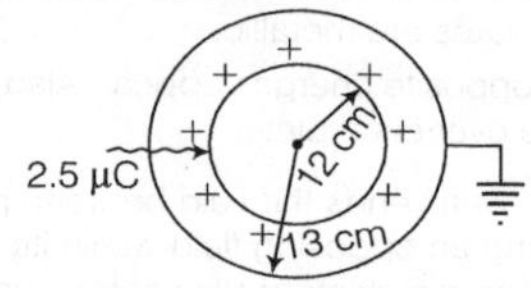

$$C = 4\pi\varepsilon_0\left(\frac{ab}{b-a}\right)$$

$$= \frac{1}{9 \times 10^9}\left(\frac{12 \times 10^{-2} \times 13 \times 10^{-2}}{(13-12) \times 10^{-2}}\right)$$

$$= \frac{12 \times 13 \times 10^{-2}}{9 \times 10^9} = 16.6 \times 10^{-11}$$

$$= 0.166 \times 10^{-9}\text{F} = 0.16\ \text{nF}$$

Also, potential of inner sphere

$$= K\left(\frac{Q_1}{R_1} - \frac{Q_1}{R_2}\right) = 9 \times 10^9\left(\frac{2.5 \times 10^{-6}}{12 \times 10^{-2}} - \frac{2.5 \times 10^{-6}}{13 \times 10^{-2}}\right)$$

$$= \frac{9 \times 10^9 \times 10^{-6} \times 2.5}{10^{-2}}\left(\frac{1}{12} - \frac{1}{13}\right)$$

$$= 9 \times 2.5 \times 10^5\left(\frac{13-12}{12 \times 13}\right)$$

$$= 9 \times 2.5 \times 10^5 \times \frac{1}{12 \times 13}$$

$$= 0.57 \times 10^5 = 5.7 \times 10^4\ \text{V}$$

18. (b,d) During charging induction occurs between plates of capacitor. Also, as $C = \dfrac{Q}{V}$; if V is reduced, then C is increased.

19. (a,b) $C_{air} = \dfrac{\varepsilon_0 A}{d} = \dfrac{8.85 \times 10^{-12} \times 0.2}{0.01}$

$= 1.77 \times 10^{-10}$ F

$K = \dfrac{C}{C_0} = \dfrac{Q/V}{Q/V_0} = \dfrac{V_0}{V}$

$= \dfrac{3000}{1000} = 3$

$C = C_0 K = 1.77 \times 10^{-10} \times 3$

$= 5.31 \times 10^{-10}$ F

$\varepsilon = \varepsilon_0 K = 8.85 \times 10^{-12} \times 3$

$= 2.65 \times 10^{-11}$ F/m

20. (a,d) Capacitor is a vital component of all transmission circuits. It is used to create oscillations in L-C oscillatory circuit.

Also, capacitors are used as memory banks as they store a sequence of charge with an imprint of a charging process.

21. (8) For an elemental length dx capacity of lower and upper elements are

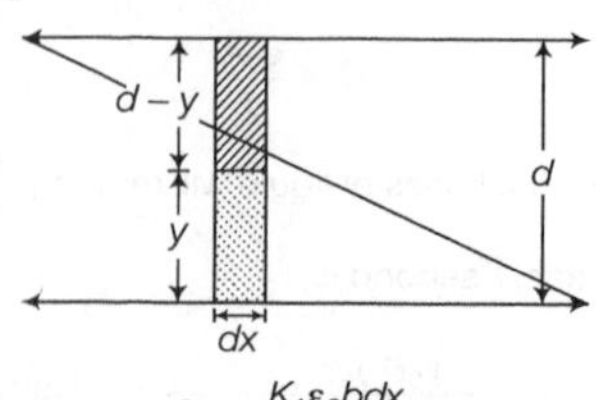

$$dC_1 = \frac{K_1\varepsilon_0 bdx}{y}$$

$$dC_2 = \frac{K_2\varepsilon_0 bdx}{d-y}$$

b = Plate width

As two elements are in series,

$$\frac{1}{dC} = \frac{1}{dC_1} + \frac{1}{dC_2}$$

Using, $$\frac{1}{C_{eq}} = \frac{1}{C_1} + \frac{1}{C_2}$$

$\Rightarrow$ $$dC = \frac{K_1K_2\varepsilon_0 b dx}{K_1 d + (K_2 - K_1)y}$$

Integrating,

$$C = \frac{K_1K_2\varepsilon_0 A}{(K_2 - K_1)d}\log\left(\frac{K_2}{K_1}\right)$$

As $K_2 = 8, K_1 = 4$, we get

$$C = \frac{8 \times 4\ \varepsilon_0 A}{(8-4)d}\log\left(\frac{8}{4}\right) = 8\frac{\varepsilon_0 A}{d}\log 2$$

So, $N = 8$

22. (2) Consider the system,

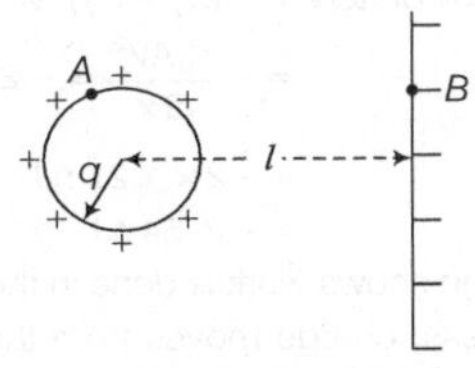

Potential difference between points A and B

$$V_B = 0,\ V_A = \frac{q}{4\pi\varepsilon_0 a}$$

$\Rightarrow$ $$\frac{q}{V} = C = 4\pi\varepsilon_0 a = \frac{2}{9} \times 10^7$$

So, $$9C \times 10^{-7} = 2$$

23. (6) Consider system of wires,

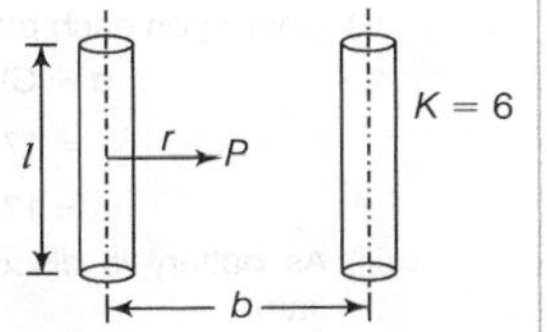

Electric field at $E = \frac{2 \times K\lambda}{r}$

where, $$k = \frac{1}{4\pi\varepsilon_0 K}$$

Now, using $$E = -\frac{dV}{dr},$$

$$V = -\int_a^{b-a} \frac{\lambda}{\pi\varepsilon_0 Kr}\cdot dr$$

$\Rightarrow$ $$V = \frac{q}{C} = \frac{\lambda}{\pi\varepsilon_0 K}\cdot\log_e\left(\frac{b}{a}\right)$$

or $$C = \frac{K(\pi\varepsilon_0 l)}{\log(b/a)}$$

or $\frac{C}{l}$ = Capacitance per metre length of wires $= K\left(\frac{\pi\varepsilon_0}{\log(b/a)}\right)$

Given, $K = 6$

24. (d)

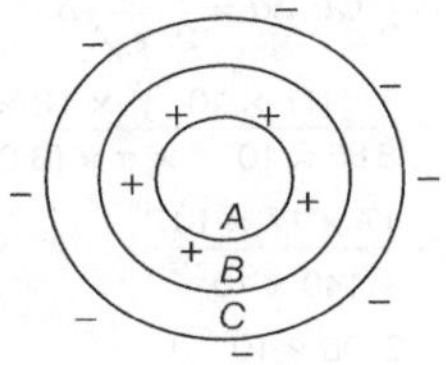

Potential on the surface of A is due to charge on A and due to charge on C.

$$V_A = \frac{q}{4\pi\varepsilon_0 a} + \frac{-q}{4\pi\varepsilon_0 C} = \frac{q}{4\pi\varepsilon_0}\left(\frac{1}{a} - \frac{1}{C}\right)$$

25. (d) Potential of shell C is

$$V_C = \frac{q}{4\pi\varepsilon_0 c} - \frac{q}{4\pi\varepsilon_0 c} = 0$$

$\therefore$ $$\Delta V = V_A - V_C = \frac{q}{4\pi\varepsilon_0}\left(\frac{1}{a} - \frac{1}{c}\right)$$

26. (c) Capacitance of the system is

$$C = \frac{q}{\Delta V} = 4\pi\varepsilon_0\left(\frac{ac}{c-a}\right)$$

27. (a) After inserting the dielectric, electric field from A to C is formed by using Gauss' law.

The field,

$$E(r) = \frac{q}{4\pi\varepsilon_0 Kr^2},\ a \le r \le b$$

and $$E(r) = \frac{q}{4\pi\varepsilon_0 r^2} \text{ for } b \le r \le c$$

So, potential difference between A and C is

$$\Delta V = V_A - V_C$$

$$= -\int_a^b \frac{q}{4\pi\varepsilon_0 Kr^2}dr - \int_b^c \frac{q}{4\pi\varepsilon_0 r^2}dr$$

$$= \frac{q}{4\pi\varepsilon_0}\left\{\frac{1}{K}\left(\frac{1}{a} - \frac{1}{b}\right) + \left(\frac{1}{b} - \frac{1}{c}\right)\right\}$$

$$= \frac{q}{4\pi\varepsilon_0}\left(\frac{b-a}{kab} + \frac{c-b}{bc}\right)$$

$$= \frac{q}{4\pi\varepsilon_0 K abc}[c(b-a) + Ka(c-b)]$$

28. (d) Capacitance of system is

$$C = \left[\frac{q}{\Delta V}\right] = \frac{4\pi\varepsilon_0 K abc}{Ka(c-b) + c(b-a)}$$

29. (a) (i) Capacity of capacitor between 2 and 3 $= \frac{\varepsilon_0 A}{2d}$

Potential difference between 2 and 3 is V.

$\therefore$ Charge on capacitor formed by plates 2 and 3 is

$$Q = CV = \frac{\varepsilon_0 A}{2d}V$$

(ii) Capacity of capacitor between 4 and 5 is

$$C = \frac{\varepsilon_0 A}{d}$$

$\therefore$ Charge on plate 5 is

$$Q = \frac{\varepsilon_0 A}{d}V$$

(iii) Between 2 and 3 battery of V volts is connected.

$\therefore$ Potential difference between 2 and 3 is V.

(iv) Plates 2 and 5 are connected to same terminal of battery, so potential difference between them is zero.

(i) → (r), (ii) → (s), (iii) → (q), (iv) → (p)

DPP-3 Energy of a Charged Capacitor, Effect of Dielectric and Force between Plates of a Capacitor

1. (i) Energy stored by capacitor

$$= \frac{1}{2}CV^2 = \frac{1}{2}\frac{\varepsilon_0 A}{d}\cdot V^2$$

$$= \frac{8.85 \times 10^{-12} \times 40 \times 10^{-4} \times (600)^2}{2 \times 1 \times 10^{-3}}$$

$$= 6.37 \times 10^{-6}\text{ J}$$

$$= 6.37\ \mu\text{J}$$

(ii) Electric field between plates

$$E = \frac{V}{d} = \frac{600}{1 \times 10^{-3}}$$

$$= 6 \times 10^5 \left(\frac{V}{m}\right)$$

(iii) Let q be the charge delivered by the cell.

Work done by cell $= qV$

Energy stored by capacitor is

$$\frac{1}{2}CV^2 = \frac{1}{2}qV$$

So, energy lost $= qV - \frac{1}{2}qV$

$$= \frac{1}{2}qV = \frac{1}{2}CV^2$$

$$= 6.37\ \mu\text{J}$$

(iv) Loss of energy is due to resistance of connecting wires, internal resistance and due to sparking (production of electro magnetic waves).

(v) Force between plates of charged capacitor is

$$F = \frac{1}{2}QE = \frac{1}{2}CV\cdot\frac{V}{d}$$

$$= \frac{1}{2} \times \frac{\varepsilon_0 A}{d} \times \frac{V^2}{d}$$

$$= \frac{8.85 \times 10^{-12} \times 40 \times 10^{-4} \times (600)^2}{(1 \times 10^{-3})^2}$$

$$= 1.27 \times 10^{-2}\text{ N}$$

2. (i) Force between plates,

$$F = \frac{1}{2}QE = \frac{1}{2}Q\cdot\frac{\sigma}{\varepsilon_0}$$

$$= \frac{1}{2}\left(\frac{Q^2}{\varepsilon_0 A}\right) = \frac{1}{2}\frac{C^2V^2}{(\varepsilon_0 A)}$$

$$= \frac{1}{2}\left(\frac{\varepsilon_0 A}{d}\right)^2 \times \frac{V^2}{\varepsilon_0 A}$$

$$= \frac{1}{2}\frac{\varepsilon_0 AV^2}{x^2},$$

Let $x = d$ = separation between plates.

Then, work done when plates are moved apart by distance dx is

$$dW = Fdx$$

So, total work done when plates are moved from separation x to $2x$ is

$$W = \int_x^{2x} dW = \int_x^{2x} \frac{1}{2}\varepsilon_0 AV^2\cdot\frac{dx}{x^2} = \frac{1}{2}\varepsilon_0 AV^2\cdot\left(\frac{-1}{x}\right)_x^{2x}$$

$$= \frac{1}{2}\varepsilon_0 AV^2\left(\frac{-1}{2x} + \frac{1}{x}\right) = \frac{1}{2}\varepsilon_0 AV^2\cdot\frac{1}{2x} = \frac{\varepsilon_0 AV^2}{4x}$$

$$= \frac{8.85 \times 10^{-12} \times 40 \times 10^{-4} \times (12)^2}{4 \times 1 \times 10^{-3}}$$

$$= 1.27 \times 10^{-9}\text{ J} = 1.27\text{ nJ}$$

(ii) Initial charge on plates with separation x is

$$Q_i = \frac{\varepsilon_0 AV}{x} = C_i\cdot V$$

Final charge on plates with separation $2x$ is

$$Q_f = C_f V = \frac{\varepsilon_0 AV}{2x}$$

$\therefore$ Charge transferred back to battery $\Delta Q = Q_i - Q_f$

$$\Rightarrow \Delta Q = \frac{\varepsilon_0 AV}{x} - \frac{\varepsilon_0 AV}{2x} = \frac{\varepsilon_0 AV}{2x}$$

$$= \frac{8.85 \times 10^{-12} \times 40 \times 10^{-4} \times 12}{2 \times 1 \times 10^{-3}}$$

$$= 2.124 \times 10^{-10}\text{ C}$$

(iii) Work done by battery $= -(Q_i - Q_f)\cdot V$

$$= -\frac{\varepsilon_0 AV^2}{2x} = -2\left(\frac{\varepsilon_0 AV^2}{4x}\right)$$

$$= -2 \times 1.27\text{ nJ}$$

$$= -2.54\text{ nJ}$$

Negative sign shows work is done in the battery.

(iv) In the process, charge moves from the terminal to negative terminal of cell and so the battery is charged in the process.

3. (i) $C = \frac{\varepsilon_0 A}{d}$, where $A = \pi r^2$

$$\therefore \quad C = \frac{8.85 \times 10^{-12} \times \pi \times (8.2 \times 10^{-2})^2}{1.3 \times 10^{-3}}$$

$$= 1.4 \times 10^{-10}\text{ F} = 140\text{ pF}$$

(ii) Charge on each plate is

$$q = CV = 1.4 \times 10^{-10} \times 120$$

$$= 1.7 \times 10^{-8}\text{ C}$$

$$= 17\text{ nC}$$

(iii) As battery is disconnected, charge on each plate remains same.

Capacitance of capacitor will be

$$C' = \frac{\varepsilon_0 A}{2d} = \frac{140}{2} = 70\text{ pF}$$

So, now the potential difference becomes

$$V' = \frac{q}{C'} = \frac{17 \times 10^{-9}}{70 \times 10^{-12}}\text{ V}$$

$$= 0.243 \times 10^3 = 243\text{ V}$$

(iv) Work done in the process $= F \times \Delta d$

(constant force results or charge is now same on each plate)

$$= \frac{1}{2}QE\,\Delta d = \frac{1}{2}\cdot\frac{Q^2}{\varepsilon_0 A}\Delta d$$

$$= \frac{(17 \times 10^{-9})^2 \times 1.3 \times 10^{-3}}{8.85 \times 10^{-12} \times \pi \times (8.2 \times 10^{-2})^2}$$

$$= \frac{17 \times 17 \times 10^{-18}}{140 \times 10^{-12}}$$

$$= 2.06 \times 10^{-6}\text{ J}$$

$$= 2.06\ \mu\text{J}$$

(v) Force between plates still remains same or the field between plates does not depends on the distance between plates and the charge on each plate is still same.

(vi) Increase in energy of capacitor occurs due to the work done by external energy.

Also, increase in energy = external work done = $2.06\ \mu$J

4. (i) Each dipole aligned itself with external field. The contribution of these dipoles reduces the net field. As each dipole falls into a configuration of lower potential energy it can contribute to increasing the internal energy of the material.

(ii) In water, there is difference in electro negativity of oxygen and hydrogen. So, the O—H pair has its own dipole moment that points towards the oxygen. In molecules like CO_2, the molecules are structured so that each CO pair has a dipole moment, but since dipole moments have same magnitude and opposite direction, the entire molecule has no dipole moment.

(iii) Heating increases disorientation of dipoles and the dielectric constant of sample is reduced.

So, $C = \frac{\varepsilon_0 KA}{d}$, capacity is reduced.

5. (d) As battery remains connected to the potential difference remains same.

$$W = U_f - U_i = \frac{1}{2}CV^2 - \frac{1}{2}CV^2 = 0$$

6. (d) Battery supplies energy,

$$U_1 = Q\Delta V$$

Energy stored in capacitor,

$$U_2 = \frac{1}{2}Q\Delta V$$

$\therefore$ $$\text{Ratio} = \frac{\frac{1}{2}Q\Delta V}{Q\Delta V} = \frac{1}{2}$$

7 (c) Charge on capacitor A, before joining with an uncharged capacitor $q_A = CV = 100 \times 3\,\mu C = 300\,\mu C$. Similarly, charge on capacitor $B q_B = 180 \times 2\,\mu C = 360\,\mu C$

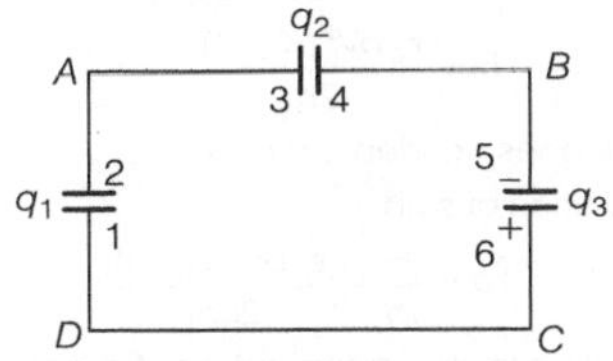

Let q_1, q_2 and q_3 be the charges on the three capacitors after joining them.

From conservation of charge, net charge on plates 2 and 3 before joining = Net charge after joining

$$300 = q_1 + q_2 \qquad \text{...(i)}$$

Similarly, net charge on plates 4 and 5 before joining = net charge after joining

$$-360 = -q_2 - q_3$$
$$360 = q_2 + q_3 \qquad \text{...(ii)}$$

Applying Kirchhoff's law in closed loop $ABCDA$

$$\frac{q_1}{3} - \frac{q_2}{2} + \frac{q_3}{2} = 0$$

or $$2q_1 - 3q_2 + 3q_3 = 0 \qquad \text{...(iii)}$$

Solving Eqs. (i), (ii) and (iii), we get

$$q_1 = 90\,\mu C$$
$$q_2 = 210\,\mu C$$

and $$q_3 = 150\,\mu C$$

8. (c) ΔU = Decrease in PE $= U_i - U_f$

$$= \frac{1}{2}C(V_1^2 + V_2^2) - \frac{1}{2}(2C)\left(\frac{V_1 + V_2}{2}\right)^2 = \frac{1}{4}C(V_1 - V_2)^2$$

9. (a) First, we required distance between plates :

As, $$V = Ed$$
$$4 \times 10^3 = 18 \times 10^6\, d$$
$$d = 4 \times 10^3 / 18 \times 10^6$$
$\Rightarrow$ $$d = \frac{4 \times 10^3}{18 \times 10^6}\text{ m}$$

Let A = plate area, then

$$C = \frac{\varepsilon_0 KA}{d}$$
$$7 \times 10^{-2} \times 10^{-6} = \frac{8.85 \times 10^{-12} \times 2.8 \times A}{4 \times 10^3 / 18 \times 10^6}$$
$\Rightarrow$ $$A = \frac{7 \times 10^{-2} \times 10^{-6} \times 4 \times 10^3}{18 \times 10^6 \times 8.85 \times 10^{-2} \times 2.8} = 63 \times 10^{-12}\text{ m}^2$$

10. (d) Ratio $\frac{\mathbf{P}}{\mathbf{E}} = x_e$ = Susceptibility (dimensionless)

As, **P** and **E** have same dimensions.

11. (d) Induced charge over surface of dielectric

$$= \sigma_{in} = \left(\frac{K-1}{K}\right)\sigma$$
$$= \left(\frac{5.4-1}{5.4}\right)\sigma = \frac{54}{44}\sigma = \frac{22}{27}\sigma$$

12. (c) Energy density in air gap,

$$U_0 = \frac{1}{2}\varepsilon_0 E_0^2$$

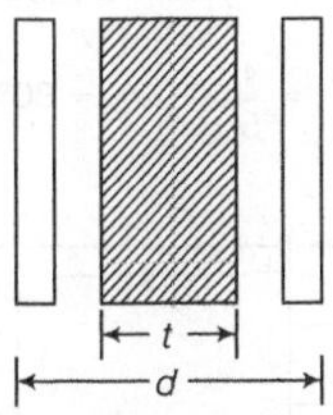

Energy density in the dielectric

$$U = \frac{1}{2}\varepsilon_0 KE^2 = \frac{1}{2}\varepsilon_0 K\left(\frac{E_0}{K}\right)^2 = \frac{1}{K}\cdot\frac{1}{2}\varepsilon_0 E_0^2 = \frac{1}{K}U_0$$

Now, U = stored energy $= U \times$ volume

$\therefore$ U_0 = energy stored in air gap
$= U_0 \times$ volume of air gap
$= U_0 \times A(d - t)$
$= \frac{1}{2}\varepsilon_0 E_0^2 A(d - t)$

Smilarly energy stored in dielectric

$U_D = U \times$ volume of dielectric slab
$= \frac{1}{K}\frac{1}{2}\varepsilon_0 E_0^2 \times A \times t$

Hence, ratio $$\frac{U_D}{U_{total}} = \frac{U_D}{U_D + U_0} = \frac{t}{(d-t)K + t}$$

Now, if $\frac{U_D}{U_{total}} = 0.5$, then

$$\frac{t}{(d-t)K + t} = 0.5$$

As, $$K = 3, \text{ so, } t = \frac{3d}{4}$$

13. (c) $\frac{U_f}{U_i} = \frac{\frac{Q^2}{2C_f}}{\frac{Q^2}{2C_i}} = \frac{C_i}{C_f} = \frac{\frac{\varepsilon_0 A}{d}}{\frac{kA\varepsilon_0}{d/2}} = \frac{1}{2K}$

$\therefore \quad U_f = \frac{1}{2K} U_i$

Energy is reduced due to negative work done by the field in pulling the dielectric slab.

14. (c) $Q_{\text{initial}} = C_{\text{initial}} \times V_{\text{initial}}$

When mica slab is inserted, capacity changes to $C_{\text{final}} = kC_{\text{initial}}$

As battery remains connected, $V_{\text{initial}} = V_{\text{final}}$

So, final charge on capacitor

$$Q_{\text{final}} = C_{\text{final}} \times V_{\text{final}} = k_{\text{mica}} C_{\text{initial}} \times V_{\text{final}}$$
$$= 6\,(3.5 \times 10^{-9}) \times 32 = 6.7 \times 10^{-7}\ \text{C}$$

15. (d) $q_i = C_i V = 2V = q$

This charge will remains constant after switch S is shifted from position 1 to position 2.

$$U_i = \frac{1}{2}\frac{q^2}{C_i} = \frac{q^2}{2 \times 2} = \frac{q^2}{4}$$

$$U_f = \frac{1}{2}\frac{q^2}{C_f} = \frac{q^2}{2 \times 10} = \frac{q^2}{20}$$

Energy dissipated $= U_i - U_f = \frac{q^2}{5}$

The energy dissipated is 80% of the initial stored energy as % decrease in energy

$$= \frac{U_i - U_f}{U_i} \times 100\% = \frac{q^2/5}{q^2/4} \times 100\%$$

$$= \frac{4}{5} \times 100\% = 80\%$$

16. (c) Given curve is

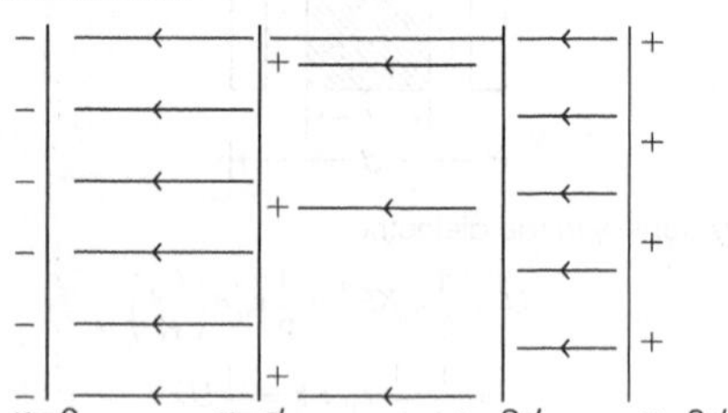

Variation of electric field is

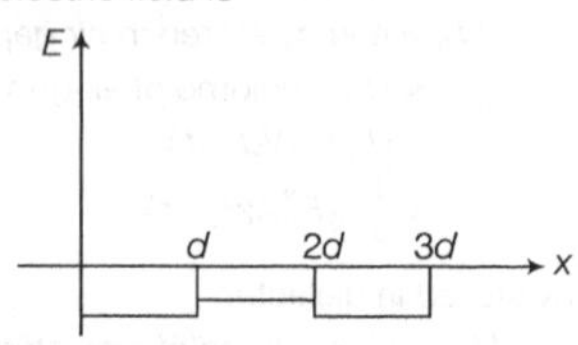

So, potential varies as follows :

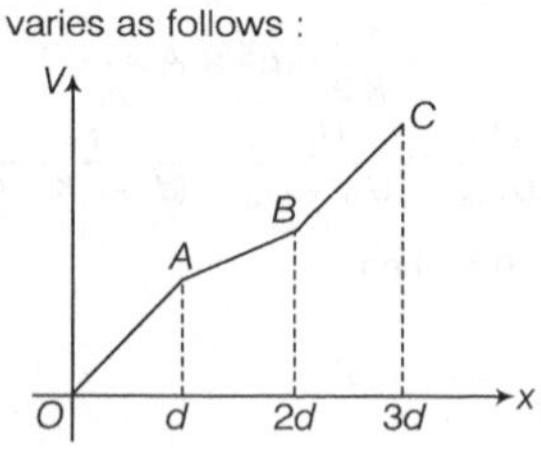

Slope OA = Slope BC

$$E_{o-d} = E_{2d-3d}$$

As, $\quad E_{o-d} > E_{d-2d}$

17. (a,c) Capacity with partially inserted slab (upto depth x)

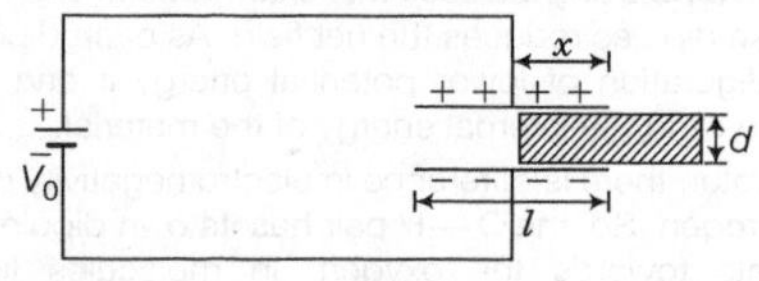

= Capacity of air capacitor + Capacity of dielectric field capacitor

$$= \frac{\varepsilon_0 b(l - x)}{d} + \frac{K\varepsilon_0 bx}{d}$$

b = plate width

$\Rightarrow \quad C = \frac{\varepsilon_0 b}{d}[l + x(K - 1)]$

Energy of capacitor, $U = \frac{1}{2}CV^2$

$\Rightarrow \quad dU = \frac{1}{2}V^2(dC)$

and $\quad (dC) = \frac{\varepsilon_0 b}{d}(K - 1)\,dx$

$\therefore \quad dU = \frac{\varepsilon_0 b}{2d}(K - 1)V^2 \cdot dx$

Work done by force = Fdx

Increase in charge = dq

$\Rightarrow \quad dq = VdC = \frac{\varepsilon_0 bV}{d}(K - 1)\,dx$

This charge is transformed from negative terminal to the positive terminal inside battery.

Hence, work done by battery,

$$dW = -dq.V = \frac{-\varepsilon_0 bV^2(K - 1)dx}{d}$$

As, $\quad dW = -dU$

Hence, $\quad Fdx = \frac{\varepsilon_0 bV^2(K - 1)dx}{2d}$

$\Rightarrow \quad F = \frac{\varepsilon_0 bV^2(K - 1)}{2d}$

Clearly, force does not depend on x.

Also, acceleration of slab

$$a = \frac{F}{m} = \frac{\varepsilon_0 bV^2(K - 1)}{2dm}$$

and acceleration is also independent of dimension x.

$\therefore$ Time period of oscillation is also independent of x.

18. (a,d) Capacity and charge both increases by K times.

$\therefore$ Both (a) and (d) are correct.

(Refer to Subjective Q. No. 2)

19. (a,c) Field inside the dielectric $= E - E_{\text{induced}}$

$$E_{\text{induced}} = \frac{\sigma_{\text{induced}}}{\varepsilon_0}$$

$\therefore \quad E(\text{inside dielectric}) = \frac{\sigma - \sigma_p}{\varepsilon_0} = \frac{\sigma}{\varepsilon_0 K}$

$\therefore \quad \%\,\Delta E = \frac{E_{\text{air}} - E_{\text{dielectric}}}{E_{\text{air}}} \times 100$

$$= \frac{\frac{\sigma}{\varepsilon_0} - \frac{\sigma}{\varepsilon_0 K}}{\frac{\sigma}{\varepsilon_0}} \times 100$$

$$= \left(1 - \frac{1}{K}\right) \times 100 = \left(1 - \frac{1}{5}\right) \times 100$$

$$= \frac{4}{5} \times 100 = 80\%$$

$\Rightarrow E$ (inside dielectric) is 20% less than electric field of air.

Alternate Method

$$E_{\text{dielectric}} = \frac{1}{5}E_{\text{air}} = \frac{100}{5} = 20\% \text{ of } E_{\text{air}}$$

Also, σ_p = charge density on the surface of dielectric

$$= \sigma\left(1 - \frac{1}{K}\right) = \sigma\left(1 - \frac{1}{5}\right) = \sigma\left(\frac{4}{5}\right)$$

$\therefore$ 20% less than on plates of cap.

20. (a,b) During charging of capacitor,

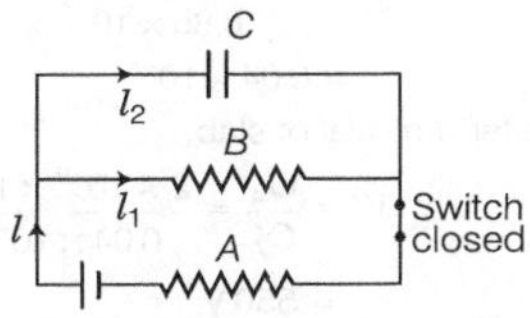

Heat developed in resistor $B = I_1^2 B$

Heat developed in resistor, $A = I^2 A$ as $I > I_1$

Clearly, $I^2 A > I^2 B$

In steady state, capacitor does not conduct.

So, $I_2 = 0$ and $I_1 = I$

Clearly, $I^2 A = I^2 B$ ($\because$ resistance of both is same)

Heat produced at same rate in A and B.

21. (a,d) When dielectric plate is inserted after disconnecting the battery, then

$$C' = CK,\ V' = \frac{V}{K'} Q' = Q$$

$$U' = \frac{1}{2}C'V'^2 = \frac{1}{2}CK \times \frac{V^2}{K^2} = \frac{1}{K}U$$

where C, V, Q and U are initial values and C', V', Q' and U' are final values.

Finally, energy of capacitor after inserting dielectric slab

$$U' = \frac{U}{K}$$

Now, we have a capacitor which has a potential difference $\frac{V}{K}$.

Now, if we consider this capacitor as air capacitor.

Then, energy U_2 (say)

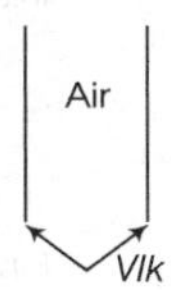

$$= \frac{1}{2}C \cdot \left(\frac{V}{k}\right)^2 = \frac{U}{k^2}$$

Also, with the dielectric energy of capacitor is

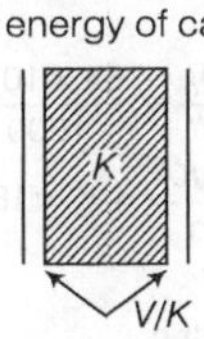

$$U' = \frac{U}{K}$$

So, energy with dielectric slab is higher and so energy stored in dielectric is

$$U' - U_2 = U\left(\frac{1}{K} - \frac{1}{K^2}\right)$$

$$= U\left(\frac{K-1}{K^2}\right) = U\left(\frac{3.5-1}{3.5 \times 3.5}\right)$$

$$= U\left(\frac{2.5}{3.5 \times 3.5}\right)$$

$$= U\left(\frac{25 \times 10}{35 \times 35}\right) = U\left(\frac{10}{49}\right)$$

Clearly, energy stored in dielectric is around 22%. So, 78% of energy is stored in air gap.

22. (5) When switch is closed, force on one spring is

$$F = \frac{Q^2}{2\varepsilon_0 A} = \frac{4C^2(\Delta V)^2}{2\varepsilon_0 A}$$

$$= \frac{2C^2(\Delta V)^2}{\left(\frac{\varepsilon_0 A}{d}\right) \cdot d} = \frac{2C(\Delta V)^2}{d}$$

On spring stretches by distance, $x = \frac{d}{4}$

So, $$k = \frac{F}{x} = \frac{2C(\Delta V)^2}{d} \times \frac{4}{d} = \frac{8C(\Delta V)^2}{d^2}$$

$$= \frac{8 \times 2 \times 10^{-6} \times (100)^2}{(8 \times 10^{-3})^2}$$

$$= 2.50 \text{ kN/m}$$

So, sum of spring constant = 2.5 + 2.5 = 5 kN/m

23. (1)

Electrostatic pressure = Electrostatic force per unit area

= Potential energy per unit volume

$$\left\{\text{Note } \left|\frac{dU}{dx}\right| = F,\ \frac{dU}{Adx} = \frac{F}{A}\right\}$$

Initially,

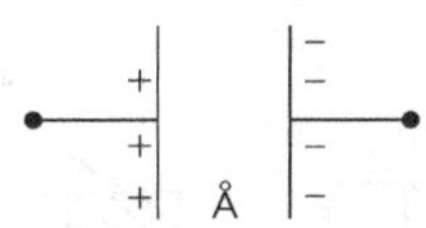

Electrostatic pressure at A

$$p_1 = \frac{U}{V} = u_E = \frac{1}{2}\varepsilon_0 E^2$$

When water fills up to point A, electrostatic pressure will be

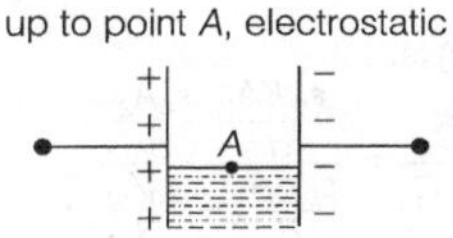

$$p_2 = u_E = \frac{1}{2}\varepsilon\varepsilon_0 E^2$$

Change in pressure $= p_2 - p_1 = \frac{1}{2}\varepsilon_0(\varepsilon - 1)\frac{V^2}{d^2}$

$$p_2 - p_1 = \varepsilon_0 \times 10^{13}$$

Hence, $K = 1$

24. (6) Electric field at a distance r from the centre is

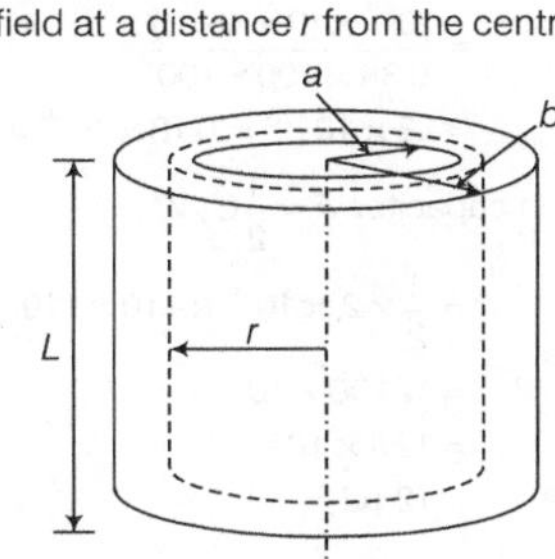

$$E = \frac{q}{2\pi\varepsilon_0 Lr}$$

Energy density of field is

$$u_E = \frac{1}{2}\varepsilon_0 E^2$$

$$\Rightarrow \quad = \frac{q^2}{8\pi^2\varepsilon_0 L^2 r^2}$$

Energy in the cylinder is volume integral of energy density.

$$U_R = \int u dV$$

$$dV = 2\pi r L \cdot dr$$

$$\Rightarrow \quad U_R = \int_0^R \frac{q^2}{8\pi^2 \varepsilon_0 L^2 r^2} \cdot 2\pi r L \cdot dr$$

$$= \frac{q^2}{4\pi\varepsilon_0 L} \cdot \int_0^R \frac{dr}{r}$$

$$= \frac{q^2}{4\pi\varepsilon_0 L} \log\left(\frac{R}{a}\right)$$

Total stored energy

$$U_b = \frac{q^2}{4\pi\varepsilon_0 L} \cdot \log\left(\frac{b}{a}\right)$$

For $\frac{U_R}{U_b} = \frac{1}{2}$, we have

$$\Rightarrow \quad \frac{1}{2}\log\left(\frac{b}{a}\right) = \log\left(\frac{R}{a}\right)$$

$$\Rightarrow \quad R = \sqrt{ab} = \sqrt{4 \times 9}$$

$$= 6 \text{ cm}$$

25. (a) Capacitor A can be viewed or a parallel combination of two capacitors :

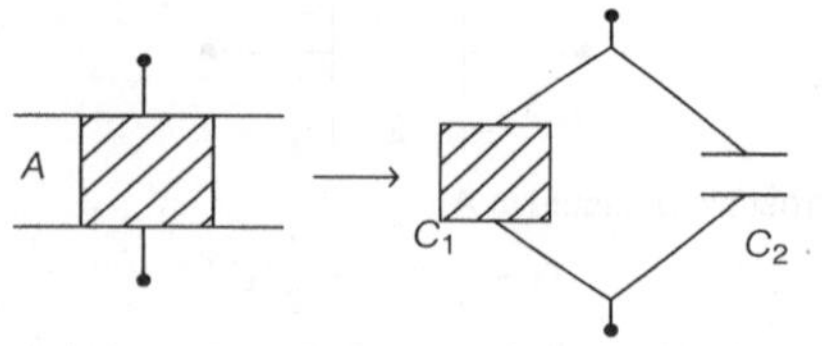

So, capacity of A is

$$C_A = \frac{GC_2}{G + C_2} = \frac{\frac{\varepsilon_0 KA_1}{d} \cdot \frac{\varepsilon_0 A_2}{d}}{\frac{\varepsilon_0 KA_1}{d} + \frac{\varepsilon_0 A_2}{d}}$$

Here, $A_1 = 0.04 \text{ m}^2$, $A_2 = 0.02 \text{ m}^2$

$$\therefore \quad C_A = \frac{\frac{\varepsilon_0}{d}(K A_1 A_2)}{(K A_1 + A_2)} = \frac{\frac{8.85 \times 10^{-12}}{8.85 \times 10^{-4}}(9 \times 0.04 \times 0.02)}{(9 \times 0.04 + 0.02)}$$

$$= \frac{10^{-8}(9 \times 0.04 \times 0.02)}{0.38}$$

$$= \frac{10^{-8} \times (9 \times 4 \times 2)}{0.38 \times 100 \times 100}$$

$$= 1.8 \times 10^{-10} = 0.18 \times 10^{-9} = 2 \text{ nF}$$

Energy stored in capacitor $A = \frac{1}{2} C_A V^2$

$$= \frac{1}{2} \times 2 \times 10^{-9} \times 110 \times 110$$

$$= 12100 \times 10^{-9}$$

$$= 12.1 \times 10^{-6}$$

$$= 12 \text{ μJ}$$

26. (c) When dielectric slab is removed from A, as battery is disconnected charge on A remains same and its capacity reduces k times.

$$\therefore \quad Q_A = C_A V_A = 2 \times 10^{-9} \times 110$$

and capacity after slab removed

$$= C'_A = \frac{\varepsilon_0 A}{d}$$

$$= \frac{8.85 \times 10^{-12} \times 0.04}{8.85 \times 10^{-4}}$$

$$= 0.04 \times 10^{-8}$$

So, potential after removal of slab,

$$V'_A = \frac{Q_A}{C'_A} = \frac{2 \times 10^{-9} \times 110}{0.04 \times 10^{-8}}$$

$$= 550 \text{ V}$$

Energy stored after removed of slab

$$\frac{1}{2} C'_A (V'_A)^2 = \frac{1}{2} \times 0.04 \times 10^{-8} \times (550)^2$$

$$= 60.5 \text{ μJ}$$

$\therefore$ Work done in process $= U_f - U_i$

$$= 60.5 - 12 = 48.5 \text{ μJ}$$

27. (d) Common potential of system of capacitors

$$V = \frac{\text{Total charge}}{\text{Total capacity}}$$

$$= \frac{2 \times 10^{-9} \times 110}{0.04 \times 10^{-8} + 9 \times 0.02 \times 10^{-8}}$$

$$= 100 \text{ V}$$

$$\text{Energy stored} = \frac{1}{2}(C_A + C_B)V^2 = 11 \text{ μJ}$$

28. (a) Original capacitance,

$$C_0 = \frac{\varepsilon_0 A}{d}$$

$$= \frac{9 \times 10^{-12} \times 0.2}{10^{-2}} = 180 \text{ pF}$$

Initial potential difference, $V_0 = 3000$ V

Initial charge, $Q_0 = C_0 V_0$

$$= 180 \times 10^{-12} \times 3000$$

$$= 18 \times 3 \times 10^{-8}$$

$$= 54 \times 10^{-8} \text{ C}$$

Final value of capacity

$$= \frac{Q_0}{V'} = \frac{54 \times 10^{-8}}{1000} = 540 \text{ pF}$$

$$\text{Final capacity} = \frac{\varepsilon_0 KA}{d} = K \times 180 = 540$$

$$\Rightarrow \quad K = 3$$

So, permittivity of dielectric

$$= \varepsilon_0 K = 3 \times 9 \times 10^{-12}$$

$$= 27 \times 10^{-12} \text{ C}^2\text{N}^{-1}\text{m}^{-2}$$

(i) $\rightarrow$ (t), (ii) $\rightarrow$ (s), (iii) $\rightarrow$ (r), (iv) $\rightarrow$ (q), (v) $\rightarrow$ (p)

DPP-4 Series and Parallel Combinations of Capacitors

1. Let $\quad C_1 = C_2 = \dfrac{\varepsilon_0 A}{d}$

When capacitors are connected in series,

$$C_1 = \frac{\varepsilon_0 A}{d/2} = \frac{2\varepsilon_0 A}{d} \quad \text{and} \quad C_2 = \frac{\varepsilon_0 A}{2d}$$

So, $$\frac{1}{C_{\underset{\text{(series)}}{\text{eq}}}} = \frac{1}{C_1} + \frac{1}{C_2} = \frac{d}{2\varepsilon_0 A} + \frac{2d}{\varepsilon_0 A}$$

$$= \frac{d}{\varepsilon_0 A}\cdot\left(\frac{1}{2} + 2\right) = \frac{5}{2}\frac{d}{\varepsilon_0 A}$$

$\Rightarrow$ $$C_{\underset{\text{(series)}}{\text{eq}}} = \frac{2\varepsilon_0 A}{5d}$$

When capacitors are connected in parallel, then

$$C_{\underset{\text{(parallel)}}{\text{eq}}} = C_1 + C_2 = \frac{2\varepsilon_0 A}{d} + \frac{\varepsilon_0 A}{2d}$$

$$= \frac{\varepsilon_0 A}{d}\left(2 + \frac{1}{2}\right) = \frac{5}{2}\frac{\varepsilon_0 A}{d}$$

$\therefore$ Ratio of equivalent capacitances in both cases

$$\frac{C_{\text{series}}}{C_{\text{parallel}}} = \frac{\frac{2}{5}\cdot\left(\frac{\varepsilon_0 A}{d}\right)}{\frac{5}{2}\cdot\left(\frac{\varepsilon_0 A}{d}\right)} = \frac{2}{5} \times \frac{2}{5} = \frac{4}{25}$$

2. System is equivalent to two capacitors C_1 and C_2 which are connected in series,

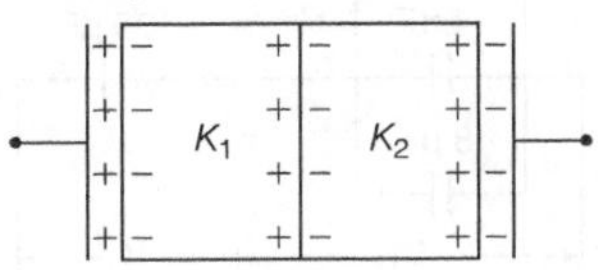

$$C_1 = \frac{\varepsilon_0 K_1 A}{d/2}, \quad C_2 = \frac{\varepsilon_0 K_2 A}{d/2}$$

$$\frac{1}{C_{eq}} = \frac{1}{C_1} + \frac{1}{C_2}$$

$\Rightarrow$ $$C_{eq} = \frac{C_1 C_2}{C_1 + C_2}$$

So, equivalent capacity is

$$C_{eq} = \frac{\frac{2\varepsilon_0 K_1 A}{d} \times \frac{2\varepsilon_0 K_2 A}{d}}{\frac{2\varepsilon_0 K_1 A}{d} + \frac{2\varepsilon_0 K_2 A}{d}} = \frac{2K_1K_2}{K_1 + K_2}\cdot\left(\frac{\varepsilon_0 A}{d}\right)$$

So, $$K_{eq} = \frac{2K_1K_2}{K_1 + K_2}$$

$\Rightarrow$ $$\frac{1}{K_1} + \frac{1}{K_2} = \frac{2}{K_{eq}}$$

So, K_1, K_{eq} and K_2 are in harmonic progression.

3. $\left(\dfrac{K_1 + K_2}{2}\right)$ System is equivalent to two capacitors in parallel.

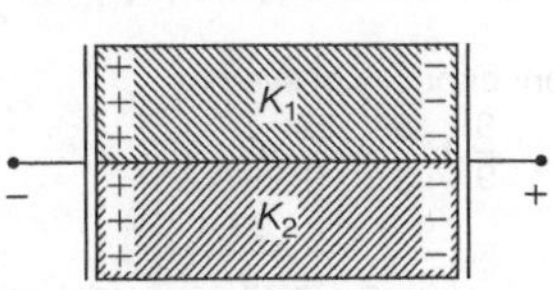

$$C_{eq} = C_1 + C_2 = \frac{\varepsilon_0 K_1 A/2}{d} + \frac{\varepsilon_0 K_2 A/2}{d}$$

$\Rightarrow$ $$C_{eq} = \left(\frac{K_1 + K_2}{2}\right)\cdot\frac{\varepsilon_0 A}{d}$$

So, $$K_{eq} = \frac{K_1 + K_2}{2}$$

4. (i) For C_1, capacitance $= \dfrac{\varepsilon_0 A}{d}$

(ii) For C_2, capacitance $= \dfrac{4\varepsilon_0 A}{d}$

(iii) For series, $C_s = \dfrac{C_1 C_2}{C_1 + C_2} = \dfrac{4}{5}\dfrac{\varepsilon_0 A}{d}$

(iv) For parallel, $C_p = C_1 + C_2 = \dfrac{5\varepsilon_0 A}{d}$

Clearly, if we take B for C_1, then slope indicates capacity as

$$C = \frac{Q}{V} \text{ or } Q = CV$$

Hence, graph C is for C_2 and graph D is for C_{parallel} and graph A is for C_{series}.

5. (a) In series the arrangement will be as follows

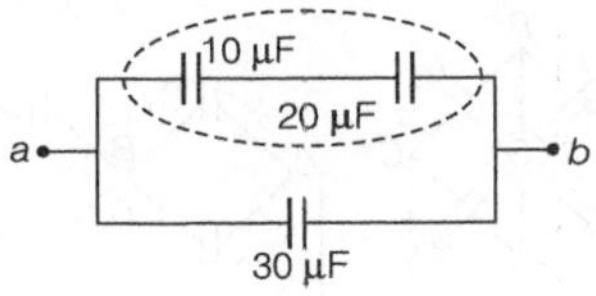

$$\frac{1}{C_{eq}} = \frac{1}{10} + \frac{1}{20} = \frac{2 + 1}{20}$$

$\Rightarrow$ $$C_{eq} = \frac{20}{3}\,\mu\text{F}$$

In parallel the arrangement will be as follows

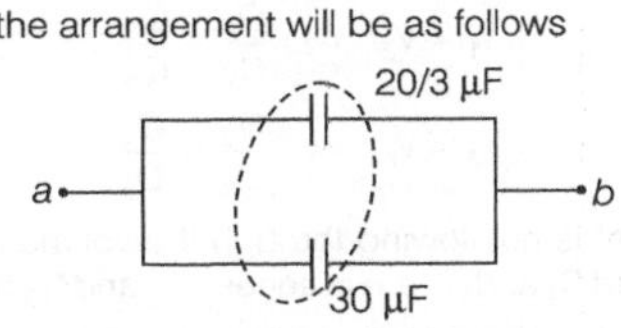

$\Rightarrow$ $$C_{eq} = C_1 + C_2$$
$$= \frac{20}{3} + 30 = \frac{110}{3}\,\mu\text{F}$$

So, equivalent circuit is

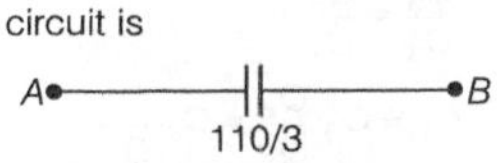

As, $$V_{ab} = 50\text{ V}$$
$$C = \frac{Q}{V}$$

$\Rightarrow$ $$Q = CV = \frac{110}{3} \times 50\,\mu\text{C} = \frac{5500}{3}\,\mu\text{C}$$

6. (b) Initially, K_1 is closed, K_2 is open, circuit is

$$V_1 + V_2 = 9\text{ V}$$

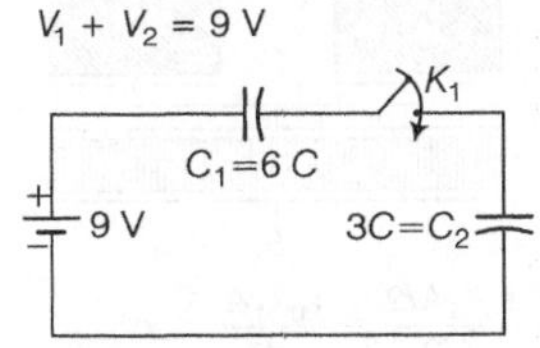

$$Q_1 = Q_2$$

$\Rightarrow$ $C_1V_1 = C_2V_2$ (as C_1 and C_2 are in series)

$\Rightarrow$ $6\,CV_1 = 3CV_2$

$\Rightarrow$ $2V_1 = V_2$

so, $V_1 + 2V_1 = 9\text{ V} = 3V_1 = 9\text{V}$

$\Rightarrow$ $V_1 = 3\text{V}$

and $V_2 = 3 \times 2 = 6\text{ V}$

Hence, charge on capacitor C_2 is

$$Q_2 = C_2V_2 = 3 \times 1\,\mu\text{F} \times 6\text{ V} = 18\,\mu\text{C}$$

Finally, K_1 is open and K_2 is closed

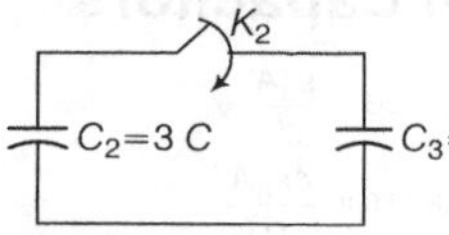

Common potential of capacitors

$$V_{\text{common}} = \frac{C_2V_2 + C_3V_3}{C_2 + C_3} = \frac{3 \times 6 + 0}{6}$$

$$\Rightarrow \quad V_{\text{common}} = 3\text{ V}$$

So, final value of charge on capacitor 2 is Q_2'

$$= C_2V_2' = C_2V_{\text{common}} = 3 \times 3 = 9\,\mu\text{C}$$

So, $\quad \dfrac{Q_2}{Q_2'} = \dfrac{18}{9} = 2:1$

7. (c) As there is no potential difference between points B and D, so given circuit can be visualised as :

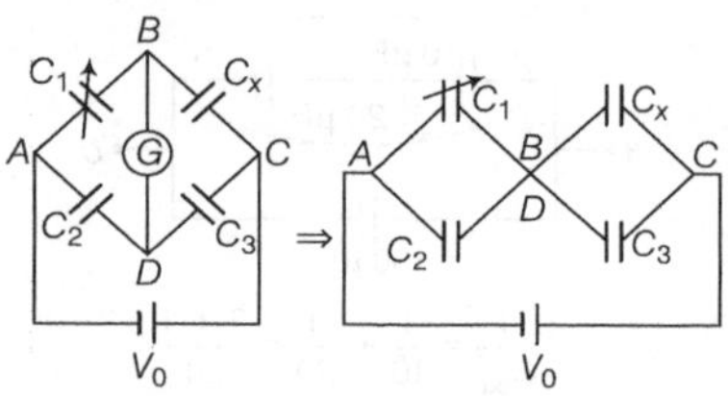

So, capacitors C_1 and C_2 are in parallel and C_x and C_3 are also in parallel and have the same voltage

$$V_1 = V_2 \Rightarrow \frac{Q_1}{C_1} = \frac{Q_2}{C_2}$$

$$V_3 = V_x \Rightarrow \frac{Q_3}{C_3} = \frac{Q_x}{C_x}$$

Here, current is not flowing through the voltmeter, we could also remove it and C_1 and C_x are in series. C_2 and C_3 are also in series. As in series, they have same charge

So, $\quad Q_1 = Q_x$ and $Q_2 = Q_3$

Solving these systems for C_x, we get

$$\frac{Q_3}{C_3} = \frac{Q_x}{C_x}$$

$$\Rightarrow \quad C_x = C_3\frac{Q_x}{Q_3} = C_3\frac{Q_1}{Q_2}$$

$$= C_3\left(\frac{C_1}{C_2}\right) = 4.8\left(\frac{8.9}{18}\right) = 2.4\,\mu\text{F}$$

8. (b) System can be viewed as a combination of 3 capacitors.

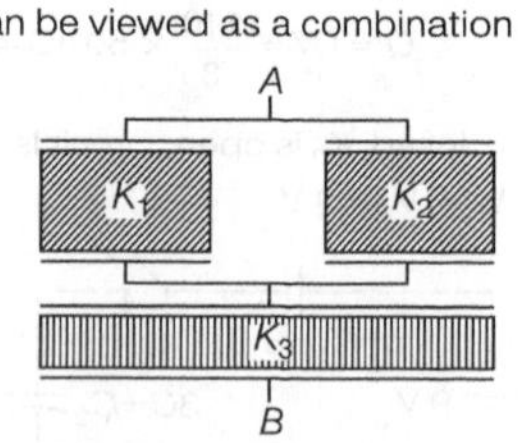

Clearly, $\quad C_1 = \dfrac{\varepsilon_0K_1A/2}{d/2} = \dfrac{\varepsilon_0K_1A}{d}$

$$C_2 = \frac{\varepsilon_0K_2A/2}{d/2} = \frac{\varepsilon_0K_2A}{d}$$

and $\quad C_3 = \dfrac{\varepsilon_0K_3A}{d/2} = \dfrac{2\varepsilon_0K_3A}{d}$

Capacitors C_1 and C_2 are in parallel and C_3 is in series with combination of C_1 and C_2. So equivalent capacitance of system is

$$\frac{1}{C_{eq}} = \frac{1}{C_1 + C_2} + \frac{1}{C_3}$$

$$\Rightarrow \quad C_{eq} = \frac{C_3(C_1 + C_2)}{C_1 + C_2 + C_3} = \frac{\dfrac{2\varepsilon_0K_3A}{d}\left(\dfrac{\varepsilon_0K_1A}{d} + \dfrac{\varepsilon_0K_2A}{d}\right)}{\dfrac{\varepsilon_0K_1A}{d} + \dfrac{\varepsilon_0K_2A}{d} + \dfrac{2\varepsilon_0K_3A}{d}}$$

$$= \frac{2K_3(K_1 + K_2)}{K_1 + K_2 + 2K_3}\left(\frac{\varepsilon_0A}{d}\right) = K_4\left(\frac{\varepsilon_0A}{d}\right)$$

So, $\quad K_4 = \dfrac{2K_3(K_1 + K_2)}{K_1 + K_2 + 2K_3}$

$$\frac{1}{K_4} = \frac{K_1 + K_2 + 2K_3}{2K_3(K_1 + K_2)} = \frac{K_1 + K_2}{2K_3(K_1 + K_2)} + \frac{2K_3}{2K_3(K_1 + K_2)}$$

$$\frac{1}{K_4} = \frac{1}{2K_3} + \frac{1}{K_1 + K_2}$$

9. (d) At $t = 0$, current goes through the capacitor.

$$\therefore \quad i = \frac{V}{R_{\text{eff}}} = \frac{2}{1000} = 2 \times 10^{-3} = 2\text{ mA}$$

When capacitor is charged,

$$R_{\text{eff}} = 2000\,\Omega \quad \text{and} \quad i = \frac{2}{2000} = 1\text{ mA}$$

10. (c) The circuit can be configured as

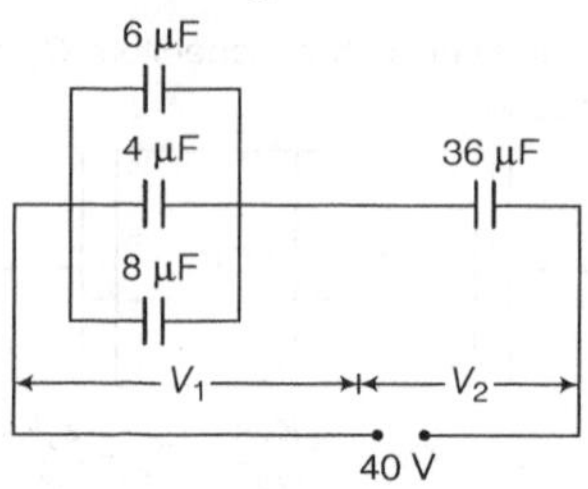

$$V_1 + V_2 = 40\text{ V} \quad \text{and} \quad \frac{V_1}{V_2} = \frac{36}{18} = 2$$

$$\Rightarrow \quad V_1 = \frac{80}{3}\text{ V} \quad \text{and} \quad V_2 = \frac{40}{3}\text{ V}$$

Charge on capacitor of 8 µF

$$= 8 \times \frac{80}{3} = 213.3\,\mu\text{F} = 214\,\mu\text{C}$$

11. (d) 12 µF and 6 µF capacitors are in series and their combination is in parallel with 4 µF capacitior

$$\therefore \quad C_{eq} = \frac{12 \times 6}{12 + 6} + 4 = 8\,\mu\text{F}$$

This system is in series with 1 µF capacitors. Its equivalent capacity

$$= \frac{8 \times 1}{8 + 1} = \frac{8}{9}\,\mu\text{F}$$

Capacity equivalent of 8 µF, 2 µF and 2 µF $= \dfrac{4 \times 8}{4 + 8} = \dfrac{32}{12} = \dfrac{8}{3}\,\mu\text{F}$

Final equivalent capacity of system

$$\frac{8}{9} + \frac{8}{3} = \frac{32}{9}\,\mu\text{F}$$

Also, $\quad C_{eq} = 1 = \dfrac{\dfrac{32}{9} \times C}{\dfrac{32}{9} + C} \Rightarrow C = \dfrac{32}{23}\,\mu\text{F}$

12. (b)

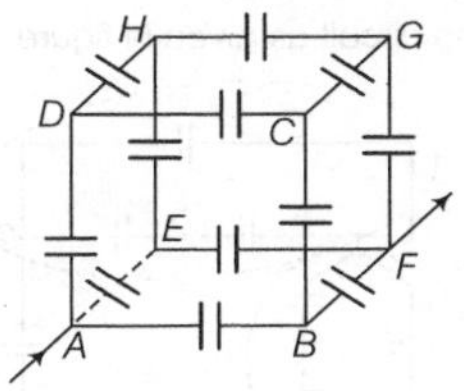

Points B and C, E and H are at the same potential. The equivalent circuit will be

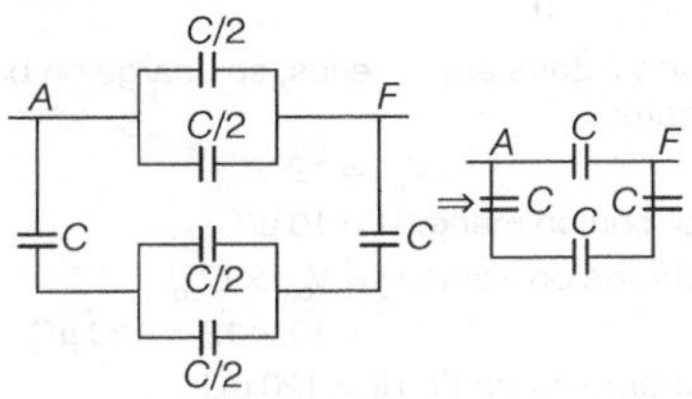

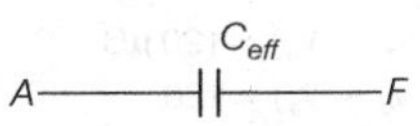

$$C_{eff} = \frac{4C}{3}$$

13. (a) $C_p = C_1 + C_2, \frac{1}{C_s} = \frac{1}{C_1} + \frac{1}{C_2}$

On substituting, $C_2 = C_p - C_1$

$$\frac{1}{C_s} = \frac{1}{C_1} + \frac{1}{C_p - C_1} = \frac{C_p - C_1 + C_1}{C_1(C_p - C_1)}$$

On simplifying, $C_1^2 - C_1C_p + C_pC_s = 0$

$$C_1 = \frac{C_p \pm \sqrt{C_p^2 - 4C_pC_s}}{2}$$

$$= \frac{1}{2}C_p \pm \sqrt{\frac{1}{4}C_p^2 - C_pC_s}$$

$$C_1 = \frac{1}{2}C_p + \sqrt{\frac{1}{4}C_p^2 - C_pC_s}$$

$$= \frac{1}{2}(9\text{ pF}) + \sqrt{\frac{1}{4}(9\text{ pF})^2 - (9\text{ pF})(2\text{ pF})}$$

$$= 6\text{ pF}$$

$$C_2 = C_p - C_1 = \frac{1}{2}C_p - \sqrt{\frac{1}{4}C_p^2 - C_pC_s}$$

$$= \frac{1}{2}(9\text{ pF}) - 1.50\text{ pF} = 3\text{ pF}$$

14. (c) Here, C and C are in parallel sum up to $2C$

$$U_i = \frac{1}{2}(2C)V^2 = CV^2$$

When the dielectric is introduced, the capacitance of each capacitor becomes $3C$. The potential of A remains constant at V and the charge on B remains fixed.

$$\therefore \quad U_f = \frac{1}{2} \times 3CV^2 + \frac{1}{2}\frac{(VC)^2}{3C}$$

$$= \frac{3}{2}CV^2 + \frac{1}{6}CV^2 = \frac{5}{3}CV^2$$

$$\therefore \quad \frac{U_i}{U_f} = \frac{CV^2}{\frac{5}{3}CV^2} = \frac{3}{5}$$

15. (b,d) Pressing S_1, charge on upper plate of C_1 is $+2CV_0$. After pressing S_2, this charge equally distributes into two capacitors. Therefore, charge on upper plates of both capacitors will be $+CV_0$. When S_2 is released and S_3 is pressed, charge on upper plate of C_1 remains unchanged $(= +CV_0)$ but charge on upper plate of C_2 is according to new battery $(= -CV_0)$.

16. (d) For minimum capacity, capacitors are connected in series,

$$\frac{1}{C_{eq}} = \frac{1}{6} + \frac{1}{6} + \frac{1}{6} \Rightarrow C_{eq} = \frac{6}{3} = 2\ \mu\text{F}$$

For maximum capacity, they are connected in parallel

$$\therefore \quad C_{eq} = 6 + 6 + 6 = 18\ \mu\text{F}$$

17. (a) For both capacitors when they are adjusted in parallel,

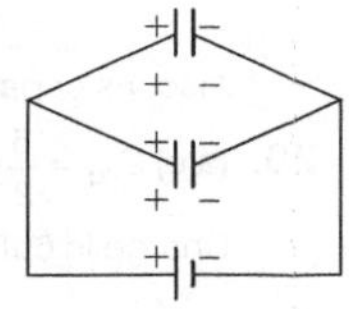

$$V_1 = V_2 \Rightarrow \frac{Q_1}{C_1} = \frac{Q_2}{C_2}$$

$$\therefore \quad C = \frac{Q}{V}$$

$$\Rightarrow \quad Q_1C_2 = Q_2C_1$$

$$\therefore \quad \text{Ratio of charges} = \frac{Q_1}{Q_2} = \frac{C_1}{C_2}$$

They are at same potential, no charge transfer takes place and no loss of energy is found. Also, a parallel combination has maximum stored energy.

18. (c) Both capacitors are in parallel. The capacitors have their high voltage plates at the same potential (the middle plate), and both capacitors have their low voltage plates at the same potential. (the outer plates, which are connected).

Let $\quad l = d_1 + d_2 = \text{constant}$

Then, $\quad C = \frac{\varepsilon_0 Al}{d_1d_2} = \frac{\varepsilon_0 Al}{d_1(l - d_1)}$

$$C \to \infty \text{ as } d_1 \to 0$$

which is $d \to 0$

To find the minimum capacitance, set $\frac{dC}{d(d_1)} = 0.$

For d_1, $\quad \frac{dC}{d(d_1)} = \frac{d}{d(d_1)}\left[\frac{\varepsilon_0 Al}{d_1 l - d_1^2}\right] = \varepsilon_0 Al \frac{(l - 2d_1)}{(d_1 l - d_1^2)^2} = 0$

$$d_1 = \frac{1}{2}l = d_2$$

$$C_{min} = \varepsilon_0 A\left(\frac{d_1 + d_2}{d_1d_2}\right)_{d_1 = \frac{l}{2}} = \varepsilon_0 A\left(\frac{l}{\left(\frac{1}{2}l\right)\left(\frac{1}{2}l\right)}\right)$$

$$= \varepsilon_0 A\left(\frac{4}{l}\right) = \varepsilon_0 A\left(\frac{4}{d_1 + d_2}\right)$$

$$C_{min} = \frac{4\varepsilon_0 A}{d_1 + d_2}$$

Charge on $C_{eq} = \varepsilon\frac{C(C + C_0)}{2C + C_0}$

By the charge distribution principle,

$$q_3 = C_{eq}\,\varepsilon = \frac{\varepsilon C(C + C_0)}{2C + C_0}$$

$$q_2 = \frac{\varepsilon C(C + C_0)C_0}{(2C + C_0)(C + C_0)} = \frac{\varepsilon CC_0}{2C + C_0}$$

$$q_1 = \frac{\varepsilon C(C + C_0)C}{(2C + C_0)(C + C_0)} = \frac{C^2\varepsilon}{(2C + C_0)}$$

In position b, $\quad q_3' = \frac{C^2\varepsilon}{2C + C_0}$

$$q_2' = q_2$$

$$q_1' = \frac{\varepsilon C(C + C_0)}{2C + C_0}$$

Heat produced = Loss in stored energy + Extra energy drawn from the battery.

Here, loss in stored energy is zero, because C_{eq} is the same in both the positions of the key.

Energy drawn from the battery $= \varepsilon\,\Delta q = \varepsilon(q_1' - q_1)$ or $\varepsilon(q_3 - q_3')$

$$\Rightarrow \text{Heat produced} = \varepsilon\left(\frac{\varepsilon C(C + C_0)}{2C + C_0} - \frac{\varepsilon C^2}{2C + C_0}\right) = \frac{\varepsilon^2 CC_0}{2C + C_0} = Q$$

19. (b,c) In combination, $Q_1 = Q_2$ (as they are in series)

$$\therefore \quad C_1V_1 = C_2V_2$$

$$\Rightarrow \quad C_1(V_A - V_D) = C_2(V_D - V_B)$$

$$\Rightarrow \quad C_1(V_1 - V_D) = C_2(V_D - V_2)$$

$$\therefore \quad V_D = \frac{C_1V_1 + C_2V_2}{C_1 + C_2}$$

Also, as capacitors are in series they have equal charges.

20. (a,c) $C_{eq} = \frac{5}{2}\mu F, Q = CV = 250\,\mu C$

Charge in $6\mu F$ branch $V_C = 150\,\mu C$

$$V_{AB} = \frac{150}{6} = 25\,V$$

and $\quad V_{BC} = 100 - 25 = 75\,V$

21. (a) As cylinder can be considered as two cylinders in parallel, their positive plates are joined as well as negative plates also.

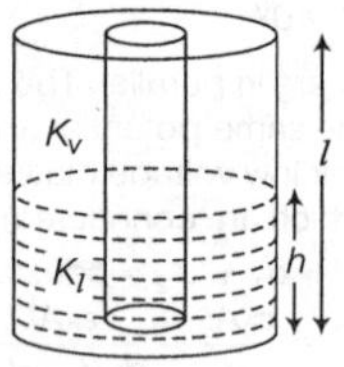

$$\therefore \quad C_{eq} = C_{liquid} + C_{vapour}$$

$$= \frac{2\pi\varepsilon_0 K_l h}{\ln(R_0/R_i)} + \frac{2\pi\varepsilon_0 K_v(l - h)}{\ln(R_0/R_i)}$$

$$= \frac{2\pi\varepsilon_0 l}{\ln(R_0/R_i)}\left\{(K_l - K_v)\frac{h}{l} + K_v\right\}$$

22. (d) When the tank is full,

$$C = \frac{2\pi\varepsilon_0 l k_l}{\ln(R_0/R_i)} = \frac{4\pi\varepsilon_0 l k_l}{2\ln(R_0/R_i)} = \frac{2 \times 1.4}{9 \times 10^9 \times 2 \times \ln(5/4.5)}$$

$$= 1.5 \times 10^{-9}\,F$$

23. (d) When the tank is empty,

$$C = \frac{2\pi\varepsilon_0 l\, K_l}{\ln(R_0/R_i)} = 1.1 \times 10^{-9}\,F$$

24. (b) Considering branch ac,

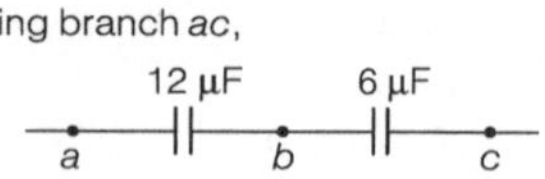

$$V_{ab} = 4V \quad \text{(given)}$$

$$\Rightarrow \quad Q_{ab} = C_{ab}V_{ab} = 12\,\mu F \times 4\,V = 48\,\mu C$$

As, $12\,\mu F$ and $6\,\mu F$ capacitors are in series

$\therefore$ Charge on both must be same.

$$\Rightarrow \quad Q_{bc} = 48\,\mu C$$

As, $\quad C_{bc} = 6\,\mu F$

$$V_{bc} = \frac{48}{6}V = 8\,V$$

Hence, $\quad V_{ac} = V_a - V_c = V_a - V_b + V_b - V_c$

$$= 4 + 8 = 12\,V$$

25. (c) In given circuit, there are two portions marked I and II as shown

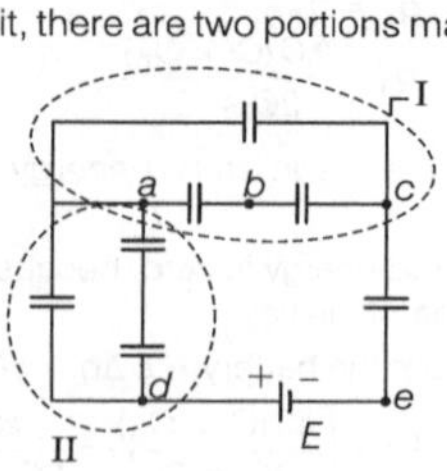

Now, considering circuit as given in figure.

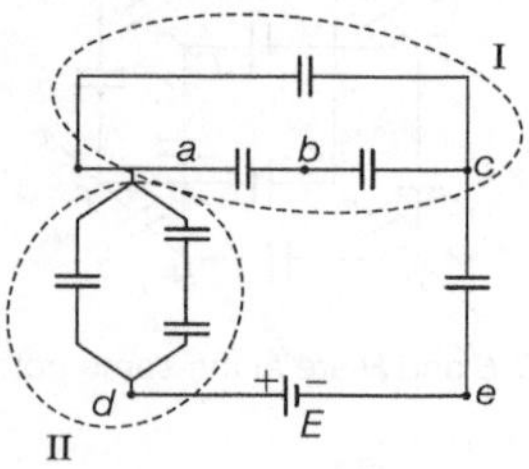

Clearly I and II parts are in series, so charge on both portions must be same.

As, $\quad V_{ac} = 12\,V$

And C_{eq} of portion marked I $= 10\,\mu F$

So $\quad$ charge on portion I $= V_{eq} \times C_{eq}$

$$= 10 \times 12 = 120\,\mu C$$

$\therefore \quad$ Charge on portion II $= 120\,\mu C$

$$\Rightarrow \quad C_{ad} \times (V_d - V_a) = 120\,\mu C$$

$$\Rightarrow \quad 12 \times (V_d - V_a) = 120 \quad \{\because C_{ad} = 12\,\mu F\}$$

$$\Rightarrow \quad V_d - V_a = 10\,V$$

26. (a) We can consider the given circuit as

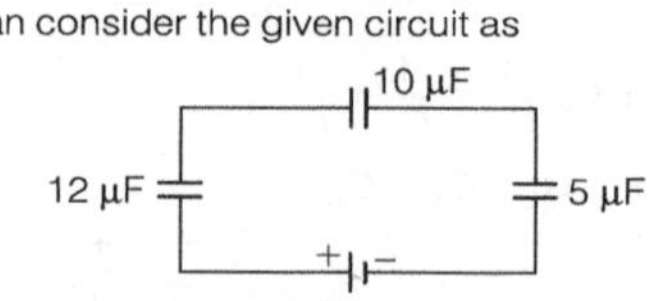

Charge on $5\,\mu F$ capacitor $= 120\,\mu C$

$$\therefore \quad V_c - V_e = \frac{120}{5} = 24\,V$$

By applying Kirchoff's law in shown loop,

$$E - V_{da} - V_{cd} - V_{ec} = 0$$

$$E = 12 + 10 + 24 = 46\,V$$

27. (a) Initially capacity of system is $\frac{C}{2}$ and when switch is closed one of capacitors is short circuited and capacity finally becomes C.

Initially potential drop across each capacitor is $\frac{E}{2}$ and finally potential drop across first capacitor is E and across second is zero.

Initially,

Capacity of system $= \frac{C}{2}$

Potential drop of each capacitor $= \frac{E}{2}$

Charge stored $= Q_1 = \frac{C}{2} \times E$

Finally charge stored $= Q_2 = C \times E$

Charge flow through cell $= CE - \frac{CE}{2} = \frac{CE}{2}$

Initially energy stored, $U_i = \frac{1}{2}\left(\frac{C}{2}\right) \cdot E^2 = \frac{CE^2}{4}$

Finally energy stored, $U_f = \frac{1}{2}(C)(E^2)$

$\therefore$ Change in energy stored $= U_f - U_i = \frac{1}{4}CE^2$

$\therefore$ Work done by cell when S is closed $= \frac{1}{2}CE^2$

Heat developed = Loss of energy of second capacitor $= \frac{1}{4}CE^2$

(i) → (q), (ii) → (p), (iii) → (r), (iv) → (r)

DPP-5 Charging of a Capacitor and *R-C* Circuits

1. (i) Consider an R-C circuit with switch S,

At $t = 0$, switch S is closed and electrons will flow from negative end of battery and accumulate over plate of capacitor. These electrons induce a positive charge over other plate of capacitor and electrons repelled will flow into positive terminal of battery.

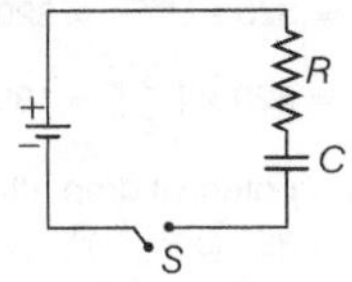

As charge accumulates on the capacitor, the potential difference across it increases ($V_{cap} = Q/C$) and the current reduces. Current becomes zero when V_{cap} will be equal to emf of battery.

Now, emf of battery = potential drop across capacitor + potential drop across resistor

$$E = IR + \frac{Q}{C}$$

(ii) Kirchhoff's loop rule,

$$\Rightarrow \quad E = R\frac{dQ}{dt} + \frac{Q}{C} \quad \ldots\text{(i)}$$

Where, R = total circuit resistance

$$\therefore \quad I = \frac{dQ}{dt}$$

$$\Rightarrow \quad \frac{dQ}{CE - Q} = \frac{dt}{RC}$$

Now, at $t = 0$, $Q = 0$ and let time $t = t$ s,

charge, $Q = Q$

$$\Rightarrow \quad \int_0^Q \frac{dQ}{CE - Q} = \int_0^t \frac{dt}{RC}$$

$$\Rightarrow \quad [-\log_e (CE - Q)]_0^Q = \left[\frac{t}{RC}\right]_0^t$$

$$\Rightarrow \quad -\log_e (CE - Q) - (-\log_e CE) = \frac{t}{RC}$$

$$\Rightarrow \quad \log_e (CE - Q) - \log_e CE = \frac{-t}{RC}$$

$$\Rightarrow \quad \log_e \left(1 - \frac{Q}{CE}\right) = -\frac{t}{RC}$$

Taking exponents on both sides

$$1 - \frac{Q}{CE} = e^{-t/RC} \quad \text{or} \quad Q = CE\,(1 - e^{-t/RC}) \quad \ldots\text{(ii)}$$

(iii) As potential difference across capacitor,

$$V = \frac{Q}{C}$$

$$\Rightarrow \quad V = \frac{Q}{C} = E\,(1 - e^{-t/RC}) \quad \ldots\text{(iii)}$$

(iv) At $t = 0$, current is maximum $I = \frac{E}{R}$ and at $t = RC$ (time constant), current drops to $\frac{1}{e}$ of its initial value

(or $0.37 \Rightarrow 37\%$ of I_{max})

(v) The quantity R-C is called time constant of the circuit.

If we take $t = RC$,

Then, $Q = CE\,(1 - e^{-1}) = 0.63\,CE \quad (\because Q_{max} = CE)$

$= 0.63\,Q_{max}$

So, we define time constant as time interval in which capacitor reach 0.63 or 63% its full charge and voltage. A circuit with higher value of time constant $\tau = RC$ reaches its maximum charge and voltage much quickly.

$$\text{As,} \quad \text{current} = \frac{dQ}{dt} \quad \ldots\text{(iv)}$$

So, current in the circuit at any instant of time t,

$$I = \frac{dQ}{dt} = \frac{d}{dt} CE\,(1 - e^{-t/RC})$$

$$\Rightarrow \quad I = \frac{E}{R} \cdot e^{-t/RC}$$

2. (i) In case, a charged capacitor with charge Q and potential V is connected across a capacitor C.

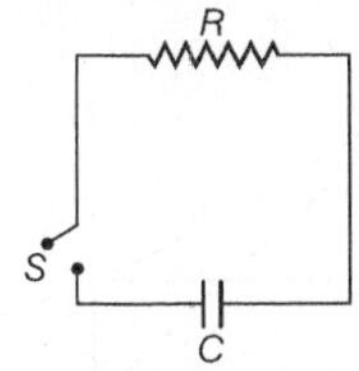

When switch S is closed charge begins to flow through resistor R from one side of capacitor towards the other side until the capacitor is fully discharged.

At any instant,

$$V_R = V_C$$

$$\Rightarrow \quad IR = \frac{Q}{C}$$

$$\text{Also,} \quad I = -\frac{dQ}{dt}$$

(negative sign as capacitor is discharging)

$$\Rightarrow \quad -\frac{dQ}{dt} \cdot R = \frac{Q}{C}$$

$$\Rightarrow \quad \frac{dQ}{Q} = -\frac{dt}{RC}$$

On integrating, we get

$$[\log_e Q]_{Q_0}^{Q} = \left[\frac{-t}{RC}\right]_0^t$$

$\therefore$ At $t = 0$, charge $= Q_0$

At $t = t$ s, charge $= Q$

$$\Rightarrow \quad \log_e \frac{Q}{Q_0} = \frac{-t}{RC}$$

$$\Rightarrow \quad Q = Q_0\, e^{-t/RC} \quad \ldots\text{(i)}$$

(ii) Potential drop across capacitor.

$$V_C = \frac{Q}{C} = \frac{Q_0 e^{-t/RC}}{C}$$

$$\Rightarrow \quad V_C = V_0 e^{-t/RC} \quad \ldots\text{(ii)}$$

$$\text{and} \quad I = \frac{-dQ}{dt}$$

$$\Rightarrow \quad I = I_0 e^{-t/RC} \quad \ldots\text{(iii)}$$

(iii) when $t = RC$, the current reduces to $\frac{1}{e}$ or 37% of its original value is one time constant, $t = \tau = RC$.

3. In an R-C circuit energy is dissipated in the form of heat through the resistor.

By Conservation law,

Energy given by battery = Energy stored in capacitor + Energy lost by resistor

Also, rate of energy loss = I^2R

4. (i) Time constant

$$\tau = RC = 10^6 (10^{-6})\text{ s} = 1\text{s}$$

Voltage as a function of time

$$V(t) = V_0 e^{-t/RC} \quad \text{for } t > 0$$

$\because \quad V(t) = 10\, e^{-t},$

$\therefore \quad V(5) = 10\, e^{-5} = 0.067 \text{ V}$

Current, $\quad i(t) = \dfrac{V_0}{R} \cdot e^{-t/RC}$

$\Rightarrow \quad i(t) = \dfrac{10}{1 \times 10^6} e^{-t}$

$\Rightarrow \quad i(t) = 10^{-5}\, e^{-t} \text{ A}$

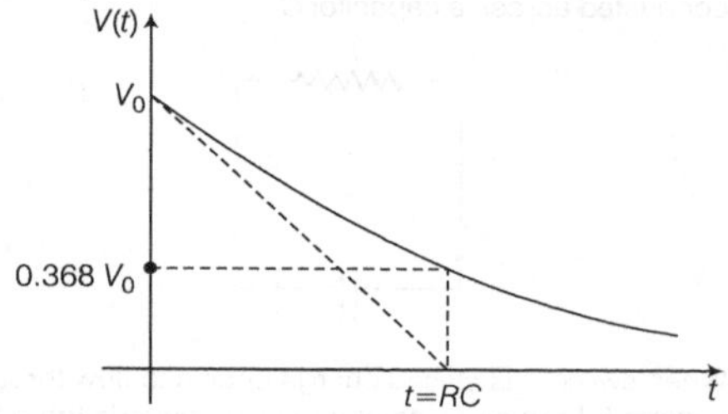

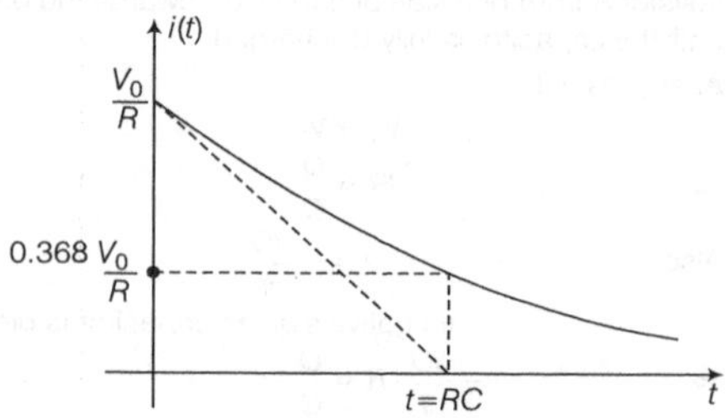

(ii) At $t = 0$, an uncharged

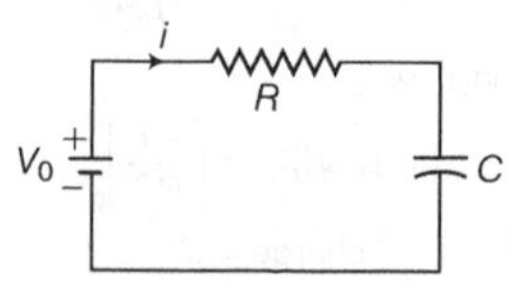

Capacitor is connected to battery, then

$$\left.\frac{dV}{dt} + \frac{1}{RC} V = \frac{1}{RC} \cdot V_0 \right\} \text{KVL}$$

$\Rightarrow \quad V = V_0 + Ae^{-t/RC}$

Now, in given circuit, time constant, $\tau = RC = 0.02$ s

$\Rightarrow \quad V(t) = 12 + Ae^{-t/RC} = 12 + Ae^{-50t} \quad \ldots\text{(i)}$

From initial conditions, at $t = 0$,

$$V(0) = 2 \text{ V}$$

$\therefore \quad V(t) = 12 + Ae^{-50t} \quad [\because t = 0]$

$\Rightarrow \quad 2 = 12 + A$

$\Rightarrow \quad A = -10$

Hence, $V(t) = 12 - 10\, e^{-50t}$ and current

$$i(t) = \frac{12 - V}{5000}$$

$$= 2 \times 10^{-3}\, e^{-50t}$$

$$= 2 \times 10^{-3}\, e^{-50t}$$

$\Rightarrow \quad (t) = 2 \times 10^{-3}\, e^{-50t}$

5. (b) During discharging of capacitor,

$$V = V_0\, e^{-t/RC}$$

After 1 s,

$$V_1 = 320\, e^{-1/RC} = 240 \text{ V}$$

$\Rightarrow \quad e^{-1/RC} = \dfrac{240}{320} = \dfrac{3}{4}$

After 2 s,

$$V_2 = 320\, e^{-2/RC} = 320\, (e^{-1/RC})^2$$

$\Rightarrow \quad V_2 = 320 \times \left(\dfrac{3}{4}\right)^2 = 180 \text{ V}$

and $\quad V_3$ = potential drop after 3 s

$$= 320\, (e^{-t/RC})^3$$

$$= 320 \times \left(\frac{3}{4}\right)^3 = 135 \text{ V}$$

6. (a) $V_0 = V_R + V_C = 4V_R$

$\Rightarrow \quad V_R = \dfrac{V_0}{4}$

$$V_R = \frac{V_0}{4} = V_0\, e^{-t/RC}$$

$\Rightarrow \quad \dfrac{1}{4} = e^{-t/RC} = e^{-t/10}$

$\Rightarrow \quad 4 = e^{t/10}$

$\Rightarrow \quad \log_e 4 = \dfrac{t}{10}$

$\Rightarrow \quad t = 10 \log_e 4$

$\Rightarrow \quad t = 10 \times \log_e 2^2$

$$= 2.0 \log_e 2 = 13.86 \text{ s}$$

7. (a) As, $Q = CE\, (1 - e^{-t/RC})$

$$V = E\, (1 - e^{-t/RC})$$

$\Rightarrow \quad C = \dfrac{Q}{V}$ is a straight line

8. (c) During discharging $I = I_0 e^{-t/RC}$

$\Rightarrow \quad \log I = \log I_0\, e^{-t/RC}$

$\Rightarrow \quad \log I = \log I_0 + \log e^{-t/RC}$

$\Rightarrow \quad = -\dfrac{t}{RC} + \log I_0$

In graph of $\log I$ *versus* t,

when C is increased slope becomes a small negative number. Thus, slope increases.

9. (a) During discharging, $Q = Q_0 e^{-t/RC}$

Hence, charge decays rapidly, if resistance is high.

So, R_3 is largest.

10. (d) At, $t = 0$

$$I = \text{initial current} = \frac{V}{R}$$

$$= \frac{10}{50} = 0.2 \text{ A}$$

and final charge $\quad Q = CV$

$\Rightarrow \quad Q = 2 \times 10^{-6} \times 10$

$$= 20 \times 10^{-6} \text{ C}$$

$$= 20\, \mu\text{C}$$

11. (d) In given circuit, total resistance across capacitor is (Note that AB is capacitor)

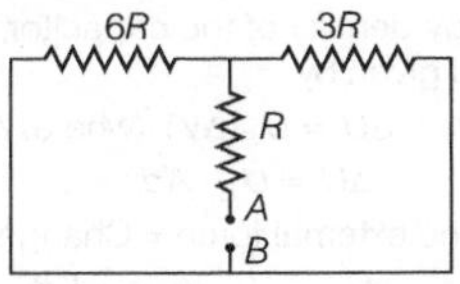

Resistance R is in series with combination of $6R$ and $3R$ in parallel.

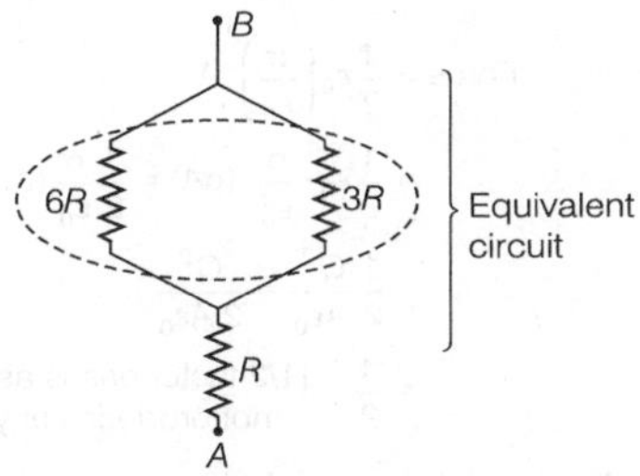

Hence, equivalent resistance of circuit is $3R$.

So, time constant is $\tau = 3RC$

12. (c) During charging circuit current drops exponentially.

13. (c) During discharging of a capacitor current decreases exponentially.

14. (a) RC has dimensions of time.

$\therefore \frac{t}{RC}$ is dimensionless.

15. (a) Final charge on capacitor is $q = q_0 e^{-\frac{t}{RC}}$

where, q_0 = charge on capacitor at $t = 0$

RC = time constant of the circuit

Putting $\quad q_0 = CV_0$

$\therefore \quad q = CV_0 e^{-\frac{t}{RC}}$

Given $\quad C = 2\text{ F}, \quad V_0 = 5\text{ V}$

$R = 6\,\Omega, t = 12\text{ s}$

Hence, $\quad q = (2 \times 5) e^{-\frac{12}{6 \times 2}} = \frac{10}{e}\text{ C}$

16. (a,d) As two capacitors of same capacity and charge can store more energy than single one hence, heat produced in (ii) is more than that of (i). Also for circuit (i),

τ = time constant = RC

For circuit (ii), $\quad \tau$ = time constant = $2R \cdot \frac{C}{2} = RC$

Hence, both capacitors will be discharged in same time.

17. (a,d) $\quad \tau = RC \quad \Rightarrow \quad C = \frac{\tau}{R}$

or $\quad C = \frac{24 \times 10^{-6}}{15 \times 10^3} = 1.6 \times 10^{-9}\text{ F}$

and $\quad V_C = \varepsilon(1 - e^{-t/\tau})$

$\Rightarrow \quad = -24 \times 10^{-6} \cdot \ln\left(1 - \frac{8}{24}\right)$

$= 9.73 \times 10^{-6}\text{ s}$

18. (a) A practical flasher circuit is

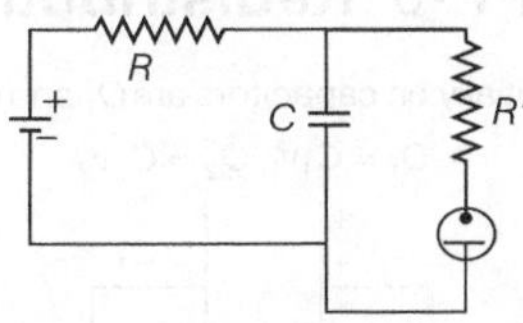

When discharging is finished the discharge tube wil not conducts longer and charging of capacitor begins repeating the cycle.

19. (d) A capacitor is not charged instantly as every practical circuit has some resistance and it is a R-C circuit.

20. (10) Time constant = Time to cost 63% of initial charge

$= \tau = RC = 5\text{ ms}$

when resistance is make $2R$, then

$\tau = 2RC = 2 \times 5$

$= 10\text{ ms}$

21. (2) Given,

$R = 400\,\Omega, C = 40\,\mu\text{F}$

$\Rightarrow \quad \frac{1}{RC} = 62.5\ (\text{s}^{-1})$

Also, $\quad V(0) = 100\text{ V}$

Hence, $\quad V(R) = V(c) = 100\, e^{-6.25t}$

and $\quad i = \frac{V_R}{R} = 0.25\, e^{-6.25t}$

$\therefore \quad$ Power, $P = V_R i$

$\Rightarrow \quad P = 25e^{-125t}$

So, energy in resistor

$W(R) = \int_0^t P dt = \int_0^t 25e^{-125t} dt$

$= 0.20\,(1 - e^{-125t})$

$= 2 \times 10^{-1}\,(1 - e^{-125t})$

So, $\quad k = 2$

22. (2) Given,

$P = 360\, e^{-t \times 10^{-5}}$

and $\quad R = 10\,\Omega$

Comparing with

$P = P_0 e^{-2t/RC}$, we get

$\frac{2}{RC} = 10^5$ or $C = 2\,\mu\text{F}$

So, $\quad W(R) = \int_0^t P \cdot dt = \int_0^t 360\, e^{-t \times 10^{-5}} \cdot dt$

$= 3.6\,(1 - e^{-t \times 10^{-5}})\text{ MJ}$

So, $W(R)$ at $\quad t = \infty$ is 3.6 MJ

$\Rightarrow \quad \frac{Q_0^2}{2C} = 3.6$

$\Rightarrow \quad Q_0 = 120\,\mu\text{C}$

So $\quad K = 2$

23. (b) Time constant of a RC circuit is $\tau = RC$, larger the value of time constant, the slower rate of growth of charge.

Also, larger the value of time constant RC, the smaller the rate of decay of charge.

(i) → (p,r), (ii) → (q,s), (iii) → (p,r), (iv) → (q,s)

DPP-6 Redistribution of Charges and Van de Graaff Generator

1. Let charges initially on capacitors are Q_1 and Q_2

$$Q_1 = C_1V_1, Q_2 = C_2V_2$$

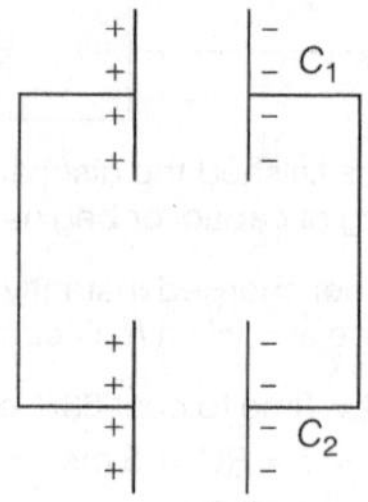

Common potential reached after stabilisation is V.

$$V = \frac{\text{Total charge}}{\text{Total capacity}}$$

$$\Rightarrow \quad V = \frac{Q_1 + Q_2}{C_1 + C_2} = \frac{C_1V_1 + C_2V_2}{C_1 + C_2}$$

Initially stored energy is

$$U_i = U_1 + U_2$$

$$= \frac{1}{2}C_1V_1^2 + \frac{1}{2}C_2V_2^2$$

Final energy is

$$U_f = \frac{1}{2}(C_1 + C_2)V^2$$

$$= \frac{1}{2}(C_1 + C_2)\left(\frac{C_1V_1 + C_2V_2}{C_1 + C_2}\right)^2$$

$$= \frac{1}{2}\frac{(C_1V_1 + C_2V_2)^2}{C_1 + C_2}$$

So, energy loss ΔU is

$$\Delta U = U_i - U_f$$

$$= \frac{1}{2}C_1V_1^2 + \frac{1}{2}C_2V_2^2 - \frac{1}{2}\frac{(C_1V_1 + C_2V_2)^2}{C_1 + C_2}$$

$$= \frac{(C_1V_1^2 + C_1V_2^2)(C_1 + C_2) - (C_1V_1 + C_2V_2)^2}{2(C_1 + C_2)}$$

$$= \frac{1}{2(C_1 + C_2)}\{C_1C_2(V_1^2 + V_2^2 - 2V_1V_2)\}$$

$$= \frac{C_1C_2}{2(C_1 + C_2)} \cdot (V_1 - V_2)^2$$

As, $(V_1 - V_2)^2 > 0$

$\therefore \quad \Delta U > 0$

$U_i > U_f$

The balance of energy ΔU appears less as a part of wire is connecting the capacitors (heating in wire) and partly radiated as electromagnetic waves.

2. Consider a charged capacitor and let a force (just equals to force between plates of capacitor) pulls the plates apart by a small distance dx, then

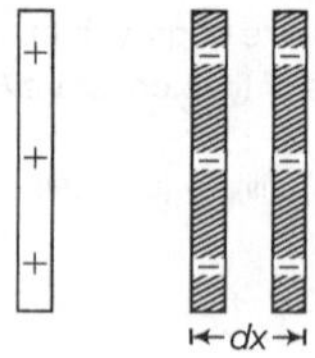

Work done by force $= F dx$

If U is the energy density of the capacitor, then change in energy of capacitor will given by

$$\Delta U = U_E(\Delta V) \quad \text{(where, } \Delta V = \text{change in volume)}$$

or $\quad \Delta U = U_E \cdot Adx$

Work done by the external force = Change in PE of the capacitor

$$\Delta U = U_E \cdot Adx = Fdx$$

$$\Rightarrow \quad F = U_E A = \frac{1}{2}\varepsilon_0 E^2 A \qquad \left(\text{as } U_E = \frac{1}{2}\varepsilon_0 E^2\right)$$

$$\therefore \quad \text{Force} = \frac{1}{2}\varepsilon_0\left(\frac{\sigma}{\varepsilon_0}\right)^2 A$$

$$= \frac{1}{2}\varepsilon_0 \cdot \frac{\sigma}{\varepsilon_0^2} \cdot (\sigma A) = \frac{1}{2}\frac{\sigma}{\varepsilon_0}Q$$

$$= \frac{1}{2}\frac{Q^2}{A\varepsilon_0} = \frac{Q^2}{2A\varepsilon_0}$$

$= \frac{1}{2}$ [1/2 factor arises as charges does not produces any force on itself]

3. Let radius of inner shell is r with charge q and radius of outer shell is R with charge Q.

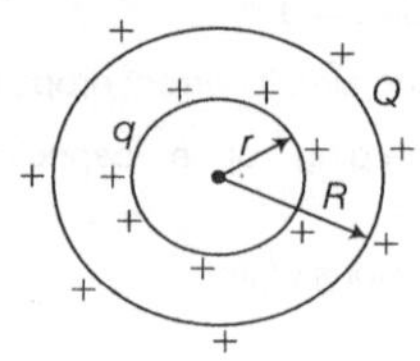

V_1 = Potential on the surface of inner sphere

= Potential due to q charge + potential due to Q charge

$$= \frac{kq}{r} + \frac{kQ}{R}$$

V_2 = Potential on the surface of outer sphere

= Potential due to q charge + Potential due to Q charge

$$= \frac{kq}{R} + \frac{kQ}{R}$$

Now, $\quad V_1 - V_2 = \left(\frac{kq}{r} + \frac{kQ}{R}\right) - \left(\frac{kq}{R} + \frac{kQ}{R}\right)$

$$= kq\left(\frac{1}{r} - \frac{1}{R}\right)$$

As, $\quad r < R \Rightarrow \frac{1}{r} > \frac{1}{R}$

So, $V_1 - V_2$ is always positive or $V_1 > V_2$.

4. (b) Only charge is conserved following conservation law of charges.

5. (c) When spheres are connected,

$$V_1 = V_2$$

$$\Rightarrow \quad \frac{Q_1}{25} = \frac{Q_2}{20}$$

$$\Rightarrow \quad \frac{Q_1}{Q_2} = \frac{25}{20}$$

$$\Rightarrow \quad \frac{Q_1}{Q_2} > 1 \quad \text{or} \quad Q_1 > Q_2$$

6. (c) Volume of small drops = Volume of big drop

$$\Rightarrow \quad 8 \times \frac{4}{3}\pi r^3 = \frac{4}{3}\pi R^3$$

$$\Rightarrow \quad R = 2r$$

As, capacity $\propto r$

So, capacity increases by a factor of 2.

7. (b) With both plates, $F = qE$

$\Rightarrow \quad F = q\dfrac{\sigma}{\varepsilon_0}$

With only one plate, $F' = qE' = q\dfrac{\sigma'}{2\varepsilon_0} = \dfrac{F}{2}$

8. (a) Since C_1 and C_2 are parallel, so their equivalent capacitance will be $(C_1 + C_2)$. Now $(C_1 + C_2)$ and C_3 are in series, so the net equivalent capacitance of circuit

$$\frac{1}{C} = \frac{1}{C_3} + \frac{1}{C_1 + C_2} = \frac{C_1 + C_2 + C_3}{C_3(C_1 + C_2)}$$

$$C = \frac{(C_1 + C_2)C_3}{C_1 + C_2 + C_3}$$

Since V is the voltage of battery, so charge on their system.

$$q = CV$$

$$q = \frac{(C_1 + C_2)C_3V}{C_1 + C_2 + C_3}$$

If capacitor C_3 breaks down then total equivalent capacitance

$$C' = C_1 + C_2$$

New stored charge, $q' = C'V$

$$q' = (C_1 + C_2)V$$

Change in total charge, $\Delta q = q' - q$

$$= (C_1 + C_2)V - \frac{(C_1 + C_2)C_3V}{C_1 + C_2 + C_3}$$

$$\Delta q = (C_1 + C_2)V\left[1 - \frac{C_3}{C_1 + C_2 + C_3}\right]$$

9. (b) Before contact, $Q_1 = \sigma 4\pi R^2$

$$Q_2 = \sigma 4\pi(2R)^2 = 4Q_1$$

After contact,

$$Q_1' + Q_2' = Q_1 + Q_2 = 5Q_1$$
$$= 5\sigma 4\pi R^2$$

The surface potentials of both spheres are same,

$$\frac{Q_1'}{R} = \frac{Q_2'}{2R}$$

$\Rightarrow \quad Q_2' = 2Q_1'$

$\Rightarrow \quad 3Q_1' = 5\sigma 4\pi R^2$

$$Q_1' = \frac{5}{3}\sigma 4\pi R^2, \; Q_2' = \frac{10}{3}\sigma 4\pi R^2$$

$\Rightarrow \quad \sigma_1' = \dfrac{5}{3}\sigma$ and $\sigma_2' = \dfrac{5}{6}\sigma$

So, $\dfrac{\sigma_1'}{\sigma_2'} = \dfrac{5}{3} \times \dfrac{6}{5} = 2$

10. (d) Before opening the switch potential difference across both the capacitors is V, as they are in parallel.

Energy stored $U_i = \dfrac{1}{2}CV^2 + \dfrac{1}{2}CV^2 = CV^2$

After opening the switch, potential difference across capacitor A is V and capacitance is $3C$.

For capacitor B,

charge $q = CV$ and

capacity $= 3C$

So, $U_f = \dfrac{q^2}{2(3C)} + \dfrac{3CV^2}{2} = \dfrac{CV^2}{6} + \dfrac{3CV^2}{2}$

$$= \frac{10}{6}CV^2 = \frac{5}{3}CV^2$$

So, $\dfrac{U_i}{U_f} = \dfrac{CV^2}{\frac{5}{3}CV^2} = \dfrac{3}{5}$

11. (a) Potential of inner sphere is independent of charge on outer sphere. So, it is still V. (Refer solution of Q. no. 3).

12. (d) For sphere 1,

$$C_1 = 4\pi\varepsilon_0 R_1$$
$$V_1 = \frac{1}{4\pi\varepsilon_0}\frac{Q_1}{R_1}$$

and for sphere 2,

$$C_2 = 4\pi\varepsilon_0 R_2$$
$$V_2 = \frac{1}{4\pi\varepsilon_0}\frac{Q_2}{R_2}$$

If $V_1 = V_2$, there is no transfer of charge and hence, no loss of energy occurs

$$V_1 = V_2$$

$\Rightarrow \quad \dfrac{Q_1}{R_1} = \dfrac{Q_2}{R_2}$

or $\quad Q_1R_2 = Q_2R_1$

Also, in all other cases whether $V_1 > V_2$ or $V_2 > V_1$, charge transfer and hence, loss of energy will takes place.

13. (a) Common potential after redistribution of charges

$$V = \frac{Q_1 + Q_2}{C_1 + C_2} = \frac{C_1V_1 + C_2V_2}{C_1 + C_2}$$

$$= \frac{4 \times 10^{-6} \times 80 + 6 \times 10^{-6} \times 30}{4 \times 10^{-6} + 6 \times 10^{-6}}$$

$$= \frac{320 + 180}{10}$$

$$= \frac{500}{10} = 50 \text{ V}$$

So, final potential energy of 4 μF capacitor

$$U_f = \frac{1}{2}CV^2 = \frac{1}{2} \times 4 \times 10^{-6} \times (50)^2$$

Initial potential energy of 4 μF capacitor

$$U_i = \frac{1}{2}CV^2 = \frac{1}{2} \times 4 \times 10^{-6} \times (80)^2$$

$\therefore$ Loss of energy $= U_i - U_f$

$\therefore \quad \Delta U = \dfrac{1}{2} \times 4 \times 10^{-6} \times (80)^2 - \dfrac{1}{2} \times 4 \times 10^{-6} \times (50)^2$

$$= \frac{1}{2} \times 4 \times 10^{-6} \times (80^2 - 50^2)$$

$$= 2 \times 10^{-6} \times 30 \times 130$$

$$= 78 \times 10^{-4} \text{ J} = 7.8 \text{ mJ}$$

14. (a,b,c,d) $E_{max} = 3 \times 10^6 = \dfrac{kQ}{r}$

$\Rightarrow \quad Q_{max} = \dfrac{E_{max} \cdot r^2}{k}$

$$= \frac{3 \times 10^6 \times (0.15)^2}{9 \times 10^9}$$

$$= 7.51 \, \mu\text{C}$$

$$V_{max} = rE_{max} = 3 \times 10^6 \times 0.15$$

$$= 450 \text{ kV}$$

Also maximum charge hold by dome is increased, when it is placed in vacuum as there will not be any breakdown.

A pointed object nearby has an induced opposite charge promoting the breakdown.

15. (a,d) By reducing the radius of dome, its potential reduces (and it also stores less charge) and so chances of discharge by dielectric breakdown decreases. Also, when pressure of a gas is increased, its ionisation tendency is suppressed.

16. (b,d) Altered capacitance $C' = \frac{C}{2}$

As charge remains same,

$$C(\Delta V) + (\Delta V) = C(\Delta V') + \frac{C}{2}(\Delta V')$$

$$\Rightarrow \qquad \Delta V' = \frac{4}{3}\Delta V$$

Also,

$$U' = \frac{1}{2}C\left(\frac{4\Delta V}{3}\right)^2 + \frac{1}{2}\left(\frac{C}{2}\right)\left(\frac{4\pi V}{3}\right)^2$$

$$= \frac{4}{3}C(\Delta V)^2$$

$$= \frac{4}{3}U$$

17. (a,d) V_f = final voltage value.

Final values of charges on capacitors are

$$Q_1 = CV_f, Q_2 = \varepsilon V = \alpha CV_f^2$$

Initial charge on capacitor is

$$Q_0 = CV_i$$

Equating charge values,

$$Q_0 = Q_1 + Q_2$$

$$CV_i = CV_f - \alpha CV_f^2 = 0$$

$$\Rightarrow \qquad \alpha V_f^2 + V_f - V_i = 0$$

$$V_f = \frac{1 \pm \sqrt{1 + 4\alpha V_i}}{2\alpha}$$

$$= \frac{-1 \pm \sqrt{1 + 624}}{4}$$

$$= \frac{24}{4} = 6\text{ V}$$

18. (a) Capacitor A is a combination of two capacitors

$$C_A = C_K + C_0 = \frac{K\varepsilon_0 A}{d} + \frac{\varepsilon_0 A}{d}$$

$$= (K+1)\frac{\varepsilon_0 A}{d}$$

$$\therefore \qquad C_A = \frac{(9+1)(8.85 \times 10^{-12})(0.02)}{8.85 \times 10^{-4}}$$

$$= 2 \times 10^{-9}\text{F}$$

$$U_A = \frac{1}{2}C_A V^2$$

$$= \frac{1}{2} \times 2 \times 10^{-9} \times (110)^2$$

$$= 1.21 \times 10^{-5}\text{ J}$$

19. (b) Charge on $A = q_A = C_A V_A = 2 \times 10^{-9} \times 110$

$$= 2.2 \times 10^{-7}\text{ C}$$

When dielectric slab is removed, the capacity of A is reduced.

$$C'_A = \frac{\varepsilon_0(2A)}{d}$$

$$= \frac{8.85 \times 10^{-12} \times 0.04}{8.85 \times 10^{-4}}$$

$$= 0.04 \times 10^{-9}\text{F}$$

Energy stored in A,

$$U'_A = \frac{1}{2}\frac{(q_A)^2}{C'_A}$$

$$= 6.05 \times 10^{-5}\text{ J}$$

20. (c) Work done is removing slab

$$W = U'_A - U_A$$

$$= (6.05 - 1.21) \times 10^{-5}\text{J}$$

$$= 4.84 \times 10^{-5}\text{ J}$$

21. (c) As capacitors are of same capacities, common potential of the combination is

$$V_{com} = \frac{\text{Total charge}}{\text{Total capacity}} = \frac{C_A V_A}{C_A + C_B}$$

$$= \frac{600 \times 200}{600 + 600} = 100\text{ V}$$

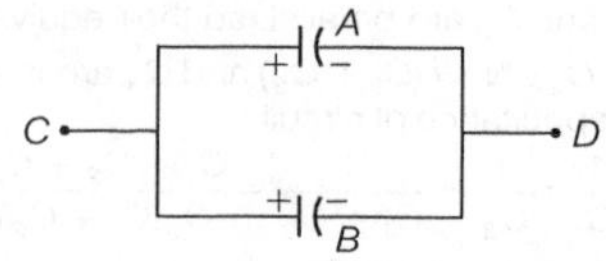

So, energy of A and $B = \frac{1}{2}(C_A + C_B)V^2$

$$= \frac{1}{2}(1200 \times 10^{-12}) \times (100)^2$$

$$= 6 \times 10^{-6}\text{ J}$$

Now, if plates of B are pulled apart, new capacity of B is

$$C'_B = \frac{\varepsilon_0 A}{2d} = \frac{C_B}{2}$$

Equivalent capacity of system

$$C_{eq} = C_A + C'_B$$

$$= \left(600 + \frac{600}{2}\right)\text{pF}$$

$$= 900\text{ pF} = 9 \times 10^{-10}\text{ F}$$

Total charge on system remains same.

$$\therefore \qquad Q = 600 \times 10^{-12} \times 200 = 12 \times 10^{-8}\text{ C}$$

Thus, potential drop across combination of capacitors (V_{CD}) is increased.

Potential drop across combination after separation of plates

$$V_{CD} = \frac{Q}{C} = \frac{12 \times 10^{-8}}{9 \times 10^{-10}}$$

$$= \frac{400}{3} = 133.33\text{ V}$$

Energy stored in combination initially $= U_i = \frac{1}{2}CV^2$

$$= \frac{1}{2}(600 \times 10^{-12} + 600 \times 10^{-12}) \times (100)^2$$

$$= 6 \times 10^{-6}\text{ J}$$

Energy stored in combination after separation of plates

$$U_f = \frac{1}{2}C_f \cdot V_f^2$$

$$= \frac{1}{2}(600 \times 10^{-12} + 300 \times 10^{-12})\left(\frac{400}{3}\right)^2$$

$$= 8 \times 10^{-6}\text{ J}$$

$$\therefore \qquad \text{Energy gain} = U_f - U_i = 2 \times 10^{-6}\text{ J}$$

Percentage gain of energy

$$= \frac{\Delta U}{U_i} \times 100 = \frac{2 \times 10^{-6}}{6 \times 10^{-6}} \times 100 = 33.33\%$$

22. (c) Energy stored in $A = \frac{1}{2}C_A \cdot V_f^2$

$$= \frac{1}{2} \times 600 \times 10^{-12} \times \left(\frac{400}{3}\right)^2$$

Energy stored in $B = \frac{1}{2}C'_B \cdot V_f^2$

$$= \frac{1}{2} \times 300 \times 10^{-12} \times \left(\frac{400}{3}\right)^2$$

So, $\qquad \text{ratio} = \frac{U'_A}{U'_B} = 2:1$

23. (c) Charge on A, $Q_A' = C_A V_A' = 600 \times 10^{-12} \times \frac{400}{3}$

Charge on B, $Q_B' = C_B V_B' = 300 \times 10^{-12} \times \frac{400}{3}$

$\therefore$ Ratio of charges $= \frac{Q_A'}{Q_B'} = \frac{2}{1}$

24. (d) Lower comb charges and sprays positively charged air molecules on the belt. Then, they transfer their charge to the belt.

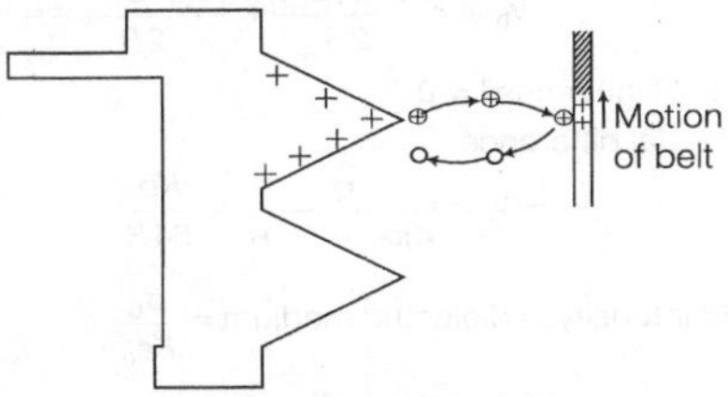

At upper comb, 'electrons' are drawn by the charged belt by motion of air molecules. There electrons neutralised the belt and lack of electrons make the shell positively charged.

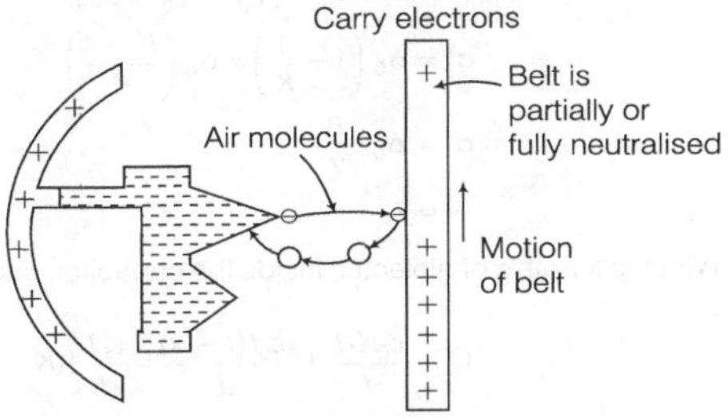

25. (b) Both Statement I and Statement II are correct but charge flows from inner to outer sphere because inner sphere is always at a higher potential.

26. (b) Both Statement I and Statement II are correct but where sphere is not kept in a vacuum chamber, dielectric breakdown of medium causes discharging of sphere.

27. (a)
Let C_A and C_B are capacities of sphere A and B. Then,

$$\frac{C_A}{C_B} = \frac{r}{R}$$

After contact with A, charge on B is q, then

$$\frac{q_1}{Q - q_1} = \frac{C_B}{C_A} = \frac{R}{r}$$

And charge on A is $Q \to q_1$.
In IInd contact A again loss Q charge, now total charge on A and B is $Q + q_1$

$$= Q\left(1 + \frac{R}{R + r}\right)$$

So, charge on B after IInd contact is

$$q_2 = Q\left(1 + \frac{R}{R + r}\right)\left(\frac{R}{R + r}\right)$$

Similarly,

$$q_n = Q\left[\left(\frac{R}{R + r} + \left(\frac{R}{R + r}\right)^2 + \dots + \left(\frac{R}{R + r}\right)^n\right)\right]$$

Series in brackets is a GP

$$\Rightarrow \quad q_n = Q\left[\frac{R}{r}\left(1 - \left(\frac{R}{R + r}\right)^n\right)\right]$$

So, electrostatic energy of B after 'n' such contacts is

$$U_n = \frac{q_n^2}{8\pi\varepsilon_0 R}$$

And when $n \to \infty$, $S_\infty = \frac{a}{1 - r}$

$$q_\infty = \frac{QR}{R + r}\left(\frac{1}{1 - \frac{R}{R + r}}\right) = Q\frac{R}{r}$$

And $\quad U_\infty = \frac{Q^2 R}{8\pi\varepsilon_0 r^2}$

(i) $\to$ (r), (ii) $\to$ (s), (iii) $\to$ (q), (iv) $\to$ (p)

Revisal Problems (JEE Main)

1. (c) Let C = capacity without slab.
Then, before the slab is removed,

$$C_1 = C \quad \text{and} \quad C_2 = KC$$

$$\Rightarrow \quad C_{\text{net}} = \frac{K}{K + 1} C$$

$$\Rightarrow \quad q_1 = q_2 = \left(\frac{KC}{K + 1}\right)E$$

E = emf of cell

When slab is removed net capacity is $C/2$.

$$\therefore \quad q_1' = q_2' = \frac{CE}{2}$$

$$\Rightarrow \quad \frac{q_2'}{q_2} = \frac{K + 1}{2K}$$

2. (b) Let charge on parallel plate capacitor $+q_1$ and that on shells be $q - q_1$, such that the potential difference across the shells would be same as potential difference across the plates.
Potentials of shells are

$$\phi_{\text{inner shell}} = \frac{q - q_1}{4\pi\varepsilon_0(2d)} + \frac{-(q - q_1)}{4\pi\varepsilon_0 d} = \frac{-(q - q_1)}{4\pi\varepsilon_0 2d}$$

$$\phi_{\text{outer shell}} = \frac{q - q_1}{4\pi\varepsilon_0(2d)} + \frac{-(q - q_1)}{4\pi\varepsilon_0 2d} = 0$$

$$\therefore \quad \Delta V = \text{Potential difference} = \phi_{\text{outer}} - \phi_{\text{inner}} = \frac{q - q_1}{2d\,(4\pi\varepsilon_0)}$$

For parallel plates, $V_{\text{plates}} = \frac{q_1}{4\pi\varepsilon_0 d}$

For parallel plates, $V_{plates} = \frac{q_1}{4\pi\varepsilon_0 d}$

Under steady state, $V_{shell} = V_{plates}$

$$\Rightarrow \quad \frac{q - q_1}{4\pi\varepsilon_0 2d} = \frac{q_1}{4\pi\varepsilon_0 d}$$

$$\Rightarrow \quad q_1 = q/3$$

$\therefore$ Charge on outer shell $= \frac{2q}{3}$

Charge on inner shell $= \frac{-2q}{3}$

Energy left in IInd plate capacitor $= \frac{1}{2}\frac{q_1^2}{C} = \frac{q^2}{72\,\pi\varepsilon_0 d}$

So, $$\frac{U_i}{U_f} = \frac{\frac{1}{2}q^2/C}{\frac{1}{2}\frac{(q/3)^2}{C}} = 9:1$$

3. (b) Let dielectric moves through a distance dx inside the plates. This increases capacitance to $C + dC$. As potential difference remains constant at V, the battery supply more charge to capacitor.

and $\quad dQ = dC \cdot V$

So, work done by battery $dW_b = VdQ = dC \cdot V^2$

and work done by force $dW_F = -Fdx$ (F = force on slab)

Total work done = change in energy of capacitor

$\Rightarrow dC \cdot V^2 - Fdx = \frac{1}{2}(dC)V^2$

$$\Rightarrow \quad F = \frac{1}{2}V^2 \cdot \frac{dC}{dx} = \frac{\varepsilon_0 bV^2(k-1)}{2d}$$

$$\therefore \quad F = \frac{\varepsilon_0 \times 4 \times 10^{-2} \times (12)^2 \times (7-1)}{2 \times 4 \times 10^{-2}}$$

$$= \frac{\varepsilon_0 \times 144 \times 6}{2} = 432\ \varepsilon_0$$

4. (d) Required work = $-$PE of system

$$= -\left[\frac{12q(-q)}{4\pi\varepsilon_0 a} + \frac{6qq}{4\pi\varepsilon_0\sqrt{2}q} + \frac{6(-q)(-q)}{4\pi\,\varepsilon_0\sqrt{2}a} + \frac{4(q)(-q)}{4\pi\varepsilon_0\sqrt{3}a}\right]$$

$$= \frac{q^2}{4\pi\varepsilon_0 a}\left(12 - 6\sqrt{2} + \frac{4}{\sqrt{3}}\right)\ \text{J}$$

5. (c) Loss of PE = Gain in KE

$$U_1 - U_2 = \Delta K$$

$$KQq\left(\frac{1}{r_1} - \frac{1}{r_2}\right) = \frac{1}{2}mv^2 - 0$$

$$v = \sqrt{\frac{2KQq}{m}\left(\frac{r_2 - r_1}{r_1 r_2}\right)}$$

$$= \sqrt{3}\,\text{m/s}$$

6. (a) When K_1 is closed for first time, outer sphere is earthed and its potential becomes zero. Let the charge on it be q_1'.

V_1' = potential due to charge on inner sphere and that due to charge on outer sphere

$$= 0 = \frac{1}{4\pi\varepsilon_0}\left(\frac{q}{2R} + \frac{q_1'}{2R}\right)$$

$$\Rightarrow \quad q_1' = -q$$

When K_2 is closed for first time the potential V_2' on inner sphere becomes zero as it is earthed.

Let the new charge on inner sphere be q_2'

$$0 = \frac{1}{4\pi\varepsilon_0}\frac{q_2'}{R} + \frac{1}{4\pi\varepsilon_0}\cdot\frac{-q}{(2R)}$$

$$\Rightarrow \quad q_2' = q/2$$

Now, when K_1 is closed for IInd time charge on outer sphere will be

$$-q_2' = -q/2$$

$\therefore$ After closing key K_1 for n times, charge on outer sphere is $+q/2^{n-1}$.

After closing key K_2 for n times, charge on inner sphere is $\frac{-q}{2^n}$.

$\therefore$ Potential of outer shell becomes

$$V_0 = K\frac{(-q/2^n)}{2R} + \frac{K(q/2^{n-1})}{2R} = \frac{Kq}{2^{n+1}R}$$

Potential of inner shell = 0

So, potential difference

$$V_0 - V_i = \frac{q}{4\pi\varepsilon_0 2^{n+1}R} = \frac{Kq}{64R} \qquad (\text{for } n = 5)$$

7. (b) Field intensity in dielectric medium $= \frac{\sigma_0}{K\varepsilon_0}$

Field intensity of free charge is $\frac{\sigma_0}{\varepsilon_0}$.

Resultant field intensity $= E_0 - E' = \frac{\sigma_0}{\varepsilon_0} - \frac{\sigma'}{\varepsilon_0} = \frac{\sigma_0}{K\varepsilon_0}$

$$\Rightarrow \quad \sigma' = \sigma_0\left(1 - \frac{1}{K}\right) = \sigma_0\left(\frac{K-1}{K}\right)$$

$$\Rightarrow \quad \sigma' = \sigma_0 \cdot \frac{6}{7}$$

$$\therefore \quad k = 7.$$

8. (c) With the length x of dielectric inside the capacitor, capacitance is

$$C = \frac{\varepsilon_0 Kxl}{d} + \frac{\varepsilon_0 l(l-x)}{d} = \frac{\varepsilon_0 l}{d}[(K-1)x + l]$$

Charge on capacitor, $Q = CV = \frac{\varepsilon_0}{d}l[(K-1)x + l] \cdot V$

$$\text{Current } I = \frac{dQ}{dt} = \frac{\varepsilon_0 l}{d}(K-1)V \cdot \frac{dx}{dt}$$

Here, $l = 10$ cm, $d = 5$ mm, $k = 7$, $V = 12$ V, $dx/dt = 1$ m/s

Hence current,

$$I = \frac{\varepsilon_0 \times 10 \times 10^{-2} \times 6 \times 12 \times 1}{5 \times 10^{-3}} = 1440\ \varepsilon_0\ \text{A}$$

9. (c) Force on dielectric is

$$F = -\frac{\varepsilon_0}{2}(K-1)E^2 \cdot \frac{b}{d}$$

For two blocks,

$$-T + mg = ma$$

$$mg - T = \frac{\varepsilon_0}{2}(K-1)E^2\frac{b}{d} = ma$$

$$\therefore \quad a = \frac{1}{2m}\left(mg - \frac{\varepsilon_0}{2}(K-1)E^2\frac{b}{d}\right)$$

Substituting value of E (given), we get

$$a = g/4$$

10. (b) Charge on all capacitors is same initially and during discharging the potential reduces as

$$V_A = V_0 e^{-\frac{t}{R_A C}} = V_0/e^{t/R_A C}$$

$$V_B = V_0 e^{-\frac{t}{R_B C}}$$

$$V_C = V_0 e^{-\frac{t}{R_C C}}$$

As, $\quad R_A > R_B > R_C$

So, potential drops much rapidly when capacitor discharges through a large resistance.

$\Rightarrow V_A$ decay fastest and V_C slowest.

$\therefore \quad 1 \to A,\ 2 \to B,\ 3 \to C$

11. (c) Given, circuit is

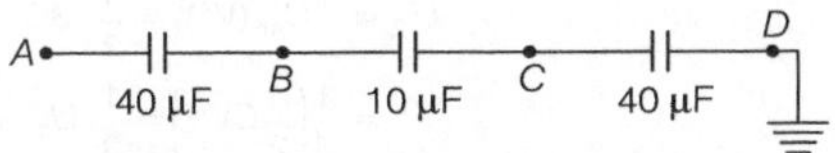

If Q charge is given to combination at point A, then all the capacitors have same charge as they are in series.

$\Rightarrow \quad Q = 40\, V_{AB} = 10\, V_{BC} = 40\, V_{CD}$

$\Rightarrow \quad 4\, V_{AB} = V_{BC} = 4\, V_{CD}$

As, centre capacitor is of least capacity, so it breakdown first,

$\Rightarrow \quad V_{BC} = 15\text{ V}$

Now, $\quad V_{AD} = V_{AB} + V_{BC} + V_{CD}$

$$= \frac{15}{4} + 15 + \frac{15}{4} = 22.5\text{ V}$$

12. (c) Dipole moment of dumbell

$$\mathbf{p} = q\,(\mathbf{d}) = 3.5 \times 10^{-9} \times (-2.6\hat{} + 2.4\hat{}) \times 10^{-3}$$

$\Rightarrow \quad |\mathbf{p}| = 12.4 \times 10^{-12}\text{ cm}$

Also, $\quad |\mathbf{E}| = \sqrt{(7.8)^2 + (4.9)^2} \times 10^3 = 9.21 \times 10^3\text{ N/C}$

So, $\quad U_{max} = |\mathbf{p}|\,|\mathbf{E}| = 114\text{ nJ}$

$U_{min} = -|\mathbf{p}|\,|\mathbf{E}| = -114\text{ nJ}$

$\Rightarrow \quad \Delta U = U_{max} - U_{min} = 228 \times 10^{-9}\text{J} = 228\text{ nJ}$

13. (b) $\quad V_{AB} = 100 = \dfrac{Q}{2C} + \dfrac{Q}{C} + \dfrac{Q}{2C}$

Also, $\quad V_{CD} = 50\text{ V}$

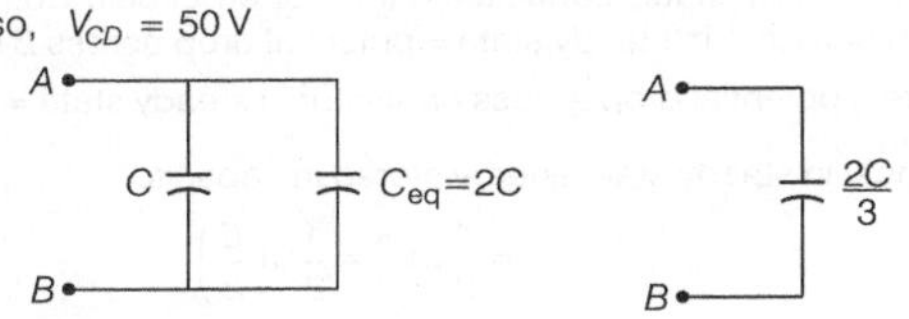

14. (d) As field is uniform field lines are parallel.

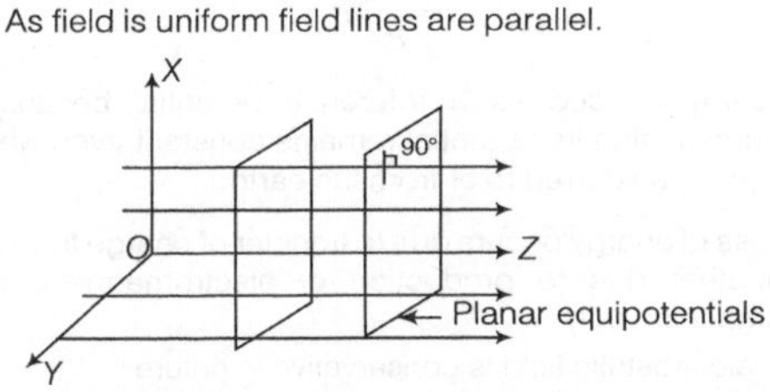

So, equipotentials are planes parallel to XOY- plane.

15. (b) In equilibrium, let x_0 = spring elongation, then

$$\frac{1}{4\pi\varepsilon_0}\left(\frac{q^2}{(x_0)^2}\right) = Kx_0$$

When a further elongation x is given to spring then, net force acting on displaced mass will be

$$F_{net} = -\left(K(x + x_0) - \frac{1}{4\pi\varepsilon_0} \cdot \frac{q^2}{(x + x_0)^2}\right)$$

$$= -\left(K(x + x_0) - \frac{q^2}{4\pi\,\varepsilon_0 x_0^2} \cdot \frac{1}{\left(1 + \dfrac{x}{x_0}\right)^2}\right)$$

$$= -\left(K(x + x_0) - \frac{q^2}{4\pi\varepsilon_0 x_0^2}\left(1 + \frac{x}{x_0}\right)^{-2}\right)$$

Now, taking approximation by using binomial expansion,

$$\left(1 + \frac{x}{x_0}\right)^{-2} = 1 - 2\frac{x}{x_0} + \ldots$$

$\therefore \quad F_{net} = -\left(K(x + x_0) - \dfrac{q^2}{4\pi\varepsilon_0 x_0^2}\left(1 - \dfrac{2x}{x_0}\right)\right)$

$$= -\left(Kx + Kx_0 - \frac{q^2}{4\pi\varepsilon_0 x_0^2} + \frac{2q^2 x}{4\pi\varepsilon_0 x_0^3}\right)$$

$$= -\left(Kx + \frac{2q^2}{4\pi\varepsilon_0 x_0^2} \cdot \left(\frac{x}{x_0}\right)\right)$$

$$= -\left(Kx + 2kx_0 \cdot \frac{x}{x_0}\right) = -3Kx$$

$\Rightarrow$ Acceleration, $a = \dfrac{-3K}{m} . x$

Comparing with equation of SHM, $a = -\omega^2 x$

$$\omega^2 = \frac{3K}{m}$$

$$T = \text{Time period of oscillation} = \frac{2\pi}{\omega} = \frac{2\pi}{\sqrt{\dfrac{3K}{m}}} = 2\pi\sqrt{\frac{m}{3K}}$$

Also, for the oscillating uncharged masses, time period of oscillations is

$$T_0 = 2\pi\sqrt{\frac{m}{2K}}$$

$\Rightarrow \quad \dfrac{T}{T_0} = \sqrt{\dfrac{2}{3}}$

$\Rightarrow \quad T_0 > T$

16. (b) Potential energy of system of charges is contribution of 2 charges taken at a time. So, total number of pairs $= {}^8C_2 = \dfrac{8!}{2!6!}$

$$= 28\text{ pairs}$$

Out of these 28 pairs, 12 unlike charge pairs at a separation a.

$\therefore \quad \text{Potential energy} = 12\left(\dfrac{-Kq^2}{a}\right)$

12 like charge pairs at separation $\sqrt{2}\,a$.

$\therefore \quad \text{Potential energy} = 12\left(\dfrac{Kq^2}{\sqrt{2}a}\right)$

4 unlike charges at separation $\sqrt{3}a$.

$\therefore \quad \text{Potential energy} = \dfrac{-4Kq^2}{\sqrt{3}a}$

17. (a) Kinetic energy of dipole = Decrease in electrostatic patential energy of dipole

$\therefore$ Kinetic energy of dipole at distance d from the origin

$$= U_i - U_f = 0 - (-\mathbf{p}\cdot\mathbf{E}) = \mathbf{p}\cdot\mathbf{E}$$

$$= (p\hat{\mathbf{i}})\cdot\left(\frac{1}{4\pi\varepsilon_0}\frac{q}{d^2}\hat{\mathbf{i}}\right) = \frac{qp}{4\pi\varepsilon_0 d^2}$$

18. (c) Initially, $V = \dfrac{Kq}{a}$

$\Rightarrow \quad q = \dfrac{Va}{K}$

Let after collapsing, the radius of droplet becomes R, then equating the volume.

We have,

$$4\pi a^2 t = \frac{4}{3}\pi R^3$$

$\Rightarrow \quad R = (3a^2 t)^{1/3}$

Potential of droplet will be

$$V' = \frac{Kq}{R} = \frac{K\left(\dfrac{Va}{K}\right)}{(3a^2 t)^{1/3}} = V.\left(\frac{a}{3t}\right)^{1/3}$$

19. (b) Ratio of capacities of spheres S_1 and S_2 is $\frac{C_1}{C_2}=\frac{r}{R}=\frac{r}{2r}=\frac{1}{2}$.

When they are kept in contact charges are distributed in the ratio of their capacities. After first contact, let charge on $S_2=q$, then ratio of charges on S_2 and S_1 after first contact is

$$\frac{q_1}{Q-q_1}=\frac{C_2}{C_1}=\frac{R}{r}\Leftrightarrow q_1=Q\left(\frac{R}{R+r}\right)$$

In second contact S_1 again acquires charge Q.

∴ Total charge in S_1 and S_2 will be

$$Q+q_1=Q\left(1+\frac{R}{R+r}\right)$$

This charge is again distributed in the same ratio. So, the charge on S_2 in second contact is

$$q_2=Q\left(1+\frac{R}{r+R}\right)\left(\frac{R}{R+r}\right)=Q\left[\frac{R}{R+r}+\left(\frac{R}{R+r}\right)^2\right]$$

∴ Charge on S_2 after third contact will be

$$q_3=Q\left[\frac{R}{R+r}+\left(\frac{R}{R+r}\right)^2+\left(\frac{R}{R+r}\right)^3\right]$$

Hence, charge on S_2 after nth contact will be

$$q_n=Q\left(\frac{R}{(R+r)}+\frac{R^2}{(R+r)^2}+\frac{R^3}{(R+r)^3}+\ldots+\frac{R^n}{(R+r)^n}\right)$$

or $$q_n=Q\frac{R}{r}\left[1-\left(\frac{R}{R+r}\right)^n\right]$$

when $n\to\infty, q_n=Q\left(\frac{R}{r}\right)$

and $$U_\infty=\frac{q^2}{2C}=\frac{Q^2R^2}{8\pi\varepsilon_0 r^2R}$$

$$U_\infty=\frac{Q^2(R)}{8\pi\varepsilon_0 r^2}=\frac{Q^2(2r)}{8\pi\varepsilon_0 r^2}=\frac{Q^2}{4\pi\varepsilon_0 r}\text{ J}$$

20. (c) Initially S_1 and S_2 are open.

Charge on $A=q_A=C_AV=100\times 3\,\mu C=300\,\mu C$

Charge on $B=q_B=(180)(2\,\mu C)=360\,\mu C$

Let q_1, q_2 and q_3 are the charges on capacitors A, B and C, respectively after closing the switches S_1 and S_2. By charge conservation, net charge on plates 2 and 3 before joining is equal to net charge after joining.

$$300=q_1+q_2 \quad \ldots\text{(i)}$$

Similarly, net charge of plates 4 and 5 remains conserved.

$$\Rightarrow \quad -360=-q_2-q_3$$

$$\Rightarrow \quad q_2+q_3=360 \quad \ldots\text{(ii)}$$

Now, by applying Kirchhoff's loop rule in closed loop formed, we have

$$\frac{q_1}{3}-\frac{q_2}{2}+\frac{q_3}{2}=0 \quad \ldots\text{(iii)}$$

Solving Eqs. (i), (ii) and (iii), we have

$$q_1=90\,\mu C, q_2=210\,\mu C$$

and $$q_3=150\,\mu C$$

21. (c) In position 1,

Energy stored, $$U_1=\frac{1}{2}CV^2$$

When switch is shifted to position 2, charge on the combination remains same

∴ $$Q=Q'$$

⇒ $$CV=C_{eq}V'$$

⇒ $$CV=3CV'$$

$$V'=\frac{V}{3}$$

So, energy stored in the capacitors after switching to position 2 is

$$U_2=\frac{1}{2}C_{eq}(V')^2=\frac{1}{2}\cdot 3C\cdot\frac{V^2}{9}$$

$$=\frac{1}{3}\left(\frac{1}{2}CV^2\right)=\frac{1}{3}\cdot U_1$$

So, energy dissipated

$$\Delta U=U_1-U_2$$

$$=U_1-\frac{1}{3}U_1=\frac{2}{3}U_1$$

Percentage of energy dissipated $=\frac{\Delta U}{U_1}\times 100$

$$=\frac{\frac{2}{3}U_1}{U_1}\times 100$$

$$=\frac{2\times 100}{3}=66.66\%$$

22. (d) Since, the capacitor plates are directly connected to the battery, it will take no time for charging.

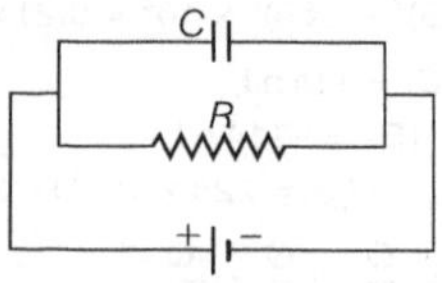

23. (c) In steady state, capacitor will be open circuit. So, potential drop across A in steady state = potential drop across B in steady state = potential drop across capacitor in steady state $=\frac{E}{2}$.

Hence, in steady state energy stored in capacitor

$$=\frac{1}{2}CV^2=\frac{1}{2}C\left(\frac{E}{2}\right)^2$$

$$=\frac{1}{8}CE^2$$

24. (b) Earth is used as a reference potential because during experimentation its potential remains constant even when a little charge is transferred to or from the earth.

25. (c) Loss of energy occurs due to transfer of charge from one body to another due to production of electromagnetic waves or resistance of path.

Also, electrostatic field is conservative in nature.

26. (c) Energy from cell = QV

Energy stored $=\frac{1}{2}CV^2=\frac{1}{2}QV$

∴ Energy loss $=\frac{1}{2}QV$

Energy is lost not only due to resistance of wires but also due to production of electromagnetic waves due to acceleration of charge.

27. (a) When plates come closer capacity $\left[C'=\frac{\varepsilon_0 A}{d'}\right]$ will increases.

As, charge remains same.

$$Q'=Q$$

⇒ $$V'=\frac{Q'}{C'}=\frac{Q}{C'}$$

Hence, potential difference between plates decreases.

So, energy will be

$$U'=\frac{1}{2}C'V'^2=\frac{1}{2}Q'V'=\frac{1}{2}QV'$$

⇒ Energy decreases and part of energy is liberated as heat.

Revisal Problems (JEE Advanced)

1. (d) For maximum heat liberation, capacitor must be discharged fully. That is possible when it is connected with a wire. So, capacity of C' must be that of a conducting wire

$$C' = \infty$$

2. (b) Using $Q = CV$,

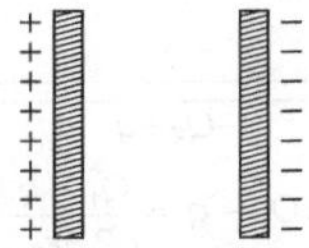

$Q = 50\ \mu C$ $\qquad$ $Q = -50\ \mu C$

Initially, $Q = 5\ \mu F \times 10\ V = 50\ \mu C$

Finally, $\qquad C = \dfrac{\varepsilon_0 A}{d} = 5\ \mu C$

$+50\ \mu C$ 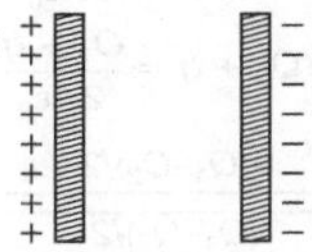 $-40\ \mu C$

Field between plates $= \dfrac{50}{2A\varepsilon_0} + \dfrac{40}{2A\varepsilon_0} = \dfrac{45}{A\varepsilon_0}$

and $\qquad V = Ed = \dfrac{45d}{A\varepsilon_0} = \dfrac{45}{\left(\dfrac{\varepsilon_0 A}{d}\right)} = \dfrac{45}{5} = 9\ V$

3. (d) When oil drop is balanced,

$$qE = mg \qquad \text{...(i)}$$

When drop falls with a terminal velocity

$$mg = 6\pi\eta r v \qquad \text{...(ii)}$$

Also, $\qquad m = \dfrac{4}{3}\pi r^3 \cdot \rho$

On solving, we get $\qquad q = 8 \times 10^{-19}$ C

4. (c) When plate is inserted between plates of capacitor, or potential is equally divided by the inserted plate, its potential is $\dfrac{Q}{2C}$.

$V = \dfrac{Q}{2C}$

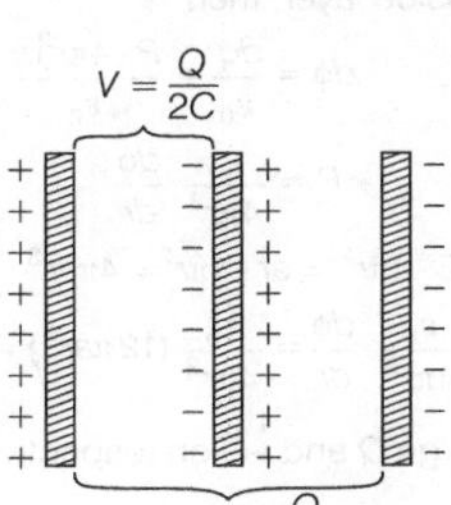

$V = \dfrac{Q}{C}$

As sphere is connected with the plate. Its potential is same as that of plate

$\therefore \qquad V_{sphere} = \dfrac{Q}{2C}$

So, $\qquad$ potential difference $= \dfrac{Q}{C} - \dfrac{Q}{2C} = \dfrac{Q}{2C}$

5. (b) $V_{surface}$ = Potential at surface

$$= \frac{1}{4\pi\varepsilon_0} \frac{\rho \dfrac{4}{3}\pi (r_2^3 - r_1^3)}{2v_0}$$

$$= \frac{7}{6}\left(\frac{\rho r_0^2}{\varepsilon_0}\right)$$

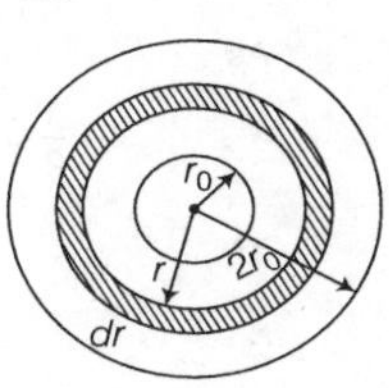

At distance r from centre,

$$E = K\frac{4}{3}\pi\left(\frac{r^3 - r_0^3}{r^2}\right)$$

Using Gauss' law,

$$\int_{V_s}^{V} dV = -\int_{2r_0}^{3r_0/2} \frac{Kq_{enclosed}}{r^2}\, dr$$

$$\Rightarrow \qquad V - V_s = -\frac{1}{4\pi\varepsilon_0}\int_{2r_0}^{\frac{3r_0}{2}} \frac{\rho \dfrac{4}{3}\pi(r^3 - r_0^3)}{r^2}\, dr$$

On solving, we get

$$V = \frac{101}{72}\left(\frac{\rho r_0^2}{\varepsilon_0}\right) V$$

6. (c) Let, x = coordinate of negative charge then

$x + 2a\cos\theta$ = coordinate of positive charge

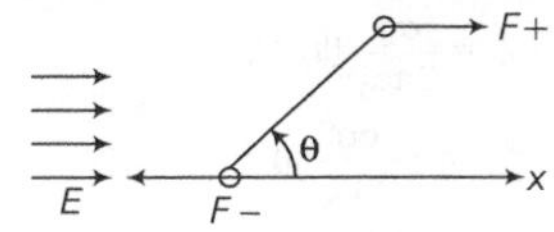

Forces on charges are

$$F_- = qE(x)\,\hat{i}$$

$$F_+ = qE(x + 2a\cos\theta)\,\hat{i}$$

$$\approx qE(x)\,\hat{i} + q\frac{dE}{dx} 2a\cos\theta\,\hat{i}$$

$$\therefore \qquad F_{net} = F_- + F_+$$

$$= \left(-qE(x) + qE(x) + q\frac{dE}{dx}2a\cos\theta\right)\hat{i}$$

$$= P\frac{dE}{dx}\cos\theta\,\hat{i}$$

7. (a) Field of wire is radial and of magnitude

$$E = \frac{\lambda}{2\pi\varepsilon_0 r}$$

$\therefore$ Potential difference between wires (due to only positive charge)

$$\Delta V_1 = -\int E\,dr = -\int_{D-d}^{d} \frac{\lambda}{2\pi\varepsilon_0 r}\, dr$$

$$= \frac{\lambda}{2\pi\varepsilon_0}\log\left(\frac{D-d}{d}\right)$$

and due to presence of negative charge on another wire.

$\Delta V = 2\Delta V_1$ = potential difference between charged wires

$$= \frac{\lambda}{\pi\varepsilon_0}\log\left(\frac{D-d}{d}\right)$$

$\therefore$ Capacitance per unit length of 2 wire system

$$C = \frac{Q}{\Delta V} = \frac{\lambda l}{\Delta V}$$

$$\Rightarrow \qquad \frac{C}{l} = \frac{\pi\varepsilon_0}{\log\left(\dfrac{D-d}{d}\right)}$$

8. (c) $h\operatorname{cosec} 30° = \dfrac{1}{2}\left[\left(\dfrac{g+a}{2}\right) - \dfrac{qE}{m}\cos 30°\right]t^2$

$$\Rightarrow \qquad h \cdot 2 = \frac{1}{2}\left[\frac{g+a}{2} - \frac{qE}{m}\frac{\sqrt{3}}{2}\right]t^2$$

$$\Rightarrow \qquad 8h = \left[(g+a) - \frac{qE}{m}\sqrt{3}\right]t^2$$

$$\Rightarrow \qquad t = 2\sqrt{\frac{2h}{(g+a) - \dfrac{qE}{m}\sqrt{3}}}$$

9. (c) $E_x = -\frac{dV}{dx}$ and $E_y = -\frac{dV}{dy}$

If V is constant (= 0)

$$\int_{V_0}^{V_A} dV = -\int_0^x E_x\, dx$$

$$V_p - V_0 = E_x(x)$$

where x is constant

$$\int_{V_A}^{V_B} dV = -\int_0^y E_y\, dy$$

$$V_B - V_A = -E_y(y)$$

Adding, $$V_B = V_0 - xE_x - yE_y$$

10. (d) As the electric field is conservative force field, thus

$$W = q(V_C - V_B) + q(V_B - V_A)$$

$$= q\left(\int_\infty^r E_C\, dr - \int_\infty^{3r} E_B\, dr\right) + 0$$

$$= \int_{3r}^r \frac{\sigma q}{2\pi\varepsilon_0}\frac{dr}{r}$$

$$= \frac{\sigma q}{2\pi\varepsilon_0}[\ln r]_{3r}^r$$

$$= -\frac{\sigma q}{2\pi\varepsilon_0}\ln 3$$

$$= \frac{-\sigma \ln 3}{2\pi\varepsilon_0} \qquad (\because q = 1)$$

11. (a,b,c) The potential at B due to the charge q on $A = \frac{q}{4\pi\varepsilon_0 b}$

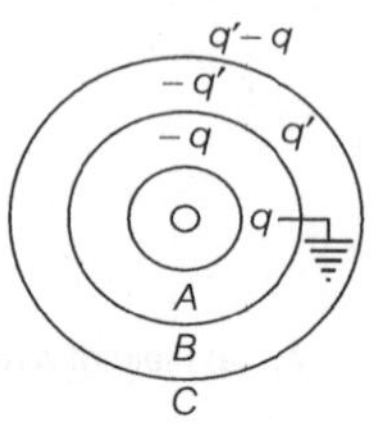

Potential due to the charge $-q$ on inner surface of $B = \frac{-q}{4\pi\varepsilon_0 b}$

Potential due to charge q' on outer surface of $B = \frac{q'}{4\pi\varepsilon_0 b}$

Potential due to charge $-q'$ on outer surface of $C = \frac{-q'}{4\pi\varepsilon_0 b}$

and Potential due to charge $q' - q$ on outer surface of C

$$= \frac{q' - q}{4\pi\varepsilon_0\, b}$$

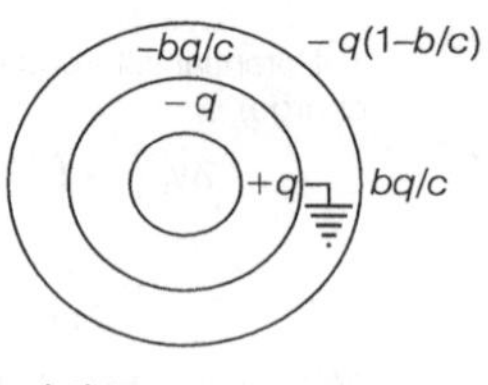

The net potential is

$$V_B = \frac{q'}{4\pi\varepsilon_0 b} - \frac{q}{4\pi\varepsilon_0 c}$$

This should be zero as shell B is earthed, thus

$$q' = \frac{b}{c}q$$

12. (b,c,d) The electric field at the either side of plates is equal and opposite, hence net electric field is zero. Hence, (b) is correct.

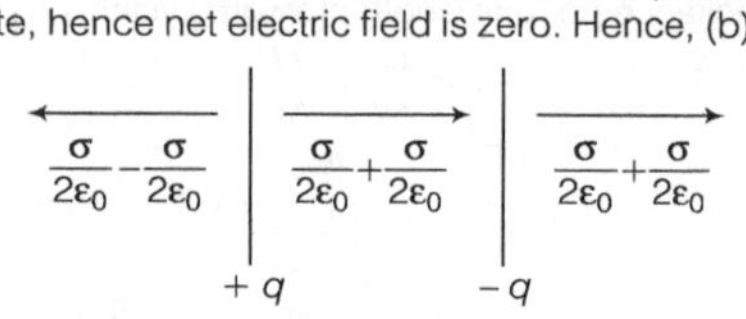

The electric field due to both plates are in the same direction, hence they add to become $\frac{\sigma}{\varepsilon_0}$

The force between the plates is given by

$$F = -QE_t$$

E_t (the electric field due to positive plate)

$$\frac{\sigma}{2\varepsilon_0} = \frac{Q}{2A\varepsilon_0}$$

$$F = \frac{-Q^2}{2A\varepsilon_0}$$

13. (a,b,c,d) The distribution should be like the one shown in figure. To find the value of q, consider the field at point P inside plate A, suppose the surface area of the plate is A. Using the equation, $E = \frac{q}{2\varepsilon_0}$, the electric field at P.

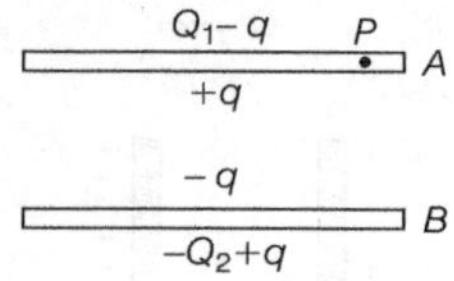

Due to charge $Q_1 - q = \frac{Q_1 - q}{2A\varepsilon_0}$ (downward)

Due to charge $+q = \frac{q}{2A\varepsilon_0}$ (upward)

Due to charge $-q = \frac{q}{2A\varepsilon_0}$ (downward)

and Due to $Q_2 + q = \frac{Q_2 + q}{2A\varepsilon_0}$ (upward)

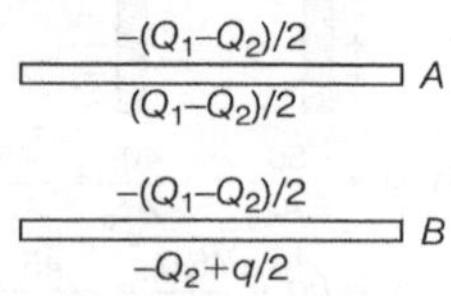

The net electric field at P due to all the four charged surfaces is

$$\frac{Q_1 - q}{2A\varepsilon_0} - \frac{q}{2A\varepsilon_0} + \frac{q}{2A\varepsilon_0} - \frac{Q_2 + q}{2A\varepsilon_0}$$

$$Q_1 - q - Q_2 - q = 0$$

$$q = \frac{Q_1 - Q_2}{2}$$

$$Q_1 - q = \frac{Q_1 + Q_2}{2}$$

$$Q_2 - q = \frac{Q_1 + Q_2}{2}$$

14. (a,b) For a layer dr of radius r,

If dq = charge inside layer, then

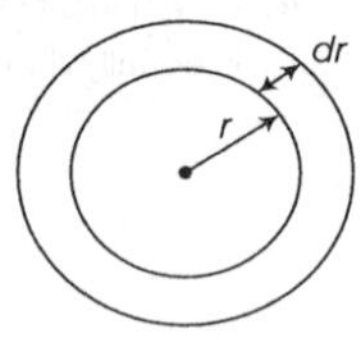

$$d\phi = \frac{dq}{\varepsilon_0} = \frac{P \cdot 4\pi r^2 dr}{\varepsilon_0}$$

$$\Rightarrow \quad P = \frac{\varepsilon_0}{4\pi r^2}\cdot\frac{d\phi}{dr}$$

Also, $\phi = E \cdot 4\pi r^2 = ar \cdot 4\pi r^2 = 4\pi a r^3$

Hence, $\rho = \frac{\varepsilon_0}{4\pi r^2}\cdot\frac{d\phi}{dr} = \frac{\varepsilon_0}{4\pi r^2}(12\pi a r^2) = 3a\varepsilon_0$

15. (a,c) Due to charge Q and $-Q$ on respective plates force between plates is

$$F = \frac{Q^2}{2\varepsilon_0 A}$$

Initially, $C = \frac{\varepsilon_0 A}{d}$, $d = 6$ mm

Equilibrium separation is 2 mm.

$\therefore$ Final value of capacitance is

$$C' = 3C$$

As, U = PE stored in equilibrium $= 7.5 \times 10^{-4}$ J

$\therefore \quad \frac{1}{2}C'V^2 = 7.5 \times 10^{-4}$

$\Rightarrow \quad C' = \frac{7.5 \times 10^{-4} \times 2}{(25)^2}$

$\therefore \quad C' = 2.4 \times 10^{-6}\text{ F} = 2.4\ \mu\text{F}$

and $C = \frac{C'}{3} = 0.8\ \mu\text{F}$

$$Q = C'V = 2.4 \times 10^{-6} \times 25$$
$$= 60 \times 10^{-6}\ \text{C} = 60\ \mu\text{C}$$

Also, $$F = \frac{Q^2}{2\varepsilon_0 A} = \frac{(60 \times 10^{-6})^2}{2 \times 0.8 \times 10^{-6} \times 6 \times 10^{-3}}$$
$$= 0.375\ \text{N}$$

For equilibrium of plate A,

Plate A is connected to spring of $2k$ spring constant

$\therefore$ $$2kx_1 = 0.375$$

where, x_1 = extension of spring

and for plate B,
$$kx_2 = 0.375$$

$\therefore$ $$2kx_1 = kx_2$$

$\Rightarrow$ $$2x_1 = x_2$$

and $$x_1 + x_2 = 4\ \text{mm}$$

$\therefore$ $$x_1 = \frac{4}{3} \times 10^{-3}\ \text{m}$$

and $$x_2 = \frac{8}{3} \times 10^{-3}\ \text{m}$$

and $$2kx_1 = 0.375$$

$\Rightarrow$ $$2k = 281.25\ \text{N/m}$$

Mechanical PE of springs $= \frac{1}{2} 2kx_1^2 + \frac{1}{2} kx_2^2$
$$= 750 \times 10^{-6}\ \text{J} = 7.5 \times 10^{-4}\ \text{J}$$

Hence, Total energy = Electrical energy + Mechanical energy
$$= 1.5 \times 10^{-3}\ \text{J}$$

16. (a,b,c,d) $V_{AB} = 100\text{V} = \frac{Q}{2C} + \frac{Q}{C} + \frac{Q}{2C}$

Also, charges on capacitors which are in series is same.

17. (b,c,d) As **E** is uniform, equipotentials exists parallel to each other and are parallel planes which are parallel to XOY plane.

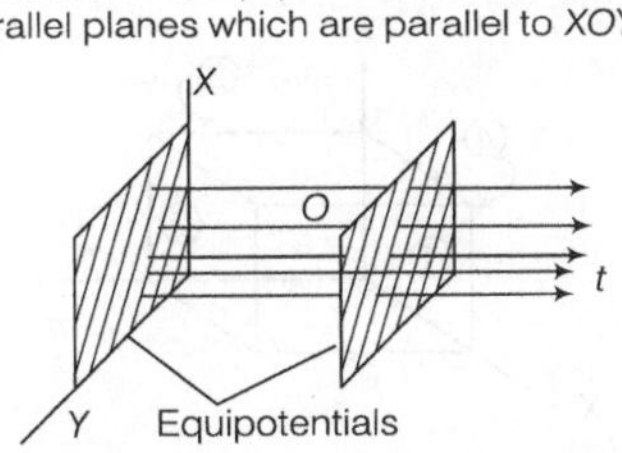

18. (a,b,c) Centre plate makes an arrangement equivalent to system shown in figure.

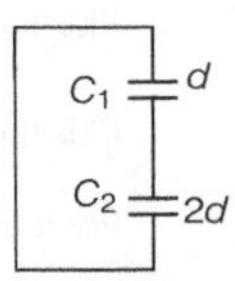

So, we have to capacitors with same potential difference V and total charge Q.

Also, $$C_1 = \frac{\varepsilon_0 A}{d},\ C_2 = \frac{\varepsilon_0 A}{2d}$$

Charge flows from ground onto each of outside plates, so that $Q_1 + Q_2 = Q$ and $\Delta V_1 = \Delta V_2 = V$

Then, $$\frac{Q_1}{C_1} = \frac{Q_2}{C_2} = \frac{Q_1 d}{\varepsilon_0 A} = \frac{Q_2 d}{\varepsilon_0 A}$$
$$Q_1 = 2Q_2;\ 2Q_2 + Q_2 = Q$$

So, $$Q_2 = \frac{Q}{3}$$

$\Rightarrow$ Charge on lower plate is $-\frac{Q}{3}$.

and $Q_1 = \frac{2Q}{3}$, charge on upper plate is $-\frac{2Q}{3}$.

Also, $$\Delta V = \frac{Q_1}{C_1} = \frac{2Qd}{3\varepsilon_0 A}$$

19. (a,b) In equilibrium, if x_0 = spring elongation then,
$$k\frac{q^2}{(l_0 + x_0)^2} = k\,x_0$$

When a further elongation x is given to spring then,
$$F = -\left\{k(x + x_0) - \frac{1}{4\pi\varepsilon_0}\frac{q^2}{(l_0 + x_0 + x)^2}\right\}$$

$\Rightarrow$ Acceleration, $a = -\frac{2k}{m}\cdot x$

As, $$F = \mu\, a$$

where $$u = \frac{m \times m}{m + m} = \frac{m}{2} = \text{reduced mass}$$

$\Rightarrow$ $$a = -\omega^2 x$$

Hence $$\omega = \sqrt{\frac{2k}{m}} \Rightarrow T = 2\pi\sqrt{\frac{m}{2k}}$$

In absence of charge, $T_0 = 2\pi\sqrt{\frac{m}{2k}}$

$\therefore$ $$\frac{T}{T_0} = 1$$

20. (b,c) From $Q = Q_0(1 - e^{-t/R_c})$ and $i = i_0 e^{-t/R_c}$

As charge and current are time dependent, readings of voltmeter and ammeter are changing. Also when $t = RC\log_e 2$ and readings of voltmeter and ammeter are equal.

21. (a,b,d) Just switch is closed, capacitor is short circuited.

and $$i = \frac{36}{R_{eq}} = 8.75\text{A}$$

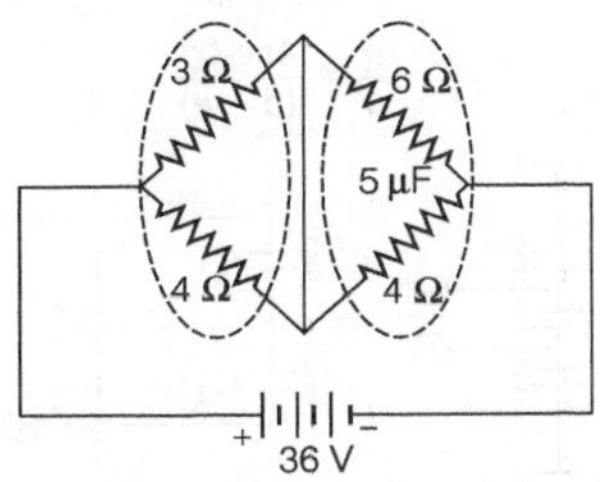

Time constant, $\tau = R_{eq}C = 4 \times 5 \times 10^{-6} = 20\ \mu\text{s}$

At steady state, capacitor is open circuit; $V_{BD} = 6\text{V}$.
$$q = CV_{BD} = 30\ \mu\text{C}.$$

22. (a,b,c) Using conservation of charge, distribute the charge on all the surfaces of the plates a, b, c, d, e, and f. Then are the concept electric field inside the plates is zero.

23. (b,c,d) Options B, C, D are true. In steady state, potential difference across capacitor C_1 is $\frac{\varepsilon}{2}$.

$\therefore$ Energy stored in capacitor C is $\frac{1}{2}C\cdot\left(\frac{\varepsilon}{2}\right)^2 = \frac{C\varepsilon^2}{8}$

In steady state current through resistors A and B will be same. So, heat will be produced at same rate in both A and B as their resistances are same.

24. (d) A and B are connected by a wire.

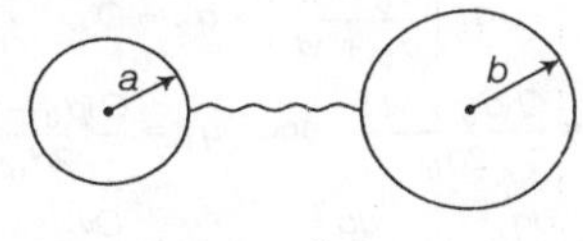

$\therefore$ They are connected in parallel.

$\therefore$ $$C_{eq} = C_A + C_B = 4\pi\varepsilon_0(a + b)$$

As, B is earthed, $V_B = 0$

$$\Rightarrow \quad \frac{Kq}{4\pi\varepsilon_0 b} + \frac{Kq'}{4\pi\varepsilon_0 b} = 0$$

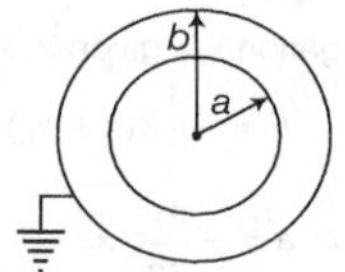

$$\Rightarrow \quad q' = -q$$

$$\therefore \quad V_A - V_B = \frac{q}{4\pi\varepsilon_0}\left(\frac{1}{a} - \frac{1}{b}\right)$$

$$\therefore \quad C = \frac{q}{V} = \frac{4\pi\varepsilon_0(ab)}{b-a}$$

As A is earthed, $V_A = 0$

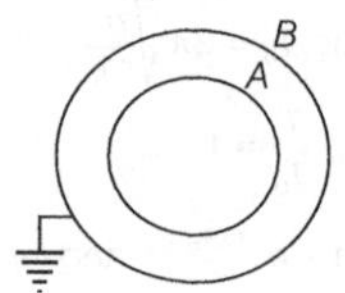

$$\Rightarrow \quad \frac{Kq'}{4\pi\varepsilon_0 a} + \frac{Kq}{4\pi\varepsilon_0 b} = 0$$

$$\Rightarrow \quad q' = -\frac{a}{b}q$$

$$\therefore \quad \Delta V = \frac{q(b-a)}{4\pi\varepsilon_0 b^2}$$

$$\therefore \quad C = \frac{q}{V} = \frac{4\pi\varepsilon_0 b^2}{(b-a)}$$

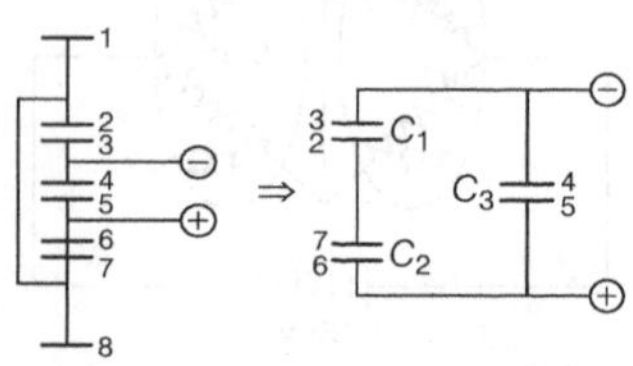

In last case, given combination is equivalent to

$$C_{eq} = (C_1 \text{ and } C_2) \,||\, C_3$$

$$= \varepsilon_0 A\left(\frac{b+2a}{2ab}\right)$$

(i) → (r), (ii) → (q), (iii) → (s), (iv) → (p)

25. (b) Let q_1 and q_2 are charges on capacitors, $V_1 = V_2$.

As, capacitors are connected in parallel.

$$\therefore \quad \frac{q_1}{C_1} = \frac{q_2}{C_2}, \; C_1 = \frac{\varepsilon_0 A}{d + vt}$$

$$C_2 = \frac{\varepsilon_0 A}{d - vt}$$

$$\therefore \quad \frac{q_1}{q_2} = \frac{C_1}{C_2} = \frac{d_0 - vt}{d_0 + vt}$$

and $\quad q_1 + q_2 = Q$

So, $\quad q_2\left(\frac{d_0 - vt}{d_0 + vt}\right) + q_2 = Q$

$$\Rightarrow \quad q_2 = \frac{Q(d_0 + vt)}{2d_0} \text{ and } q_1 = \frac{Q(d_0 - vt)}{2d_0}$$

$$i = -\frac{dq_1}{dt} \text{ or } \frac{dq^2}{dt} \Rightarrow i = \frac{Qv}{2d_0}$$

Also, $\quad u_i = \dfrac{Q^2 d_0}{2\varepsilon_0 A} \times 2$

$$u_f = \frac{1}{2} \times \frac{Q^2(d_0 - vt)(d_0 + vt)}{4d_0^2 \times \varepsilon_0 A} + \frac{1}{2}\frac{Q^2(d_0 + vt)^2}{4d_0^2} \times \frac{d_0 - vt}{\varepsilon_0 A}$$

$$= \frac{Q^2(d_0^2 - v^2t^2)}{2\varepsilon_0 A(2d_0)}$$

(i) → (r), (ii) → (q), (iii) → (s), (iv) → (p)

26. (b) As battery is disconnected, Q remains same.

C increases, $C' = \dfrac{\varepsilon_0 KA}{d} = KC$

Potential difference decreases as $= V' = \dfrac{Q'}{C'} = \dfrac{Q}{KC} = \dfrac{V}{K}$

And potential energy is also reduced as

$$U' = \frac{1}{2}C'V'^2 = \frac{1}{2}KC \times \frac{V^2}{K^2} = \frac{U}{K}$$

(i) → (p), (ii) → (q), (iii) → (s), (iv) → (r)

27. (c) $E = \dfrac{1}{2}m_p v^2 + \dfrac{eQ}{4\pi\varepsilon_0 r}$

28. (d) Protons are initially absorbed by the metal sphere, thus increasing charge on the sphere which in turn introduces a repulsive force between a proton and the metal sphere. Eventually, the repulsive force become great enough to deflect the proton beam away from the sphere, such that no further charge accumulates on the metal sphere. Thus the metal sphere maintains a constant electrostatic potential.

29. (b) Angular momentum remains conserved

$$m_p(v_0)\frac{R}{2} = m_p vR$$

$$\therefore \quad v = \left(\frac{v_0}{2}\right)$$

Explanations for Q. Nos. (30-32)

For face 1 and 3, area vectors are $d^2\hat{\mathbf{i}}$ and $-d^2\hat{\mathbf{j}}$

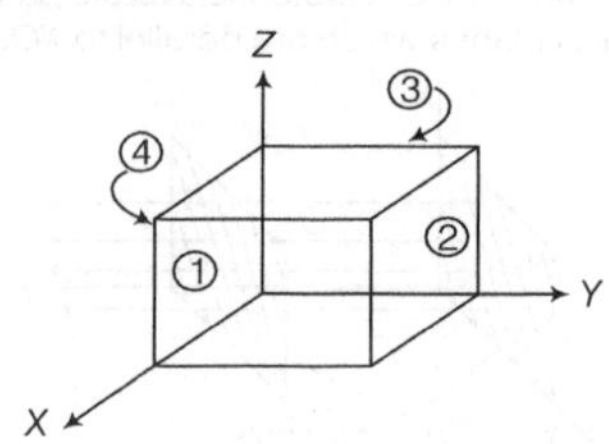

Flux through surface 1 is $= \mathbf{E}\cdot\mathbf{A} = E_0\left(1 + \dfrac{Z}{a}\right)d^2$

Flux through surface Z is $= -E_0\left(1 + \dfrac{Z}{a}\right)d^2$

For faces 2 and 4 $\mathbf{A}$ is $d^2\hat{\mathbf{j}}$ and $-d^2\hat{\mathbf{j}}$

$\therefore$ Flux through surface 2 is $= \mathbf{E}\cdot\mathbf{A} = E_0\dfrac{Z}{a}d^2$

and flux through surface 4 $= -E_0\dfrac{Z}{a}d^2$

For faces parallel to XOY-plane, area vectors are along $+Z$ and $-Z$ axis.

$\therefore$ Flux through there faces is zero.

And total flux,

$$\phi = E_0\left(1 + \frac{Z}{a}\right)d^2 - E_0\left(1 + \frac{Z}{a}\right)d^2 + E_0\frac{Z}{a}d^2 - E_0\frac{Z}{a}d^2 = 0$$

As, total flux is zero so, net charge enclosed by the cubical volume must be zero.

30. (a) **31.** (c)

32. (d)

33. (b) When field is applied, electrostatic force on A is zero but electrostatic force on B is

$F_B = q_B E$ (in negative x direction)

Acceleration of block B will be

$$a_B = \frac{F_B}{m_B} = \frac{q_B E}{m_B} = \frac{1 \times 10}{1} = 10 \text{ m/s}^2$$

Just before collision, speed of block B will be

$$v_B = \sqrt{2a_B s} \qquad \text{(using } v^2 - u^2 = 2)$$

$$= \sqrt{2 \times 10 \times 1.8} = 6 \text{ m/s}$$

In the collision, momentum is conserved. So,

$$(m_A + m_B)v = m_B v_B$$

$$\Rightarrow \quad v = \frac{m_B v_B}{m_A + m_B} = \frac{1 \times 6}{2} = 3 \text{ m/s}$$

34. (c) At equilibrium position, net force on system of blocks will be zero.

So, $\quad F_e = kx_0 \quad$ (where, k = spring constant)

$\Rightarrow \quad 10 = 18\, x_0$

$\Rightarrow \quad x_0 = \frac{5}{9}$ m

Which occurs during compression of spring.

Hence, $x_0 = \frac{-5}{9}$ m

35. (d) Angular frequency of

SHM is $\quad \omega = \sqrt{\frac{k}{(m_A + m_B)}} = \sqrt{\frac{18}{2}} = 3$ rad/s

At $x = \frac{5}{9}$ m, speed is 3 m/s

$\therefore \quad v = \omega \sqrt{A^2 - x^2}$

$3 = 3\sqrt{A^2 - x^2} \qquad$ (gives)

$\Rightarrow \quad A^2 = 1 + x^2 = 1 + \frac{25}{81} = \frac{106}{81}$

$\Rightarrow \quad A = \sqrt{\frac{106}{9}}$ m

36. (a) Given, capacitor can be viewed as a parallel combination of two-capacitors.

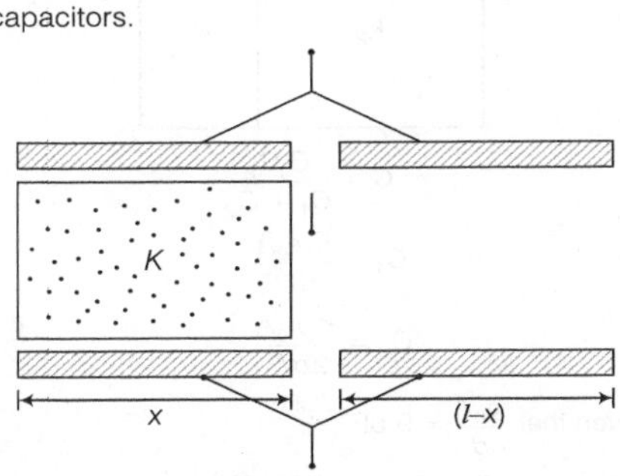

So, capacity of given arrangement is

$$C = \frac{\varepsilon_0 K x l}{d} + \frac{\varepsilon_0 l (l - x)}{d} = \frac{\varepsilon_0 l}{d}(Kx + l - x)$$

$$= \frac{\varepsilon_0 l}{d}(l + (K - 1)x)$$

So, energy stored in the capacitor is

$$U = \frac{1}{2} C \cdot V^2 = \frac{\varepsilon_0 l}{2d}(l + (K - 1)x) \cdot V^2 = 61 \,\mu\text{J}$$

37. (d) Force on the dielectric slab is $F = -\left(\frac{dU}{dx}\right)\hat{\mathbf{i}}$

$\Rightarrow \quad F = -\frac{\varepsilon_0 l}{2d}(K - 1)V^2 \hat{\mathbf{i}}$

$= -1.55 \times 10^{-3}\, \hat{\mathbf{i}}$ N

38. (a) Rate of change of energy

$$= \frac{dU}{dt} = \frac{dU}{dx} \cdot \frac{dx}{dt} = \frac{\varepsilon_0 l}{2d}(k - 1)V^2 \cdot v$$

$$= 1.55 \times 10^{-3} \times 2 \times 10^{-2} = 3.1 \times 10^{-5} \text{ J/s}$$

39. (2) For entire motion, $S = ut + \frac{1}{2}at^2$

$\Rightarrow \quad 0 = xt + \frac{1}{2}at^2$

$\Rightarrow \quad a = -\frac{Zu}{t}$

Also, $\quad \Sigma F = -mg - qE$

$\Rightarrow \quad \Sigma F = -ma$

$\therefore \quad -mg - qE = m\left(\frac{-Zu}{t}\right)$

$\Rightarrow \quad E = \frac{m}{q}\left(\frac{Zu}{t} - g\right)$

For upward flight, $\quad v^2 = u^2 + 2as$

$0 = u^2 + 2\left(\frac{-Zu}{t}\right) \cdot h$

$\Rightarrow \quad h = \frac{1}{4}ut.$

Also, $\quad \Delta V = -\int_0^h E dh = \frac{m}{q}\left(\frac{2u}{t} - g\right) \cdot h = \frac{m}{q}\left(\frac{2u}{t} - g\right)\left(\frac{1}{4}ut\right)$

$\therefore \quad \Delta V = \frac{2}{5 \times 10^{-6}}\left(\frac{2 \times 20.1}{4.10} - 9.8\right)\left(\frac{1}{4} \times 20.1 \times 4.10\right)$

$= 40.2 \text{ kV} \approx 40 \text{ kV}$

So, $\quad 40 = 20\,k, k = 2$

40. (1) Potential at the centre of a ring will be due to charge on both the rings. Also, a symmetrical distribution does not effects the value of potential at centre as each elemental charge is at same distance from the centre.

and $\quad V_2 = \frac{1}{4\pi\varepsilon_0}\left\{\frac{q_2}{R_2} + \frac{q_1}{\sqrt{R_1^2 + x^2}}\right\}$

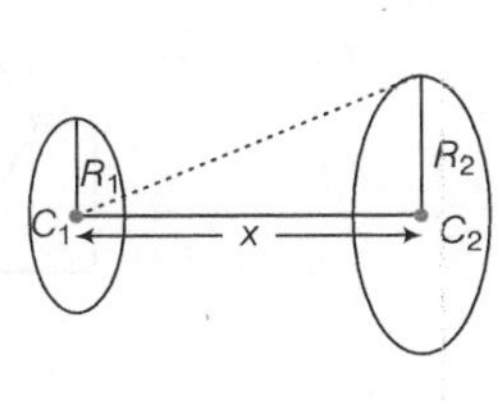

Substituting values, we get

$V_1 = 2.4 \times 10^5$ V,

$V_2 = 1.69 \times 10^5$ V

So, $\quad V_1 = \frac{1}{4\pi\varepsilon_0}\left\{\frac{q_1}{R_1} + \frac{q_2}{\sqrt{R_2^2 + x^2}}\right\}$

$\therefore \quad V_1 - V_2 = 71$ kV

So, $\quad N \times 71 = 71$ or $N = 1$

41. (3) Potential at P at distance due to strip is

$$V_p = \frac{\sigma q}{4\pi\varepsilon_0} \log_e\left(1 + \frac{l}{d}\right)$$

and at point Q at distance $d/2$ due to strip is

$$V_Q = \frac{\sigma q}{4\pi\varepsilon_0} \log_e\left(1 + \frac{2l}{d}\right)$$

$$\Rightarrow \quad V_Q - V_p = \frac{\sigma q}{4\pi\varepsilon_0} \log_e\left(\frac{d + 2l}{d + l}\right)$$

By conservation of energy,

$$\frac{1}{2}mv^2 = \frac{\sigma q}{4\pi\varepsilon_0} \log_e\left(\frac{d + 2l}{d + l}\right)$$

$$\Rightarrow \quad V^2 = \frac{\sigma q}{4\pi\varepsilon_0 m} \cdot \log_e\left(\frac{3}{2}\right); \text{ as } d = l$$

or $\quad v^2 \propto \frac{\sigma q}{4\pi\varepsilon_0 m}$

Hence, proportionality constant in $\log\left(\frac{3}{2}\right)$

$\Rightarrow \quad 2N = 3$

JEE Main & AIEEE Archive

1. (a) Electric field inside dielectric, $\dfrac{\sigma}{K\varepsilon_0} = 3 \times 10^4$

$\Rightarrow$ $\sigma = 2.2 \times 8.58 \times 10^{-12} \times 3 \times 10^4$

$= 6.6 \times 8.85 \times 10^{-8} = 5.841 \times 10^{-7}$

$= 6 \times 10^{-7}$ C/m^2

2. (b)

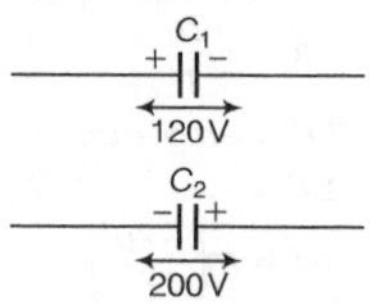

For potential to be made zero after connection, the charge of both capacitors are equal.

$\therefore$ $q_1 = q_2$

$\Rightarrow$ $C_1V_1 = C_2V_2$

$\Rightarrow$ $120\,C_1 = 200\,C_2$

$\Rightarrow$ $3C_1 = 5C_2$

3. (d) If C_e is the effective capacitance, then

$$V_C = \frac{1}{2}V_0$$

$$\frac{q}{C_e} = \frac{q_0}{2C_e}$$

As $q = q_0\left(1 - e^{\frac{-t}{RC_e}}\right)$

$\Rightarrow$ $q_0\,(1 - e^{-t/RC_e}) = \dfrac{q_0}{2}$

$\Rightarrow$ $t = RC_e \ln 2$

For parallel grouping,

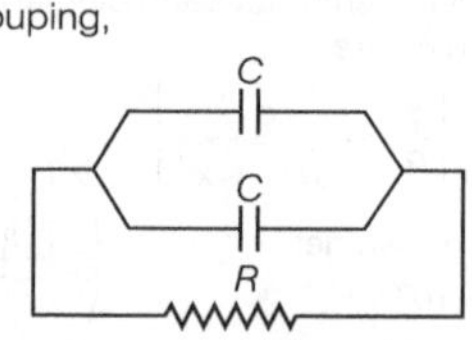

$$C_e = 2C$$

$\therefore$ $t_2 = 2\,RC\ln 2$

For series grouping,

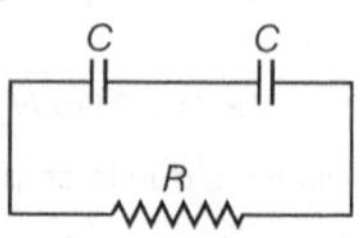

$$C_e = \frac{C}{2}$$

$\therefore$ $t_1 = \dfrac{RC}{2}\ln 2$

$\therefore$ $\dfrac{t_2}{t_1} = \dfrac{1}{4}$

$\Rightarrow$ $t_2 = 2.5$ s

4. (b) Neon bulb is filled with gas, so its resistance is infinite, hence no current flows through it.

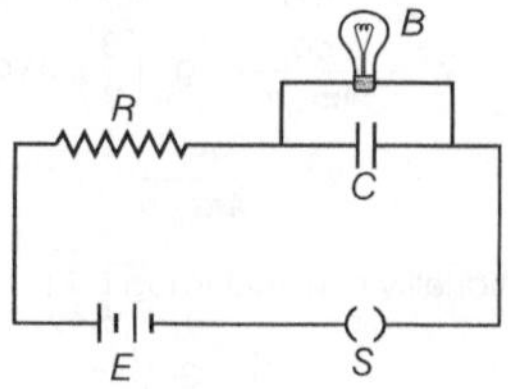

Now, $V_c = E\,(1 - e^{-t/RC})$

$\Rightarrow$ $120 = 200\,(1 - e^{-t/RC})$

$\Rightarrow$ $e^{-t/RC} = \dfrac{2}{5}$

$\Rightarrow$ $t = RC\,\ln 2.5$

$\Rightarrow$ $R = \dfrac{t}{C\ln 2.5}$

$= \dfrac{t}{2.303C\,\log 2.5}$

$= 2.7 \times 10^6\,\Omega$

5. (c) Energy stored in capacitor,

$$U = \frac{1}{2}\frac{q^2}{C} = \frac{1}{2C}(q_0e^{-t/\tau})^2$$

$$= \frac{q_0^2}{2C}e^{-2t/\tau} \quad \text{(where, } \tau = CR\text{)}$$

$$U = U_ie^{-2t/\tau}$$

$$\frac{1}{2}U_i = U_ie^{-2t_1/\tau}$$

$$\frac{1}{2} = e^{-2t_1/\tau}$$

$\Rightarrow$ $t_1 = \dfrac{\tau}{2}\ln 2$

Now, $q = q_0e^{-t/\tau}$

$$\frac{1}{4}q_0 = q_0e^{-t_2/\tau}$$

$$t_2 = \tau\ln 4 = 2\tau\ln 2$$

$\therefore$ $\dfrac{t_1}{t_2} = \dfrac{1}{4}$

6. (c) As both capacitors are connected in series.

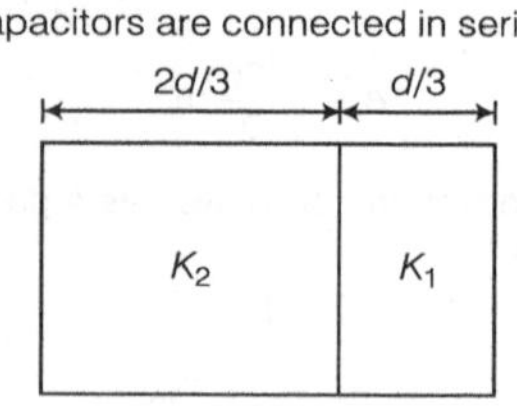

$\therefore$ $C = \dfrac{C_1C_2}{C_1 + C_2}$ ...(i)

where, $C_1 = \dfrac{K_1\varepsilon_0A}{d/3}$ ...(ii)

and $C_2 = \dfrac{K_2\varepsilon_0A}{2d/3}$ ...(iii)

It is given that, $\dfrac{\varepsilon_0A}{d} = 9$ pF

On substituting Eqs. (ii) and (iii) in Eq. (i), we get the result $C = 40.5$ pF.

7. (d) On introduction and removal and again on introduction, the capacity and potential remain same. So, net work done by the system in this process,

$$W = U_f - U_i$$

$$= \frac{1}{2}CV^2 - \frac{1}{2}CV^2$$

$$= 0$$

8. (d) Ratio of energy stored in the capacitor and the work done by the battery

$$= \frac{\frac{1}{2}qV}{qV} = \frac{1}{2}$$

9. (a) Each plate is taking part in the formation of two capacitors except the plates at the ends.

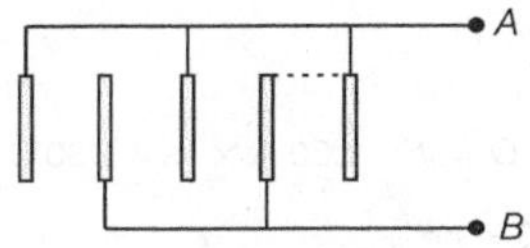

These capacitors are in parallel and n plates form $(n-1)$ capacitors.

Thus, equivalent capacitance between A and $B = (n-1)C$

10. (d) $$E = \left(\frac{1}{2}\right)CV^2 \quad \ldots(i)$$

The energy stored in capacitor is lost in form of heat energy.

$$\therefore \quad H = ms\,\Delta T \quad \ldots(ii)$$

From Eqs. (i) and (ii), we have

$$ms\,\Delta T = \left(\frac{1}{2}\right)CV^2 \quad \Rightarrow \quad V = \sqrt{\frac{2\,ms\,\Delta T}{C}}$$

11. (b) Aluminium is a metal, so when we insert an aluminium foil, equal and opposite charges appear on its two surfaces. Since, it is of negligible thickness, it will not affect the capacitance.

Alternate Method

From the formula, $C = \dfrac{\varepsilon_0 A}{d - t + \dfrac{t}{K}}$

Here, $K = \infty$ and $t \to 0$

So, $$C = \frac{\varepsilon_0 A}{d+0} = \frac{\varepsilon_0 A}{d} = C_0$$

12. (d) $W = \dfrac{1}{2}\dfrac{q^2}{C}$

$$= \frac{1}{2} \times \frac{(8 \times 10^{-18})^2}{100 \times 10^{-6}}$$

$$= \frac{1}{2} \times \frac{64 \times 10^{-36}}{100 \times 10^{-6}}$$

$$= 32 \times 10^{-32}\ \text{J}$$

13. (b) Energy stored by any system of capacitors

$$= \frac{1}{2}C_{net}V^2$$

where, V is source voltage.

Thus, n capacitors are connected in parallel.

So, $$C_{net} = nC$$

$$\therefore \quad E_{net} = \frac{1}{2}nCV^2$$

14. (a) Capacitance of spherical conductor

$$= 4\pi\varepsilon_0 a$$

where, a is radius of conductor.

Therefore, $$C = \frac{1}{9 \times 10^9} \times 1$$

$$= \frac{1}{9} \times 10^{-9}$$

$$= 0.11 \times 10^{-9}\ \text{F}$$

$$= 1.1 \times 10^{-10}\ \text{F}$$

JEE Advanced & IIT JEE Archive

1. (a, d) $C = C_1 + C_2$

$$C_1 = \frac{K\varepsilon_0 A/3}{d}$$

$$C_2 = \frac{\varepsilon_0 2A/3}{d}$$

$$\Rightarrow \quad C = \frac{(K+2)\varepsilon_0 A}{3d}$$

$$\Rightarrow \quad \frac{C}{C_1} = \frac{K+2}{K}$$

Also, $E_1 = E_2 = \dfrac{V}{d}$, where V is potential difference between the plates.

2. (b, d) After pressing S_1 charge on upper plate of C_1 is $+2CV_0$.

After pressing S_2 this charge equally distributes in two capacitors. Therefore charge an upper plates of both capacitors will be $+CV_0$

When S_2 is released and S_3 is pressed, charge on upper plate of C_1 remains unchanged $(= +CV_0)$ but charge on upper plate of C_2 is according to new battery $(= -CV_0)$.

3. (d) $$q_i = C_i V = 2V = q \quad \text{(say)}$$

This charge will remain constant after switch is shifted from position 1 to position 2.

$$U_i = \frac{1}{2}\frac{q^2}{C_i} = \frac{q^2}{2 \times 2} = \frac{q^2}{4}$$

$$U_f = \frac{1}{2}\frac{q^2}{C_f} = \frac{q^2}{2 \times 10} = \frac{q^2}{20}$$

$\therefore$ Energy dissipated $= U_i - U_f = \dfrac{q^2}{5}$

This energy dissipated $\left(= \dfrac{q^2}{5}\right)$ is 80% of the initial stored energy $\left(= \dfrac{q^2}{4}\right)$.

4. (a) After time t, thickness of liquid will remain $\left(\dfrac{d}{3} - vt\right)$.

Now, time constant as function of time :

$$\tau_c = CR$$

$$= \frac{\varepsilon_0(1) \cdot R}{\left(d - \dfrac{d}{3} + vt\right) + \dfrac{d/3 - vt}{2}}$$

$$\left(\text{Applying } C = \frac{\varepsilon_0 A}{d - t + \dfrac{t}{k}}\right)$$

$$= \frac{6\varepsilon_0 R}{5d + 3vt}$$

3. Current Electricity

DPP-1 Electric Current

1. Volume of silver deposited, V = Area × Thickness

$$= 700 \times 10^{-4} \times 0.133 \times 10^{-3}$$
$$= 9.31 \times 10^{-6}\ \text{m}^3$$

Mass of silver deposited, m = Volume × Density

$$= 9.31 \times 10^{-6} \times 10.5 \times 10^3$$
$$= 9.78 \times 10^{-2}\ \text{kg}$$

Number of silver atoms deposited

$$N = \text{Number of moles} \times 6.022 \times 10^{23}\ \text{atoms}$$
$$= \frac{9.78 \times 10^{-2} \times 10^3 (\text{g})}{107.9\ (\text{g})} \times 6.022 \times 10^{23}$$
$$= 5.45 \times 10^{23}\ \text{atoms}$$

Current in circuit $= 5\ \text{A}$

$$I = \frac{\Delta Q}{\Delta t}$$

$$\Rightarrow \quad \Delta t = \text{time} = \frac{\Delta Q}{I} = \frac{Ne}{I}$$
$$= \frac{5.45 \times 10^{23} \times 1.6 \times 10^{-19}}{5}$$
$$= 1.74 \times 10^4\ \text{s} = 4.83\ \text{h}$$

2. (i) No, electrons move randomly in a conductor colliding with other electrons and ions. The path of an individual electron is *zig-zag*.

(ii) Yes, drift velocity is an average of velocities of all electrons which they have under an applied potential difference. So, when all velocities are averaged, they give a value possessed by an electron (on an average) both in direction and magnitude without under going any collision.

(iii) When current flows through a conductor, then electrons are accelerated by the field and they suffer many collisions. In each collision, a part of energy is liberated in the form of heat making the conductor hot.

(iv) Since, electric field causes drift velocity.

$$\therefore \quad E \propto v_d \text{ and } \frac{v_d}{E} = \text{constant for the conductor.}$$

This ratio is called electron mobility, it is a scalar quantity.

(v) Current in wire does not change when the wire is bent (in any shape), i.e. current in wire does not follow vector addition laws and hence, it is treated as a scalar quantity.

3. (i) As, current at both ends is same and diameter of wire is also same at both ends

$$I = nev_d A$$

This means that drift speed is also same at both ends of the wire.

(ii) Potential energy keeps on falling as they travel across the resistor.

So, PE at the exit is less than the PE at the entrance.

4. Charge on a proton $= +1.6 \times 10^{-19}$ C

Charge on an electron $= -1.6 \times 10^{-19}$ C

(i) The drifting of electrons in a certain direction means the flow of positive charge in the opposite direction. Therefore, the rate of flow of positive charge in the direction of proton drifting, i.e. the current in the tube is

$$i = (1 \times 10^{18})(1.6 \times 10^{-19}) + (2.7 \times 10^{18})(1.6 \times 10^{-19})$$
$$= 3.7 \times 10^{18} \times 1.6 \times 10^{-19} = 5.92\ \text{A}$$

(ii) The resistance of the tube

$$R = \frac{V}{I} = \frac{230}{5.92} = 388\ \Omega$$

5. (a) $I = \dfrac{Q}{t} \Rightarrow Q = It = 1.30\ \text{A} \times 1\text{s} = 1.30\ \text{C}$

Also, $Q = ne$

$$\Rightarrow \quad n = \frac{1.30}{1.6 \times 10^{-19}}$$
$$n = \frac{1.30 \times 10^{19}}{1.6} = 0.8125 \times 10^{19}$$
$$= 8.125 \times 10^{18}\ \text{electrons per sec}$$

6. (b) As, $q = I \times t = \int_{t_1}^{t_2} I\,dt$

$\therefore$ Charge = Area under graph of $(I - t)$

$$\Rightarrow \quad q_1 = \text{Area of triangle} = \frac{1}{2} \times 7.5 \times 6$$
$$q_2 = \text{Area of rectangle} = (15 - 7.5) \times 6 = 7.5 \times 6$$
$$\Rightarrow \quad \text{Ratio, } \frac{q_1}{q_2} = \frac{\frac{1}{2} \times 7.5 \times 6}{7.5 \times 6} = \frac{1}{2}$$

7. (c) Charge that flows in a wire in time interval t_1 to t_2 sec is

$$Q = \int_{t_1}^{t_2} I\,dt = \int_0^{12} (I_0 + \alpha t)\,dt$$
$$\Rightarrow \quad Q = \left[I_0 t + \frac{\alpha t^2}{2} \right]_0^{12}$$
$$= I_0 \times 12 + \frac{\alpha}{2}(12^2 - 0^2)$$
$$= 12 I_0 + 72\alpha$$
$$= 12 \times 15 + 72 \times 6$$
$$= 180 + 432 = 612\ \text{C}$$

8. (d) $I = \text{current} = \dfrac{dQ}{dt}$

$$= \frac{d}{dt}(2t^2 - 4t + 3) \times 10^{-3}$$
$$= (4t - 4) \times 10^{-3}$$

At $t = 4$ s,

$$I = (4 \times 4 - 4) \times 10^{-3}\ \text{A}$$
$$= 12\ \text{mA}$$

9. (d) An electric field exists inside the wire which accelerates electrons till they move with an average velocity v_d and $I = nev_d A$.

10. (b) Due to a large number densities, we are able to get high current.

11. (b) A moving charge produces both electric and magnetic fields. As, the observer is not moving with respect to the charge, so observer only experiences an electric field.

12. (c) Due to current flowing, electrons travel through a conductor and they collide one another resulting in a loss of energy in the form of heat.

13. (c) In case of a varying section conductor, current flowing through any section is same as energy section remains electrically neutral even when current flows through it.

So, number of electrons entering any section

= number of electrons leaving that section

Hence, I is same throughout the conductor.

14. (a) Total charge crossing any cross-section of tube in 1 s

$$= 3.13 \times 10^{15} + 3.12 \times 10^{15} = 6.25 \times 10^{15}$$

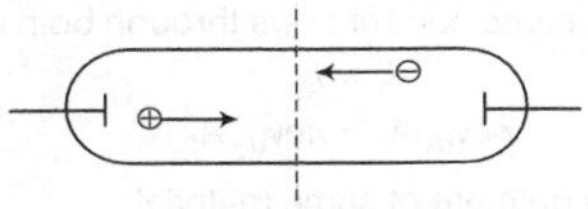

$\therefore$ Charge crossing the cross-section in 1s

$$= 6.25 \times 10^{15} \times 1.6 \times 10^{-19}$$
$$= 10 \times 10^{-4} \text{ C in 1 s}$$

$\therefore \quad I = 1 \text{ mA}$ (from left to right)

15. (a,b) $\because \quad I = I_0 e^{-t/x}$

At $t = 0$, $\quad I = I_0 e^0 = I_0$

Charge passing through the wire in τ interval

$$Q = \int_{t_1}^{t_2} I dt = \int_0^{\tau} I_0 e^{-t/\tau} dt = I_0 \int_0^{\tau} e^{-t/\tau} dt$$

$$\because \quad \int e^{ax} dx = \frac{1}{a} e^{ax} = -I_0 \tau [e^{-t/\tau}]_0^{\tau}$$
$$= I_0 \tau \{e^{-t/\tau}\}_{\tau}^{0} = I_0 \tau \{e^0 - e^{-1}\}$$
$$= I_0 \tau \left(1 - \frac{1}{e}\right) = 0.632 I_0 \tau$$

16. (a,d) Speed of each deuteron

$$v = \sqrt{\frac{2K}{m}} = 1.38 \times 10^7 \text{ m/s}$$

As, kinetic energy, $K = \frac{1}{2} mv^2$

$$m = 2 \times 1.67 \times 10^{-27} \text{kg}$$
$$1 \text{ eV} = 1.6 \times 10^{-19} \text{ J}$$

Time between 2 deuterons, $t = \frac{q}{I}$

$$\because \quad I = \frac{q}{t}$$
$$\therefore \quad v = \frac{d}{t}$$

and distance between 2 deuterons

$$d = vt = v \cdot \frac{q}{I} = 2.21 \times 10^{-2} \text{ m}$$

Also, repulsion is very less comparing to accelerating potential of 2 mV.

17. (b) In metals, only free electrons carry current.

In gases, electrons and positive ions carry current.

In a cell, positive and negative ions are charge carriers.

In a semiconductor, electrons and holes (vacancy for an electron) are charge carriers.

(i) $\to$ (r), (ii) $\to$ (p), (iii) $\to$ (q), (iv) $\to$ (s)

18. (b) Current = 175 mA and time = 150 ns

$$Q = It$$
$$= 175 \times 10^{-3} \times 150 \times 10^{-9} \text{ C}$$
$$= 2.625 \times 10^{-8} \text{ C}$$

As, $Q = ne$

$$n = \frac{Q}{e} = \frac{2.625 \times 10^{-8}}{1.6 \times 10^{-19}}$$

or $\quad n = 1.64 \times 10^{11}$

19. (d) A pulse is completed in 5 ms.

So, average current

$$I_{av} = \frac{\text{Total charge}}{\text{Total time}} = \frac{2.625 \times 10^{-8}}{5 \times 10^{-3}}$$
$$= 5.25 \times 10^{-3} \text{ mA}$$

20. (c) Pulse time, $\Delta t = 5$ ms

Number of pulses in 1 s

$$= \frac{1}{5 \times 10^{-3}} = 200$$

So, total charge delivered in 1 s

$$= 200 \times \text{charge in pulse}$$
$$= 200 \times 2.625 \times 10^{-8}$$
$$= 5.25 \text{ }\mu\text{C}$$

21. (b) Energy delivered per second

= Energy delivered per pulse × Number of pulses per sec

$$= 1.64 \times 10^{11} \times 50 \text{ MeV} \times 200$$
$$= 1.64 \times 10^{11} \times 50 \times 1.6 \times 10^{-13} \times 200$$
$$= 262.4 \text{ J}$$

So, energy absorbed by the absorber

$$= \frac{80}{100} \times 262.4 = 210 \text{ J}$$

22. (1) As, when an electron approaches a junction, in addition to a uniform field (that it normally faces which keeps v_d fixed), there is a accumulation of charges on wire surface at the junction. These produce electric field. These fields alter the direction of momentum. But till A, momentum is conserved, so $N = 1$.

23. (4) Change in the Kinetic Energy of an ion is

$$\Delta K = K_f - K_i$$
$$= 8.6 \times 10^{-17} - 5.4 \times 10^{-17}$$
$$= 3.2 \times 10^{-17} \text{ J}$$

This change is produced by accelerating potential.

So, potential energy provided by accelerating potential

= Gain in kinetic energy of an ion

$$\Rightarrow \quad q\Delta V = \Delta K$$
$$\Rightarrow \quad q\,(5) = 3.2 \times 10^{-17}$$
$$\Rightarrow \quad q = \frac{3.2 \times 10^{-17}}{5}$$
$$= \frac{320 \times 10^{-19}}{5}$$
$$= 64 \times 10^{-19} \text{ C}$$

Using $q = ne$, $\quad n = \frac{q}{e} = \frac{64 \times 10^{-19}}{1.6 \times 10^{-19}}$

$$\Rightarrow \quad n = 40$$
$$\therefore \quad N = 4$$

24. (5) In equilibrium, $\Sigma F = 0$

$$\Rightarrow \quad -F_{\text{spring}} + F_{\text{electrostatic}} = 0$$
$$\Rightarrow \quad kx = QE$$
$$\Rightarrow \quad x = Q$$
$$\Rightarrow \quad 8 \times 10^{-15} = n \times e$$
$$\Rightarrow \quad n = \frac{8 \times 10^{-15}}{1.6 \times 10^{-19}}$$
$$= 50000 \text{ electrons}$$
$$\therefore \quad N = 5$$

DPP-2 Relation among I, v_d, j, η, ρ and τ

1. (i) Heat as well as electrical energy is carried mostly by free electrons.

A good electrical conductor has many free electrons.

Thus, it is also a good conductor of heat.

(ii) Resistance occurs because a travelling electron collides with other electrons/ions. If we consider no resistance that means no collision.

Thus, electron will have very high velocity.

(iii) Drift velocity of electrons is very less, but as soon as an electron enters the conductor, some other electrons from other end are expelled. In this way, current reaches other end of conductor almost instantly.

(iv) Conductor in a circuit is not in electrostatic equilibrium.

The battery maintains a potential difference across the conductor and an electric field in the conductor. The steady current is due to the response of the electrons in the conductor due to the electric field.

2. Since, n = number density of charge carriers

= number of electrons per unit volume

= number of atoms per unit volume

(as 1 atom denotes 1 electron)

Also, mass per atom

$$= \frac{\text{Simple mass}}{\text{Avogadro number}} = \frac{27}{N_A}$$

$$= 4.49 \times 10^{-23} \text{ g/atom}$$

$$\Rightarrow \quad n = \frac{\text{Density of aluminium}}{\text{Mass per atom}} = \frac{2.7 \text{ g/cm}^3}{4.49 \times 10^{-23} \text{ g/atom}}$$

$$= 6.02 \times 10^{28} \text{ atoms/m}^3$$

$$\Rightarrow \quad v_d = \frac{I}{ne_A} = 0.130 \text{ mm/s}$$

3. Find area $A = \frac{\pi d^2}{4}$

Then, use $I = nev_d A$

4. (i) Relaxation time mainly depends on velocities of electrons. As electric field affects speed only at the order of 1 mm/s and the change in temperature affects velocities at the order of 10^2 m/s.

So, effect of field is insignificant comparing to the effect of temperature.

(ii) Conductivity, $\sigma = \frac{ne^2\tau}{m} \Rightarrow \frac{\sigma}{n} = \frac{e^2\tau}{m}$

$$\Rightarrow \quad \frac{\sigma}{n\tau} = \text{a constant}$$

$$\Rightarrow \quad \frac{\sigma_1}{n_1\tau_1} = \frac{\sigma_2}{n_2\tau_2} = \frac{\sigma_3}{n_3\tau_3}$$

$$\Rightarrow \quad \frac{\sigma}{n\tau_1} = \frac{2\sigma}{2n\tau_2} = \frac{2\sigma}{n\tau_3}$$

$$\Rightarrow \quad \frac{1}{\tau_1} = \frac{1}{\tau_2} = \frac{2}{\tau_3}$$

$$\Rightarrow \quad \tau_1 = \tau_2 = \frac{\tau_3}{2}$$

$\therefore$ τ_3 is maximum.

5. (b) Drift speed of electrons is of order of 10^{-4} m/s or 10^{-2} cm/s.

6. (a) Mean relaxation time decreases when number of collisions increases. Number of collisions increases when temperature increases due to increased lattice vibrations.

$$\therefore \quad \tau \propto \frac{1}{T}$$

7. (c) In series, same current flows through both wires

$$\therefore \quad I_1 = I_2$$

$$\Rightarrow \quad nev_{d_1}A_1 = nev_{d_2}A_2$$

n is same as both are of same material.

$$\Rightarrow \quad \frac{v_{d_1}}{v_{d_2}} = \frac{A_2}{A_1} \quad \text{...(i)}$$

Now, given $\frac{R_1}{R_2} = \frac{1}{2} \Rightarrow \frac{\rho \frac{l_1}{A_1}}{\rho \frac{l_2}{A_2}} = \frac{1}{2}$

$$\Rightarrow \quad \frac{l_1}{l_2} \cdot \frac{A_2}{A_1} = \frac{1}{2}$$

$$\Rightarrow \quad \frac{A_2}{A_1} = \frac{1}{2} \times \frac{l_2}{l_1} = \frac{1}{2} \times \frac{2}{1} = 1:1 \quad \text{...(ii)}$$

So, from Eqs. (i) and (ii), we get

$$\frac{v_{d_1}}{v_{d_2}} = 1:1$$

8. (d) As, volume of conductor remains constant

$$\therefore \quad l_i A_i = l_f A_f$$

Given, $l_f = 3l_i \Rightarrow A_f = \frac{l_i}{l_f} \cdot A_i = \frac{A_i}{3}$

Also, $R_i = \rho \frac{l_i}{A_i}$

and $R_f = \rho \frac{l_f}{A_f} = \rho \frac{3l_i}{A_i/3} = 9R_i$

As, $E = IR \Rightarrow I = \frac{E}{R}$

$$\therefore \quad I_i = \frac{E}{R_i}, \quad I_f = \frac{E}{9R_i} = \frac{I_i}{9}$$

$$I_i = nev_{d_i}A_i$$

$$I_f = nev_{d_f}A_f \Rightarrow \frac{I_i}{I_f} = 9 = \frac{nev_{d_i}A_i}{nev_{d_f}A_f}$$

$$\Rightarrow \quad 9 = \frac{v_{d_i}}{v_{d_f}} \cdot 3$$

$$\Rightarrow \quad v_{d_f} = \frac{1}{3} v_{d_f}$$

9. (a) Current through any cross-section remains same.

10. (b) $I = nev_d A$

$$\Rightarrow \quad v_d = \frac{I}{neA}$$

$$\Rightarrow \quad v_d \propto \frac{1}{A}$$

$$\Rightarrow \quad v_d A = \text{a constant}$$

$$\therefore \quad v_{d_A} > v_{d_B}$$

11. (a) $v_d = \frac{I}{neA} = \frac{1.344}{10^{-6} \times 1.6 \times 10^{-19} \times 8.4 \times 10^{22}}$

$= 0.11$ cm/s $= 0.1$ mm/s

12. (b) Without any applied potential difference for an electron

$$v_{rms} = \sqrt{\frac{3K_BT}{M}}$$

$$\therefore \quad v_{rms} \propto \sqrt{T}$$

13. (a) Number of protons crossing any cross-section of beam

$= 1.43 \times 10^{13}$ protons

14. (c) $v_d A$ = a constant

$$\Rightarrow \quad v_{d_1}\left(\pi\,\frac{10^2}{4}\right) = v_{d_2}\left(\pi\,\frac{14^2}{4}\right)$$

$$\Rightarrow \quad \frac{v_{d_1}}{v_{d_2}} = \frac{14\times 14}{10\times 10} = \frac{7\times 7}{5\times 5} = \frac{49}{25}$$

15. (d) Due to randomly moving electrons drift velocity is very less as

$$v_d = \frac{v_1 + v_2 + \cdots + v_n}{N}$$

16. (d) Current flowing through both the bars is equal.
Now, the heat produced is given by

$$H = I^2Rt \text{ or } H \propto R$$

or $$\frac{H_{AB}}{H_{BC}} = \frac{R_{AB}}{R_{BC}} = \frac{(1/2r)^2}{(1/r)^2} \quad \left(\text{as } R \propto \frac{1}{A} \propto \frac{1}{r^2}\right)$$

$$= \frac{1}{4}$$

or $$H_{BC} = 4H_{AB}$$

17. (a,b) When temperature increases, then number of free charge carriers (electrons in metals, ions in gases and electrons and hole pairs in semiconductors) increases.
With temperature, lattice vibration increases and hence, number of collisions increases.

18. (b,c,d) $I = \frac{6}{6} = nev_d A$

$$\Rightarrow \quad v_d = \frac{1}{10^{29}\times 1.6\times 10^{-18}\times 10^{-6}} = \frac{10^{-4}}{1.6}\text{ m/s}$$

$$\text{KE} = \frac{1}{2}mv_d^2 \times nAl$$

$$= \frac{1}{2}\times 9.1\times 10^{-31}\times \frac{1}{2.56}\times 10^{-8}\times 10^{29}\times 10^{-6}\times 10^{-1}$$

$$= 2\times 10^{-17}\text{ J}$$

Ohmic loss = I^2R = 6 J/s

$\therefore$ All kinetic energy of the electrons will be lost in $\frac{2\times 10^{-17}}{6}$ s

$\approx 10^{-17}$ s

19. (a, c) $v_d = \frac{I}{neA} = \frac{4I_m}{N\rho_D e\pi d^2}$

Substituting values, we get

$$v_d = 5.1\times 10^{-10}\text{ m/s and}$$

$$E = \rho j = 1.2\times 10^{-7}\text{ V/m}$$

20. (a, d) Use, $R = \frac{\rho l}{A}$ to find resistivity

$$\rho = \frac{RA}{l} = 1.6\times 10^{-8}\ \Omega\text{-m}$$

Find current density using $j = \frac{I}{A} = 2.4\times 10^5$ A/m^2

Then, find number density of charge carriers as

$$n = \frac{j}{v_d e} = 8.8\times 10^{28}\text{ m}^{-3}$$

21. (5) Mean relaxation time is given by

$$\tau = \frac{m}{ne^2\rho} = \frac{9.1\times 10^{-31}}{8.5\times 10^{28}\times(1.6\times 10^{-19})^2\times 1.7\times 10^{-8}}$$

$$= 2.5\times 10^{-14}\text{ s}$$

As, free electrons have only thermal velocities

$$v_{\text{rms}} = \sqrt{\frac{3KT}{m}} = \sqrt{\frac{3\times 1.38\times 10^{-23}\times 300}{9.1\times 10^{-31}}} = 10^5\text{ m/s}$$

So, mean free path = $v_{\text{rms}}\times\tau = 10^5\times 2.5\times 10^{-14}$

$$= 2.5\times 10^{-10}\text{ m} = 25\text{ Å}$$

So, $N = 5$

22. (4) Current through outer portion is $i = \int \hat{\mathbf{j}}\cdot d\mathbf{A}$

$$= \int_{R/2}^{R}\hat{\mathbf{j}}\cdot d\mathbf{A} \qquad \{\text{As, } \hat{\mathbf{j}} \parallel d\mathbf{A}\}$$

$$= \int_{R/2}^{R} ar^2 2\pi r\cdot dr \qquad \{\because d\mathbf{A} = 2\pi r dr\}$$

$$= 2\pi a\int_{R/2}^{R} r^3 dr = 2\pi a\left(\frac{r^4}{4}\right)_{R/2}^{R}$$

$$= \frac{15}{32}\pi a R^4$$

So, $i \propto R^4$.

23. (9) As, each conduction electron is given by 10^6 atoms.
$\therefore$ Number of conduction electrons per unit volume is

$$n = \frac{\text{Atoms per unit volume}}{10^6}$$

$$= \frac{(\text{Atoms / mole})\times(\text{moles / mass})\times(\text{mass / vol})}{10^6}$$

$$= N_A\times\frac{1}{M}\times\rho\times\frac{1}{10^6}$$

$$= \frac{6.02\times 10^{23}\times 9\times 10^3}{6.02\times 10^{-2}\times 10^6} = 9\times 10^{22}$$

$\therefore$ $N = 9$

24. (a) For metals, drift speed is of order 10^{-4} m/s, speed of electron in atomic orbit is of order 10^6 m/s and thermal speed of electrons is of order 10^5 m/s.
(i) $\to$ (t), (ii) $\to$ (s), (iii) $\to$ (r), (iv) $\to$ (q), (v) $\to$ (p)

25. (c) $R = \frac{\rho l}{A} = \frac{4\times 10^{-8}\times 0.5}{\pi(1\times 10^{-4})} = 0.637\ \Omega$

26. (c) $I = \frac{\Delta V}{R} = \frac{4}{0.637} = 6.28$ A

27. (b) $\mathbf{j} = \frac{I}{A}\hat{\mathbf{i}} = \frac{6.28}{\pi(1\times 10^{-4})}\hat{\mathbf{i}} = 2\times 10^8\hat{\mathbf{i}} = 200\hat{\mathbf{i}}$ mA/m^2

28. (a) Charge that strikes the surface in time Δt is

$$q = i\Delta t$$

As charge on each α-particle is $2e$, we have

$$q = n(2e) = i\ \Delta t$$

or $$n = \frac{i\ (\Delta t)}{2e} = \frac{0.25\times 10^{-6}\times 1.6}{2\times 1.6\times 10^{-19}}$$

$$= \left(\frac{0.25}{2}\right)\times 10^{13}$$

$$= 1.5\times 10^{12}\text{ particles}$$

29. (b) Kinetic energy of α-particles is

$$= \frac{1}{2}mv^2 = k$$

$$\Rightarrow \quad v = \sqrt{\frac{2K}{mx}}$$

$$= \sqrt{\frac{20\times 10^6\times 1.6\times 10^{-19}\times 2}{6.68\times 10^{-27}}}$$

$$= 3.1\times 10^7\text{ m/s}$$

and number of particles in a beam of length L is

$$N = \frac{iL}{2ev} = \frac{0.25\times 10^{-6}\times 10\times 10^{-2}}{2\times 1.6\times 10^{-19}\times 3.1\times 10^7}$$

$$= 2.5\times 10^3$$

30. (c) Using energy conservation,

$$\text{KE} = V \Rightarrow \text{KE} = qV = 2eV$$

$$\Rightarrow \quad V = \frac{\text{KE}}{2e} = \frac{3.2\times 10^{-12}}{2\times 1.6\times 10^{-19}}$$

$$= 10\times 10^6\text{ V}$$

DPP-3 Resistance, Resistivity, Temperature Dependence and Ohm's Law

1. As resistance varies, we first find the variation of resistance with respect to time.

As graph is parabolic, we assume that

$$R = at^2 + bt + c \quad \text{(let)}$$

For $t = 0$ s, $R = 9\,\Omega$

$\therefore$ $c = 9$ ...(i)

For $t = 1$ s, $R = 14\,\Omega$

$\Rightarrow$ $14 = a + b + 9$ ...(ii)

For $t = 2$ s, $R = 17\,\Omega$

$\Rightarrow$ $17 = 4a + 2b + 9$ ...(iii)

From Eqs.(i), (ii) and (iii), we get

$$R = (-t^2 + 6t + 9)\Omega$$

So, R_{av} = average value of resistance in first 3 s is

$$= \frac{1}{(b-a)}\int_a^b R\,dt = \frac{1}{(3-0)}\cdot\int_0^3 (-t^2 + 6t + 9)dt = 15\,\Omega$$

So, average current using Ohm's law will be

$$I_{av} = \frac{V}{R_{av}} = \frac{3}{15} = 0.2\text{ A}$$

2. Saturation current, $I = neAv_d$

$$n = 2 \times \frac{10^9}{10^{-6}} = 2 \times 10^{15}\text{ m}^{-3}$$

$$v = v_d = \frac{6 \times 10^{-2}}{1} = 6 \times 10^{-2}\text{ m/s}$$

$\therefore$ $I = 2 \times 10^{15} \times 1.6 \times 10^{-19} \times 100 \times 10^{-4} \times 6 \times 10^{-2}$

$= 0.192\ \mu\text{A}$

Electric field, $E = \dfrac{v}{\text{Mobility}} = \dfrac{I}{\text{Mobility}} = \dfrac{6 \times 10^{-2}}{3.5 \times 10^{-4}} = 171.4\text{ Vm}^{-1}$

3. A more general definition of temperature coefficient of resistance is $\alpha = \dfrac{1}{\rho}\dfrac{d\rho}{dT}$, where ρ = resistivity, T = temperature.

$\Rightarrow$ $\dfrac{d\rho}{\rho} = \alpha\,dT \quad \Rightarrow \quad \int_{\rho_0}^{\rho}\dfrac{d\rho}{\rho} = \alpha\int_{T_0}^{T} dT$

$\Rightarrow$ $\log\dfrac{\rho}{\rho_0} = \alpha(T - T_0)$

Integrating on both sides, we get

$\Rightarrow$ $\rho = \rho_0 e^{\alpha(T - T_0)}$

Taking antilog

Using approximation, $e^x = 1 + x$, we have

$$\rho = \rho_0[1 + \alpha(T - T_0)]$$

4. Resistance of dielectric block is

$$R = \rho\frac{l}{A} = \frac{d}{\sigma A} \quad \left[\therefore\ l = D \text{ and } \frac{1}{\rho} = \sigma\right]$$

Capacitance of capacitor is

$$C = \frac{k\varepsilon_0 A}{d}$$

Then, $RC = \dfrac{d}{\sigma A}\cdot\dfrac{k\varepsilon_0 A}{d} = \dfrac{k\varepsilon_0}{\sigma}$

which is a characteristic of the material only.

5. (d) Using $R_t = R_0(1 + \alpha t)$

Taking base temperature as 0°C

$\Rightarrow$ $\dfrac{R_{T_1}}{R_{T_2}} = \dfrac{1 + \alpha t_1}{1 + \alpha t_2} \quad \Rightarrow \quad \dfrac{20}{60} = \dfrac{1 + 20\alpha}{1 + 500\alpha}$

$\Rightarrow$ $\alpha = \dfrac{1}{220}$

Now, again

$$\frac{20}{25} = \frac{1 + \left(\frac{1}{220} \times 20\right)}{1 + \left(\frac{1}{220} \times t\right)} \Rightarrow t = 80°\text{C}$$

6. (c) Ohm's law is true for metallic conductors under the condition that temperature of conductor remains constant.

7. (a) If length is increased by 100%, then final length becomes $2l$.

Now, equating initial and final volumes

$$l_1A_1 = l_2A_2 \Rightarrow lA = 2l \cdot A_2$$

$\Rightarrow$ $A_2 = \dfrac{A}{2}$

So, final resistance becomes

$$R_2 = \rho\frac{l_2}{A_2} = \rho\cdot\frac{2l}{A/2} = 4\rho\frac{l}{A} = 4R_1$$

$\therefore$ Increase = 300%

8. (d) Alloys have low temperature coefficient of resistance, so they show a little increase in resistance when temperature is increased.

9. (a) Using $R = R_0(1 + \alpha\,\Delta T)$, we get

$$1.15R_0 = R_0(1 + \alpha\,\Delta T)$$

$\Rightarrow$ $1 + \alpha\,\Delta T = 1.15$

$\Rightarrow$ $\alpha\,\Delta T = 0.15$

$\Rightarrow$ $\Delta T = \dfrac{0.15}{\alpha} = \dfrac{0.15}{0.0068}$

$\Rightarrow$ $\Delta T = 22°\text{C}$

So, the temperature rise of 22°C is required.

10. (c) $R_x = \dfrac{\rho l_x}{A_{yz}} = \dfrac{3 \times 10^{-5} \times 1 \times 10^{-2}}{2 \times 10^{-2} \times 4 \times 10^{-2}} = 3.75 \times 10^{-4}\,\Omega$

$R_y = \dfrac{\rho l_y}{A_{xz}} = \dfrac{3 \times 10^{-5} \times 2 \times 10^{-2}}{1 \times 10^{-2} \times 4 \times 10^{-2}} = 1.5 \times 10^{-3}\,\Omega$

$R_z = \dfrac{\rho l_z}{A_{xy}} = \dfrac{3 \times 10^{-5} \times 4 \times 10^{-2}}{1 \times 10^{-2} \times 2 \times 10^{-2}} = 6 \times 10^{-3}\,\Omega$

$\Rightarrow$ $\dfrac{R_z}{R_y} = \dfrac{6 \times 10^{-3}}{1.5 \times 10^{-3}} = \dfrac{60}{15} = 4$

11. (c) All the conductors have same length and same material.

Now, area of Ist conductor $A = (\sqrt{3}a)^2 - (\sqrt{2}a)^2 = a^2$

Area of cross-section of conductor $B = (\sqrt{2}\,a)^2 - a^2 = a^2$

Area of cross-section of conductor $C = a^2$

As, area of conductors is also same.

$\therefore$ $R_A = R_B = R_C$

12. (a) $R = \rho\dfrac{l}{A}$

$\Rightarrow$ $\dfrac{R_A}{R_B} = \dfrac{A_B}{A_A}$ [ρ and l are same for both rods]

$\Rightarrow$ $\dfrac{A_B}{A_A} = \dfrac{R_A}{R_B} = \dfrac{1}{2}$

When a body is dipped in water.

$$\text{Loss of weight} = V\sigma_L g = Al\sigma_L g$$

$\therefore$ Loss of weight $\propto$ Area

$\Rightarrow$ $\dfrac{\text{Loss of weight of } A}{\text{Loss of weight of } B} = \dfrac{A_A}{A_B} = 2$

$\therefore$ Loss of weight of A is more than that of B.

13. (b) $V = IR$ comparing with $Y = mX$, we have

Slope = Resistance of wire

In given graph,

$$\text{Slope} = \tan(\text{angle with } X\text{-axis})$$
$$= \tan(90^\circ - 60^\circ) = \tan 30^\circ = \frac{1}{\sqrt{3}}$$

14. (c) As, resistance $\propto$ temperature

$\Rightarrow \quad R_1 \propto T_1 \Rightarrow \tan\theta \propto T_1$

$\Rightarrow \quad \tan\theta = KT_1$

and $\quad R_2 \propto T_2 \Rightarrow \tan(90^\circ - \theta) \propto T_2$

$\Rightarrow \quad \cot\theta = KT_2$

$$\Rightarrow \quad T_2 - T_1 = \left(\frac{\cos\theta}{\sin\theta} - \frac{\sin\theta}{\cos\theta}\right) = \frac{\cos^2\theta - \sin^2\theta}{\sin\theta\cos\theta}$$
$$= \frac{2\cos 2\theta}{\sin 2\theta} = 2\cot 2\theta$$

$\Rightarrow \quad (T_2 - T_1) \propto \cot 2\theta$

15. (c) We will require a voltmeter, an ammeter, a test resistor and a variable battery to varify Ohm's law.

Voltmeter which is made by connecting a high resistance with a galvanometer is connected in parallel with the test resistor.

Further, an ammeter which is formed by connecting a low resistance in parallel with galvanometer is required to measure the current through test resistor.

16. (a,b) R = resistance of combination = $R_N + R_C$

As, resistance does not change with temperature

$$R_N + R_C = R_C(1 + \alpha_C \Delta T) + R_N(1 + \alpha_N \Delta T)$$
$$\Rightarrow \quad R_C\alpha_C\Delta T + R_N\alpha_N\Delta T = 0 \quad \text{or} \quad R_C = -R_N\frac{\alpha_N}{\alpha_C}$$

$\therefore$ Total resistance,

$$\Rightarrow \quad R = -R_N\frac{\alpha_N}{\alpha_C} + R_N$$

So, $\quad R_N = R\left(1 - \dfrac{\alpha_N}{\alpha_C}\right)^{-1}$ and $R_C = R\left(1 - \dfrac{\alpha_C}{\alpha_N}\right)^{-1}$

$$R_N = 10\left(1 - \frac{0.4 \times 10^{-3}}{-0.5 \times 10^{-3}}\right)^{-1} = 5.56\text{ k}\Omega$$

Similarly, $\quad R_C = 4.44\text{ k}\Omega$

17. (a,c) Initially, stored energy $= \dfrac{1}{2}\dfrac{Q^2}{C}$

When switch S is closed, charge Q distributes over the plates of capacitors.

Total capacitance of system = $4C$

Final potential difference between plates of both capacitors is

$$V_f = \frac{Q}{4C}$$

Charge on small capacitor after stabilisation $= Q'_1 = CV_f$

$$= \frac{Q}{4C} \cdot C = \frac{Q}{4}$$

and so larger capacitor has charge, $Q'_2 = \dfrac{3Q}{4}$

Hence, final energy stored in a smaller capacitor

$$= \frac{1}{2}C\left(\frac{Q}{4C}\right)^2 = \frac{Q^2}{32C}$$

Final energy stored in a large capacitor $= \dfrac{1}{2}\dfrac{Q_2'^2}{3C}$

$$\text{Energy dissipated} = \frac{Q^2}{2C} - \frac{Q^2}{8C} = \frac{3Q^2}{8C}$$

18. (a, d) Germanium is a semiconductor, whereas sodium is a metal. So, with temperature the conductivity of germanium (σ_1) increases and that of sodium (σ_2) decreases.

19. (b) Resistivity of semiconductor decreases with temperature because number of electrons thermally agitated to conduction bend increases with temperature. At higher temperature, the number density of charge carriers is more as compared to that at low temperature.

20. (c) Third pin is connected to metal box covering the instrument or appliance and the wall socket connects third pin to ground. In case of a leakage current or short circuit, current is conducted to the earth and instrument does not give any shock to user.

21. (b) Due to bending, there is no change in the length of the wire and resistance $R = \rho\dfrac{l}{A}$ does not change.

Also, resistance depends on the resistivity of material of a conductor.

22. (a) Resistivity of metal increases with temperature. Resistivity of a semiconductor decreases with temperature due to formation of new electrons and holes due to thermal agitation. Resistivity of metal alloys is nearly constant with temperature. Resistivity of a diamagnetic substance falls to zero or it exhibits superconductivity.

(i) $\rightarrow$ (q), (ii) $\rightarrow$ (p), (iii) $\rightarrow$ (r), (iv) $\rightarrow$ (s)

23. (3) Mean relaxation time

$$\tau = \frac{m}{ne^2\rho} = \frac{9.11 \times 10^{-31}}{(8.49 \times 10^{28}) \times (1.6 \times 10^{-19})^2 \times (1.7 \times 10^{-8})}$$
$$= 2.5 \times 10^{-14}\text{ s}$$

Given $v = 1.6 \times 10^6$ m/s

Mean free path (on an average) travelled by an electron

$$= l = v\tau = 2.5 \times 10^{-14} \times 1.6 \times 10^6$$
$$= 4 \times 10^{-8}\text{ m} = 40\text{ nm}$$

Average atomic spacing = 0.2 nm

Number of atomic spacings covered by electron between collisions $= \dfrac{0.2}{40} = 200$

$\therefore N = 200$

24. $\left(\dfrac{6}{5}\alpha\right) N = 3$. Resistance of wires are

$$R_A = \frac{\rho l_A}{A_A} = \frac{\rho 2l}{A/2} = 4\frac{\rho l}{A}$$
$$R_B = \frac{\rho l}{A}$$

$\therefore$ Temperature coefficient of resistance of composite conductor is given by

$$\alpha_{AB} = \frac{R_A\alpha_A + R_B\alpha_B}{R_A + R_B} = \frac{4R \cdot \alpha_1 + R \cdot 2\alpha_1}{4R + R} = \frac{6}{5}\alpha$$

25. (4) Resistance, $r = \dfrac{\rho l}{a}$

Volume, $x = a \times l$

So, $\quad r = \dfrac{\rho l^2}{al} = \dfrac{\rho l^2}{x} \Rightarrow l = \left(\dfrac{r}{\rho}\right)^{1/2}$

Also, $\quad a = \text{area} = \dfrac{\pi d^2}{4} \Rightarrow r = \dfrac{\rho l \times a}{a \times a} = \dfrac{\rho x}{a^2}$

$$\Rightarrow \quad r = \frac{\rho x}{\left(\frac{\pi}{4}d^2\right)^2} \Rightarrow d^4 = \frac{16\rho x}{\pi^2 r}$$

$$\Rightarrow \quad d = \left(\frac{16\rho x}{\pi^2 r}\right)^{1/4}$$

DPP-4 **Series and Parallel Combination of Resistances**

1. (i) In series combination, current through each resistor is same.

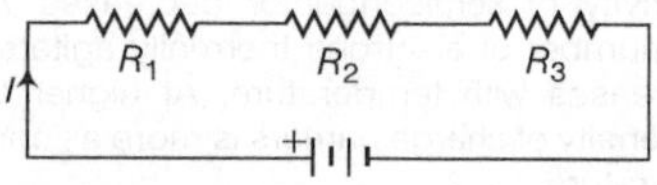

But, potential difference and power used across each resistor is different.

Potential differences are IR_1, IR_2 and IR_3.

Powers are I^2R_1, I^2R_2, I^2R_3.

(ii) In parallel combination,

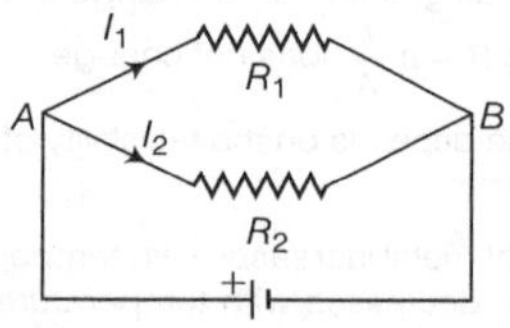

Potential difference across each resistor $= V_A - V_B = V_{AB}$

But, currents are different

$\therefore \quad I_1 = \dfrac{V_{AB}}{R_1}$ and $I_2 = \dfrac{V_{AB}}{R_2}$

and powers used are also different.

$$P_1 = \frac{V_{AB}^2}{R_1} \text{ and } P_2 = \frac{V_{AB}^2}{R_2}$$

2. (i) (a) A voltage divider is a set of series connected resistors.

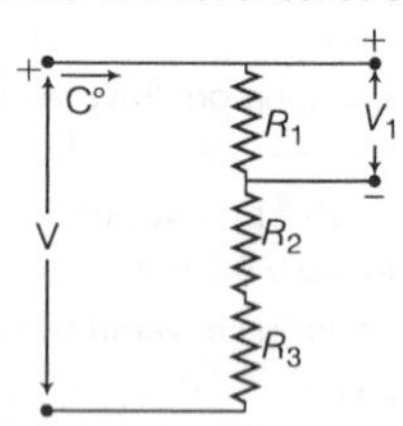

To obtain a particular output V_1, the values of resistance can be suitably adjusted.

$V_1 = iR$ and $V = i(R_1 + R_2 + R_3)$, we get

$$V_1 = \frac{VR_1}{(R_1 + R_2 + R_3)}$$

(b) A current divider is an arrangement of parallel resistances. The current through each branch can be adjusted by putting resistors of suitable values.

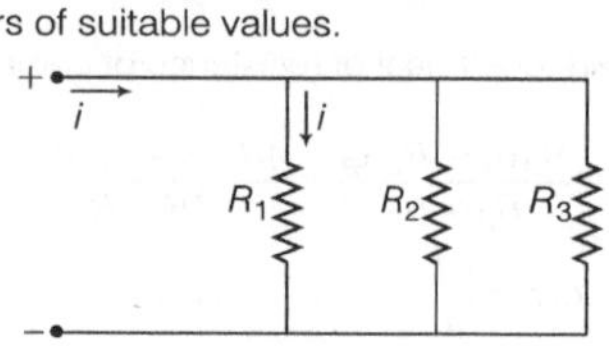

$$i = \frac{V}{R_1} + \frac{V}{R_2} + \frac{V}{R_3} \text{ and } i_1 = \frac{V}{R_1}$$

$$\Rightarrow \quad \frac{i_1}{i} = \frac{R_2R_3}{R_1R_2 + R_1R_3 + R_2R_3}$$

$$\Rightarrow \quad i_1 = \frac{R_2R_3 i}{R_1R_2 + R_1R_3 + R_2R_3}$$

(ii) Here, $\dfrac{V_1}{V} = 10\% = 0.1$

$\Rightarrow \quad 0.1 = \dfrac{R_1}{R_1 + R_2} = \dfrac{R_1}{50}$

$\Rightarrow \quad R_1 = 50 \times 0.1 = 5\,\Omega$

and $\quad R_2 = 45\,\Omega$

(iii) Here, $\dfrac{20\text{ mA}}{30\text{ mA}} = \dfrac{R_2}{R_1 + R_2}$ and $\dfrac{10\text{ mA}}{30\text{ mA}} = \dfrac{R_1}{R_1 + R_2}$

also, $\dfrac{R_1R_2}{R_1 + R_2} \geq 10\,\Omega$

Solving these, we have $R_1 \geq 15\,\Omega$ and $R_2 \geq 30\,\Omega$

3. (i) As, maximum power of bulb, i.e. wattage $W = \dfrac{V_s^2}{R}$

where, V_s is specified voltage on the bulb.

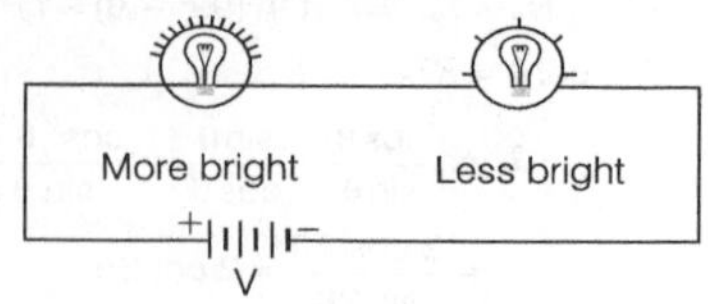

So, resistance of filament of bulb is

$$R = \frac{V_s^2}{W}$$

So, $\quad R \propto \dfrac{1}{W}$

When bulbs are connected in series, I is a constant.

Power consumed, $P = I^2R$

$$P \propto R \propto \frac{1}{W}$$

So, less wattage bulb will consume more power so, give more brightness.

As, first bulb gives more brightness, then wattage of first bulb is less than wattage of second bulb. Hence, claim of student is not correct.

(ii) A typical household circuit is

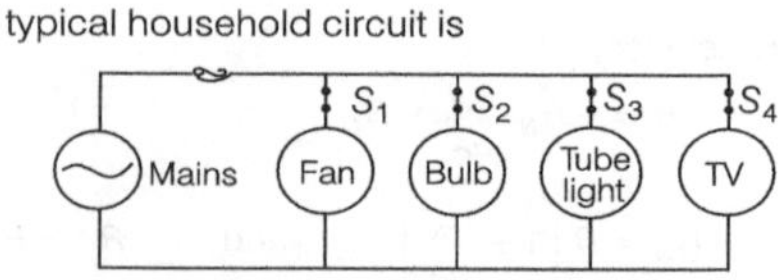

In parallel connection, if one of the appliances is switched OFF (or ON) or is fused, there is no effect on the power supplied to other appliances. Hence, appliances can work independently and even if one of the appliance is removed from the circuit even, then circuit works as efficiently as before.

(iii) As, $V = IR \quad \Rightarrow \quad I = \dfrac{V}{R}$

and on comparing above equation with $y = mx$, we get

$$\text{Slope of line} \propto \frac{1}{R}$$

Now, $\quad$ slope of line $B = \tan\theta = \tan 60° = \sqrt{3}$

and $\quad$ slope of line $A = \tan 30° = \dfrac{1}{\sqrt{3}}$

So, resistance of series combination

$$= R_1 + R_2 = \tan 60° = \sqrt{3}$$

Resistance of parallel combination

$$= \frac{R_1R_2}{R_1 + R_2} = \tan 30° = \frac{1}{\sqrt{3}}$$

So, $\quad \text{ratio} = \dfrac{\sqrt{3}}{1/\sqrt{3}} = 3:1$

4. (10.5. 3.8) Amount of heat absorbed by water

= amount of heat generated by coil

$$= I^2RT = \frac{V^2}{R}t$$

For Ist coil, $H = I^2R_1\,(6 \times 60) = \dfrac{V^2}{R_1}(6 \times 60)$

For IInd coil, $H = I^2R_2\,(8 \times 60) = \dfrac{V^2}{R_2}(8 \times 60)$

As, amount of heat required to heat water is same in both cases

$\therefore \quad I^2R_1\,(6 \times 60) = I^2R_2\,(8 \times 60)$

$\Rightarrow \quad 6R_1 = 8R_2 \Rightarrow R_1 = \dfrac{4}{3}R_2$

(i) Now, when the both coils are connected in series,

$$R_{eq} = R_1 + R_2 = R_1 + \frac{3}{4}R_1 = \frac{7}{4}R_1$$

$\Rightarrow$ Heat produced in time $\Delta t = H$

= amount of heat required for boiling

$= I^2(R_1 + R_2)\,\Delta t$

$= \dfrac{V^2}{(R_1 + R_2)}\Delta t = \dfrac{V^2}{R}(6 \times 60)$

(as, heat required is still same)

$$\Rightarrow \Delta t = \left(\frac{R_1 + R_2}{R_1}\right) \times 6 \times 60$$

$$\Rightarrow \Delta t = \frac{\frac{7}{4}R_1}{R_1} \times 6 \times 60 = \frac{7}{4} \times 6 \times 60 = \frac{7}{2} \times 3 \times 60 \text{ s}$$

$= 3.5 \times 3 \text{ min} = 10.5 \text{ min}$

(ii) When resistors are connected in parallel,

$$R_{eq} = \frac{R_1R_2}{R_1 + R_2} = \frac{\frac{7}{4}R_1 \times R_1}{R_1 + \frac{7}{4}R_1} = \frac{\frac{7}{4}}{\frac{11}{4}}R_1 = \frac{7}{11}R_1$$

$\therefore$ Heat produced in time Δt,

$$\Rightarrow \frac{V^2}{R_{eq}} \cdot \Delta t = \frac{V^2}{R_1}(6 \times 60)$$

$$\Rightarrow \frac{V^2}{\frac{7}{11}R_1}\Delta t = \frac{V^2}{R_1} \times 6 \times 60$$

$$\Rightarrow \Delta t = \frac{7}{11} \times 6 \text{ min} = \frac{42}{11} \approx 3.8 \text{ min}$$

5. (d) Circuit can be redrawn as follows

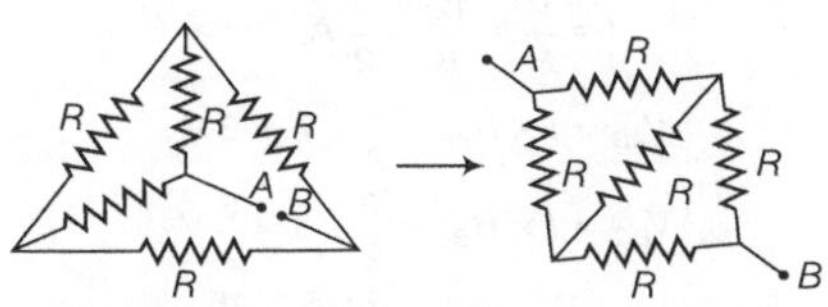

So, given figure is a balanced Wheatstone's bridge and hence, above circuit is equivalent to

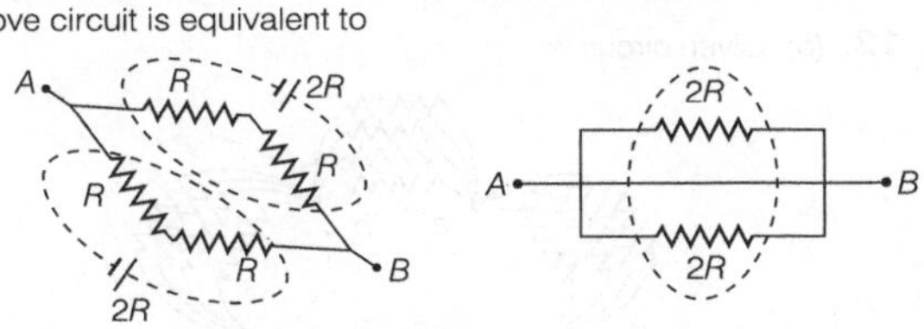

$\Rightarrow \quad R_{eq} = \dfrac{2R \times 2R}{2R + 2R}$

$\Rightarrow \quad = R$

So, using $V = IR$,

$I = \dfrac{12}{2} = 6$ A

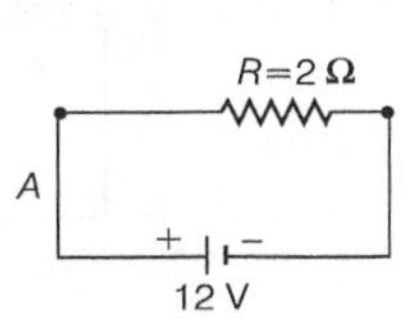

6. (d) Given circuit is a combination of a balanced Wheatstone's bridge and a resistance of $R\ \Omega$.

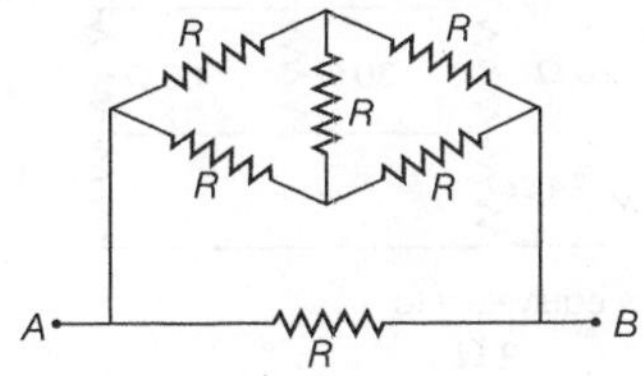

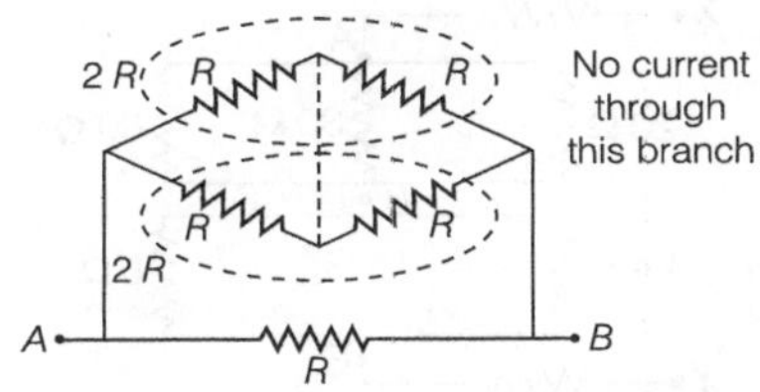

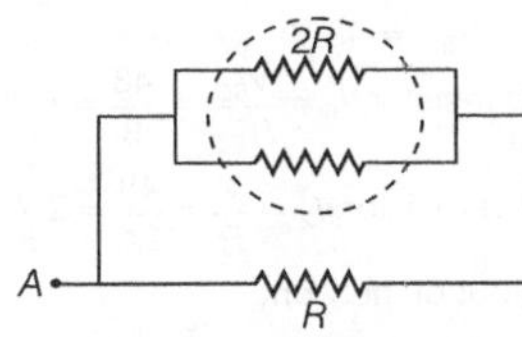

$R_{eq} = R$ using formula for parallel combination

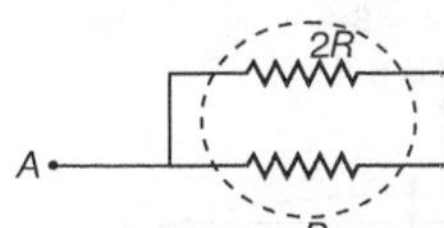

Parallel combination

$R_{eq} = \dfrac{R \times R}{R + R} = \dfrac{R}{2}\ \Omega$

7. (b) Given circuit is a network equivalent to

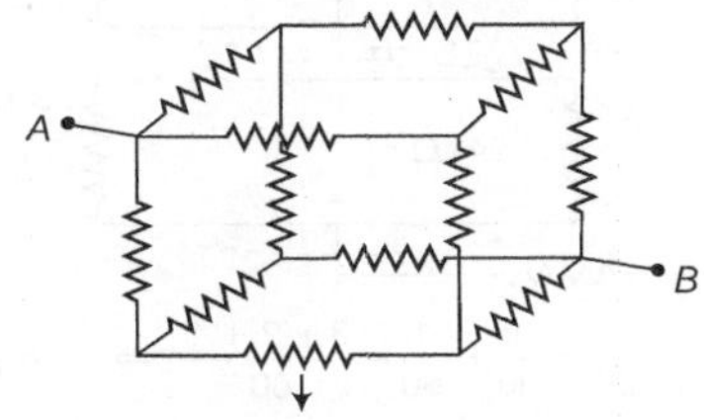

To redraw circuit, we join resistors across points which are same potentials. This can be judged on the basis of symmetry of circuit.

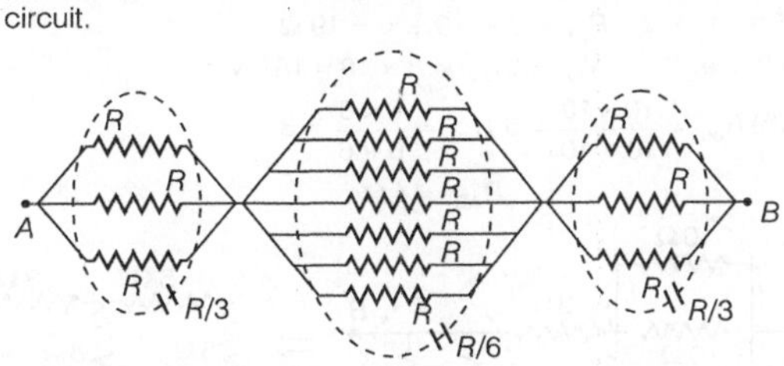

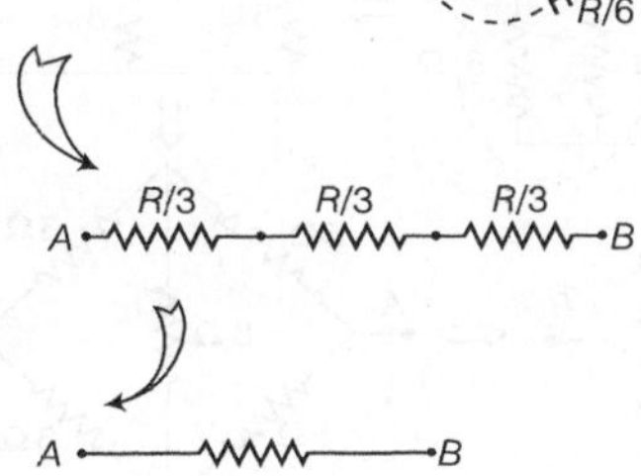

$R_{eq} = \dfrac{R}{3} + \dfrac{R}{6} + \dfrac{R}{3}$

$= \dfrac{2R + R + 2R}{6}$

$= \dfrac{5R}{6}\ \Omega$

8. (c) Given circuit is

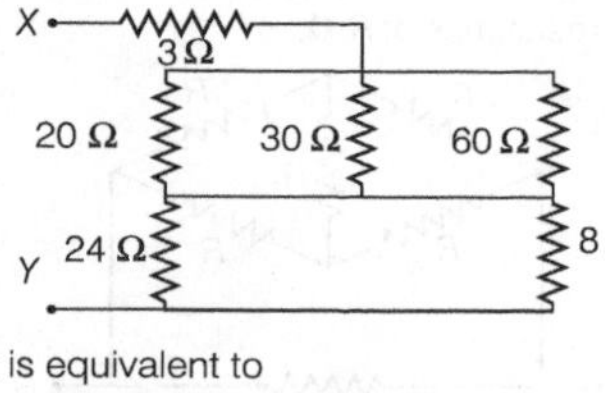

The circuit is equivalent to

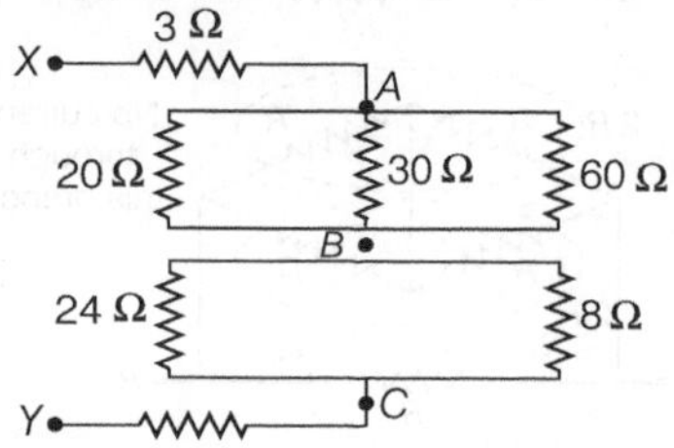

Now, given $V_{BC} = 48$ V

So, current through 8 Ω resistor, $I_1 = \frac{V_{BC}}{R_{BC}} = \frac{48}{8} = 6$ A

and current through 24 Ω resistor, $I_2 = \frac{V}{R} = \frac{48}{24} = 2$ A

Hence, current flowing out of network

$$I_1 + I_2 = 6 + 2 = 8 \text{ A}$$

∴ Current flowing through XY is 8 A.

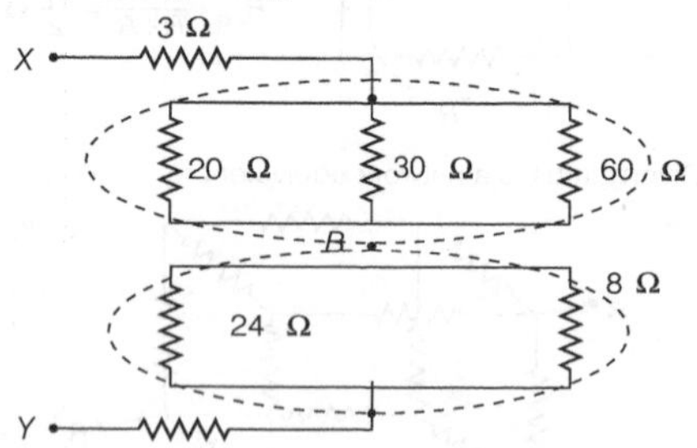

Now, $\frac{1}{R_{eq}} = \frac{1}{20} + \frac{1}{30} + \frac{1}{60} = \frac{3+2+1}{60} \Rightarrow R_{eq} = \frac{60}{6} = 10\,\Omega$

$\frac{1}{R_{eq}} = \frac{1}{24} + \frac{1}{8} = \frac{1+3}{24} \Rightarrow R_{eq} = \frac{24}{4} = 6\,\Omega$

So, $R_{XY} = 3 + 10 + 6 = 19\,\Omega$

Hence, $V_{XY} = IR_{XY} = 8 \times 19 = 152$ V

9. (b) $R_{eq} = \frac{10 \times 10}{10 + 10} = 5, R_{eq} = \frac{6 \times 6}{6+6} = 3$

∴ $R_{AB} = 4\,\Omega$

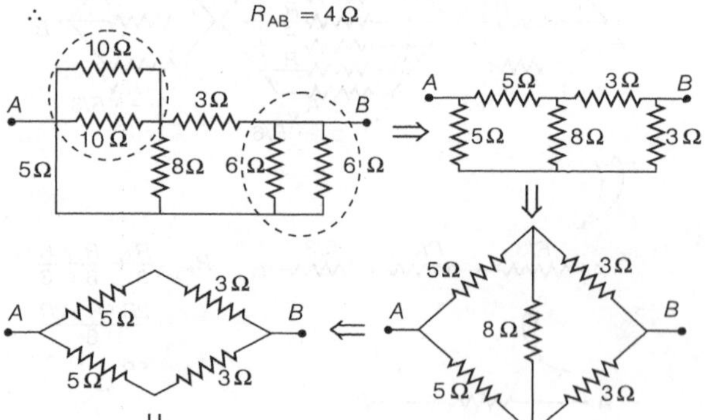

Balance Wheatstone's bridge

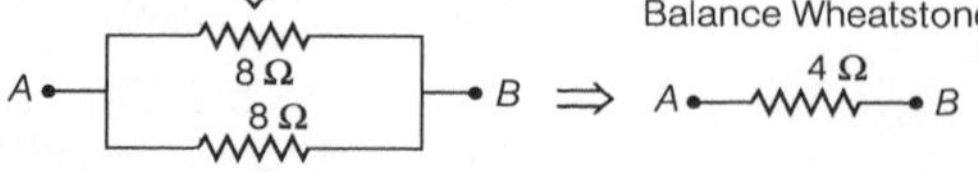

10. (c) Given circuit can be simplified as

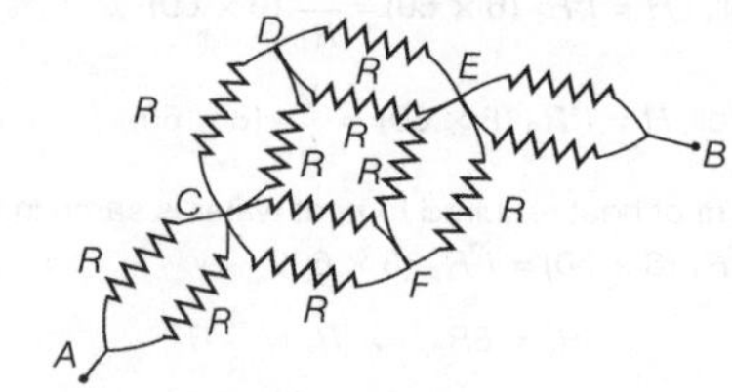

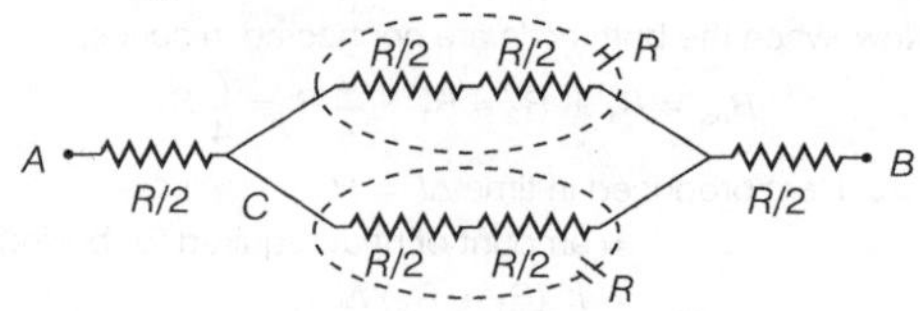

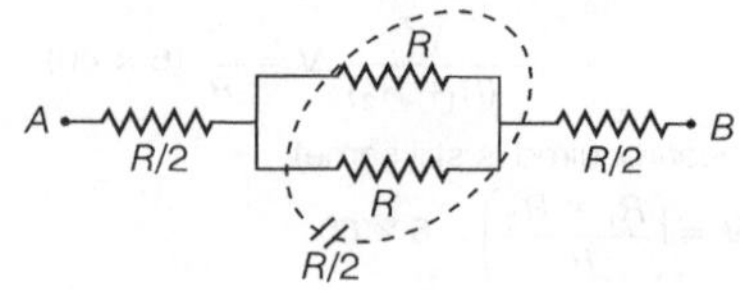

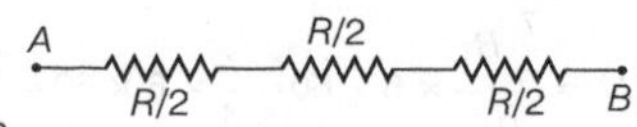

$\Rightarrow R_{AB} = \frac{3}{2}R$

11. (a)

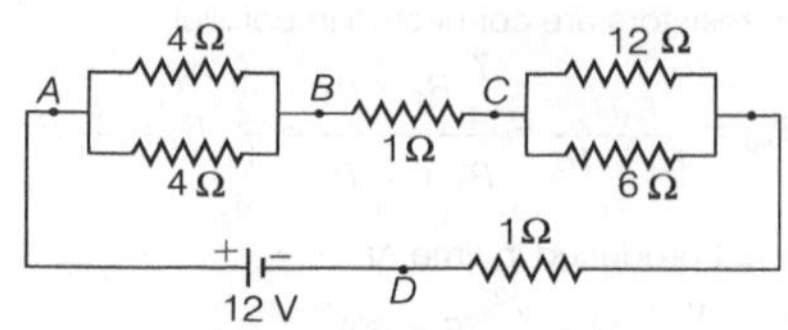

Total resistance of circuit

= resistance of two of 4 Ω resistance in parallel + resistance of 1 Ω + resistance of 12 Ω and 6 Ω in parallel + 1 Ω resistance

$$= 2 + 1 + \frac{12 \times 6}{12 + 6} + 1 = 8\,\Omega$$

So, current through cell

$$I = \frac{V}{R} = \frac{12}{8} = \frac{3}{2} \text{ A}$$

Now, $V_{AB} = I \times R_{AB} = \frac{3}{2} \times 2 = 3$ V

$V_{BC} = I \times R_{BC} = \frac{3}{2} \times 1 = \frac{3}{2}$ V

$V_{CD} = I \times R_{CD} = \frac{3 \times 5}{2} = \frac{15}{2} = 7.5$ V

12. (a) Given circuit is

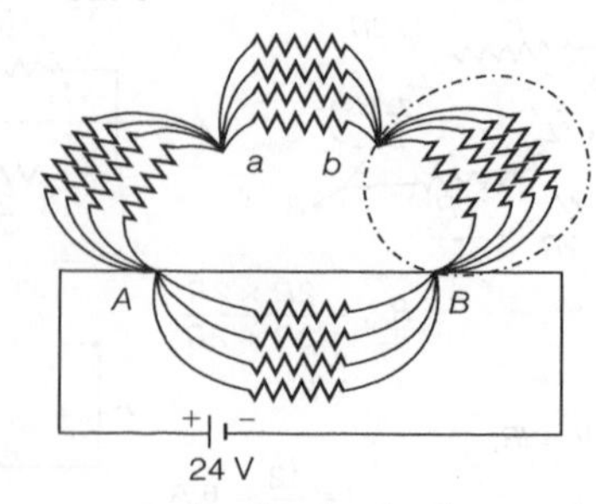

$$\frac{1}{R} = \frac{1}{5} + \frac{1}{5} + \frac{1}{5} + \frac{1}{5}$$

$$R = \frac{5}{4}\,\Omega$$

So, circuit can be redrawn as

$$R = 3\times\frac{5}{4} = \frac{15}{4}\text{ V}$$

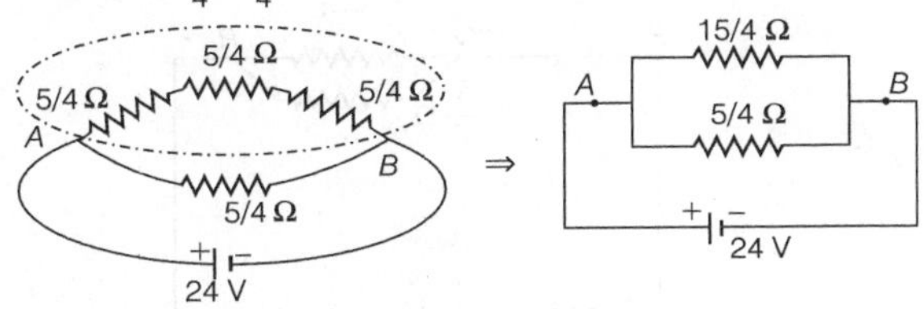

$$R = 3\times\frac{5}{4} = \frac{15}{4}\text{ V}$$

$$\Rightarrow \quad R_{\text{total}} = R_{AB} = \frac{\frac{15}{4}\times\frac{5}{4}}{\frac{15}{4}+\frac{5}{4}} = \frac{15\times5\times4}{20\times16} = \frac{15}{4\times4} = \frac{15}{16}\,\Omega$$

Now, as $\quad V_{AB} = 24\text{ V}$

Current through arm of resistance $\frac{15}{4}\,\Omega$

$$= I_1 = \frac{V_{AB}}{R} = \frac{24}{15/4} = \frac{24\times4}{15}\text{ A}$$

Now, $\quad I' = \frac{1}{5}\text{ of } I_1 = \frac{1}{5}\times\frac{24\times4}{15} = \frac{96}{75}\text{ A}$

13. (b) Circuit can be redrawn as

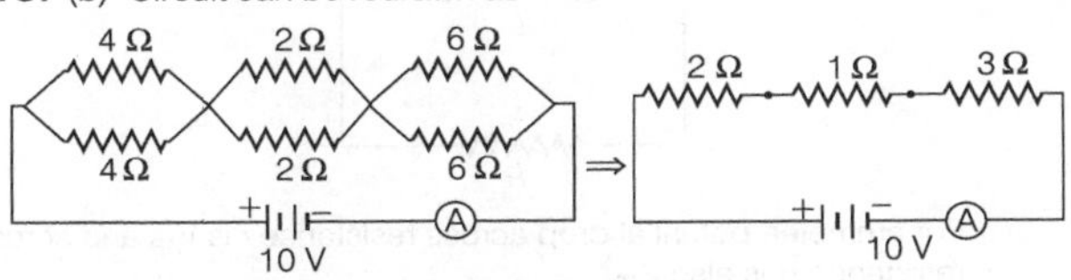

Clearly, $\quad R_{\text{eq}} = 2+1+3 = 6\,\Omega$

So, ammeter reading $I = \frac{V}{R} = \frac{10}{6} = \frac{5}{3}\text{ A}$

14. (c) Current I can be independent of R_6 only when R_1, R_2, R_3, R_4 and R_6 form a balanced Wheatstone's bridge

Therefore, $\quad \frac{R_1}{R_2} = \frac{R_3}{R_4}$ or $R_1R_4 = R_2R_3$

15. (a) The circuit can be redrawn as follows

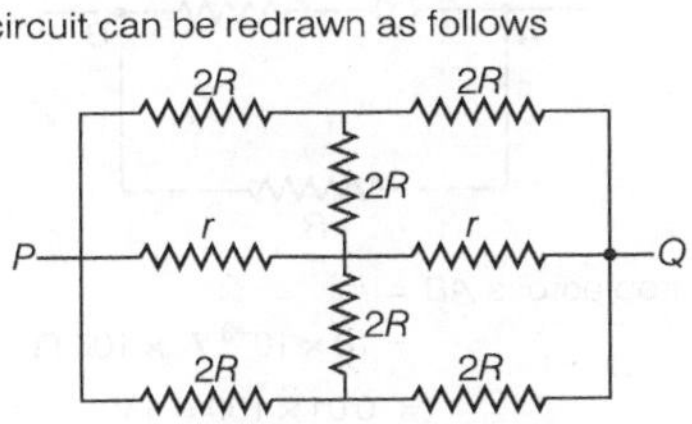

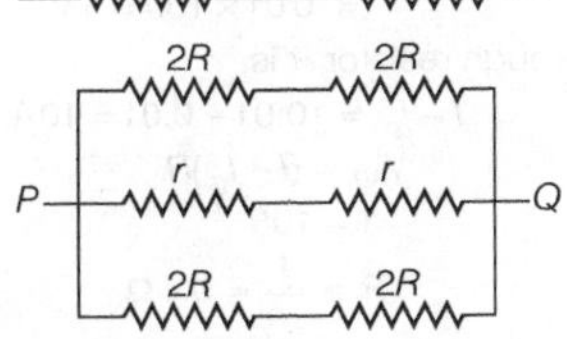

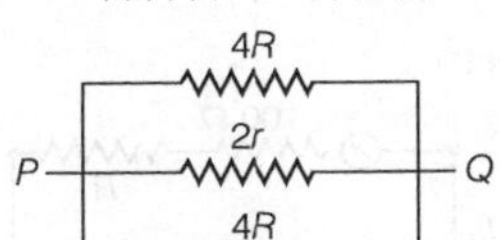

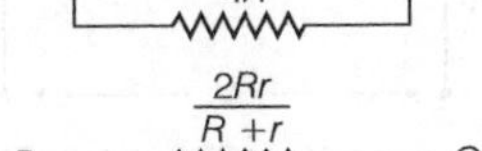

Thus, effective resistance between P and Q is $\frac{2Rr}{R+r}$.

16. (a,d) Given circuit can be redrawn as

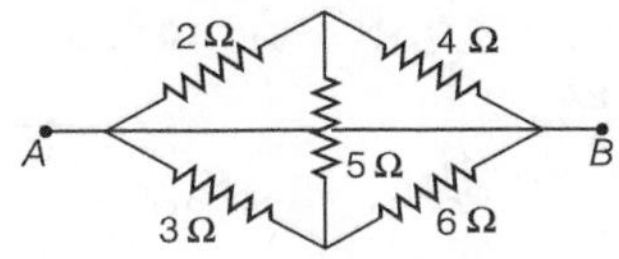

and as $\quad \frac{2}{3} = \frac{4}{6}$

So, as this is a case of Wheatstone's bridge.

So, $5\,\Omega$ resistor does not participate in the circuit,

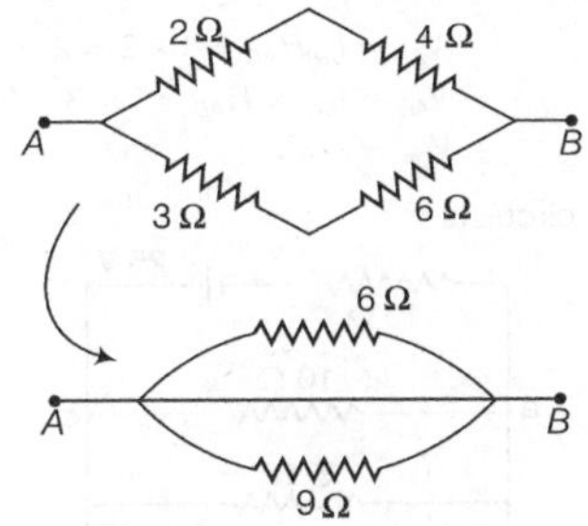

$$\Rightarrow \quad R_{AB} = \frac{9\times6}{9+6} = \frac{9\times6}{15} = \frac{18}{5}\,\Omega$$

17. (b,d) We can simplify the given circuit as

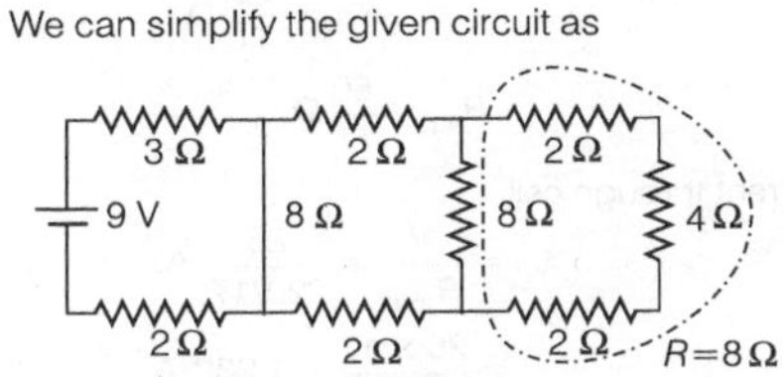

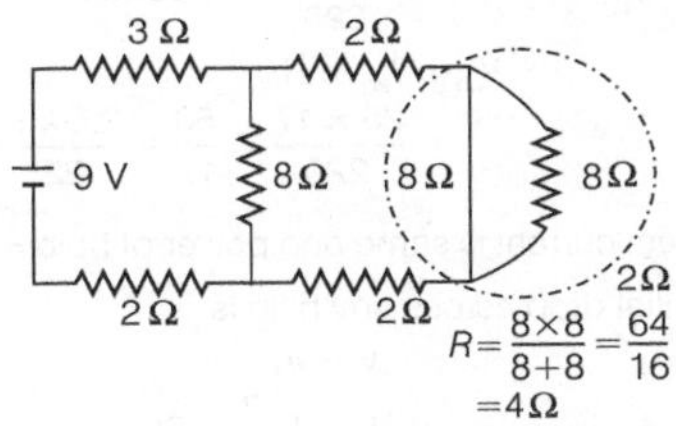

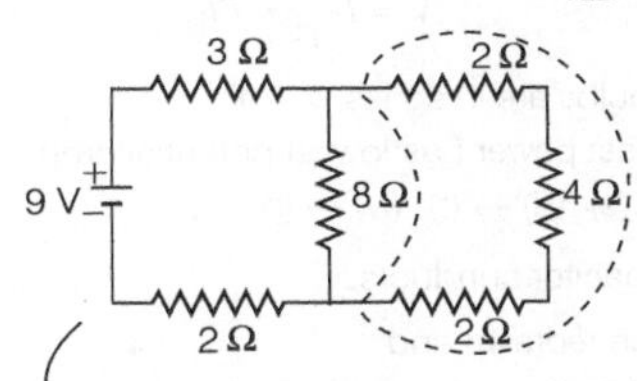

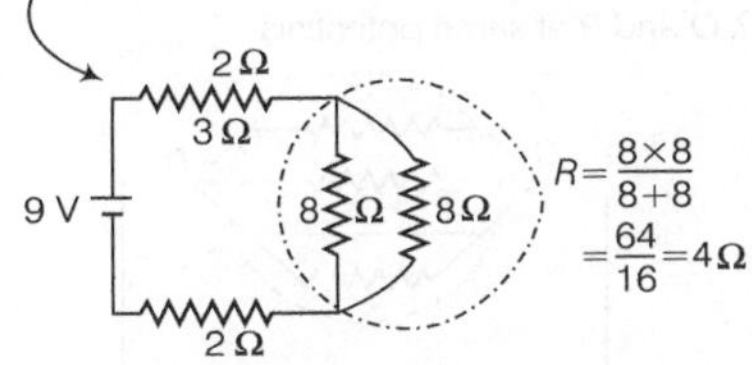

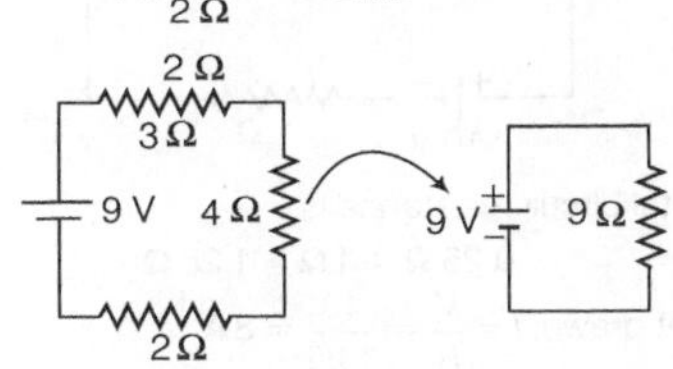

So, current through cell $= I = \frac{V}{R} \Rightarrow I = \frac{9}{9} = 1\text{ A}$

18. (a,b)

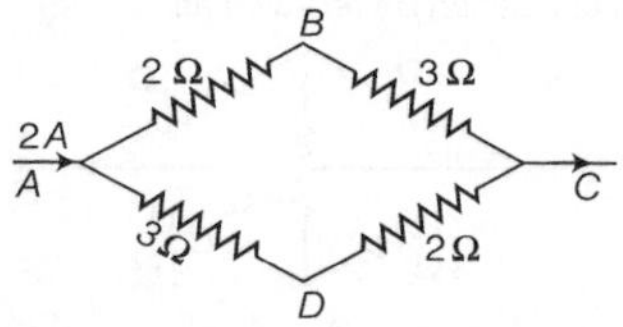

As, resistance of branch ABC and ADC are equal.

$\therefore$ Same current flows through both branches and

$$I_{ABC} = I_{ADC} = \frac{2}{2} = 1\text{ A}$$

$\therefore$ $$V_{AB} = I_{AB}R_{AB} = 1 \times 2 = 2\text{ V}$$

and $$V_{AD} = I_{AD} \times R_{AD} = 1 \times 3 = 3\text{ V}$$

$\Rightarrow$ $$V_{DA} = -3\text{V}$$

19. (a,c) Given circuit is

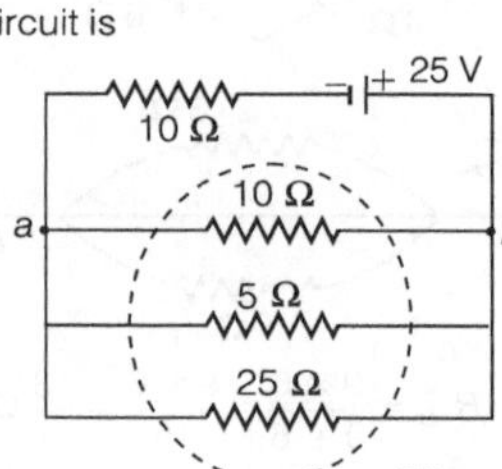

$R_{eq} = \frac{50}{17}\ \Omega$

$$R_{eq} = \frac{50}{17}\ \Omega$$

So, current through cell

$$I = \frac{V}{R_{total}} = \frac{25}{220/17}\text{ A}$$

$$= \frac{25 \times 17}{220} = 1.932\text{ A}$$

Also, $$V_{ab} = I_{ab} \times R_{ab}$$

$$= \frac{25 \times 17}{220} \times \frac{50}{17} = \frac{2.5 \times 5}{22} = 5.68\text{ V}$$

20. (b) In series, current is same and power of bulb $= I^2R = P$

and potential drop across any bulb is

$$V = IR$$

$\Rightarrow$ $$V = I \cdot \frac{P}{I^2} = PI$$

Clearly, when bulbs are in series, $V \propto P$.

So, bulb of lowest power has lowest potential drop.

(i) $\rightarrow$ (q), (ii) $\rightarrow$ (s), (iii) $\rightarrow$ (r), (iv) $\rightarrow$ (p)

21. (b) At $t = 0$, capacitor conducts.

$\therefore$ Circuit can be redrawn and

points P, Q and R at same potentials.

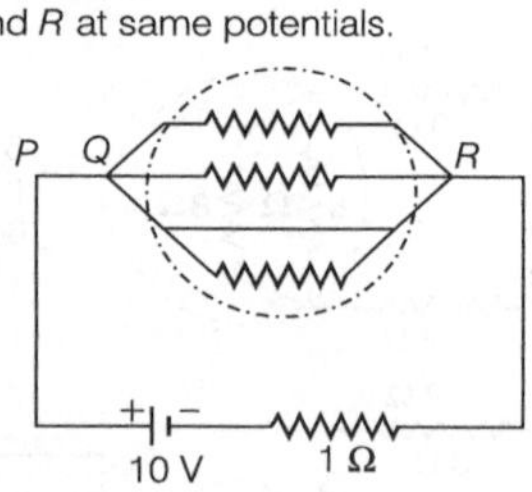

Clearly, total circuit resistance is

$$0.25\ \Omega + 1\ \Omega = 1.25\ \Omega$$

So, current drawn, $I = \frac{V}{R} = \frac{10}{1.25} = 8\text{A}$

22. (c) When steady state is reached capacitor acts as an open switch. So, circuit is equivalent to

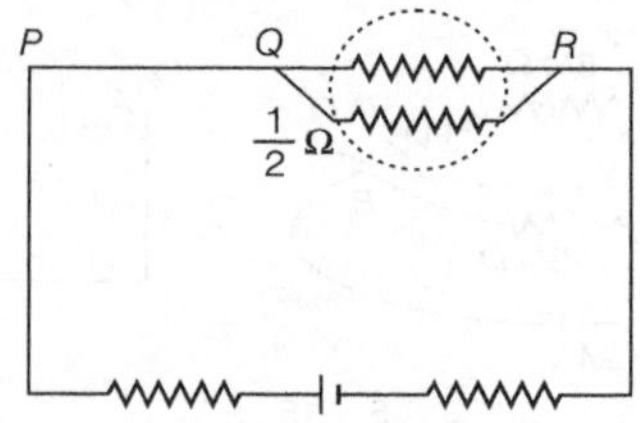

So, $$R_{eq} = 2.5\ \Omega$$

$$I = \frac{V}{R} = \frac{10}{2.5} = 4\text{ A}$$

23. (a) In steady state,

$$V_{AB} = 4\text{ A} \times 1\ \Omega = 4\text{ V}$$

Charge on capacitor, $A = 0$

Charge on capacitor, $B = q_B = 4\ \mu\text{C}$

Charge on capacitor, $C = q_C = 4\ \mu\text{C}$

$$\Delta H = \text{work done by battery} - \text{energy stored}$$
$$= 8 \times 10 - 24 = 56\ \mu\text{J}$$

24. (d)

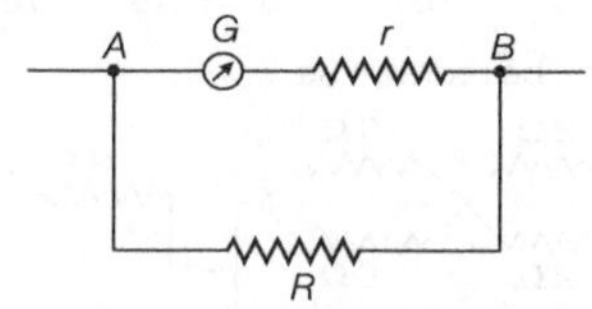

For ammeter, potential drop across resistance r is V_{AB} and across resistance R is also V_{AB}.

$\therefore V_r = V_R$ for ammeter or both are in parallel.

For voltmeter, r and R are in series and $R > r$. As, current is same in series,

$$V_R = iR \text{ and } v_r = ir$$

$\Rightarrow$ $$V_R > v_r$$

25. (a) An ammeter circuit is

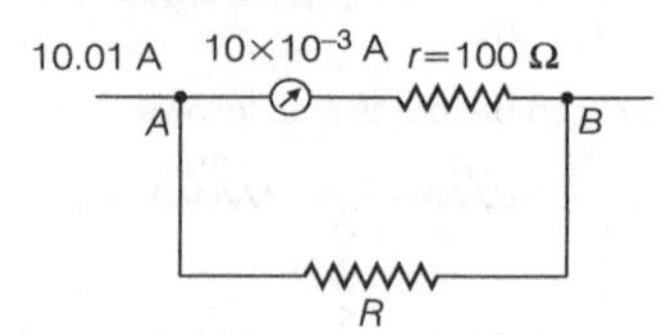

Potential drop across $AB = ir$

$$= 10 \times 10^{-3}\text{ A} \times 100\ \Omega$$

$$= 0.01 \times 100 = 1\text{V}$$

As, current through resistor R is,

$$I - I_g = 10.01 - 0.01 = 10A$$

So, $$V_{AB} = (I - I_g)R$$

$$1 = 10R$$

$$R = \frac{1}{10} = 0.1\ \Omega$$

26. (c) For a voltmeter,

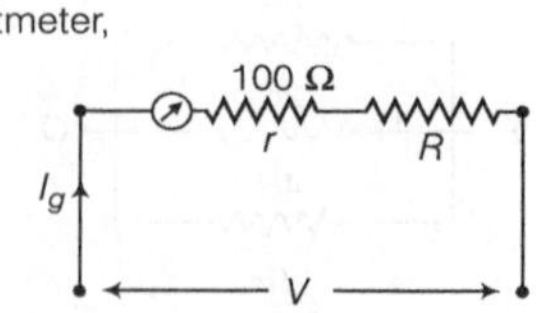

$$V = 100 = I_g(r + R), \quad \text{where } I_g = 0.01\text{ A}$$

So, $$R = 9900\ \Omega$$

27. (5) From $P = \frac{V^2}{R}$,

Resistance of heater $\quad R = \frac{V^2}{P} \quad \Rightarrow R = \frac{(100)^2}{1000} = 10\,\Omega$

From power, $P = i^2R$,

Current across heater for a power output of 62.5 W is

$$i = \sqrt{\frac{P}{R}} = \sqrt{\frac{62.5}{10}} = 2.5\,\text{A}$$

Mains current in circuit is

$$I = \frac{100}{10 + \frac{10R}{10+R}} = \frac{10(10+R)}{10+2R}$$

This current will distribute in inverse ratio of resistance between heater and R

$$\therefore \quad i = \left(\frac{R}{10+R}\right) I$$

$$\Rightarrow \quad 2.5 = \left(\frac{R}{10+R}\right)\left[\frac{10(10+R)}{10+2R}\right]$$

By solving, we get $\quad R = 5\,\Omega$

28. (5) For an ammeter, a parallel resistance is connected as shown in figure

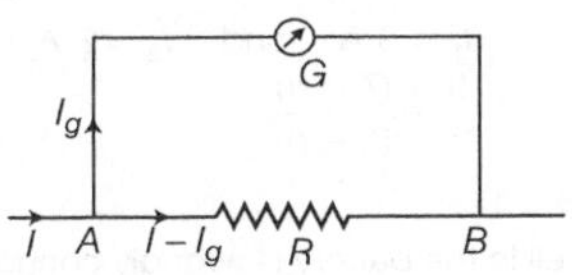

Also, $\quad V_{AB} = I_g G = (I - I_g)R$

Now, when ammeter is OP,

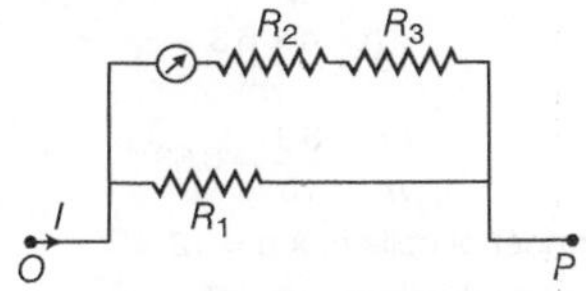

$$I_g(R_2 + R_3 + G) = (I - I_g)R_1$$

$$\Rightarrow \quad I_g(R_2 + R_3 + 25) = (10 - I_g)R_1 \quad \ldots\text{(i)}$$

When ammeter is OQ,

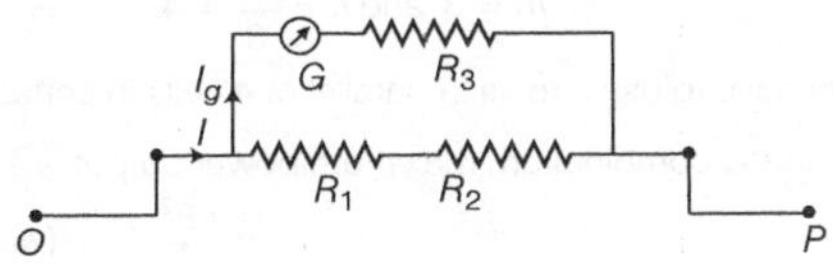

$$I_g(R_3 + 25) = (1 - I_g)(R_1 + R_2) \quad \ldots\text{(ii)}$$

and when ammeter is OR,

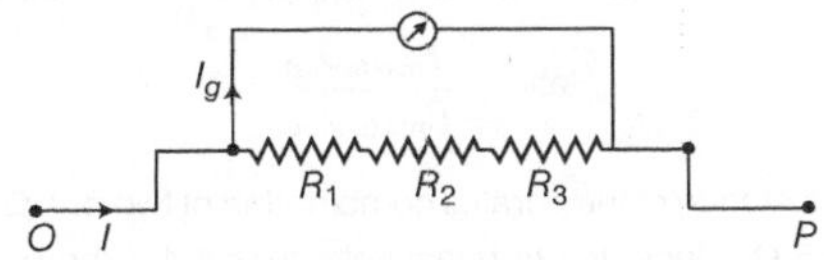

$$I_g(25) = (0.1 - I_g)(R_1 + R_2 + R_3) \quad \ldots\text{(iii)}$$

Solving for R_3 using Eqs.(i), (ii) and (iii), we get

$$2R_3 = 5\,\Omega$$

DPP- 5 Cells and Combination of Cells

1. (i) Let I current flows in the circuit,

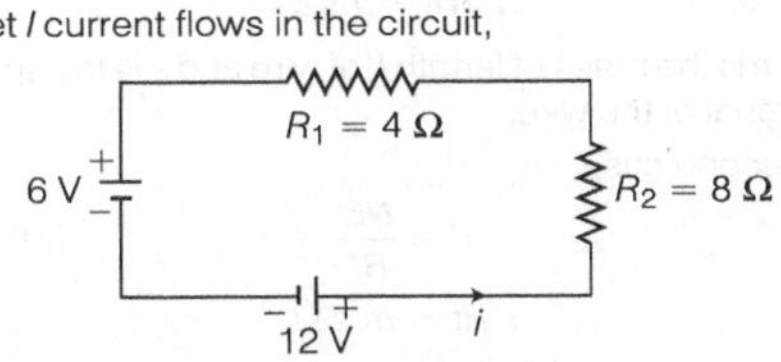

As batteries are oppositely placed, net emf of circuit is $12 - 6 = 6$ V

Also, equivalent resistance of circuit

$$= 4 + 8 = 12\,\Omega$$

So, current in circuit is $\quad I = \frac{V}{R} = \frac{6}{12}\,\text{A} = 0.5\,\text{A}$

(ii) Power dissipated in the resistor R_1 is

$$P_1 = I^2R_1$$
$$= (0.5)^2(4) = 1\,\text{W}$$

Power dissipated in the resistor R_2 is

$$P_2 = I^2R_2$$
$$= (0.5)^2(8) = 2\,\text{W}$$

(iii) If I current flows through a battery of emf E, rate of energy supplied by battery is

$$P = IE$$

Now, for battery of 6 V,

$$\text{Power} = EI = 0.5 \times 6 = 3\,\text{W}$$

and for battery of 12 V,

$$\text{Power} = EI = 0.5 \times 12 = 6\,\text{W}$$

By conservation of energy, power output of 12 V battery is used by 6 V battery and 2 resistors.

Hence, only 12 V battery is expensing power of 6 W in the circuit.

2. (i) During charging for a battery,

$$V = E + Ir \quad \text{(In discharging it is } E - Ir = V\text{)}$$

So, V = potential difference across terminals of cell

$$= 12 + 50 \times 0.04$$
$$= 12 + 2 = 14\,\text{V}$$

(ii) Part of energy supplied to cell is dissipated in the form of heat due to internal resistance of cell.

Rate of energy dissipation $= I^2r = (50)^2 \times (0.04)$

$$= 100\,\text{W}$$

(iii) Total energy supplied to battery by the source which is being used to charge the battery is

$$VI = 14 \times 50 = 700\,\text{W}$$

By using energy conservation.

Energy stored as chemical energy = Energy supplied − Energy converted into heat

$\Rightarrow$ Rate of increase of chemical energy $= 700 - 100$
$= 600\,\text{W}$

3. Maximum power (and hence current) is transferred to a load when load resistance is equal to the internal resistance of source.

For series grouping,

Battery resistance $= nr = 12 \times 6.3 = 75.6\,\Omega$

For parallel grouping,

Battery resistance $= r/m = \frac{6.3}{12} = 0.525\,\Omega$

So, these grouping cannot be provided. (as, $R_{ext} = 10\,\Omega$)

Hence, we have to go for a mixed grouping.

Let, n = number of cells in series in each row

and m = number of rows in parallel.

Then, internal resistance of battery so formed is $\frac{nr}{m}$.

Now, for maximum power output

$$R = \frac{nr}{m}$$

$$\Rightarrow \quad 10 = \frac{n \times 6.3}{m}$$

$$\Rightarrow \quad \frac{m}{n} = \frac{6.3}{10} = 0.63$$

Also, total number of cells $m \times n = 12$

Multiplying these equations, we get

$$m^2 = 12 \times 0.63 \approx 9 \quad \text{(approximately as } m \text{ cannot be a fraction)}$$

$$\Rightarrow \quad m = 3 \text{ and } n = \frac{12}{3} = 4.$$

So, we have to use 3 rows in parallel of 4 cells in series.

4. (d) In series combination, maximum power output

$$= \frac{nE^2}{4r} \quad (\text{when, } R = nr)$$

In parallel combination,

$$\text{Maximum power output} = \frac{nE^2}{4r} \quad \left(\text{when, } R = \frac{r}{n}\right)$$

$$\therefore \quad \text{Ratio} = \frac{P_{\max\,(\text{series})}}{P_{\max\,(\text{parallel})}} = 1$$

5. (d) Resistance of the parallel combination of two of 7 Ω resistors is $R = 3.5\ \Omega$. Since, the terminal voltage is 1.4 V the current in the circuit is $I = 1.4/3.5 = 0.4$ A. Therefore, current through each cell $= I/2 = 0.2$ A.

Let r be the internal resistance of each cell. The voltage drop across r is

$$1.5 - 1.4 = 0.1 \text{ V}$$

Therefore, $\quad 0.2\, r = 0.1$ or $r = 0.5\ \Omega$

6. (c) Since, the cells are connected in series, the total emf is $2E$. Therefore, the current in the circuit is

$$I = \frac{2E}{R + r_1 + r_2} \quad \ldots\text{(i)}$$

The potential difference across the cell of internal resistance r_1 is

$$E = Ir_1 = 0$$

i.e. $\quad E = Ir_1 \quad$ [using this in Eq. (i)]

$$I = \frac{2Ir_1}{R + r_1 + r_2}$$

which gives $\quad R = r_1 - r_2$

7. (b) Since, the two cells are in opposition, the effective voltage $6 - 4 = 2$ V. The current in the circuit is

$$I = \frac{2}{2 + 8} = 0.2 \text{ A}$$

∴ Terminal voltage of 6 V cell $= 6 - 2 \times 0.2 = 5.6$ V

Terminal voltage of 4 V cell $= 4 + 8 \times 0.2 = 5.6$ V

Therefore, the potential difference across terminals X and Y is 5.6 V.

8. (d) The emf of cells connected in reverse polarity cancel each other. Hence, cells marked 2, 3 and 4 together cancel the effect of cells marked 5, 6 and 7, the circuit reduces to that shown in the figure.

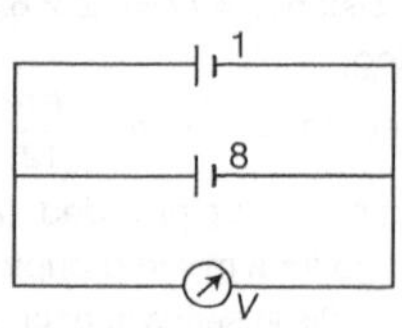

Now, cells 1 and 8 are in reverse polarity.

Hence, voltmeter reading $= 5 - 5 = 0$

9. (b) Since, the cells are in opposition, the effective emf $= 6 - 2 = 4$ V. Since, the current is taken to flow from the positive to the negative terminal of the battery, a current of $\frac{4}{5+3} = 0.5$ A flows from B to C.

10. (a) Let E be the emf of each cell and let n be the number of cells connected wrongly. Then, the net emf of the battery is equal to the emf of correctly connected cells – emf of wrongly connected cells.

$$= (12 - n)E - nE = (12 - 2n)E$$

When the two cells aid the battery, the net emf is

$$E_1 = (12 - 2n)E + 2E = (14 - 2n)E$$

If R is the resistance in the circuit, the current is

$$I_1 = \frac{E_1}{R} = \frac{(14 - 2n)E}{R} \quad \ldots\text{(i)}$$

When the two cells oppose the battery, the net emf is

$$E_2 = (12 - 2n)E - 2E = (10 - 2n)E \quad \ldots\text{(ii)}$$

Now, the current in the circuit is

$$I_2 = \frac{(10 - 2n)E}{R}$$

Dividing Eqs. (i) by Eq. (ii), we get

$$\frac{I_1}{I_2} = \frac{(14 - 2n)}{(10 - 2n)} = \frac{(7 - n)}{(5 - n)}$$

Given, $\quad I_1 = 3$ A and $I_2 = 2$ A

Thus, $\quad \frac{3}{2} = \frac{(7 - n)}{(5 - n)}$

Which gives $n = 1$.

Hence, one cell in the battery is wrongly connected.

11. (b) In the first case, $i = \frac{3E}{R}$

where, E is the emf of each cell and $R(\propto L)$ is the resistance of the wire.

Also, $\quad i^2RT = ms\Delta T \quad \ldots$(i)

where, m is the mass of length L of wire and s is the specific heat of the material of the wire.

In the second case,

$$i' = \frac{NE}{R'} \quad (\text{where, } R' \propto 2L)$$

$$i^2Rt = m's\Delta T$$

Dividing Eq. (ii) by Eq. (i), we get

$$\left(\frac{N^2}{9}\right)\cdot\left(\frac{R}{R'}\right) = \left(\frac{m'}{m}\right) \Rightarrow \left(\frac{N^2}{9}\right)\cdot\left(\frac{1}{2}\right) = 2$$

$$N = 6$$

12. (d) $V_{AB} = 4 = \frac{5x + 2 \times 10}{x + 10} = 20\ \Omega$

13. (b) Given circuit can be redrawn as

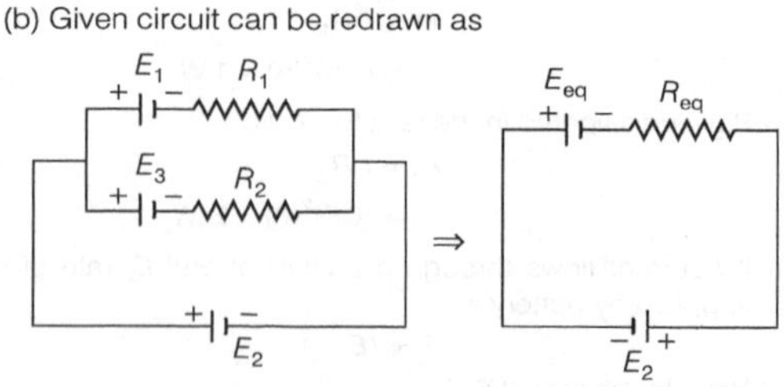

$$\because \quad E_{eq} = \frac{E_1R_2 + E_2R_2}{R_1 + R_2} = \frac{2 \times 4 + 2 \times 4}{4 + 4} = 2 \text{ V}$$

and $\quad R_{eq} = \frac{4}{2} = 2\ \Omega$

∴ Current, $i = \frac{2 + 2}{2} = 2$ A

This current is from A to B.

14. (c) Strength of a cell is total charge capacity

$= I \times t = 5\ A \times 18\ h = 90\ Ah$

15. (a,c,d) The emf of the combination is E and not E/n.

16. (a,b) Let m be the number of cells in each row and let the number of rows connected in parallel be n. Then, $mn = 24$. Since, the internal resistance of each cell is 2 Ω, the resistance of each row $= 2m\Omega$

The total resistance of n such rows connected in parallel $= \dfrac{2m}{n}\ \Omega$

The current through the external resistance of the cells = external resistance, i.e. if

$$\frac{2m}{n} = 12 \quad \text{or} \quad m = 6n$$

But $\quad mn = 24$

$\therefore \quad m = 12$ and $n = 2$

$$\text{Current in the circuit} = \frac{\text{Total emf}}{\text{Total resistance}} = \frac{1.5 \times 12}{12 + 12} = 0.75\ A$$

As, the two rows have the same resistance, then current in each row 0.375 A = current in each cell.

17. (a,b) $\dfrac{1}{R_p} = \dfrac{1}{R_1} + \dfrac{1}{R_2} + \cdots\cdots + \dfrac{1}{R_{min}} + \dfrac{1}{R_n}$

$$\therefore \quad \frac{R_{min}}{R_p} = \frac{R_{min}}{R_1} + \frac{R_{min}}{R_2} + \cdots\cdots + 1 + \frac{R_{min}}{R_n}$$

$$\therefore \quad \frac{R_{min}}{R_p} > 1$$

$\Rightarrow \quad R_{min} > R_p$ or $R_p < R_{min}$

Also, $\quad R_s = R_1 + R_2 + \cdots\cdots + R_{max} + R_n$

So, $\quad R_s \geq R_{max}$ {equality exists when $R_1 + R_2 \cdots R_n = 0$}

So, $\quad R_s > R_{max}$

18. (a,c) Current in circuit, $I = \dfrac{6}{6} = 1\ A$

$$I = nev_d\ A$$

$$\Rightarrow \quad v_d = \frac{I}{neA} = \frac{1}{10^{29} \times 1.6 \times 10^{-19} \times 10^{-6}} = \frac{1}{1.6} \times 10^{-4}\ m/s$$

$$\text{KE of electrons} = \frac{1}{2} mv^2 dnAl$$

$$= \frac{1}{2} \times 9.1 \times 10^{-31} \times \frac{1}{2.56} \times 10^{-8} \times 10^{29} \times 10^{-6} \times 10^{-1}$$

$$= 2 \times 10^{-17}\ J$$

Ohmic loss, $I^2R = 6 \times 1^2 = 6\ J/s$

$\therefore$ Time to loss all kinetic energy

$$= \frac{2 \times 10^{-17}}{6} = \frac{20}{6} \times 10^{-18} = 2 \times \frac{10}{6} \times 10^{-18} = 3.4 \times 10^{-18}\ s$$

19. (10) $I = \dfrac{E}{R + nR}$ and $\dfrac{E}{R + \dfrac{R}{n}} = 10\ I$

$$\Rightarrow \quad \frac{1+n}{1+\dfrac{1}{n}} = 10 = \frac{1+n}{n+1} \cdot n = n$$

$$\therefore \quad n = 10$$

20. (2) $I = \dfrac{E + E}{R + r_1 + r_2} = \dfrac{2E}{R + r_1 + r_2}$

$$V_1 = E - Ir_1 = E - \frac{2E}{r_1 + r_2 + R} \cdot r_1$$

Given, $\quad V_1 = 0$

$$\Rightarrow \quad E = \frac{2Er_1}{r_1 + r_2 + R}$$

$$\Rightarrow \quad r_1 + r_2 + R = 2r_1$$

$$\Rightarrow \quad R = r_1 - r_2 = 3 - 1 = 2\ \Omega$$

Explanations for Q. Nos. (21-24)

Equivalent emf of cells is

$$E_{eq} = \frac{\Sigma(E/r)}{\Sigma(1/r)} = \frac{3/1 + 2/1 + 1/1}{1/1 + 1/1 + 1/1} = 2\ V$$

Equivalent circuit is

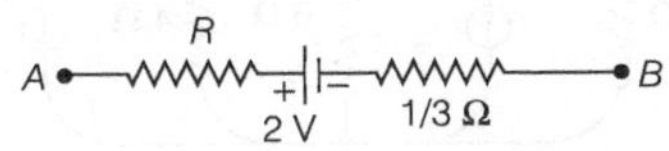

By using $V = E - ir$,

$$i_1 = \frac{V_B - V_A + E_1}{r_1} = \frac{-2 + 3}{1} = 1\ A$$

$$i_2 = \frac{V_B - V_A + E_2}{r_2} = 0$$

$$i_3 = \frac{V_B - V_A + E_3}{r_3} = -1\ A$$

When r_2 is short-circuited and A and B are connected,

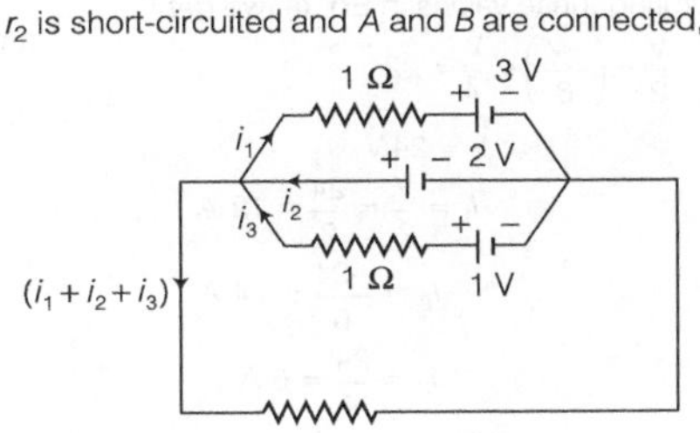

By KVL in 3 loops,

Then,

$$3 - i_1 - (i_1 + i_2 + i_3) = 0 \quad \text{...(i)}$$

$$2 - (i_1 + i_2 + i_3) = 0 \quad \text{...(ii)}$$

$$1 - i_3 - (i_1 + i_2 + i_3) = 0 \quad \text{...(iii)}$$

$$i_2 = 2\ A$$

21. (a) **22.** (b)

23. (c) **24.** (d)

25. (b) Power transferred to load is maximum when, $r = R$

(i) → (p), (ii) → (q), (iii) → (s), (iv) → (r)

DPP-6 Network Analysis and Kirchhoff's Rules

1. (i) Power dissipated in 5 Ω resistor is 20 W.

$$\therefore \quad \frac{V^2_{DC}}{5} = 20$$

$$\Rightarrow \quad V_{DC} = 10 \text{ V}$$

Applying KVL to loop *acdba*, we have

$$V_{ac} + V_{cd} + V_{db} + V_{ba} = 0$$

$$0 + 10 + 0 + V_{ba} = 0$$

$$\Rightarrow \quad V_{ba} = -10 \text{ V}$$

An analog meter shows 10 V and a digital meter shows −10 V.

(ii) Applying KVL to loop *cefge*, we have

$$-V_{ce} + V_{ef} - V_{fg} - V_{ge} = 0$$

$$-17 \times 2 + 90 - 6 \times 2 - V_{gc} = 0$$

$$\Rightarrow \quad V_{gc} = 44 \text{ V}$$

2. We assign direction of current in each branch as

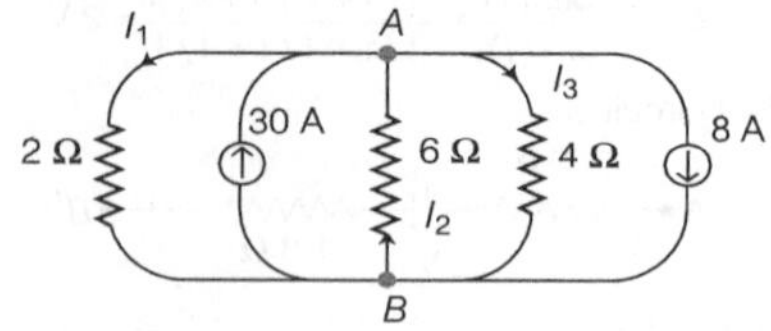

At junction *A* by KCL,

$$\Sigma I = 0$$

$$\Rightarrow \quad I_2 + 30 = I_1 + I_3 + 8$$

$$\Rightarrow \quad I_1 - I_2 + I_3 = 22 \qquad \ldots \text{(i)}$$

By Ohm's law, $\quad I_1 = \frac{V}{2}, I_3 = \frac{V}{4}, I_2 = \frac{-V}{6}$

Substituting these values in Eq. (i), we get

$$\frac{V}{2} - \left(\frac{-V}{6}\right) + \frac{V}{4} = 22$$

or $\quad V = 24 \text{ V}$

$$\therefore \quad I_1 = \frac{V}{2} = \frac{24}{2} = 12 \text{ A}$$

$$I_2 = \frac{-24}{6} = -4 \text{ A}$$

and $\quad I_3 = \frac{24}{4} = 6 \text{ A}$

Negative sign of current means that it flows in opposite sense.

3. Sign convention for junction rule

Current entering the junction is taken as positive while leaving the junction is taken negative.

or

Current entering the junction is taken as negative while leaving the junction is taken positive.

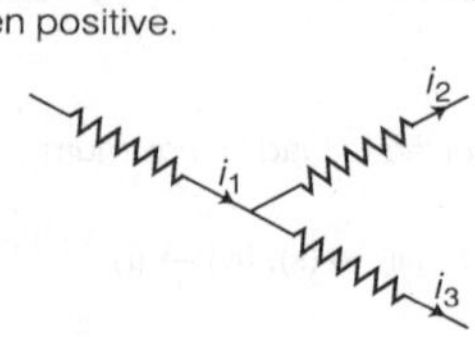

$$i_1 - i_2 - i_3 = 0$$

Sign convention for loop rule

For loop rule, one of the best method is to assign the + ve and − ve signs to every component and then the sign of each component which occures later in the path is taken, e.g., in given circuit,

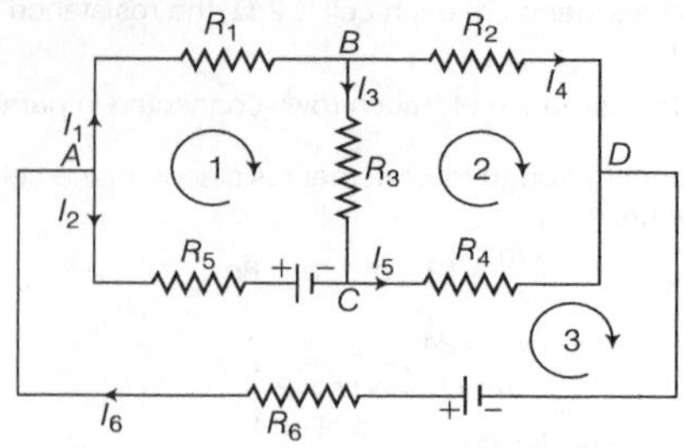

In loop 1 *ABCA*,

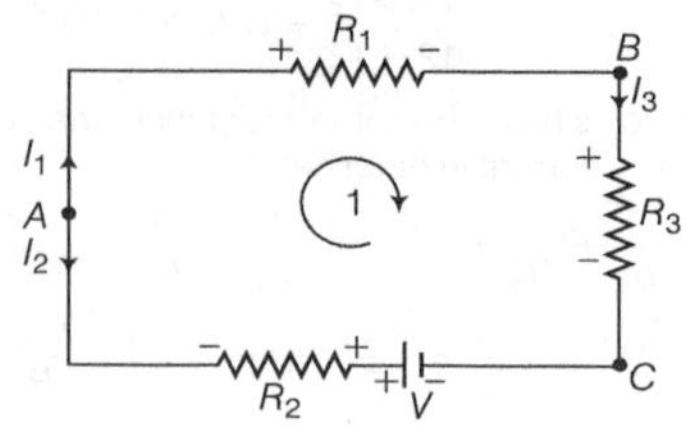

Step 1 Assign current to each branch.

Step 2 Assign signs + ve and − ve keeping in mind direction of flow.

Step 3 Put $\Sigma V = 0$, taking sign of potential drop or later sign occuring across each component.

In loop 1, $\quad \Sigma V = 0$

$$-I_1R_1 - R_3I_3 + V - I_2R_2 = 0$$

or $\quad V = I_1R_1 + I_2R_2 + I_3R_3$

The following gives a better aid to remember.

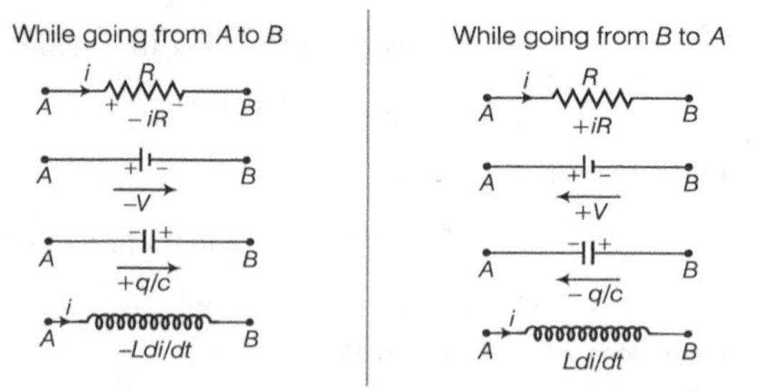

4. (i) $E_{eq} = \frac{\Sigma(E/r)}{\Sigma(1/r)} = \frac{(3/1 + 2/1 + 1/1)}{(1/1 + 1/1 + 1/1)} = 2 \text{ V}$

Further r_1, r_2 and r_3 each are of 1 Ω. Therefore, internal resistance of the equivalent battery will be $\frac{1}{3}$ Ω as all three are in parallel.

The equivalent circuit is therefore shown in the figure.

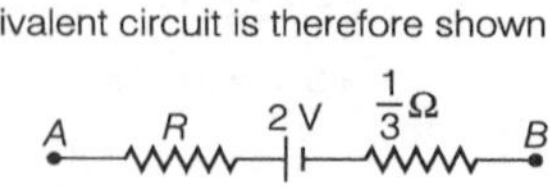

Since, no current is taken from the battery.

$$V_{AB} = 2\text{ V (From } V = E - ir)$$

Further, $V_{AB} + V_A - V_B = E_1 - i_1 r_1$

$\therefore$ $i_1 = \dfrac{V_B - V_A + E_1}{r_1} = \dfrac{-2+3}{1} = 1\text{ A}$

Similarly, $i_2 = \dfrac{V_B - V_A + E_2}{r_2} = \dfrac{-2+2}{1} = 0\text{ A}$

and $i_3 = \dfrac{V_B - V_A + E_3}{r_3} = \dfrac{-2+1}{1} = -1\text{ A}$

(ii) r_2 is short-circuited that means the resistance of this branch becomes zero. Making a closed circuit with a battery and resistance R. Applying Kirchhoff's second law in three loops so formed.

$$3 - i_1 - (i_1 + i_2 + i_3) = 0 \quad \text{...(i)}$$

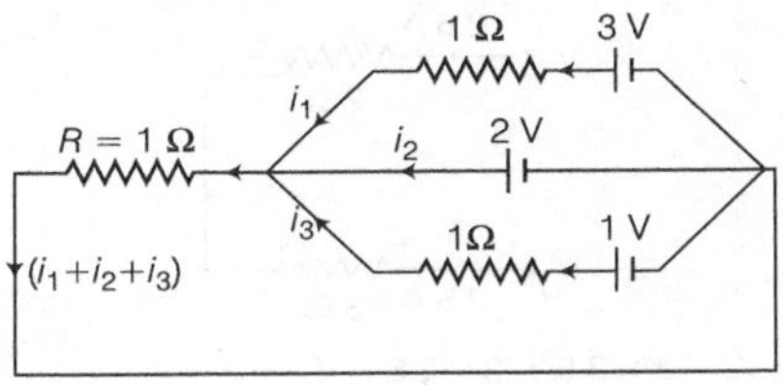

$$2 - (i_1 + i_2 + i_3) = 0 \quad \text{...(ii)}$$

$$1 - i_3 - (i_1 + i_2 + i_3) = 0 \quad \text{...(iii)}$$

From Eq. (ii), we get $i_1 + i_2 + i_3 = 2\text{ A}$

$\therefore$ Substituting in Eq. (i), we get $i_1 = 1\text{ A}$

Substituting in Eq. (iii), we get $i_3 = -1\text{ A}$

$\therefore$ $i_2 = 2\text{ A}$

5. Let at any time t charge on capacitor C be Q and currents are as shown. Since, charge Q will be increased with time t. Therefore,

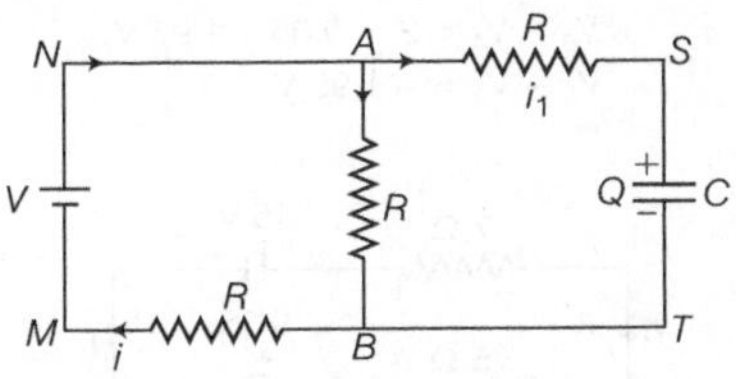

(i) Applying Kirchhoff's second law in loop $MNABM$,

$$V = (i - i_1)R + iR$$

or $$V = 2iR - i_1 R \quad \text{...(i)}$$

Similarly, applying Kirchhoff's second law in loop $MNSTM$, we have

$$V = i_1 R + \frac{Q}{C} + iR \quad \text{...(ii)}$$

Eliminating i from Eqs. (i) and (ii), we get

$$V = 3i_1 R + \frac{2Q}{C} \quad \text{or} \quad 3i_1 R = V - \frac{2Q}{C}$$

or $$i_1 = \frac{1}{3R}\left(V - \frac{2Q}{C}\right) \quad \text{or} \quad \frac{dQ}{dt} = \frac{1}{3R}\left(V - \frac{2Q}{C}\right)$$

or $$\frac{dQ}{V - \frac{2Q}{C}} = \frac{dt}{3R} \quad \text{or} \quad \int_0^Q \frac{dQ}{V - \frac{2Q}{C}} = \int_0^t \frac{dt}{3R}$$

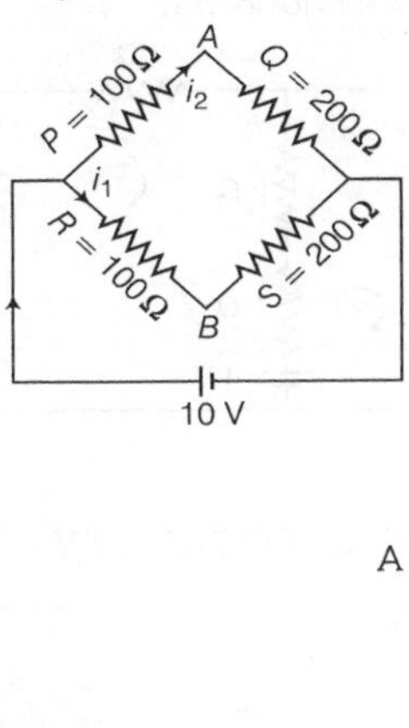

This equation gives

$$Q = \frac{CV}{2}(1 - e^{-2t/3RC})$$

(ii) $$i_1 = \frac{dQ}{dt} = \frac{V}{3R} e^{-2t/3RC}$$

From Eq. (i), $i = \dfrac{V + i_1 R}{2R} = \dfrac{V + \frac{V}{3} e^{-2t/3RC}}{2R}$

$\therefore$ Current through AB,

$$i_2 = i - i_1 = \frac{V + \frac{V}{3} e^{-2t/3RC}}{2R} - \frac{V}{3R} e^{-2t/3RC}$$

$$i_2 = \frac{V}{2R} - \frac{V}{6R} e^{-2t/3RC}$$

$$i_2 = \frac{V}{2R} \quad \text{as } t \to \infty$$

6. (a) At junction A, $i_{AB} = 2 + 2 = 4\text{ A}$

At junction B, $i_{BC} = i_{AB} - 1 \Rightarrow i_{BC} = 4 - 1 = 3\text{ A}$

At junction C, $i = i_{BC} - 1.3 = 3 - 1.3 = 1.7\text{ A}$

7. (b) $i_{CD} = i_{BC} + i_{AC} = i_2 + i_3$

8. (a) As per marked currents, by Kirchhoff's junction rule,

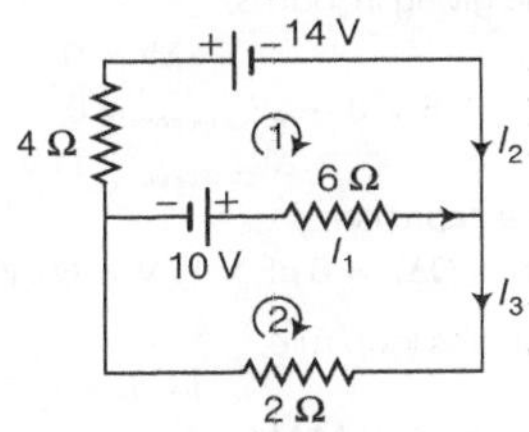

$$I_1 + I_2 = I_3 \quad \text{...(i)}$$

and by Kirchhoff's loop rule in loop 1,

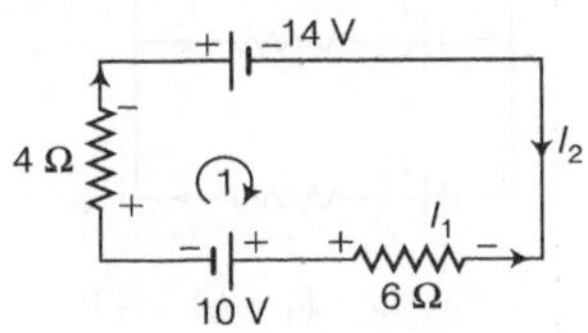

$$\Sigma V = 0$$

$\Rightarrow$ $$-14 + 6I_1 - 10 - 4I_2 = 0$$

$\Rightarrow$ $$-24 + 6I_1 - 4I_2 = 0 \quad \text{...(ii)}$$

In loop 2, $$\Sigma V = 0$$

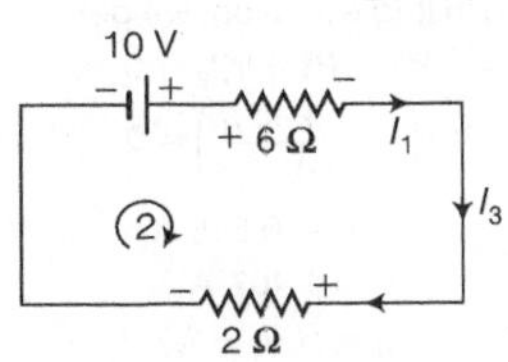

$\Rightarrow$ $$10 - 6I_1 - 2I_3 = 0 \quad \text{...(iii)}$$

From Eqs. (i) and (iii), we have

$$10 - 6I_1 - 2(I_1 + I_2) = 0$$

$\Rightarrow$ $$10 - 8I_1 - 2I_2 = 0$$

$\Rightarrow$ $$5 - 4I_1 - I_2 = 0 \quad \text{...(iv)}$$

Solving Eqs. (iv) and (ii) simultaneously, we get

$$I_1 = 2\text{ A and } I_2 = -3\text{ A}$$

($\therefore$ current is flowing in the opposite direction)

and substituting these values in Eq. (i), we get

$$I_3 = -1.0\text{ A}$$

(So, I_3 is also flowing in opposite direction to be assumed 1).

9. (c) Current distribution in the given circuit is taken as

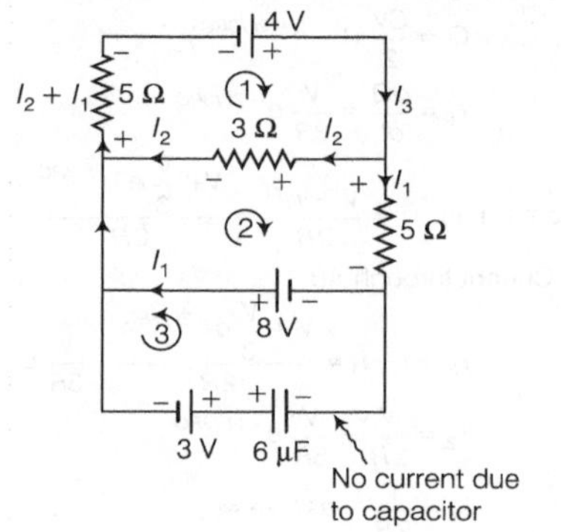

Using equations of Kirchhoff's law are given as below

$$I_1 + I_2 = I_3$$

$$4 - 3I_2 - 5I_3 = 0 \quad \text{(loop 1)}$$

$$3I_2 - 5I_1 + 8 = 0 \quad \text{(loop 2)}$$

On solving these equations, we get

$$I_2 = -0.364\text{ A},\quad I_1 = 1.38\text{ A},\quad I_3 = 1.02\text{ A}$$

Now, clockwise giving in loop 3,

$$\Sigma V = 0$$

$$\Rightarrow \quad 8 + 3 - \Delta V_{\text{capacitor}} = 0$$

$$\Rightarrow \quad \Delta V_{\text{capacitor}} = 11\text{ V}$$

So, charge on a capacitor

$$C = Q\Delta V = 6\,\mu\text{F} \times 11\text{ V} = 66\,\mu\text{C}$$

10. (b) From Kirchhoff's loop rule,

$$I = I_1 + I_2$$

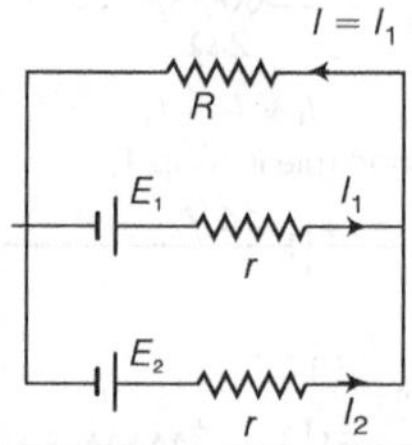

$$E_1 - I_1r - (I_1 + I_2)R = 0 \quad \text{...(i)}$$

and

$$E_2 - I_2r - (I_1 + I_2)R = 0 \quad \text{...(ii)}$$

On solving Ist equation for I_2, we get

$$I_2 = \frac{E_1 - (R + r)I_1}{R} = \frac{2 - 4.45\,I_1}{4}$$

and substituting in IInd equation, we get

$$E_2 - RI_1 - (R + r)I_2 = 0 \quad \text{...(iii)}$$

$$\Rightarrow 3 - 4I_1 - (4.45)\left(\frac{2 - 4.45\,I_1}{4}\right) = 0$$

$$\Rightarrow \quad I_1 = -0.815\text{ A}$$

$$I_2 = 1.407\text{ A}$$

$$\Rightarrow \quad V_R = R(I_1 + I_2)$$

$$= 4(-0.815 + 1.407) = 2.37\text{ V}$$

11. (d) From Kirchhoff's loop rule,

$$-I(2) + 18 - I(6.6) - 12 - I(1) = 0$$

$$\Rightarrow \quad I = \frac{6}{9.6}\text{ A}$$

Terminal voltage of 18 V battery is

$$V = E - Ir_1 = 18 - I \times 2 = 18 - 0.625 \times 2$$

$$= 16.75 \approx 17\text{ V}$$

For 12 V battery,

$$V = E + Ir_2 = 12 + 1 \times 0.625$$

$$= 12.625 \approx 13\text{ V}$$

12. (d) $R_{eq} = \frac{5}{2} + 1.5 = 4$ A

$$\text{Current through cell} = \frac{V}{R} = \frac{20}{4} = 5\text{ A}$$

So, current through each branch $= \frac{i}{2} = 2.5$ A

Now, considering loop

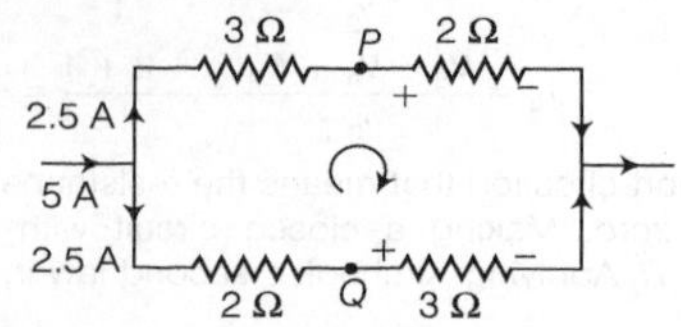

Now, applying Kirchhoff's rule to part loop P to Q

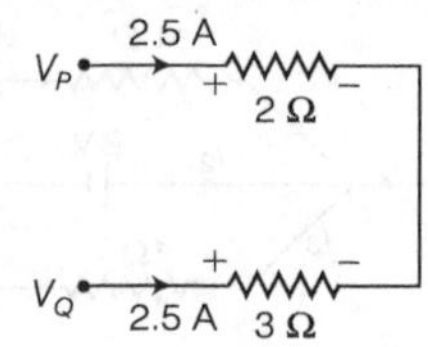

$$V_P - 2 \times 2.5 + 3 \times 2.5 - V_Q = 0$$

$$\Rightarrow \quad V_P - V_Q = -2.5\text{ V}$$

13. (b) As current flows from A to B,

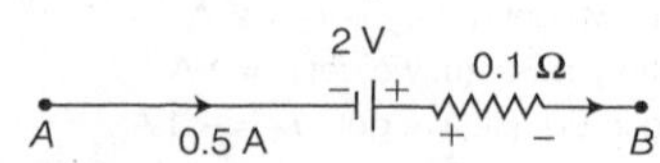

To calculate potential difference between points A and B, we start at A and add each individual potential difference until reaching B.

So, $\quad V_{AB} = V_B - V_A = 2 - 0.1 \times 0.5$

$$\Rightarrow \quad V_B - V_A = 2 - 0.05 = 1.95\text{ V}$$

$$\Rightarrow \quad V_A - V_B = -1.95\text{ V}$$

14. (d) Given,

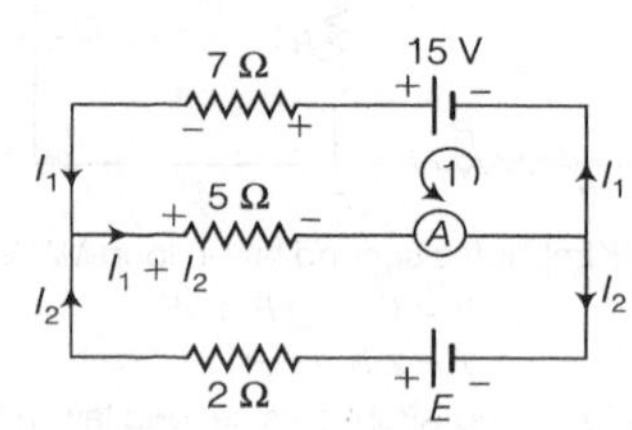

Given, $\quad I_1 + I_2 = 2$ A

In loop 1, by Kirchhoff's rule $\Sigma V = 0$

$$15 - 7I_1 - 2 \times 5 = 0$$

$$\Rightarrow \quad I_1 = \frac{5}{7}\text{ A}$$

$$\Rightarrow \quad I_1 = 0.714\text{ A}$$

Also, $\quad I_1 + I_2 = 2\text{ A} \Rightarrow I_2 = 2 - \frac{5}{7} = \frac{9}{7}$ A

$$\Rightarrow \quad I_2 = 1.29\text{ A}$$

15. (a) By Kirchhoff's rule for loop 1,

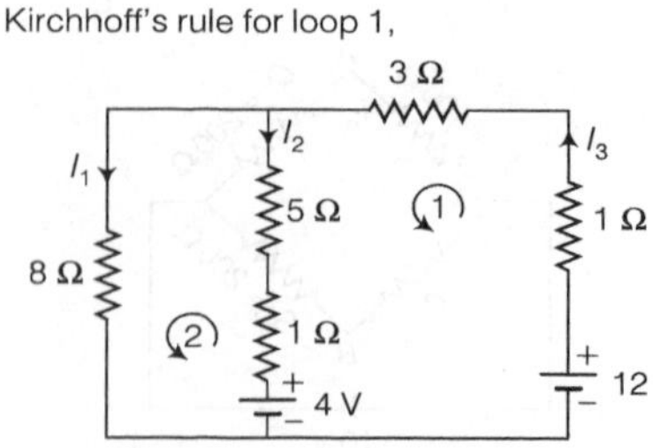

By Kirchhoff's rule for loop1,

$$12 - 4I_3 - 6I_2 - 4 = 0$$

$$\Rightarrow \quad 8 = 4I_3 + 6I_2$$

and from loop, $2 - 6I_2 - 4 + 8I_1 = 0$

On solving, we get $\quad I_1 = 846 \text{ mA}$

and $\quad I_2 = 462 \text{ mA}$

Hence, current through 8 Ω resistance is 846 mA.

16. (b) Let currents in branches are I_1, I_2 and I_3.

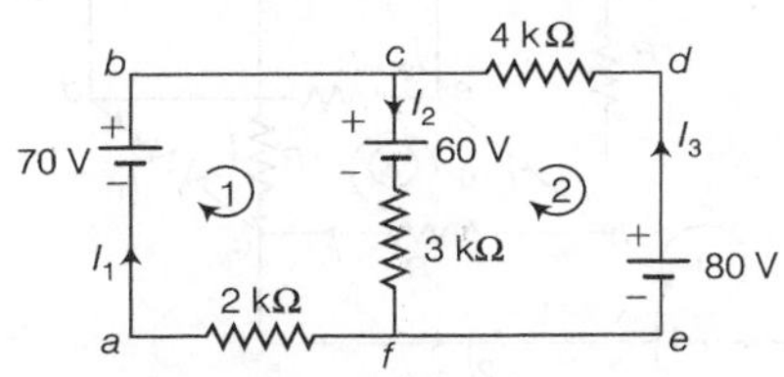

Then, from loops 1 and 2, we have

$$70 - 60 - I_2 \times 3 - I_1 \times 2 = 0 \qquad \ldots\text{(i)}$$

$$80 - I_3 \times 4 - 60 - I_2 \times 3 = 0 \qquad \ldots\text{(ii)}$$

and $\quad I_2 = I_1 + I_3 \qquad \ldots\text{(iii)}$

On solving, we get

$$I_1 = 0.385 \text{ mA}$$

$$I_2 = 3.08 \text{ mA}$$

$$I_3 = 2.69 \text{ mA}$$

and $\quad V_{cf} = -60 - 3.08 \times 3$

i.e. $\quad V_c - V_f = -69.2 \text{ V}$

Point c is at higher potential.

17. (c) In loop 1,

$$I_1 = \frac{5}{10} = \frac{1}{2} \text{ A}$$

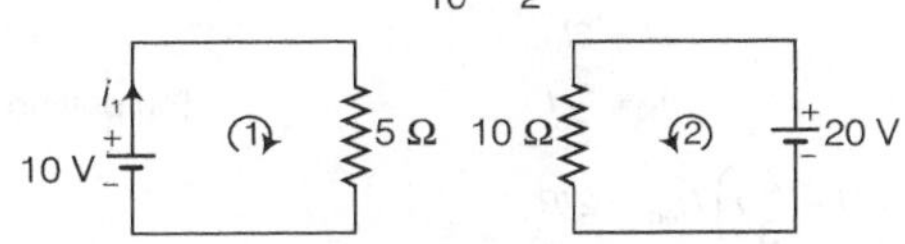

In loop 2, $\quad I_2 = \frac{20}{10} = 2 \text{ A}$

But, both currents are confined to separate loops and so, current through 2 Ω resistor is zero.

18. (a) By Kirchhoff's laws,

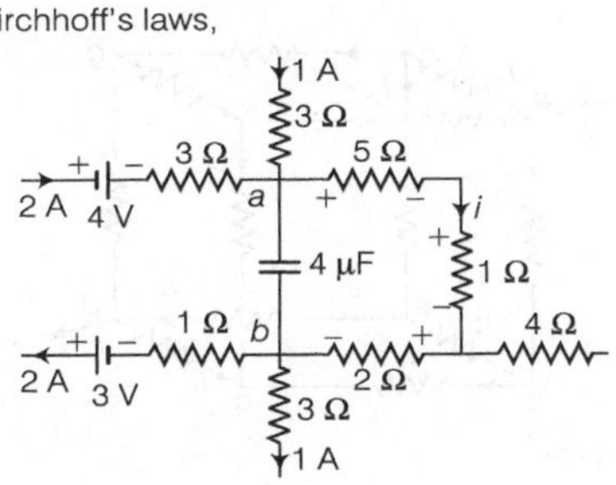

At junction a and b,

$$i = 2\text{A} + 1\text{ A} = 3 \text{ A}$$

and also current through corner 4 Ω resistor is zero.

Now, for path ab

$$-3 \times 5 - 3 \times 1 + 3 \times 2 = V_{ab} = V_b - V_a$$

$$\Rightarrow \quad V_{ab} = V_b - V_a = -12 \text{ V}$$

So, potential drop across capacitor = 12 V

Hence, energy stored $= \frac{1}{2} CV^2 = \frac{1}{2} \times 4 \times 10^{-6} \times 12^2$ J

$= 0.288 \text{ mJ} \approx 0.3 \text{ mJ}$

19. (a,c)

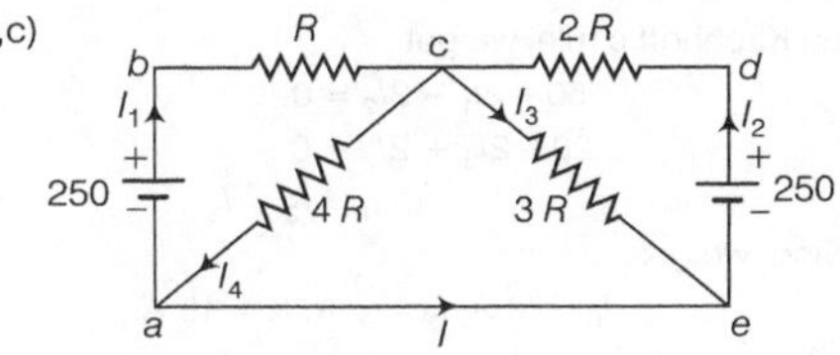

Given circuit can be modified as

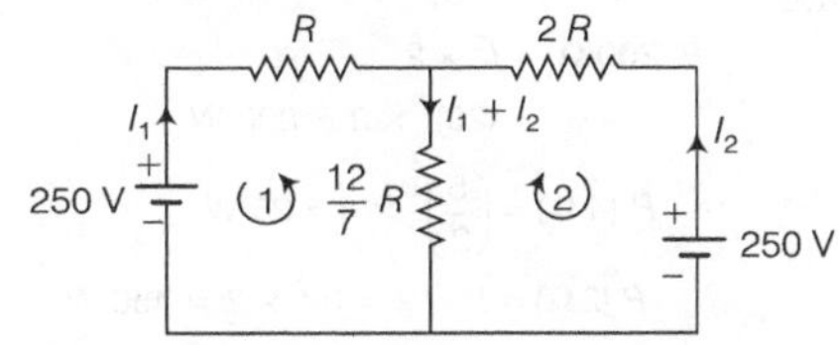

Now, applying Kirchhoff's rules to both loops, we have

$$\left(\frac{12}{7} R\right)(I_1 + I_2) + I_1 R = 250 \qquad \ldots\text{(i)}$$

$$\frac{12}{7} R\,(I_1 + I_2) + I_2\, 2R = 250 \qquad \ldots\text{(ii)}$$

On substituting, $R = 1000\ \Omega$, we get

$$I_1 = 10 \text{ mA}, I_2 = 130 \text{ mA and } I_4 = 60 \text{ mA}$$

As, $\quad I = I_4 - I_1 \Rightarrow I = 50 \text{ mA}$

20. (a,c) Let currents in loops are named as w, x and z.

$$w + x + z = y$$

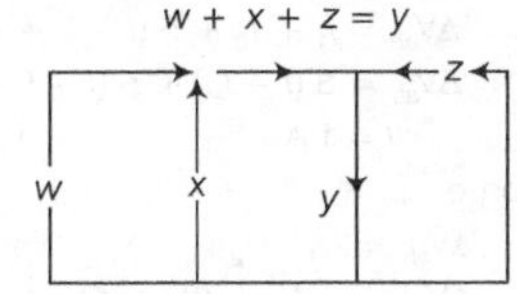

Then, equations are

$$-200\,w - 40 + 80x = 0$$

$$-80x + 40 + 360 - 20 = 0$$

$$360 - 20y - 70\,z + 80 = 0$$

Eliminating y by substitution, we get

$$x = 2.5\,w + 0.5$$

$$400 - 100x - 20w - 20z = 0$$

$$440 - 20w - 20x - 90z = 0$$

Eliminating x, $\quad 350 - 270w - 20z = 0$

$$430 - 70w - 20x - 90z = 0$$

Eliminating z, $430 - 70w - 1575 + 1215w = 0$

$$w = \frac{1145}{1145} = 1 \text{ A}$$

$\therefore \quad z = 4 \text{ A}, x = 3 \text{ A}, y = 8 \text{ A}$

and $\Delta V = IR = 1 \times 200 = 200$ V

21. (a,b,c) Given circuit is slightly modified as

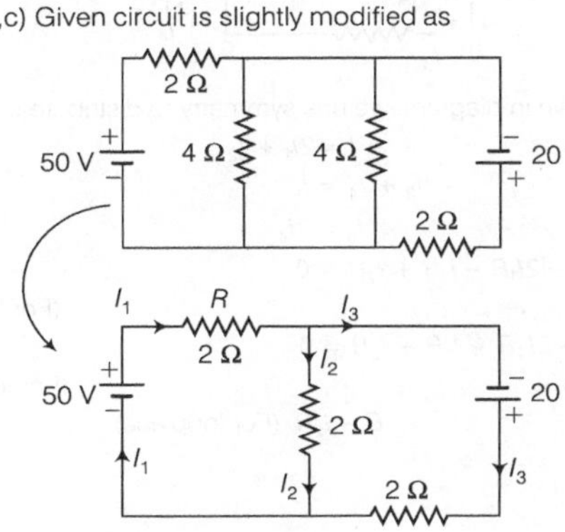

Applying Kirchhoff's rule, we get

$$50 - 2I_1 - 2I_2 = 0 \quad \text{...(i)}$$

$$20 - 2I_3 + 2I_2 = 0 \quad \text{...(ii)}$$

$$I_1 = I_2 + I_3 \quad \text{...(iii)}$$

On solving, we get

$$I_1 = 20 \text{ A}, I_2 = 5 \text{ A}, I_3 = 15 \text{ A}$$

Using $P = I^2R$,

We have

$$P\,(200\,\Omega) = I_1^2 \times 2$$

$$= (20)^2 \times 2 = 800 \text{ W}$$

$$P\,(4\,\Omega) = \left(\frac{5}{2}\right)^2 \times 4 = 25 \text{ W}$$

$$P\,(2\Omega) = I_3^2 \times 2 = 15^2 \times 2 = 450 \text{ W}$$

22. (a,c)

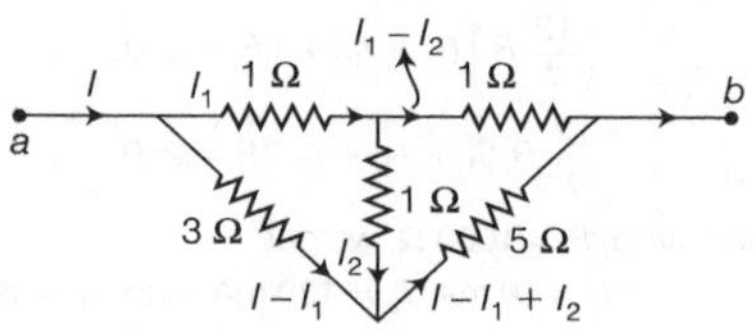

Equations are

$$\Delta V_{ab} = I_1 + 1\,(I_1 - I_2)$$

$$\Delta V_{ab} = I_1 + I_2 + 5\,(I - I_1 + I_2)$$

$$\Delta V_{ab} = 3\,(I - I_1) + 5\,(I - I_1 + I_2)$$

As, $I = 1$ A

Equations become

$$\Delta V_{ab} = 2I_1 - I_2 \quad \text{...(i)}$$

$$\Delta V_{ab} = -\,4I_1 + 6I_2 + 5 \quad \text{...(ii)}$$

$$\Delta V_{ab} = 8 - 8I_1 + 5I_2 \quad \text{...(iii)}$$

On solving these equations, we get

$$\Delta V_{ab} = \frac{27}{17} \text{ V}$$

$$R_{ab} = \frac{\Delta V_{ab}}{I} = \frac{27}{17}\,\Omega$$

23. (d)

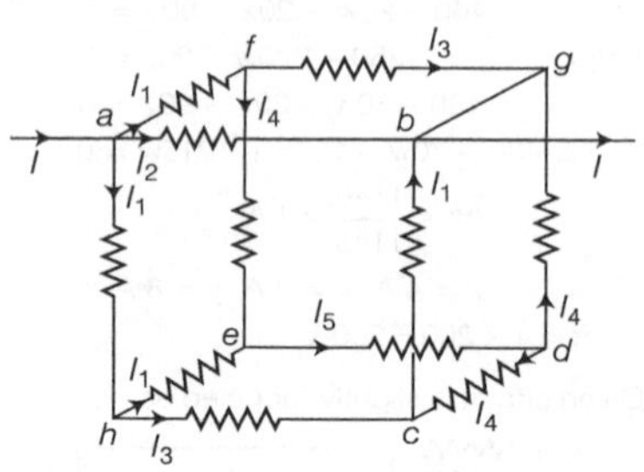

As shown in diagram, we use symmetry to distribute the current

$$I = 2I_1 + I_2 \quad \text{...(i)}$$

$$I_3 + I_4 = I_1 \quad \text{...(ii)}$$

$$I_5 = 2I_4 \quad \text{...(iii)}$$

$$-\,2I_1R - I_3R + I_2R = 0 \quad \text{...(iv)}$$

(For loop *afgba*)

$$-\,2I_4R - I_5R + I_3R = 0 \quad \text{...(v)}$$

(For loop *hedch*)

$$E - I_2R \text{ (For loop } aba) \quad \text{...(vi)}$$

On solving, we get

$$R_{eq} = \frac{E}{I} = \frac{7}{12}R$$

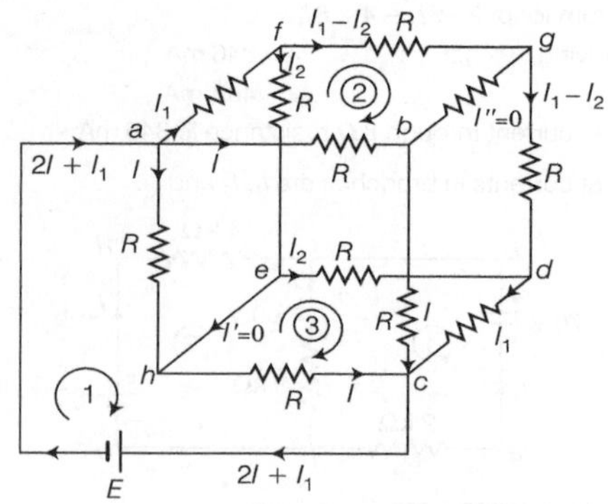

Loop 1 $\quad E = (2I + I_1)\,R_{eq} = IR + IR$

$$(2I + I_1)\,R_{eq} = 2IR \quad \text{...(i)}$$

Loop 2, *afgba*

$$\Sigma E = \Sigma\, IR$$

$$0 = I_2R + (I_1 - I_2)\,R - IR$$

$$0 = 2I_1 - I_2 - I$$

$$I = 2I_1 - I_2 \quad \text{...(ii)}$$

Loop 3, *edche*

$$\Sigma E = \Sigma\, IR$$

$$0 = I_2R + I_1R - IR$$

$$I = I_1 + I_2 \quad \text{...(iii)}$$

$$I = 2I_1 - I_2 \quad \text{...(iv)}$$

Adding Eqs. (iii) and (iv), we get

$$2I = 3I_1$$

$$\Rightarrow \quad I_1 = \frac{2}{3}I \qquad \text{[on putting in Eq. (i)]}$$

$$\Rightarrow \quad \left(2I + \frac{2}{3}I\right)R_{eq} = 2IR$$

$$\Rightarrow \quad R_{eq} = \frac{3}{4}R$$

In case of resistance between ends of a diagonal, we use symmetry to reduce number of currents to just three.

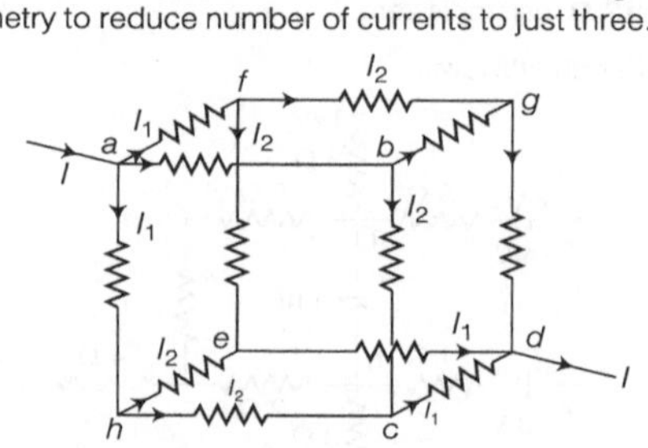

Equations are

$$I = 3I_1 \quad \text{...(i)}$$

$$I_1 = 2I_3 \quad \text{...(ii)}$$

Loop *abgda*,

$$-\,2I_1R - I_2R + E = 0 \quad \text{...(iii)}$$

On solving, we get

$$R_{eq} = \frac{5}{6}R$$

(i) → (q), (ii) → (p), (iii) → (r)

Explanations for Q. Nos. (24-27)

First we mark branch currents and node voltages as

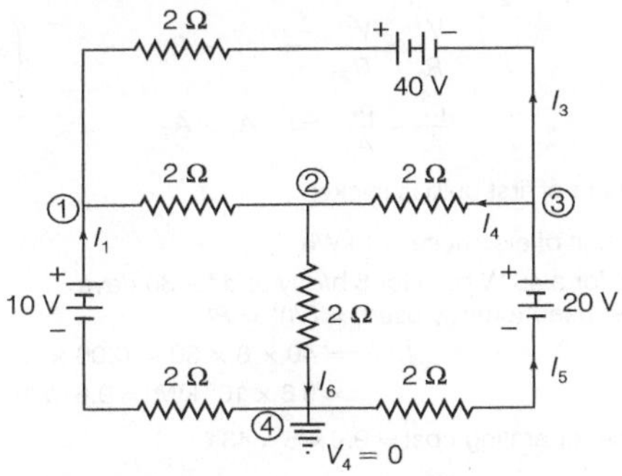

At node 1, by KCL we have

$$I_1 + I_3 = I_2$$

$$\frac{0 + 10 - V_1}{2} + \frac{V_3 + 40 - V_1}{2} = \frac{V_1 - V_2}{2}$$

$$\Rightarrow \quad 3V_1 - V_2 - V_3 = 50 \quad \text{...(i)}$$

At node 2,

$$I_2 + I_4 = I_6$$

$$\frac{V_1 - V_2}{2} + \frac{V_3 - V_2}{2} = \frac{V_2 - 0}{2}$$

$$V_1 - 3V_2 + V_3 = 0 \quad \text{...(ii)}$$

At node 3,

$$I_5 = I_3 + I_4$$

$$\frac{0 + 20 - V_3}{2} + \frac{V_3 + 40 - V_1}{2} = \frac{V_3 - V_2}{2}$$

$$V_1 + V_2 - 3V_3 = 0 \quad \text{...(iii)}$$

On solving Eqs. (i), (ii) and (iii), we get

$$V_1 = 20\,\text{V}$$
$$V_2 = 7.5\,\text{V}$$
$$V_3 = 2.5\,\text{V}$$

Also, current in branches 2, 3 is

$$I = I_4 = \frac{V_3 - V_2}{2} = \frac{2.5 - 7.5}{2} = \frac{-5}{2}$$

$\therefore \quad I = -2.5\,\text{A}$ (current flows from point 2 to 3)

24. (a) **25.** (b)

26. (c) **27.** (b)

DPP-7 Electrical Energy

1. (0.32 %) Power consumption = 10 units

In 5 h = 2 units

$$2\,\frac{\text{kW}}{\text{h}} = 2000\,\text{Js}^{-1}$$

Current in circuit, $I = \frac{P}{V} = \frac{2000}{220} \approx 9\,\text{A}$

Power loss in wires $= I^2R\,\text{Js}^{-1} = \rho\frac{l}{A}\cdot I^2$

$$= 1.7 \times 10^{-8} \times \frac{10}{\pi \times 10^{-6}} \times 81 \approx 4\,\text{Js}^{-1}$$

Hence, power loss in wires $= \frac{4}{2000} \times 100 = 0.2\%$

With aluminium wires,

$$\text{Power loss} = \rho\frac{l}{A}\cdot I^2 = 4 \times \frac{\rho_{Al}}{\rho_{Cu}}$$

$$= 6.4\,\text{Js}^{-1} = 0.32\%$$

2. (i) $(2 \times 10^{-17}, 10^{-17})$ $I = \frac{V}{R} = \frac{6}{6} = 1\,\text{A}$

$$I = nev_d A$$

$$\Rightarrow \quad v_d = \frac{I}{neA} = \frac{1}{10^{29} \times 1.6 \times 10^{-19} \times 10^{-6}}$$

$$= \frac{1}{1.6} \times 10^{-4}\,\text{m/s}$$

$$= 6.25 \times 10^{-5}\,\text{m/s}$$

So, KE of electrons $= \left(\frac{1}{2}mv_d^2\right) \times$ total number of electrons

$$= \frac{1}{2}mv_d^2 \times nAl$$

$$= \frac{1}{2} \times 9.1 \times 10^{-31} \times (6.25 \times 10^{-5})^2 \times 10^{29} \times 10^{-6} \times 10^{-1}$$

$$= 2 \times 10^{-17}\,\text{J}$$

(ii) Ohmic loss $= I^2R$

$$= 6 \times (1^2) = 6\,\text{Js}^{-1}$$

$\therefore$ All KE of electrons will be lost in $\frac{2 \times 10^{-17}}{6} \approx 10^{-17}\,\text{s}$

3. Rated power and rated voltage are the voltage (or emf) at which that equipment is designed to operate and that equipment gives rated output power at rated voltage. It is also indicated with the specifications of that appliance, e.g. if it is a bulb, rated power and rated voltage are printed over it.

As, electricity from power stations is supplied at a constant emf, current in circuit depends on the load and hence, there is no rated current value.

4. We have $\quad P_{out} = IV_{AB} = I^2R$

Power generated by source

$$P_{gen} = EI$$

$$\text{Efficiency} = \eta = \frac{P_{out}}{P_{gen}} = \frac{IV_{AB}}{IE} = \frac{V_{AB}}{E} = \frac{IR}{I(R + r)}$$

$$= \frac{R}{R + r} = \frac{1}{1 + (r/R)}$$

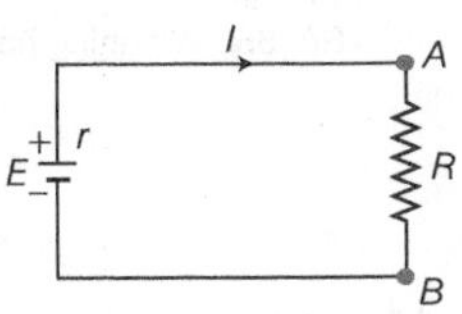

$\therefore \quad \eta \to 1$ or $R \to \infty$

At $\quad R = r, \eta = 50\%$

Also, $\quad V_{AB} = IR = \frac{E}{R + r}\cdot R$

and $\quad I_{AB} = \frac{E}{R + r}$

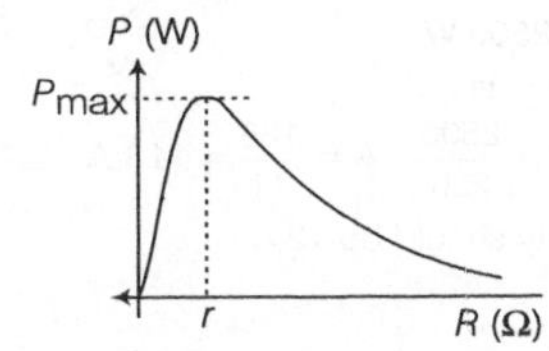

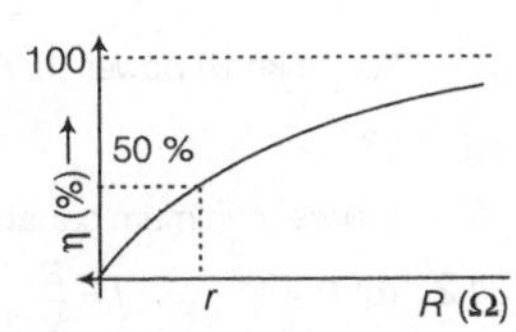

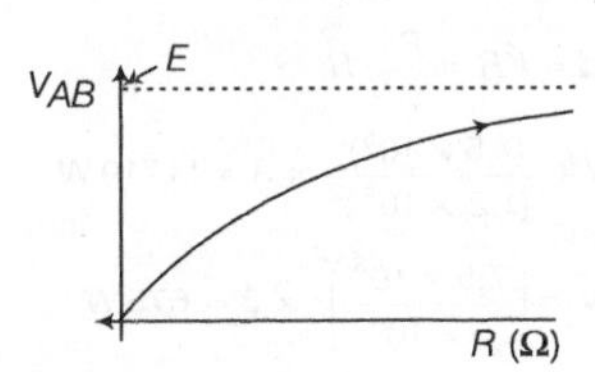

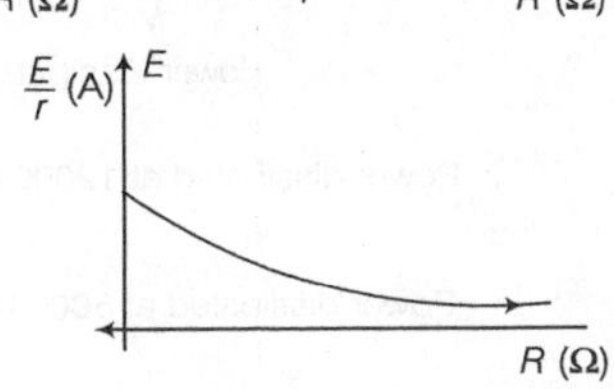

5. (b) Heat produced, $H = \frac{V^2 t}{R}$

$$= \frac{(120)^2 \times 10 \times 60}{6}$$

$$= 14.4 \times 10^5 \text{J}$$

6. (d) Power consumed by bulb $= \frac{V^2}{V_R^2} \times P_R$

$$= \left(\frac{110}{220}\right)^2 \times 100 = 25 \text{ W}$$

7. (c) Power consumed $= \left(\frac{\text{Supply voltage}}{\text{Rate voltage}}\right)^2 \times$ Rated power

$$= \left(\frac{110}{115}\right)^2 \times 500 = 456.6 \text{ W}$$

$\therefore$ % drop in heat output $= \frac{P_{\text{actual}} - P_{\text{consumed}}}{P_{\text{actual}}} \times 100$

$$= \left(\frac{500 - 456.6}{500}\right) \times 100 = 8.6\%$$

8. (b) Ratio of currents $= \frac{i_1}{i_2} = \frac{10}{5} = 2$

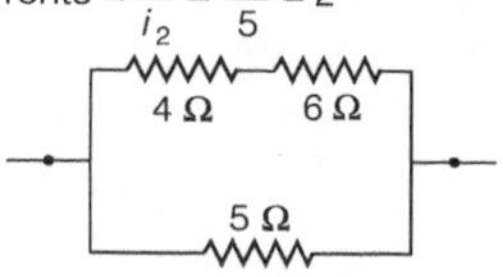

and ratio of heat produced $= \frac{H_1}{H_2} = \frac{i_1^2 R_1 t_1}{i_2^2 R_2 t_1} = \frac{i_1^2 R_1}{i_2^2 R_2} = \frac{2^2}{1^2} \times \frac{5}{4}$

$\therefore$ $H_2 = 2$ J/s

9. (d) Rated power of bulb is given 75 W but rated voltage is not given, so it cannot be concluded that bulb's output is 75 W at 110 V (or at some other supply voltage value)

So, data is insufficient to reach any conclusion.

10. (a) Since, $P = \frac{V^2}{R}$

$\Rightarrow$ $R = \frac{V^2}{P}$

Resistance is inversely proportional to power for a constant voltage.

So, 850 W setting has higher resistance.

$$R = \frac{V^2}{P} = \frac{(120)^2}{850} = 17 \ \Omega$$

$$R = \frac{V^2}{P} = \frac{(120)^2}{1250} = 12 \ \Omega$$

11. (c) Total power (P) consumed

$$= (15 \times 40) + (5 \times 100) + (5 \times 80) + (1 \times 1000)$$

$$= 2500 \text{ W}$$

As we know power i.e. $P = VI$

$\Rightarrow$ $I = \frac{2500}{220} \text{ A} = \frac{125}{11} = 11.3 \text{ A}$

Hence, minimum capacity should be 12 A.

12. (b) $P = VI \Rightarrow I = \frac{P}{V}$

Power dissipated $= I^2 R = \frac{P^2}{V^2} \cdot R$

$\therefore$ Power dissipated at 12000 V $= \frac{(7.5 \times 10^5)^2}{(1.2 \times 10^4)^2} \times 3 = 11719$ W

Power dissipated at 50000 V $= \left(\frac{7.5 \times 10^5}{5 \times 10^4}\right)^2 \times 3 = 675$ W

So, difference is $11719 - 675 = 1.1 \times 10^4$ W

13. (a) As, bulbs can be compared only when they are to be used with same source, so we take V = constant for both.

Now, $P_1 > P_2$

$$\frac{V^2}{R_1} > \frac{V^2}{R_2} \Rightarrow R_1 < R_2 \qquad \left(\text{as, } R = \frac{\rho l}{A}\right)$$

$$\Rightarrow \frac{\rho l}{A_1} < \frac{\rho l}{A_2} \Rightarrow A_1 > A_2$$

Filament of first bulb is thicker.

14. (a) 1 unit of electricity is 1 kWh.

Now, for a 40 W bulb for 8 h/day and for 30 days, power used (energy used) $= (VI)t = Pt$

$$= 40 \times 8 \times 30 = 1200 \times 8 = 9600 \text{ Wh}$$

$$= 9.6 \times 10^3 \text{ kWh} = 9.6 \text{ kWh}$$

Hence, operating cost = 9.6× 5 = 48 ₹

15. (d) $P = \frac{V^2}{R}$

$$\Rightarrow \frac{P_{200}}{P_{800}} = \frac{V^2 / R_{200}}{V^2 / R_{800}} = \frac{R_{800}}{R_{200}} = \frac{R_{200}(1 + \alpha \Delta T)}{R_{200}}$$

$$\Rightarrow P_{200} = P_{800} \times (1 + 4 \times 10^{-4} \times 600)$$

$$= 500\,(1 + 4 \times 10^{-4} \times 600) = 620 \text{ W}$$

So, most close option is 611W, [disperency arises because base [temperature is taken at 200°C.]

16. (c) Let water requires H joule of heat to boil.

Then, $H = \frac{V^2}{R} t$ ...(i)

where, $t = 15$ min

If length of heating element is made $\frac{2}{3}$ of initial length, then

$$R' = \text{New resistance} = \frac{2}{3} R$$

With same voltage, heat produced will be

$$H' = \frac{V^2}{\frac{2}{3}R} \cdot t' = \frac{3V^2}{2R} \cdot t'$$

As, $H' = H$ for boiling of water,

$$\frac{V^2}{R} t = \frac{3}{2} \frac{V^2}{R} t' \Rightarrow t' = \frac{2}{3} t = \frac{2}{3} \times 15 \text{ min} = 10 \text{ min}$$

17. (d) $R_{\text{total}} = 2 + \frac{6 \times 1.5}{6 + 1.5} = 3.2 \text{ k}\Omega$

(a) $I = \frac{24 \text{ V}}{3.2 \text{ k}\Omega} = 7.5 \text{ mA} = I_{R_1}$, $\quad I_{R_2} = \left(\frac{R_L}{R_L + R_2}\right) I$

$$I = \frac{1.5}{7.5} \times 7.5 = 1.5 \text{ mA}$$

$$I_{R_L} = 6 \text{ mA}$$

(b) $V_{R_L} = (I_{R_L})(R_L) = 9$ V

(c) $\frac{P_{R_1}}{P_{R_2}} = \frac{(I_{R_1}^2)R_1}{(I_{R_2}^2)R_2} = \frac{(7.5)^2(2)}{(1.5)^2(6)} = \frac{25}{3}$

(d) When R_1 and R_2 are interchanged, then

$$\frac{R_2 R_L}{R_2 + R_L} = \frac{2 \times 1.5}{3.5} = \frac{6}{7} \text{ k}\Omega$$

Now, potential difference across R_L will be

$$V_L = 24 \left[\frac{6/7}{6 + 6/7}\right] = 3 \text{ V}$$

Earlier, it was 9 V

Since, $P = \frac{V^2}{R}$ or $P \propto V^2$

In new situation, potential difference has been decreased three times. Therefore, power dissipated will decrease by a factor of 9.

18. (c,d) As, bulbs are in parallel V remains same and so current will increase. As, bulbs are identical current becomes twice.

19. (b,c) $P = \frac{V^2}{R} \Rightarrow R = \frac{V^2}{P}$

Also, $P = VI \Rightarrow I = \frac{P}{V}$

Now, for the bulb,

V_R = rated voltage = 3 V and P_R = rated power = 2 W

$\therefore \quad R = \frac{V_R^2}{P} = \frac{3 \times 3}{2} = 4.5\,\Omega$

To use bulb with a 9 V supply, a series resistance is used so that rated power and power of bulb have same value.

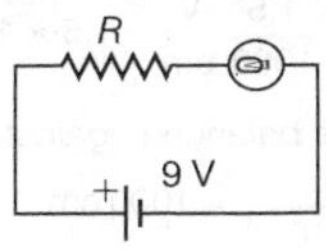

For above circuit,

$$E = I(R + R_{bulb})$$

where, $I = \left(\frac{P_{bulb}}{V_{bulb}}\right)_{\text{rated value}}$

$\therefore \quad R = \frac{V_{bulb}}{P_{bulb}}(E - V_{bulb}) = \frac{3}{2} \times (9 - 3) = 9\,\Omega$

20. (a,b) 7 resistors each of 460 Ω in series has an equivalent resistance of

$$= 7 \times 460 = 3220\,\Omega \approx 3.2\,\text{k}\Omega$$

and power $= \frac{1}{2} \times 7 = 3.5\,\text{W}$

Also, seven resistors of 22.4 kΩ in parallel has an equivalent resistance of $R = \frac{22.4}{7}\,\text{k}\Omega \approx 3.2\,\text{k}\Omega$

In parallel, $P_{net} = 7 \times \frac{1}{2} = 3.5\,\text{W}$

21. (c)

(i) $n = 0$ switch closed, effective circuit resistance is 16 kΩ.

$$I = \frac{16\text{ V}}{16\text{ k}\Omega} = 1\text{ mA}$$

Voltage across resistance R is $V = IR = 1\text{ mA} \times 1\,\Omega = 1\text{ mV}$

(ii) $n = 1$ switch closed, $R_{effective} = 8\text{ k}\Omega$

$$I = \frac{16\text{ V}}{8\text{ k}\Omega} = 2\text{ mA}$$

$$V = 2\text{mA} \times 1\,\Omega = 2\text{mV}$$

(iii) $n = 2$ switch closed, $R_{effective} = 4\text{ k}\Omega$

$$I = \frac{16\text{ V}}{4\text{ k}\Omega} = 4\text{ mA}$$

$$V = 4\text{ mA} \times 1\,\Omega = 4\text{ mV}$$

(iv) $n = 3$ and $n = 1$ switches closed, $R_{effective} = 1.6\text{ k}\Omega$

$I = 10$ mA

$V = 10 \times 10^{-3} \times 1 = 10$ mV

(i) → (s), (ii) → (r), (iii) → (q), (iv) → (p)

22. (a) When resistors are in series, then

$$R_s = R_1 + R_2 + \cdots + R_n$$

$$\frac{V^2}{P_s} = \frac{V_1^2}{P_1} + \frac{V_2^2}{P_2} + \cdots + \frac{V_n^2}{P_n}$$

When bulbs are of equal power, then

$$\frac{V^2}{P_s} = \frac{nV^2}{P} \Rightarrow P_s = \frac{P}{n}$$

23. (c) When resistors are in parallel, then

$$\frac{1}{R_p} = \frac{1}{R_1} + \frac{1}{R_2} + \cdots + \frac{1}{R_n} \quad \text{or} \quad \frac{P_p}{V^2} = \frac{P_1}{V^2} + \frac{P_2}{V^2} + \cdots + \frac{P_n}{V^2}$$

$$P_p = P_1 + P_2 + \cdots + P_n \quad \text{or} \quad P_p = nP$$

24. (b) Power consumed in series, $P_s = V^2 / nR$

and in parallel, $P_p = V^2 / R / n$

$\therefore \quad P_p = n^2 P_s$

DPP-8 **Measuring Instruments**

1. (i) For a meter bridge in balanced condition,

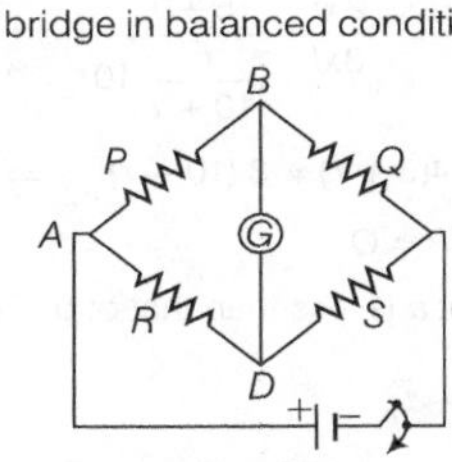

$$\frac{P}{Q} = \frac{R}{S}$$

⇒ Ratio of resistances is equal.

$\Rightarrow \quad \frac{P}{R} = \frac{Q}{S}$

Now, if galvanometer and cell are interchanged,

As, $\frac{P}{R} = \frac{Q}{S}$

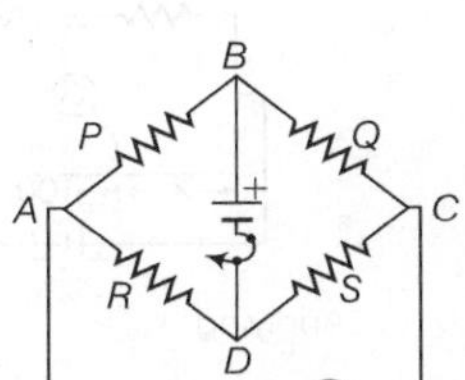

Galvanometer is still in balanced condition. So, there is still no deflection in the galvanometer.

(ii) End error is a error due to non-measured length of wire at ends which is clamped on the connecting strips.

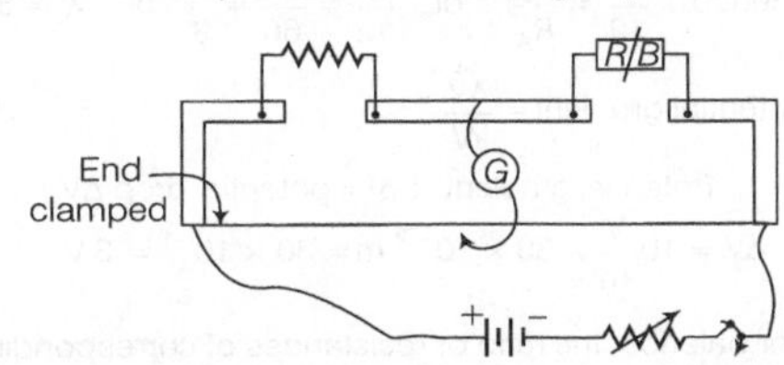

To minimise or eradicate end error, unknown resistance value is measured in putting resistance initially in right gap and then in left gap. The average of both values is finally taken.

2. Sensitivity of a potentiometer is ability to measure (balance) a very small potential difference against certain length of potentiometer wire.

Potential gradient is fall of potential per unit length, potential gradient $= \frac{V_{AB}}{l_{AB}}$

AB is potentiometer wire.

If potential gradient of a potentiometer is small, then there is a small potential drop over a large length. So, sensitivity of a potentiometer is proportional to reciprocal of potential gradient.

$$\Rightarrow \quad \text{Sensitivity} \propto \frac{1}{\text{Potential gradient}}$$

To increase sensitivity, we have to decrease potential gradient. Hence, sensitivity is increased if

(i) length of potentiometer wire is increased.

Or

(ii) V_{AB} – potential drop across potentiometer wire is decreased either by connecting a resistance in series with wire or by setting rheostat to a larger value.

3. (i) So that potential drop is uniform in each part of wire.

(ii) A voltmeter draws a small current from the cell, so measured emf value by a voltmeter is $E' = E - Ir$

$\therefore$ Voltmeter reading has a error

$$E - E' = Ir$$

whereas, potentiometer does not draw any current.

So, Emf measured is accurate.

(iii) Emf of driver cell is kept large so that potential drop across a part of potentiometer wire is equal to emf of slave cell and so balance can be reached. If emf of driver cell is smaller, then there is no null deflection point-over-entire length of potentiometer wire.

(iv) Metal alloys have high resistivity and low temperature coefficient. So, alloy wires produce a large potential drop and their resistance stays nearly same with slight increase in temperature during experiment.

4. (i) There are no positive and negative terminals on the galvanometer because only zero deflection is needed.

(ii)

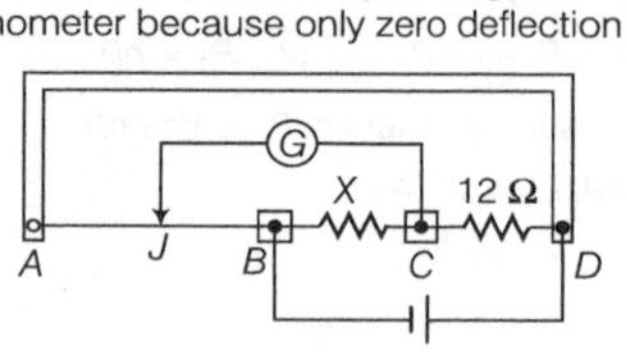

(iii) $AJ = 60\,\text{cm}$

$\therefore$ $BJ = 40\,\text{cm}$

If no deflection is taking place. Then, Wheatstone bridge is said to be balanced.

Hence, $\frac{X}{12} = \frac{R_{BJ}}{R_{AJ}}$ or $\frac{X}{12} = \frac{40}{60} = \frac{2}{3}$ or $X = 8\,\Omega$

5. (a) Potential gradient $= \frac{\Delta V}{\Delta l}$

$\Rightarrow$ Potential gradient $\times \Delta l$ = potential drop ΔV

$\Rightarrow$ $\Delta V = 10 \frac{\text{V}}{\text{m}} \times 30 \times 10^{-2}\,\text{m} = 30 \times 10^{-1} = 3\,\text{V}$

6. (c) For balance, the ratio of resistances of corresponding arms is same.

$\therefore$ $\frac{P}{Q} = \frac{R}{S} \Rightarrow S = \frac{RQ}{P}$

$\Rightarrow$ $S = \frac{4 \times 11}{9} = \frac{44}{9}$

Now, S is given $6\,\Omega$ if another resistance of $R\,\Omega$ is put in parallel then, equivalent resistance of combination is $\frac{6 \times R}{R + 6}$

So, $\frac{6 \times R}{R + 6} = \frac{44}{9}$ (for balance)

$\Rightarrow$ $54R = 44R + 6 \times 44$

$\Rightarrow$ $10R = 6 \times 44$

$$R = \frac{6 \times 44}{10} = 26.4\,\Omega$$

7. (c)

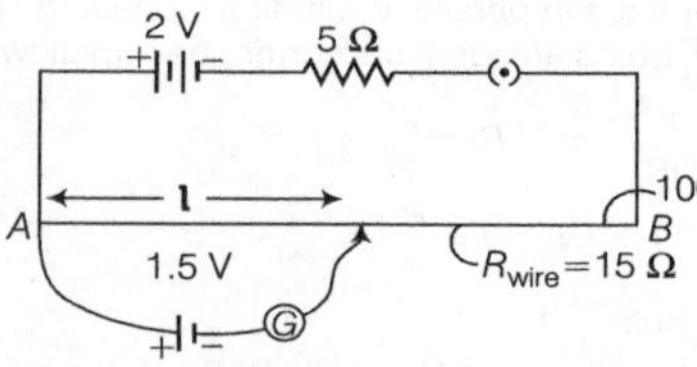

V_{AB} = potential difference across ends of wire

$$= I_{AB} R_{AB} = \frac{E}{R_{\text{total}}} \cdot R_{AB} = \frac{2}{20} \times 15 = 1.5\,\text{V}$$

So, potential gradient of potentiometer wire

$$= \frac{V_{AB}}{l_{AB}} = \frac{1.5\ \text{V}}{1000\ \text{cm}} = 1.5 \times 10^{-3}\ \frac{\text{V}}{\text{cm}}$$

$\therefore$ Potential of 1.5 V is balanced against a length

$$l' = \frac{1.5}{1.5 \times 10^{-3}} = 1000\,\text{cm}$$

8. (b)

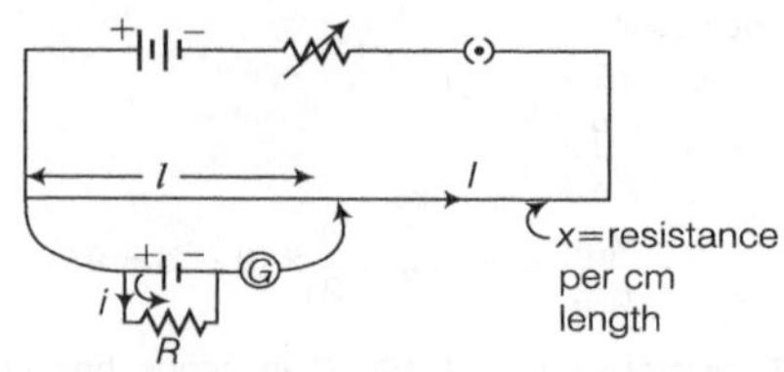

When cell is shunted by a resistance R, then potential drop across terminals of cell is balanced against potentiometer wire.

Now, terminal potential difference of cell, $V = E - ir = iR$

E = emf, i = current in cell circuit

r = internal resistance of cell and $i = \frac{E}{R + r}$

So, in Ist case, $V = l_1 x l = \frac{E}{R + r} . R$

$\Rightarrow$ $2xl = \frac{E}{5 + r} . 5$ ...(i)

and in IInd case, $3xl = \frac{E}{10 + r} . 10$ ...(ii)

Dividing both equations,

$$\frac{2xl}{3xl} = \frac{\frac{E}{5 + r} . 5}{\frac{E}{10 + r} . 10} \Rightarrow \frac{2}{3} = \frac{10 + r}{10} \times \frac{5}{5 + r}$$

$\Rightarrow$ $4(5 + r) = 3(10 + r)$ $\Rightarrow$ $r = 10\,\Omega$

9. (b) As, $X > Y$ and $P = Q$

$\therefore$ Potential of point a is less than that of b. Current flows from b to a.

10. (a) $R > 2\,\Omega$

$\therefore$ $100 - x > x$

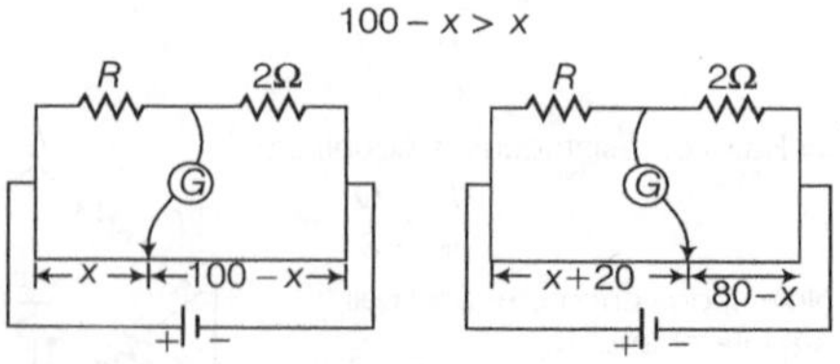

Applying $\frac{P}{Q} = \frac{R}{S}$

We have $\frac{2}{R} = \frac{x}{100 - x}$ ...(i)

$\frac{R}{2} = \frac{x + 20}{80 - x}$ ...(ii)

On solving Eqs. (i) and (ii), we get

$$R = 3\,\Omega$$

11. (d) As, cells are in parallel, emf of combination = emf of a single cell = E

So, in balance condition emf E is balanced against 300 cm length.

$$\Rightarrow \quad E = 300xl$$

Now, $$x = \frac{5\,\Omega}{5\,\text{m}} = \frac{1}{100}\,\Omega\text{cm}^{-1} = 10^{-2}\,\Omega/\text{cm}$$

and $$I = \frac{E_{\text{driver}}}{\text{Total resistance}} = \frac{10}{4+5} = \frac{10}{9}\,\text{A}$$

Hence, $$E = 300xI = 300 \times 10^{-2} \times \frac{10}{9}$$

$$= 3 \times \frac{10}{9} = \frac{10}{3} = 3.33\,\text{V}$$

12. (b) Using the concept of balanced Wheatstone bridge, we get

$$\frac{P}{Q} = \frac{R}{S}$$

$$\Rightarrow \quad \frac{X}{(52+1)} = \frac{10}{(48+2)}$$

$$\therefore \quad X = \frac{10 \times 53}{50} = 10.6\,\Omega$$

13. (b) At null deflection position, equivalent emf of cell combination (of 2 cells in parallel) is balanced against potential drop of AN length of potentiometer wire.

Now, equivalent emf of combination is

$$E_{\text{eq}} = \frac{E_1 r_2 + E_2 r_1}{r_1 + r_2} = \frac{2 \times 6 + 4 \times 2}{2+6} = \frac{12+8}{8}$$

$$\therefore \quad E_{\text{eq}} = \frac{20}{8} = 2.5\,\text{V}$$

and potential drop across AN length = potential drop per unit length × length AN.

Now, current in potentiometer wire

$$= I_{AB} = \frac{E}{R_{\text{total}}} = \frac{12}{4+4} = \frac{12}{8}\,\text{A} = 1.5\,\text{A}$$

So, V_{AB} = potential drop across AB length of wire

$$= I_{AB}\,R_{AB} = 1.5 \times 4 = 6\,\text{V}$$

Hence, potential drop across AN length $= \dfrac{V_{AB}}{l_{AB}} \times AN = \dfrac{6}{4} \times AN$

So, $$\frac{6}{4}AN = E_{\text{eq}}$$

As, N is null deflection point

$$\Rightarrow \quad \frac{6}{4} \times AN = 2.5 \Rightarrow AN = \frac{2.5 \times 4}{6} = \frac{5}{3}\,\text{m}$$

14. (d) Given network of resistances is equivalent to two balanced Wheatstone bridge

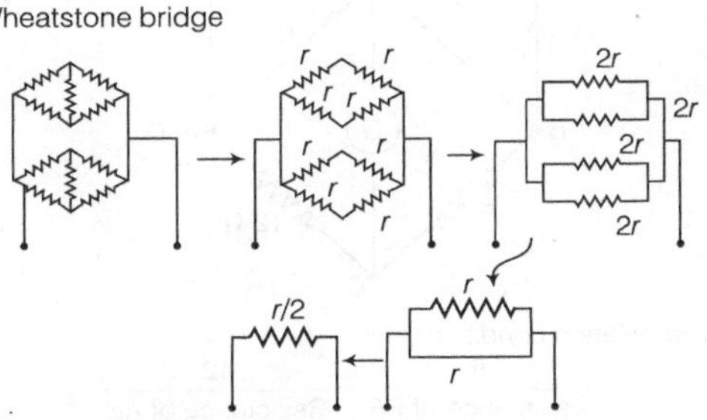

As, meter bridge is balanced,

$$\therefore \quad \frac{10}{30} = \frac{r/2}{60} \Rightarrow \frac{10 \times 60}{30} \times 2 = r \quad \text{or} \quad r = 40\,\Omega$$

15. (d) A meter bridge is most sensitive (gives most accurate reading) when balance is obtained near mid-point.

So, $R = X$ gives most accurate value of X.

16. (c) For balanced meter bridge,

$$\frac{X}{R} = \frac{l}{(100-l)} \quad \text{(where, } R = 90\,\Omega)$$

$$\therefore \quad \frac{X}{90} = \frac{40}{100-40}$$

$$\therefore \quad X = 60\,\Omega$$

$$X = R\frac{l}{(100-l)}$$

$$\frac{\Delta X}{X} = \frac{\Delta l}{l} + \frac{\Delta l}{100-l} = \frac{0.1}{40} + \frac{0.1}{60}$$

$$\Delta X = 0.25$$

So, $$X = (60 \pm 0.25)\,\Omega$$

17. (a,c) Let cell delivers current i and i_1 is apart of current i that goes through R_3.

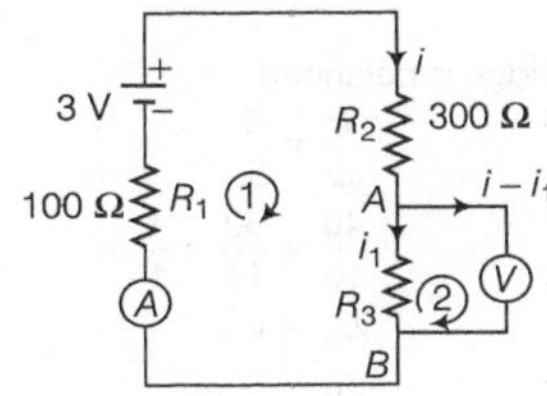

So, by loop rule

$$E - iR_2 - i_1R_3 - iR_1 = 0 \quad \text{(loop 1)} \ldots \text{(i)}$$

$$i_1R_3 - (i - i_1)R_V = 0 \quad \text{(loop 2)} \ldots \text{(ii)}$$

On solving for i_1, Eq. (ii) yields

$$i = \frac{R_3 + R_V}{R_V}.i_1$$

Substituting this into Eq. (i), we get

$$E - \left(\frac{R_3 + R_V}{R_V}\right).R_2.i_1 - i_1R_3 - \left(\frac{R_3 + R_V}{R_V}\right)R_1i_1 = 0$$

$$\Rightarrow \quad E = i_1\left\{\left(\frac{R_3 + R_V}{R_V}\right)R_2 - R_3 - \left(\frac{R_3 + R_V}{R_V}\right)R_1\right\}$$

$$= i_1\left\{\frac{(R_3 + R_V)R_2 - R_VR_3 - (R_3 + R_V)R_1}{R_V}\right\}$$

So, $$i_1 = \frac{ER_V}{(R_3 + R_V)R_2 - R_VR_3 - (R_3 + R_V)R_1}$$

$$= \frac{3 \times 5 \times 10^3}{\left[(250 + 5 \times 10^3)\,300 - 5 \times 10^3 \times 250 - (250 + 5 \times 10^3) \times 100\right]}$$

and potential drop across resistance $R_3 = i_1R_3 = 1.12$ V

Current in the absence of voltmeter is

$$i' = \frac{E}{R_1 + R_2 + R_3}$$

∴ True potential drop across

$$R_3 = i'R_3 = \frac{3 \times 250}{250 + 300 + 100} = 1.15\,\text{V}$$

So, fractional error is

$$\frac{1.15 - 1.12}{1.15} = 0.030$$

This is about 3%.

18. (a,c) Balance point is not obtained if

(i) polarity of slave cell is reversed.

(ii) emf of driver cell is less than that of slave cell.

19. (b,d) In balance condition,

$$\frac{2R}{R} = \frac{2R}{RRO}$$

$$\Rightarrow \quad R = RRO = 22 \times 10^3\,\Omega$$

(using colour code for resistors)

If X and $2R$ are interchanged, then

for balance, $\frac{2R}{Y} = \frac{X}{2R}$

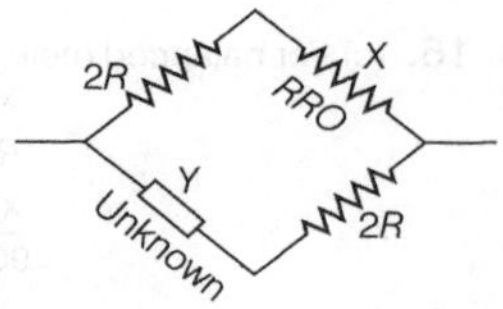

$$\Rightarrow \quad Y = \frac{2R \times 2R}{X} = \frac{2R \times 2R}{R}$$

$$= 4R = 4 \times 22 \times 10^3$$

$$= 88 \times 10^3 \ \Omega$$

So, colour code is GGY.

20. (b) Unknown resistance is put in left gap, then in right gap and average value is taken to eradicate net error.

Statement II is correct but it is not the explanation of Statement I.

21. (a) Sensitivity of a potentiometer increases with decrease in potential gradient and potential gradient decreases with increase in length of potentiometer wire.

22. (a)

Wheatstone bridge is balanced,

As, $$\frac{P}{Q} = \frac{R}{S}$$

$$\Rightarrow \quad \frac{40}{10} = \frac{60}{15} = \frac{4}{1}$$

$$\Rightarrow \quad V_{AB} = V_{AD}$$

So, $$40 I_1 = 60 I_2$$

$$I_1 = 1.5 I_2$$

Heat produced in $AB = I_1^2 Rt = (1.5 I_2)^2 \times 40 \times t = 90\, I_2^2 t$

Heat produced in $BC = I_1^2 Rt = (1.5 I_2)^2 \times 10 \times t = 22.5\, I_2^2 t$

Heat produced in $AD = I_2^2 Rt = I_2^2 \times 60 \times t = 60 I_2^2 t$

Heat produced in $DC = I_2^2 Rt = I_2^2 \times 15 \times t = 15 I_2^2 t$

(i) → (s), (ii) → (r), (iii) → (q), (iv) → (p).

23. (b) Emf of thermocouple = Potential drop occurring across 400 mm of potentiometer wire.

$$= I(R) = I(x)(l) \quad \ldots \text{(i)}$$

where, x = resistance per metre length = $\frac{10\ \Omega}{1\ \text{m}}$

I = current in potentiometer wire

$$= \frac{E}{R} = \frac{2}{(990 + 10)} = 2 \times 10^{-3}\ \text{A}$$

Substituting values in Eq. (i), we get

E = emf of thermocouple

$$= 2 \times 10^{-3} \times \frac{10}{1} \times 400 \times 10^{-3} = 8 \times 10^{-3}\ \text{V}$$

24. (a) Balance length for emf E_0 is l_0.

$$\therefore \quad E_0 = l_0 x$$

$$x = \frac{E_0}{l_0}$$

If i current flows through 1 Ω resistance giving potential difference as

$$V' = i'(1) = x l_1$$

where, l_1 = balancing length,

So, error $\Delta i = i - x l_1 = i - \frac{E_0}{l_0} l_1$

where, i = measured value from ammeter.

25. (b) Let E_0 is balanced against l_0 length.

$$\Rightarrow \quad x = \frac{E_0}{l_0}$$

Balancing length for unknown potential difference $V' = x l_1 = \frac{E_0}{l_0} l_1$

$$\therefore \ \text{Error} = V - V' = V - \frac{E_0}{l_0} l_1$$

26. (3) In given case, circuit is

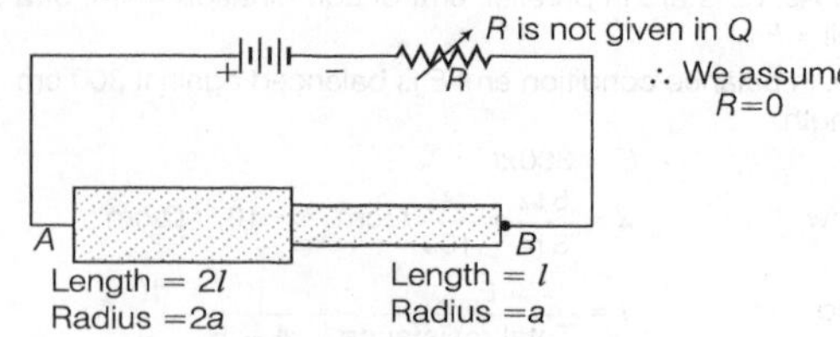

Resistance of wire of length $3l$

= Resistance of thick part + Resistance of thin part

$$= \frac{\rho 2l}{\pi(2a)^2} + \frac{\rho l}{\pi(a^2)} = \frac{\rho l}{\pi a^2}\left(\frac{1}{2} + 1\right) = \frac{3}{2} \cdot \frac{\rho l}{\pi a^2}$$

Total resistance of circuit

= Resistance of potentiometer wire

+ Internal resistance of battery

$$= \frac{3}{2}\frac{\rho l}{\pi a^2} + \frac{\rho l}{2\pi a^2} = \frac{2\rho l}{\pi a^2}$$

So, current in potentiometer wire $= \frac{\text{Emf of supply}}{\text{Total resistance}} = \frac{\varepsilon}{\left(\frac{2\rho l}{\pi a^2}\right)}$

Maximum voltage which can be balanced against potentiometer wire.

V = Potential drop in potentiometer wire

$= I \times R$ (resistance of potentiometer wire)

$$= \frac{E}{\frac{2\rho l}{\pi a^2}} \times \frac{3}{2}\left(\frac{\rho l}{\pi a^2}\right) = \frac{3}{4} E$$

27. (8) To measure resistance of an unknown resistor, the circuit used is

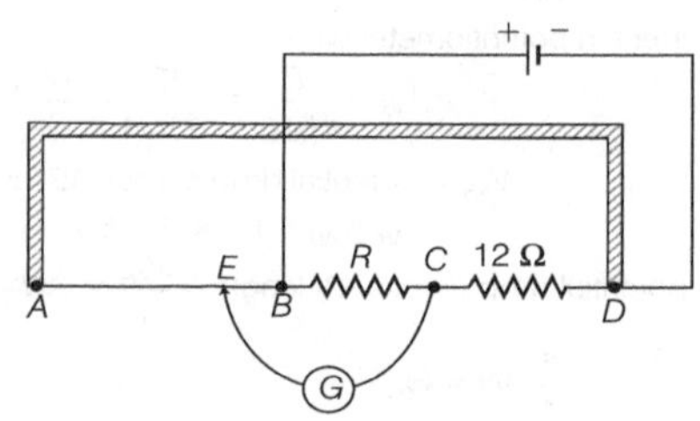

Null deflection length AE = 60 cm above circuit is equivalent to

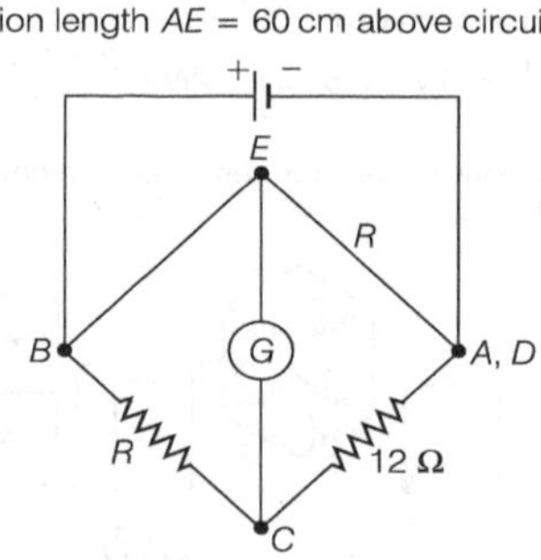

So, in balance condition,

$$\frac{R}{\text{Resistance of } BE} = \frac{12}{\text{Resistance of } AE}$$

$$\frac{R}{l_{BE} \times \text{resistance per cm length}} = \frac{12}{l_{AE} \times \text{resistance per cm length}}$$

$$\frac{R}{40} = \frac{12}{60}$$

$$\Rightarrow \quad R = \frac{12 \times 40}{60} = 8\ \Omega$$

DPP-9 DC Circuits, Power and Energy Considerations

1. Output power $= I^2R = \dfrac{\varepsilon^2 R}{(R+r)^2}$

$$\frac{dP}{dR} = 0$$

$\Rightarrow$ $R = r$

This is maximum at $R = r$

and $P_{max}(\text{output}) = \dfrac{\varepsilon^2}{4r}$

2. In series, $i = \dfrac{2E}{2+R}$

$\therefore$ $J_1 = i^2R = \left(\dfrac{2E}{2+R}\right)^2 R$

In parallel, $i = \dfrac{E}{0.5+R}$

$\therefore$ $J_2 = i^2R = \left(\dfrac{E}{0.5+R}\right)^2 R$

$$\frac{J_1}{J_2} = 2.25 = \frac{4(0.5+R)^2}{(2+R)^2}$$

or $1.5 = \dfrac{2(0.5+R)}{(2+R)}$

On solving, we get $R = 4\,\Omega$

3. For a cell,

$$V = E - Ir$$

r = internal resistance

When there is no current drawn from circuit, $I = 0$ and $V = E$.

In case of charging of a cell direction of current is from positive to negative terminal of cell.

$\therefore$ $V = E + Ir$

or $V > E$

In case of discharging,

$$V = E - Ir \quad \text{or} \quad V < E$$

4. (c) Total power consumed

$$= 60 + 1000 + 40 = 1100\,\text{W}$$

As, power $= \dfrac{V^2}{R}$

$\Rightarrow$ $R = \dfrac{V^2}{P} = \dfrac{220 \times 220}{1100} = 44\,\Omega$

Circuit current,

$$I = \frac{V}{R} = \frac{220}{44} = 5\,\text{A}$$

5. (d) $P = \dfrac{V^2}{R} \Rightarrow R = \dfrac{V^2}{P}$

$$R_1 = \frac{V^2}{100}, \quad R_2 = R_3 = \frac{V^2}{60}$$

Now, $W_1 = \dfrac{(250)^2}{(R_1+R_2)^2} \cdot R_1$

$$W_2 = \frac{(250)^2}{(R_1+R_2)^2} \cdot R_2$$

$$W_3 = \frac{(250)^2}{R_3}$$

$\therefore$ $W_1 : W_2 : W_3 :: 15 : 24 : 64$

or $W_1 < W_2 < W_3$

6. (c) When switch S is closed, resistance of circuit decreases and current increases.

So, A becomes more bright and brightness of B remains same.

7. (a) $P = 100\,\text{MW} = 100 \times 10^6\,\text{W}$

Let R be the resistance of the cable. At 20000 V, the current in the cable is

$$I_1 = \frac{100 \times 10^6}{20000} = 5000\,\text{A}$$

$\therefore$ Power dissipated is

$$P_1 = I_1^2 R = (5000)^2 R = 2.5 \times 10^7 R$$

At 200 V, the current in the cable is

$$I_2 = \frac{100 \times 10^6}{200} = 5 \times 10^5\,\text{A}$$

$\therefore$ Power dissipated is

$$P_2 = I_2^2 R = 2.5 \times 10^{11} R$$

$\therefore$ $\dfrac{P_2}{P_1} = \dfrac{2.5 \times 10^{11}}{2.5 \times 10^7} = 10^4$

8. (c) The heat transmitted per second through the walls of the closed box is given by

$$\frac{Q}{t} = \frac{kA(Q_2 - Q_1)}{d}$$

where, A is the total surface area of the box. Since, a cube has 6 faces, the total surface area is

$$= 6 \times \text{surface area of one face}$$
$$= 6 \times (0.5\,\text{m} \times 0.5\,\text{m}) = 1.5\,\text{m}^2$$

$\therefore$ $\dfrac{Q}{t} = \dfrac{1.68 \times 10^{-1} \times 1.5 \times (100)}{1 \times 10^{-3}}$

$$= 2.52 \times 10^4\,\text{Js}^{-1}$$

In order to maintain the temperature difference between the inside and outside. The heat lost must be compensated by the production of heat through electric current in the coil. The heat produced per second is given by

$$H = \frac{V^2}{R} = 2.52 \times 10^4$$

or $R = \dfrac{420 \times 420}{2.52 \times 10^4} = 7\,\Omega$

9. (a) $P = i^2R$

Current is same, so $P \propto R$

In the first case, it is $3r$, in second case it is $\dfrac{2}{3}r$, in third case it is $\dfrac{r}{3}$ and in fourth case the net resistance is $\dfrac{3r}{2}$.

$$R_{III} < R_{II} < R_{IV} < R_I$$

$\therefore$ $P_{III} < P_{II} < P_{IV} < P_I$

10. (c) Applying, $P = \dfrac{V}{R}$, $R_1 = 1\,\Omega$, $R_2 = 0.5\,\Omega$ and $R_3 = 2\,\Omega$

$$V_1 = V_1 = V_3 = 3\,\text{V}$$

$\therefore$ $P_1 = \dfrac{(3)^2}{1} = 9\,\text{W}$, $P_2 = \dfrac{(3)^2}{0.5} = 18\,\text{W}$

and $P_3 = \dfrac{(3)^2}{2} = 4.5\,\text{W}$

$\therefore$ $P_2 > P_1 > P_3$

11. (c) Power dissipated in resistor is large and amount of heat generated per second is more than the amount of heat dissipated per second (rated power) and resistor may be damaged.

12. (b) As ribbon B, nichrome ribbon has lesser resistance, $(R = \rho l / A)$.

$\therefore$ Rate of heat production in nichrome ribbon is large and so, it melts very quickly when current in the circuit rises.

13. (c) Temperature of a wire increases when heat produced per second is larger than heat lost per second.

$\Rightarrow \quad I^2R = h = 2\pi r l$

h = heat lost per second, per unit area

$\Rightarrow \quad \dfrac{I^2\rho l}{\pi r^2} = h \times 2\pi r l$

So, we have $I^2 \propto r^3$

and steady state temperature is independent of length of wire. So, both wires melt for same value of current.

14. (c) As, $\quad I^2 \propto r^3$

and $\quad A = \pi r^2$

$$r = \left(\frac{A}{\pi}\right)^{1/2}$$

$\Rightarrow \quad I^2 \propto A^{3/2}$

$\Rightarrow \quad \dfrac{I_2^2}{I_1^2} = \dfrac{A_2^{3/2}}{A_1^{3/2}}$

$\Rightarrow \quad I_2^2 = I_1^2\left(\dfrac{A_2}{A_1}\right)^{3/2}$

$$I_2^2 = I_1^2\left(\frac{2A_1}{A_1}\right)^{3/2}$$

$\Rightarrow \quad I_2 = I_1 \cdot 2^{3/4}$

15. (b) Steady state temperature of resistor depends not only on the power but also on surface area, emissivity, etc. Bulb is of less power but acquires more temperature (~ 1600°C).

16. (c) Resistance of bulb,

$$R = \frac{V_R^2}{P_R} = \frac{230 \times 230}{100}\,\Omega$$

and heat produced in 20 min when bulb is connected across 110 V supply.

= Power of bulb in 110 V socket × time

$$= \frac{V^2}{R} \times \text{time} = \frac{110 \times 110 \times 100 \times (20 \times 60)}{230 \times 230}\,\text{J}$$

$$= \frac{110 \times 110 \times 20 \times 60}{23 \times 23} \approx 25\,\text{kJ}$$

17. (b,d) The mass of a wire of length l, cross-sectional area A and density d is given by

$$m = Ald$$

Since, A and d are the same for the two wires and since, their masses are in the ratio of 1 : 2, their lengths and hence, their resistances are in the ratio of 1 : 2, when the wires are connected in series, the same current I flows in each. Since, the heat H produced in time t is given by

$$H = I^2Rt \quad \text{or} \quad H \propto R$$

On the other hand, if the wires are connected in parallel, the potential difference V is the same across the two wires. Since, H is also given by

$$H = \frac{V^2t}{R}$$

or $\quad H \propto \dfrac{1}{R}$

18. (b,c,d) Resistance of 40 W bulb $= \dfrac{V^2}{W} = \dfrac{200 \times 200}{40} = 1000\,\Omega$

Resistance of 60 W bulb $= \dfrac{200 \times 200}{60} = 666.7\,\Omega$

Resistance of 100 W bulb $= \dfrac{200 \times 200}{100} = 400\,\Omega$

Hence, option (c) is correct. Since, the bulbs are connected in series, the current in each is the same. Hence, option (b) is incorrect.

The total resistance = 1000 + 666.7 + 400 = 2066.7 Ω

The current in the circuit is

$$I = \frac{200}{2066.7} = 0.097\text{ A}$$

The option (d) is correct.

Since, the 40 W bulb has the highest resistance and the current in each bulb is same, the potential difference across the 40 W bulb is highest.

19. (a,b,c)

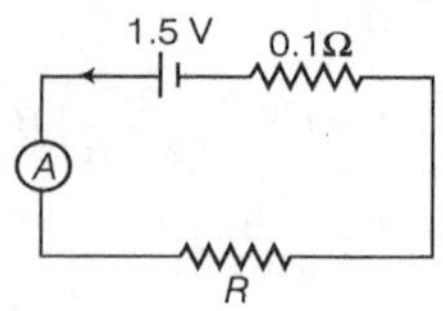

As ammeter shows 2 A, an external resistance of R ohm is connected in the circuit.

$$I = \frac{V}{r + R}$$

$$2 \times (0.1 + R) = 1.5$$

$$0.2 + 2R = 1.5$$

$$2R = 1.3$$

$$R = \frac{1.3}{2} = 0.65\,\Omega$$

Rate of chemical energy consumption

$$\varepsilon\frac{dQ}{dt} = \varepsilon I = 1.5 \times 2 = 3\text{W}$$

Rate of energy dissipation in the cell $= I^2r = 4 \times 0.1 = 0.4\text{W}$

Rate of energy dissipation inside the resistor

$$= I^2R = 4 \times 0.65 = 2.60\text{W}$$

20. (a)

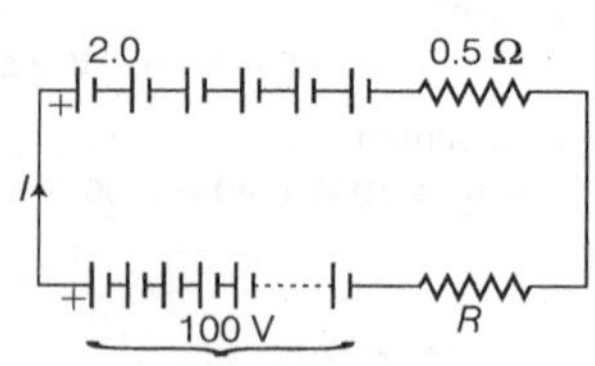

For above circuit,

$$E_{\text{net}} = E_1 - E_2 = 100 - 2 \times 6$$
$$= 100 - 12 = 88\text{ V}$$

As, supply is of much higher emf, so there must be an external resistance in the circuit which limit the current.

$$\text{Current}, I = \frac{E}{R + r}$$

$\Rightarrow \quad 8 = \dfrac{88}{R + 0.5}$

$\Rightarrow \quad R = 11 + 0.5 = 11.5\,\Omega$

Power supplied by source

$$= EI = 100 \times 8 = 800\text{ W}$$

21. (a) 252000 J

22. (a) 201600 J

Obtain power dissipated as heat the remaining power is the sum of the mechanical power yielded by the motor and the chemical power stored in the battery.

Explanations for Q. Nos. (23-25)

We use Kirchhoff's laws to find current through each branch of circuit.

It is found that in every branch, current flows in the directions shown as below.

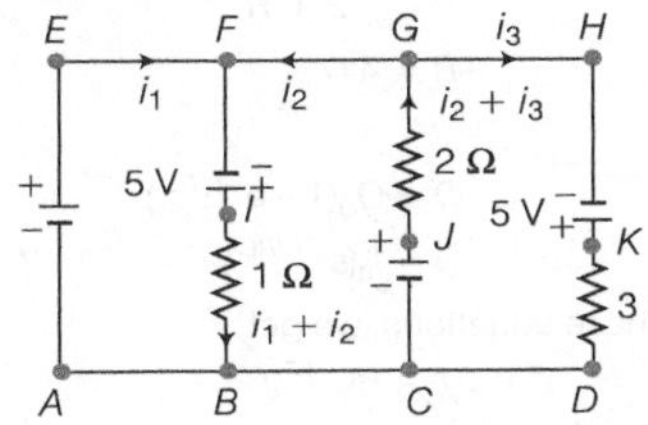

In every branch, current is from the +ve to – ve terminal.
So, batteries are discharging, so all are acting like source.

Power consumed by 1 Ω resistor is maximum and power output by battery in AE branch is maximum.

23. (d)

24. (a)

25. (a)

26. (a) (i) → (q), (ii) → (r), (iii) → (s), (iv) → (p).

In case A, current through each bulb is same. So, power output

$$= I^2R = I^2\left(\frac{V^2}{P}\right)$$

∴ Power output or brightness

$$\propto P_{out} \propto \frac{1}{P_{rated}}$$

Most bright bulbs is of least rated power. Hence, C is glowing with maximum intensity.

In case C, voltage across each bulb is same. So, power output

$$= \frac{V^2}{R} = \frac{V^2}{\frac{V^2_{rated}}{P_{rated}}} = \frac{V^2 P_{rated}}{V^2_{rated}}$$

So, power output or brightness $\propto P_{rated}$

Most bright bulbs are of largest power. Hence, A is most bright.

27. (8) Rate of energy developed $= I^2R$

So, energy developed in

$$10 \text{ s} = I^2Rt = 400$$

$$\Rightarrow \quad R = \frac{400}{2 \times 2 \times 10} = 10\,\Omega$$

When current is

increased to 4 A, energy dissipated will be

$$E = I^2Rt = 4^4 \times 10 \times 20$$

$$= 3200 \text{ joules} = 8 \times 400 \text{ joules}$$

$$\therefore \quad k = 8$$

28. (8) Amount of heat delivered by heater = Amount of heat required by water in phase change

$\Rightarrow$ power × time = mass × latent heat

$\Rightarrow$ 1000×60 (watt × s) $= m(\text{g}) \times 540$ (cal × g)

As 1 cal = 4.2 J

$\Rightarrow$ $60000 = m(\text{g}) \times 540 \times 4.2$ J/g

$$\Rightarrow \quad \frac{60000}{540 \times 4.2} = m$$

$$\Rightarrow \quad m \text{ (in grams)} = \frac{80}{3} \text{ g}$$

$$\therefore \quad N = 8$$

Revisal Problems (JEE Main)

1. (a) Let τ = Relaxation time

$$= \frac{m}{ne^2\rho}$$

$$= \frac{9.1 \times 10^{-31}}{(8.5 \times 10^2)(1.6 \times 10^{-19})^2(1.7 \times 10^{-8})}$$

$$= 2.5 \times 10^{14} \text{ s}$$

$$v_{rms} = \sqrt{\frac{3k_BT}{m}}$$

$$= \sqrt{\frac{3 \times 1.38 \times 10^{-23} \times 300}{9.1 \times 10^{-31}}} = 10^5 \text{ m/s}$$

∴ Mean free path, $\lambda = v_{rms} \times \tau$

$$= 10^5 \times (2.5 \times 10^{-14})$$

$$= 25 \text{ Å}$$

2. (b) Current through conductor,

$$I = \frac{V}{R} = \frac{30}{0.5} = 60 \text{ A}$$

Magnitude of linear momentum of all electrons

= Number of electrons × momentum of 1 electrons

∴ p = sum of momentum of all electrons

$$= nAl \times mv_d$$

$$= (nev_dA) \cdot \frac{lm}{e} \qquad (\text{as, } I = neAv_d)$$

$$= \frac{Ilm}{e} = \frac{60 \times 100 \times 9.1 \times 10^{-31}}{1.6 \times 10^{-19}}$$

$$= 3.4 \times 10^{-8} \text{ kg-m/s}$$

3. (c) We use

$$J = \sigma E, \; E = \rho J, \; \sigma = \frac{1}{\rho}$$

$$E = \frac{dV}{dx}, \; J = \frac{I}{\pi(b^2 - a^2)}$$

When we consider longitudinal resistance,

$$\therefore \quad \frac{dV}{dx} = \rho \frac{I}{\pi(b^2 - a^2)}$$

$$\Rightarrow \quad -\int_{V_1}^{V_2} dV = \int_0^L \frac{\rho I \, dx}{\pi(b^2 - a^2)}$$

$$\Rightarrow \quad V = V_1 - V_2 = \frac{\rho LI}{\pi(b^2 - a^2)}$$

$$\text{So,} \quad R = \frac{V}{I} = \frac{\rho L}{\pi(b^2 - a^2)}$$

Also, when radial resistance is considered,

$$E = -\frac{dV}{dr}, \; J = \frac{I}{2\pi rL}$$

$$\therefore \quad -\frac{dV}{dr} = \frac{\rho I}{2\pi rL}$$

$$\therefore \quad -\int_{V_1}^{V_2} dV = \int_a^b \frac{\rho I \, dr}{2\pi rL} \quad \text{or} \quad V = V_1 - V_2 = \frac{\rho I}{2\pi L} \log_e\left(\frac{b}{a}\right)$$

$$\text{Hence,} \quad R = \frac{V}{I} = \frac{\rho}{2\pi L} \cdot \log_e\left(\frac{b}{a}\right)$$

∴ The ratio of resistances is

$$\frac{R_e}{R_r} = \frac{\frac{\rho L}{\pi(b^2 - a^2)}}{\frac{\rho}{2\pi L} \log_e\left(\frac{b}{a}\right)} = \frac{2L^2}{(b^2 - a^2)\log_e\left(\frac{b}{a}\right)}$$

4. (d) In series combination $P_{max} = \frac{E^2}{2r}, R = 2r$

In parallel combination $P_{max} = \frac{E^2}{2r}, R = \frac{r}{2}$

$\therefore$ The ratio of power output = 1 : 1

5. (d) Circuit can be redrawn as

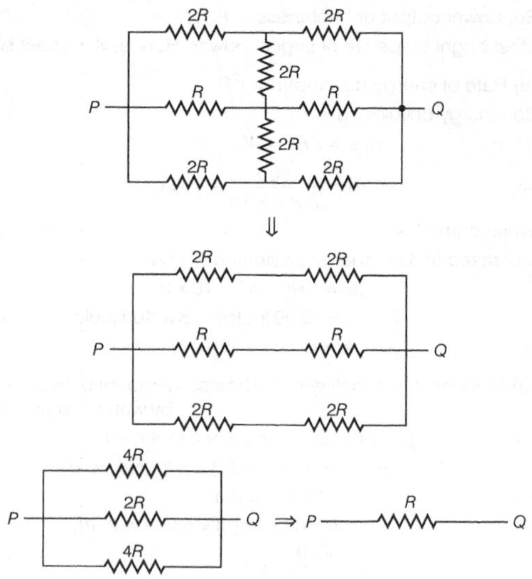

$\therefore \qquad R_{eq} = 5\,\Omega$

6. (c) Let Q_0 = steady state charge on the capacitor

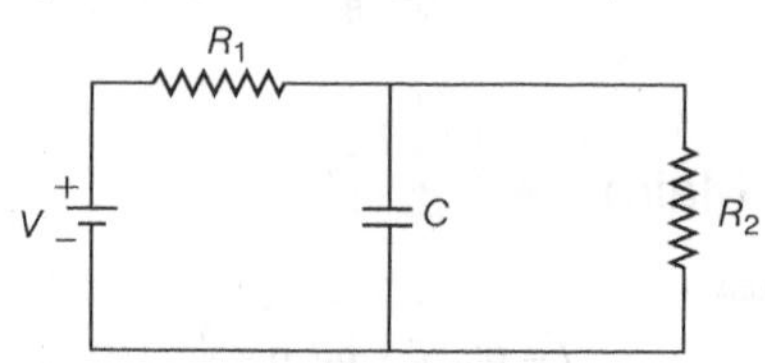

$= C$ (potential drop across C in steady state)

$= C\,(IR_2) = C\left(\frac{V R_2}{R_1 + R_2}\right)$

$\alpha = \frac{1}{f}$ or $\frac{1}{CR_{eq}} = \frac{1}{C\left(\frac{R_1 R_2}{R_1 + R_2}\right)}$

$\therefore \qquad \alpha = \frac{R_1 + R_2}{CR_1R_2} = \frac{1000}{4 \times 10^{-6} \times \frac{1000}{4}}$

$= \frac{10^3}{10^{-6}} = 10^9$

7. (c) In series combination, $i = \frac{2E}{2 + R}$

$\Rightarrow \qquad P_1 = i^2R = \left(\frac{2E}{2 + R}\right)^2 R$

In parallel combination $i = \frac{E}{0.5 + R}$

$P_2 = i^2R = \left(\frac{E}{0.5 + R}\right)^2 R$

Given that, $\qquad P_2 = \frac{P_1}{2.25}$

$\Rightarrow \qquad \frac{P_1}{P_2} = 2.25 = \frac{4(0.5 + R)^2}{(2 + R)^2}$

or $\qquad 1.5 = \frac{2(0.5 + R)}{2 + R}$

or $\qquad R = 4\,\Omega$

8. (c) We use

$$Q = Q_0(1 - e^{-t/RC})$$

$$I = I_0(e^{-t/RC})$$

On solving these equations, we get

$$t = RC \log 2$$

9. (b) Potential drop across R_3 = Potential drop across R_2

$$i_3R_3 = i_2R_2$$

$\Rightarrow \qquad i_2 = i_3\frac{R_3}{R_2} = i \times \frac{6}{3} = 2i$

$\therefore$ Potential drop across $R_2 = 3i\ R_3 = 6i$

Potential drop across,

$$R_4 = i_4R_4$$

$\therefore \qquad i_4R_4 = 6i + 6i$

$\Rightarrow \qquad i_4 = i$

10. (a) By using Kirchhoff's law,

$$V_C + 1 - 2 = V_E$$

$\therefore \qquad V_C - V_E = 1\text{ V}$

11. (c) To minimise error due to uneven damping of ends. Unknown resistance is switched between gaps and then mean value is taken.

12. (d) As deflection is increasing towards end B, battery of box must be reversed or polarity of driver cell must be reversed.

13. (c) At balance point, no current flows through the galvanometer, hence increasing or decreasing R_2 has no effect on balance length.

14. (a) Both the Wheatstone bridges are given balanced.

$\therefore \qquad \frac{3}{R_1} = \frac{9}{6}$

and $\qquad \frac{4}{8} = \frac{R_2}{16}$

$\Rightarrow \qquad R_1 = \frac{6 \times 3}{9} = 2$

$R_2 = \frac{4 \times 16}{8} = 8$

Hence, $\qquad \frac{R_1}{R_2} = \frac{2}{8} = \frac{1}{4}$

So, $\qquad k = 4$

15. (b) $R \times 0.1 = ki_1$

$$\frac{R}{R + x} = \frac{i_1}{i_2}$$

On solving, we get

$$R = 10\,\Omega$$

$$x = 15\,\Omega$$

So, $\qquad \frac{R}{x} = \frac{10}{15} = \frac{2}{3}$

So, $\qquad k = 2$

16. (c) Consider any section of given network.

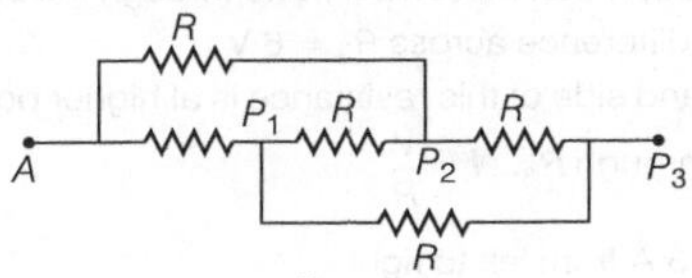

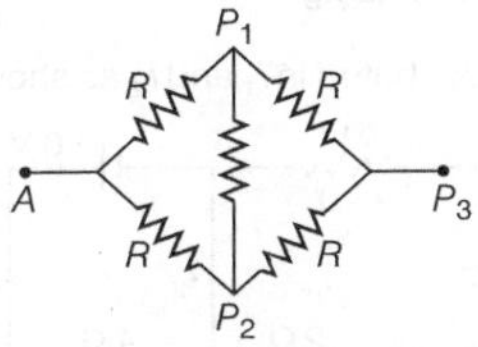

This is a balanced Wheatstone bridge.

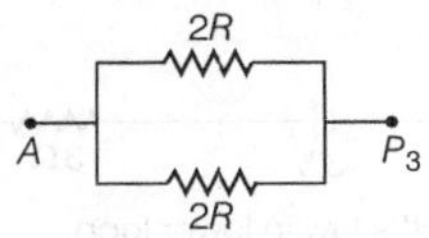

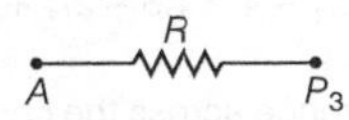

Which reduces given circuit to

$$R_{eq} = R\left(1 + \frac{1}{2} + \frac{1}{4} + \frac{1}{8} + \ldots\ldots\right)$$

Infinite GP sum $= \frac{a}{1-r} = \frac{1}{1-\frac{1}{2}} = 2$

$\therefore \quad R_{eq} = 2R$

17. (b) Given,

***Case* I** $i = \frac{4E}{R}$

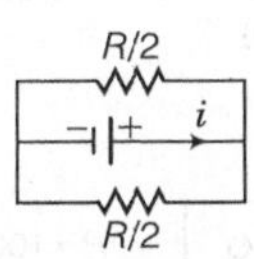

***Case* II** $i = \frac{IbE}{3R}$

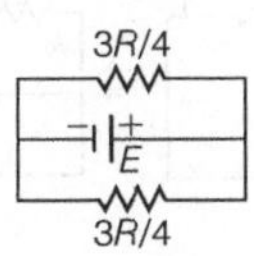

18. (a) $I = \dfrac{1.4}{2 + \frac{4}{3} + \frac{100R}{100 + R}} = \frac{2}{100} = 200\,\Omega$

$$V = I \times 200 = \left(\frac{2}{100} - I\right) \times 100$$

$\Rightarrow \quad I = \frac{2}{300}$

So, $\quad V = \frac{2}{300} \times 200 = 1.33\text{ V}$

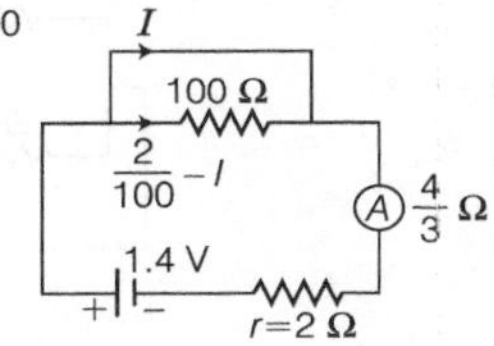

So, error in measurement of voltage

$= 1.33 - 1.0 = 0.33$ V

19. (b) When S is open, R_1 and R_2 are in series, total resistance

$$R_1 = 4000 + 6000 = 10000\,\Omega$$

Voltmeters are also in series, then resistance of voltmeters

$$R_2 = 6000 + 4000 = 10000\,\Omega$$

So, $\quad R_{eq} = \frac{R_1 R_2}{R_1 + R_2} = 5000\,\Omega$

$$I = \frac{E}{R_{eq}} = \frac{250}{5000} = \frac{1}{20}\text{ A}$$

Current i_1 (in voltmeters) $= \frac{1}{2} \times \frac{1}{20} = \frac{1}{40}$ A

$\therefore$ Potential drop across, $V_1 = \frac{1}{40} \times 6000 = 150$ V

and potential drop across, $V_2 = \frac{1}{40} \times 4000 = 100$ V

So, $\quad \frac{V_1}{V_2} = \frac{150}{100} = 1.5$

and when switch is closed,

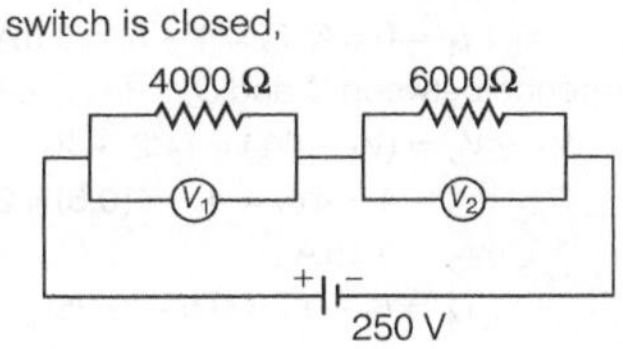

$V_1 = 125$ V, $V_2 = 125$ V

$\therefore \quad \frac{V_1}{V_2} = 1$

20. (b) $I^2RT = ms\Delta\theta$

$$i^2\rho\frac{l}{A}\cdot t = Al\rho \cdot s\Delta\theta$$

On solving, we get $\quad t = 55.5$ s

Also, time is not length dependent.

Revisal Problems (JEE Advanced)

1. (a) $\tau_{C_1} = \tau_{C_2} = 2CR = \tau_C$

$$q_1 = CE(1 - e^{-\frac{t}{\tau_C}}),\quad q_2 = EC(1 - e^{-\frac{t}{\tau_C}})$$

$$\frac{q_1}{q_2} = \frac{1}{2} \qquad \text{(at anytime)}$$

The ratio of steady state charges is also 1 : 2.

$$\frac{dq_1}{dt} = \frac{EC}{\tau_C}e^{-\frac{t}{\tau_C}}$$

$$\frac{dq_2}{dt} = \frac{2EC}{\tau_C}e^{-\frac{t}{\tau_C}}$$

$$\therefore \quad \left(\frac{dq_1}{dt}\right) \neq \left(\frac{dq_2}{dt}\right)$$

2. (a) Under steady state power developed = power lost

or $\quad i^2R = \lambda(\theta - \theta_0)$

$$\theta - \theta_0 = \frac{i^2R}{\lambda}$$

This is also the increase in temperature of the resistance.

$$\Delta l = l \propto \Delta\theta$$

$$\text{Strain} = \frac{\Delta l}{l} = \alpha\Delta\theta = \alpha(\theta - \theta_0) = \frac{i^2R\alpha}{\lambda}$$

3. (b) Given circuit is a balanced wheatstone bridge, so no current flows through $6R$ resistor and equivalent circuit is

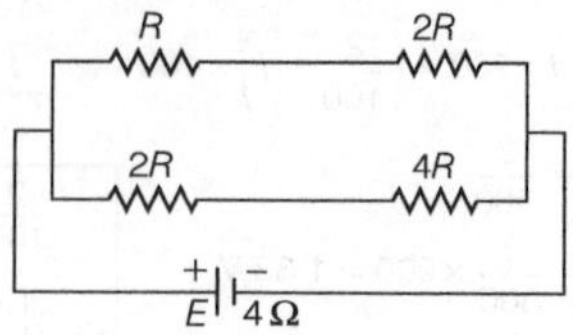

So, $R_{eq} = \left(\frac{1}{3R} + \frac{1}{6R}\right)^{-1} = 2R$

For maximum power output, internal resistance of cell = external resistance

$\Rightarrow \quad 4 = 2R$

or $\quad R = 4/2 = 2\,\Omega$

4. (d) Net resistance of a circuit is $9\,\Omega$. So, current from cell $i = \frac{9}{9} = 1\text{ A}$ = current through $3\,\Omega$ resistor.

Potential difference between points A and B is

$$V_A - V_B = 9 - 1(3+2) = 4\text{ V} = 8i$$

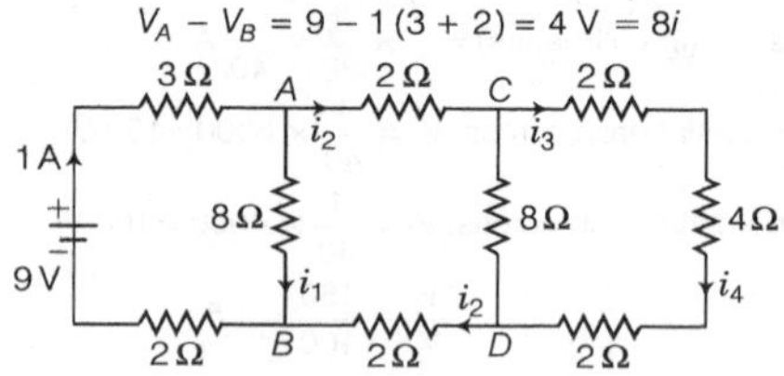

$\therefore \quad i_1 = 0.5\text{A}, i_2 = 1 - i_1 = 0.5\text{A}$

Potential difference between C and D,

$$V_C - V_D = (V_A - V_B) = i_2(2+2)$$
$$= 4 - 4i_2 = 4 - 4(0.5) = 2\text{ V} = 8i_3$$

$\Rightarrow \quad i_3 = 0.25\text{A}$

$\Rightarrow \quad i_4 = i_2 - i_3 = 0.5 - 0.25$

$\therefore \quad i_4 = 0.25\text{ A}$

5. (c) Equivalent circuit is

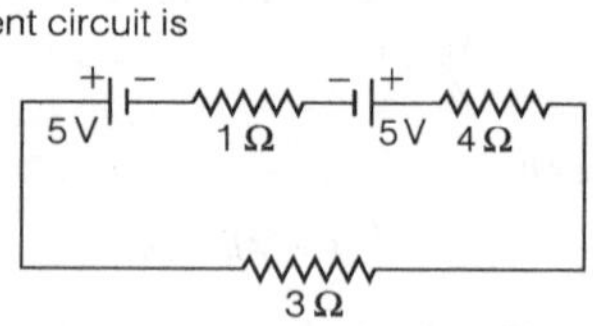

As, net emf of circuit = 0

$\therefore \quad I = 0$

Also, current through 6 V battery is

$$I' = \frac{6-4}{2} = \frac{1}{2}\text{ A}$$

6. (d) At 20 cm mark, reading of voltmeter is zero.

At some distance x from centre,

$$V_P = 4\left(\frac{20+x}{40}\right)$$

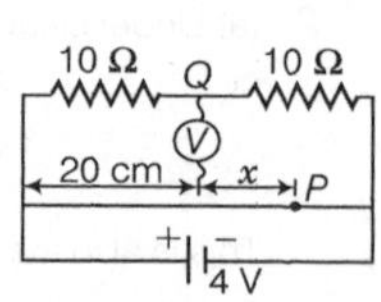

and $\quad V_Q = 2\text{ V}$

$\therefore \quad V_P - V_Q = \frac{x}{10}$ volt

$\Rightarrow \quad 2\sin \pi t = \frac{x}{10}$

As, $\quad x = 20\sin \pi t$

$\therefore \quad \frac{dx}{dt}$ = velocity

$= 20\pi \cos \pi t$ cm/s

7. (a) In steady state, no current flows through the capacitor.

Potential difference across $R_3 = 6$ V.

As, left hand side of this resistance is at higher potential,

Current through R_3, : $I = \frac{V}{R}$

$i = \frac{6}{4} = 1.5$ A from left to right.

This current I is distributed in i_1 and i_2 as shown

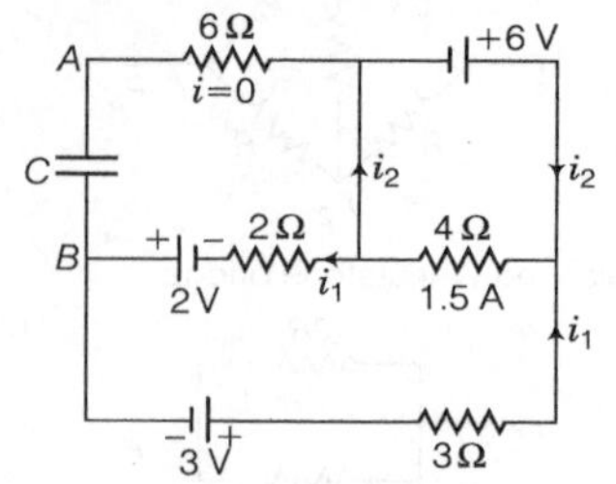

Applying Kirchhoff's law in lower loop,

$$3 - 3i_1 - 4 \times 1.5 - 2i_1 + 2 = 0$$

$\Rightarrow \quad i_1 = -0.2\text{ A}$

Now, potential difference across the capacitor V_{AB} is

$$V_A - 2i_1 + 2 = V_B$$
$$V_A - V_B = 2i_1 - 2 = -2.4\text{V}$$

or $\quad V_{BA} = 2.4\text{V}$

So, energy stored in capacitor

$$U = \frac{1}{2}CV_{AB}^2$$
$$= \frac{1}{2} \times 5 \times 10^{-6} \times (2.4)^2$$
$$= 1.44 \times 10^{-5}\text{ J}$$

8. (c) Given circuit is a balanced Wheatstone bridge (including the voltmeter's resistance)

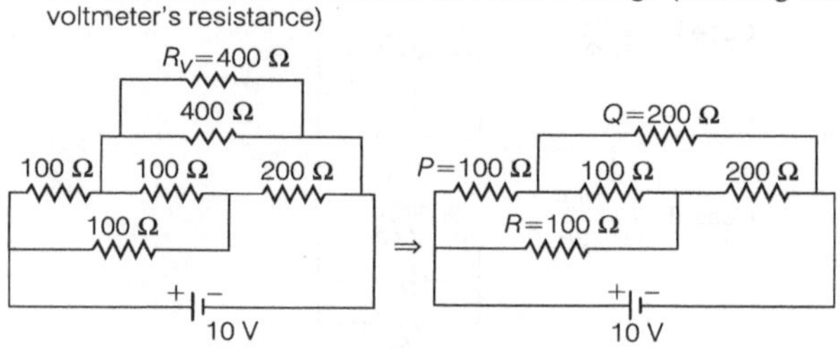

$\therefore \quad \frac{P}{Q} = \frac{R}{S}$ and voltmeter will read the potential difference across resistance Q.

$$\text{Current, } i_1 = i_2 = \frac{10}{100+200} = \frac{1}{30}\text{A}$$

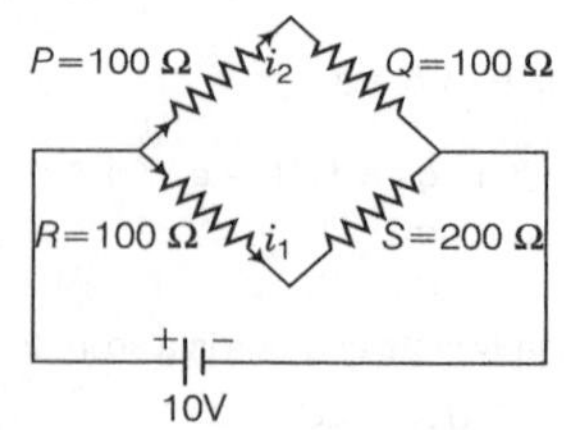

$\therefore$ Potential difference across voltmeter $= Qi_1 = 200 \times \frac{1}{30}$

$$= \frac{20}{3}\text{ V}$$

9. (d) Points H and F are at same potentials.

Points I and E are at same potentials.

Points J and D are at same potentials.

Points K and C are at same potentials.

So, circuit can be redrawn by joining points of same potentials as

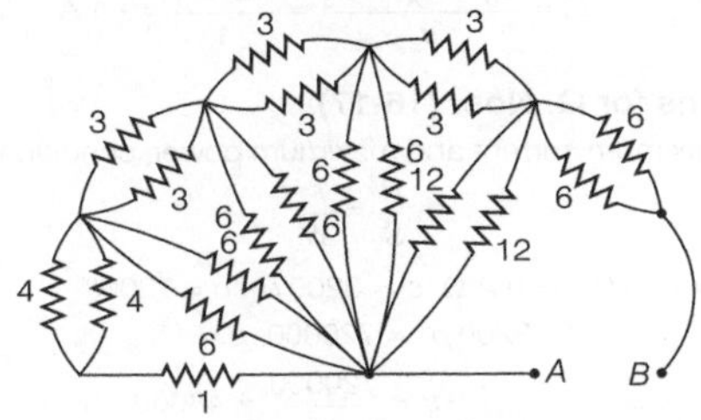

Or

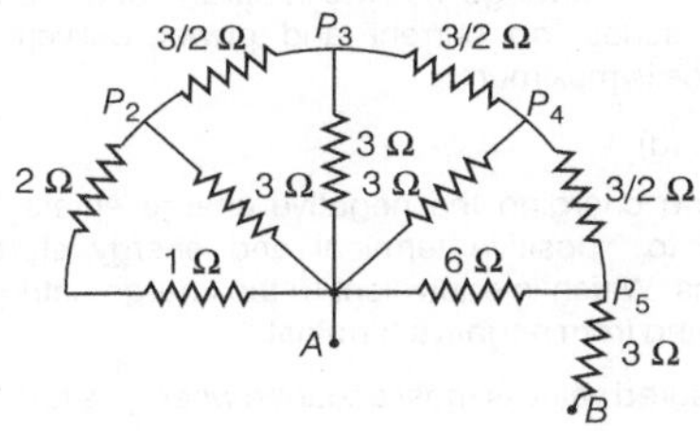

Or

$R_{eq} = H = 3\,\Omega$

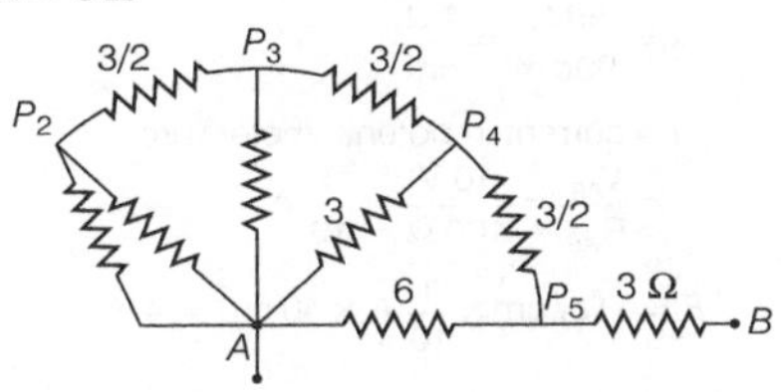

$$R_{eq} = \frac{3 \times 3}{3+3} = \frac{3}{2}\ \Omega$$

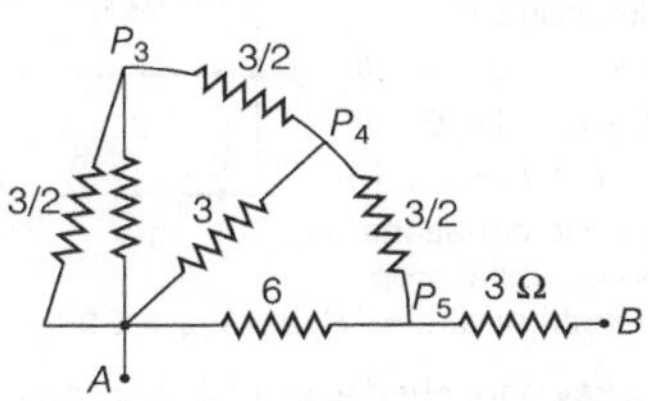

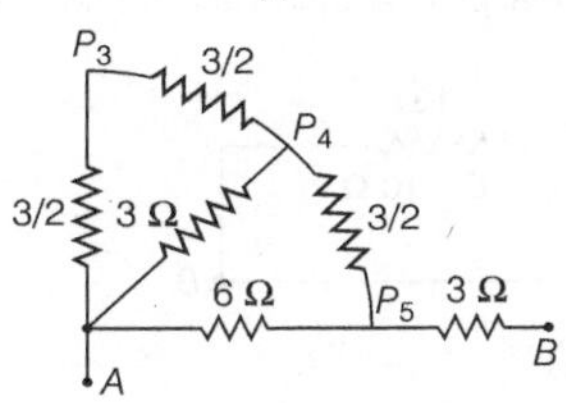

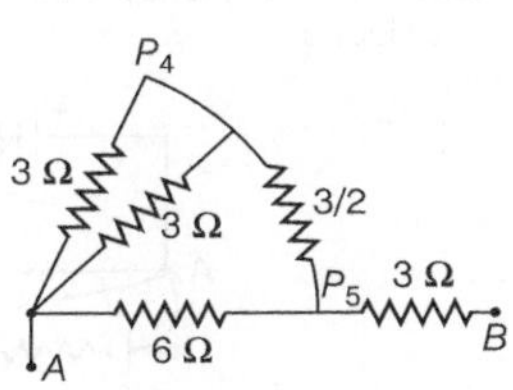

10. (d) Due to internal resistance of driving cell, V_{CD} = potential drop across CD length is less than ε (emf of driver cell)

$$\Rightarrow \qquad V_{CD} < \varepsilon \Rightarrow \frac{V_{CD}}{2} < \frac{\varepsilon}{2}$$

$\Rightarrow$ Potential drop of 50 cm of length $< \frac{\varepsilon}{2}$.

Now, when a cell of emf $\varepsilon/2$ has to be balanced against potential drop of potentiometer wire, as no current flows through slave cell.

Potential drop across balance length = $\varepsilon_1/2$.

$\therefore$ Balance length is more than 50 cm.

11. (b,c) As, current enters through A and distributes in infinite network. So, by symmetry current divides equally in 4 equal parts

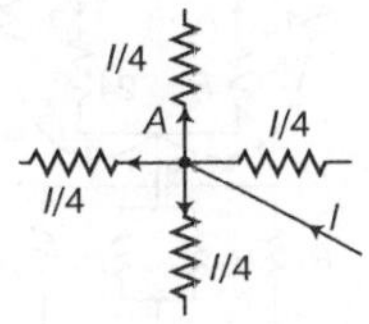

and currents from circuit are directed towards B,

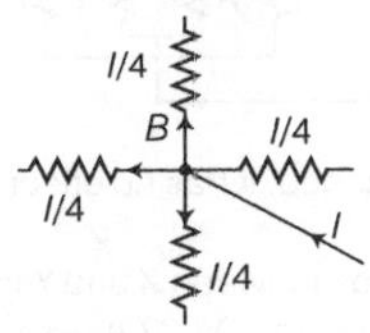

When we consider A and B as

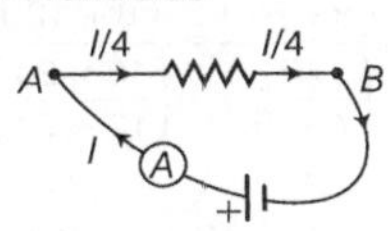

So, currents in AB branch superimposes and $I_{AB} = \frac{I}{2}$.

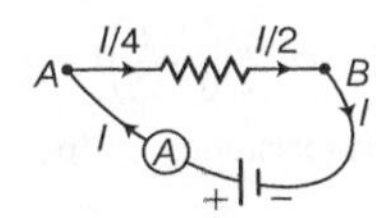

and so we consider remaining circuit as R'_{AB}.

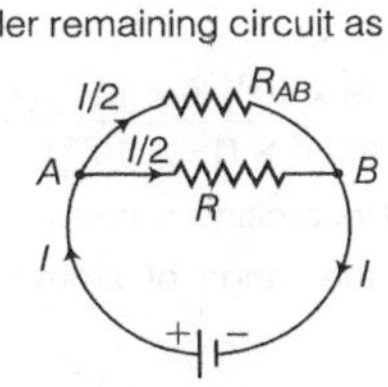

Now, $\quad V_{AB} = \frac{I}{2}R = \frac{I}{2}R_{eq} \Rightarrow R'_{eq} = \frac{R}{2}$

So, equivalent resistance is $\frac{R}{2}$.

Also, $\qquad V = IR_{eq}$

$$\Rightarrow \qquad I = \frac{V}{R_{eq}} = \frac{V}{R/2}$$

$$\Rightarrow \qquad \frac{1}{4/2} = \frac{1}{2} = 0.5\ \text{A}$$

12. (a,c) Given circuit can be redrawn as,

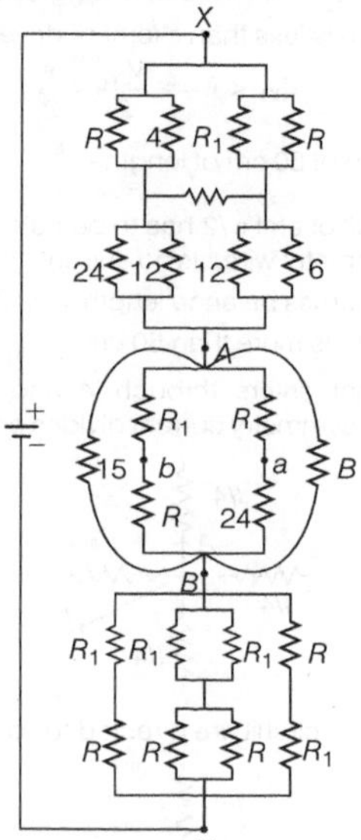

If value of R_2 is changed, it has no effect on equivalent resistance of circuit.

∴ Wheatstone bridge between X and Y is balanced.

So, $$\frac{\left(\frac{RR_1}{R+R_1}\right)}{\left(\frac{4R}{4+R}\right)} = \frac{\left(\frac{6\times 12}{6+12}\right)}{\left(\frac{12\times 24}{12+24}\right)}$$

On solving, we get

$$R_1 = \frac{4R}{4+2R} = \frac{2R}{2+R}$$

13. (b,d) $I^2Rt = ms\Delta Q \Rightarrow I^2\left(\rho_1 \frac{l}{A}\right)t = Al\rho_2 s\Delta Q$

ρ_1 = specific resistance, ρ_2 = density

$$\therefore \quad t = \frac{A^2\rho_2 s\Delta Q}{I^2\rho_1} \quad \text{...(i)}$$

$$= \frac{(0.5\times 10^{-6})^2\,(9\times 10^3)\,(9\times 10^{-2})\,(4.18\times 1050)}{(10)^2\times(1.6\times 10^{-8})} = 55.5\text{ s}$$

Also, time required for melting is independent of length of wire.

14. (b,c) To increase the range of ammeter, a 1Ω resistance is required which is

$$S = \left(\frac{i_g}{\rho - i_g}\right)G$$

From option (c), $$S = \frac{50\times 10^{-6}\times 100}{5\times 10^{-3} - 50\times 10^{-6}}$$

$\Rightarrow$ $S = 1\Omega$

To change ammeter into voltmeter, a high resistance in species must be used and it is given by

$$R = \frac{V}{i_g} - G$$

From option (b), $$R = \frac{10}{5\times 10^{-6}} - 100 = 200\text{ k}\Omega$$

15. (a,d) Equivalent emf of three cells is $E_{eq} = \frac{\Sigma(E/r)}{\Sigma(1/r)} = 2\text{ V}$

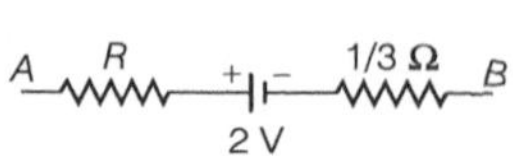

Equivalent circuit is

Since, no current is drawn from cell,

$V_{AB} = 2\text{ V}$ and $V_{AB} = V_A - V_B = E_1 - i_1 r_1$

$$\therefore \quad i_1 = \frac{V_B - V_A + E_1}{r_1} = \frac{-2+3}{1} = 1\text{A}$$

$$i_2 = \frac{V_B - V_A + E_2}{r_2} = \frac{-2+2}{1} = 0$$

$$i_3 = \frac{V_B - V_A + E_3}{r_3} = \frac{-2+1}{1} = -1\text{ A}$$

Explanations for Q. Nos. (16-17)

For maximum current and maximum power, condition is same

$$\frac{R}{S} = \frac{r}{p},$$

$R = 640\,\Omega$, $r = 0.2\,\Omega$, $s = 3200\,p$, $sp = 72000$

$$\therefore \quad 3200\,p^2 = 720000,\ p = 15$$

$$s = \frac{720000}{15} = 48000$$

So, we should arrange 15 rows in parallel each containing 48000 cells in series, so current and power delivered to external resistance is maximum.

16. (a) **17.** (d)

18. (b) During charging the negative charge enters from negative terminal to positive terminal and energy stored in battery increases. Which is equivalent to the charge entring the terminal and leaving from negative terminal.

19. (b) Measured value is most accurate when $\frac{P}{Q}$ is nearly equals to 1.

20. (a) The emf of cell = potential drop of balance length = $xl l_{(AC)}$

Here, $$x = \frac{99\,\Omega}{99\text{cm}} = \frac{1\,\Omega}{\text{cm}}$$

I = current in potentiometer wire

$$= \frac{V_{AB}}{R_{AB}} = \frac{10\text{ V}}{100\,\Omega} = \frac{1}{10}\text{ A}$$

$$\therefore \quad E = 1\,\Omega/\text{cm}\times\frac{1}{10}\text{ A}\times 40\text{ cm} = 4\text{ V}$$

21. (a) When jockey is touched at point B, the circuit will be applying Kirchhoff's law.

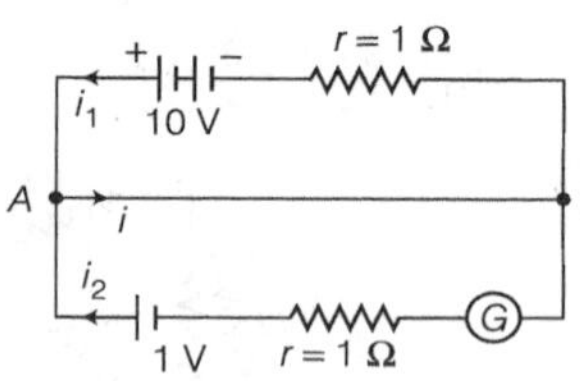

$-99i - i_1 + 10 = 0$ ...(i)

$-99i - i_2 + 1 = 0$ ...(ii)

and $i_1 + i_2 = i$ ...(iii)

On solving these equations, we get i and potential drop across AB length of wire $= i\,R_{AB} = V_{AB} = 6.93\text{ V}$

22. (d) When jockey is touched at point A, then there are two separate loops.

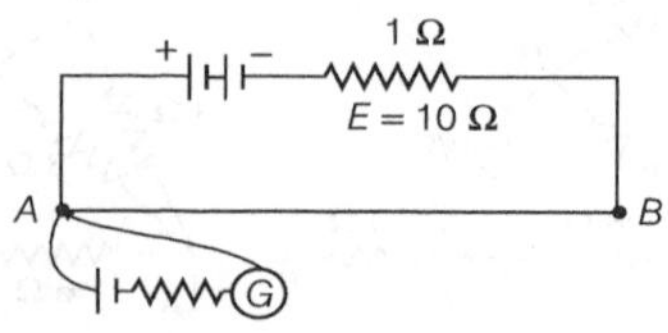

23. (a) When jockey is touched at point A, then potential difference across the terminals of driver cell is given

$$V = E - Ir = 9.9\text{ V}$$

24. (b) (i) → (p,q,r,s), (ii) → (s), (iii) → (p,q), (iv) → (r)

All are Wheat stone bridge. Only S is balanced as

$$\frac{P}{Q} = \frac{R}{S}$$

To get direction of current, find V_B and V_D.

JEE Main & AIEEE Archive

1. (c) Total power (P) consumed

$$= (15\times 40) + (5\times 100) + (5\times 80) + (1\times 1000) = 2500 \text{ W}$$

As, we know that, $P = VI$

$$\Rightarrow \quad I = \frac{2500}{200} = \frac{125}{11} = 11.3 \text{ A}$$

Hence, minimum capacity should be 12 A.

2. (d) $P = \dfrac{V^2}{R}$

Resistance of the bulb,

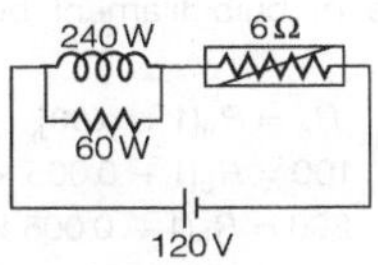

Initially with only bulb in circuit,

$$R = \frac{120\times 120}{60} = 240\,\Omega$$

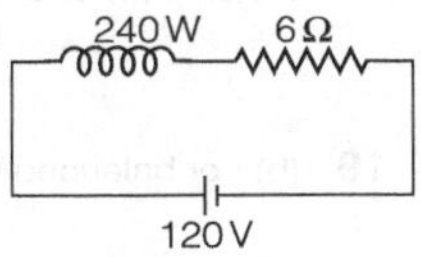

$$R_{eq} = 240 + 6 = 246\,\Omega$$

$$\Rightarrow \quad I_1 = \frac{V}{R_{eq}} = \frac{120}{246}$$

$$V_1 = I_1R_1 = \frac{120}{246}\times 240$$

$$= 117.073 \text{ V}$$

Resistance of the heater $= \dfrac{V^2}{P} = \dfrac{120\times 120}{240} = 60\,\Omega$

As, bulb and heater are connected in parallel.

Net resistance $= \dfrac{240\times 60}{300} = 48\,\Omega$

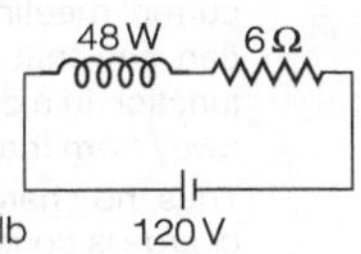

Total resistance, $R_2 = 48 + 6 = 54\,\Omega$

Total current, $I_2 = V/R_2 = 120/54$

Potential across heater = Potential across bulb

$$V_2 = \frac{120}{54}\times 48 = 106.66 \text{ V}$$

$$V_1 - V_2 = 117.073 - 106.66 = 10.04 \text{ V}$$

3. (d) To increase the range of ammeter, resistance should be decreased (so additional shunt is connected in parallel).

4. (c) As, the rated power of 25 W is less than 100 W, it implies that 25 W bulb has higher resistance. As in series connection, current through both the bulbs is same but heating in 25 W bulb is more than that of 100 W bulb. So, 25 W bulb will get fused.

5. (a) From Ohm's law, $R = V/I$

By error method, $\dfrac{\Delta R}{R} = \dfrac{\Delta V}{V} + \dfrac{\Delta I}{I} = 3\% + 3\% = 6\%$

6. (b) Resistance of combination, $R_e = 4R$

$$\Rightarrow \quad \frac{\Delta R_e}{R_e} = \frac{\Delta R}{R} = \frac{5\times 100}{100} = 5\%$$

7. (d) Potential gradient of a potentiometer,

$$K = \frac{V}{l} = \frac{IR}{l} = \frac{I\rho l}{lA}$$

$$K = \frac{I\rho}{A} = \frac{0.2\times 4\times 10^{-7}}{8\times 10^{-7}}$$

$$= 0.1 \text{ V/m}$$

8. (d)

$$R = \frac{\rho l}{A} = \frac{\rho l^2}{V} \qquad [\because V = \text{volume}]$$

$$\therefore \quad \frac{\Delta R}{R} = 2\frac{\Delta l}{l} = +\,0.2\%$$

P and V are constants.

9. (b) A moving conductor is equivalent to a battery of emf $= VBl$ [motion emf]

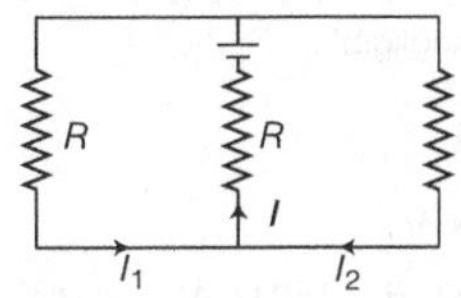

Equivalent circuit, $\quad I = I_1 + I_2 \quad$ ...(i)

Applying Kirchhoff's law, $I_1R + IR - VBl = 0 \quad$ ...(ii)

$I_2R + IR - VBl = 0 \quad$ ...(iii)

Adding Eqs. (ii) and (iii), we get

$$I_1R + IR - VBl + I_2R + IR - VBl = 0$$

$$\Rightarrow \quad I_1R + I_2R + 2IR - 2VBl = 0$$

$$\Rightarrow \quad (I_1 + I_2)R + 2IR - 2VBl = 0$$

$$\Rightarrow \quad 2IR + IR = 2VBl \qquad \text{[From Eq. (i)]}$$

$$\Rightarrow \quad I = \frac{2VBl}{3R}$$

Subtracting Eq.(iii) from Eq.(ii), we get

$$I_1R + IR - VBl = 0$$

$$\underset{-}{}I_2R \underset{-}{+} IR \underset{+}{-} VBl \underset{-}{=} 0$$

$$I_1R - I_2R = 0$$

$$\Rightarrow \quad R(I_1 - I_2) = 0 \quad \Rightarrow \quad I_1 - I_2 = 0 \qquad [\because R \neq 0]$$

$$\Rightarrow \quad I_1 = I_2 \qquad \text{...(iv)}$$

From Eq. (i), we get

$$I = 2I_1 \qquad (\because I_1 = I_2)$$

$$\Rightarrow \quad 2I_1 = I \quad \Rightarrow \quad I_1 = \frac{I}{2} = \frac{2VBl}{2\times 3R} = \frac{VBl}{3R}$$

$$\therefore \quad I_1 = I_2 = \frac{VBl}{3R}$$

10. (b) At $t = 0$, inductor behaves like an infinite resistance.

So, at $t = 0$, $I = \dfrac{V}{R_2}$

and at $t = \infty$, inductor behaves like a conducting wire.

$$\therefore \quad I = \frac{V}{R_{eq}} = \frac{V(R_1 + R_2)}{R_1R_2}$$

11. (d) Let R_0 be the initial resistance of both conductors.

$\therefore$ At temperature θ, their resistances will be

$$R_1 = R_0(1 + \alpha_1\theta)$$

and

$$R_2 = R_0(1 + \alpha_2\theta)$$

For series combination,

$$R_s = R_1 + R_2$$

$$R_{s0}(1 + \alpha_s\theta) = R_0(1 + \alpha_1\theta) + R_0(1 + \alpha_2\theta)$$

where, $\quad R_{s0} = R_0 + R_0 = 2R_0$

$$\therefore \quad 2R_0(1 + \alpha_s\theta) = 2R_0 + R_0\theta(\alpha_1 + \alpha_2)$$

or

$$\alpha_s = \frac{\alpha_1 + \alpha_2}{2}$$

For parallel combination,

$$R_p = \frac{R_1R_2}{R_1 + R_2}$$

$$R_{p0}(1 + \alpha_p\theta) = \frac{R_0(1 + \alpha_1\theta)R_0(1 + \alpha_2\theta)}{R_0(1 + \alpha_1\theta) + R_0(1 + \alpha_2\theta)}$$

where, $\quad R_{p0} = \dfrac{R_0R_0}{R_0 + R_0} = \dfrac{R_0}{2}$

$$\therefore \quad \frac{R_0}{2}(1 + \alpha_p\theta) = \frac{R_0^2(1 + \alpha_1\theta + \alpha_2\theta + \alpha_1\alpha_2\theta^2)}{R_0(2 + \alpha_1\theta + \alpha_2\theta)}$$

As α_1 and α_2 are small quantities.

$\therefore$ $\alpha_1 \alpha_2$ is negligible.

or $$\alpha_p = \frac{\alpha_1 + \alpha_2}{2 + (\alpha_1 + \alpha_2)\theta} = \frac{\alpha_1 + \alpha_2}{2}\left[1 - \left(\frac{\alpha_1 + \alpha_2}{2}\right)\theta\right]$$

As $(\alpha_1 + \alpha_2)^2$ is negligible.

$$\therefore \quad \alpha_p = \frac{\alpha_1 + \alpha_2}{2}$$

12. (d) $R = R_t\ (1 + \alpha\, \Delta t)$

Given, $R_t = 100\ \Omega$, $R = 150\ \Omega$, $\Delta t = 200°$ C

$\therefore$ $150 = 100\,(1 + \alpha\, 200) \Rightarrow 0.5 = 200\,\alpha$

$\Rightarrow$ $\alpha = \dfrac{0.5}{200} = \dfrac{1}{400} = 25 \times 10^{-4}/°C$

13. (b) When bridge is balanced, ratio of resistance in opposite arms is equal.

$$\frac{P}{Q} = \frac{R}{S}$$

Given, $P = 55\ \Omega$, $Q = R$

As $R \propto l$

$R \propto l_1$, where $l_1 = 20$ cm

$S \propto l_2$, where $l_2 = 100 - 20 = 80$ cm

$$\frac{R}{S} = \frac{l_1}{l_2} = \frac{20}{80}$$

$\therefore$ $\dfrac{55}{R} = \dfrac{20}{80}$ or $R = 220\ \Omega$

14. (c) Let potential at P_1 be 0 V and potential at P_2 be V_0.
Now apply KCL at P_2,

$$\Sigma I = 0$$

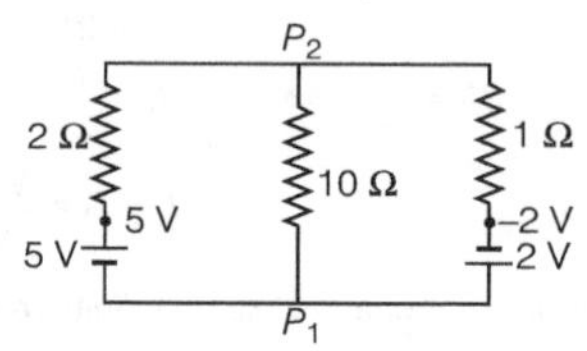

$$\frac{V_0 - 5}{2} + \frac{V_0 - 0}{10} + \frac{V_0 - (-2)}{1} = 0 \quad \text{or} \quad V_0 = \frac{5}{16}$$

So, current through 10 Ω resistor is $\dfrac{V_0}{10}$ from P_2 to P_1.

15. (c) From $R_t = R_0(1 + \alpha t)$

$\therefore$ $5 = R_0(1 + 50\alpha)$ ...(i)

and $6 = R_0(1 + 100\alpha)$ ...(ii)

$\therefore$ $\dfrac{5}{6} = \dfrac{1 + 50\alpha}{1 + 100\alpha}$ or $\alpha = \dfrac{1}{200}$

Putting the value of α in Eq. (i), we get

$$5 = R_0\left(1 + 50 \times \frac{1}{200}\right) \text{ or } R_0 = 4\ \Omega$$

16. (d) Let (ρ_A, l_A, r_A, A_A) and (ρ_B, l_B, r_B, A_B) be specific resistances, lengths, radii and areas of wires A and B, respectively.

Resistance of $A = R_A = \dfrac{\rho_A l_A}{A_A} = \dfrac{\rho_A l_A}{\pi r_A^2}$

Resistance of $B = R_B = \dfrac{\rho_B l_B}{A_B} = \dfrac{\rho_B l_B}{\pi r_B^2}$

From given information,

$$\rho_B = 2\rho_A,\ r_B = 2r_A \text{ and } R_A = R_B$$

$\therefore$ $\dfrac{\rho_A l_A}{\pi r_A^2} = \dfrac{\rho_B l_B}{\pi r_B^2}$ or $\dfrac{\rho_A l_A}{\pi r_A^2} = \dfrac{2\rho_A \times l_B}{\pi (2r_A)^2}$

or $\dfrac{l_B}{l_A} = \dfrac{2}{1} = 2:1$

17. (c) Resistance of electric bulb, $R = \dfrac{V^2}{P}$, where subscripts denote for rated parameters.

$\Rightarrow$ $R = \dfrac{(220)^2}{100}$

Power consumed at 110 V,

$$P_{\text{consumed}} = \frac{V^2}{R}$$

$\therefore$ $P_{\text{consumed}} = \dfrac{(110)^2}{(220)^2/100} = 25$ W

18. (b) Let resistance of bulb filament be R_0 at 0°C, then from expression,

$$R_\theta = R_0[1 + \alpha\Delta\theta]$$

We have, $100 = R_0[1 + 0.005 \times 100]$

and $200 = R_0[1 + 0.005 \times x]$

where, x is temperature in °C at which resistance become 200 Ω.

Dividing the above two equations, we get

$$\frac{200}{100} = \frac{1 + 0.005x}{1 + 0.005 \times 100} \Rightarrow x = 400°C$$

19. (b) For balanced Wheatstone bridge,

$$\frac{P}{Q} = \frac{R}{S}$$

Here, $S = S_1 \,||\, S_2 = \dfrac{S_1 S_2}{S_1 + S_2}$

$\Rightarrow$ $\dfrac{P}{Q} = \dfrac{R(S_1 + S_2)}{S_1 S_2}$

20. (d) Kirchhoff's first law or KCL states that the algebraic sum of current meeting at any junction is equal to zero. In other words, we can say that "The sum of all the currents directed towards a junction in a circuit is equal to the sum of all the currents directed away from that junction."

Thus, no charge has been accumulated at any junction, i.e. charge is conserved and hence we can say that KCL ($\Sigma i = 0$) is based on conservation of charge.

Kirchhoff's IInd law or KVL states that algebraic sum of changes in potential around any closed resistor loop must be zero. In other words, "Around any closed loop, voltage drops are equal to voltage rises".

No energy is gained or lost in circulating a charge around a loop, thus we can say that KVL is based on conservation of energy.

21. (b) The given circuit can be redrawn as

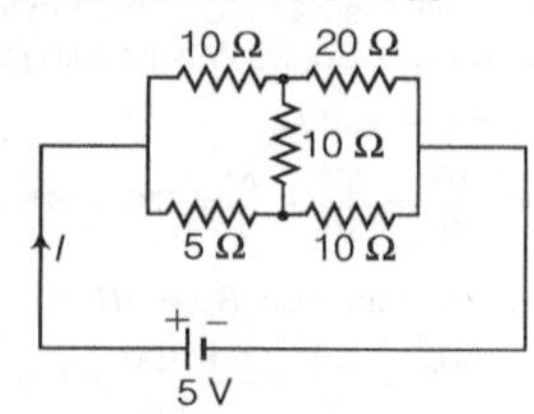

Which is a balanced Wheatstone bridge and hence no current flows in the middle resistor, so equivalent circuit would be as shown below:

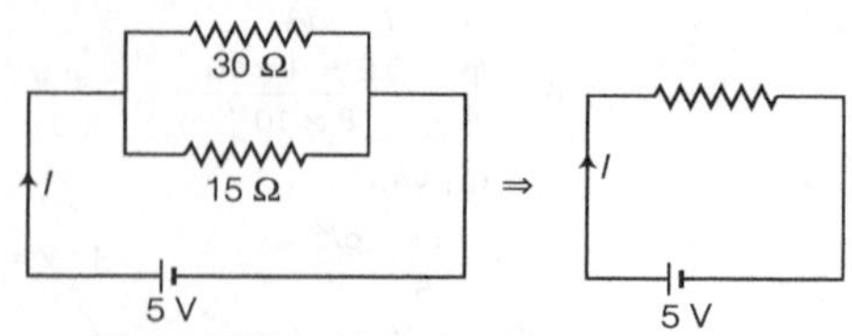

$$30\ \Omega \,||\, 15\ \Omega = 10\ \Omega$$

$$I = \frac{V}{R} = \frac{5}{10} = 0.5 \text{ A}$$

22. (b) The galvanometer shows zero deflection, i.e. current through XY is zero.

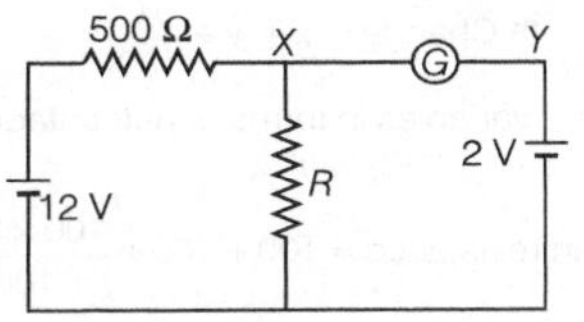

As a result potential drop across R is 2 V. Circuit can be redrawn as

$$I = \frac{12}{500 + R}$$

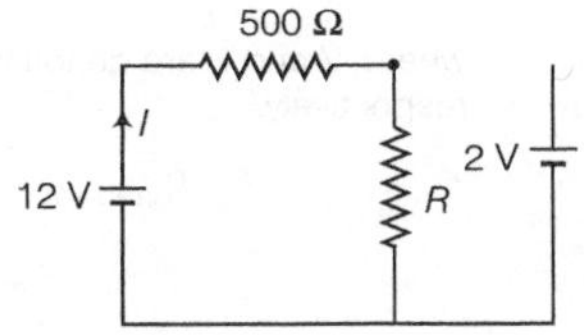

Voltage across R, $V = IR$

$$\Rightarrow \quad 2 = \frac{12}{500 + R} \times R$$

or $1000 + 2R = 12R$

or $\quad R = 100\,\Omega$

23. (*b*) As R_1, R_2 and R in series.

$$R_{eq} = R_1 + R_2 + R$$

$\therefore$ Net current, $I = \dfrac{2E}{R_1 + R_2 + R}$

According to the question,

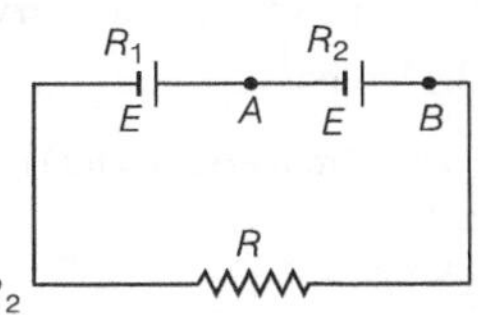

$$-(V_A - V_B) = E - IR_2$$

$\therefore \quad 0 = E - IR_2$

or $\quad E = IR_2$

or $\quad E = \dfrac{2E}{R_1 + R_2 + R} R_2$

or $\quad R_1 + R_2 + R = 2R_2$ or $R = R_2 - R_1$

24. (a) Is heat generated

$$H_1 = \frac{V^2}{R}t \quad \text{and} \quad H_2 = \frac{V^2}{(R/2)}t$$

$\therefore \quad \dfrac{H_2}{H_1} = 2$ or $H_2 = 2H_1$

25. (c) $\quad I = \dfrac{E}{R + r}$

$$I = \frac{E}{R} = \text{constant}$$

where, R = external resistance, r = internal resistance = 0

26. (d) The internal resistance of the cell,

$$r = \left(\frac{l_1 - l_2}{l_2}\right) R = \left(\frac{240 - 120}{120}\right) \times 2 = 2\,\Omega$$

27. (a) $\quad P = V^2/R$

$\therefore \quad R_{hot} = \dfrac{V^2}{P} = \dfrac{200 \times 200}{100} = 400\,\Omega$

$$R_{cold} = \frac{400}{10} = 40\,\Omega$$

28. (c)The equivalent circuit can be drawn as 6 Ω and 2 Ω are in parallel.

$$R' = \frac{2 \times 6}{8} = 1.5\,\Omega$$

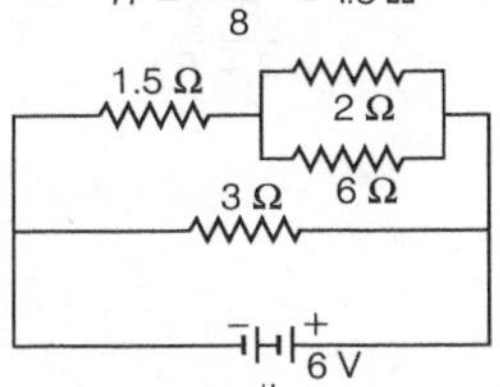

⇓

As 1.5 Ω and 1.5 Ω are in series.

$$R'' = 1.5 + 1.5 = 3\,\Omega$$

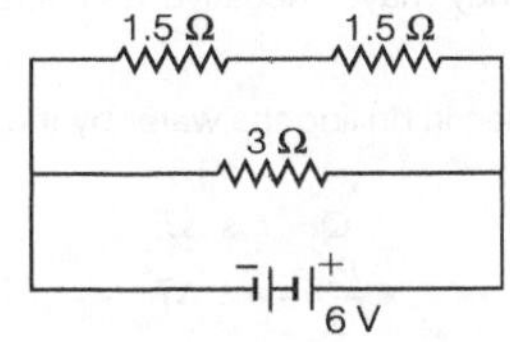

As 3 Ω and 3 Ω are in parallel.

$$R_{eq} = \frac{3 \times 3}{6} = 1.5\,\Omega$$

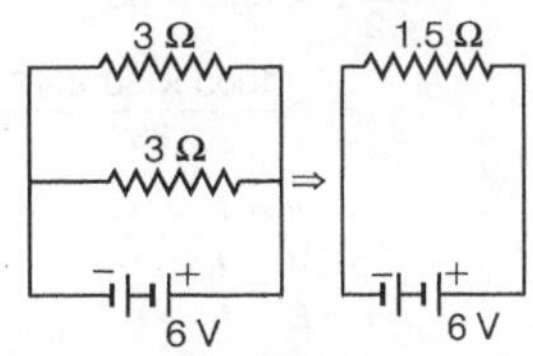

Hence, current supplied by the battery is

$$I = \frac{V}{R_{eq}} = \frac{6}{1.5} = 4\text{ A}$$

29. (a) Let resistances be R_1 and R_2.

Then, $S = R_1 + R_2$ and $P = \dfrac{R_1 R_2}{R_1 + R_2}$

$\therefore \quad (R_1 + R_2) = \dfrac{n \times R_1 R_2}{R_1 + R_2} \quad [\because S = nP]$

or $\quad (R_1 + R_2)^2 = nR_1R_2$

$$\Rightarrow \quad n = \left[\frac{R_1^2 + R_2^2 + 2R_1R_2}{R_1R_2}\right] = \left[\frac{R_1}{R_2} + \frac{R_2}{R_1} + 2\right]$$

We know, Arithmetic Mean ≥ Geometric Mean

$$\frac{\frac{R_1}{R_2} + \frac{R_2}{R_1}}{2} \geq \sqrt{\frac{R_1}{R_2} \times \frac{R_2}{R_1}} \quad \Rightarrow \quad \frac{R_1}{R_2} + \frac{R_2}{R_1} \geq 2$$

So, n (minimum value) = 2 + 2 = 4

30. (a) Meter bridge is an arrangement which works on Wheatstone's principle, so the balancing condition is

$$\frac{R}{S} = \frac{l_1}{l_2}$$

where, $\quad l_2 = 100 - l_1$

***Case* I** $\quad R = X, S = Y, l_1 = 20$ cm,

$l_2 = 100 - 20 = 80$ cm

$$\therefore \quad \frac{X}{Y} = \frac{20}{80} \quad \ldots\text{(i)}$$

***Case* II** Let the position of null point be obtained at a distance l from same end.

$\therefore \quad R = 4X, \; S = Y,$

$l_1 = l, \; l_2 = 100 - l$

So, from Eq. (i), we get

$$\frac{4X}{Y} = \frac{l}{100 - l}$$

$$\frac{X}{Y} = \frac{l}{4(100 - l)} \quad \ldots\text{(ii)}$$

Therefore, from Eqs. (i) and (ii), we get

$$\frac{l}{4(100 - l)} = \frac{20}{80} \quad \Rightarrow \quad \frac{l}{4(100 - l)} = \frac{1}{4}$$

or $\quad l = 100 - l$

or $\quad 2l = 100$

Hence, $\quad l = 50$ cm

31. (c) Thermistors are the resistors made up of metal oxides whose resistance decreases with the increase in temperature. This implies that they have negative temperature coefficient of resistivity.

32. (c) Let time taken in boiling the water by the heater be t second. Then,

$$Q = ms\,\Delta T$$

$\Rightarrow$ $$\frac{Pt}{J} = ms\,\Delta T$$

$\therefore$ $$\frac{836}{4.2}t = 1 \times 1000(40 - 10)$$

or $$\frac{836}{4.2}t = 1000 \times 30$$

or $$t = \frac{1000 \times 30 \times 4.2}{836} = 150\text{ s}$$

33. (d) $$V \propto l$$

$\therefore$ $$\frac{V}{E} = \frac{l}{L}$$

where, l = balance point distance

L = length of potentiometer wire

or $$V = \frac{l}{L}E$$

or $$V = \frac{30 \times E}{100} = \frac{30}{100}E$$

34. (b) Resistance in the arms AC and BC are in series.

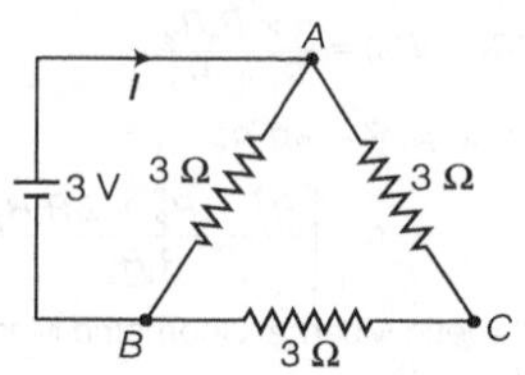

$\therefore$ $$R' = 3 + 3 = 6\,\Omega$$

Now, R' and $3\,\Omega$ are in parallel.

$\therefore$ $$R_{eq} = \frac{6 \times 3}{6 + 3} = 2\,\Omega$$

Now, $V = IR$

$\Rightarrow$ $$I = \frac{3}{2} = 1.5\text{ A}$$

35. (d) Given, $l' = l + 100\%\,l = 2l$

Initial volume = Final volume

i.e. $$\pi r^2 l = \pi r'^2 l'$$

or $$r'^2 = \frac{r^2 l}{l'} = r^2 \times \frac{l}{2l}$$

or $$r'^2 = \frac{r^2}{2}$$

$\therefore$ $$R' = \rho\frac{l'}{A'} = \rho\frac{2l}{\pi r'^2} \quad \left(\because R = \frac{\rho l}{A}\right)$$

$$= \frac{\rho \cdot 4l}{\pi r^2} = 4R$$

Thus, $$\Delta R = R' - R = 4R - R = 3R$$

$\therefore$ $$\%\Delta R = \frac{3R}{R} \times 100\% = 300\%$$

Alternate Method

By the shortcut formula,

$$\%\text{ Change} = x + y + \frac{xy}{100}$$

Valid only for 2 variables in terms of percentage.

As $$R \propto l^2\%$$

$$\text{Increase in resistance} = 100 + 100 + \frac{100 \times 100}{100}$$

$$= 300\%$$

36. (c) $$R = \frac{V^2}{P} = \frac{(220)^2}{1000}$$

where, V and P are denoting rated voltage and power, respectively.

$\therefore$ $$P_{consumed} = \frac{V^2_{applied}}{R}$$

$$= \frac{110 \times 110}{220 \times 220} \times 1000$$

$$= 250\text{ W}$$

37. (b) ***Case I***

Using the formula, $$P = \frac{V^2}{R} \quad \ldots\text{(i)}$$

where, R is resistance of wire, V is voltage across wire and P is power dissipation in wire

and $$R = \frac{\rho l}{A} \quad \ldots\text{(ii)}$$

From Eqs. (i) and (ii), we have

$$P_1 = \frac{V^2}{\rho l/A}$$

or $$P_1 = \frac{V^2}{\rho l} \cdot A \quad \ldots\text{(iii)}$$

Case II

Let R_2 be net resistance of two wires in parallel, then

$$R_2 = \frac{R \times R}{R + R} = \frac{R}{2}$$

where, R is the resistance of half wire.

$\Rightarrow$ $$R_2 = \frac{\rho \cdot \left(\frac{l}{2}\right)}{A \cdot 2} = \frac{\rho l}{4A}$$

or $$P_2 = \frac{V^2}{\rho l} \cdot 4A \quad \ldots\text{(iv)}$$

Hence, from Eqs. (iii) and (iv), we get

$$\frac{P_1}{P_2} = \frac{1}{4} \Rightarrow \frac{P_2}{P_1} = \frac{4}{1}$$

38. (b) $P = \dfrac{V^2}{R_{net}}$

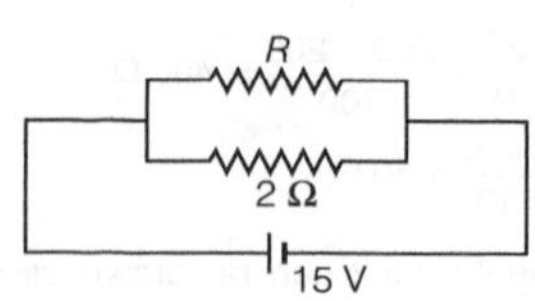

$\therefore$ $$150 = \frac{(15)^2}{\frac{R \cdot 2}{2 + R}} \Rightarrow R = 6\,\Omega$$

JEE Advanced & IIT JEE Archive

1. (c) $R = \dfrac{x}{100-x} \times 90 = \dfrac{40 \times 90}{100-40} = \dfrac{3600}{60}$

$\Rightarrow \quad R = 60\,\Omega$

$$\frac{dR}{R} = \frac{100}{x(100-x)}dx$$

$\therefore \quad dR = \dfrac{100}{40 \times 60} \times 0.1 \times 60 = 0.25\,\Omega$

The unkown resistance = $(60 \pm 0.25)\,\Omega$

2. (5) $i_g(G + 4990) = V$

$\Rightarrow \quad \dfrac{6}{1000}(G + 4990) = 30$

$\Rightarrow \quad G + 4990 = \dfrac{30{,}000}{6} = 5000 \Rightarrow G = 10\,\Omega$

$$V_{ab} = V_{cd}$$

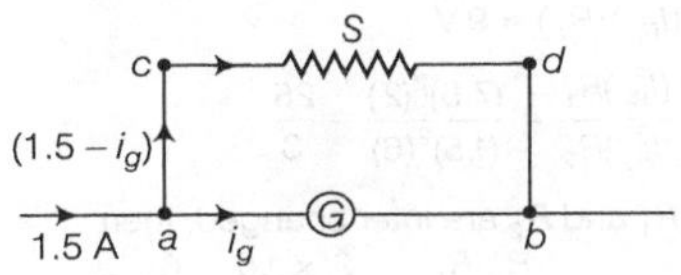

$\Rightarrow \quad i_g G = (1.5 - i_g)S$

$\Rightarrow \quad \dfrac{6}{1000} \times 10 = \left(1.5 - \dfrac{6}{1000}\right)S$

$\Rightarrow \quad S = \dfrac{60}{1494} = \dfrac{2n}{249}$

$\Rightarrow \quad n = \dfrac{249 \times 30}{1494} = \dfrac{2490}{498} = 5$

3. (5) $B_R = B_T - B_C$

R = Remaining portion, T = Total portion and C = Cavity

$$B_R = \frac{\mu_0 I_T}{2a\pi} - \frac{\mu_0 I_C}{2\,(3a/2)\pi} \qquad \ldots(i)$$

$$I_T = J\,(\pi a^2)$$

$$I_C = J\left(\frac{\pi a^2}{4}\right)$$

Substituting the values in Eq. (i), we have

$$B_R = \frac{\mu_0}{a\pi}\left[\frac{I_T}{2} - \frac{I_C}{3}\right] = \frac{\mu_0}{a\pi}\left[\frac{\pi a^2 J}{2} - \frac{\pi a^2 J}{12}\right] = \frac{5\,\mu_0 aJ}{12}$$

$\therefore \quad N = 5$

4. (a,c)

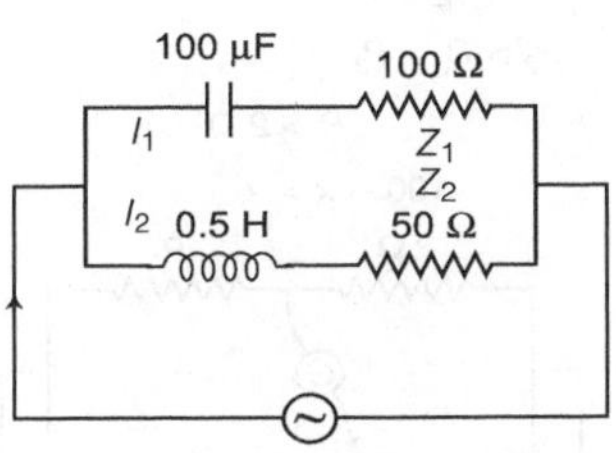

Circuit 1 $\quad X_C = \dfrac{1}{\omega C} = 100\,\Omega$

$\therefore \quad Z_1 = \sqrt{(100)^2 + (100)^2} = 100\sqrt{2}\,\Omega$

$$\phi_1 = \cos^{-1}\left(\frac{R_1}{Z_1}\right) = 45^\circ$$

In this circuit current leads the voltage.

$$I_1 = \frac{V}{Z_1} = \frac{20}{100\sqrt{2}} = \frac{1}{5\sqrt{2}}\text{ A}$$

$$V_{100\,\Omega} = (100)\,I_1 = (100)\,\frac{1}{5\sqrt{2}}\text{ V} = 10\sqrt{2}\text{ V}$$

Circuit 2 $\quad X_L = \omega L = (100)\,(0.5) = 50\,\Omega$

$$Z_2 = \sqrt{(50)^2 + (50)^2} = 50\sqrt{2}\,\Omega$$

$$\phi_2 = \cos^{-1}\left(\frac{R_2}{Z_2}\right) = 45^\circ$$

In this circuit voltage leads the current.

$$I_2 = \frac{V}{Z_2} = \frac{20}{50\sqrt{2}} = \frac{\sqrt{2}}{5}\text{ A}$$

$$V_{50\,\Omega} = (50)\,I_2 = 50\left(\frac{\sqrt{2}}{5}\right) = 10\sqrt{2}\text{V}$$

Further, I_1 and I_2 have a mutual phase difference of 90°.

$\therefore \quad I = \sqrt{I_1^2 + I_2^2} = \sqrt{\dfrac{1}{50} + \dfrac{4}{50}}$

$$= \frac{1}{\sqrt{10}}\text{ A} \approx 0.3\text{ A}$$

5. (c) Between 3 μF and 2 μF (in parallel) total charge of 80 μC will distribute in direct ratio of capacity.

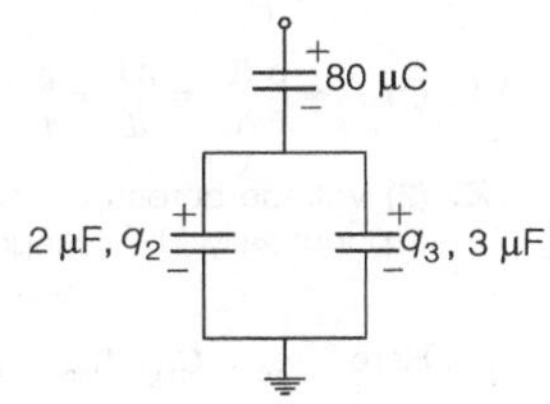

$\therefore \quad \dfrac{q_3}{q_2} = \dfrac{3}{2}$

or $\quad q_3 = \left(\dfrac{3}{3+2}\right)(80)$

$$= 48\,\mu\text{C}$$

6. (d) r = distance of a point from centre.

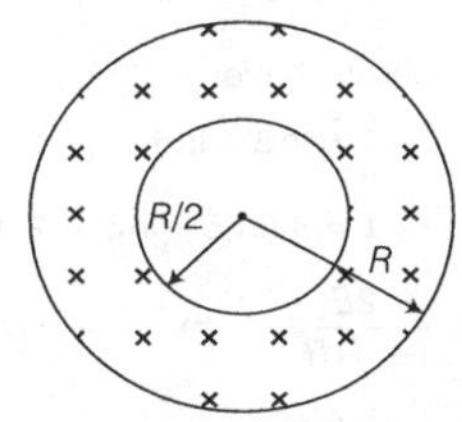

For $r \le R/2$

Using Ampere's circuital law,

$$\oint \mathbf{B} \cdot d\mathbf{l} \text{ or } Bl = \mu_0\,(I_{in}) \text{ or } B\,(2\pi r) = \mu_0\,(I_{in})$$

or $\quad B = \dfrac{\mu_0}{2\pi}\dfrac{I_{in}}{r} \qquad \ldots(i)$

Since, $\quad I_{in} = 0$

$\therefore \quad B = 0$

For $\dfrac{R}{2} \le r \le R$, $\quad I_{in} = \left[\pi r^2 - \pi\left(\dfrac{R}{2}\right)^2\right]\sigma$

Here, σ = current per unit area.

Substituting in Eq. (i), we have

$$B = \frac{\mu_0}{2\pi}\,\frac{\left[\pi r^2 - \pi\dfrac{R^2}{4}\right]\sigma}{r} = \frac{\mu_0\sigma}{2r}\left(r^2 - \frac{R^2}{4}\right)$$

At $r = \dfrac{R}{2}$, $\quad B = 0$

At $r = R$, $\quad B = \dfrac{3\mu_0\sigma R}{8}$

For $r \ge R$

$$I_{in} = I_{Total} = I \text{ (say)}$$

Therefore, substituting in Eq. (i), we have

$$B = \frac{\mu_0}{2\pi}\cdot\frac{I}{r} \text{ or } B \propto \frac{1}{r}$$

7. (b) Using the concept of balanced Wheat stone bridge, we have,

$$\frac{P}{Q}=\frac{R}{S} \Rightarrow \frac{X}{(52+1)}=\frac{10}{(48+2)}$$

$$\therefore \quad X=\frac{10\times 53}{50}=10.6\,\Omega$$

8. (5) V_{AB} = Equivalent emf of two batteries in parallel.

$$=\frac{E_1/r_1+E_2/r_2}{1/r_1+1/r_2}=\frac{(6/1)+(3/2)}{(1/1)+(1/2)}=5\text{ V}$$

9. (c) We will require a voltmeter, an ammeter, a test resistor and a variable battery to verify Ohm's law.

Voltmeter which is made by connecting a high resistance with a galvanometer is connected in parallel with the test resistor.

Further, an ammeter which is formed by connecting a low resistance in parallel with galvanometer is required to measure the current through test resistor.

10. (d) $R=\frac{V^2}{P}$ or $R\propto\frac{1}{P} \Rightarrow \frac{1}{R_{100}}>\frac{1}{R_{60}}>\frac{1}{R_{40}}$

11. (c) $R=\frac{\rho\,(L)}{A}=\frac{\rho L}{tL}=\frac{\rho}{t}$ i.e., R is independent of L.

12. (2) Voltage across the capacitors will increase from 0 to 10 V exponentially. The voltage at time t will be given by

$$V=10\,(1-e^{-t/\tau_C})$$

Here $\tau_c=C_{net}\,R_{net}=(1\times10^6)\,(4\times10^{-6})=4\text{s}$

$$\therefore \quad V=10\,(1-e^{-t/4})$$

Substituting $V=4$ volt, we have

$$4=10\,(1-e^{-t/4}) \quad \text{or} \quad e^{-t/4}=0.6=\frac{3}{5}$$

Taking log both sides we have,

$$-\frac{t}{4}=\ln 3-\ln 5$$

or $\quad t=4\,(\ln 5-\ln 3)=2\text{ s}$

13. (4) In series, $\quad i=\frac{2E}{2+R} \Rightarrow J_1=i^2R=\left(\frac{2E}{2+R}\right)^2\cdot R$

In parallel, $\quad i=\frac{E}{0.5+R} \Rightarrow J_2=i^2R=\left(\frac{E}{0.5+R}\right)^2\cdot R$

$$\frac{J_1}{J_2}=2.25=\frac{4\,(0.5+R)^2}{(2+R)^2} \quad \text{or} \quad 1.5=\frac{2\,(0.5+R)}{(2+R)}$$

On solving, we get,

$$R=4\,\Omega$$

14. (a,b,d) Due to symmetry on upper side and lower side, points P and Q are at same potentials. Similarly, points S and T are at same potentials. Therefore, the simple circuit can be drawn as shown below

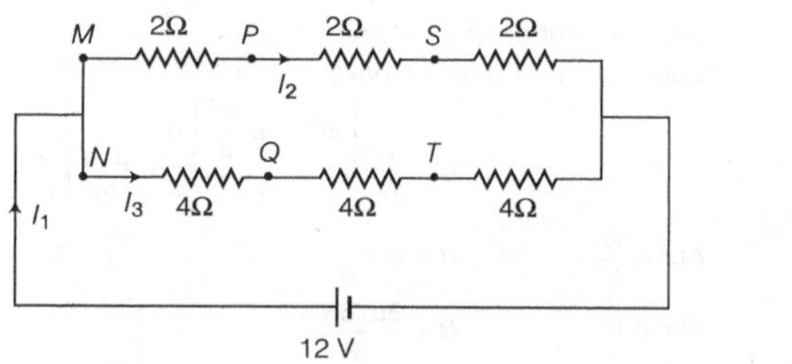

$$I_2=\frac{12}{2+2+2}=2\text{ A} \Rightarrow I_3=\frac{12}{4+4+4}=1\text{ A}$$

$$\therefore \quad I_1=I_2+I_3=3\text{ A} \Rightarrow I_{PQ}=0$$

because $V_P=V_Q$

Potential drop (from left to right) across each resistance is

$$\frac{12}{3}=4\text{ V}$$

$\therefore \quad V_{MS}=2\times4=8\text{ V}$

$\Rightarrow \quad V_{NQ}=1\times4=4\text{ V}$

or $\quad V_S<V_Q$

15. (a,d) $R_{\text{total}}=2+\frac{6\times1.5}{6+1.5}=3.2\text{ k}\Omega$

(a) $\quad I=\frac{24\text{V}}{3.2\text{ k}\Omega}=7.5\text{mA}=I_{R_1}$

$$I_{R_2}=\left(\frac{R_L}{R_L+R_2}\right)I$$

$\Rightarrow \quad I=\frac{1.5}{7.5}\times7.5=1.5\text{mA}$

$$I_{R_L}=6\text{mA}$$

(b) $V_{R_L}=(I_{R_L})(R_L)=9\text{ V}$

(c) $\frac{P_{R_1}}{P_{R_2}}=\frac{(I^2_{R_1})R_1}{(I^2_{R_2})R_2}=\frac{(7.5)^2(2)}{(1.5)^2(6)}=\frac{25}{3}$

(d) When R_1 and R_2 are inter changed, then

$$\frac{R_2\,R_L}{R_2+R_L}=\frac{2\times1.5}{3.5}=\frac{6}{7}\text{k}\Omega$$

Now, potential difference across R_L will be

$$V_L=24\left[\frac{6/7}{6+6/7}\right]=3\text{ V}$$

Earlier it was 9 V

Since, $\quad P=\frac{V^2}{R}$ or $P\propto V^2$

In new situation potential difference has been decreased three times. Therefore, power dissipated will decrease by a factor of 9.

16. (c) Applying $P=\frac{V^2}{R}$

$$R_1=1\Omega,\ R_2=0.5\Omega \text{ and } R_3=2\Omega$$

$$V_1=V_2=V_3=3\text{ V}$$

$\therefore \quad P_1=\frac{(3)^2}{1}=9\text{ W}$

$$P_2=\frac{(3)^2}{0.5}=18\text{ W}$$

and $\quad P_3=\frac{(3)^2}{2}=4.5\text{ W}$

$\therefore \quad P_2>P_1>P_3$

17. (a) $\quad R>2\,\Omega$

$\therefore \quad 100-x>x$

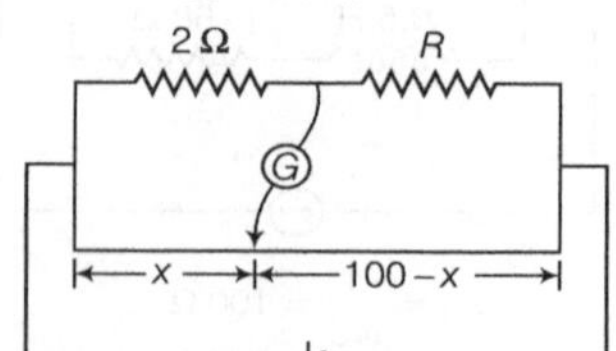

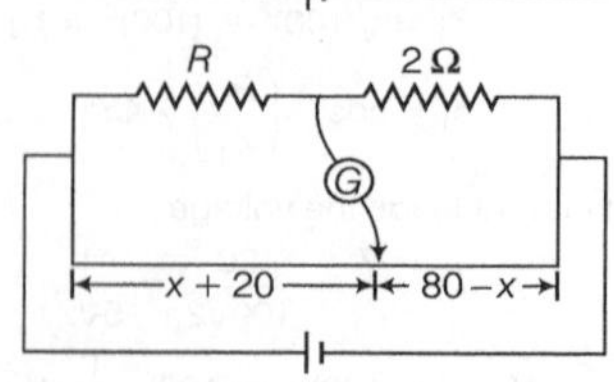

Applying $\frac{P}{Q} = \frac{R}{S}$

we have, $\frac{2}{R} = \frac{x}{100 - x}$ ...(i)

$\frac{R}{2} = \frac{x + 20}{80 - x}$ ...(ii)

On solving Eqs. (i) and (ii), we get, $R = 3\,\Omega$

18. (c) From Y to X charge flows to plates a and b.

$(q_a + q_b)_i = 0,\ (q_a + q_b)_f = 27\ \mu$

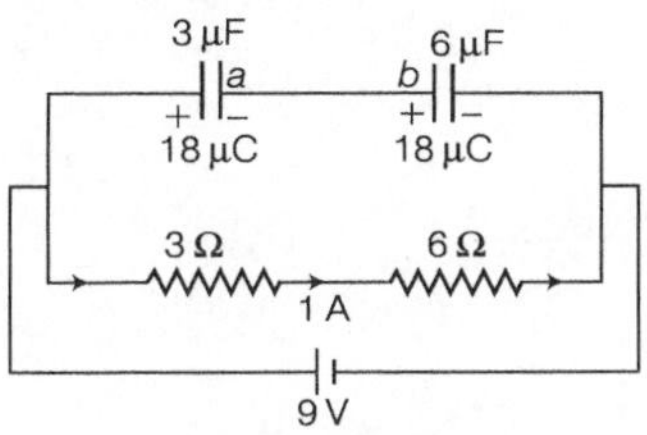

Initial Figure
(when switch was open)

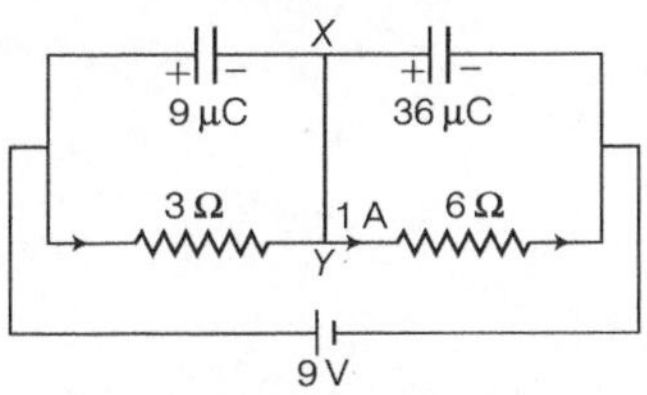

Final Figure
(when switch is closed)

$\therefore$ 27 μC charge flows from Y to X.

19. (b) $\tau = CR$

$$\tau_1 = (C_1 + C_2)(R_1 + R_2) = 18\ \mu s,$$

$$\tau_2 = \left(\frac{C_1C_2}{C_1 + C_2}\right)\left(\frac{R_1R_2}{R_1 + R_2}\right) = \frac{8}{6} \times \frac{2}{3} = \frac{8}{9}\ \mu s$$

$$\text{and } \tau_3 = (C_1 + C_2)\left(\frac{R_1R_2}{R_1 + R_2}\right) = (6)\left(\frac{2}{3}\right) = 4\ \mu s$$

20. (a) Current flowing through both the bars is equal.

Now, the heat produced is given by

$$H = I^2Rt \text{ or } H \propto R$$

or $\frac{H_{AB}}{H_{BC}} = \frac{R_{AB}}{R_{BC}} = \frac{(1/2r)^2}{(1/r)^2}$ $\left(\text{as } R \propto \frac{1}{A} \propto \frac{1}{r^2}\right)$

$= \frac{1}{4}$

or $H_{BC} = 4\,H_{AB}$

21. (a) Given, $V_C = 3V_R = 3(V - V_C)$

here, V is the applied potential.

$\therefore$ $V_C = \frac{3}{4}V$

or $V(1 - e^{-t/\tau_c}) = \frac{3}{4}V$

$\therefore$ $e^{-t/\tau_c} = \frac{1}{4}$ ...(i)

Here, $\tau_c = CR = 10\ s$

Substituting this value of τ_c in Eq. (i) and solving for t, we get

$$t = 13.86\,s$$

22. (c) Current in the respective loop will remain confined in the loop itself. Therefore, current through 2 Ω resistance is zero

23. (d) $W = 0$. Therefore, from first law of thermodynamics,

$$\Delta U = \Delta Q = i^2Rt = (1)^2\,(100)\,(5 \times 60)\ J = 30\ kJ$$

24. (a) $V_{ab} = i_g.G = (i - i_g)S \Rightarrow i = \left(1 + \frac{G}{S}\right)i_g$

Substituting the values, we get

$$i = 100.1\ mA$$

25. (b) Slide wire bridge is most sensitive when the resistance of all the four arms of bridge is same.

26.

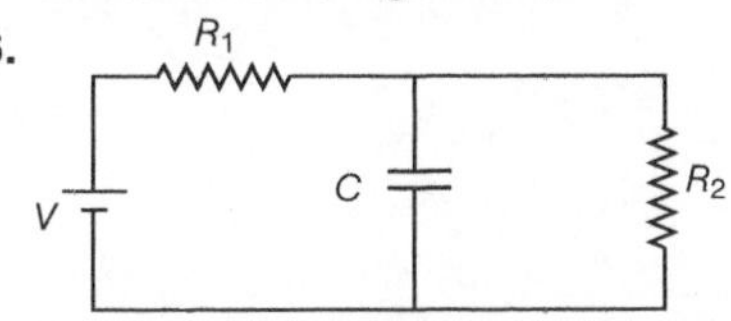

Q_0 is the steady state charge stored in the capacitor.

$Q_0 = C$ [PD across capacitor in steady state]

$= C$ [steady state current through R_2] $(R_2) = C\left(\frac{V}{R_1 + R_2}\right) \cdot R_2$

$$\therefore\quad Q_0 = \frac{C\,V\,R_2}{R_1 + R_2}$$

α is $\frac{1}{\tau_c}$ or $\frac{1}{C\,R_{net}}$

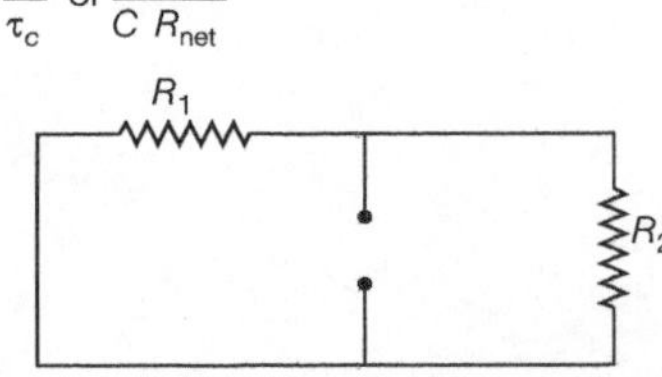

Here, R_{net} is equivalent resistance across capacitor after short circuiting the battery. Thus,

$$R_{net} = \frac{R_1R_2}{R_1 + R_2} \quad (\text{As } R_1 \text{ and } R_2 \text{ are in parallel})$$

$$\alpha = \frac{1}{C\left(\frac{R_1R_2}{R_1 + R_2}\right)} = \frac{R_1 + R_2}{C\,R_1R_2}$$

27. (d) With increase in temperature, the value of unknown resistance will increase.

In balanced Wheat stone bridge condition,

$$\frac{R}{X} = \frac{l_1}{l_2}$$

Where, R = value of standard resistance,
X = value of unknown resistance.

To take null point at same point or $\frac{l_1}{l_2}$ to remain unchanged, $\frac{R}{X}$ should also remain unchanged.

Therefore, if X is increasing R, should also increase.

Last 3 Years' Questions
JEE Main & Advanced
(2016-2018)

JEE Main

2016

1. The region between two concentric spheres of radii a and b, respectively (see the figure), has volume charge density $\rho = \frac{A}{r}$, where, A is a constant and r is the distance from the centre. At the centre of the spheres is a point charge Q. The value of A, such that the electric field in the region between the spheres will be constant, is **[Electrostatics]**

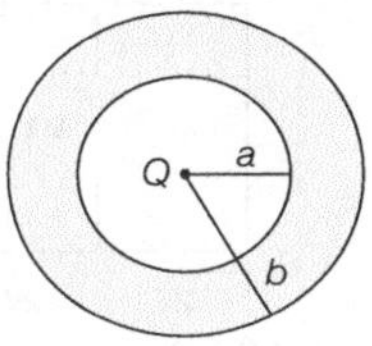

(a) $\frac{Q}{2\pi a^2}$ (b) $\frac{Q}{2\pi(b^2 - a^2)}$

(c) $\frac{2Q}{\pi(a^2 - b^2)}$ (d) $\frac{2Q}{\pi a^2}$

2. A combination of capacitors is set-up as shown in the figure. The magnitude of the electric field, due to a point charge Q (having a charge equal to the sum of the charges on the 4 μF and 9μF capacitors), at a point distant 30 m from it, would equal to **[Capacitance]**

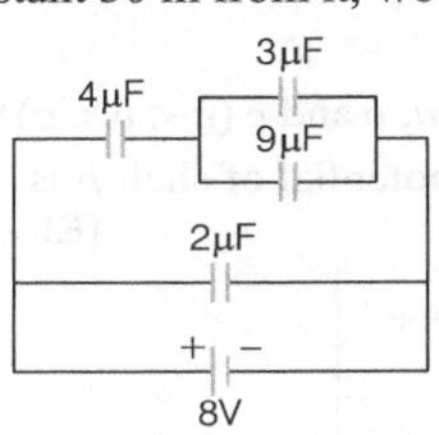

(a) 240 N/C (b) 360 N/C

(c) 420 N/C (d) 480 N/C

3. The temperature dependence of resistances of Cu and undoped Si in the temperature range 300-400 K, is best described by **[Current Electricity]**

(a) linear increase for Cu, linear increase for Si

(b) linear increase for Cu, exponential increase for Si

(c) linear increase for Cu, exponential decrease for Si

(d) linear decrease for Cu, linear decrease for Si

2017

4. Which of the following statements is false? **[Current Electricity]**

(a) In a balanced Wheatstone bridge, if the cell and the galvanometer are exchanged, the null point is disturbed

(b) A rheostat can be used as a potential divider

(c) Kirchhoff's second law represents energy conservation

(d) Wheatstone bridge is the most sensitive when all the four resistances are of the same order of magnitude

5. A capacitance of 2 μF is required in an electrical circuit across a potential difference of 1kV. A large number of 1 μF capacitors are available which can withstand a potential difference of not more than 300 V. The minimum number of capacitors required to achieve this is **[Capacitance]**

(a) 16 (b) 24 (c) 32 (d) 2

6. In the given circuit diagram, when the current reaches steady state in the circuit, the charge on the capacitor of capacitance C will be **[Capacitance]**

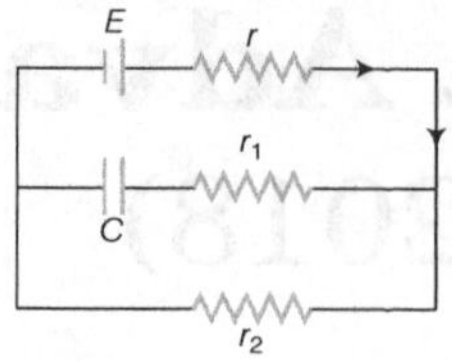

(a) $CE\dfrac{r_1}{(r_2 + r)}$ (b) $CE\dfrac{r_2}{(r + r_2)}$ (c) $CE\dfrac{r_1}{(r_1 + r)}$ (d) CE

7. In the below circuit, the current in each resistance is **[Current Electricity]**

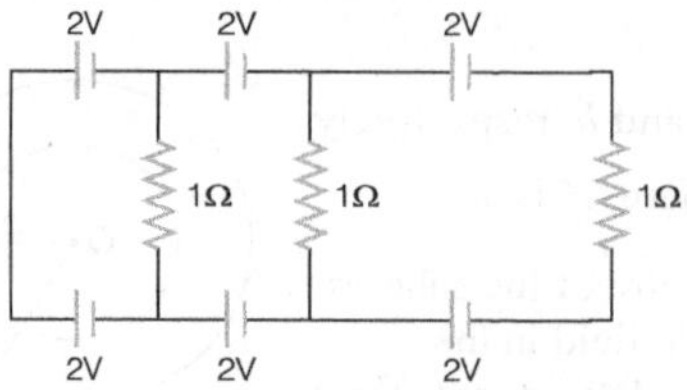

(a) 0.25 A (b) 0.5 A (c) 0 A (d) 1 A

8. An electric dipole has a fixed dipole moment **p**, which makes angle θ with respect to X-axis. When subjected to an electric field $\mathbf{E}_1 = E\hat{\mathbf{i}}$, it experiences a torque $\mathbf{T}_1 = \tau\hat{\mathbf{k}}$. When subjected to another electric field $\mathbf{E}_2 = \sqrt{3}E_1\hat{\mathbf{j}}$, it experiences a torque $\mathbf{T}_2 = -\mathbf{T}_1$. The angle θ is **[Electrostatics]**

(a) 45° (b) 60° (c) 90° (d) 30°

2018

9. Three concentric metal shells A, B and C of respective radii a, b and c $(a < b < c)$ have surface charge densities $+\sigma, -\sigma$ and $+\sigma$, respectively. The potential of shell B is **[Electrostatics]**

(a) $\dfrac{\sigma}{\varepsilon_0}\left[\dfrac{a^2 - b^2}{a} + c\right]$ (b) $\dfrac{\sigma}{\varepsilon_0}\left[\dfrac{a^2 - b^2}{b} + c\right]$

(c) $\dfrac{\sigma}{\varepsilon_0}\left[\dfrac{b^2 - c^2}{b} + a\right]$ (d) $\dfrac{\sigma}{\varepsilon_0}\left[\dfrac{b^2 - c^2}{c} + a\right]$

10. A parallel plate capacitor of capacitance 90 pF is connected to a battery of emf 20 V. If a dielectric material of dielectric constant $K = 5/3$ is inserted between the plates, the magnitude of the induced charge will be **[Capacitance]**

(a) 1.2 nC (b) 0.3 nC (c) 2.4 nC (d) 0.9 nC

11. Two batteries with emf 12 V and 13 V are connected in parallel across a load resistor of 10 Ω. The internal resistances of the two batteries are 1 Ω and 2 Ω, respectively. The voltage across the load lies between **[Current Electricity]**

(a) 11.6 V and 11.7 V (b) 11.5 V and 11.6 V

(c) 11.4 V and 11.5 V (d) 11.7 V and 11.8 V

12. In a potentiometer experiment, it is found that no current passes through the galvanometer when the terminals of the cell are connected across 52 cm of the potentiometer wire. If the cell is shunted by a resistance of 5 Ω, a balance is found when the cell is connected across 40 cm of the wire. Find the internal resistance of the cell. **[Current Electricity]**

(a) 1 Ω (b) 1.5 Ω (c) 2 Ω (d) 2.5 Ω

13. On interchanging the resistances, the balance point of a meter bridge shifts to the left by 10 cm. The resistance of their series combination is 1 kΩ. How much was the resistance on the left slot before interchanging the resistances? **[Current Electricity]**

(a) 990 Ω (b) 505 Ω (c) 550Ω (d) 910 Ω

JEE Advanced

2016

1. An infinite line charge of uniform electric charge density λ lies along the axis of an electrically conducting infinite cylindrical shell of radius R. At time $t = 0$, the space inside the cylinder is filled with a material of permittivity ε and electrical conductivity σ. The electrical conduction in the material follows Ohm's law. Which one of the following graphs best describes the subsequent variation of the magnitude of current density J (t) at any point in the material? **[Electrostatics, Type 1]**

(a)
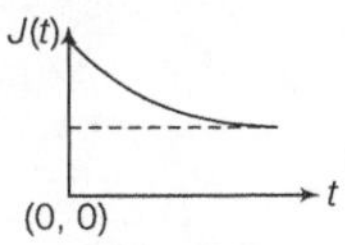

(b)
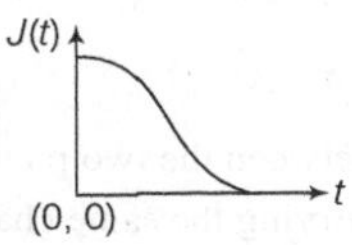

(c)
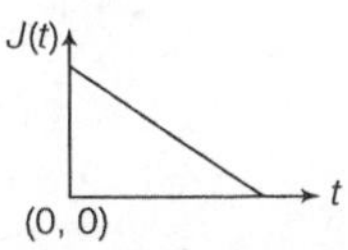

(d)
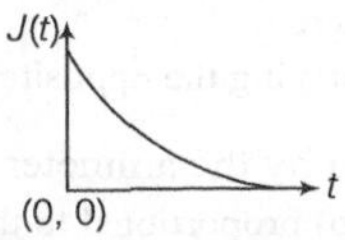

2. An incandescent bulb has a thin filament of tungsten that is heated to high temperature by passing an electric current. The hot filament emits black-body radiation. The filament is observed to break up at random locations after a sufficiently long time of operation due to non-uniform evaporation of tungsten from the filament. If the bulb is powered at constant voltage, which of the following statement(s) is (are) true? **[Current Electricity, Type 2]**

(a) The temperature distribution over the filament is uniform
(b) The resistance over small sections of the filament decreases with time
(c) The filament emits more light at higher band of frequencies before it breaks up
(d) The filament consumes less electrical power towards the end of the life of the bulb

3. Consider two identical galvanometers and two identical resistors with resistance R. If the internal resistance of the galvanometers $R_c < R/2$, which of the following statement(s) about anyone of the galvanometers is (are) true? **[Current Electricity, Type 2]**

(a) The maximum voltage range is obtained when all the components are connected in series
(b) The maximum voltage range is obtained when the two resistors and one galvanometer are connected in series, and the second galvanometer is connected in parallel to the first galvanometer
(c) The maximum current range is obtained when all the components are connected in parallel
(d) The maximum current range is obtained when the two galvanometers are connected in series, and the combination is connected in parallel with both the resistors

4. In the circuit shown below, the key is pressed at time $t = 0$. Which of the following statement(s) is (are) true? **[Current Electricity, Type 2]**

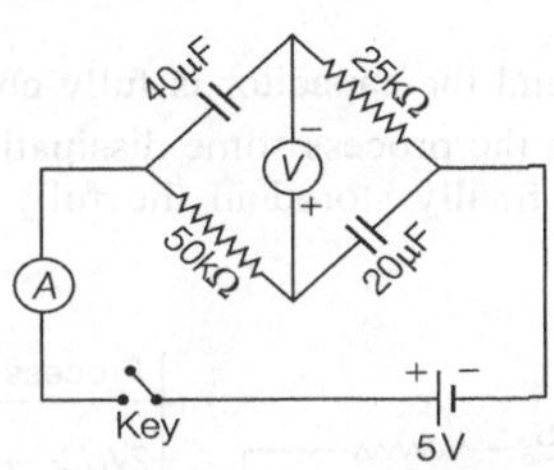

(a) The voltmeter display -5 V as soon as the key is pressed and displays $+5$ V after a long time
(b) The voltmeter will display 0 V at time $t = \ln 2$ seconds
(c) The current in the ammeter becomes $1/e$ of the initial value after 1 second
(d) The current in the ammeter becomes zero after a long time

Paragraph 1 (Q.Nos. 5 & 6)

Consider an evacuated cylindrical chamber of height h having rigid conducting plates at the ends and an insulating curved surface as shown in the figure. A number of spherical balls made of a light weight and soft material and coated with a conducting material are placed on the

***Note** **Type-1** Only one correct option, **Type-2** More then one correct option, **Type-3** Single digit Answer, **Type-4** Numerical value based*

bottom plate. The balls have a radius $r << h$. Now, a high voltage source (HV) connected across the conducting plates such that the bottom plate is at $+V_0$ and the top plate at $-V_0$. Due to their conducting surface, the balls will get charge, will become equipotential with the plate and are repelled by it. The balls will eventually collide with the top plate, where the coefficient of restitution can be taken to be zero due to te soft nature of the material of the balls. The electric field in the chamber can be considered to be that of a parallel plate capacitor. Assume that there are no collisions between the balls and the interaction between them is negligible. (Ignore gravity) **[Current Electricity]**

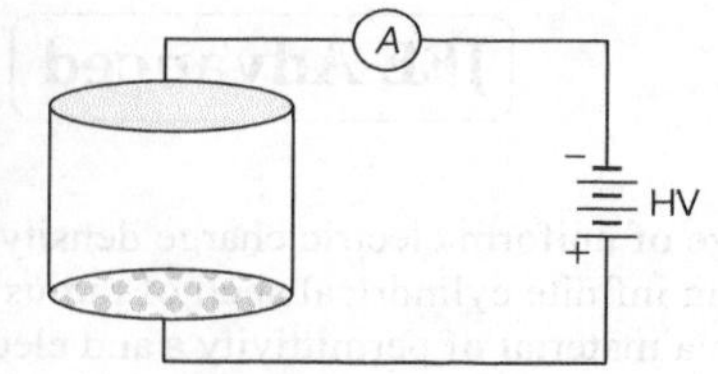

5. Which one of the following statement is correct?

(a) The balls will execute simple harmonic motion between the two plates

(b) The balls will bounce back to the bottom plate carrying the same charge they went up with

(c) The balls will stick to the top plate and remain there

(d) The balls will bounce back to the bottom plate carrying the opposite charge they went up with

6. The average current in the steady state registered by the ammeter in the circuit will be

(a) proportional to V_0^2 (b) proportional to the potential V_0

(c) zero (d) proportions to $V_0^{1/2}$

2017

7. A point charge $+Q$ is placed just outside an imaginary hemispherical surface of radius R as shown in the figure. Which of the following statements is/are correct? **[Electrostatics, Type 2]**

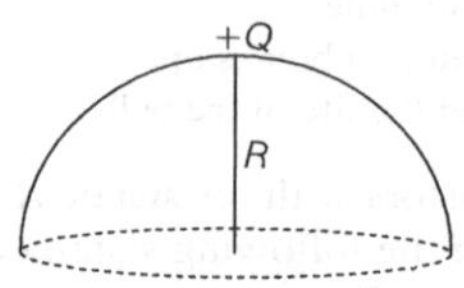

(a) The electric flux passing through the curved surface of the hemisphere is $-\frac{Q}{2\varepsilon_0}\left(1-\frac{1}{\sqrt{2}}\right)$.

(b) The component of the electric field normal to the flat surface is constant over the surface

(c) Total flux through the curved and the flat surfaces is $\frac{Q}{\varepsilon_0}$

(d) The circumference of the flat surface is an equipotential

Paragraph 2 (Q.Nos. 8 & 9)

Consider a simple RC circuit as shown in Figure 1.

Process 1 In the circuit the switch S is closed at $t = 0$ and the capacitor is fully charged to voltage V_0 (i.e. charging continues for time $T >> RC$). In the process some dissipation (E_D) occurs across the resistance R. The amount of energy finally stored in the fully charged capacitor is E_c.

Process 2 In a different process the voltage is first set to $\frac{V_0}{3}$ and maintained for a charging time $T >> RC$. Then, the voltage is raised to $\frac{2V_0}{3}$ without discharging the capacitor and again maintained for a time $T >> RC$. The process is repeated one more time by raising the voltage to V_0 and the capacitor is charged to the same final voltage V_0 as in Process 1. **[Capacitance]**

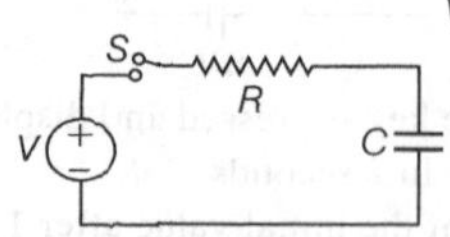

Figure 1

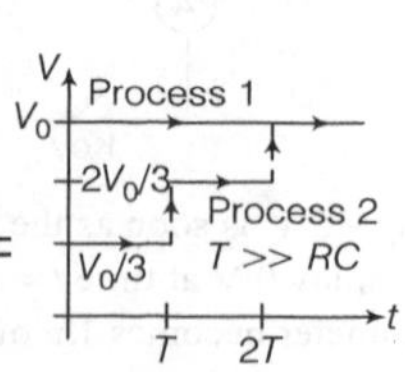

Figure 2

These two processes are depicted in Figure 2.

8. In Process 1, the energy stored in the capacitor E_C and heat dissipated across resistance E_D are related by

(a) $E_C = E_D \ln 2$ (b) $E_C = E_D$ (c) $E_C = 2E_D$ (d) $E_C = \frac{1}{2}E_D$

9. In Process 2, total energy dissipated across the resistance E_D is

(a) $E_D = \frac{1}{3}\left(\frac{1}{2}CV_0^2\right)$ (b) $E_D = 3\left(\frac{1}{2}CV_0^2\right)$ (c) $E_D = 3CV_0^2$ (d) $E_D = \frac{1}{2}CV_0^2$

2018

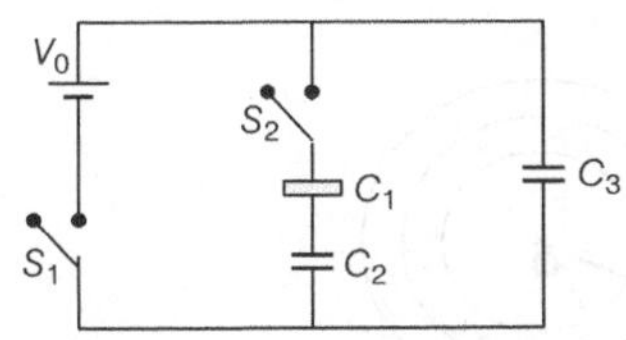

10. Three identical capacitors C_1, C_2 and C_3 have a capacitance of 1.0μF each and they are uncharged initially. They are connected in a circuit as shown in the figure and C_1 is then filled completely with a dielectric material of relative permittivity ε_r. The cell electromotive force (emf) $V_0 = 8V$. First the switch S_1 is closed while the switch S_2 is kept open. When the capacitor C_3 is fully charged, S_1 is opened and S_2 is closed simultaneously. When all the capacitors reach equilibrium, the charge on C_3 is found to be 5μC. The value of ε_r =............. **[Capacitance, Type 4]**

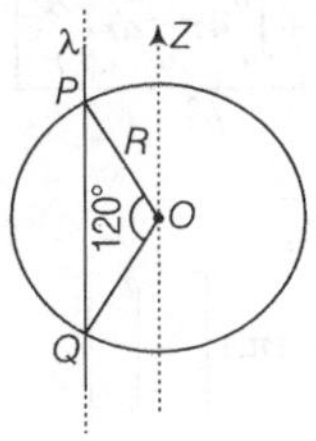

11. An infinitely long thin non-conducting wire is parallel to the Z-axis and carries a uniform line charge density λ. It pierces a thin non-conducting spherical shell of radius R in such a way that the arc PQ subtends an angle 120° at the centre O of the spherical shell, as shown in the figure. The permittivity of free space is ε_0. Which of the following statements is (are) true? **[Electrostatics, Type 2]**

(a) The electric flux through the shell is $\sqrt{3}\,R\lambda/\varepsilon_0$

(b) The z-component of the electric field is zero at all the points on the surface of the shell

(c) The electric flux through the shell is $\sqrt{2}\,R\lambda/\varepsilon_0$

(d) The electric field is normal to the surface of the shell at all points

12. The electric field E is measured at a point $P\,(0, 0, d)$ generated due to various charge distributions and the dependence of E on d is found to be different for different charge distributions. List-I contains different relations between E and d. List-II describes different electric charge distributions, along with their locations. Match the functions in List-I with the related charge distributions in List-II.

List-I		List-II
P. E is independent of d	1.	A point charge Q at the origin
Q. $E \propto \frac{1}{d}$	2.	A small dipole with point charges Q at $(0, 0, l)$ and $-Q$ at $(0, 0, -1)$. (Take, $2l << d$)
R. $E \propto \frac{1}{d^2}$	3.	An infinite line charge coincident with the X-axis, with uniform linear charge density λ.
S. $E \propto \frac{1}{d^3}$	4.	Two infinite wires carrying a uniform linear charge density parallel to the X- axis. The one along $(y = 0, z = l)$ has a charge density $+\lambda$ and the one along $(y = 0, z = -l)$ has a charge density $-\lambda$. (Take, $2l << d$).
	5.	Infinite plane charge coincident with the xy-plane with uniform surface charge density.

[Electrostatics]

(a) P → 5; Q → 3,4; R → 1; S → 2
(b) P → 5; Q → 3; R → 1, 4; S → 2
(c) P → 5; Q → 3; R → 1, 2; S → 4
(d) P → 4; Q → 2, 3; R → 1; S → 5

13. A particle of mass 10^{-3} kg and charge 1.0 C is initially at rest. At time $t = 0$, the particle comes under the influence of an electric field $\mathbf{E}(t) = E_0 \sin \omega t\,\hat{\mathbf{i}}$, where $E_0 = 1.0\,NC^{-1}$ and $\omega = 10^3$ rad s^{-1}. Consider the effect of only the electrical force on the particle. Then, the maximum speed in m s^{-1}, attained by the particle at subsequent times is .... **[Electrostatics, Type 4]**

Answer *with* Explanations

JEE Main

1. *(a)* As E is constant,

Hence, $E_a = E_b$

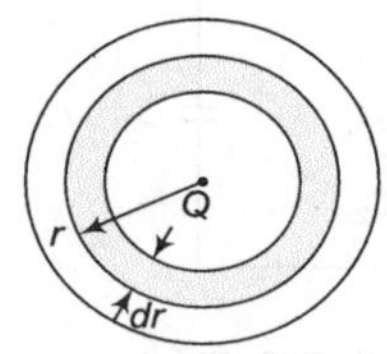

As per Guass theorem, only Q_{in} contributes in electric field.

$$\therefore \quad \frac{kQ}{a^2} = \frac{k\left[Q+\int_a^b 4\pi r^2 dr \cdot \frac{A}{r}\right]}{b^2}$$

Here, $k = \frac{1}{4\pi\varepsilon_0}$

$$\Rightarrow \quad Q\frac{b^2}{a^2} = Q + 4\pi A\left[\frac{r^2}{2}\right]_a^b$$

$$= Q + 4\pi A \cdot \left(\frac{b^2 - a^2}{2}\right)$$

$$\Rightarrow \quad Q\left(\frac{b^2}{a^2}\right) = Q + 2\pi A\,(b^2 - a^2)$$

$$\Rightarrow \quad Q\left(\frac{b^2 - a^2}{a^2}\right) = 2\pi A\,(b^2 - a^2)$$

$$\Rightarrow \quad A = \frac{Q}{2\pi a^2}$$

2. *(c)* 3 μF and 9μF = 12μF

$$4\mu\text{F and } 12\mu\text{F} = \frac{4\times 12}{4+12} = 3\mu\text{F}$$

$Q = CV = 3\times 8 = 24\mu\text{C}$ (on 4μF and 3μF)

Now, this 24μC distributes in direct ratio of capacity between 3μF and 9μF.

Therefore, $Q_{9\mu\text{F}} = 18\mu\text{C}$

$\therefore \quad Q_{4\mu\text{F}} + Q_{9\mu\text{F}} = 24 + 18 = 42\mu\text{C} = Q$ (say)

$$E = \frac{kQ}{R^2} = \frac{9\times 10^9 \times 42\times 10^{-6}}{30^2} = 420\text{ N/C}$$

3. *(c)* As, we know Cu is conductor, so increase in temperature, resistance will increase. Then, Si is semiconductor, so with increase in temperature, resistance will decreases.

4. *(a)* In a balanced Wheatstone bridge, there is no effect on position of null point, if we exchange the battery and galvanometer. So, option (a) is incorrect.

5. *(c)* As each capacitors cannot withstand more than 300 V, so there should be four capacitors in each row became in this condition 1 kV i.e. 1000 V will be divided by 4 (i.e. 250 not more than 300 V).

Now, equivalent capacitance of one row

$$= \frac{1}{4}\times 1\mu F = 0.25\mu\text{F}$$

$$\left[\because \text{in series combination, } C_{eq} = \frac{c}{n}\right]$$

Now, we need equivalent of 2μF, so let we need n such rows

$$\therefore \quad n\times 0.25 = 2\mu\text{F}$$

$$[\because \text{in parallel combination } C_{eq} = nc]$$

$$n = \frac{2}{0.25} = 8$$

$\therefore$ Total number of capacitors

= number of rows × number of capacitors in each row

$= 8\times 4 = 32$

6. *(b)* In steady state, no current flows through the capacitor. So, resistance r_p becomes ineffective.

So, the current in circuit, $I = \frac{E}{r + r_2 \text{ (Total Resistance)}}$

$\because$ Potential drop across capacitor

$$= \text{Potential dropacross } r_2 = Ir_2 = \frac{Er_2}{r + r_2}$$

$\therefore$ Stored charge of capacitor, $Q = CV = CE\frac{r_2}{r + r_2}$

7. *(c)* Each resistance is converted with two cells combined in opposite direction, so potential drop across each resistor is zero. Hence the current through each of resistor is zero.

8. *(b)*

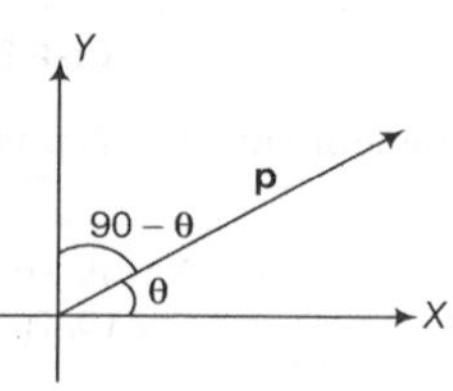

Torque applied on a dipole $\tau = pE\sin\theta$

where θ = angle between axis of dipole and electric field.

For electric field $E_1 = E\hat{\mathbf{i}}$

it means field is directed along positive X direction, so angle between dipole and field will remain θ, therefore torque in this direction $E_1 = pE_1\sin\theta$

In electric field $E_2 = \sqrt{3}\,E\hat{\mathbf{j}}$, it means field is directed along positive Y-axis, so angle between dipole and field will be $90 - \theta$.

Torque in this direction $T_2 = pE\sin(90 - \theta)$

$$= p\sqrt{3}\,E_1\cos\theta$$

According to question $\tau_2 = -\tau_1 \Rightarrow |\tau_2| = |\tau_1|$

$$\therefore \quad pE_1\sin\theta = p\sqrt{3}\,E_1\cos\theta$$

$$\tan\theta = \sqrt{3}$$

$$\Rightarrow \quad \tan\theta = \tan 60^\circ$$

$$\therefore \quad \theta = 60^\circ$$

9. *(b)* Key Idea *Potential of B = Potential due to charge on A + Potential due to charge on B + Potential due to charge on C.*

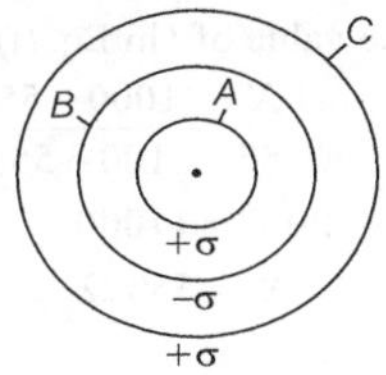

$$\therefore \quad V_B = \frac{k(Q_A + Q_B)}{b} + \frac{kQ_C}{c}$$

$$= \frac{1}{4\pi\varepsilon_0}\left[\frac{\sigma 4\pi a^2}{b} - \frac{\sigma 4\pi b^2}{b} + \frac{\sigma 4\pi c^2}{c}\right]$$

$$= \frac{\sigma\varepsilon}{\varepsilon_0}\left(\frac{a^2 - b^2}{b} + \frac{c^2}{c}\right) = \frac{\sigma}{\varepsilon_0}\left(\frac{a^2 - b^2}{b} + c\right)$$

$$V_B = \frac{\sigma}{\varepsilon_0}\left(\frac{a^2 - b^2}{b} + c\right)$$

10. *(a)* Magnitude of induced charge is given by $Q' = (K - 1)CV_0$

$$= \left(\frac{5}{3} - 1\right) 90 \times 10^{-12} \times 20$$

$$= 1.2 \times 10^{-9}\,\text{C}$$

$$\Rightarrow \quad Q' = 1.2\text{nC}$$

11. *(b)*

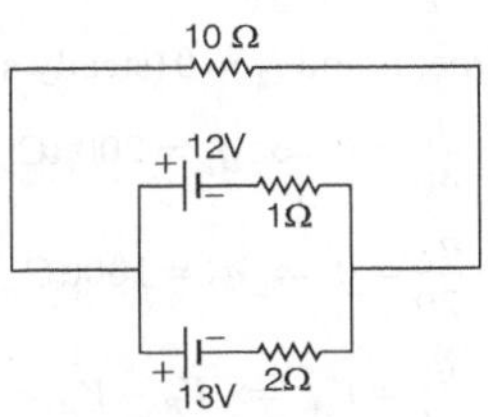

For parallel combination of cells,

$$E_{eq} = \frac{\dfrac{E_1}{r_1} + \dfrac{E_2}{r_2}}{\dfrac{1}{r_1} + \dfrac{1}{r_2}}$$

$$\therefore \quad E_{eq} = \frac{\dfrac{12}{1} + \dfrac{13}{2}}{\dfrac{1}{1} + \dfrac{1}{2}} = \frac{37}{3}\,\text{V}$$

Potential drop across 10 Ω resistance,

$$V = \left(\frac{E}{R_{total}}\right) \times 10 = \frac{\dfrac{37}{3}}{\left(10 + \dfrac{2}{3}\right)} \times 10 = 11.56\text{V}$$

$$\therefore \quad V = 11.56\,\text{V}$$

Alternative Method

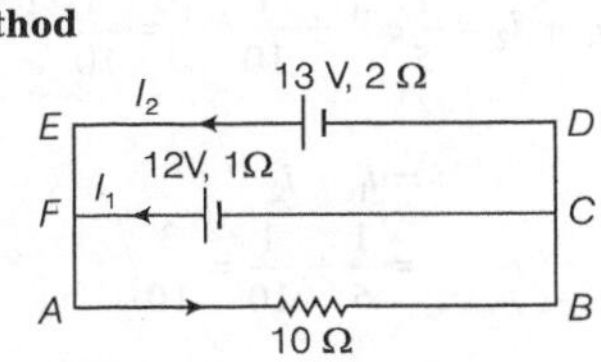

Applying KVL, in loop $ABCFA$,

$$-12 + 10(I_1 + I_2) + 1 \times I_1 = 0$$

$$\Rightarrow \quad 12 = 11I_1 + 10I_2 \qquad \text{...(i)}$$

Similarly, in loop $ABDEA$,

$$-13 + 10(I_1 + I_2) + 2 \times I_2 = 0$$

$$\Rightarrow \quad 13 = 10I_1 + 12I_2 \qquad \text{...(ii)}$$

Solving Eqs. (i) and (ii), we get

$$I_1 = \frac{7}{16}\text{A}, \; I_2 = \frac{23}{32}\text{A}$$

∴ Voltage drop across 10 Ω resistance is,

$$V = 10\left(\frac{7}{16} + \frac{23}{32}\right) = 11.56\,\text{V}$$

12. *(b)* With only the cell,

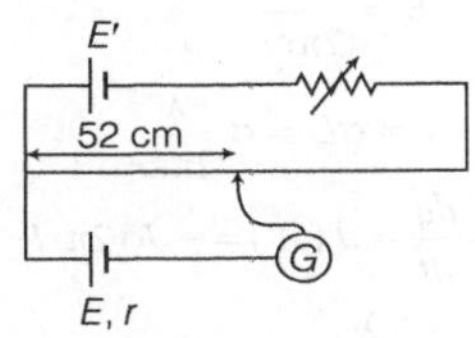

On balancing, $\quad E = 52 \times x \qquad$ ...(i)

where, x is the potential gradient of the wire.

When the cell is shunted,

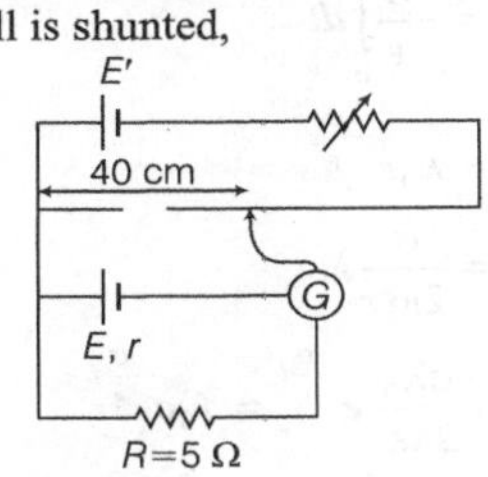

Similarly, on balancing,

$$V = E - \frac{Er}{(R + r)} = 40 \times x \qquad \text{...(ii)}$$

Solving Eqs. (i) and (ii), we get

$$\frac{E}{V} = \frac{1}{1 - \dfrac{r}{R + r}} = \frac{52}{40}$$

$$\Rightarrow \quad \frac{E}{V} = \frac{R + r}{R} = \frac{52}{40}$$

$$\Rightarrow \quad \frac{5 + r}{5} = \frac{52}{40} \Rightarrow r = \frac{3}{2}\,\Omega \Rightarrow r = 1.5\,\Omega$$

13. *(c)* We have, $\quad X + Y = 1000\,\Omega$

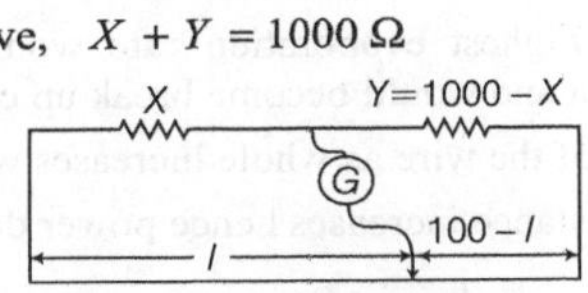

$$\text{Initially,} \quad \frac{X}{l} = \frac{1000 - X}{100 - l} \qquad \text{...(i)}$$

When X and Y are interchanged, then

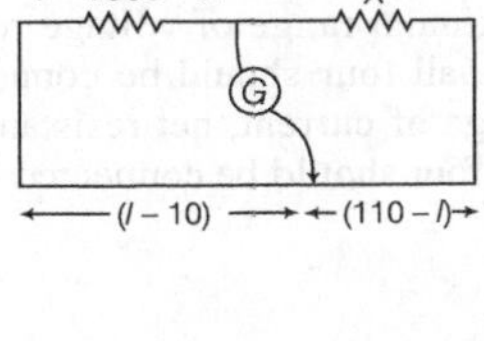

$$\frac{1000-X}{l-10}=\frac{X}{100-(l-10)}$$

or $$\frac{1000-X}{l-10}=\frac{X}{110-l} \quad \ldots(\text{ii})$$

From Eqs. (i) and (ii), we get

$$\frac{100-l}{l}=\frac{l-10}{110-l}$$

$$(100-l)(110-l)=(l-10)l$$

$$11000-100l-110l+l^2=l^2-10l$$

$$\Rightarrow \quad 11000=200l$$

$$\therefore \quad l=55\text{ cm}$$

Substituting the value of l in Eq. (i), we get

$$\frac{X}{55}=\frac{1000-55}{100-55}$$

$$\Rightarrow \quad 20X=11000$$

$$\therefore \quad X=550\,\Omega$$

JEE Advanced

1. *(d)* Suppose charger per unit length at any instant is λ. Initial value of λ is suppose λ_0. Electric field at a distance r at any instant is

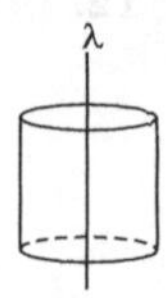

$$E=\frac{\lambda}{2\pi\varepsilon r}$$

$$J=\sigma E=\sigma\frac{\lambda}{2\pi\varepsilon r}$$

$$i=\frac{dq}{dt}=J(A)=-J\sigma 2\pi rl$$

$$\frac{d\lambda l}{dt}=-\frac{\lambda}{2\pi\varepsilon r}\times\sigma 2\pi rl \quad (q=\lambda l)$$

$$\int_{\lambda_o}^{\lambda}\frac{d\lambda}{\lambda}=-\frac{\sigma}{\varepsilon}\int_{o}^{t}dt$$

$$\Rightarrow \quad \lambda=\lambda_0 e^{-\frac{\sigma}{\varepsilon}t}$$

$$J=\frac{\sigma}{2\pi\varepsilon r}\lambda$$

$$=\frac{\sigma\lambda_0}{2\pi\varepsilon r}e^{-\frac{\sigma}{\varepsilon}t}=J_0 e^{-\frac{\sigma}{\varepsilon}t}$$

Here, $$J_0=\frac{\sigma\lambda_0}{2\pi\varepsilon r}$$

$\therefore$ $J(t)$ decreases exponentially as shown in figure below.

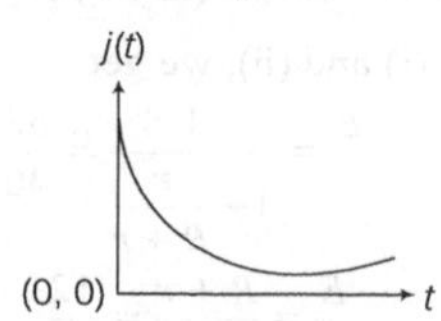

2. *(c,d)* Because of non-uniform evaporation at different section, area of cross-section would be different at different sections.

   Region of highest evaporation rate would have rapidly reduced area and would become break up cross-section.

   Resistance of the wire as whole increases with time.

   Overall resistance increases hence power decreases.

   $\left(P=\frac{V^2}{R} \text{ or } P\propto\frac{1}{R} \text{ as } V \text{ is constant}\right)$. At break up junction temperature would be highest, thus light of highest band frequency would be emitted at those cross-section.

3. *(a,c)* For maximum range of voltage resistance should be maximum. So, all four should be connected in series. For maximum range of current, net resistance should be least. Therefore, all four should be connected in parallel.

4. *(a,b,c,d)*

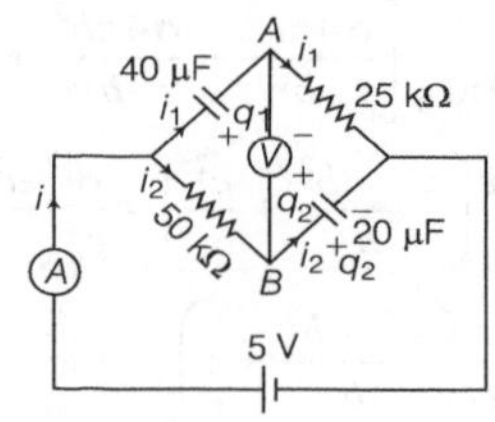

Just after pressing key,

$$5-25000i_1=0$$

$$5-50000i_2=0$$

(As charge in both capacitors = 0)

$$\Rightarrow \quad i_1=0.2\text{mA} \Rightarrow i_2=0.1\text{ mA}$$

And $V_B+25000i_1=V_A$

$$\Rightarrow \quad V_B-V_A=-5\text{V}$$

After a long time, i_1 and $i_2=0$ (steady state)

$$\Rightarrow \quad 5-\frac{q_1}{40}=0 \Rightarrow q_1=200\mu\text{C}$$

and $$5-\frac{q_2}{20}=0 \Rightarrow q_2=100\mu\text{C}$$

$$V_B-\frac{q_2}{20}=V_A \Rightarrow V_B-V_A=+5\text{V}$$

$\Rightarrow$ (a) is correct.

For capacitor 1,

$$q_1=200[1-e^{-t/1}]\mu\text{C} \Rightarrow i_1=\frac{1}{5}e^{-t/1}\text{mA}$$

For capacitor 2,

$$q_2=100[1-e^{-t/1}]\mu\text{C} \Rightarrow i_2=\frac{1}{10}e^{-t/1}\text{mA}$$

$$\Rightarrow \quad V_B-\frac{q_2}{20}+i_1\times 25=V_A$$

$$\Rightarrow \quad V_B-V_A=5[1-e^{-t}]-5e^{-t}=-5[1-2e^{-t}]$$

At $t=\ln 2$,

$$V_B-V_A=5[1-1]=0$$

$\Rightarrow$ (b) is correct.

At $t=1$,

$$i=i_1+i_2=\frac{1}{5}e^{-1}+\frac{1}{10}e^{-1}=\frac{3}{10}\cdot\frac{1}{e}$$

At $t=0$,

$$i=i_1+i_2$$

$$=\frac{1}{5}+\frac{1}{10}=\frac{3}{10}$$

$\Rightarrow$ (c) is correct.

After a long time, $i_1 = i_2 = 0$

$\Rightarrow$ (d) is correct.

5. *(d)* Balls will gain positive charge and hence move towards negative plate.

On reaching negative plate, balls will attain negative charge and come back to positive plate.

And so on, balls will keep oscillating.

But oscillation is not SHM, as force on balls is not $\propto x$.

$\Rightarrow$ (d) is correct.

6. *(a)* As the balls keep on carrying charge form one plate to another, current will keep on flowing even in steady state. When at bottom plate, if all balls attain charge q,

$$\frac{kq}{r} = V_0\left(k = \frac{1}{4\pi\varepsilon_0}\right) \Rightarrow q = \frac{V_0 r}{k}$$

Inside cylinder, electric field,

$$E = [V_0 - (-V_0)]h = 2V_0 h.$$

$\Rightarrow$ Acceleration of each ball,

$$a = \frac{qE}{m} = \frac{2hr}{km}\cdot V_0^2$$

$\Rightarrow$ Time taken by balls to reach other plate,

$$t = \sqrt{\frac{2h}{a}} = \sqrt{\frac{2h.km}{2hrV_0^2}} = \frac{1}{V_0}\sqrt{\frac{km}{r}}$$

If there are n balls, then Average current,

$$i_{av} = \frac{nq}{t} = n \times \frac{V_0 r}{k} \times V_0\sqrt{\frac{r}{km}}$$

$$\Rightarrow \quad i_{av} \propto V_0^2$$

7. *(a,d)*

(a) $\Omega = 2\pi(1-\cos\theta)$; $\theta = 45°$

$$\phi = -\frac{\Omega}{4\pi}\times\frac{Q}{\varepsilon_0} = -\frac{2\pi(1-\cos\theta)}{4\pi}\frac{Q}{\varepsilon_0} = -\frac{Q}{2\varepsilon_0}\left(1-\frac{1}{\sqrt{2}}\right)$$

(b) The component of the electric field perpendicular to the flat surface will decrease so we move away from the centre as the distance increases (magnitude of electric field decreases) as well as the angle between the normal and electric field will increase. Hence, the component of the electric field normal to the flat surface is not constant.

Alternate solution

$$x = \frac{R}{\cos\theta}$$

$$E = \frac{KQ}{x^2} = \frac{KQ\cos^2\theta}{R^2}$$

$$\Rightarrow \quad E_\perp = \frac{KQ\cos^3\theta}{R^2}$$

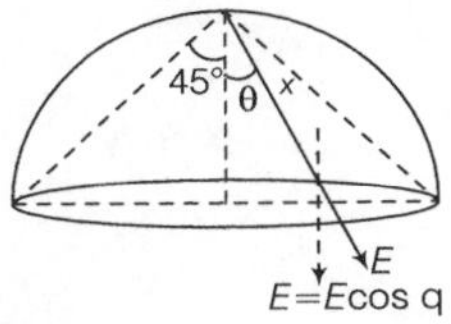

As we move away from centre

$\theta \uparrow \cos\theta \downarrow$ so $E_\perp \downarrow$

(c) Total flux ϕ due to charge Q is $\frac{Q}{\varepsilon_0}$.

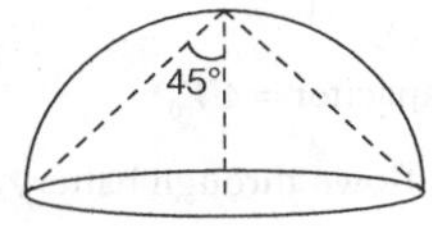

So, ϕ through the curved and flat surface will be less than $\frac{Q}{\varepsilon_0}$.

(d) Since, the circumference is equidistant from Q it will be equipotential $V = \frac{KQ}{\sqrt{2}R}$.

8. *(b)* When switch is closed for a very long time capacitor will get fully charged and charge on capacitor will be $q = CV$

Energy stored in capacitor,

$$E_C = \frac{1}{2}CV^2 \quad \text{...(i)}$$

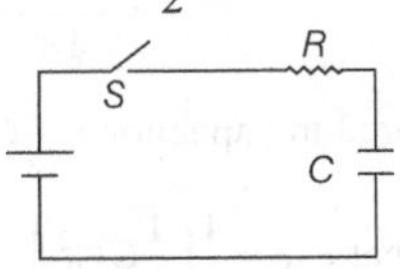

Work done by a battery, $W = Vq = VCV = CV^2$

Energy dissipated across resistance

$$E_D = \text{(work done by battery)} - \text{(energy stored)}$$

$$E_D = CV^2 - \frac{1}{2}CV^2 = \frac{1}{2}CV^2 \quad \text{...(ii)}$$

From Eqs. (i) and (ii)

$$E_D = E_C$$

9. *(a)* **For process (1)**

Charge on capacitor $= \frac{CV_0}{3}$

Energy stored in capacitor $= \frac{1}{2}C\frac{V_0^2}{9} = \frac{CV_0^2}{18}$

Work done by battery $= \frac{CV_0}{3}\times\frac{V}{3} = \frac{CV_0^2}{9}$

Heat loss $= \frac{CV_0^2}{9} - \frac{CV_0^2}{18} = \frac{CV_0^2}{18}$

For process (2)

Charge on capacitor $= \frac{2CV_0}{3}$

Extra charge flow through battery $= \frac{CV_0}{3}$

Work done by battery $= \frac{CV_0}{3}\cdot\frac{2V_0}{3} = \frac{2CV_0^2}{9}$

Final energy stored in capacitor $= \frac{1}{2}C\left(\frac{2V_0}{3}\right)^2 = \frac{4CV_0^2}{18}$

Energy stored in process

$$2 = \frac{4CV_0^2}{18} - \frac{CV_0^2}{18} = \frac{3CV_0^2}{18}$$

Heat loss in process (2) = work done by battery in process (2) − energy store in capacitor process (2)

$$= \frac{2CV_0^2}{9} - \frac{3CV_0^2}{18} = \frac{CV_0^2}{18}$$

For process (3)

Charge on capacitor = CV_0

Extra charge flown through battery = $CV_0 - \frac{2CV_0}{3} = \frac{CV_0}{3}$

Work done by battery in this process = $\left(\frac{CV_0}{3}\right)(V_0) = \frac{CV_0^2}{3}$

Final energy stored in capacitor = $\frac{1}{2}CV_0^2$

Energy stored in this process = $\frac{1}{2}CV_0^2 - \frac{4CV_0^2}{18} = \frac{5CV_0^2}{18}$

Heat loss in process (3) = $\frac{CV_0^2}{3} - \frac{5CV_0^2}{18} = \frac{CV_0^2}{18}$

Now, total heat loss (E_D)

$$= \frac{CV_0^2}{18} + \frac{CV_0^2}{18} + \frac{CV_0^2}{18} = \frac{CV_0^2}{6}$$

Final energy stored in capacitor = $\frac{1}{2}CV_0^2$

So we can say that $E_D = \frac{1}{3}\left(\frac{1}{2}CV_0^2\right)$

10. *(1.50)*

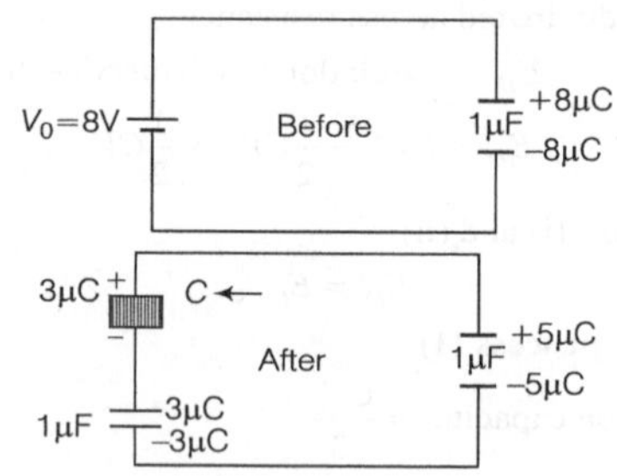

$C = \varepsilon_r C_1 = (\varepsilon_r)\mu F$

Applying loop rule,

$$\frac{5}{1} - \frac{3}{\varepsilon_r} - \frac{3}{1} = 0 \Rightarrow \frac{3}{\varepsilon_r} = 2$$

$$\varepsilon_r = 1.50$$

11. *(a, b)* $PQ = (2)R\sin 60°$

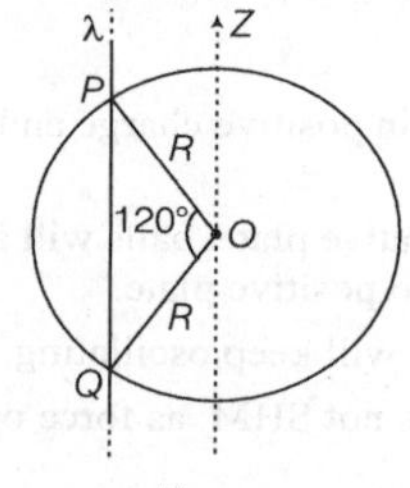

$$= (2R)\frac{\sqrt{3}}{2} = (\sqrt{3}R)$$

$$q_{\text{enclosed}} = \lambda(\sqrt{3}R)$$

We have, $\phi = \frac{q_{\text{enclosed}}}{\varepsilon_0} \Rightarrow \phi = \left(\frac{\sqrt{3}\lambda R}{\varepsilon_0}\right)$

Also, electric field is perpendicular to wire, so Z-component will be zero.

12. *(b)* **List-II**

(1) $E = \frac{1}{4\pi\varepsilon_0}\frac{Q}{d^2} \Rightarrow E \propto \frac{1}{d^2}$

(2) $E_{\text{axis}} = \frac{1}{4\pi\varepsilon_0}\frac{2Q(2l)}{d^3} \Rightarrow E \propto \frac{1}{d^3}$

(3) $E = \frac{\lambda}{2\pi\varepsilon_0 d} \Rightarrow E \propto \frac{1}{d}$

(4) $E = \frac{\lambda}{2\pi\varepsilon_0(d-l)} - \frac{\lambda}{2\pi\varepsilon_0(d+l)} = \frac{\lambda(2l)}{2\pi\varepsilon_0 d^2} \Rightarrow E \propto \frac{1}{d^2}$

(5) $E = \frac{\sigma}{2\varepsilon_0} \Rightarrow$ E is independent of d

13. *(2)* $a = \frac{F}{m} = \frac{qE}{m} = 10^3\sin(10^3 t)$

$$\frac{dv}{dt} = 10^3\sin(10^3 t)$$

$$\Rightarrow \int_0^v dv = \int_0^t 10^3\sin(10^3 t)\,dt$$

$$\therefore \quad v = \frac{10^3}{10^3}[1 - \cos(10^3 t)]$$

Velocity will be maximum when $\cos(10^3 t) = -1$

$v_{\max} = 2\text{ m/s}$